Prealgebra

Fifth Edition

Jamie Blair

Orange Coast College
Costa Mesa, California

John Tobey

North Shore Community College
Danvers, Massachusetts

Jeffrey Slater

North Shore Community College
Danvers, Massachusetts

Jennifer Crawford

Normandale Community College
Bloomington, Minnesota

PEARSON

Boston   Columbus   Indianapolis   New York   San Francisco   Upper Saddle River
Amsterdam   Cape Town   Dubai   London   Madrid   Milan   Munich   Paris   Montréal   Toronto
Delhi   Mexico City   São Paulo   Sydney   Hong Kong   Seoul   Singapore   Taipei   Tokyo

Editorial Director, Mathematics: *Christine Hoag*
Editor-in-Chief: *Paul Murphy*
Acquisitions Editor: *Dawn Giovanniello*
Executive Content Editor: *Kari Heen*
Senior Content Editor: *Lauren Morse*
Editorial Assistant: *Chelsea Pingree*
Vice President, Executive Director of Development: *Carol Trueheart*
Senior Development Editor: *Elaine Page*
Senior Managing Editor: *Karen Wernholm*
Senior Production Supervisor: *Ron Hampton*
Design Manager: *Andrea Nix*
Interior Design: *Tamara Newnam*
Senior Design Specialist: *Barbara Atkinson*
Digital Assets Manager: *Marianne Groth*
Supplements Production Project Manager: *Katherine Roz*
Executive Manager, Course Production: *Peter Silvia*
Media Producers: *Audra Walsh and Vicki Dreyfus*
Content Development Manager: *Rebecca E. Williams*
Senior Content Developer: *Mary Durnwald*
Executive Marketing Manager: *Michelle Renda*
Marketing Manager: *Rachel Ross*
Marketing Assistant: *Ashley Bryan*
Senior Author Support/Technology Specialist: *Joe Vetere*
Procurement Manager/Boston: *Evelyn M. Beaton*
Procurement Specialist: *Debbie Rossi*
Senior Media Buyer: *Ginny Michaud*
Permissions Project Supervisor: *Michael Joyce*
Production Management, Composition, and Answer Art: *Integra*
Text Art: *Scientific Illustrators*
Cover Images: *Illustration by Amy DeVoog*

Many of the designations used by manufacturers and sellers to distinguish their products are claimed as trademarks. Where those designations appear in this book, and Pearson Education was aware of a trademark claim, the designations have been printed in initial caps or all caps.

Library of Congress Cataloging-in-Publication Data

Prealgebra / Jamie Blair ... [et al].—5th ed.
   p. cm.
 Includes index.
 978-0-321-75645-9
1. Mathematics—Textbooks. I. Blair, Jamie.
  QA39.3.B43 2011
  510—dc23   2011023720

1 2 3 4 5 6 7 8 9 10—DOW—16 15 14 13 12

ISBN-10: 0-321-75645-2 (paperback)
ISBN-13: 978-0-321-75645-9 (paperback)

PEARSON

pearsonhighered.com

This book is dedicated to my husband, Jerry Blair,
and my two children, Joe and Wendy. Their support
and patience during the production of the text
are greatly appreciated.

Contents

Preface

TO THE INSTRUCTOR

This book was developed to bridge the gap between arithmetic and algebra topics, covering all the key arithmetic topics and introducing basic algebra concepts and topics. The approach used in the text integrates algebra rules and concepts with those of arithmetic, teaches "why," not memorization, and emphasizes translation skills (the language of mathematics to the English language). The text spirals topics and teaches students the specific study skills necessary to accommodate their individual learning styles. The text offers students an effective and proven learning program suitable for a variety of course formats—including lecture-based classes; discussion-oriented classes; modular, self-paced courses; distance learning; mathematics laboratories; and computer-supported centers.

We have visited and listened to teachers across the country and have incorporated a number of suggestions into this edition to help you with the particular learning delivery system at your school.

What's New in the Fifth Edition?

- **Chapter Organizers** have been updated to include a You Try It column that provides additional opportunity for students to practice relevant chapter topics and procedures.

- A solid correlation has been made between the material on the **How Am I Doing? Chapter Test** and the examples, exercises, Chapter Review, and Cumulative Review. Each Chapter Test problem has at least 1 example, 2 Chapter Review exercises, and 2 Cumulative Review exercises that represent the same problem type. New assessment check boxes allow students to tally their answers and gauge their preparedness for the actual test.

- Following each Chapter Test, the new **Math Coach** provides students with a personal office hour experience by walking them through some helpful hints to keep them from making common errors on test problems. For additional help, students can also watch the authors work through these problems on the accompanying Math Coach videos, available on YouTube and in MyMathLab.

- Select **Examples and Student Practice** problems, representing some of the most difficult concepts for students to master in a chapter, have been placed side by side to encourage students to work through each step of these problems to gain further understanding. These concepts are also covered on the Chapter Test and in the Math Coach.

- **Enhanced emphasis on Steps to Success boxes** (formerly Developing Your Study Skills) have been integrated throughout the text to provide students with more guided techniques for improving their study skills and succeeding in math.

- The **Use Math to Save Money** features are now assignable so that students can apply this new knowledge to their everyday lives. All of the topics have been chosen based on a student survey of over 1000 developmental math college students. These give practical, realistic examples of how students can use math to cut costs and spend less.

- Ten percent of the exercises throughout the text have been refreshed.

- All real-world application problems have been updated.

- *New* **The Lecture Series on DVD** have been completely revised to provide students with extra help for each section of the textbook. The videos include:

  - **Interactive Lectures** that highlight key examples and exercises from every section of the textbook. A new interface allows for easy navigation to sections, objectives, and examples.

  - **Math Coach Videos**, featuring the text authors (Jamie Blair, John Tobey, Jeffrey Slater, and Jennifer Crawford), coach students in avoiding the most commonly made mistakes in a particular problem when students need the most help: the night before an exam.

  - **Chapter Test Prep Videos** provide step-by-step video solutions to every problem in each How Am I Doing? Chapter Test in the textbook.

Student and Instructor Resources

Worksheets with the Math Coach

Provides extra vocabulary and practice exercises for every section of the text. Each chapter also includes the Math Coach problems with ample space for students to show their work. The worksheets can be packaged with the textbook or with the MyMathLab access kit.

Student Solutions Manual

Provides worked-out solutions to all odd-numbered section exercises, even and odd exercises in the Quick Quiz, mid-chapter reviews, chapter reviews, chapter tests, Math Coach, and cumulative reviews.

Lecture Series on DVD Featuring Math Coach and Chapter Test Prep Videos

Provides students with extra help for each section of the textbook. The videos include

- A complete lecture for each section of the textbook. The new interface allows easy navigation to objectives and examples.

- Math Coach videos that coach students in avoiding the most commonly made mistakes in a particular problem.

- Step-by-step video solutions to every problem in each How Am I Doing? Chapter Test.

Math Coach and Chapter Test Videos are also available in MyMathLab and on YouTube.

All Student Resources are available for purchase at www.mypearsonstore.com.

Annotated Instructor's Edition

Contains all of the content found in the student edition, plus the following:

- Answers to all practice problems, section exercises, mid-chapter reviews, chapter reviews, chapter tests, cumulative tests, and practice final exam

- Teaching Tips placed in the margin at key points where students historically need extra help

- Teaching Examples placed in the margins to accompany each example

Instructor's Solutions Manual

- Detailed, step-by-step solutions to the even-numbered section exercises

- Solutions to every exercise (odd and even) in the Classroom Quiz, mid-chapter reviews, chapter reviews, chapter tests, cumulative tests, and practice final exam

(Available for download from the Instructor's Resource Center)

Instructor's Resource Manual with Tests and Mini-Lectures

- Mini-lecture for each text section

- Two short group activities per chapter

- Three forms of additional practice exercises

- Two pretests per chapter—free response and multiple choice

- Six tests per chapter—free response and multiple choice

- Two cumulative tests per even-numbered chapter—free response and multiple choice

- Two Final Exams—free response and multiple choice

- Answers to all items

(Available for download from the Instructor's Resource Center)

MyMathLab® Online Course (access code required)

MathXL® Online Course (access code required)

TestGen® (Available for download from the Instructor's Resource Center)

Should I buy or lease a car? How do I balance my checkbook with the bank's statement? The mathematics you learn in this chapter will help you to answer these kinds of questions.

Whole Numbers and Introduction to Algebra

1.1 Understanding Whole Numbers

Student Learning Objectives

After studying this section, you will be able to:

① Understand place values of whole numbers.

② Write whole numbers in expanded notation.

③ Write word names for whole numbers.

④ Use inequality symbols with whole numbers.

⑤ Round whole numbers.

Often we learn a new concept in stages. First comes learning the new *terms* and basic assumptions. Then we have to master the *reasoning*, or logic, behind the new concept. This often goes hand in hand with learning a *method* for using the idea. Finally, we can move quickly with a *shortcut*.

For example, in the study of stock investments, before tackling the question "What is my profit from this stock transaction?" you must learn the meaning of such terms as *stock, profit, loss,* and *commission*. Next, you must understand how stocks work (reasoning/logic) so that you can learn the method for calculating your profit. After you master this concept, you can quickly answer many similar questions using shortcuts.

In this book, watch your understanding of mathematics grow through this same process. In the first chapter we review the whole numbers, emphasizing *concepts,* not shortcuts. Do not skip this review even if you feel you have mastered the material since understanding each stage of the concepts is crucial to learning algebra. With a little patience in looking at the terms, reasoning, and step-by-step methods, you'll find that your understanding of whole numbers has deepened, preparing you to learn algebra.

① Understanding Place Values of Whole Numbers

We use a set of numbers called **whole numbers** to count a number of objects.

> The whole numbers are as follows:
> 0, 1, 2, 3, 4, 5, 6, 7, 8, 9, 10, 11, 12, 13, 14, 15, 16, . . .

There is no largest whole number. The three dots . . . indicate that the set of whole numbers goes on forever. The numbers 0, 1, 2, 3, 4, 5, 6, 7, 8, and 9 are called **digits.** The *position* or *placement* of the digit in a number tells the *value* of the digit. For this reason, our number system is called a **place-value system.** For example, look at the following three numbers.

632 The "6" means 6 hundreds (600).

61 The "6" means 6 tens (60).

6 The "6" means 6 ones (6).

To illustrate the values of the digits in a number, we can use the following place-value chart. Consider the number 847,632, which is entered on the chart.

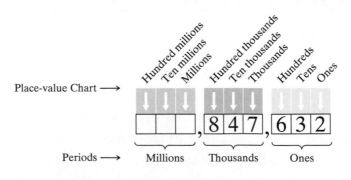

The digit 8 is in the hundred thousands place.

The digit 4 is in the ten thousands place.

The digit 7 is in the thousands place.

The digit 6 is in the hundreds place.

The digit 3 is in the tens place.

The digit 2 is in the ones place.

When we write very large numbers, we place a comma after every group of three digits, moving from right to left. These three-digit groups are called **periods.** It is usually agreed that four-digit numbers do not have a comma, but numbers with five or more digits do.

EXAMPLE 1   In the number 573,025:

(a) In what place is the digit 7?   **(b)** In what place is the digit 0?

Solution

(a) 5 7 3,025   **(b)** 573, 0 25
　　　↑　　　　　　　　　　　　　　　　　　　↑
　　ten thousands   hundreds

Student Practice 1   In the number 3,502,781:

(a) In what place is the digit 5?

(b) In what place is the digit 0?

NOTE TO STUDENT: *Fully worked-out solutions to all of the Student Practice problems can be found at the back of the text starting at page SP-1.*

② Writing Whole Numbers in Expanded Notation

We sometimes write numbers in **expanded notation** to emphasize place value. The number 47,632 can be written in expanded notation as follows:

40,000	+	7000	+	600	+	30	+	2
4 ten	+	7	+	6	+	3	+	2
thousands		thousands		hundreds		tens		ones

EXAMPLE 2   Write 1,340,765 in expanded notation.

Solution

We write 1 followed by a zero for each of the remaining digits.
↓

We write 1 ,340,765 as 1 ,000,000 + 3 00,000 + 4 0,000 + 7 00 + 6 0 + 5
　　　　　　　　　　　　　　We continue in this manner for each digit.

Since there is a zero in the thousands place, we do not write it as part of the sum.

Student Practice 2   Write 2,507,235 in expanded notation.

EXAMPLE 3   Jon withdraws $493 from his account. He requests the minimum number of bills in one-, ten-, and hundred-dollar bills. Describe the quantity of each denomination of bills the teller must give Jon.

Solution   If we write $493 in expanded notation, we can easily describe the denominations needed.

400	+	90	+	3
4		**9**		**3**
hundred-dollar		ten-dollar		one-dollar
bills		bills		bills

Student Practice 3   Christina withdraws $582 from her account. She requests the minimum number of bills in one-, ten-, and hundred-dollar bills. Describe the quantity of each denomination of bills the teller must give Christina.

Understanding the Concept

The Number Zero   Not all number systems have a zero. The Roman numeral system does not. In our place-value system the zero is necessary so that we can write a number such as 308. By putting a zero in the tens place, we indicate that there are zero tens. Without a zero symbol we would not be able to indicate this. For example, 38 has a different value than 308. The number 38 means three *tens* and eight ones, while 308 means three *hundreds* and eight ones. In this case, we use *zero* as a **placeholder.** It holds a position and shows that there is no other digit in that place.

③ Writing Word Names for Whole Numbers

Sixteen, *twenty-one*, and *four hundred five* are **word names** for the numbers 16, 21, and 405. We use a hyphen between words when we write a two-digit number greater than twenty. To write a word name, start from the left. Name the number in each period, followed by the name of the period, and a comma. The last period name, "ones," is not used.

EXAMPLE 4   Write a word name for each number.

(a) 2135   **(b)** 300,460

Solution   Look at the place-value chart on page 2 if you need help identifying the period.

(a) 2135   The number begins with 2 in the *thousands* place. The word name is
two *thousand*, one hundred thirty-five.

                              ↑
              We use a hyphen here.

(b) 300,460   The number begins with 3 in the *hundred thousands* place.

     Three *hundred thousand*, four hundred sixty

                            ↑
We place a comma here to match the comma in the number.

Student Practice 4   Write a word name for each number.

(a) 4006   **(b)** 1,220,032

CAUTION: We should not use the word *and* in the word names for whole numbers. Although we may hear the phrase "three hundred and two" for the number 302, it is not technically correct. As we will see later in the book, we use the word *and* for the decimal point when using decimal notation.

④ Using Inequality Symbols with Whole Numbers

It is often helpful to draw pictures and graphs to help us visualize a mathematical concept. A **number line** is often used for whole numbers. The following number line has a point matched with zero and with each whole number. Each number is equally spaced, and the "→" arrow at the right end indicates that the numbers go on forever. The numbers on the line increase from left to right.

If one number lies to the *right* of a second number on the number line, it *is greater than* that number.

4 lies to the *right* of 2 on the number line because 4 *is greater than* 2.

A number *is less than* a given number if it lies to the *left* of that number on the number line.

3 lies to the *left* of 5 on the number line because 3 *is less than* 5.

The symbol > means *is greater than*, and the symbol < means *is less than*. Thus we can write

             $4 > 2$ $3 < 5$
              ↓   ↓
     4 *is greater than* 2.   3 *is less than* 5.

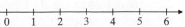

The symbols < and > are called **inequality symbols.** The statements $4 > 2$ and $2 < 4$ are both correct. Note that the inequality symbol always points to the smaller number.

EXAMPLE 5   Replace each question mark with the inequality symbol < or >.

(a) 1 ? 6   **(b)** 8 ? 7   **(c)** 4 ? 9   **(d)** 9 ? 4

Solution

(a) 1 < 6   **(b)** 8 > 7   **(c)** 4 < 9   **(d)** 9 > 4

   ↓   ↓   ↓   ↓

1 is less than 6.   8 is greater than 7.   4 is less than 9.   9 is greater than 4.

Student Practice 5   Replace each question mark with the inequality symbol < or >.

(a) 3 ? 2   **(b)** 6 ? 8   **(c)** 1 ? 7   **(d)** 7 ? 1

EXAMPLE 6   Rewrite using numbers and an inequality symbol.

(a) Five is less than eight.   **(b)** Nine is greater than four.

Solution

(a) Five *is less than* eight.   **(b)** Nine *is greater than* four.

  ↓   ↓   ↓   ↓   ↓   ↓

  5   <   8   9   >   4

Remember, the inequality symbol always points to the smaller number.

Student Practice 6   Rewrite using numbers and an inequality symbol.

(a) Seven is greater than two.   **(b)** Three is less than four.

---

⑤ Rounding Whole Numbers

We often approximate the values of numbers when it is not necessary to know the exact values. These approximations are easier to use and remember. For example, if our hotel bill was $82.00, we might say that we spent about $80. If a car cost $14,792, we would probably say that it cost approximately $15,000.

    Why did we approximate the price of the car at $15,000 and not $14,000? To understand why, let's look at the number line.

The number 14,792 is closer to 15,000 than to 14,000, so we approximate the cost of the car at $15,000.

    It would also be correct to approximate the cost at $14,800 or $14,790, since each of these values is close to 14,792 on the number line. How do we know which approximation to use? We specify how accurate we would like our approximation to be. **Rounding** is a process that approximates a number to a specific **round-off place** (ones, tens, hundreds, . . .). *Thus the value obtained when rounding depends on how accurate we would like our approximation to be.* To illustrate, we round the price of the car discussed above to the thousands and to the hundreds place.

    14,792 rounded to the nearest *thousand* is 15,000.   The *round-off place* is thousands.

    14,792 rounded to the nearest *hundred* is 14,800.   The *round-off place* is hundreds.

    We can use the following set of rules instead of a number line to round whole numbers.

> **PROCEDURE TO ROUND A WHOLE NUMBER**
> 1. Identify the round-off place digit.
> 2. If the digit to the *right* of the round-off place digit is:
>   (a) *Less than 5,* do not change the round-off place digit.
>   (b) *5 or more,* increase the round-off place digit by 1.
> 3. Replace all digits to the *right* of the round-off place digit with zeros.

EXAMPLE 7   Round 57,441 to the nearest thousand.

Solution   The round-off place digit is in the thousands place.

5⑦,441   **1.** Identify the round-off place digit 7.
             2. The digit to the right is less than 5.

Do not change the
round-off place digit.

57,**000**

             3. Replace all digits to the right with zeros.

We have rounded 57,441 to the nearest thousand: **57,000.** This means that 57,441 is closer to 57,000 than to 58,000.

Student Practice 7   Round 34,627 to the nearest hundred.

EXAMPLE 8   Round 4,254,423 to the nearest hundred thousand.

Solution   The round-off place digit is in the hundred thousands place.

4, ②54,423   **1.** Identify the round-off place digit 2.
                 2. The digit to the right is 5 or more.

CAUTION: The round-off place digit either stays the same or increases by 1. It never decreases.

Increase the round-off
place digit by 1.

4,300,000

                 3. Replace all digits to the right with zeros.

We have rounded 4,254,423 to the nearest hundred thousand: **4,300,000.**

Student Practice 8   Round 1,335,627 to the nearest ten thousand.

👣 STEPS TO SUCCESS Finding a Study Partner

Attempt to make a friend in your class and become study partners. You may find that you enjoy sitting together and drawing support and encouragement from each other. You must not depend on a friend or fellow student to tutor you, do your work for you, or in any way be responsible for your learning. However, you will learn from each other as you seek to master the course. Studying with a friend and comparing notes, methods, and solutions can be very helpful. And it makes learning mathematics a lot more fun!

Making it personal:

1. Exchange contact information with someone in class so you can contact each other whenever you are having difficulty with your studying.

Name of study partner: _____

Phone number: _____

E-mail address: _____

2. Set up convenient times to study together on a regular basis, to do homework, and to review for exams.

Day and time you and your partner will meet:

Day ____   Time ____

Verbal and Writing Skills, Exercises 1–8

1. Write the word name for
   (a) 8002.
   (b) 802.
   (c) 82.
   (d) What is the place value of the digit 0 in the number eight hundred twenty?

2. Write in words.
   (a) $2 < 5$
   (b) $5 > 2$
   (c) What can you say about parts **(a)** and **(b)**?

3. In the number 9865:
   (a) In what place is the digit 8?
   (b) In what place is the digit 5?

4. In the number 23,981:
   (a) In what place is the digit 2?
   (b) In what place is the digit 9?

5. In the number 754,310:
   (a) In what place is the digit 4?
   (b) In what place is the digit 7?

6. In the number 913,728:
   (a) In what place is the digit 9?
   (b) In what place is the digit 1?

7. In the number 1,284,073:
   (a) In what place is the digit 1?
   (b) In what place is the digit 0?

8. In the number 3,098,269:
   (a) In what place is the digit 0?
   (b) In what place is the digit 8?

Write in expanded notation.

9. 5876

10. 7632

11. 4921

12. 3562

13. 867,301

14. 913,045

15. Damian withdraws $562 from his account. He requests the minimum number of bills in one-, ten-, and hundred-dollar bills. Describe the quantity of each denomination of bills the teller must give Damian.

16. Erin withdraws $274 from her account. She requests the minimum number of bills in one-, ten-, and hundred-dollar bills. Describe the quantity of each denomination of bills the teller must give Erin.

17. Describe the denominations of bills for $46:
   (a) Using only ten- and one-dollar bills.

   (b) Using tens, fives, and only 1 one-dollar bill.

18. Describe the denominations of bills for $96:
   (a) Using only ten- and one-dollar bills.

   (b) Using tens, fives, and only 1 one-dollar bill.

Write a word name for each number.

19. 6079

20. 4032

21. 86,491

22. 33,224

23. Fill in the check with the amount $672.

```
James Hunt
4 Platt St.                                      2824
Mapleville, RI  02839
                                DATE_____ 20____
PAY to the
ORDER of  Hampton Apartments          $ [        ]
_____ DOLLARS

Mason Bank
California

MEMO _____  _____
⑆5800520⑆ 55202205⑈ 2824
```

24. Fill in the check with the amount $379.

```
Ellen Font
22 Rose Place                                    2520
Garden Grove, CA  92641
                                DATE_____ 20____
PAY to the
ORDER of  Atlas Insurance             $ [        ]
_____ DOLLARS

Mason Bank
California

MEMO _____  _____
⑆5800520⑆ 55202205⑈ 2520
```

Replace each question mark with the inequality symbol < or >.

25. 5 ? 7   **26.** 3 ? 1   **27.** 6 ? 8   **28.** 9 ? 6

29. 13 ? 10   **30.** 9 ? 11   **31.** 9 ? 0   **32.** 0 ? 9

33. 2131 ? 1909   **34.** 3010 ? 3210   **35.** 52,647 ? 616,000   **36.** 101,351 ? 101,251

Rewrite using numbers and an inequality symbol.

37. Five is greater than two.   **38.** Seven is less than ten.

39. Two is less than five.   **40.** Six is greater than four.

Round to the nearest ten.

41. 45   **42.** 85   **43.** 661   **44.** 123

Round to the nearest hundred.

45. 63,854   **46.** 12,790   **47.** 823,042   **48.** 701,529

Round to the nearest thousand.

49. 38,431   **50.** 56,212   **51.** 143,526   **52.** 312,540

Round to the nearest hundred thousand.

53. 5,254,423   **54.** 1,395,999   **55.** 9,007,601   **56.** 3,116,201

57. *The Sun* The diameter of the sun is approximately 865,000 miles. Round this figure to the nearest ten thousand.

58. *Inches and Miles* There are 3,484,800 inches in 55 miles. Round 3,484,800 to the nearest ten thousand.

Mixed Practice *Automobile Prices The table lists the 2010 sticker prices on some popular vehicles. Use this table to answer exercises 59–62.*

Type of Automobile	2010 MSRP
Ford Expedition XLT	$35,010
Ford Supercab XLT	$29,605
Dodge Charger Rallye	$27,395
Dodge Caravan Crew	$29,195

Source: www.dodge.com; www.fordvehicles.com

Replace the question mark with an inequality symbol to indicate the relationship between the prices of the vehicles.

59. Ford Expedition XLT ? Ford Supercab XLT

60. Dodge Caravan Crew ? Dodge Charger Rallye

Round each vehicle's MSRP to the nearest thousand.

61. Dodge Charger Rallye

62. Ford Supercab XLT

One Step Further *Round to the nearest hundred.*

63. 16,962

64. 44,972

Very large numbers are used in some disciplines to measure quantities, such as distance in astronomy and the national debt in macroeconomics. We can extend the place-value chart to include these large numbers.

65. Write 5,311,192,809,000 using the word name.

66. Round 5,311,192,809,000 to the nearest million.

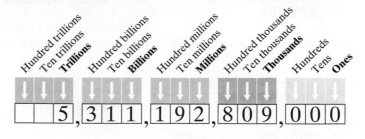

To Think About *Sometimes to get an approximation we must round to the nearest unit, such as a foot, yard, hour, or minute.*

67. *Train Travel Time* A train takes 3 hours and 50 minutes to reach its destination. Approximately how many hours does the trip take?

68. *Automobile Travel Time* An automobile trip takes 5 hours and 40 minutes. Approximately how many hours does the drive take?

69. *Fence Measurements* The Nguyens' backyard has a fence around it that measures 123 feet 5 inches. Approximately how many feet of fencing do the Nguyens have?

70. *Yardage Measurements* Jessica has 15 yards 4 inches of material. Approximately how many yards of material does Jessica have?

Quick Quiz 1.1

1. Write 6402 in expanded notation.

2. Replace each question mark with the appropriate symbol $<$ or $>$.
  (a) 0 ? 10   **(b)** 15 ? 10

3. Round 154,572 to
  (a) the nearest ten thousand
  (b) the nearest hundred

4. **Concept Check** Explain how to round 8937 to the nearest hundred.

👣 STEPS TO SUCCESS Preparing to Learn Algebra

People often learn arithmetic by memorizing facts and properties without understanding why the facts are true or what the properties mean. **Learning strictly by memorization can cause problems.** For example:

- Many of the shortcuts in arithmetic do not work in algebra.
- Memorizing does not help one develop reasoning and logic skills, which are essential to understanding algebra concepts.
- Memorization can eventually cause *memory overload*. Trying to remember a collection of unrelated facts can cause you to become *anxious* and *discouraged*.

Helpful Hints to Learn Concepts and Avoid Memory Overload

Do not skip familiar arithmetic topics. The explanation may be different because it emphasizes teaching *why*, not *memorization*.

Be patient and keep trying. If you don't understand a concept the first time, you may see why it works once you work through the exercises.

Read all the Understanding the Concept boxes in the book. The information and exercises will help you learn concepts and avoid memory overload.

Making it personal: Write in words why the commutative property of addition reduces the amount of memorization necessary to learn addition facts.

1.2 Adding Whole Number Expressions

Student Learning Objectives

After studying this section, you will be able to:

1. Use symbols and key words for expressing addition.
2. Use properties of addition to rewrite algebraic expressions.
3. Evaluate algebraic expressions involving addition.
4. Add whole numbers when carrying is needed.
5. Find the perimeters of geometric figures.

① Using Symbols and Key Words for Expressing Addition

What is *addition*? We perform **addition** when we group items together. Consider the following illustration involving the sale of bikes.

Bikes sold Saturday		Bikes sold Sunday		Total bikes sold
🚲🚲🚲🚲	+	🚲🚲🚲	=	🚲🚲🚲🚲🚲🚲🚲
4	+	3	=	7 bikes

is equal to

We see that the number 7 is the total of 4 and 3. That is, $4 + 3 = 7$ is an addition fact. The numbers being added are called **addends.** The result is called the **sum.**

$$4 \quad + \quad 3 \quad = \quad 7$$
$$\text{addend} \quad \text{addend} \quad \text{sum}$$

In mathematics we use symbols such as "+" in place of the words *sum* or *plus*. The English phrase "five plus two" written using symbols is "$5 + 2$." Writing English phrases using math symbols is like translating between languages such as Spanish and French.

There are several English phrases that describe the operation of addition. The following table gives some of them and their translated equivalents written using mathematical symbols.

English Phrase	Translation into Symbols
Six *more than* nine	$9 + 6$
The *sum of* some number and seven	$x + 7$
Four *increased by* two	$4 + 2$
Three *added* to a number	$n + 3$
One *plus* a number	$1 + x$

When we do not know the value of a number, we use a letter, such as x, to represent that number. A letter that represents a number is called a **variable.** Notice that the variables used in the table above are different. We can choose any letter as a variable. Thus we can represent "a number plus seven" by $x + 7, a + 7, n + 7, y + 7$, and so on. Combinations of variables and numbers such as $x + 7$ and $a + 7$ are called **algebraic expressions** or **variable expressions.**

EXAMPLE 1 Translate each English phrase using numbers and symbols.

(a) The sum of six and eight   **(b)** A number increased by four

Solution

(a) The *sum of* six and eight

$$6 + 8$$

(b) A number *increased by* four

$$x \quad + \quad 4$$

Although we used the variable x to represent the unknown quantity in part (b), any letter could have been used.

Student Practice 1 Translate each English phrase using numbers and symbols.

(a) Five added to some number   **(b)** Four more than five

NOTE TO STUDENT: Fully worked-out solutions to all of the Student Practice problems can be found at the back of the text starting at page SP-1.

② **Using Properties of Addition to Rewrite Algebraic Expressions**

Most of us memorized some basic addition facts. Yet if we study these sums, we observe that there are only a few addition facts for each one-digit number that we must memorize. For example, we can easily see that when 0 items are added to any number of items, we end up with the same number of items: $5 + 0 = 5, 0 + 8 = 8$, and so on. This illustrates the **identity property of zero:** $a + 0 = a$ and $0 + a = a$.

EXAMPLE 2   Express 4 as the sum of two whole numbers. Write all possibilities. How many addition facts must we memorize? Why?

Solution   Starting with $4 + 0$, we write all the sums equal to 4 and observe any patterns.

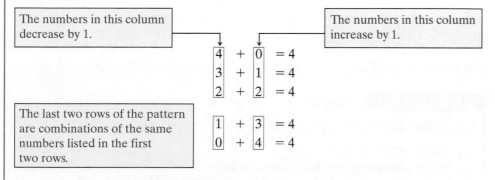

| The numbers in this column decrease by 1. | The numbers in this column increase by 1. |

$$4 + 0 = 4$$
$$3 + 1 = 4$$
$$2 + 2 = 4$$

The last two rows of the pattern are combinations of the same numbers listed in the first two rows.

$$1 + 3 = 4$$
$$0 + 4 = 4$$

We need to learn only **two addition facts** for the number four: $3 + 1$ and $2 + 2$. The remaining facts are either a repeat of these or use the fact that when 0 is added to any number, the sum is that number.

Student Practice 2   Express 8 as the sum of two whole numbers. Write all possibilities. How many addition facts must we memorize? Why?

In Example 2 we saw that the order in which we add numbers doesn't affect the sum. That is, $3 + 1 = 4$ and $1 + 3 = 4$. This is true for all numbers and leads us to a property called the **commutative property of addition.**

> **COMMUTATIVE PROPERTY OF ADDITION**
> $$a + b = b + a \qquad 4 + 9 = 9 + 4$$
> $$13 = 13$$
> Two numbers can be added in either order with the same result.

EXAMPLE 3   Use the commutative property of addition to rewrite each sum.

(a) $8 + 2$ **(b)** $7 + n$ **(c)** $x + 3$

Solution

(a) $8 + 2 = 2 + 8$ **(b)** $7 + n = n + 7$ **(c)** $x + 3 = 3 + x$

Notice that we applied the commutative property of addition to the expressions with variables n and x. That is because variables represent numbers, even though they are unknown numbers.

Student Practice 3   Use the commutative property of addition to rewrite each sum.

(a) $x + 3$ **(b)** $9 + w$ **(c)** $4 + 0$

EXAMPLE 4   If $2566 + 159 = 2725$, then $159 + 2566 = \,?$

Solution

$159 + 2566 = 2725$ Why? The commutative property states that the order in which we add numbers doesn't affect the sum.

Student Practice 4   If $x + y = 6075$, then
$$y + x = \,?$$

To **simplify** an expression like $8 + 1 + x$, we find the sum of 8 and 1.

$$8 + 1 + x = 9 + x \ \text{ or } \ x + 9$$

Simplifying $8 + 1 + x$ is similar to rewriting the English phrase "8 plus 1 plus some number" as the simpler phrase "9 plus some number." Since addition is commutative, we can write this simplification as either $9 + x$ or $x + 9$. We choose to write this sum as $x + 9$, since it is standard to *write the variable in the expression first*.

EXAMPLE 5   Simplify. $3 + 2 + n$

Solution   To simplify, we find the sum of the known numbers.
$$3 + 2 + n = 5 + n \text{ or } n + 5$$

We cannot add the variable n and the number 5 because n represents an unknown quantity; we have no way of knowing what quantity to add to the number 5.

Student Practice 5   Simplify. $6 + 3 + x$

Addition of more than two numbers may be performed in more than one manner. To add $5 + 2 + 1$ we can first add the 5 and 2, or we can add the 2 and 1 first. We indicate which sum we add first by using parentheses. *We perform the operation inside the parentheses first*.

$$5 + 2 + 1 = (5 + 2) + 1 = 7 + 1 = 8$$
$$5 + 2 + 1 = 5 + (2 + 1) = 5 + 3 = 8$$

In both cases the order of the numbers 5, 2, and 1 remains unchanged and the sums are the same. This illustrates the **associative property of addition.**

ASSOCIATIVE PROPERTY OF ADDITION

$$(a + b) + c = a + (b + c) \qquad (4 + 9) + 1 = 4 + (9 + 1)$$
$$13 + 1 = 4 + 10$$
$$14 = 14$$

When we add three or more numbers, the addition may be grouped in any way.

EXAMPLE 6   Use the associative property of addition to rewrite the sum and then simplify. $(x + 3) + 6$

Solution

$$(x + 3) + 6 = x + (3 + 6) \quad \text{The associative property allows us to regroup.}$$
$$= x + 9 \qquad \text{Simplify: } 3 + 6 = 9.$$

Student Practice 6   Use the associative property of addition to rewrite the sum and then simplify. $(w + 1) + 4$

Sometimes we must use both the associative and commutative properties of addition to rewrite a sum and simplify. In other words, we can *change the order in which we add* (commutative property) and *regroup the addition* (associative property) to simplify an expression.

EXAMPLE 7   Use the associative and/or commutative property as necessary to simplify the expression. $5 + (n + 7)$

Solution

$$5 + (n + 7) = 5 + (7 + n)$$ The commutative property allows us to change the order of addition.

$$= (5 + 7) + n$$ Regroup the sum using the associative property.

$$= 12 + n$$ Simplify.

$$5 + (n + 7) = n + 12$$ Write $12 + n$ as $n + 12$.

Student Practice 7   Use the associative and/or commutative property as necessary to simplify each expression.

(a) $(2 + x) + 8$ **(b)** $(4 + x + 3) + 1$

Understanding the Concept

Addition Facts Made Simple   There are many methods that can be used to add one-digit numbers. For example, if you can't remember that $7 + 8 = 15$ but can remember that $7 + 7 = 14$, just add 1 to 14 to get 15.

$$7 + 8 = 7 + (7 + 1)$$
$$= (7 + 7) + 1$$
$$= 14 + 1 = 15$$

Another quick way to add is to use the sum $5 + 5 = 10$, since it is easy to remember. Let's use this to add $7 + 5$.

$$7 + 5 = (2 + 5) + 5$$
$$= 2 + (5 + 5)$$
$$= 2 + 10 = 12$$

Exercises

1. Use the fact that $5 + 5 = 10$ to add $8 + 5$.
2. Use the fact that $6 + 6 = 12$ to add $6 + 8$.

③ Evaluating Algebraic Expressions Involving Addition

We have already learned that when we do not know the value of a number, we designate the number by a letter. We call this letter a *variable*. We use a variable to represent an unknown number until such time as its value can be determined. For example, if 6 is added to a number but we do not know the number, we could write

$$n + 6 \quad \text{where } n \text{ is the unknown number.}$$

If we were told that n has the value 9, we could *replace n* with 9 and then simplify.

$$n + 6$$
$$9 + 6 \quad \text{Replace } n \text{ with 9.}$$
$$15 \quad \text{Simplify by adding.}$$

Thus $n + 6$ has the value 15 when n is replaced by 9. This is called evaluating the expression $n + 6$ if n is equal to 9.

> To **evaluate** an algebraic expression, we replace the variables in the expression with their corresponding values and simplify.

An algebraic expression has different values depending on the values we use to replace the variable.

EXAMPLE 8   Evaluate $x + y + 3$ for the given values of x and y.

(a) x is equal to 6 and y is equal to 1   **(b)** x is equal to 4 and y is equal to 2

Solution

(a) $x + y + 3$ ⎫ Replace x with
$\quad 6 + 1 + 3$ ⎭ 6 and y with 1.

$\qquad 10 \qquad$ Simplify.

When x is equal to 6 and y is equal to 1, $x + y + 3$ is equal to 10.

(b) $x + y + 3$ ⎫ Replace x with
$\quad 4 + 2 + 3$ ⎭ 4 and y with 2.

$\qquad 9 \qquad$ Simplify.

When x is equal to 4 and y is equal to 2, $x + y + 3$ is equal to 9.

Student Practice 8   Evaluate $x + y + 6$ for the given values of x and y.

(a) x is equal to 9 and y is equal to 3   **(b)** x is equal to 1 and y is equal to 7

④ Adding Whole Numbers When Carrying Is Needed

Of course, we are often required to add numbers that have more than a single digit. In such cases we must:

1. Arrange the numbers vertically, lining up the digits according to place value.

2. Add first the digits in the ones column, then the digits in the tens column, then those in the hundreds column, and so on, moving from *right to left*.

Sometimes the sum of a column is a several-digit number—that is, a number larger than 9. When this happens we evaluate the place values of the digits to find the sum.

EXAMPLE 9   Add. $68 + 25$

Solution   We arrange numbers vertically and add the digits in the ones column first, then the digits in the tens column.

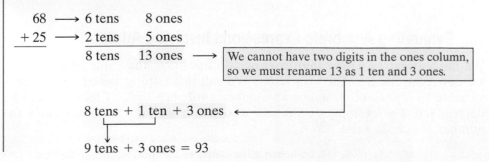

A shorter way to do this problem involves a process called "carrying." Instead of rewriting 13 ones as *1 ten and 3 ones*, we would carry the *1 ten* to the tens column by placing a 1 above the 6 and writing the 3 in the ones column of the sum.

$$\begin{array}{r} \overset{1}{6}8 \\ +\,25 \\ \hline 3 \end{array}$$ 8 ones + 5 ones = 13 ones

$$\begin{array}{r} \overset{1}{6}8 \\ +\,25 \\ \hline 93 \end{array}$$ Add 1 ten + 6 tens + 2 tens.

Student Practice 9   Add. 247 + 38

Often you must *carry* several times, by bringing the left digit into the next column to the left.

EXAMPLE 10   A market research company surveyed 1870 people to determine the type of beverage they order most often at a restaurant. The results of the survey are shown in the table. Find the total number of people whose responses were iced tea, soda, or coffee.

Solution   We add whenever we must find the "total" amount.

$$\begin{array}{r} \overset{21}{3}57 \quad \text{Iced tea} \\ 577 \quad \text{Soda} \\ +\,84 \quad \text{Coffee} \\ \hline 1018 \end{array}$$

We add 7 + 7 + 4 = 18. Since 18 equals 1 ten and 8 ones, we carry 1 ten placing a 1 at the top of the tens column.

We add 1 + 5 + 7 + 8 = 21. Since 21 tens equals 2 hundreds and 1 ten, we carry 2 hundreds placing a 2 at the top of the hundreds column.

We add 2 + 3 + 5 = 10. Since 10 hundreds equals 1 thousand and zero hundreds, we write 0 in the hundreds column and 1 in the thousands column.

$$357 + 577 + 84 = 1018$$

A total of 1018 people responded iced tea, soda, or coffee.

Type of Beverage	Number of Responses
Soda	577
Orange juice	475
Coffee	84
Iced tea	357
Milk	286
Other	91

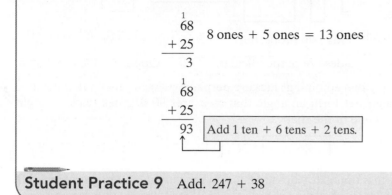

Student Practice 10   Use the survey results from Example 10 to answer the following: Find the total number of people whose responses were milk, orange juice, or other.

⑤ Finding the Perimeters of Geometric Figures

Geometry has a visual aspect that many students find helpful to their learning. Numbers and abstract quantities may be hard to visualize, but we can take pen in hand and draw a picture of a rectangle that represents a room with certain dimensions. We can easily visualize problems such as "What is the distance around the outside edges of the room (perimeter)?" In this section we study rectangles, squares, triangles, and other complex shapes that are made up of these figures.

A **rectangle** is a four-sided figure like the ones shown here.

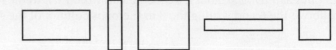

A rectangle has the following two properties:

1. Any two adjoining sides are perpendicular.   **2.**   Opposite sides are equal.

When we say that any two adjoining sides are **perpendicular,** we mean that any two sides that join at a corner form an angle that measures 90 degrees (called a **right angle**) and thus form one of the shapes.

When we say that opposite sides are equal, we mean that the measure of a side is equal to the measure of the side across from it. When all sides of a rectangle are the same length, we call the rectangle a **square.**

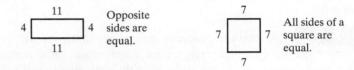

A **triangle** is a three-sided figure with three angles.

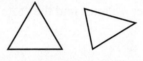

The distance around an object (such as a rectangle or triangle) is called the **perimeter**. To find the perimeter of an object, add the lengths of all its sides.

EXAMPLE 11   Find the perimeter of the triangle. (The abbreviation "ft" means feet.)

Solution   We add the lengths of the sides to find the perimeter.

$$5 \text{ ft} + 5 \text{ ft} + 7 \text{ ft} = 17 \text{ ft} \quad \text{The perimeter is 17 ft.}$$

Student Practice 11

Student Practice 11   Find the perimeter of the square.

If you are unfamilar with the value, meaning, and abbreviations for the metric and U.S. units of measure, refer to Appendix B, which contains a brief summary of this information.

EXAMPLE 12   Find the perimeter of the shape consisting of a rectangle and a square.

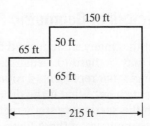

Solution   We want to find the distance around the figure. We look only at the *outside* edges since dashed lines indicate *inside* lengths.

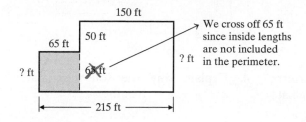

Now we must find the lengths of the unlabeled sides. The shaded figure is a square since the length and width have the same measure. Thus each side of the shaded figure has a measure of 65 ft.

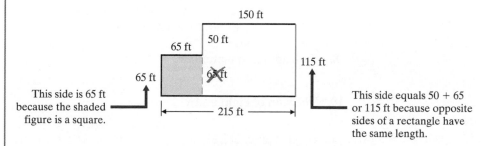

This side is 65 ft because the shaded figure is a square.

This side equals 50 + 65 or 115 ft because opposite sides of a rectangle have the same length.

Next, we add the length of the six sides to find the perimeter.

$$150 \text{ ft} + 115 \text{ ft} + 215 \text{ ft} + 65 \text{ ft} + 65 \text{ ft} + 50 \text{ ft} = 660 \text{ ft}$$

The perimeter is 660 ft.

Student Practice 12   Find the perimeter of the shape consisting of a rectangle and a square.

Student Practice 12

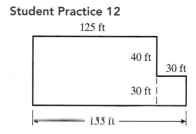

Understanding the Concept

Using Inductive Reasoning to Reach a Conclusion   When we reach a conclusion based on specific observations, we are using **inductive reasoning.** Much of our early learning is based on simple cases of inductive reasoning. If a child touches a hot stove or other appliance several times and each time gets burned, he is likely to conclude, "If I touch something that is hot, I will get burned." This is inductive reasoning. The child has thought about several actions and their outcomes and has made a conclusion or generalization.

The following is an illustration of how we use inductive reasoning in mathematics.

Find the next number in the sequence 10, 13, 16, 19, 22, 25, 28, . . .

We observe a pattern that each number is 3 more than the preceding number: $10 + 3 = 13; 13 + 3 = 16$, and so on. Therefore, if we add 3 to 28, we conclude that the next number in the sequence is 31.

Exercise

1. For each of the following find the next number by identifying the pattern.
   (a) 8, 14, 20, 26, 32, 38, . . .
   (b) 17, 28, 39, 50, 61, . . .

For more practice, complete exercises 89–94 on page 21.

Verbal and Writing Skills, Exercises 1–6

Write in words.

1. $10 + x$

2. $n + 4$

3. Write in your *own words* the steps you must perform to find the answer to the following problem. Evaluate $x + 6$ if x is equal to 9.

4. Explain why the following statement is true. If $x + y + z = 105$, then $z + y + x = 105$.

State what property is represented in each mathematical statement.

5. $(2 + 3) + 4 = 2 + (3 + 4)$

6. $4 + (x + 3) = 4 + (3 + x)$

Translate using numbers and symbols.

7. A number plus two

8. Two added to a number

9. The sum of five and y

10. The sum of eight and x

11. Some number added to twelve

12. Twelve more than a number

13. A number increased by seven

14. A number plus four

Use the commutative property of addition to rewrite each sum.

15. $5 + a$

16. $y + 6$

17. $3 + x$

18. $5 + x$

19. If $3542 + 216 = 3758$, then $216 + 3542 = ?$

20. If $8790 + 157 = 8947$, then $157 + 8790 = ?$

21. If $5 + n = 12$, then $n + 5 = ?$

22. If $8 + x = 31$, then $x + 8 = ?$

Simplify.

23. $x + 4 + 2$

24. $a + 6 + 2$

25. $9 + 3 + n$

26. $4 + 4 + y$

27. $x + 0 + 2$

28. $x + 3 + 0$

Use the associative property of addition to rewrite each sum, then simplify.

29. $(x + 2) + 1$

30. $(x + 5) + 1$

31. $9 + (3 + n)$

32. $3 + (4 + x)$

33. $(n + 3) + 8$

34. $(a + 3) + 7$

Use the associative and/or commutative property as necessary to simplify each expression.

35. $(x + 4) + 11$

36. $(y + 1) + 4$

37. $(2 + n) + 5$

38. $(4 + x) + 5$

39. $8 + (1 + x)$

40. $5 + (3 + a)$

41. $2 + (3 + n) + 4$

42. $3 + (n + 2) + 1$

43. $(3 + a + 2) + 8$

44. $(6 + x + 4) + 4$

45. $(5 + x + 7) + 4$

46. $(2 + n + 8) + 5$

47. Evaluate $y + 7$ for the given values of y.
  (a) y is equal to 3
  (b) y is equal to 8

48. Evaluate $n + 8$ for the given values of n.
  (a) n is equal to 4
  (b) n is equal to 7

49. Evaluate $x + y$ if x is 6 and y is 13.

50. Evaluate $a + b$ if a is 5 and b is 10.

51. Evaluate $a + b + c$ if a is 9, b is 15, and c is 12.

52. Evaluate $x + y + z$ if x is 11, y is 18, and z is 15.

53. Evaluate $n + m + 13$ if n is 26 and m is 44.

54. Evaluate $x + y + 21$ if x is 33 and y is 43.

Yearly Bonus Pay For exercises 55 and 56, use the table and the formula **Bonus $= x + y + 250$** to calculate the yearly bonus for MJ Industry employees.

Bonus $= x + y + 250$

x represents the number of productivity units earned.

y represents the number of years of employment.

Employee Name	Employee Number	Years of Employment	Productivity Units Earned
Julio Sanchez	00315	15	150
Mary McCab	00316	12	180
Jamal March	00317	18	125
Leo J. Cornell	00318	10	175

55. Calculate the yearly bonus for
  (a) Mary McCab.
  (b) Leo J. Cornell.

56. Calculate the yearly bonus for
  (a) Julio Sanchez.
  (b) Jamal March.

Add. For more practice, refer to Appendix D.

57. 15
      $+\,23$

58. 71
      $+\,12$

59. 236
      $+\,43$

60. 331
      $+\,57$

61. 32
      11
      20
      $+\,7$

62. 33
      11
      6
      $+\,4$

63. 105
      8
      133
      $+\,98$

64. 308
      7
      245
      $+\,75$

65. $236 + 467 + 26$

66. $531 + 217 + 18$

67. $281 + 64 + 539$

68. $562 + 65 + 133$

69. $7287 + 732 + 423$

70. $3366 + 152 + 485$

71. $922{,}876 + 54 + 1287 + 5000$

72. $836{,}147 + 99 + 2413 + 4000$

73. $3107 + 9063 + 54 + 379{,}626$

74. $2902 + 9050 + 12 + 986{,}100$

Applications, Exercises 75–78 *Answer each question.*

75. *Checking Account* Angelica's check register indicates the deposits and debits (checks written or ATM withdrawals) for a 1-month period.

Date	Deposits	Debits
12/3/09	$159	
12/9/09		$63
12/13/09	$241	
12/15/09		$121
12/22/09		$44

(a) What is the total of the deposits made to Angelica's checking account?

(b) What is the total of the debits made to Angelica's checking account?

76. *Checking Account* The bookkeeper for the Spaulding Appliance Company examined the following record from the company account for the month of March.

Date	Deposits	Debits
3/6/08	$3477	
3/9/08		$120
3/13/08		$3500
3/15/08	$4614	
3/22/08		$1388

(a) What is the total of the deposits in this time period?

(b) What is the total of the debits in this time period?

77. *Apartment Expenses* The rent on an apartment was $875 per month. To move in, Charles and Vincent were required to pay the first and last months' rent, a security deposit of $500, a connection fee with the utility company of $24, and a cable T.V. installation fee of $35. How much money did they need to move into the apartment?

78. *Car Expenses* Shawnee found that for a 6-month period, in addition to gasoline, she had the following car expenses: insurance, $562; repair to brakes, $276; and new tires, $142. If gasoline for her car cost $495 for this time period, what was the total amount she spent on her car?

Find the perimeter of each rectangle.

79.
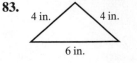
13 in.

5 in.

80.
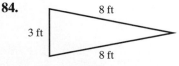
7 in.

1 in.

Find the perimeter of each square.

81.   3 ft

82.   8 ft

Find the perimeter of each triangle.

83.
4 in.   4 in.

6 in.

84.

8 ft

3 ft

8 ft

Find the perimeters of the shapes made of rectangles.

85.
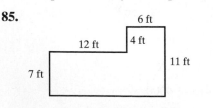
6 ft

12 ft   4 ft

11 ft

7 ft

86.
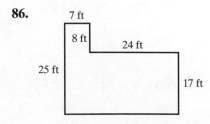
7 ft

8 ft   24 ft

25 ft

17 ft

87.

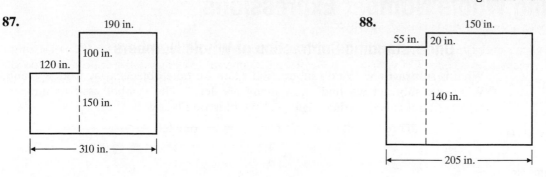

88.

To Think About *For each of the following, find the next number in the sequence by identifying the pattern.*

89. 1, 3, 5, 7, 9, 11, 13,…

90. 2, 4, 6, 8, 10, 12,…

91. 0, 5, 10, 15, 20, 25,…

92. 24, 31, 38, 45, 52, 59, 66,…

93. 7, 16, 25, 34, 43,…

94. 12, 25, 38, 51, 64,…

Quick Quiz 1.2

1. Use the associative and/or commutative property as necessary to simplify each expression.

　(a) $(4 + a) + 9$

　(b) $2 + (1 + x + 7)$

3. Find the perimeter of the shape consisting of a rectangle and a square.

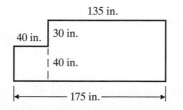

2. Evaluate $m + n + 13$ if m is 25 and n is 8.

4. Concept Check

$$\begin{array}{r} \overset{11}{395} \\ +\ 28 \\ \hline 423 \end{array}$$

　(a) When we carry, what is the value of the 1 that is placed above the 9?

　(b) When we carry, what is the value of the 1 that is placed above the 3?

1.3 Subtracting Whole Number Expressions

Student Learning Objectives

After studying this section, you will be able to:

① Understand subtraction of whole numbers.

② Use symbols and key words for expressing subtraction.

③ Evaluate algebraic expressions involving subtraction.

④ Subtract whole numbers with two or more digits.

⑤ Solve applied problems involving subtraction of whole numbers.

① Understanding Subtraction of Whole Numbers

What is **subtraction**? We do subtraction when we take objects away from a group. When we subtract we find "how many are left." The symbol used to indicate subtraction is called a **minus sign** "−." We illustrate below.

Three   take away   two   =   one left

$3 \quad - \quad 2 \quad = \quad 1$

$\square \, \boxtimes \, \boxtimes = \square \quad 3 - 2 = 1$

There are three parts to a subtraction problem: **minuend, subtrahend**, and **difference**.

$$3 \quad - \quad 2 \quad = \quad 1$$

minuend   subtrahend   difference

Subtraction is defined in terms of addition. Thus each subtraction sentence has a related addition sentence. For example, to find the value of $3 - 2$ we think of the number that when added to 2 gives 3.

$3 - 2 = 1$ Subtraction sentence

$\square \, \boxtimes \, \boxtimes = \square$

$3 = 1 + 2$ Related addition sentence

$\square \, \square \, \square = \square + \square \, \square$

Therefore we can use related addition sentences to help with subtraction. To subtract $12 - 8 = ?$, we can think $12 = ? + 8$.

EXAMPLE 1   Subtract.

(a) $9 - 5$ **(b)** $7 - 2$ **(c)** $15 - 0$ **(d)** $15 - 15$

Solution

(a) $9 - 5 = 4$ **(b)** $7 - 2 = 5$ **(c)** $15 - 0 = 15$ **(d)** $15 - 15 = 0$

Student Practice 1   Subtract.

(a) $5 - 2$ **(b)** $6 - 3$ **(c)** $18 - 0$ **(d)** $18 - 18$

NOTE TO STUDENT: *Fully worked-out solutions to all of the Student Practice problems can be found at the back of the text starting at page SP-1.*

Observe the pattern in the following subtraction problems: $6 - 0 = 6$; $6 - 1 = 5$; $6 - 2 = 4$; $6 - 3 = 3$. Each time you subtract the next larger whole number, the result decreases by 1. We can use this subtraction pattern to subtract mentally.

EXAMPLE 2 $800 - 50 = 750$. Use this fact to find $800 - 53$.

Solution   Since we know $800 - 50 = 750$, we can use subtraction patterns to find $800 - 53$.

Increase numbers in this column by 1.		Decrease numbers in this column by 1.
	$800 - 50 = 750$	
	$800 - 51 = 749$	
	$800 - 52 = 748$	
	$800 - 53 = 747$	

Student Practice 2 $600 - 50 = 550$. Use this fact to find $600 - 54$.

② Using Symbols and Key Words for Expressing Subtraction

There are several English phrases to describe the operation of subtraction. The following table presents some English phrases and their translated equivalents written using mathematical symbols.

English Phrase	Translation into Symbols
The *difference* of three and x	$3 - x$
Eight *minus* a number	$8 - n$
Two *subtracted from* seven	$7 - 2$
A number *decreased by* four	$n - 4$
Five *less than* nine	$9 - 5$

CAUTION: Math symbols are not always written in the same order as the words in the English phrase. Notice that when we translate the phrases "less than" or "subtracted from," the math symbols are not written in the same order as they are read in the statement.

EXAMPLE 3   Translate using numbers and symbols.

(a) The difference between five and x **(b)** Four less than seven

Solution

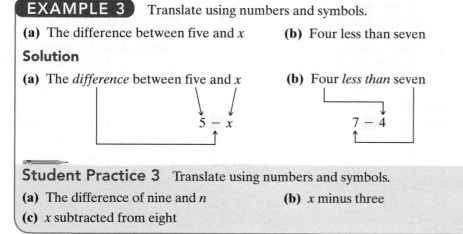

(a) The *difference* between five and x

$$5 - x$$

(b) Four *less than* seven

$$7 - 4$$

Student Practice 3   Translate using numbers and symbols.

(a) The difference of nine and n **(b)** x minus three

(c) x subtracted from eight

CAUTION: Note that the symbol $<$ means "is less than" while the symbol means "less than." Therefore in Example 3b we use the minus symbol $-$, not the inequality symbol $<$.

When we use the phrases "less than" and "subtracted from," the order in which we write the numbers in the subtraction is *reversed*. It is important to write these numbers in the correct order because, in general, *subtraction is not commutative*. In other words, $30 - 20$ is not the same as $20 - 30$. To show this, let's see what happens when we change the order of the numbers in subtraction.

$30 - $20   You have $30 in your checking account and write a check for $20; your balance will be $10.

$20 - $30   You have $20 in your checking account and write a check for $30; you will be overdrawn!

Obviously, the results are *not* the same. We summarize as follows.

SUBTRACTION IS NOT COMMUTATIVE

If a and b are not the same number, then

  $a - b$ does not equal $b - a$. $30 - 20$ does not equal $20 - 30$.

---

③ Evaluating Algebraic Expressions Involving Subtraction

Recall from Section 1.2 that to evaluate an expression we replace the variables in the expression with the given values and simplify.

EXAMPLE 4   Evaluate $7 - x$ for the given values of x.

(a) x is equal to 2   **(b)** x is equal to 4

Solution

(a) $7 - x$
    $7 - 2$ Replace x with 2.
      5 Simplify.

When x is equal to 2,
$7 - x$ is equal to 5.

(b) $7 - x$
    $7 - 4$ Replace x with 4.
      3 Simplify.

When x is equal to 4, $7 - x$ is
equal to 3.

Student Practice 4   Evaluate $8 - n$ for the given values of n.

(a) n is equal to 3   **(b)** n is equal to 6

④ Subtracting Whole Numbers with Two or More Digits

Often, we cannot subtract mentally, especially if the numbers being subtracted involve more than one digit. In this case we follow the same procedure as we did in addition, except we subtract digits instead of adding them. Therefore, we must:

1. Arrange the numbers vertically.

2. Subtract the digits in the ones column first, then the digits in the tens column, then those in the hundreds column, and so on, moving from *right to left*.

Many times, however, a digit in the lower number (subtrahend) is greater than the digit in the upper number (minuend) for a particular place value, as illustrated below.

$$\begin{array}{r} 7\ \boxed{2} \\ -3\ \boxed{8} \end{array} \quad 8 > 2$$

When this happens we must *rename* 72 using place values so we can subtract.

EXAMPLE 5   Subtract. $72 - 38$

Solution

We cannot subtract 8 ones from 2 ones, so we rewrite 7 tens as "tens and ones."

$$\begin{array}{r} 7\boxed{2} \\ -\ 3\boxed{8} \end{array} \qquad \begin{array}{rl} & \overset{6\ \text{tens}}{\cancel{7\ \text{tens}}} \quad \overset{10\ \text{ones}}{2\ \text{ones}} \\ - & 3\ \text{tens} \quad 8\ \text{ones} \end{array}$$

7 tens → 6 tens   2 ones → 10 ones   2 ones

$$\begin{array}{rl} 72 & \quad 6\ \text{tens} \quad 12\ \text{ones} \\ -\ 38 & -3\ \text{tens} \quad\ \ 8\ \text{ones} \\ \hline 34 & \quad 3\ \text{tens} \quad\ \ 4\ \text{ones} \end{array}$$

10 ones + 2 ones = 12 ones

12 ones − 8 ones = 4 ones;
6 tens − 3 tens = 3 tens

Thus $72 - 38 = 34$.

A shorter way to do this is called **borrowing.** Instead of rewriting 7 *tens* and 2 *ones* as 6 *tens* and 12 *ones*, we would borrow 1 ten from the 7 tens by crossing out the 7 and placing 6 above the 7. Then we would cross out the 2 and place 12 above the 2.

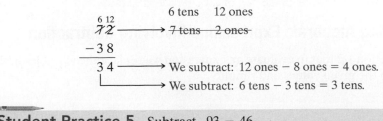

6 tens   12 ones
$\cancel{7\ \text{tens}}$ $\cancel{2\ \text{ones}}$

We subtract: 12 ones − 8 ones = 4 ones.
We subtract: 6 tens − 3 tens = 3 tens.

Student Practice 5   Subtract. $93 - 46$

Sometimes we cannot borrow from the digit directly to the left because this digit is 0. In this case we borrow from the next nonzero digit to the left of the 0, as illustrated in the next example.

EXAMPLE 6   Subtract. 304 − 146

Solution

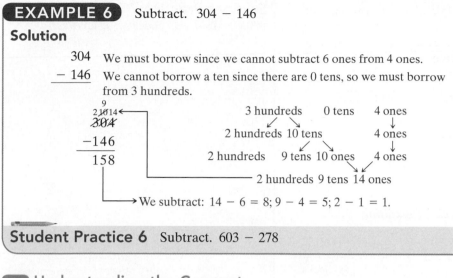

304   We must borrow since we cannot subtract 6 ones from 4 ones.

− 146   We cannot borrow a ten since there are 0 tens, so we must borrow from 3 hundreds.

3 hundreds	0 tens	4 ones
2 hundreds	10 tens	4 ones
2 hundreds	9 tens	10 ones	4 ones

2 hundreds 9 tens 14 ones

We subtract: 14 − 6 = 8; 9 − 4 = 5; 2 − 1 = 1.

$$\begin{array}{r} 304 \\ -146 \\ \hline 158 \end{array}$$

Student Practice 6   Subtract. 603 − 278

Understanding the Concept

Money and Borrowing   Converting money (changing $100 bills to $10 and $1 bills) illustrates the process of borrowing. To see this, let's look at the following: A cashier in a gift shop must give a customer $11 change for a purchase. Since the cashier is out of small bills and has only 3 hundred-dollar bills in the register, she must ask another cashier to convert a hundred-dollar bill to tens and ones.

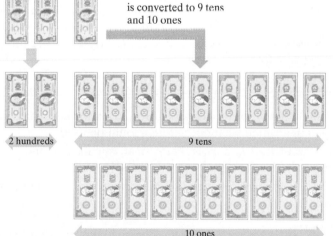

3 hundreds

1 hundred-dollar bill is converted to 9 tens and 10 ones

2 hundreds   9 tens

10 ones

The cashier now has 2 hundreds, 9 tens, and 10 ones and can give the customer $11 change.

$$\begin{array}{r} \$300 \\ -11 \end{array}$$

two hundreds   nine tens   ten ones

$$\begin{array}{r} 2\ 9\ 10 \\ 300 \\ -11 \\ \hline 289 \end{array}$$

Exercises

1. What happens when we must borrow from 0? That is, when subtracting 400 − 68, why must we change the middle 0 to 9, then borrow 1 from the first nonzero whole number to the left of it? Explain.

2. Explain why changing 1 ten-dollar bill to 10 one-dollar bills is similar to borrowing in subtraction.

We can check our subtraction problems using the related addition problems. For example, to check that $7 - 2 = 5$, we verify that $7 = 5 + 2$.

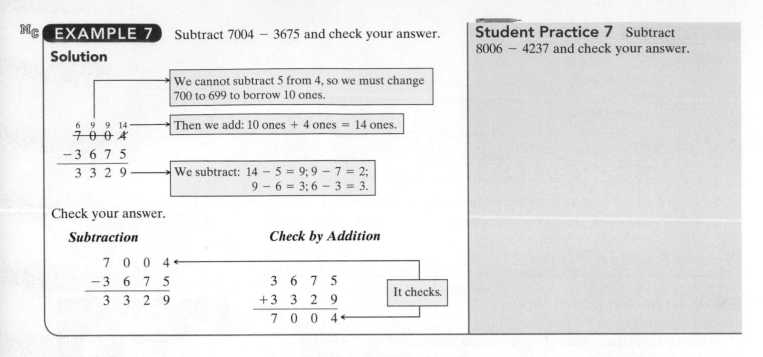

Mc EXAMPLE 7   Subtract $7004 - 3675$ and check your answer.

Solution

$$\begin{array}{r} \overset{6}{\cancel{7}}\ \overset{9}{\cancel{0}}\ \overset{9}{\cancel{0}}\ \overset{14}{\cancel{4}} \\ -3\ 6\ 7\ 5 \\ \hline 3\ 3\ 2\ 9 \end{array}$$

We cannot subtract 5 from 4, so we must change 700 to 699 to borrow 10 ones.

Then we add: 10 ones + 4 ones = 14 ones.

We subtract: $14 - 5 = 9$; $9 - 7 = 2$; $9 - 6 = 3$; $6 - 3 = 3$.

Check your answer.

Subtraction		*Check by Addition*

$$\begin{array}{r} 7\ 0\ 0\ 4 \\ -3\ 6\ 7\ 5 \\ \hline 3\ 3\ 2\ 9 \end{array} \qquad \begin{array}{r} 3\ 6\ 7\ 5 \\ +3\ 3\ 2\ 9 \\ \hline 7\ 0\ 0\ 4 \end{array}$$

It checks.

Student Practice 7   Subtract $8006 - 4237$ and check your answer.

⑤ Solving Applied Problems Involving Subtraction of Whole Numbers

Key words and phrases found in applied problems often help determine which operations should be used for computations. Subtraction is often used in real-life problems when we are comparing more than one amount. Often we want to know *how much more* or *how much less* one amount is than another. Subtraction is also necessary when we want to know *how much is left* or when the problem uses the key words or phrases for subtraction, such as *difference, minus, subtracted from, decreased by*, or *less than*. When we solve applied problems it is a good idea to use the following three steps in the problem-solving process.

Step 1. *Understand the problem.* Draw pictures. Look for key words and phrases to help you determine what operations should be used.

Step 2. *Calculate and state the answer.* Perform all calculations and answer the question asked in the problem.

Step 3. *Check your answer.* You may use a different method to find the answer, or you may estimate to see if your answer is reasonable.

EXAMPLE 8   Fish counts of calico bass caught out of a local sportfishing wharf on the last three days of May are given in the table.

	Number of Calico Bass Caught
May 29	232
May 30	311
May 31	133

How many more calico bass were caught off the wharf on May 30 than on May 31?

Solution *Understand the problem.* The key phrase "how many more" indicates that we subtract.

Calculate and state the answer. We subtract the number of calico bass caught on May 31 from the number of calico bass caught on May 30: 311 − 133

$$
\begin{array}{r}
\overset{\cancel{0}\;11}{\cancel{3}\,\cancel{1}\,\cancel{1}} \\
-1\,3\,3 \\
\hline
8
\end{array}
$$
→ We borrow 1 ten.

→ We cannot subtract 3 tens from 0 tens so we borrow again.

$$
\begin{array}{r}
\overset{2\;\;\;10}{\underset{0\;\;11}{\cancel{3}\,\cancel{1}\,\cancel{1}}} \\
-1\,3\,3 \\
\hline
1\,7\,8
\end{array}
$$
→ We borrow: write *3 hundreds* as *2 hundreds and 10 tens*.

Thus, 178 more fish were caught on May 30 than on May 31.

We leave the check to the student.

Student Practice 8   Use the information in Example 8 to answer the following question. How many fewer fish were caught off the wharf on May 31 than on May 29?

EXAMPLE 9   Find the perimeter of the shape consisting of rectangles.

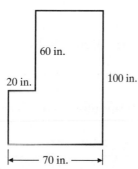

Solution To find the perimeter we must find the distance around the figure. Therefore, we must find the measures of the unlabeled sides.

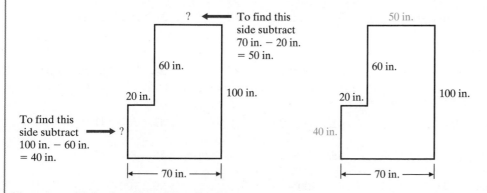

To find this side subtract
70 in. − 20 in.
= 50 in.

To find this side subtract
100 in. − 60 in.
= 40 in.

Next we add the lengths of the six sides.

50 in. + 100 in. + 70 in. + 40 in. + 20 in. + 60 in. = 340 in.

The perimeter is 340 in.

Student Practice 9   Find the perimeter of the shape consisting of rect-angles.

Student Practice 9

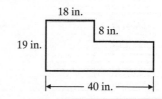

Verbal and Writing Skills, Exercises 1–4

Write using words.

1. $6 - x$

2. $10 - 2$

Fill in the blank.

3. The key phrase "how many more" indicates the operation _____.

Answer true or false.

4. The English phrase "five less than x" written using symbols is $5 - x$.

Subtract.

5. $7 - 4$ **6.** $7 - 5$ **7.** $6 - 2$ **8.** $8 - 4$

9. $9 - 3$ **10.** $9 - 6$ **11.** $8 - 7$ **12.** $6 - 5$

13. $15 - 0$ **14.** $14 - 0$ **15.** $20 - 20$ **16.** $12 - 12$

17. If $700 - 600 = 100$, find $700 - 603$ using subtraction patterns.

18. If $900 - 800 = 100$, find $900 - 806$ using subtraction patterns.

19. If $300 - 200 = 100$, find $300 - 205$ using subtraction patterns.

20. If $800 - 700 = 100$, find $800 - 705$ using subtraction patterns.

Translate using symbols.

21. Nine minus two

22. Three decreased by a number

23. The difference of eight and y

24. The difference of three and a number

25. Ten subtracted from seventeen

26. Seven subtracted from a number

27. A number decreased by one

28. Eight minus two

29. Two less than some number

30. Nine less than twelve

Evaluate $9 - n$ for the given values of n.

31. If n is equal to 4   **32.** If n is equal to 6   **33.** If n is equal to 9   **34.** If n is equal to 1

Evaluate $x - 2$ for the given values of x.

35. If x is equal to 9   **36.** If x is equal to 5   **37.** If x is equal to 3   **38.** If x is equal to 10

Subtract and check. For more practice, refer to Appendix D.

39. $97 - 35$ **40.** $98 - 25$ **41.** $56 - 23$ **42.** $76 - 41$

43. $83 - 67$ **44.** $57 - 38$ **45.** $72 - 18$ **46.** $73 - 35$

47. $873 - 195$ **48.** $764 - 545$ **49.** $500 - 43$ **50.** $700 - 29$

51. $8912 - 3847$ **52.** $8711 - 644$ **53.** $5301 - 185$ **54.** $8801 - 4583$

55. $\begin{array}{r} 15{,}107 \\ -\ 6{,}428 \end{array}$ **56.** $\begin{array}{r} 29{,}002 \\ -\ 3{,}667 \end{array}$ **57.** $\begin{array}{r} 164{,}300 \\ -\ 58{,}923 \end{array}$ **58.** $\begin{array}{r} 796{,}020 \\ -\ 68{,}431 \end{array}$

Find the perimeter of each shape consisting of rectangles.

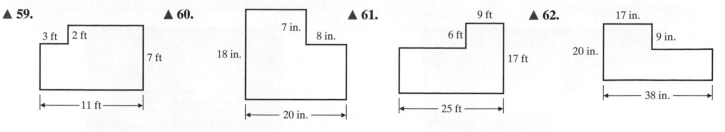

▲ **59.**   ▲ **60.**   ▲ **61.**   ▲ **62.**

Applications

Roller Coasters *When a roller coaster descends at high speeds, the force exerted on a rider's body by the roller coaster becomes less than that of gravity, producing a sensation of weightlessness. Then when the roller coaster hits the bottom and either shoots up or turns sharply, a g-force is exerted on the rider's body for a fraction of a second. This g-force can be stronger than the one felt by astronauts during a space shuttle launch.*

For exercises 63–66, refer to the bar graphs, which display the top speeds and maximum drops for some of the most popular roller coaster rides.

63. How much faster is the top speed of Superman the Escape than that of Goliath?

64. How much slower is the top speed of Ghostrider than that of Superman the Escape?

65. How much less is the maximum drop of Magnum XL-200 than that of Millennium Force?

66. How much greater is the maximum drop of Superman the Escape than that of Colossus?

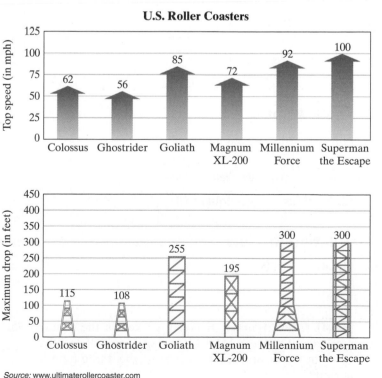

Source: www.ultimaterollercoaster.com

67. *Sun vs. Moon Diameter* The moon is about 400 times smaller than the sun.

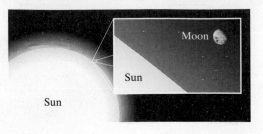

The diameter of the sun is approximately 865,000 miles, and the diameter of the moon is approximately 2160 miles. How many more miles is the diameter of the sun than that of the moon?

68. *Earth vs. Moon Diameter* If the moon were next to Earth it would be like a tennis ball next to a basketball.

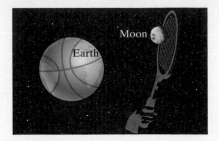

The approximate polar diameter of Earth (distance through Earth from North Pole to South Pole) is 7900 miles. The diameter of the moon is approximately 2160 miles. Find the difference in the diameters of Earth and the moon.

69. *Checking Account* Fill in the balances in Pedro's check register.

Check Number	Amount	Balance $1364
# 123	$238	
# 124	$137	
# 125	$ 69	
# 126	$ 98	
# 127	$369	

70. *Whale Population Decline* Although the International Whaling Commission has banned commercial whaling since 1987, several countries still hunt whales. As a result, the number of whales continues to decline, as shown in the following chart.

(a) Which species has had the largest decline in population?

(b) What is the decline in the total whale population?

Species	Approximate Population Years Earlier	Population in the World's Oceans 2000
Blue	275,000	5,000
Bowhead	60,000	8,500
Humpback	150,000	20,000

Source: Orange County Register

To Think About

71. For what value(s) of x and y will $x - y = y - x$?

Translate using symbols, then evaluate.

72. Eight minus y, if y is equal to 3

Cumulative Review *Replace each question mark with an inequality symbol.*

73. **[1.1.4]** 5,117,206 ? 13,842

74. **[1.1.4]** 2,386,702 ? 117,401

Add.

75. **[1.2.4]** *Hours Worked* Edward worked in the supermarket 120 hours in May, 135 in June, and 105 in July. How many hours did he work in the three-month period?

76. **[1.2.4]** *Pet Supply Purchases* Drew bought a dog for $430. He returned to the store the next day to purchase the following items for the dog: bed, $32; leash, $12; dog food, $28; and dog treats, $6. How much did Drew pay for the dog and all the supplies?

Quick Quiz 1.3

1. Translate the following using numbers and symbols.
   (a) Five subtracted from a number
   (b) A number decreased by 7
   (c) Eight less than a number

2. Subtract and check.
   (a) $14{,}062 - 7283$
   (b) $601{,}307 - 192{,}512$

3. Jose's salary was $2860 per month at his former job. His new job pays a salary of $3270 per month. How much more per month will Jose earn at his new job?

4. **Concept Check** Explain why when we subtract $800 - 35$, we change 8 to 7 in the borrowing process.

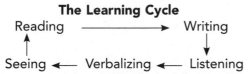 STEPS TO SUCCESS Class Attendance and the Learning Cycle

Did you know that an important part of the learning process happens in the classroom? People learn by *reading, writing, listening, verbalizing,* and *seeing.* These activities are all part of the **learning cycle** and always occur in class

- *Listening and seeing:* hearing and watching the instructor's lecture
- *Reading:* reading the information on the board and in handouts
- *Verbalizing:* asking questions and participating in class discussions
- *Writing:* taking notes and working problems assigned in class

The Learning Cycle

Reading $\longrightarrow$ Writing
$\uparrow$ $\downarrow$
Seeing $\longleftarrow$ Verbalizing $\longleftarrow$ Listening

Attendance in class completes the entire learning cycle once. Completing the following activates the entire learning cycle one more time.

Making it personal:

Place a ✓ *in the blank as you complete each of the following.*

Notebook for Class Notes

_____ 1. Shortly after class, read your class notes and the text for the section covered in class. Add any facts that you recall from class that you may have missed, or other facts from the text.

_____ 2. Rewrite your notes if they are too messy to understand.

_____ 3. Place a ? beside any notes from class that you don't understand, then ask a classmate or the instructor for clarification.

_____ 4. Place these notes in a notebook so you can refer to them when needed.

Keep in mind that you must pay attention and participate to learn. Just being there is not enough.

1.4 Multiplying Whole Number Expressions

Student Learning Objectives

After studying this section, you will be able to:

① Understand multiplication of whole numbers.

② Use symbols and key words for expressing multiplication.

③ Use multiplication properties to simplify numerical and algebraic expressions.

④ Multiply two several-digit numbers.

⑤ Solve applied problems involving multiplication of whole numbers.

① Understanding Multiplication of Whole Numbers

Multiplication of whole numbers can be thought of as repeated addition. For example, suppose that a small parking lot has 4 rows of parking spaces with 8 spaces in each row. How many parking spaces are in the lot?

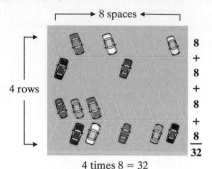

4 times 8 = 32

To get the total we add 8 four times, $8 + 8 + 8 + 8 = 32$, or we can use a shortcut: 4 rows of 8 is the same as 4 times 8, which equals 32. This is multiplication, a shortcut for repeated addition. When numbers are large, multiplication is easier than addition, but for smaller numbers, you can—if you are stuck—do a multiplication problem by working the equivalent addition problem.

The illustration of the parking lot is an example of an **array,** a rectangular figure that consists of rows and columns. Since the parking lot has 4 rows and 8 columns, it is a 4 by 8 array (always write the rows first). We can use dots, squares, or any figure to represent the elements of an array.

EXAMPLE 1 Draw two arrays that represent the multiplication 3 times 4.

Solution There are two arrays consisting of twelve items that represent the multiplication 3 times 4. One array has 4 rows and 3 columns, and the other one has 3 rows and 4 columns.

$$
\begin{array}{cc}
 & \star\ \star\ \star \\
\star\ \star\ \star\ \star & \star\ \star\ \star \\
3\ \star\ \star\ \star\ \star & 4\ \star\ \star\ \star \\
\star\ \star\ \star\ \star & \star\ \star\ \star \\
4 & 3 \\
\text{12 items} & \text{12 items}
\end{array}
$$

Student Practice 1 Draw two arrays that represent the multiplication 5 times 3.

NOTE TO STUDENT: Fully worked-out solutions to all of the Student Practice problems can be found at the back of the text starting at page SP-1.

It is often helpful to use arrays for real-life multiplication problems.

EXAMPLE 2 L&M's Print Shop makes business cards in 3 colors: white, beige, and light blue. The shop has 4 types of print to choose from: boldface, italic, fine line, and Roman.

(a) Set up an array that describes all possible business cards that can be made.

(b) Determine how many different types of cards can be made.

Solution

(a) We set up a 4 by 3 array where *each row corresponds to a type of print* and *each column corresponds to a color*. Each item in the array represents one possible business card.

	White	Beige	Light blue
Boldface	**Jesse Willettes** **Sales Manager** **(312) 123-5462**	**Jesse Willettes** **Sales Manager** **(312) 123-5462**	**Jesse Willettes** **Sales Manager** **(312) 123-5462**
Italic	*Jesse Willettes* *Sales Manager* *(312) 123-5462*	*Jesse Willettes* *Sales Manager* *(312) 123-5462*	*Jesse Willettes* *Sales Manager* *(312) 123-5462*
Fine line	Jesse Willettes Sales Manager (312) 123-5462	Jesse Willettes Sales Manager (312) 123-5462	Jesse Willettes Sales Manager (312) 123-5462
Roman	Jesse Willettes Sales Manager (312) 123-5462	Jesse Willettes Sales Manager (312) 123-5462	Jesse Willettes Sales Manager (312) 123-5462

(b) We have a 4 by 3 array that corresponds to the multiplication 4 times 3, or 12 business cards.

Student Practice 2 A manufacturer makes 3 different types of bikes: dirt, racer, and road. Each type comes in 5 different colors: red, blue, green, pink, and black.

(a) Set up an array that describes all possible bikes that can be made.

(b) Determine how many different bikes can be made.

② Using Symbols and Key Words for Expressing Multiplication

In mathematics there are several ways of indicating multiplication. We write the multiplication problem *4 times 5* as illustrated in the margin.

   If two variables a and b are multiplied, we indicate this by writing ab, with *no symbol between the a and b*. If a number is multiplied by a variable, we write the number first with *no symbol* between the number and the variable. Thus $6a$ indicates "six times a number."

   The numbers or variables we multiply are called **factors.** The *result* of the multiplication is called the **product.**

$$4(5) \qquad (4)5 \qquad 4 \cdot 5$$
$$4 \times 5 \qquad (4)(5) \qquad 4 * 5$$

ab means $a \cdot b$

$6a$ means $6 \cdot a$

$$\underset{\text{factor}}{6} \; \cdot \; \underset{\text{factor}}{8} \; = \; \underset{\text{product}}{48}$$

EXAMPLE 3   Identify the product and the factors in each equation.

(a) $5(4) = 20$ 　　　　　　　　　**(b)** $3x = 12$

Solution

(a) 5 and 4 are the factors and 20 is the product.

(b) 3 and x are the factors and 12 is the product.

Student Practice 3 Identify the product and the factors in each equation.

(a) $9 \cdot 7 = 63$ 　　　　　　　　　**(b)** $xy = z$

   The word *product* is also used to indicate the operation of multiplication. There are several other English phrases used to describe multiplication. The following table gives some English phrases and their translated equivalents written using mathematical symbols.

English Phrase	Translation into Symbols
The *product* of two and three	2(3) or $2 \cdot 3$
The *product* of x and y	xy
Six *times* a number	$6x$
Double a number	$2x$
Twice a number	$2x$
Triple a number	$3x$

EXAMPLE 4   Translate using numbers and symbols.

(a) The product of four and a number   (b) Triple a number

Solution

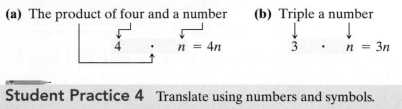

(a) The product of four and a number   (b) Triple a number

$$4 \cdot n = 4n \qquad 3 \cdot n = 3n$$

Student Practice 4   Translate using numbers and symbols.

(a) Double a number   (b) Two times a number

③ Using Multiplication Properties to Simplify Numerical and Algebraic Expressions

Like addition, multiplication is **commutative.** By this we mean that the order in which we multiply factors does not change the product. We use an array to illustrate this fact.

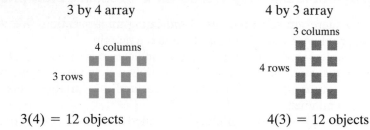

3 by 4 array   4 by 3 array

4 columns   3 columns

3 rows   4 rows

$3(4) = 12$ objects $4(3) = 12$ objects

Both arrays represent multiplication of 3 and 4; $3(4) = 12$ and $4(3) = 12$, illustrating that multiplication is commutative.

Multiplication is also **associative,** meaning that we can regroup the factors when multiplying and the product does not change.

We state these properties as follows.

COMMUTATIVE PROPERTY OF MULTIPLICATION $ab = ba$

Changing the order of factors does not $5(6) = 6(5)$
change the product. $30 = 30$

ASSOCIATIVE PROPERTY OF MULTIPLICATION $(ab)c = a(bc)$

Changing the grouping of factors does not $(7 \cdot 3) \cdot 2 = 7 \cdot (3 \cdot 2)$
change the product. $21(2) = 7(6)$
                                                    $42 = 42$

In addition to these properties there are two other properties of multiplication. The **identity property of 1** states that when any number is multiplied by 1, the product is that number: $a \cdot 1 = a; 2 \cdot 1 = 2$. The **multiplication property of 0** states that when any number is multiplied by 0, the product is 0: $a \cdot 0 = 0; 2 \cdot 0 = 0$.

We list a few other facts that can help us with multiplication.

1. Multiplying by 2 is the same as doubling a number.

2. Multiplying by 5 is the same as repeatedly adding 5, which is easy since all the numbers end with 0 or 5: 5, 10, 15, 20, 25, ...

3. Multiplying any number by 10 can be done simply by attaching a 0 to the end of that number.

$$3(10) = 30 \qquad 4(10) = 40 \qquad 5(10) = 50$$

We can use these properties and facts to make multiplication of several numbers easier.

EXAMPLE 5   Multiply. $4 \cdot 2 \cdot 4 \cdot 5$

Solution

$\begin{aligned} &= \quad 4 \cdot 2 \cdot 4 \cdot 5 &&\text{Use the commutative property to change the order of factors so} \\ &&&\text{that one factor is 10.} \end{aligned}$

$\begin{aligned} &= (4 \cdot 4) \cdot (2 \cdot 5) &&4 \cdot 4 = 16; \quad 2 \cdot 5 = 10 \\ &= \quad 16 \cdot 10 &&\text{To multiply 16(10), write 16 and attach a zero at the end.} \\ &= \quad 160 \end{aligned}$

Student Practice 5   Multiply.

(a) $2 \cdot 6 \cdot 0 \cdot 3$ **(b)** $2 \cdot 3 \cdot 1 \cdot 5$

We follow the same process with algebraic expressions.

EXAMPLE 6   Simplify. $2(3)(n \cdot 7)$

Solution   It may help to rewrite expressions using familiar notation: the multiplication symbol $\cdot$.

$\begin{aligned} 2(3)(n \cdot 7) &= 2 \cdot 3 \cdot (n \cdot 7) &&\text{Rewrite using familiar notation.} \\ &= 6 \cdot (n \cdot 7) &&\text{Multiply } 2 \cdot 3 = 6. \\ &= 6 \cdot (7 \cdot n) &&\text{Change the order of factors.} \\ &= (6 \cdot 7) \cdot n &&\text{Regroup.} \\ 2(3)(n \cdot 7) &= 42n &&\text{Multiply and write in standard notation: } 42 \cdot n = 42n. \end{aligned}$

Student Practice 6   Simplify.

(a) $4(x \cdot 3)$ **(b)** $2(4)(n \cdot 5)$

● Understanding the Concept

Memorizing Multiplication Facts   If we think of multiplication as repeated addition, very little memorization is needed to learn the multiplication facts. Once we know the 2, 5, and 10 times tables, which are fairly easy to learn, we can get the rest using the methods that follow.

For example, from the 5 times table we can get the 4 and 6 times tables as follows. To find 4(7) we think

$$5(7) - 7 \text{ is the same as } 4(7)$$

$$(7 + 7 + 7 + 7 + 7) \qquad 7 + 7 + 7 + 7 = 28$$

$$35 - 7 = 28$$

Continued on next page

$$\overset{\text{5(7) + 7 is the same as 6(7)}}{\underbrace{(7 + 7 + 7 + 7 + 7) + 7}\quad\underbrace{7 + 7 + 7 + 7 + 7 + 7}} = 42$$

$$35 + 7 = 42$$

Similarly, from the 10 times table we can get the 9 times table, and from the 2 times table we can get the 3 times table.

Exercise

1. Use the techniques discussed to find each product.

(a) 3(7)   (b) 4(8)

(c) 6(8)   (d) 9(8)

④ Multiplying Two Several-Digit Numbers

The numbers 10, 100, 200, and 2000 have **trailing zeros** (zeros at the end). We can multiply these numbers fairly easily. For example, to find 3 times 300 we use repeated addition: $300 + 300 + 300 = 900$. We see that to find 3(300) we need only *multiply the nonzero digits* (numbers that are not equal to zero) and attach the number of trailing zeros to the right side of the product.

EXAMPLE 7   Multiply. (547)(600)

Solution   Since the number 600 has trailing zeros, we multiply the nonzero digits and attach the trailing zeros to the right side of the product.

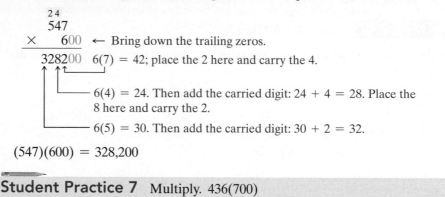

$$\begin{array}{r} \overset{2\,4}{547} \\ \times\quad 600 \\ \hline 328200 \end{array}$$

← Bring down the trailing zeros.

$6(7) = 42$; place the 2 here and carry the 4.

$6(4) = 24$. Then add the carried digit: $24 + 4 = 28$. Place the 8 here and carry the 2.

$6(5) = 30$. Then add the carried digit: $30 + 2 = 32$.

$(547)(600) = 328,200$

Student Practice 7   Multiply. 436(700)

How can we multiply numbers with several digits when there are no trailing zeros? Consider the multiplication $2 \cdot 23$. Recall that in expanded notation $23 = 20 + 3$ or $3 + 20$. Thus $2 \cdot 23 = 2(3 + 20)$. We can use the expanded notation to see how to multiply large numbers using a *condensed form*.

Expanded Notation Process   **Condensed Form**

$2 \cdot 23 = 2 \cdot (3 + 20)$ To multiply $2 \cdot (3 + 20)$, we   23

$\quad\quad = (3 + 20) + (3 + 20)$ can add $(3 + 20)$ twice. $\times\ \ 2$

$\quad\quad = (3 + 3) + (20 + 20)$ We regroup. 46 $2 \cdot 3 = 6$

$\quad\quad = 2 \cdot 3 + 2 \cdot 20$ $3 + 3 = 2 \cdot 3$; $20 + 20 = 2 \cdot 20$ $2 \cdot 20 = 40$

$\quad\quad = 6 + 40 = 46$

We see that we can multiply $2 \cdot 23$ simply by calculating $2 \cdot 3$ and $2 \cdot 20$ using the condensed form.

EXAMPLE 8   Multiply. 857(43)

Solution   To multiply 857(43), we multiply 857(3 + 40) or 857(3) + 857(40) using the condensed form.

$$
\begin{array}{r}
857 \\
\times\ 43 \\
\hline
2571 \\
34280 \\
\hline
36{,}851
\end{array}
$$

2571 ←———— Multiply: 3(857) = 2571.

34280 ←———— To find the product 40(857) = 34,280, we multiply 4(857) and add one trailing zero.

↑————— Add.

The products 2571 and 34,280 are called **partial products.**

Student Practice 8   Multiply. 936(38)

EXAMPLE 9   Multiply. 3679(102)

Solution

$$
\begin{array}{r}
3679 \\
\times\ 102 \\
\hline
7358 \\
00000 \\
367900 \\
\hline
375{,}258
\end{array}
$$

7358   Multiply: 2(3679).

00000   Multiply: 0(3679), and attach 1 trailing zero.

367900   Multiply: 100(3679), or 1(3679) and attach 2 trailing zeros.

375,258 Add.

(3679)(102) = 375,258

We can eliminate the trailing zeros in the partial products if we line up the partial products correctly.

$$
\begin{array}{r}
3679 \\
\times\ 102 \\
\hline
7358 \\
0000 \\
3679 \\
\hline
375{,}258
\end{array}
$$

7 3 5 8   Place the 8 under the 2.

00 0 0   Place the 0 under the 0.

367 9   Place the 9 under the 1.

375,2 5 8

Student Practice 9   Multiply. 203(4651)

⑤ Solving Applied Problems Involving Multiplication of Whole Numbers

One of the most important steps in solving a word problem is determining what operation(s) we must perform to find the answer. Applied problems that require the multiplication operation often state key words such as *times* and *product*, deal with *arrays* (rows and columns), or represent situations involving *repeated addition*. When reading a word problem, look for this information so that you can easily determine that you must perform the multiplication operation to solve the problem.

Remember to use the following three steps in the problem-solving process.

Step 1. *Understand the problem.*

Step 2. *Calculate and state the answer.*

Step 3. *Check your answer.*

EXAMPLE 10   Jessica drove an average speed of 60 miles per hour for 7 hours (per hour means each hour). How far did she drive?

Solution   *Understand the problem.* We draw a diagram and see that this is a situation that involves repeated addition, which indicates that we multiply.

⊢ 60 miles ⊣   ⊢ 60 miles ⊣   ⊢ 60 miles ⊣   and so on . . .

↓   ↓   ↓

1 hour   1 hour   1 hour   . . .

Calculate and state the answer.

$$\begin{array}{r} 60 \\ \times\ 7 \\ \hline 420 \end{array}$$ Miles driven each hour
    Number of hours driven
    Total miles driven

Check. From the diagram we can see that in 3 hours Jessica drove 180 miles $(60 + 60 + 60)$. Thus in 6 hours she drove 360 miles (180 miles + 180 miles). Now, since she drove 60 miles the seventh hour, we add $360 + 60 = 420$ miles.

Student Practice 10   Drew earns $9 per hour as a retail clerk. How much will he earn if he works 30 hours?

EXAMPLE 11   An apartment building is 4 stories high with 6 apartments on each floor. How many apartments are in the apartment building?

Solution   *Understand the problem.* We draw a picture and see that this situation deals with an array and thus requires that we multiply.

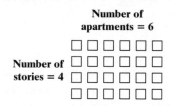

Calculate and state the answer. We have a 4 by 6 array. To find the total number of items in the array, we multiply $4 \cdot 6 = 24$.

There are 24 apartments in the building.

Check. We can use repeated addition and add 6 four times: $6 + 6 + 6 + 6 = 24$. We get the same result.

Student Practice 11   Allen is building a brick wall. The wall will be 12 bricks high and 30 bricks long. How many bricks will Allen need to build the wall?

Verbal and Writing Skills, Exercises 1–6

Translate the symbols into words.

1. **(a)** $4x$

   (b) ab

2. **(a)** $7y$

   (b) xy

Draw two arrays that represent each product.

3. 2 times 3

4. 4 times 2

State what property is represented in each mathematical statement.

5. $3(6 \cdot 5) = (3 \cdot 6)5$

6. $3(6 \cdot 5) = (6 \cdot 5) \cdot 3$

Fill in each box to complete each problem.

7. $3 \cdot 4(2y)$

   $= (3 \cdot 4 \cdot \boxed{}) \cdot y$

   $= \boxed{} y$

8. $4 \cdot 5(3x)$

   $= (4 \cdot 5 \cdot \boxed{}) \cdot x$

   $= \boxed{} x$

9. $(3a) \cdot 4 \cdot 2$

   $= 3 \cdot \boxed{} \cdot 4 \cdot 2$

   $= (3 \cdot 4 \cdot 2) \cdot \boxed{}$

   $= \boxed{} a$

10. $(4y) \cdot 3 \cdot 2$

    $= 4 \cdot \boxed{} \cdot 3 \cdot 2$

    $= (4 \cdot 3 \cdot 2) \cdot \boxed{}$

    $= \boxed{} y$

11. ***Shirt and Tie*** Anthony has 4 ties: brown, black, gray, and dark blue, and 3 shirts: white, pink, and blue.

    (a) Set up an array that shows all the possible outfits that Anthony can make.

    (b) How many different outfits are possible?

12. ***Carpet and Window Blinds*** Gerry has a choice of 4 carpet colors: beige, gray, blue, and light brown; and 3 colors of blinds: white, pale blue, and rose.

    (a) Set up an array that shows all the possible color combinations of carpet and blinds that Gerry can choose from.

    (b) How many different combinations are possible?

Ice Cream Toppings The Ice Cream Palace has 8 flavors of ice cream: vanilla, French vanilla, chocolate, strawberry, coffee, pecan, chocolate chip, and mint chip. There are 5 toppings for the ice cream: fudge, cherry, candy sprinkle, caramel, and nut.

13. How many different one-topping single-scoop ice cream dishes can you order?

14. If the Ice Cream Palace increases the number of flavors to 10, how many different one-topping ice cream dishes can you order?

Identify the factors and the product in each equation.

15. $6(3) = 18$

16. $4(7) = 28$

17. $22x = 88$

18. $7a = 49$

Translate using numbers and symbols.

19. Seven times a number

20. A number times five

21. Triple a number

22. Double a number

23. The product of six and a number

24. The product of a and b

Use what you have learned about the properties of multiplication to answer each question.

25. If $x \cdot y = 0$ and $x = 6$, then $y = ?$

26. If $a \cdot b = 0$ and $a = 10$, then $b = ?$

27. If $x(y \cdot z) = 40$, then $(x \cdot y)z = ?$

28. If $b(a \cdot c) = 30$, then $(a \cdot b) \cdot c = ?$

Multiply. See Example 5.

29. $(3)(6)(2)(5)$

30. $(4)(5)(2)(2)$

31. $(2)(3)(8)(5)$

32. $(5)(4)(3)(2)$

33. $2 \cdot 4 \cdot 6 \cdot 0$

34. $9 \cdot 0 \cdot 8 \cdot 6$

35. $4 \cdot 2 \cdot 4 \cdot 5$

36. $3 \cdot 2 \cdot 4 \cdot 5$

Simplify.

37. $8(6b)$

38. $7(5b)$

39. $5(z \cdot 8)$

40. $3(x \cdot 8)$

41. $8(a \cdot 7)$

42. $2(a \cdot 9)$

43. $2(7 \cdot c)$

44. $5(8 \cdot x)$

45. $9(2)(x \cdot 5)$

46. $2(3)(5 \cdot z)$

47. $9(2)(0 \cdot y)$

48. $0(7)(z \cdot 8)$

49. $6(3)(1 \cdot b)$

50. $4(7)(x \cdot 1)$

51. $2 \cdot 3(5y)$

52. $6 \cdot 4(3y)$

53. $(6x)3 \cdot 7$

54. $(4a)5 \cdot 2$

55. $3(5y) \cdot 6$

56. $4(3a) \cdot 5$

Multiply. For more practice, refer to Appendix D.

57. $9(637)$

58. $8(926)$

59. $7(602)$

60. $6(405)$

61. $398(300)$

62. $578(500)$

63. $793(600)$

64. $871(300)$

65. $\begin{array}{r} 76 \\ \times\, 68 \\ \hline \end{array}$

66. $\begin{array}{r} 81 \\ \times\, 34 \\ \hline \end{array}$

67. $\begin{array}{r} 32 \\ \times\, 59 \\ \hline \end{array}$

68. $\begin{array}{r} 44 \\ \times\, 68 \\ \hline \end{array}$

69. $\begin{array}{r} 847 \\ \times\, 56 \\ \hline \end{array}$

70. $\begin{array}{r} 668 \\ \times\, 95 \\ \hline \end{array}$

71. $\begin{array}{r} 455 \\ \times\, 86 \\ \hline \end{array}$

72. $\begin{array}{r} 322 \\ \times\, 74 \\ \hline \end{array}$

73. $354(702)$

74. $632(201)$

75. $409(432)$

76. $(201)631$

77. $8324(922)$

78. $4456(578)$

79. $3006(837)$

80. $9002(563)$

81. $12,107(808)$

82. $23,109(605)$

83. $61,711(1000)$

84. $86,246(2000)$

Applications

85. *Total Weekly Pay* A restaurant cook earns $8 per hour and works 40 hours per week. Calculate the cook's total pay for the week.

86. *Airplane Travel Distance* An airplane travels for 6 hours at an average speed of 450 miles per hour. How far does it travel?

87. *Orange Trees in a Grove* An orange grove has 15 rows of trees with 25 trees in each row. How many orange trees are in the grove?

88. *Flowers in a Garden* John plants 6 rows of plants in his garden. Each row contains 12 small plants. How many plants does he have?

89. *Spelling Books Purchased* East Gate Academy purchased 327 spelling workbooks at $12 per book. What was the total cost of the workbooks?

90. *Yards Rushed* A football player averages 116 yards per game rushing. At this average, how many rushing yards will be gained in a 9-game season?

91. *Hotel Curtain Purchase* A five-story hotel has 40 rooms on each floor. The owners are purchasing 25 boxes of curtains at a discount. If there are 10 sets of curtains in each box, can the owners replace one set of curtains in every room of the hotel? Why or why not?

92. *Tiles Needed* Robert will be laying tile on sections of both floors of a two-story store. He has determined that each floor will require 50 rows of tile with 35 tiles in each row. Robert ordered 46 boxes of tiles at a discount. If there are 75 tiles in each box, will Robert have enough tiles to complete the job? Why or why not?

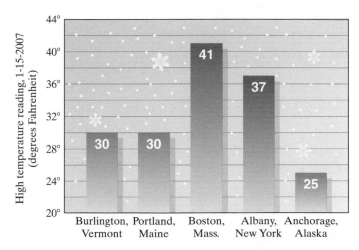

Temperatures in Various Cities *Use the bar graph to answer exercises 93 and 94.*

93. (a) What was the high temperature in Portland on January 15, 2007?

(b) If the high temperature reading in Honolulu, Hawaii, was two times the high temperature in Albany, New York, what was the high in Honolulu that day?

94. (a) What was the high temperature in Boston on January 15, 2007?

(b) On January 15, 2007, the high temperature reading in Buffalo, New York, was two times the high temperature in Burlington, Vermont. What was the high in Buffalo that day?

One Step Further *Simplify.*

95. $2a(4b)(5c)$

96. $(4x)(2y)(6z)$

97. $4(3a)(2b)(10c)$

98. $2(3x)(3y)(5z)$

99. $x(4y)7z$

100. $8a(5b)2c$

To Think About *Multiplication facts can be listed in a table such as shown here. For example, the product of 9 and 3 is placed where row 9 and column 3 meet.*

101. Fill in the multiplication table using the following step-by-step directions.

(a) Use the multiplication property of zero to fill in the second row. Now, use the commutative property to fill in the second column.

(b) Use the identity property of 1 to fill in the third row. Now, use the commutative property to fill in the third column.

(c) Complete the 2 times table: $2 \cdot 1, 2 \cdot 2, 2 \cdot 3$, and so on. Place the products in the fourth row. Now, use the commutative property to place the products in the fourth column.

(d) Complete the 5 times table: $5 \cdot 1, 5 \cdot 2, 5 \cdot 3$, and so on. Place the products in the seventh row. Now, use the commutative property to place the products in the seventh column.

(e) How many multiplication facts are blank in the table?

(f) Since the 0, 1, 2, and 5 times tables are fairly simple to learn, what does this process tell you about the amount of memorization necessary to learn all the multiplication facts?

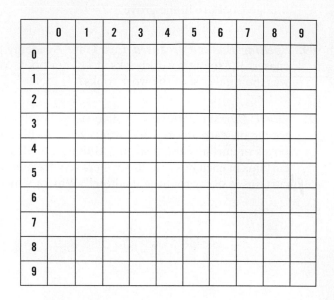

	0	1	2	3	4	5	6	7	8	9
0										
1										
2										
3										
4										
5										
6										
7										
8										
9										

Cumulative Review

102. **[1.2.4]** Add. $426,862 + 2128$

103. **[1.3.4]** Subtract. $7000 - 142$

104. **[1.1.5]** Round to the nearest thousand. 826,540

105. **[1.1.5]** Round to the nearest ten thousand. 168,406,000

106. **[1.3.5]** *Electric Bill* Julio's electric bill for April was $97. If he planned a budget that included electricity expenses of $120 a month, how much less was the bill than the budget allotment?

107. **[1.3.5]** *Distance Traveled* Mary Ann is planning to drive 920 miles to her sister's house over a two-day period. If she stays at a hotel 455 miles from her house the first night, how far must she drive the second day?

Quick Quiz 1.4

1. Translate using numbers and symbols. The product of six and a number

2. Multiply. $(1610)(105)$

3. Simplify by multiplying. $3a(2b)(5)$

4. **Concept Check** Explain what to do with the zeros when you multiply 546×2000.

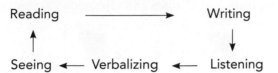 # STEPS TO SUCCESS Previewing New Material

Does the pace of the lecture seem too fast for you? Do you miss parts of the instructor's explanation? Previewing the new material can help you with these problems as well as enhance the amount of learning that happens while you are *listening* to the lecture.

The Learning Cycle

Part of your study time each day should consist of looking ahead to those sections in your text that are to be covered the following day. You do not necessarily have to learn the material on your own. Survey the concepts, terminology, diagrams, and examples, so that you are familiar with the new ideas when the instructor presents them.

To help yourself in class:

- Before the next class meeting read the section to be covered in class.

- Take note of concepts that appear confusing or difficult as you read.
- Listen carefully for your instructor's explanation of material that gave you difficulty.
- Be prepared to ask questions if you still do not understand.

Previewing new material enables you to see what is coming and prepares you to learn.

Making it personal: List below the questions you have or the concepts you do not understand.

1. _____

2. _____

3. _____

4. _____

1.5 Dividing Whole Number Expressions

Student Learning Objectives

After studying this section, you will be able to:

① Understand division of whole numbers.

② Use symbols and key words for expressing division.

③ Master basic division facts.

④ Perform long division with whole numbers.

⑤ Solve applied problems involving division of whole numbers.

When is division necessary to solve real-life problems? How do I divide whole numbers? Both these questions are answered in this section. It is just as important to know when a situation requires division as it is to know how to divide. Even if we use a calculator, we must know when the situation requires us to divide.

① Understanding Division of Whole Numbers

Suppose we wanted to display 12 roses in bouquets of 3. To determine the number of bouquets we can make, we count out 12 roses and repeatedly take out sets of 3.

$12 - 3 = 9$ 1 bouquet	$9 - 3 = 6$ 1 bouquet
$6 - 3 = 3$ 1 bouquet	$3 - 3 = 0$ 1 bouquet
4 bouquets can be made.	

By repeatedly subtracting 3, we found how many groups of 3 are in 12. In mathematics we express this as **division:**

$$12 \text{ divided by } 3 \text{ equals } 4.$$

The symbols used for division are $\overline{)}$, $\div$, $/$, $\underline{}$. We can write a division problem in any of the following ways:

$$3\overline{)12}^{\,4} \qquad 12 \div 3 = 4 \qquad 12/3 = 4 \qquad \frac{12}{3} = 4$$

$$3 \text{ divided into } 12 \text{ equals } 4 \qquad 12 \text{ divided by } 3 \text{ equals } 4$$

EXAMPLE 1 Write the division statement that corresponds to the following situation. You need not carry out the division.

180 chairs in an auditorium are arranged so that there are 12 chairs in each row. How many rows of chairs are there?

Solution We draw an array with 12 columns.

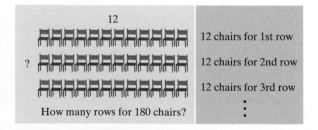

We want to know *how many groups of 12* are in 180.

The division statement that corresponds to this situation is $180 \div 12$.

NOTE TO STUDENT: Fully worked-out solutions to all of the Student Practice problems can be found at the back of the text starting at page SP-1.

Student Practice 1 Write the division statement that corresponds to the following situation. You need not carry out the division.

John has $150 to spend on paint that costs $15 per gallon. How many gallons of paint can John purchase?

We also divide when we want to split an amount equally into a certain number of parts. For example, if we split the 12 roses into 3 equal groups, how many roses would be in each group? There would be 4 roses in each group.

The division statement that represents this situation is

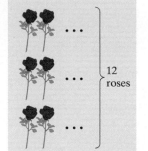

12 divided by 3 equals 4 or 12 ÷ 3 = 4.

EXAMPLE 2   Write the division statement that corresponds to the following situation. You need not carry out the division.

120 students in a band are marching in 5 rows. How many students are in each row?

Solution   We draw a picture. We want to *split 120 into 5 equal groups*.

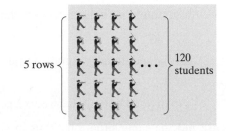

The division statement that corresponds to this situation is 120 ÷ 5.

Student Practice 2   Write the division statement that corresponds to the following situation. You need not carry out the division.

Rita would like to donate $170 to 5 charities, giving each charity an equal amount of money. How much money will each charity receive?

② Using Symbols and Key Words for Expressing Division

When referring to division we sometimes use the words *quotient, divisor,* and *dividend* to identify the three parts in a division problem.

$$\text{divisor}\overline{)\text{dividend}}^{\text{quotient}}$$

There are also several phrases to describe division. The following table gives some English phrases and their mathematical equivalents.

English Phrase	Translation into Symbols
n divided by six	$n \div 6$
The *quotient* of seven and thirty-five	$7 \div 35$
The *quotient* of thirty-five and seven	$35 \div 7$
Fifteen items *divided equally* among five groups	$15 \div 5$
Fifteen items *shared equally* among five groups	$15 \div 5$

EXAMPLE 3   Translate using numbers and symbols.

(a) The quotient of forty-six and two   **(b)** The quotient of two and forty-six

Solution

(a) The quotient of forty-six and two   **(b)** The quotient of two and forty-six

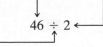

$$46 \div 2$$ $$2 \div 46$$

Continued on next page

Student Practice 3   Translate using symbols.

(a) The quotient of twenty-six and three

(b) The quotient of three and twenty-six

Understanding the Concept

The Commutative Property and Division   Example 3 illustrates that the order in which we write the numbers in the division is different when we use the phrases

"the quotient of **46** and **2**"   and   "the quotient of **2** and **46**."

$$46 \div 2 \qquad\qquad\qquad\qquad 2 \div 46$$

It is important to write these numbers in the correct order, as illustrated below.

The division statement: $2 \div 46$

The situation: \$2 divided equally among 46 people

The division statement: $46 \div 2$

The situation: \$46 divided equally between 2 people

We can see that these are **not** the same situations; thus in general, *division is not commutative:* $2 \div 46 \neq 46 \div 2$.

Exercise

1. Can you think of one case where $a \div b = b \div a$?

③ Mastering Basic Division Facts

By looking at rectangular arrays we can see how multiplication and division are related. Earlier we saw that the number of items in an array is equal to the *number of rows* × *the number of columns*. We can use this fact to find how many groups of 2 are in 6. That is, $6 \div 2 = $ **?**

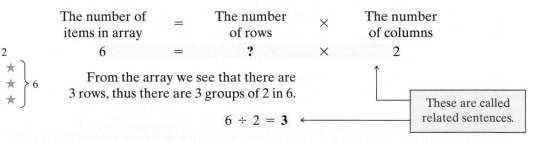

The number of items in array	=	The number of rows	×	The number of columns
6	=	**?**	×	2

From the array we see that there are 3 rows, thus there are 3 groups of 2 in 6.

$$6 \div 2 = 3$$

These are called related sentences.

We see that the answer to the division $6 \div 2$ is that number which when multiplied by 2 yields 6. We can use this fact when we divide.

To find $6 \div 2 = $ **?** , think $6 = $ **?** $\times 2$.

EXAMPLE 4   Divide. $18 \div 3$

Solution

$$18 \div 3 = ? \quad \text{Think, } 18 = ? \cdot 3.$$

$$18 \div 3 = 6 \quad 18 = 6 \cdot 3$$

Student Practice 4   Divide. $21 \div 7$

What about division by 0? Zero can be divided by any nonzero number, but division by zero is not possible. To see why, suppose that we could divide by zero. Then $7 \div 0 = $ *some number*. Let us represent "some number" by ? .

If	$7 \div 0 = ?$	
then	$7 = 0 \times ?$	The related multiplication sentence.
Which would mean	$7 = 0$	Since any number times 0 equals 0.

That is, we would obtain $7 = 0$, which we know is not true. Therefore, our assumption that $7 \div 0 = $ *some number* is wrong. Thus we conclude that we cannot divide by zero. We say division by 0 is **undefined.** It is helpful to remember the following basic concepts.

DIVISION PROBLEMS INVOLVING THE NUMBER 1 AND THE NUMBER 0

1. Any nonzero number divided by itself is 1 $\left(7 \div 7 = 1, \ \frac{7}{7} = 1, \text{ and } 7\overline{)7} \right)$.

2. Any number divided by 1 remains unchanged $\left(29 \div 1 = 29, \ \frac{29}{1} = 29, \text{ and } 1\overline{)29} \right)$.

3. Zero may be divided by any nonzero number; the result is always zero $\left(0 \div 4 = 0, \ \frac{0}{4} = 0, \text{ and } 4\overline{)0} \right)$.

4. Zero can never be the divisor in a division problem $\left(3 \div 0, \frac{3}{0}, \text{ and } 0\overline{)3} \text{ are } \textbf{undefined.} \ 0 \div 0, \frac{0}{0}, \text{ and } 0\overline{)0} \text{ are impossible to determine.} \right)$

EXAMPLE 5 Divide.

(a) $0 \div 9$ **(b)** $9 \div 0$ **(c)** $\dfrac{16}{16}$

Solution

(a) $0 \div 9 = 0$ 0 divided by any nonzero number is equal to 0.
(b) $9 \div 0$ Zero can never be the divisor in a division problem. $9 \div 0$ is undefined.
(c) $16 \div 16 = 1$ Any number divided by itself is 1.

Student Practice 5 Divide.

(a) $3 \div 3$ **(b)** $3 \div 0$ **(c)** $\dfrac{0}{3}$

---

④ Performing Long Division with Whole Numbers

Suppose that we want to split 17 items equally between 2 people.

| 8 items | 8 items | 1 item |

Each person would get 8 items with 1 *left over*. We call this 1 the **remainder** (R) and write $17 \div 2 = 8$ R1.

We use related multiplication sentences and the division symbol $\overline{)}$ when division involves large numbers, or remainders. For example,

$$\overset{?}{2)\overline{17}} \qquad 17 \div 2 = ?$$

$$\overset{?}{2)\overline{17}} \qquad \text{Think: } 2 \cdot ? = 17; \text{ two times what number is close to or equal to 17?}$$

$$\overset{8\ \text{R}}{2)\overline{17}} \qquad 2 \cdot 8 = 16, \text{ which is close to 17, so we have a remainder.}$$

$$\overset{8\ \text{R1}}{2)\overline{17}}$$
$$\underline{-16} \qquad \text{We subtract and get a remainder 1.}$$
$$1$$

Thus to divide, we *guess* the quotient and *check* by multiplying the quotient by the divisor. If the guess is too large or too small, we *adjust* it and continue the process until we get a remainder that is less than the divisor.

EXAMPLE 6 Divide and check your answer. $38 \div 6$

Solution We *guess* that 6×6 is close to 38.

$$\overset{6}{6)\overline{38}} \qquad \text{Our guess, 6, is placed here.}$$
$$\underline{-36} \qquad 6 \times 6 = 36; \textit{Check: } 36 \textit{ must be less than } 38.$$

Since $36 < 38$, we do not need to *adjust* our guess to a smaller number.

$$\overset{6\ \text{R2}}{6)\overline{38}}$$
$$\underline{-36} \qquad \text{We subtract: } 38 - 36 = 2.$$
$$2 \qquad \textit{Check: } 2 \textit{ must be less than } 6. \text{ We write R2 in the quotient.}$$

Since $2 < 6$, we do not need to *adjust* our guess to a larger number.

To verify that this is correct, we multiply the divisor by the quotient, then add the remainder:

$$\text{Multiply } 6 \times 6 = 36$$
$$\overset{6\ \text{R2}}{6)\overline{38}} \qquad \qquad \underline{+\ 2} \qquad \text{Then add the remainder.}$$
$$38$$
$$38 = 38 \ \checkmark$$

$$38 \div 6 = 6\ \text{R2}$$

Student Practice 6 Divide and check your answer. $43 \div 6$

Let's see what we do if our guess is either too large or too small.

EXAMPLE 7 Divide and check your answer. $293 \div 41$

Solution *First guess* (too large):

$$\overset{8}{41)\overline{293}} \qquad \textit{Guess: } 41 \text{ times what number is close to 293? } 8$$
$$\qquad \text{We write 8 in the quotient.}$$
$$\underline{-328} \qquad \textit{Check: } 41(8) = 328; \text{ Our guess is too large}$$
$$\qquad \text{so we must } \textit{adjust.}$$
$$\qquad \qquad \text{too large}$$

Second guess (too small):

$$\overset{6}{41)\overline{293}}$$
$$\text{too small} \qquad \underline{-246} \qquad \textit{Check: } 41(6) = 246; 246 \textit{ is less than } 293,$$
$$47 \qquad \text{but 47 is } \textbf{not} \text{ less than 41.}$$
$$\qquad \text{Our guess is too small so we must } \textit{adjust.}$$

Third guess:

$$7\text{ R}6$$
$$41\overline{)293}$$
$$-287$$
$$6$$

Guess: We try 7.

Check: $41(7) = 287$; 287 *is less than* 293, and 6 *is less than* 41. We *do not* need to *adjust* our guess, and 6 is the remainder. We write R6 in the quotient.

We verify that the answer is correct:

$$\begin{array}{ccccccc} (\text{divisor} & \cdot & \text{quotient}) & + & \text{remainder} & = & \text{dividend} \\ (41 & \cdot & 7) & + & 6 & = & 293 \end{array}$$

$$293 \div 41 = 7\text{ R}6$$

Student Practice 7 Divide and check your answer. $354 \div 36$

EXAMPLE 8 Divide and check your answer. $70\overline{)3672}$

Solution Accurate guesses can shorten the division process. If we consider only the *first digit of the divisor* and the *first two digits of the dividend,* it is easier to get accurate guesses.

First set of steps:

$$5$$
$$70\overline{)3672}$$
$$-350$$
$$17$$

Guess: We look at 7 and 36 to make our guess.

7 times what number is close to 36? 5

Check: $5(70) = 350$.

350 *is less than* 367, and

17 *is less than* 70.   We *do not adjust* our guess.

Second set of steps: We bring down the next number in the dividend: 2. Then we continue the guess, check, and adjust process until there are no more numbers in the dividend to bring down.

$$52\text{ R}32$$
$$70\overline{)3672}$$
$$-350$$
$$172$$
$$-140$$
$$32$$

Guess: We look at 7 and 17 to make our guess. We try 2.

Check: $2(70) = 140$; 140 *is less than* 172.

Check: 32 *is less than* 70.

32 is the remainder because there are no more numbers to bring down.

$$3672 \div 70 = 52\text{ R}32$$

Check:
$$\begin{array}{ccccccc} (\text{divisor} & \cdot & \text{quotient}) & + & \text{remainder} & = & \text{dividend} \\ (70 & \cdot & 52) & + & 32 & = & 3672. \end{array}$$

Student Practice 8 Divide and check your answer. $80\overline{)2611}$

McC **EXAMPLE 9** Divide and check your answer. $33{,}897 \div 56$

Solution First set of steps:

$$60$$
$$56\overline{)33897}$$
$$-336$$
$$29$$

Guess: We look at 5 and 33 to make our guess. We try 6.

Check: $6(56) = 336$; 336 *is less than* 338.

Check: 2 *is less than* 56.

We bring down the 9. Since 56 cannot be divided into 29, we write 0 in the quotient.

Student Practice 9 Divide and check your answer. $14{,}911 \div 37$

Continued on next page

Second set of steps: We bring down the 7.

$$
\begin{array}{r}
605\ \text{R}17 \\
56\overline{)33897} \\
-336 \\
\hline
297 \\
-280 \\
\hline
17
\end{array}
$$

Guess: We look at 5 and 29 to make our guess. We try 5.
Check: $5(56) = 280$; 280 *is less than* 297, and
17 *is less than* 56.

17 is the remainder because there are no more numbers to bring down.

CAUTION: In Example 9 we placed a zero in the quotient because 56 did not divide into 29. You must remember to place a zero in the quotient when this happens, otherwise you will get the wrong answer. There is a big difference between 65 and 605, so be careful.

⑤ Solving Applied Problems Involving Division of Whole Numbers

As we have seen, there are various key words, phrases, and situations that indicate when we must perform the division operation. Knowing these can help us solve real-life applications.

EXAMPLE 10   Twenty-six students in Ellis High School entered their class project in a contest sponsored by the Falls City Baseball Association. The class won first place and received 250 tickets to the baseball play-offs. The teacher gave each student in the class an equal number of tickets, then donated the extra tickets to a local boys and girls club. How many tickets were donated to the boys and girls club?

Solution   *Understand the problem.* Since we must split 250 equally among 26 students, we divide.

Calculate and state the answer.

$$
\begin{array}{r}
9\ \text{R}16 \\
26\overline{)250} \\
-234 \\
\hline
16
\end{array}
$$

Since there are 16 tickets left over, 16 tickets are donated to the boys and girls club.

Check: $(26 \cdot 9) + 16 = 250.$

Student Practice 10   Twenty-two players on a recreational basketball team won second place in a tournament sponsored by Meris and Mann 3DMax Movie Theater. The team won 100 movie passes and divided these passes equally among players on the team. The extra passes were donated to a local children's home. How many passes were donated to the children's home?

Understanding the Concept

Conclusions and Inductive Reasoning   In Section 1.2 we saw how to use inductive reasoning to find the next number in a sequence. How accurate is inductive reasoning? Do we always come to the right conclusion? Conclusions arrived at by inductive reasoning are always tentative. They may require further investigation to avoid reaching the wrong conclusion. For example, inductive reasoning can result in more than one probable next number in a list as illustrated next.

Identify 2 different patterns and find the next number for the following sequence: 1, 2, 4, . . .

Notice that $1 \cdot 2 = 2$ and $2 \cdot 2 = 4$. Using a pattern of multiplying the preceding number by 2, the next number is $4 \cdot 2 = 8$ and thus we have: 1, 2, 4, 8, . . .

For the second pattern we see that $1 + 1 = 2$ and $2 + 2 = 4$. Using a pattern of adding consecutive counting numbers, the next number is $4 + 3 = 7$ and thus we have 1, 2, 4, 7, . . .

To know for sure which answer is correct, we would need more information such as more numbers in the sequence to verify the pattern. You should always treat inductive reasoning conclusions as tentative, requiring further verification.

Exercise

1. Identify 2 different patterns and find the next number for the following sequence: 1, 1, 2, . . .

For more practice, complete exercises 55–62 on page 54.

👣 STEPS TO SUCCESS Time Management

Planning and organizing your schedule is an efficient, low-stress way to juggle school, work, family, and social activities. It allows you to set realistic goals and priorities. You will also be less likely to forget assignments or appointments. A time management schedule provides you with a road map to achieving your goals.

Making it personal:

1. Make a list of your daily activities.
2. Make a list of exam and assignment due dates.
3. Place this information in a weekly planner.
4. Plan your study time for each day. Since exam and assignment schedules vary, these activities may vary from day to day.

5. Leave space to insert last-minute things that come up or specific questions that you want to ask during class.
6. Review the planner periodically during the day so that you don't forget a task and can adjust for unexpected changes.

Think of your time management plan as a contract with yourself. You will find that by adhering to the contract, your grades will improve and you will have more free time.

Mon.	Tues.
7–9 A.M. Jogging and breakfast ⟶	
9 A.M. Preview math lecture material.	Prepare test review questions.
10 A.M. Math class ⟶	
	*Homework due
11 A.M. English class, questions for final draft of paper ⟶	
	*Term paper due
12 P.M. Lunch ⟶	
1 P.M. Review math class notes and begin homework.	Do practice test—math.
2:30–5:30 P.M. Work ⟶	
5:30–8 P.M. Dinner and social time ⟶	
8–10 P.M. Finish math homework, review term paper.	Review math and read history.

Verbal and Writing Skills, Exercises 1–4

Write the division statement that corresponds to each situation. You need not carry out the division.

1. 220 paintings are arranged in rows so that 4 paintings are in each row. How many rows of paintings are there?

2. In the school gym, 320 chairs must be arranged in rows with 16 chairs in each row. How many rows of chairs are there?

How many rows?   16 chairs

3. 225 tickets to the Dodgers' first game of the year will be distributed equally among *n* people.

4. A dinner bill totaling $*n* was split among 5 people.

5. For the division problem $15 \div 3$, which wording is correct? There may be more than one right answer.

  (a) 3 divided by 15   **(b)** 15 divided by 3

  (c) 3 divided into 15   **(d)** 15 divided into 3

6. For the division problem $24 \div 3$, which wording is correct? There may be more than one right answer.

  (a) 3 divided by 24   **(b)** 24 divided by 3

  (c) 3 divided into 24   **(d)** 24 divided into 3

Translate using numbers and symbols.

7. Twenty-seven divided by a number

8. Eight divided by a number

9. Forty-two dollars divided equally among six people

10. Sixty-three jelly beans divided equally among three children

11. The quotient of thirty-six and six

12. The quotient of forty-four and eleven

13. The quotient of three and thirty-six

14. The quotient of eleven and forty-four

Divide.

15. $42 \div 42$

16. $25 \div 25$

17. $\dfrac{0}{5}$

18. $\dfrac{0}{99}$

19. $17 \div 0$

20. $45 \div 0$

Divide and check your answer. Refer to Appendix D for more practice.

21. $58 \div 9$

22. $60 \div 9$

23. $7\overline{)2597}$

24. $6\overline{)3726}$

25. $3\overline{)1346}$

26. $6\overline{)4046}$

27. $\dfrac{1268}{30}$

28. $\dfrac{1863}{20}$

29. $30\overline{)632}$

30. $20\overline{)783}$

31. $19\overline{)5817}$

32. $32\overline{)6436}$

33. $\dfrac{1403}{29}$ **34.** $\dfrac{1301}{24}$ **35.** $1369 \div 19$ **36.** $1350 \div 16$

37. $18{,}985 \div 27$ **38.** $12{,}854 \div 42$ **39.** $11{,}571 \div 34$ **40.** $37{,}780 \div 118$

41. $113{,}317 \div 223$ **42.** $123{,}264 \div 136$ **43.** $70{,}141 \div 136$ **44.** $21{,}945 \div 29$

Applications

45. *Computer Conference Tickets* The 14 members of the Carver High School Chess Club team won first place in a tournament sponsored by the Carver Convention Center. The chess team won 60 tickets to the World-wide Computer Conference. The team decided to divide the tickets equally among all 14 team members and to donate the extra tickets to the PTA. How many tickets were donated to the PTA?

46. *Entertainment Event Tickets* The 21 members of the Laurel High School track team won first place in a tournament sponsored by the Laurel Recreation Center. The team won 75 tickets to the county fair. The team decided to divide the tickets equally among all 21 team members and to donate the extra tickets to the homeless shelter. How many tickets were donated to the shelter?

47. *Restaurant Bill* The bill for dinner, including tip, at Lido's Restaurant was $85. If 5 people split the bill evenly, how much did each person have to pay?

48. *Banquet Ticket Price* The members of the Elks Club are planning a banquet. The cost of the entire banquet will be $1105. If 65 members plan to attend, how much should the ticket price be to cover the cost of the banquet?

49. *Travel Allowance* JoAnn received a travel allowance of $1050 from her employer for food and lodging. If her business trip takes 6 days, how much money should she budget each day so that she will not go over her total travel allowance?

50. *Cow Pasture Capacity* A rancher plans to have 250 square feet of pasture for each cow on his field. If the area of the field is 156,250 square feet, how many cows should the rancher allow on the field?

51. *Photographing Deer* A photographer sets a telephoto lens so that she can be twice as far away from her subject as she would be with a regular lens. She is taking pictures of deer that are 124 feet from her camera. How far from the deer would the photographer have to be to get the same shot with a regular lens?

52. *Cross-Stitch Pattern* Janice is making a cross-stitch pattern on 14-count material. This means that there are 14 squares to the inch. If Janice's pattern is 98 squares across, how many inches wide will it be?

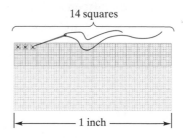

53. How many 41-cent stamps can be bought with 1300 cents?

54. A young toy-car collector has 218 miniature cars. He bought carrying cases to store these cars. Each carrying case holds 15 cars. He plans to give his younger brother any cars that won't fill up a case.

(a) How many cases can he fill completely?

(b) How many cars will he give to his brother?

To Think About *For each of the following, find the next number in the sequence by identifying the pattern.*

55. 5, 15, 45, 135, . . .

56. 4, 16, 64, 256, . . .

57. 3, 4, 7, 12, 19, 28, 39, . . .

58. 0, 2, 6, 12, 20, . . .

59. 7, 9, 10, 12, 13, 15, 16, . . .

60. 1, 6, 8, 13, 15, 20, . . .

Identify two patterns and find the next number for each of the following.

61. 0, 1, 4, . . .

62. 1, 4, 8, . . .

Complete each of the following.

63. **(a)** $(32 \div 4) \div 2$

    (b) $32 \div (4 \div 2)$

    (c) What can you say about division and the associative property?

64. **(a)** $(48 \div 6) \div 2$

    (b) $48 \div (6 \div 2)$

    (c) What can you say about division and the associative property?

Cumulative Review

65. **[1.2.1]** Translate into symbols. Seven plus x equals eleven.

66. **[1.3.4]** Subtract. $1060 - 114$

67. **[1.4.4]** Multiply. 4031 (202)

68. **[1.1.5]** Round 556,432 to the nearest thousand.

69. **[1.3.5]** *Distance Traveled* Leo wanted to make a 1389-mile trip in 3 days to visit his aunt. He drove 430 miles the first day and 495 miles the second day. How far does Leo have to drive the third day to reach his destination?

70. **[1.3.5]** *Truck Purchase Price* The total cost of the truck Ranak purchased, including tax and license, is $29,599. If the dealer gave Ranak $6200 for his car as a trade-in and Ranak put $5500 down, what is the balance owed on the truck?

Quick Quiz 1.5

1. Translate using numbers and symbols.

    (a) the quotient of fourteen and seven

    (b) the quotient of seven and fourteen

2. Divide. $15{,}916 \div 39$

3. A school district receives a grant for $5,484,000 to be distributed equally among its three junior colleges. How much does each college receive?

4. **Concept Check**

$$\begin{array}{r} 2 \\ 13\overline{)2645} \\ -26 \\ \hline 04 \end{array}$$

Explain the next 2 steps for this division problem.

1.6 Exponents and the Order of Operations

1 Writing Whole Numbers and Variables in Exponent Form

Student Learning Objectives

After studying this section, you will be able to:

1. Write whole numbers and variables in exponent form.

2. Evaluate numerical and algebraic expressions in exponent form.

3. Use symbols and key words for expressing exponents.

4. Follow the order of operations.

Recall that in the multiplication problem $3 \cdot 3 \cdot 3 \cdot 3 \cdot 3 = 243$ the number 3 is called a **factor.** We can write the repeated multiplication $3 \cdot 3 \cdot 3 \cdot 3 \cdot 3$ using a shorter notation, 3^5, because there are five factors of 3 in the repeated multiplication. We say that 3^5 is written in **exponent form.** 3^5 is read "three to the fifth power."

EXPONENT FORM

The small number 5 is called an **exponent.** Whole number exponents, except zero, tell us how many factors are in the repeated multiplication. The number 3 is called the **base.** The base is the number that is multiplied.

$$3 \cdot 3 \cdot 3 \cdot 3 \cdot 3 = 3^5 \longrightarrow \text{The exponent is 5.}$$

3 appears as a factor 5 times.   The base is 3.

We do not multiply the base 3 by the exponent 5. The 5 just tells us how many 3's are in the repeated multiplication.

If a whole number or variable does not have an exponent visible, the exponent is understood to be 1.

$$9 = 9^1 \quad \text{and} \quad x = x^1$$

EXAMPLE 1 Write in exponent form.

(a) $2 \cdot 2 \cdot 2 \cdot 2 \cdot 2 \cdot 2$ (b) $4 \cdot 4 \cdot 4 \cdot x \cdot x$ (c) 7   (d) $y \cdot y \cdot y \cdot 3 \cdot 3 \cdot 3 \cdot 3$

Solution

(a) $2 \cdot 2 \cdot 2 \cdot 2 \cdot 2 \cdot 2 = 2^6$

(c) $7 = 7^1$

(b) $4 \cdot 4 \cdot 4 \cdot x \cdot x = 4^3 \cdot x^2$ or $4^3 x^2$

(d) $y \cdot y \cdot y \cdot 3 \cdot 3 \cdot 3 \cdot 3 = y^3 \cdot 3^4$, or $3^4 y^3$
Note, it is standard to write the number before the variable in a term. Thus $y^3 3^4$ is written $3^4 y^3$.

Student Practice 1 Write in exponent form.

(a) n (b) $6 \cdot 6 \cdot y \cdot y \cdot y \cdot y$

(c) $5 \cdot 5 \cdot 5 \cdot 5 \cdot 5 \cdot 5 \cdot 5 \cdot 5$ (d) $x \cdot x \cdot 8 \cdot 8 \cdot 8$

NOTE TO STUDENT: Fully worked-out solutions to all of the Student Practice problems can be found at the back of the text starting at page SP-1.

EXAMPLE 2 Write as a repeated multiplication.

(a) n^3 (b) 6^5

Solution

(a) $n^3 = n \cdot n \cdot n$ (b) $6^5 = 6 \cdot 6 \cdot 6 \cdot 6 \cdot 6$

Student Practice 2 Write as a repeated multiplication.

(a) x^6 (b) 1^7

2 Evaluating Numerical and Algebraic Expressions in Exponent Form

To *evaluate,* or find the *value* of, an expression in exponent form, we first write the expression as repeated multiplication, then multiply the factors.

EXAMPLE 3   Evaluate each expression.

(a) 3^3 **(b)** 1^9 **(c)** 2^4

Solution

(a) $3^3 = 3 \cdot 3 \cdot 3 = 27$

(b) $1^9 = 1$

We do not need to write out this multiplication because repeated multiplication of 1 will always equal 1.

(c) $2^4 = 2 \cdot 2 \cdot 2 \cdot 2 = 16$

Student Practice 3   Evaluate each expression.

(a) 4^3 **(b)** 8^1 **(c)** 10^2

Sometimes we are asked to express an answer in *exponent form* and other times to *find the value of (evaluate)* an expression. Therefore, it is important that you read the question carefully and express the answer in the correct form.

Write $5 \cdot 5 \cdot 5$ in *exponent* form: $5 \cdot 5 \cdot 5 = 5^3$.

Evaluate 5^3: $5^3 = 5 \cdot 5 \cdot 5 = 125$.

Large numbers are often expressed using a number in exponent form that has a base of 10: 10^1, 10^2, 10^3, 10^4, and so on. Let's look for a pattern to find an easy way to evaluate an expression when the base is 10.

$10^1 = 1\,0$ $10^3 = (10)(10)(10) = 1\,000$

$10^2 = (10)(10) = 1\,00$ $10^4 = (10)(10)(10)(10) = 1\,0,000$

Notice that when the exponent is 1 there is 1 trailing zero; when the exponent is 2 there are 2 trailing zeros; when it is 3 there are 3 trailing zeros; and so on. Thus to calculate a power of 10, we write 1 and attach the number of trailing zeros named by the exponent.

EXAMPLE 4   Evaluate 10^7.

Solution

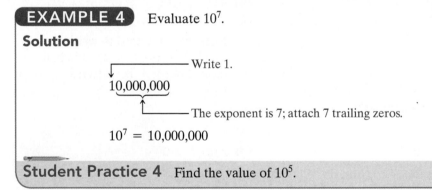

$10^7 = 10,000,000$

Student Practice 4   Find the value of 10^5.

To evaluate the expression x^2 when x *is equal to 4*, we replace the variable x with the number 4 and find the value of 4^2: $4^2 = 4 \cdot 4 = 16$. We can write the statement "x is equal to 4" using math symbols: "$x = 4$."

EXAMPLE 5   Evaluate x^3 for $x = 3$.

Solution

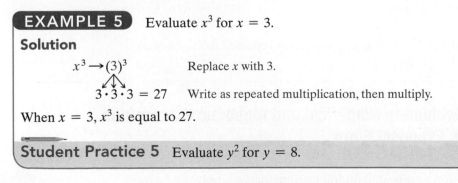

When $x = 3$, x^3 is equal to 27.

Student Practice 5   Evaluate y^2 for $y = 8$.

③ Using Symbols and Key Words for Expressing Exponents

How do you say 10^2 or 5^3? We can say "10 raised to the power 2," or "5 raised to the power 3," but the following phrases are more commonly used.

If the value of the exponent is 2, we say the base is **squared.**

6^2 is read "six squared."

If the value of the exponent is 3, we say the base is **cubed.**

6^3 is read "six cubed."

If the value of the exponent is *greater than* 3, we say that the base is raised to the **(exponent)th power.**

6^5 is read "six to the fifth power."

EXAMPLE 6   Translate using symbols.

(a) Five cubed   **(b)** Seven squared   **(c)** y to the eighth power

Solution

(a) Five cubed $= 5^3$ **(b)** Seven squared $= 7^2$ **(c)** y to the eighth power $= y^8$

Student Practice 6   Translate using symbols.

(a) Four to the sixth power   **(b)** x cubed   **(c)** Ten squared

④ Following the Order of Operations

It is often necessary to perform more than one operation to solve a problem. For example, if you bought one pair of socks for $3 and 4 undershirts for $5 each, you would multiply first and then add to find the total cost. In other words, the order in which we performed the operations (order of operations) was multiply first, then add. However, the order of operations may not be as clear when dealing with a math statement. When we see the problem written as $3 + 4(5)$, understanding what to do can be tricky. Do we add, then multiply, or do we multiply before adding? Let's work this calculation both ways.

Add First	Multiply First
$3 + 4(5) = 7(5) = 35$ Wrong!	$3 + 4(5) = 3 + 20 = 23$ Correct

Since $3 + 4(5)$ can be written $3 + (5 + 5 + 5 + 5) = 3 + 20$, 23 is correct. Thus we see that the order of operations makes a difference. The following rule tells which operations to do first: the correct **order of operations.** We call this a *list of priorities.*

ORDER OF OPERATIONS

Follow this order of operations.

Do first   **1.** Perform operations inside *parentheses.*

             2. Simplify any expression with *exponents.*

             3. *Multiply* or *divide* from left to right.

Do last   **4.** *Add or subtract* from left to right.

     parentheses → exponents → multiply or divide → add or subtract

---

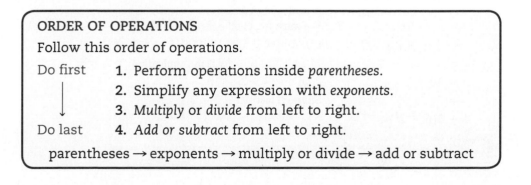

Now, following the order of operations, we can clearly see that to find $3 + 4(5)$, we multiply and then add. You will find it easier to follow the order of operations if you keep your work neat and organized, perform one operation at a time, and follow the sequence *identify, calculate, replace*.

1. *Identify* the operation that has the highest priority.
2. *Calculate* this operation.
3. *Replace* the operation with your result.

EXAMPLE 7 Evaluate. $2^3 - 6 + 4$

Solution

$$2^3 - 6 + 4 = 8 - 6 + 4$$ *Identify*: The highest priority is **exponents.**
                                     Calculate: $2 \cdot 2 \cdot 2 = 8$. *Replace*: 2^3 with 8.

$$= 8 - 6 + 4$$ *Identify*: **Subtraction** has the highest priority.
                                     Calculate: $8 - 6 = 2$. *Replace*: $8 - 6$ with 2.

$$= 2 + 4$$ *Identify*: **Addition** is last. *Calculate*: $2 + 4 = 6$.

$$2^3 - 6 + 4 = 6$$ *Replace*: $2 + 4$ with 6.

Note that addition and subtraction have equal priority. We do the operations as they appear, reading from *left* to *right*. In Example 7 the subtraction appears first, so we subtract before we add.

Student Practice 7 Evaluate. $3^2 + 2 - 5$

EXAMPLE 8 Evaluate. $2 \cdot 3^2$

Solution

$$2 \cdot 3^2 = 2 \cdot 9$$ *Identify*: The highest priority is **exponents.** *Calculate*: $3 \cdot 3 = 9$.
                                  Replace: 3^2 with 9.

$$= 2 \cdot 9$$ *Identify*: **Multiplication** is last. *Calculate*: $2 \cdot 9 = 18$.
                          Replace: $2 \cdot 9$ with 18.

$$2 \cdot 3^2 = 18$$

CAUTION: $2 \cdot 3^2$ *does not equal* 6^2! We must follow the rules for the order of operations and simplify the exponent 3^2 before we multiply; otherwise, we will get the wrong answer.

Student Practice 8 Evaluate. $4 \cdot 2^3$

EXAMPLE 9 Evaluate. $4 + 3(6 - 2^2) - 7$

Solution We always perform the calculations inside the parentheses first. Once inside the parentheses, we proceed using the order of operations.

$$4 + 3(6 - 2^2) - 7$$ Within the parentheses, **exponents** have the highest priority: $2^2 = 4$.

$$= 4 + 3(6 - 4) - 7$$ We must finish all operations inside the parentheses, so we **subtract:** $6 - 4 = 2$.
$$= 4 + 3(2) - 7$$

$$= 4 + 6 - 7$$ The highest priority is **multiplication:** $3 \cdot 2 = 6$.

$$= 10 - 7$$ **Add** first: $4 + 6 = 10$.

$$= 3$$ **Subtract** last: $10 - 7 = 3$.

$$4 + 3(6 - 2^2) - 7 = 3$$

Student Practice 9 Evaluate. $2 + 7(10 - 3 \cdot 2) - 4$

As we stated earlier, it is easier to follow the order of operations if we keep our work neat and organized, perform one operation at a time, and follow the sequence **identify, calculate, replace.**

EXAMPLE 10   Evaluate. $\dfrac{(6 + 6 \div 3)}{(5 - 1)}$

Solution   We rewrite the problem as division and then follow the order of operations.

$(6 + 6 \div 3)$	$\div$	$(5 - 1)$	We perform operations inside parentheses first.
$(6 + 2)$	$\div$	4	$6 \div 3 = 2$; $5 - 1 = 4$.
8	$\div$	$4 = 2$	Divide.

Student Practice 10   Evaluate.

$$\dfrac{(4 + 8 \div 2)}{(7 - 3)}$$

STEPS TO SUCCESS Getting the Most from Your Study Time

Did you know that there are many things you can do to increase your learning when you study? If you use the following strategies, you can improve the way you study and learn more while studying less.

Making it personal:

Place a ✔ *in the blank as you complete each of the following.*

Completing Homework

_____ **1.** Read the text and review your class notes on the *same day* that your class meets.

_____ **2.** Either the same day, or the day after your class meets, complete the **first half** of your home-work assignment.

_____ **3.** Put an * beside any problem that you get wrong or don't know how to start. For best results check your answer only **after** you complete a problem.

_____ **4.** Follow up on wrong answers (*).
   - Double-check all your work for errors.
   - Look in the book for a similar example or practice problem and rework the problem using the book's solution as a guide.
   - If you still can't solve the problem, watch a video lecture, or complete a similar problem on MyMathLab. If this does not work, ask for help.

_____ **5.** Repeat steps 3 and 4 to complete the **second half** of your homework. Do this in a separate sitting so that you are fresh.

_____ **6.** Complete the Quick Quiz at the end of each exercise set. Study the example referenced for each problem you get wrong, then rework the problem.

_____ **7.** Complete the How Am I Doing? mid-chapter test.

Studying Your Homework

_____ **8.** Look over past homework for the current chapter.

_____ **9.** Revisit * problems. Before the quiz or test, work another problem that is like each * problem. In this text, an even-numbered problem is similar to the preceding odd-numbered problem.

It is important that you realize that completing your homework assignment and studying your home-work are separate activities. Activities 2–7 describe the process of *completing your homework*, whereas 8–9 describe the process of *studying your homework*. For best results, these two activities should be done at different times.

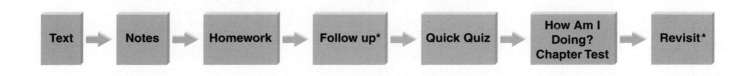

Verbal and Writing Skills, Exercises 1–2

1. Write in words the question being asked by the equation $n^2 = 16$.

2. Write in words the question being asked by the equation $x^3 = 27$.

Write each product in exponent form.

3. $2 \cdot 2 \cdot 2$

4. $4 \cdot 4$

5. $a \cdot a \cdot a \cdot a \cdot a$

6. $x \cdot x$

7. 4

8. z

9. $3 \cdot 3 \cdot 3 \cdot 3$

10. $7 \cdot 7 \cdot 7$

Write each product in exponent form.

11. $5 \cdot 5 \cdot a \cdot a \cdot a$

12. $8 \cdot 8 \cdot x \cdot x \cdot x$

13. $2 \cdot 2 \cdot z \cdot z \cdot z \cdot z \cdot z$

14. $3 \cdot 3 \cdot y \cdot y \cdot y \cdot y$

15. $5 \cdot 5 \cdot 5 \cdot y \cdot y \cdot x \cdot x$

16. $6 \cdot 6 \cdot x \cdot y \cdot y$

17. $n \cdot n \cdot n \cdot n \cdot n \cdot 9 \cdot 9$

18. $x \cdot x \cdot x \cdot x \cdot x \cdot 7 \cdot 7$

Write as a repeated multiplication.

19. **(a)** 7^3 **(b)** y^5

20. **(a)** 7^6 **(b)** x^2

Evaluate.

21. 2^3

22. 3^3

23. 5^2

24. 6^2

25. 1^6

26. 1^{15}

27. 7^2

28. 3^2

29. 4^4

30. 9^3

31. 10^1

32. 8^1

33. 5^3

34. 2^4

35. 10^6

36. 10^5

37. x^2 for $x = 5$

38. y^3 for $y = 3$

39. a^4 for $a = 1$

40. b^{18} for $b = 1$

Translate using numbers and exponents.

41. Seven to the third power

42. Three cubed

43. Nine squared

44. Four to the seventh power

Evaluate.

45. $3 \cdot 4 - 7$

46. $3 \cdot 5 - 2$

47. $7^2 + 5 - 3$

48. $6^3 + 4 - 8$

49. $5 \cdot 3^2$

50. $4 \cdot 2^2$

51. $2 \cdot 2^2$

52. $4 \cdot 4^2$

53. $5^2 - 7 + 3$

54. $4^3 - 8 + 7$

55. $9 + 2 \cdot 2$

56. $5 + 3 \cdot 9$

57. $9 + (6 + 2^2)$

58. $8 + (7 + 4^3)$

59. $40 \div 5 \times 2 + 3^2$

60. $6^2 \div 6 \times 2 + 1$

61. $2 \times 15 \div 5 + 10$

62. $3 \times 12 \div 4 + 2$

63. $2^2 + 8 \div 4$

64. $3^3 + 6 \div 3$

65. $\dfrac{(8 + 4 \div 2)}{(5 - 3)}$

66. $\dfrac{(5 + 15 \div 5)}{(9 - 5)}$

67. $\dfrac{(3 + 1)}{(12 \div 6 \times 2)}$

68. $\dfrac{(16 - 4)}{(36 \div 6 \times 2)}$

69. $7 + 5(3 \cdot 4 + 7) - 2$

70. $3 + 4(5 \cdot 2 + 8) - 3$

71. $59 - 4(1 + 5 \cdot 2) + 4$

72. $88 - 3(2 + 6 \cdot 4) + 6$

73. $6 + 2(4 \cdot 5 + 9) - 11$

74. $2 + 12(3 \cdot 2 + 1) - 10$

One Step Further

75. $32 \cdot 6 - 4(4^3 - 5 \cdot 2^2) + 3$

76. $63 \cdot 4 - 5(3^2 + 4 \cdot 2^3) + 5$

77. $12 \cdot 5 - 3(3^3 - 2 \cdot 3^2) + 1$

78. $42 \cdot 5 - 3(5^2 + 2 \cdot 4^2) + 3$

To Think About

79. Fred wanted to evaluate $3 \cdot 2 + 4$. He multiplied 3 times 6 to get 18. What is wrong with his reasoning? What is the correct answer?

80. Sara wanted to evaluate $2 \cdot 4^2$. She squared 8 to get 64. What is wrong with her reasoning? What is the correct answer?

81. Multiply: $21 \cdot 10^1$; $21 \cdot 10^2$; $21 \cdot 10^3$; $21 \cdot 10^4$. Do you see a pattern that might suggest a quick way to multiply a number by a power of 10? Explain.

82. Multiply: $10^1 \cdot 10^2$; $10^1 \cdot 10^3$; $10^1 \cdot 10^4$. Do you see a pattern that might suggest a quick way to multiply 10 by a power of 10? Explain.

Cumulative Review

83. **[1.2.4]** Add. $4079 + 2762$

84. **[1.3.4]** Subtract. $8900 - 477$

85. **[1.4.4]** Multiply. $(387)(196)$

86. **[1.4.2]** Translate using symbols. The product of two and some number.

Quick Quiz 1.6

1. Write the product in exponent form.
   (a) $9 \cdot 9 \cdot 9 \cdot x \cdot x$ **(b)** $5 \cdot 5 \cdot 5 \cdot 5 \cdot 5$

2. Evaluate.
   (a) 2^4 **(b)** 1^5

3. Evaluate. $2^2 + 2(10 \div 2) - 11$

4. **Concept Check** Explain in what order you would do the steps to evaluate $50 + 3 \times 5^2 \div 25$.

Did You Know...
That Balancing Your Checkbook Can Help You Keep Track of What You Are Spending?

BALANCING YOUR FINANCES

Understanding the Problem:
One of the first steps in saving money is to determine your current spending trends. The first step in that process is learning to balance your finances.

Terry balanced his checkbook once a month when he received his bank statement. Below is a table that records the deposits Terry made for the month of May. The beginning balance for May was $300.50.

Date	Deposit
May 1	$200.00
May 3	$150.50
May 10	$120.25
May 25	$50.00
May 28	$25.00

Keeping a Record of Checks:
Terry needs to know if he is depositing enough money to cover his monthly expenses. Below is a table that records each check Terry wrote for the month of May.

Date	Check Number	Checks
May 2	102	$238.50
May 6	103	$75.00
May 12	104	$200.00
May 28	105	$28.56
May 30	106	$36.00

Finding the Facts:
Step 1: Terry must determine how much he deposits into the bank every month.
Task 1: *Determine how much Terry deposited in the bank in May.*
Step 2: Terry must determine how much he spends each month.
Task 2: *Determine the total amount of the checks Terry wrote in May.*

Task 3: *Based on the given information, will Terry be able to cover all his expenses for the month of May?*

Making a Decision:
Step 3: Terry needs to know if he can continue to spend money at the same rate, or if he needs to cut back on his spending.
Task 4: *Assuming all the checks cleared for May, what would Terry's balance be at the beginning of June?*

Task 5: *If Terry continues these spending habits, what will happen?*

Applying the Situation to Your Life:
Knowing your monthly income and spending habits can help you to save. Balance your checkbook and monitor your spending habits each month, and try to cut out unnecessary expenses. If you are charged ATM fees for making withdrawals from your checking account with an ATM card, be sure to subtract those costs from your checkbook balance. If you pay monthly fees to the bank for the cost of your checking account, be sure to subtract those costs from your checkbook balance.
Task 6: *How often do you balance your checkbook?*

Task 7: *Do you have any unnecessary expenses that can be cut out?*

How Am I Doing? Sections 1.1–1.6

How are you doing with your homework assignments in Sections 1.1 to 1.6? Do you feel you have mastered the material so far? Do you understand the concepts you have covered? Before you go further in the textbook, take some time to do each of the following problems.

1.1

1. Write in expanded notation. 9062

2. Replace the question mark with the appropriate inequality symbol < or >. 16 ? 22

3. Round 17,248,954 to the nearest hundred thousand.

1.2

4. Use the associative and/or commutative properties of addition, then simplify.

   (a) $(6 + a) + 3$ **(b)** $(6 + x + 4) + 2$

5. Evaluate $x + y$ if x is equal to 9 and y is equal to 11.

6. Add. $9532 + 251 + 322$

▲ 7. Find the perimeter of the following shape made up of two rectangles.

1.3

8. Translate using numbers and symbols. Eleven decreased by a number

9. Subtract and check. $39,204 - 5982$

1.4

10. Translate using numbers and symbols. Double a number

11. Simplify. $2(4)(y \cdot 5)$ 12. Multiply. $(2371)126$

13. A small hotel is 6 stories high with 12 rooms on each floor. How many rooms are in the hotel?

1.5

14. Translate using numbers and symbols. The quotient of 144 and x

15. Divide. $\dfrac{362,664}{721}$

1.6

16. Write the product in exponent form. $n \cdot n \cdot n \cdot n \cdot 3 \cdot 3 \cdot 3$

17. Evaluate. 4^3

Evaluate.

18. $2 \cdot 3^2$

19. $(2 + 10) + 12 \div 6 - 3^2$

Now turn to page SA-2 for the answers to each of these problems. Each answer also includes a reference to the objective in which the problem is first taught. If you missed any of these problems, you should stop and review the Examples and Student Practice problems in the referenced objective. A little review now will help you master the material in the upcoming sections of the text.

1. _____
2. _____
3. _____
4. (a) _____
   (b) _____
5. _____
6. _____
7. _____
8. _____
9. _____
10. _____
11. _____
12. _____
13. _____
14. _____
15. _____
16. _____
17. _____
18. _____
19. _____

1.7 More on Algebraic Expressions

Student Learning Objectives

After studying this section, you will be able to:

① Use symbols and key words for expressing algebraic expressions.

② Evaluate algebraic expressions involving multiplication and division.

③ Use the distributive property to simplify numerical and algebraic expressions.

In this section we will see how to translate some new types of phrases into symbols. We will also use the skills we learned in previous sections to simplify and evaluate algebraic expressions.

① Using Symbols and Key Words for Expressing Algebraic Expressions

When we translate phrases into numbers and symbols we must take care to preserve the order of operations indicated by the phrase. When a phrase contains key words for more than one operation, the phrases *sum of* or *difference of* indicate that these operations must be placed within parentheses so that they are completed first. We illustrate below.

Three *times* the *difference of* five and two
↓ ↓ ↓ ↓
$3 \quad \cdot \quad (5 - 2)$

The phrase *difference of* indicates that we must place $5 - 2$ within parentheses.

Three times five minus two
↓ ↓ ↓ ↓ ↓
$3 \quad \cdot \quad 5 \quad - \quad 2$

We must include parentheses when we see the phrases *sum of* or *difference of* or we will get the wrong answer: $3 \cdot (5 - 2) = 3 \cdot 3 = 9$ but $3 \cdot 5 - 2 = 15 - 2 = 13$.

EXAMPLE 1 Translate using numbers and symbols.

(a) Two times x plus seven

(b) Two times the sum of x and seven

Solution

(a) Two times x plus seven
↓ ↓ ↓ ↓ ↓
$2 \quad \cdot \quad x \quad + \quad 7$

$2x + 7$

(b) Two times the sum of x and seven
↓ ↓ ↓ ↓
$2 \quad \cdot \quad (x + 7)$

$2(x + 7)$

The key phrase *sum of* indicates that $x + 7$ is placed within parentheses.

Student Practice 1 Translate using numbers and symbols.

(a) Five times y plus three

(b) Three times the sum of m and two

NOTE TO STUDENT: Fully worked-out solutions to all of the Student Practice problems can be found at the back of the text starting at page SP-1.

② Evaluating Algebraic Expressions Involving Multiplication and Division

We evaluate variable expressions involving multiplication and division just as we did expressions involving addition and subtraction. For example, to evaluate $2n$ if n is equal to 5, we replace the variable in the expression with 5 and then simplify: $2n \rightarrow 2(5) = 10$.

EXAMPLE 2 Evaluate $\dfrac{(2a + 3)}{7}$ for $a = 9$.

Solution

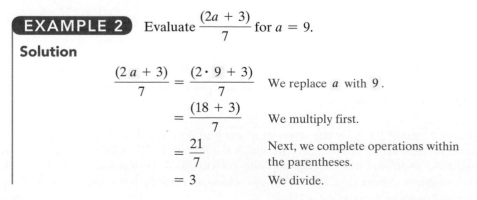

$$\frac{(2a + 3)}{7} = \frac{(2 \cdot 9 + 3)}{7} \qquad \text{We replace } a \text{ with } 9.$$

$$= \frac{(18 + 3)}{7} \qquad \text{We multiply first.}$$

$$= \frac{21}{7} \qquad \text{Next, we complete operations within the parentheses.}$$

$$= 3 \qquad \text{We divide.}$$

Student Practice 2

Evaluate $\dfrac{(5y - 4)}{3}$ for $y = 2$.

$\mathbb{M}_{\mathbb{C}}$ **EXAMPLE 3**   Evaluate.

(a) $3x + 3y + 6$ for $x = 3$ and $y = 5$ **(b)** $\dfrac{(x^2 - 2)}{y}$ for $x = 4$ and $y = 2$

Solution   We replace each variable with the indicated value and then follow the order of operations to simplify.

(a) $3x + 3y + 6$ $= 3 \cdot 3 + 3 \cdot 5 + 6$ We replace x with 3 and y with 5.

                   $= 9 + 15 + 6$ We multiply first.

                   $= 30$ We add last.

(b) $\dfrac{(x^2 - 2)}{y}$ $= \dfrac{(4^2 - 2)}{2}$ We replace x with 4 and y with 2.

            $= \dfrac{14}{2}$ We square 4 first then subtract: $4^2 = 16, 16 - 2 = 14$.

            $= 7$ We divide last: $14 \div 2 = 7$.

Student Practice 3   Evaluate.

(a) $5m - 2n + 1$ for $m = 7$ and $n = 3$

(b) $\dfrac{(a^3 - 2)}{b}$ for $a = 2$ and $b = 3$

③ Using the Distributive Property to Simplify Numerical and Algebraic Expressions

A property that is often used to simplify and multiply is the **distributive property.** This property states that we can distribute multiplication over addition or subtraction. The following example will help you understand what we mean by this.

     "4 *times* $(n + 7)$" is written $4(n + 7)$.   We can find this product using repeated addition.

$4(n + 7) = (n + 7) + (n + 7) + (n + 7) + (n + 7)$ We write $4(n + 7)$ as repeated addition.

$= (n + n + n + n) + (7 + 7 + 7 + 7)$ We change the order of addition and group the n's and 7's together.

            $\downarrow$ $\downarrow$

         $4n$ $+$ $4 \cdot 7$ We have 4 n's plus 4 7's.

$4(n + 7) = 4n + 28$

     A shorter way to do this is to **distribute** the 4 by multiplying each number or variable inside the parentheses by 4.

                      $4 \cdot n$

$4(n + 7) = 4(n + 7) = 4 \cdot n + 4 \cdot 7 = 4n + 28$

                      $4 \cdot 7$

     We can state the distributive property as follows.

DISTRIBUTIVE PROPERTY

If a, b, and c are numbers or variables, then

$$a(b + c) = ab + ac \quad \text{and} \quad a(b - c) = ab - ac$$

We distribute a over addition and subtraction by multiplying every number or variable inside the parentheses by a. Then we simplify the result.

CAUTION: We must only use the distributive property if the numbers or variables inside the parentheses are separated by a $+$ or $-$ sign. We *do not use the distributive property* when the numbers or variables inside the parentheses are *separated by multiplication or division symbols*. Thus, we can use the distributive property in Example 4 because in the expression $3(x - 2)$, the x and the 2 are separated by a $-$ sign. We *cannot use the distributive property* for the expression $3(2x)$ because the 2 and the x are not separated by a $+$ or $-$ sign. Thus $3(x - 2) = 3x - 6$ while $3(2x) = 3 \cdot 2 \cdot x = 6x$.

EXAMPLE 4   Use the distributive property to simplify. $3(x - 2)$

Solution

$$3(x - 2) = 3(x - 2) = 3 \cdot x - 3 \cdot 2 \qquad \text{Multiply 3 times } x.$$

Multiply 3 times 2.

$$3(x - 2) = 3x - 6 \qquad \text{Simplify.}$$

Student Practice 4   Use the distributive property to simplify.

(a) $2(x - 5)$ **(b)** $4(y + 3)$

EXAMPLE 5   Simplify. $2(y + 1) + 4$

Solution   First we use the distributive property and then we simplify.

$$2(y + 1) + 4 = 2 \cdot y + 2 \cdot 1 + 4 \qquad \text{We use the distributive property}$$
$$= 2y + 2 + 4 \qquad \text{to multiply } 2(y + 1).$$
$$= 2y + 6 \qquad \text{We simplify: } 2 + 4 = 6.$$

Student Practice 5   Simplify. $7(y + 3) + 2$

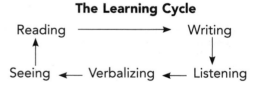 **STEPS TO SUCCESS** Reviewing for an Exam

Reviewing for an exam enables you to connect concepts you learned over several classes. Your review activities should cover all the components of the learning cycle.

The Learning Cycle

Reading ⟶ Writing
Seeing ← Verbalizing ← Listening
Seeing ↑ (from Reading)

Making it personal:

Place a ✔ in the blank as you complete each of the following.

_____ **1.** Make 3-by-5 study cards as follows.
  • Write the name of any new term or rule on the front of the card. Then write the definition of the term or the rule on the back.
  • Write a warning about a common error you make on the front side of the card. Write an example of the common error along with the correction for it on the back side of the card.

  • Periodically use these cards as flash cards and quiz yourself, or study with a classmate.

_____ **2.** Reread your notes. Study returned homework and quizzes and redo problems you got wrong.

_____ **3.** Read the Chapter Organizer at the end of the chapter, then complete the *You Try It* problems in the Organizer.

_____ **4.** Complete the Chapter Review exercises.

_____ **5.** Complete the How Am I Doing? Chapter Test and place a * beside any problem you got wrong.

_____ **6.** Refer to the Math Coach following the test. You will find hints on how to correct the problems you got wrong.

_____ **7.** Get help with the problems you do not understand.

It is not a good idea to complete all seven steps at one time. For best results, complete each step at a separate sitting and start the process early so that you are done at least three days before the exam.

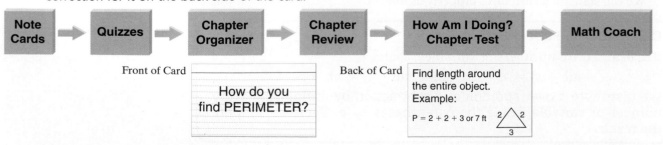

Verbal and Writing Skills, Exercises 1–4

State what property is represented in each mathematical statement.

1. $5(3 + 4) = 5 \cdot 3 + 5 \cdot 4$

2. $3(6 - 4) = 3 \cdot 6 - 3 \cdot 4$

Are the following true or false? Explain your answers.

3. **(a)** $8(3y) = 8 \cdot 3 \cdot 8 \cdot y$

(b) $8(3 + y) = 8 \cdot 3 + 8 \cdot y$

4. **(a)** $4(5x) = 4 \cdot 5 \cdot 4 \cdot x$

(b) $4(5 + x) = 4 \cdot 5 + 4 \cdot x$

Fill in each box with the correct number or variable.

5. $2(x + 1) = 2 \cdot \square + 2 \cdot \square$

6. $9(y + 2) = 9 \cdot \square + 9 \cdot \square$

7. $6(y - 3) = 6 \cdot \square + 6 \cdot \square$

8. $8(x - 1) = 8 \cdot \square - 8 \cdot \square$

Translate using numbers and symbols.

9. Six times y plus two

10. Seven times x plus three

11. Seven times four minus one

12. Eleven times five minus two

13. Four times the sum of three and nine

14. Nine times the sum of four and six

15. Triple the sum of y and six

16. Double the sum of x and one

17. Eight times the difference of four and y

18. Three times the difference of six and x

Mixed Practice, Exercises 19–24

Translate using numbers and symbols, then simplify.

19. **(a)** Four times two plus seven

(b) Four times the sum of two and seven

20. **(a)** Five times six plus one

(b) Eight times the sum of six and one

21. **(a)** Four times three minus one

(b) Four times the difference of three and one

22. **(a)** Two times seven minus one

(b) Two times the difference of seven and one

23. **(a)** Twelve times one plus three

(b) Twelve times the sum of one and three

24. **(a)** Nine times four plus one

(b) Nine times the sum of four and one

Evaluate for the given values.

25. $4a + 5b$ for $a = 2$ and $b = 6$

26. $3m + 2n$ for $m = 4$ and $n = 5$

27. $8x - 6y$ for $x = 9$ and $y = 2$

28. $9x - 2y$ for $x = 8$ and $y = 5$

29. $\dfrac{(x + 4)}{3}$ for $x = 11$

30. $\dfrac{(y + 7)}{5}$ for $y = 13$

31. $\dfrac{(a^2 - 4)}{b}$ for $a = 5$ and $b = 3$

32. $\dfrac{(m^2 - 6)}{n}$ for $m = 6$ and $n = 3$

33. $\dfrac{(x^3 + 4)}{y}$ for $x = 2$ and $y = 2$

34. $\dfrac{(x^3 + 9)}{y}$ for $x = 3$ and $y = 6$

35. $\dfrac{(a^2 + 6)}{b}$ for $a = 2$ and $b = 5$

36. $\dfrac{(n^2 + 5)}{m}$ for $n = 3$ and $m = 7$

37. $\dfrac{(y-2)}{2}$ for $y = 16$

38. $\dfrac{(y-3)}{3}$ for $y = 18$

39. $4m + 3n$ for $m = 2$ and $n = 7$

40. $5x + 4y$ for $x = 4$ and $y = 6$

41. $\dfrac{(x^2-5)}{y}$ for $x = 5$ and $y = 4$

42. $\dfrac{(x^2-3)}{y}$ for $x = 6$ and $y = 11$

Use the distributive property to simplify.

43. $4(x+1)$

44. $2(x+1)$

45. $3(n-5)$

46. $6(n-4)$

47. $3(x-6)$

48. $4(x-3)$

49. $4(x+4)$

50. $5(x+9)$

51. $2(x+6)+5$

52. $4(x+2)+6$

53. $2(y+1)+5$

54. $7(y+1)+3$

55. $4(x+3)+6$

56. $3(x+2)+5$

57. $9(y+1)-3$

58. $5(y+1)-2$

59. $3(x+1)-1$

60. $6(x+1)-3$

One Step Further

Evaluate for the given values.

61. $yx^2 - 3$ for $y = 6$ and $x = 2$

62. $ab^2 + 4$ for $a = 5$ and $b = 3$

63. $\dfrac{(a^2-3)+2^3}{b}$ for $a = 5$ and $b = 2$

64. $\dfrac{(a^3-4)-3^2}{b}$ for $a = 3$ and $b = 7$

To Think About

65. **(a)** Add $(x+2)$ four times.

(b) Multiply $4(x+2)$ using the distributive property.

(c) What do you notice about the answers in **(a)** and **(b)**?

66. **(a)** Add $(x+4)$ three times.

(b) Multiply $3(x+4)$ using the distributive property.

(c) What do you notice about the answers in **(a)** and **(b)**?

Cumulative Review

67. **[1.4.3]** Simplify. $8(2)(x \cdot 4)$

68. **[1.2.3]** Evaluate $4 + x$ if x is 2.

69. **[1.2.3]** Evaluate $x + y + 4$ if x is 1 and y is 3.

70. **[1.3.4]** Subtract. $2001 - 463$

Quick Quiz 1.7

1. Translate using numbers and symbols.
Double the sum of n and five

2. Use the distributive property to simplify.
$6(y+1)+3$

3. Evaluate.
(a) $3x + 2y$ if $x = 3$ and $y = 4$
(b) $\dfrac{(x^2-2)}{y}$ if $x = 4$ and $y = 7$

4. **Concept Check** Simplify $5(x+1)$, then evaluate $5(x+1)$ for $x = 2$. Compare results and state the difference in the process to simplify and to evaluate.

1.8 Introduction to Solving Linear Equations

① Combining Like Terms

In algebra we often deal with terms such as $4y$ or $7x$. What do we mean by *terms*?

A **term** is a number, a variable, or a product of a number and one or more variables. Terms are separated from other terms in an expression by a $+$ sign or a $-$ sign. Often a term has a number factor and a variable factor. The number factor is called the **coefficient.**

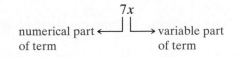

$7x$

numerical part ◄──┘ └──► variable part
of term   of term

A term that has no variable is called a **constant term,** and a term that has a variable is called a **variable term.**

$$9 + 3n + 4x$$

constant term   variable terms

What do the expressions $3n$ and $4x$ mean? $3n$ is the term that represents the sum $n + n + n$, and $4x$ is the term that represents the sum $x + x + x + x$. As we saw earlier, $3n$ and $4x$ also indicate multiplication: 3 times n and 4 times x.

$n + n + n$ We count three n's added $x + x + x + x$ We count four x's added.
   ↓   ↓
$3n$ $4x$

EXAMPLE 1 Write a term that represents each of the following.

(a) Two y's   **(b)** $a + a + a + a$

(c) Seven   **(d)** One x

Solution

(a) Two y's $= 2y$ **(b)** $a + a + a + a = 4a$

(c) Seven $= 7$ **(d)** One $x = 1x$ or x

Student Practice 1 Write a term that represents each of the following.

(a) Four n's   **(b)** $y + y + y$

(c) Eight   **(d)** One y

We see many examples of adding and subtracting quantities that are like quantities, as shown in the following example.

$$3 \text{ feet} + 7 \text{ feet} = 10 \text{ feet} \qquad 7 \text{ trucks} - 2 \text{ trucks} = 5 \text{ trucks}$$

However, we cannot combine things that are not the same:

$$7 \text{ trucks} - 4 \text{ feet} \qquad \text{(cannot be done!)}$$

Similarly, in algebra we cannot combine terms that are not like terms. **Like terms** are terms that have *identical variable parts*. For example, in the expression $8x + 6b + 2x$, the terms $8x$ and $2x$ are called like terms since they have the same variable parts. They are both counting x's.

Expression   *Like Terms*

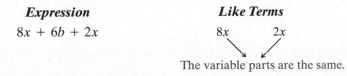

$8x + 6b + 2x$ $8x$ $2x$

The variable parts are the same.

There are no like terms for $6b$, since none of the terms have exactly the same variable part as $6b$.

Student Learning Objectives

After studying this section, you will be able to:

① **Combine like terms.**

② **Translate English statements into equations.**

③ **Solve equations using basic arithmetic facts.**

④ **Translate and solve equations.**

NOTE TO STUDENT: Fully worked-out solutions to all of the Students Practice problems can be found at the back of the text starting at page SP-1.

EXAMPLE 2 Identify the like terms. $7ab + 4a + 2ab + 3y$

Solution

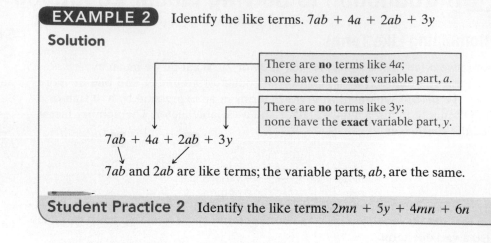

There are **no** terms like $4a$; none have the **exact** variable part, a.

There are **no** terms like $3y$; none have the **exact** variable part, y.

$7ab + 4a + 2ab + 3y$

$7ab$ and $2ab$ are like terms; the variable parts, ab, are the same.

Student Practice 2 Identify the like terms. $2mn + 5y + 4mn + 6n$

The numerical part of a term is called the **coefficient** of the term. The coefficient tells you how many you have of whatever variable follows. To combine like terms, we either add or subtract the coefficients of like terms.

> **COMBINING LIKE TERMS**
>
> To combine like terms, add or subtract the numerical coefficients of like terms. The variable parts stay the same.
>
> $$6x + 4x = 10x \qquad 8y - 2y = 6y$$

EXAMPLE 3 Identify like terms, then combine like terms.

(a) $3x + 7y + 2x$ **(b)** $9m - m - 8$ **(c)** $4xy + 8y + 2xy$

Solution

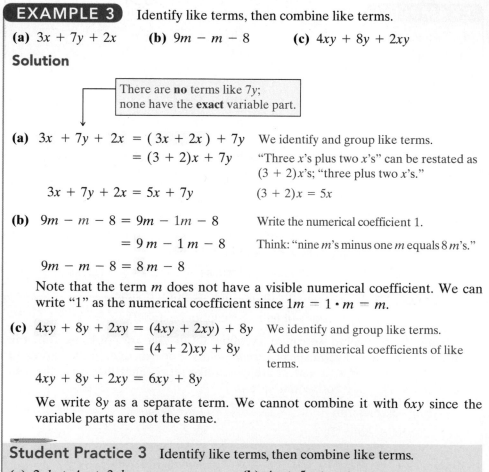

There are **no** terms like $7y$; none have the **exact** variable part.

(a) $3x + 7y + 2x = (3x + 2x) + 7y$ We identify and group like terms.

$\qquad\qquad\qquad = (3 + 2)x + 7y$ "Three x's plus two x's" can be restated as $(3 + 2)x$'s; "three plus two x's."

$\quad 3x + 7y + 2x = 5x + 7y$ $(3 + 2)x = 5x$

(b) $9m - m - 8 = 9m - 1m - 8$ Write the numerical coefficient 1.

$\qquad\qquad\qquad = 9\,m - 1\,m - 8$ Think: "nine m's minus one m equals $8\,m$'s."

$\quad 9m - m - 8 = 8\,m - 8$

Note that the term m does not have a visible numerical coefficient. We can write "1" as the numerical coefficient since $1m = 1 \cdot m = m$.

(c) $4xy + 8y + 2xy = (4xy + 2xy) + 8y$ We identify and group like terms.

$\qquad\qquad\qquad\qquad = (4 + 2)xy + 8y$ Add the numerical coefficients of like terms.

$\quad 4xy + 8y + 2xy = 6xy + 8y$

We write $8y$ as a separate term. We cannot combine it with $6xy$ since the variable parts are not the same.

Student Practice 3 Identify like terms, then combine like terms.

(a) $2ab + 4a + 3ab$ **(b)** $4y + 5x + y + x$

(c) $7x + 3y + 3z$

▲ **EXAMPLE 4**   Write the perimeter of the rectangular figure as an algebraic expression and simplify.

$4a + 7b$

$2a + 3b$

Solution   Since the figure is a rectangle, opposites sides are equal.

$2a + 3b$ $4a + 7b$ $2a + 3b$

$4a + 7b$

We add all sides to find the perimeter:

$(2a + 3b) + (4a + 7b) + (2a + 3b) + (4a + 7b)$ We must combine like terms.

$= (2a + 2a + 4a + 4a) + (3b + 3b + 7b + 7b)$ We use the associative and commutative properties to change the order of addition and regroup.

$= 12a + 20b$ We combine like terms.

The algebraic expression for the perimeter is $12a + 20b$.

▲ **Student Practice 4**   Write the perimeter of the triangular figure in the margin as an algebraic expression and simplify.

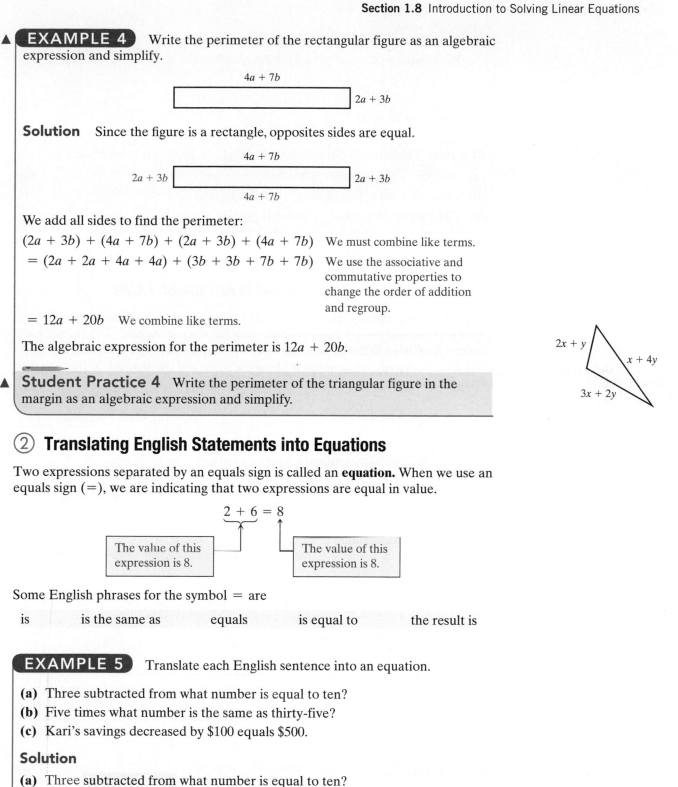

$2x + y$

$x + 4y$

$3x + 2y$

② Translating English Statements into Equations

Two expressions separated by an equals sign is called an **equation.** When we use an equals sign (=), we are indicating that two expressions are equal in value.

$$2 + 6 = 8$$

The value of this expression is 8.   The value of this expression is 8.

Some English phrases for the symbol = are

is   is the same as   equals   is equal to   the result is

EXAMPLE 5   Translate each English sentence into an equation.

(a) Three subtracted from what number is equal to ten?
(b) Five times what number is the same as thirty-five?
(c) Kari's savings decreased by $100 equals $500.

Solution

(a) Three subtracted from what number is equal to ten?

$$n - 3 = 10$$

(b) Five times what number *is the same as* thirty-five?

$$5 \cdot n = 35$$

$5n = 35$

Continued on next page

(c) We let x represent Kari's savings, the unknown value.

Kari's savings decreased by $100 *equals* $500.

$$\begin{array}{ccccc} \downarrow & \downarrow & \downarrow & \downarrow & \downarrow \\ x & - & \$100 & = & \$500 \end{array}$$

$x - \$100 = \$500$

Student Practice 5   Translate each English sentence into an equation.

(a) Four times what number is the same as seven?

(b) Three subtracted from what number is equal to nine?

(c) The number of baseball cards in a collection plus 20 new cards equals 75 cards.

---

③ Solving Equations Using Basic Arithmetic Facts

Suppose we ask the question "Three plus what number is equal to nine?" The answer to this question is 6, since *three plus six is equal to nine*. The number 6 is called the **solution** to the equation $3 + x = 9$ and is written $x = 6$: "the value of x is 6." In other words, an *equation* is like a *question,* and the *solution* is the *answer* to this question.

English Phrase	**Math Symbols**
Question	*Equation*
Three plus what number is equal to nine?	$3 + x = 9$
Answer to the Question	*Solution*
Three plus *six* is equal to nine.	$x = 6$

The solution to an equation must make the equation a true statement. For example, if 6 is a solution to $3 + x = 9$, we must get a true statement when we evaluate the equation for $x = 6$.

$$3 + x = 9$$
$$3 + 6 = 9 \qquad \text{We replace the variable with 6, then simplify.}$$
$$9 = 9 \qquad \text{We get a true statement.}$$

> To *solve* an *equation* we must find a value for the variable in the equation that makes the equation a true statement.

EXAMPLE 6   Is 2 a solution to $6 - x = 9$?

Solution   If 2 is a solution to $6 - x = 9$, when we replace x with the value $x = 2$ we will get a true statement.

$$6 - x = 9 \qquad \text{"Six minus what number equals nine?"}$$
$$6 - 2 \overset{?}{=} 9 \qquad \text{Replace the variable with 2 and simplify.}$$
$$4 \overset{?}{=} 9 \qquad \text{This is a false statement.}$$

Since $4 = 9$ is *not* a true statement, 2 is *not* a solution to $6 - x = 9$.

Student Practice 6   Is 5 a solution to $x + 8 = 11$?

EXAMPLE 7   Solve the equation $3 + n = 10$ and check your answer.

Solution   To solve the equation $3 + n = 10$, we answer this question:

"Three plus what number is equal to ten?"

Using addition facts we see that the *answer, or solution,* is 7.

To check the solution, we replace n with the value $n = 7$ and verify that we get a true statement.

Check: $3 + n = 10$ Write the equation.

$3 + 7 \overset{?}{=} 10$ Replace the variable with 7 and simplify.

$10 = 10$ ✓   Verify that we get a true statement.

Since we get a true statement, the solution to $3 + n = 10$ is 7 and is written $n = 7$.

Student Practice 7   Solve the equation $4 + n = 9$ and check your answer.

EXAMPLE 8   Solve the equation $9n = 45$ and check your answer.

Solution   To solve the equation $9n = 45$, we answer this question:

"Nine times what number equals forty-five?"

The answer or solution is 5 and is written $n = 5$.

To check the answer, we replace n with the value $n = 5$ and verify that we get a true statement.

Check: $9 n = 45$

$9(5) \overset{?}{=} 45$ Replace the variable with 5 and simplify.

$45 = 45$ ✓   Verify that this is a true statement.

Thus the solution to $9n = 45$ is 5 and is written $n = 5$.

Student Practice 8   Solve the equation $6x = 48$ and check your answer.

EXAMPLE 9   Solve the equation $\dfrac{6}{x} = 3$ and check your answer.

Solution $\dfrac{6}{x} = 3$ "Six divided by what number equals 3?"

Using division facts, we see that the answer or solution is 2 and is written $x = 2$.

Check:

$$\frac{6}{x} = 3 \rightarrow \frac{6}{2} \overset{?}{=} 3 \qquad \text{Replace the variable with 2 and simplify.}$$

$$3 = 3 \checkmark \qquad \text{Verify that this is a true statement.}$$

Thus $x = 2$ is the solution.

Student Practice 9   Solve the equation $\frac{x}{4} = 2$ and check your answer.

Sometimes, we must first use the associative and commutative properties to simplify an equation and then find the solution.

EXAMPLE 10   Simplify using the associative and commutative properties and then find the solution to the equation $(5 + n) + 1 = 9$.

Solution   First we simplify.

$$(5 + n) + 1 = 9$$

$$(n + 5) + 1 = 9 \qquad \text{Commutative property}$$

Continued on next page

$$n + (5 + 1) = 9 \quad \text{Associative property}$$
$$n + 6 = 9 \quad \text{Simplify.}$$

Next we solve $n + 6 = 9$.

$$n + 6 = 9 \quad \text{"What number plus 6 is equal to 9?"}$$
$$n = 3$$

The solution to the equation is 3 and is written $n = 3$.
   We leave the check to the student.

Student Practice 10   Simplify using the associative and commutative properties and then find the solution to the equation $(3 + x) + 1 = 7$.

We may need to combine like terms before we solve an equation.

EXAMPLE 11   Simplify by combining like terms and then find the solution to the equation $n + 5n = 18$.

Solution

$$\begin{array}{ll} n + 5n = 18 & \text{Write the equation.} \\ 1n + 5n = 18 & \text{Write } n \text{ as } 1n. \\ (1 + 5)n = 18 & \text{Add numerical coefficients of like terms.} \\ 6n = 18 & \text{Think: "Six times what number equals eighteen?" 3.} \\ n = 3 & \end{array}$$

Student Practice 11   Simplify by combining like terms and then find the solution to the equation $n + 3n = 20$.

④ Translating and Solving Equations

In many real-life applications we must translate an English statement into an equation and then solve the equation.

EXAMPLE 12   Translate, then solve. Double what number is equal to eighteen?

Solution   Double what number is equal to eighteen?

$$\begin{array}{ccccc} \downarrow & & \downarrow & \downarrow & \downarrow \\ 2 & \cdot & n & = & 18 \quad \text{Translate.} \\ & & n & = & 9 \quad \text{Use multiplication facts to find } n. \end{array}$$

Student Practice 12   Translate, then solve. What number times five is equal to twenty?

Understanding the Concept

Evaluate or Solve?   Do you know the difference between evaluating the expression $8x$ when x is 3 and solving the equation $8x = 16$?

- *Evaluate an expression.* We replace the variable in the expression with the given number and then perform the calculation(s).
  Evaluate $8x$ when x is 3. $8 \cdot 3 = 24$

- *Solve an equation.* We find the value of the variable that makes the equation a true statement—that is, the solution to the equation.
  Solve: $8x = 16$. $x = 2$

We can illustrate this idea with the following situations.

1. Evaluating
   (a) *Fact.* You are given directions to the Lido Movie Theater.
   (b) *Evaluate.* You follow these directions to the movie theater.

2. Solving
   (a) *Fact.* You know the address of the theater.
   (b) *Solve.* You must find the directions yourself.

In summary, an equation has an equals sign, and an expression does not. We find the solutions to equations, and we evaluate expressions as directed.

Exercise

1. Can you think of other real-life situations that illustrate the difference between evaluating and solving?

STEPS TO SUCCESS Improving Your Test-Taking Skills

Step 1 *Write key facts on your test.* As soon as you get your test, find a blank area to write down any important strategies, formulas, or key facts.

Step 2 *Scan the test and work problems that are easy first.* Quickly glance at each question on the test, placing an * beside the ones you feel confident you can complete. Then complete these problems first. This will help build your confidence.

Step 3 *Keep track of the time as you complete the rest of the test.* Determine how much time is left so you can plan the strategy for the rest of the test. This plan should include determining how many minutes you should spend on each of the remaining unanswered questions so that you can finish the test in the time that is left.

Step 4 *Complete the rest of the problems on the test.* Complete the remaining problems on the test, starting with the ones you feel most confident about. If you get stuck on a problem, stop working on it and move on to another one.

Step 5 *Relax periodically.* If you start to feel anxious at any time during the test, take a few moments to relax. Close your eyes, place yourself in a comfortable position in your chair, breathe deeply, and take a moment to think about something pleasant. Next, think positive thoughts such as "I will answer the question to the best of my ability and will not worry about what I have forgotten or do not understand. Instead I will show that I can master what I do understand."

You may think that taking a few minutes away from the test to relax is wasting time. This is not true. You will perform better if you are relaxed.

Step 6 *Revisit the problems you are not sure of or did not complete.* Try to rework the problems that you struggled with earlier. You may recall how to complete these problems once you have completed the majority of the test.

Step 7 *Review the entire test to check for careless errors.* Take whatever time is left to review all your work. Check for careless errors and be sure that you have followed all directions properly.

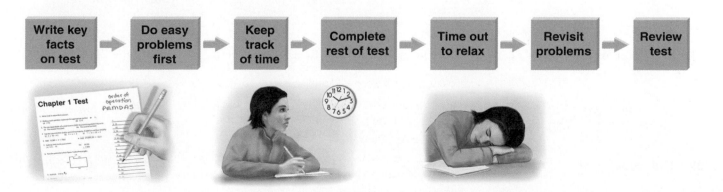

Verbal and Writing Skills, Exercises 1–12

Translate the mathematical symbols using words.

1. $7x$

2. $8x$

3. $8x = 40$

4. $6x = 30$

5. Can we add $2x + 3y$? Why or why not?

6. Can we add $4x + 2xy$? Why or why not?

Fill in the blanks.

7. In the expression $6x + 5$, $6x$ is called a _____ term and 5 is called a _____ term.

8. When two expressions are separated by an equals sign, we call it an _____.

9. The numerical part of $8x$ is ___ and is called the _____ of the term.

10. The numerical part of x is ___ and is called the _____ of the term.

11. Rewrite y with a coefficient: ___.

12. In the expression $12x + 9x$, $12x$ and $9x$ are called _____ terms.

Fill in each box to complete each problem.

13. $7x + 3\,\square = 10x$

14. $10x - 2\,\square = 8x$

15. $3xy + \square = 7xy$

16. $\square + 4ab = 6ab$

17. $3x + 5xy + \square = 7x + 5xy$

18. $7a + 2ab + \square = 7a + 4ab$

Write a term that represents each expression.

19. Three x's

20. Six y's

21. $a + a + a + a$

22. $x + x + x + x + x$

Identify like terms.

23. $5x + 3y + 2x + 8m + 7y$

24. $2m + 4b + 6m + 3x + 4b$

25. $2mn + 3y + 4mn + 6$

26. $7x + 3xy + 4 + 2xy$

Combine like terms.

27. $7x + 2x$

28. $13x + 3x$

29. $9y - y$

30. $7m - m$

31. $3x + 2x + 6x$

32. $4a + 8a + 3a$

33. $8x + 4a + 3x + a$

34. $9y + 2b + 2y + b$

35. $6xy + 4b + 3xy$

36. $7ab + 5x + 5ab$

37. $6xy + 3x + 9 + 9xy$

38. $5mn + 6m + 1 + 2mn$

39. $12ab - 5ab + 9$

40. $11xy - 2xy + 3$

41. $14xy + 4 + 3xy + 6$

42. $12ab + 6 + 5ab + 2$

Write the perimeter of each rectangular figure as an algebraic expression, then simplify.

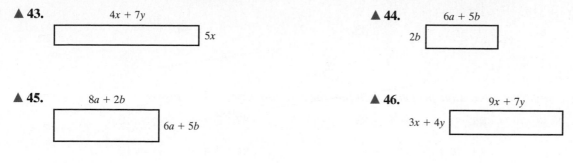

▲ **43.** $4x + 7y$ $5x$

▲ **44.** $6a + 5b$ $2b$

▲ **45.** $8a + 2b$ $6a + 5b$

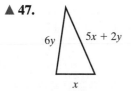

▲ **46.** $9x + 7y$ $3x + 4y$

Write the perimeter of each triangle as an algebraic expression, then simplify.

▲ **47.** $6y$ $5x + 2y$ x

▲ **48.** $3a + 2b$ $6b$ a

Translate to an equation. **Do not solve** *the equation.*

49. Five plus what number equals sixteen?

50. When twenty-four is added to a number, the result is fifty.

51. What number times three equals thirty-six?

52. What number times two is equal to forty?

53. If a number is subtracted from forty-five, the result is six.

54. If a number is subtracted from twelve, the result is two.

55. Twenty-five divided by what number is equal to five?

56. Twenty-two divided by what number is equal to eleven?

57. Let J represent James' age. James' age plus 12 years equals 25.

58. Let S represent Sherie's checking account balance. Sherie's checking account balance plus $14 equals $56.

59. Let C represent Chuong's monthly salary. Chuong's monthly salary decreased by $50 equals $1480.

60. Let P represent the price of the ticket. The price of the ticket decreased by $5 equals $16.

Answer yes or no.

61. Is 4 a solution to the equation $9 - x = 3$?

62. Is 3 a solution to the equation $5 - x = 3$?

63. Is 15 a solution to the equation $x + 4 = 19$?

64. Is 20 a solution to the equation $x + 6 = 26$?

Solve and check your answer.

65. $x + 5 = 9$

66. $x + 4 = 10$

67. $11 - n = 3$

68. $13 - n = 10$

69. $x \quad 6 = 0$

70. $x - 2 = 0$

71. $2 + x = 13$

72. $21 + x = 25$

73. $25 - x = 20$

74. $44 - n = 42$

75. $8x = 16$

76. $7y = 14$

77. $4y = 12$ **78.** $9x = 63$ **79.** $8x = 56$ **80.** $10y = 30$

81. $\dfrac{15}{y} = 1$ **82.** $\dfrac{12}{x} = 1$ **83.** $\dfrac{14}{x} = 2$ **84.** $\dfrac{20}{x} = 2$

Simplify using the associative and/or commutative property, then find the solution. Check your answer.

85. $(x + 1) + 3 = 7$ **86.** $(x + 6) + 5 = 13$ **87.** $2 + (7 + y) = 10$

88. $(3 + x) + 2 = 7$ **89.** $3 + (n + 5) = 10$ **90.** $2 + (8 + x) = 12$

Simplify by combining like terms, then find the solution. Check your answer.

91. $3n + n = 12$ **92.** $6n + n = 21$ **93.** $7x - 2x - x = 4$ **94.** $3y + y + 2y = 12$

Mixed Practice, Exercises 95–102

Solve.

95. $5x = 20$ **96.** $\dfrac{30}{x} = 15$ **97.** $16 - x = 1$ **98.** $38 - n = 34$

99. $1 + (4 + a) = 15$ **100.** $(6 + x) + 1 = 10$ **101.** $8x - 5x - x = 10$ **102.** $4y + y + 2y = 14$

For the following English sentences,

    (a) *Translate into an equation.*   **(b)** *Solve the equation.*

103. Four plus what number equals eight?

104. Three added to what number equals nine?

105. Three times what number is equal to nine?

106. Four times what number is equal to twelve?

One Step Further

▲ **107.** Find the missing side of the following triangle if the perimeter is 170 feet.

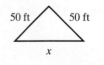

50 ft   50 ft

x

▲ **108.** Find the missing side of the following triangle if the perimeter is 110 yards.

30 yd   30 yd

x

Simplify by combining like terms.

109. $6 + (2x^2 + 5) + (7 + 3x^2) + x^2$

110. $(2 + 8x^2) + 9 + (4x^2 + 6) + x^2$

Simplify.

111. (a) $2x + 3x + 5y$
    (b) $(2x)(5y)$

112. (a) $5x + 4x + 6y$
    (b) $(5x)(6y)$

113. (a) $5a + 6y + 2a$
    (b) $(5a)(6y)$

114. (a) $6a + 7y + 3a$
    (b) $(6a)(7y)$

To Think About

Running Speeds Compared *Use the bar graph to answer exercises 115 and 116.*

115. (a) How fast can a domestic cat run if it is as fast as a grizzly bear?

(b) The speed of a lion is two times the speed of an elephant. How fast is the elephant?

116. (a) Which animal is faster, a zebra or a Cape hunting dog?

(b) The speed of a cheetah is two times the speed of a rabbit. How fast is the rabbit?

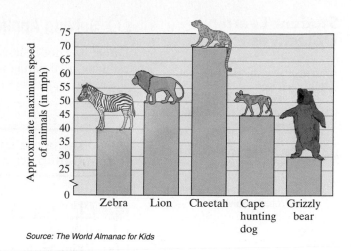

Source: The World Almanac for Kids

Cumulative Review

Match each operation described by the phrase in the right column with the appropriate operation listed in the left column. Place the correct letter in the blank space.

(a) Addition

(b) Subtraction

(c) Multiplication

(d) Division

117. [1.5.2] ____ Split equally between

118. [1.4.2] ____ Find the number of items in an array

119. [1.2.1] ____ Find the total

120. [1.3.2] ____ How much less

Quick Quiz 1.8

1. Combine like terms.

$2ab + 4a + 1 + ab$

2. Solve each equation and check your answer.

(a) $\dfrac{10}{a} = 2$

(b) $4 + (x + 7) = 12$

(c) $8y + y = 72$

3. Translate into an equation and solve.

(a) The product of three and what number is equal to eighteen?

(b) Let D represent Dave's age. Dave's age increased by seven is equal to twenty-one.

4. Concept Check Explain the difference in the process you must use to complete (a) and (b).

(a) Combine like terms. $3x + x + 2x$

(b) Solve. $3x + x + 2x = 12$

1.9 Solving Applied Problems Using Several Operations

Student Learning Objectives

After studying this section, you will be able to:

1. Solve applied problems involving estimation.

2. Solve applied problems involving charts and diagrams.

3. Use the Mathematics Blueprint for Problem Solving.

① Solving Applied Problems Involving Estimation

Often, it is not necessary to know the exact sum or difference; in this case we can estimate. Estimating is also helpful when it is necessary to do mental calculations. There are many ways to estimate, but in this book we use the following rule to make estimations.

> To estimate a sum or difference, round each number to the same round-off place and then find the sum or difference.

EXAMPLE 1 Some sample sale prices for 2010 Ford motor vehicles are listed below.

(a) Estimate the difference in the price between an F-150 SVT Raptor and a Mustang V6 Premium Coupe by rounding each price to the nearest thousand.

(b) Calculate the exact difference in cost. Is your estimate reasonable?

Manufacturers Suggested Retail Prices on 2010 Ford Vehicles

F-150 SVT Raptor	$38,020
Taurus SEL	$28,195
Focus SE Coupe	$15,895
Mustang V6 Premium Coupe	$26,695

Source: www.fordvehicles.com

Solution *Understand the problem.* The information we need to solve the problem is listed in the table.

(a) *Calculate and state the answer.* To estimate, we round each number to the thousands place.

	Exact Value		Rounded Value
The price of the F-150 SVT Raptor	38,020	→	38,000
The price of the Mustang V6 Premium Coupe:	26,695	→	27,000

We subtract the *rounded figures* to *estimate* the difference in the cost of the two vehicles.

$$38,000 - 27,000 = 11,000$$

The estimated difference in price is $11,000.

(b) We subtract the *original figures* to find the *exact* difference in the cost of the two vehicles.

$$\$38,020 - \$26,695 = \$11,325$$

The exact difference in price is $11,325.

The estimated difference in price, $11,000, is close to the exact difference, $11,325, so our estimate is reasonable. Note that if you round each number to the nearest hundred instead of to the nearest thousand, your estimate will not be wrong, just a little closer to the exact amount. When we estimate we want to make calculations with numbers that are easy to work with. In this case, it is easier to subtract numbers rounded to the thousands place than to the hundreds place; that is, subtracting $38,000 - 27,000$ is easier than subtracting $38,000 - 26,700$.

Student Practice 1 Use the sale prices listed in Example 1 to answer the following.

(a) Estimate the difference in price between an F-150 SVT Raptor and a Taurus SEL.

(b) Calculate the exact difference in cost. Is your estimate reasonable?

NOTE TO STUDENT: *Fully worked-out solutions to all of the Student Practice problems can be found at the back of the text starting at page SP-1.*

② Solving Applied Problems Involving Charts and Diagrams

How much material do I need to fence my yard? How much gasoline will I need for my trip? How much profit did my business make? One important use of mathematics is to answer these types of questions. In this section we combine problem-solving skills with the mathematical operations of addition, subtraction, multiplication, and division to solve everyday problems. We follow the three-step problem-solving process discussed earlier in the chapter when solving applied problems. We restate the three steps for your review.

Step 1. *Understand the problem.* Read the problem and organize the information. Use pictures and charts to help you see facts more clearly.

Step 2. *Calculate and state the answer.* Use arithmetic and algebra to find the answer.

Step 3. *Check the answer.* Use estimation and other techniques to test your answer.

EXAMPLE 2 The three owners of the Pizza Palace redecorated their business. The items purchased are listed on the invoice below. If the cost of these purchases excluding tax was divided equally among the owners, how much did each owner pay?

	INVOICE NO. 35513
AAA Restaurant Supply	SOLD TO Pizza Palace

QUANTITY	TYPE OF PURCHASE	PRICE PER ITEM	TOTAL COST PER ITEM
9	Tables		
54	Chairs	26	
2	Ovens	1020	
Thank You	Total Cost for all Items		

Solution *Understand the problem.* Read the problem carefully and study the invoice. Then fill in the invoice.

Calculate and state the answer. Multiply to get the total cost per item. Then place these amounts on the invoice.

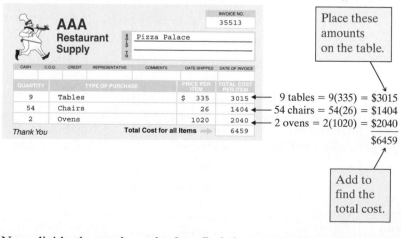

	INVOICE NO. 35513
AAA Restaurant Supply	SOLD TO Pizza Palace

QUANTITY	TYPE OF PURCHASE	PRICE PER ITEM	TOTAL COST PER ITEM
9	Tables	$ 335	3015
54	Chairs	26	1404
2	Ovens	1020	2040
Thank You	Total Cost for all Items		6459

Place these amounts on the table.

9 tables = 9(335) = $3015
54 chairs = 54(26) = $1404
2 ovens = 2(1020) = $2040
                              $6459

Add to find the total cost.

Now, divide the total cost by 3 to find the amount that each owner paid: 6459 ÷ 3 = 2153.

Each owner paid $2153.

Student Practice 2 The two owners of a Chinese restaurant redecorated their place of business. The items purchased are listed on the invoice in the margin. If the cost of these purchases was divided equally between the owners, excluding tax, how much did each owner pay?

INVOICE NO. 25104	
SOLD TO The China Palace	WHOLESALE SUPPLY STORE

QUANTITY	TYPE OF PURCHASE	PRICE PER ITEM	TOTAL COST PER ITEM
9	Tables	$ 230	
50	Chairs	25	
3	Ovens	910	
Thank You	Total Cost for all Items		

Continued on next page

Check your answer. We estimate and compare the estimate with our calculated answer.

Round:	*Number of Items*	*Price per Item*	*Find Total Cost*
	$9 \rightarrow 10$	$\$335 \rightarrow \$300$	$10 \cdot 300 = \$3000$
	$54 \rightarrow 50$	$\$26 \rightarrow \$30$	$50 \cdot 30 = 1500$
	$2 \rightarrow 2$	$\$1020 \rightarrow \$1000$	$2 \cdot 1000 = \underline{2000}$
			$\$6500$

Divide the estimated total by 3: $3\overline{)6500} \quad \dfrac{2166}{} \text{ R2} \approx \$2167$

Our estimate of $2167 per owner is close to our exact calculation of $2153. Thus our *exact answer is reasonable.*

③ Using the Mathematics Blueprint for Problem Solving

To solve a word problem that contains many facts and requires several steps, it is helpful to *organize* the information and then *plan* the process that you will use. This is very similar to what we do when we use a daily planner or date book to *organize* and *plan* our days. A Mathematics Blueprint for Problem Solving will be used to organize the information in a word problem and plan the method to solve it. From the blueprint we will be able to see clearly the three steps for solving problems in real-life situations: understand the problem; calculate and state the answer; and then check the answer.

EXAMPLE 3 A frequent-flyer program offered by many major airlines to first-class passengers awards 3 frequent-flyer mileage points for every 2 miles flown. When customers accumulate a certain number of frequent-flyer points, they can cash them in for free air travel, ticket upgrades, or other awards. How many frequent-flyer points would a customer accumulate after flying 3500 miles in first class?

Solution *Understand the problem.* Sometimes, drawing charts or pictures can help us understand the problem as well as plan our approach to solving the problem.

$$2 \text{ miles} + 2 \text{ miles} \cdots = 3500 \text{ miles} \quad \text{How many groups of 2's are in 3500?}$$
$$3 \text{ points} + 3 \text{ points} \cdots = ? \text{ points}$$

We organize the information and make our plan in the Mathematics Blueprint.

Mathematics Blueprint for Problem Solving

Gather the Facts	What Am I Asked to Do?	How Do I Proceed?	Key Points to Remember
A customer is awarded 3 frequent-flyer points for every 2 miles flown.	Determine how many frequent-flyer points a customer earns after flying 3500 miles.	1. *Divide* 3500 by 2. 2. *Multiply* 3 times the number obtained in step 1.	Frequent-flyer points are determined by the number of miles flown.

Calculate and state the answer.

Step 1. We divide to find how many groups of 2 are in 3500.
$$3500 \div 2 = 1750$$

Step 2. We multiply 1750 times 3 to find the total points earned.
$$1750 \cdot 3 = 5250 \text{ points}$$

The customer would earn 5250 points.

Check. If the customer earned 4 points (instead of 3) for every 2 miles traveled, we could just double the mileage to find the points earned.

2 miles earns 4 points (double 2).

3500 miles earns 7000 points (double 3500).

Since the customer earned a little less than 4 points, the total should be less than 7000. It is: $5250 < 7000$. The customer also earned more points than miles traveled (3 points for every 2 miles), so the total points should be more than the total miles traveled. It is: $5250 > 3500$. *Our answer is reasonable.*

Student Practice 3   Use the information in Example 3 to determine how many frequent-flyer points a customer would accumulate if she flew 4500 miles.

EXAMPLE 4   Koursh was offered two different jobs: a 40-hour-a-week store assistant management position that pays $14 per hour and an executive secretary position paying a monthly salary of $2600. Which job pays more per year?

Solution   *Understand the problem.* We organize the information in the Mathematics Blueprint.

Mathematics Blueprint for Problem Solving

Gather the Facts	What Am I Asked to Do?	How Do I Proceed?	Key Points to Remember
The store assistant management position pays $14 per hour for 40 hours. The secretary's position pays $2600 per month.	Determine which job pays a higher salary per year.	1. Calculate the assistant manager's weekly pay. 2. Multiply the result of step 1 by 52 weeks to find yearly pay. 3. Multiply the secretary's pay by 12 months to find yearly pay. 4. Compare both salaries.	I must find *yearly* pay: 12 months = 1 year 52 weeks = 1 year

Calculate and state the answer. From the information organized in the blueprint, we can write out a process to find the answer.

$14 \times 40 = $560   Pay for 1 week (management)

$560 \times 52 = $29,120   Pay for 1 year (management)

$2600 \times 12 = $31,200 Pay for 1 year (secretary)

The yearly pay is $29,120 for the management position and $31,200 for the secretary's position.

The *secretary's position* pays more per year.

Continued on next page

Check the answer. We estimate the assistant manager's pay per year by rounding $14 per hour to $10 and 52 weeks to 50 weeks.

$$\$10 \times 40 \text{ hr} = \$400 \text{ per week;} \qquad \$400 \times 50 \text{ weeks} = \$20,000 \text{ per year}$$

We estimate the secretary's pay per year by rounding 12 months per year to 10.

$$\$10 \times \$2600 = \$26,000 \text{ per year}$$

Since $\$26,000 > \$20,000$, the secretary position pays more. ✓

Student Practice 4   Emily is a salesperson for A&E Appliance. For the last two years she has averaged about 7 sales per week, and she is paid solely on commission—$55 per sale. The store manager has decided to offer all salespersons the options of accepting a salary of $1770 per month or remaining on commission. If Emily continues to maintain her past sales record, which option will earn her more money per year?

👣 STEPS TO SUCCESS The Day Before the Exam

The day before the exam is not a good time to start reviewing. Use this day to skim your chapter review, homework, quizzes, and other review material. On this day you can fine-tune what you already know and review what you are unsure of. Starting your review early reduces anxiety so that you can think clearly during the test. Often, low test scores are related to high anxiety. Plan ahead so that you can relax on the day of the exam.

If you have been following the advice in the other Steps to Success boxes, you should be almost ready for the exam. If you have not read these boxes, be sure to allow extra time to study for the exam.

Making it personal:

Place a ✔ *in the blank as you complete each of the following.*

The day before the exam complete the following.

___ 1. If you have not read the Steps to Success boxes, reading them now will provide valuable advice. Pay special attention to "Reviewing for an Exam" on page 66, and "Improving Your Test-Taking Skills" on page 75.

___ 2. Re-take the How Am I Doing? test at the end of the chapter. Take this test as if it were the real exam. Do not refer to notes or to the text while completing the test.

___ 3. Grade the test. The problems you missed on this test are the type of problems that you should review.

___ 4. Watch the Test Prep Video and read the Math Coach to review the solutions to the How Am I Doing? test.

Relax, don't stay up late, and you will be alert for the exam.

Applications

Estimate each of the following.

1. Supplies Purchased Emma purchased supplies to paint her kitchen, family room, and dining room. The following store receipt indicates the supplies she purchased.

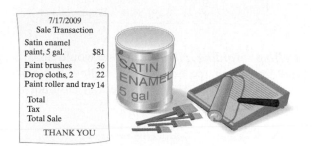

7/17/2009 Sale Transaction	
Satin enamel paint, 5 gal.	$81
Paint brushes	36
Drop cloths, 2	22
Paint roller and tray	14
Total	
Tax	
Total Sale	
THANK YOU	

(a) Excluding tax, *estimate* how much money Emma spent by rounding each amount to the nearest ten.

(b) Find the total money spent.

(c) Is your *estimate* reasonable?

2. Supplies Purchased Julio Arias bought schoolbooks and school supplies. The following store receipt indicates his purchases.

8/27/2010 Sale Transaction	
Math book	$41
History book	37
Notebooks, 2	13
Graphing calculator	89
Total	
Tax	
Total Sale	

THANK YOU FOR SHOPPING AT THE CORNER BOOK STORE

(a) Excluding tax, *estimate* how much money Julio spent by rounding each amount to the nearest ten.

(b) Find the total money spent.

(c) Is your *estimate* reasonable?

3. Mileage on Vehicle The Arismendi family took a scenic drive across the country. From Phoenix, Arizona, they drove east 597 miles the first day, 512 miles the second day, 389 miles the third day, and 310 miles the fourth day. Round each amount to the nearest hundred and then *estimate* how many more miles the Arismendi family drove the first two days than the last two days.

4. Mileage on Vehicle Mike drove his Toyota truck 15,300 miles the first year he owned it, 14,880 the second year, 9100 the third year, and 13,950 the fourth year. Round each amount to the nearest thousand and then *estimate* how many more miles Mike drove his truck the first two years than the second two years.

5. Restaurant Remodeling The 5 owners of Mei's Restaurant remodeled their business. They bought 7 tables, 20 chairs, and 2 crystal light fixtures. The cost of these purchases was divided equally among the owners. Excluding tax, how much did each owner pay?

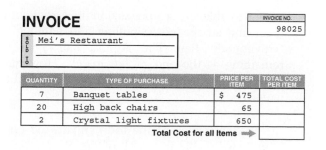

INVOICE   INVOICE NO. 98025

SOLD TO: Mei's Restaurant

QUANTITY	TYPE OF PURCHASE	PRICE PER ITEM	TOTAL COST PER ITEM
7	Banquet tables	$ 475	
20	High back chairs	65	
2	Crystal light fixtures	650	
		Total Cost for all Items ➡	

6. Appliances Purchased Last weekend, May's Appliance Store sold 10 washing machines, 5 dryers, and 20 dishwashers to a private college. The college will divide the expense for upgrades equally among the 200 students in the college apartments by charging each student a one-time assessment fee. How much will the assessment fee be for each student?

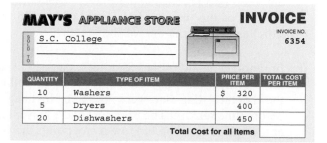

MAY'S APPLIANCE STORE   **INVOICE**   INVOICE NO. 6354

SOLD TO: S.C. College

QUANTITY	TYPE OF ITEM	PRICE PER ITEM	TOTAL COST PER ITEM
10	Washers	$ 320	
5	Dryers	400	
20	Dishwashers	450	
		Total Cost for all Items	

Solve each problem involving charts and diagrams.

7. ***Entertainment Event Tickets*** Dave and his family went to the Middletown Amusement Park. They purchased 2 adult tickets, 4 child tickets, and 1 senior citizen ticket. How much did they spend on the tickets?

Middletown Ticket Prices
Adult – $ 13
Child – $ 5 (under 12 years)
Senior citizen – $ 7 (over 55 years)

8. ***Entertainment Event Tickets*** Ranak and her friends went to an outdoor jazz concert. They purchased 5 adult, 4 student, and 3 child tickets. How much money did they spend on concert tickets?

Outdoor Jazz Concert Ticket Prices
Adult – $ 17
Child – $ 8 (under 12 years)
Student discount – $ 9 (college ID required)

▲ 9. ***Fencing Needed*** John Tulson wants to put a fence around the back and sides of his property (see the diagram). How many feet of fence must he purchase?

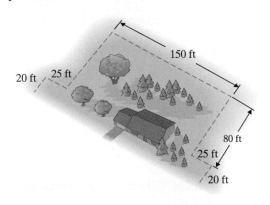

▲ 10. ***Ceiling Molding Needed*** Rosa would like to put molding along the edge of the ceiling in her kitchen (see the diagram). How many feet of molding will she need?

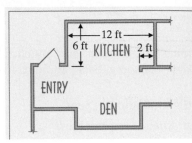

Paintball *In many paintball tournaments there are 100 points possible in each match of a round of play. There are 7 players on each team and when a player is tagged with a paintball, he is out of the match. The team with the higher point total at the end of each round of play moves on to the next round. Use the chart below to answer exercises 11 and 12.*

Point Chart	
Tagging the opposing team players with paintballs	3 points/player
Pulling the flag of the opposing team	22 points
Hanging the flag of the opposing team	50 points
Players left in the game at the end of play	1 point/player

11. **(a)** The Torches tagged 6 of the Shooters' players, and succeeded in pulling and hanging the Shooters' flag. The Torches had 5 players remaining at the end of the match. How many points did the Torches have at the end of the match?

   (b) In another match, the Shooters tagged all of the Torches' players. They succeeded in pulling the Torches' flag but were not able to hang it before time ran out. The Shooters had 1 player left at the end of the match. How many points did the Shooters have at the end of the match?

12. **(a)** In the second half of the tournament the Alpha team was able to tag 5 players on the Greyhounds' team and succeeded in pulling and hanging the Greyhounds' flag. The Alphas had 3 players remaining at the end of the match. How many points did the Alpha team have at the end of the match?

   (b) In the final match of the tournament, the Greyhounds were able to tag all of the players on the Alpha team. The Greyhounds had 5 players remaining at the end of the match, but were unable to succeed in pulling and hanging the flag of the other team before time ran out. How many points did the Greyhounds receive in this match?

13. *Hourly/Overtime Wages* A restaurant cook earns $8 per hour for the first 40 hours worked and $12 per hour for overtime (hours worked in addition to the 40 hours a week). Last week the cook worked 52 hours. Calculate the cook's total pay for that week.

    (a) Fill in the Mathematics Blueprint for Problem Solving.

    (b) Calculate and state the answer.

Mathematics Blueprint for Problem Solving

Gather the Facts	What Am I Asked to Do?	How Do I Proceed?	Key Points to Remember

14. *Apartment Expenses* Four roommates share equally the following expenses for their apartment: $920 for rent, $96 for utilities, and $56 for the telephone. How much is each roommate's monthly share?

    (a) Fill in the Mathematics Blueprint for Problem Solving.

    (b) Calculate and state the answer.

Mathematics Blueprint for Problem Solving

Gather the Facts	What Am I Asked to Do?	How Do I Proceed?	Key Points to Remember

15. *Business Profit* T. B. Etron's Company made $68,542 last year. The expenses for that year were $14,372.

    (a) How much profit did the company make?

    (b) If the two owners divided the profits equally, how much money did each owner receive?

16. *PTA Raffle* The R. L. Saunders High School PTA sold $2568 in raffle tickets. The expenses for the prizes were $1062.

    (a) How much profit did the PTA make?

    (b) If the profits were divided equally among three clubs, how much money did each club receive?

17. *Fishing Trip Expenses* Carlos and three of his friends will equally share the expenses for a 2-day fishing trip they plan to take this summer. The boat rental will be $450 per day, and they estimate that the gasoline will cost $50 per day. If the total cost of food and bait for the entire 2-day trip is $200, what will be each person's share of the total expenses?

18. *Comparing Earnings* Sara's current job as a computer technician at ComTec pays a salary of $3200 per month. BLM Accountants offered her a programmer's position that pays $16 per hour for a 40-hour week. Which job pays more per year?

19. *Bus Pass vs. Daily Rate* Round-trip bus fare is $2. Justin rides the bus 5 days a week to work and 2 nights a week to school. He can buy a pass at school that allows 6 months of unlimited bus rides for $400. If Justin only rides the bus round-trip to work and school, is it cheaper for Justin to buy the pass or to pay each time he rides the bus?

20. *Salary vs. Commission* Myra sells new memberships for a Total Flex Fitness Center chain. She is paid only on commission—$40 for each new membership. For the last three years she has signed up an average of 10 new members per week. She has been offered an alternative pay option—a salary of $1800 per month. Based on her average sales, with which pay option will she earn more money per year?

21. *Inheritance Proceeds* Glenda has inherited $6000, which will be distributed in two equal payments. She will receive the first half of the inheritance now and the remainder of the inheritance in 6 months. Glenda plans to invest the first payment in a certificate of deposit (CD) at her bank. She will distribute the second payment equally among her three children.

  (a) How much will Glenda invest in a CD?

  (b) How much will each child receive?

22. *Stamp Collection* Lester donated one-half of the 2500 stamps in his collection to the local senior citizen group. He distributed the remaining stamps in his collection equally among his five grandchildren.

  (a) How many stamps did Lester donate to the senior citizen group?

  (b) How many stamps from Lester's collection did each grandchild receive?

Grocery Store Purchases Al's Grocery Store gave customers the following incentive to shop at the store.

23. Jesse made three purchases at Al's Grocery Store during the month of June: $30, $240, and $170.

  (a) How many points did Jesse earn during the month of June?

  (b) How many discount dollars did Jesse earn?

24. Marsha made three purchases at Al's Grocery Store during the month of June: $230, $140, and $180.

  (a) How many points did Marsha earn during the month of June?

  (b) How many discount dollars did Marsha earn?

Promotional Points The L&M Clothing chain offered customers the following incentive to use their L&M charge card.

25. Alyssa made two purchases at L&M Clothing during the month of January: $170 and $260.

  (a) How many points did Alyssa earn during the month of January?

  (b) How many discount dollars did Alyssa earn?

26. Ian made three purchases at L&M Clothing during the month of January: $80, $160, and $220.

  (a) How many points did Ian earn during the month of January?

  (b) How many discount dollars did Ian earn?

January's Promotion

Earn points every time you use your *L & M* charge card.

Earn 10 points for every $50 charged during the month of January plus an additional 50 points for any single purchase over $200.

Cash in your points!

25 points earn you a $5 discount in February.

To Think About

27. *Credit Card Debt Repayment* A $5000 debt on a credit card will take 32 years to repay if only the minimum monthly payment is made. This debt will cost the borrower about $7800 in interest. The borrower could be out of debt in 3 years by paying $175 per month. Find the amount of interest paid at the end of 3 years.

Writing numbers in exponent form can often help us identify a pattern in a sequence of numbers. For example, the sequence 1, 8, 27, 64, 125, . . . can be written as $1^3, 2^3, 3^3, 4^3, 5^3$, . . . and we see that the next number would be 6^3 or 216. Identify a pattern and then find the next number in the following sequences.

28. 4, 16, 36, 64, 100, . . .

29. 9, 25, 49, 81, 121, . . .

Find the next two figures that would appear in the sequence.

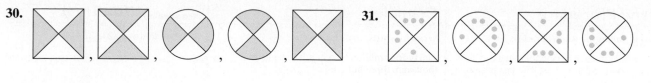

30.

31.

Cumulative Review

32. **[1.4.3]** Multiply. $4 \cdot 3 \cdot 2 \cdot 5$

33. **[1.8.3]** Solve. $6x = 30$

34. **[1.8.3]** Solve. $x + 9 = 12$

Quick Quiz 1.9

1. The Belmont High Service Club organized a charity jazz festival. Tickets to the festival were $20 for adults, $15 for students, and $7 for children under 12 years old. The service club sold 350 adult tickets, 200 student tickets, and 47 child tickets.

   (a) Find the total income from the sale of tickets.

   (b) If the expenses for the festival were $3400, how much profit did the Club make from the charity event?

2. **Dairy Cows** A dairy cow produces an average of 7 gallons of milk a day. If a farmer has a herd of 35 cows, how much milk will they produce in 1 day? In 1 week?

3. Luis is opening his new accounting office and must order furniture for the front office. He found a sale at a local store that is offering free delivery and no sales tax. Luis must order 2 filing cabinets, 1 desk, 1 office chair, 2 bookcases, 6 guest chairs, 2 end tables, and 1 coffee table. The prices are listed in the table below. Round each price to the nearest hundred and then *estimate* the total price of the furniture.

Filing Cabinet	$467	Guest Chair	$197
Desk	$765	End Table	$317
Office Chair	$255	Coffee Table	$421
Bookcase	$299		

4. **Concept Check** At the end of January, Sahara had $200 left in her vacation savings account and $1000 in her household savings account. Each month for the next six months, Sahara plans to put $100 in her vacation account and $200 in her household account. In addition, she plans to split her $900 tax return equally between both accounts. Explain how to determine if Sahara will have enough money in her vacation account at the end of six months to take a $1500 vacation.

Chapter 1 Organizer

Topic and Procedure	Examples	✏ You Try It
Writing word names for whole numbers, p. 4 Name the number in each period followed by the name of the period and a comma. The last period name, "ones," is not used. (millions) (thousands) (ones) xxx xxx xxx	Write in words. 34,218,316 Thirty-four million, two hundred eighteen thousand, three hundred sixteen	**1.** Write in words. 23,327,414

Inequality symbols, p. 4
 	
< and > are called inequality symbols. The symbol ">" means *is greater than,* and "<" means *is less than.* The inequality symbol always points to the smaller number.
	Replace ? with the inequality symbol < or >.
	
	  9 ? 3 15 ? 20
	
	  9 > 3 15 < 20
		2. Replace ? with the inequality symbol < or >.
		
		2 ? 11 17 ? 13

Rounding whole numbers, pp. 5, 6
 	
1. Identify the round-off place.
 	
2. If the digit to the right of the round-off place is
  (a) Less than 5, do not change the round-off place digit.
  (b) 5 or more, increase the round-off place digit by 1.
 	
3. In either case, replace all digits to the right of the round-off place digit with zeros.
	Round 27,468 to the nearest hundred.
	
	┌── The round-off place digit is 4.
	2 7,④6 8
	    └── The digit to the right is 5 or more.
	
	┌── Increase the round-off place digit.
	2 7,5 0 0
	   └ Replace digits to the right with zero.
		3. Round 133,442 to the nearest thousand.

Using properties of addition to simplify, pp. 11, 12 The commutative property states that we can change the order of numbers when adding. The associative property states that we can regroup numbers when adding. We use both of these properties to simplify expressions.	Use the commutative and/or associative property as necessary to simplify. $3 + (n + 2)$ Change the order $3 + (n + 2)$ of addition. $= 3 + (2 + n)$ Regroup. $= (3 + 2) + n$ Add. $= 5 + n$ or $n + 5$	**4.** Use the commutative and/or associative property as necessary to simplify. $(x + 6) + 8$
Adding whole numbers, p. 14 Starting with the far right column, add each column separately. If a two-digit sum occurs, carry the first digit over to the next column to the left.	Add. 382 + 156 + 73 + 5 $$\begin{array}{r} \overset{2\,1}{3}82 \\ 156 \\ 73 \\ +\ \ 5 \\ \hline 616 \end{array}$$	**5.** Add. 121 + 46 + 592 + 3

Finding the perimeter, pp. 15, 16
 	
The perimeter is the distance around an object. We add the lengths of all sides to find the perimeter.
	Find the perimeter of the shape consisting of rectangles.
	
	1. We find the length of the unlabeled sides.
	
	  3 m + 6 m = 9 m; unlabeled vertical side
	  7 m + 8 m = 15 m; unlabeled horizontal side
	
	2. We add the lengths of all sides.
	
	  6 + 7 + 3 + 8 + 9 + 15 = 48 m
		6. Find the perimeter of the shape consisting of rectangles.

Subtracting whole numbers, pp. 24, 25 Starting with the right column, subtract each column separately. If necessary, borrow a unit from the column to the left and bring it to the right as a "10."	Subtract. 26,581 − 4832 $$\begin{array}{r} {}^{5\ 15\ 7\,11}{2\cancel{6},\cancel{5}\cancel{8}\cancel{1}} \\ -\ \ 4,832 \\ \hline 2\,1,749 \end{array}$$	**7.** Substract. 47,621 − 5,935

Topic and Procedure	Examples	✏ You Try It

Key words for multiplication, p. 33
 	
The key words or phrases that represent multiplication are *times, product of, double,* and *triple*.
	Translate using numbers and symbols.
	
	Double a number: Triple a number:
	    2x **3x**
	
	The product of Six times x:
	two and three: **2·3** **6x**
		8. Translate using numbers and symbols.
		
		  (a) Twice a number:
		
		  (b) Five times a number:
		
		  (c) A number times eight:
		
		  (d) The product of four and two:

Properties of multiplication, p. 34 We can regroup and change the order of numbers when we multiply since multiplication is associative and commutative.	Simplify. $2(x \cdot 3)$ Change the order of multiplication and regroup. $(2 \cdot 3) \cdot x$ Multiply. $(2 \cdot 3) \cdot x = 6x$	9. Simplify. $4(y \cdot 5)$
Multiplying whole numbers, p. 37 Multiply the top factor by the ones digit, then by the tens digit, then by the hundreds digit of the lower factor. Add the partial products.	Multiply. $567 \cdot 238$ $$\begin{array}{r} 567 \\ \times\ 238 \\ \hline 4536 \\ 1701 \\ 1134 \\ \hline 134{,}946 \end{array}$$	10. Multiply. $468 \cdot 251$

Key words for division, p. 45
 	
The key words or phrases that represent division are *divided by, shared equally, divided equally,* and *quotient*.
	Translate using numbers and symbols.
	  Nine divided by n: $9 \div n$
	The quotient of fifteen and five: $15 \div 5$
	The quotient of five and fifteen: $5 \div 15$
		11. Translate using numbers and symbols.
		
		  (a) The quotient of six and x:
		
		  (b) The quotient of x and six:
		
		  (c) A number divded by 3:

Long division, p. 47 We *guess* the quotient and *check* by multiplying the quotient by the divisor. We *adjust* our guess if it is too large or too small and continue the process until we get a remainder less than the divisor.	Divide. $1278 \div 25$ $$\begin{array}{r} 51\ \text{R3} \\ 25\overline{)1278} \\ -125 \\ \hline 28 \\ -\ 25 \\ \hline 3 \end{array}$$	12. Divide. $988 \div 21$
Exponents, p. 55 2^3 is written in exponent form. The exponent is 3, and the base is 2. Thus 2^3 is read "two to the third power" and means that there are 3 factors of 2.	Write in exponent form. $4 \cdot 4 \cdot 4 \cdot 4 \cdot x \cdot x$ $$4^4 x^2$$ Find the value of 3^3 and 7^2. $$3^3 = 3 \cdot 3 \cdot 3 = 27 \qquad 7^2 = 7 \cdot 7 = 49$$	13. **(a)** Write in exponent form. $$8 \cdot 8 \cdot 8 \cdot n \cdot n$$ **(b)** Find the value of 2^4.

Order of operations, p. 57
 	
1. Perform the operations inside parentheses.
2. Then simplify any expressions with exponents.
3. Then do multiplication and division in order from left to right.
4. Then do addition and subtraction in order from left to right.
	Evaluate. $2^3 + 16 \div 4^2 \cdot 5 - 3$
	$8 + 16 \div \mathbf{16} \cdot 5 - 3$ Raise to a power first.
	$8 + \mathbf{1} \cdot 5 - 3$ Then multiply and
	           divide from left to right.
	$8 + \mathbf{5} - 3$
	$\mathbf{13} - 3 = 10$ Then add and subtract
	           from left to right.
		14. Evaluate. $4 + 8 \div 2^2 \cdot 5 - 3^2$

Using symbols and key words for expressing algebraic expressions, p. 64
 	
When a phrase contains key words for more than one operation, the phrases *sum of* or *difference of* indicate that these operations must be placed within parentheses.
	Translate using numbers and symbols.
	Six times x minus two.
	$$6x - 2$$
	Six times the difference of x and two.
	$$6(x - 2)$$
		15. Translate using numbers and symbols.
		
		  (a) Four times the difference of x and 5:
		
		  (b) Four times x minus five:

Topic and Procedure	Examples	✏ You Try It
Distributive property, p. 65 To multiply $a(b + c)$ and $a(b - c)$ we distribute the a by multiplying every number or variable inside the parentheses by a and then simplifying.	Simplify. $4(x + 2)$ $$4 \cdot x + 4 \cdot 2 = 4x + 8$$	16. Simplify. $7(n - 3)$
Combining like terms, p. 69 We either add or subtract numerical parts of like terms. The variable part stays the same.	Combine like terms. $2xy + 3x + 5xy$ $$7xy + 3x$$	17. Combine like terms. $4mn + 2n + 6mn$
Solving equations, p. 72 We solve simple equations using basic arithmetic facts. To check, we replace the variable with the number to see if it makes the equation a true statement.	Solve. **(a)** $6n = 42 \rightarrow 6 \cdot 7 = 42$ $\qquad\qquad\qquad n = 7$ **(b)** $\dfrac{27}{x} - 9 \rightarrow \dfrac{27}{3} = 9$ $\qquad\qquad\qquad x = 3$	18. Solve. **(a)** $4n = 24$ **(b)** $\dfrac{35}{x} = 7$
Simplifying and solving equations, pp. 73, 74 We simplify using the commutative and associative properties before we solve equations.	Solve. $$2 + (n + 7) = 11$$ $$2 + (7 + n) = 11$$ $$(2 + 7) + n = 11$$ $$9 + n = 11$$ $$n = 2$$	19. Solve. $3 + (x + 2) = 15$
Translating and solving equations, p. 74 We translate the English sentence into math symbols and then solve for the variable.	What number subtracted from twelve is equal to four? $$12 - x = 4$$ $$12 - 8 = 4 \rightarrow x = 8$$	20. What number subtracted from ten is equal to two?
Evaluating expressions, p. 74 **Addition, pp. 13, 14** **Subtraction, pp. 23, 24** **Multiplication, pp. 64, 65** **Division, pp. 64, 65** To evaluate an expression, we replace the variable in the expression with the given value and then simplify.	Evaluate. **(a)** $3a + 2b$ for $\quad$ **(b)** $\dfrac{(x - 1)}{2}$ for $x = 5$ $a = 2$ and $b = 4$ $3 \cdot 2 + 2 \cdot 4 \qquad \dfrac{(5 - 1)}{2} = \dfrac{4}{2} = 2$ $= 6 + 8 = 14$	21. Evaluate. **(a)** $\quad 5x + 3y \quad$ for $x = 3$ and $y = 2$ **(b)** $\dfrac{(x - 4)}{3} \quad$ for $x = 10$
Estimating, p. 80 We round each number to the same round-off place before using the number in calculations.	A Ford dealership reduced the price of a Ford Taurus from \$23,995 to \$22,950 for the Labor Day weekend sale. Estimate the savings on the Taurus. We round \$23,995 to \$24,000 and the sale price of \$22,950 to \$23,000. To estimate the savings, we subtract. $\$24,000 - \$23,000 = \$1000$ savings	22. Jason bought a flat screen TV in a department store for \$2499. Sara shopped online and bought the same TV for \$2130. Estimate how much money Sara saved by shopping online.

Procedure for Solving Applied Problems

Using the Mathematics Blueprint for Problem Solving, p. 82

When solving problems, students often find it helpful to complete the following steps. You will not use all the steps all the time. Choose the steps that best fit the condition of the problem.

Understand the problem.

(a) Read the problem carefully and then draw a picture or chart.

(b) Think about the facts you are given and what you are asked for.

(c) Use the Mathematics Blueprint for Problem Solving to organize your work.

Calculate and state the answer. Perform the necessary calculations and state the answer, including the units of measure.

Check.

(a) Estimate your answer and check it with the value calculated to see if your answer is reasonable, or

(b) Repeat the calculations working the problem a different way.

EXAMPLE The Austin Department of Housing purchased several computer stations for a total cost of $7850 and bought software to update their system for $1055. The department had $9450 in their supply budget for the quarter. After this purchase, how much money was left for supplies?

Understand the problem.

Mathematics Blueprint for Problem Solving

Gather the Facts	What Am I Asked to Do?	How Do I Proceed?	Key Points to Remember

Cost of purchases:
 	
  Computers: $7850
 	
  Software: $1055
 	
Money available: $9450
	Find the amount of money left after the purchases.	1. Find the total cost of the computers and software.
		
		2. Subtract this amount from the money available in the supply budget.
			Be sure to copy all the facts correctly.

Calculate and state the answer. Calculate the cost of the computers and the software.

$$\begin{array}{r} \$7850 \\ +\ \ 1055 \\ \hline \end{array}$$

Cost of purchases: $8905

$$\begin{array}{r} \$9450 \\ -\ \ 8905 \\ \hline \end{array}$$

Money left: $ 545

Check. We estimate the amount of money left in the budget.

$7850 rounds to: $7900
$1055 rounds to: + 1100
Estimated cost of purchases: $9000

$9450 rounds to: $9500
                  − 9000
Estimate of money left: $ 500

The estimated balance of $500 is close to the answer, $545. We determine that the answer is reasonable.

Chapter 1 Review Problems

Vocabulary *Write the definition for each word.*

1. (1.2) Rectangle: _____

2. (1.2) Square: _____

3. (1.2) Right angle: _____

4. (1.2) Triangle: _____

5. (1.2) Perimeter: _____

6. (1.4) Factors: _____

7. (1.8) Term: _____

8. (1.8) Constant term: _____

9. (1.8) Coefficient: _____

10. (1.8) Like terms: _____

11. (1.8) Equation: _____

Section 1.1

12. In the number 175,493:

    (a) In what place is the digit 7?

    (b) In what place is the digit 5?

Fill in the check for the amount indicated.

13. $187

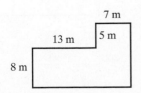

Write each number in expanded notation.

14. 7694

15. 5831

Replace each question mark with the inequality symbol < or >.

16. 2 ? 8

17. 12 ? 0

Rewrite using numbers and an inequality symbol.

18. Six is greater than one.

19. Three is less than five.

Round to the nearest hundred.

20. 61,269

21. 382,240

Round to the nearest hundred thousand.

22. 6,365,534

23. 8,118,701

Section 1.2

Translate using symbols.

24. Seven more than a number

25. The sum of some number and five

Use whichever properties (associative and/or commutative) are necessary to simplify each expression.

26. $7 + (9 + x)$

27. $(2 + n) + 9$

28. $5 + (n + 2)$

29. $(5 + x + 3) + 2$

Add.

30. $8398 + 372 + 255$

31. $17,456 + 213 + 982$

Answer the following.

32. ***College Enrollment*** A private college has 1434 freshmen, 1596 sophomores, 1423 juniors, and 1565 seniors. How many students are attending the college?

▲ 33. Find the perimeter of the shape made up of two rectangles (measured in meters).

Section 1.3

Translate using numbers and symbols.

34. Eight decreased by a number   **35.** The difference of a number and six   **36.** Ten subtracted from a number

Evaluate.

37. $8 - x$ if x is equal to 3

38. $y - 9$ if y is equal to 15

Subtract and check your answer.

39. $8502 - 2957$

40. $9021 - 5862$

41. $29{,}104 - 4988$

Golf Championships *A professional golf player's earnings from winning six golf championships are listed on the bar graph. Use the graph to answer exercises 42 and 43.*

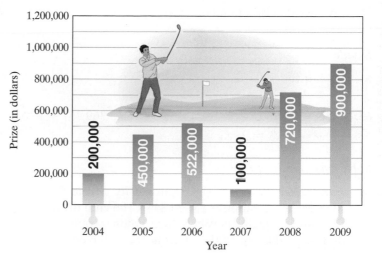

42. How much more did the player earn in the 2009 championship than in the 2006 championship?

43. How much less did the player earn in the 2005 championship than in the 2008 championship?

Section 1.4

Identify the factors.

44. $4x = 32$

Translate using numbers and symbols.

45. Triple a number

Translate the symbols to words.

46. $7y = 63$

Multiply.

47. $7 \cdot 2 \cdot 3 \cdot 0$

48. $5 \cdot 3 \cdot 2 \cdot 2$

Simplify.

49. $6(y \cdot 7)$

50. $3(5)(x \cdot 2)$

51. $3(2)(x \cdot 4)$

Multiply.

52. $(416)(2000)$

53. $(4251)352$

54. $\begin{array}{r} 6424 \\ \times\ 903 \\ \hline \end{array}$

Solve each applied problem.

55. *Miles Per Gallon* Lisa has a truck that averages 17 miles per gallon on the highway. Approximately how far can she travel if she has 18 gallons of gas in her tank?

56. *Doors Replaced* J&R Doors & Windows is replacing all the interior doors in a 6-unit apartment complex that has 21 apartments in each unit. If each apartment has 4 interior doors, how many doors will need to be replaced?

Section 1.5

Write the division that corresponds to each situation. You need not carry out the division.

57. 300 desks are arranged so that 20 desks are in each row. How many rows are there?

58. A $500 prize is divided equally between n people. How much will each person receive?

Translate each phrase using numbers and symbols.

59. Five divided by a number

60. The quotient of a number and thirteen

Divide.

61. $10 \div 0$

62. $33 \div 33$

63. $1456 \div 29$

64. $369{,}757 \div 922$

65. $\dfrac{510{,}144}{846}$

Solve each applied problem.

66. *Fundraiser Proceeds* The Dalton City Music Club fund-raising committee raised $447. The club divided the funds equally between four youth groups and deposited the rest of the funds in their club account. How much money did the club deposit in their club account?

67. *Loan Payments* Lisa wishes to pay off a loan of $3528 in 24 months. How large will her monthly payments be?

Section 1.6

Write each product in exponent form.

68. $2 \cdot 2 \cdot 2 \cdot n \cdot n$

69. $z \cdot z \cdot z \cdot z \cdot 5 \cdot 5 \cdot 5$

Write as a repeated multiplication.

70. x^3

71. 6^5

Evaluate.

72. 10^3

73. 2^4

Translate each phrase using numbers, symbols, and exponents.

74. Six cubed

75. x to the fifth power

Evaluate.

76. $6 + 24 \div 8 - 2^2$

77. $(15 + 25 \div 5) \div (8 - 4)$

78. $5 \cdot 2^2$

Section 1.7

Translate using numbers and symbols.

79. (a) Three times x plus two

    (b) Three times the sum of x and two

80. (a) Four times x minus five

    (b) Four times the difference of x and 5

Translate using numbers and symbols, then simplify.

81. (a) Three times seven plus one

    (b) Three times the sum of seven and one

82. Evaluate $\dfrac{x^3 - 1}{y}$ if x is equal to 3 and y is equal to 2.

83. Evaluate $2m + 3n$ if m is equal to 8 and n is equal to 2.

Use the distributive property to simplify.

84. $5(x + 1)$

85. $4(x - 1)$

86. $3(x + 1) + 5$

Section 1.8

Combine like terms.

87. $2x + x + 6x$

88. $5x + 6y + 6x$

89. $3xy + 5y + 2xy + 8y$

▲ **90.** Find an expression that represents the perimeter.

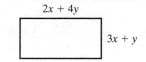

Solve and check your answer

91. $x + 2 = 9$

92. $10 - n = 6$

Simplify using the associative and/or commutative property and then find the solution.

93. $(3 + x) + 1 = 8$

94. $2 + (n + 7) = 10$

Solve each equation and check your answer.

95. $9x = 27$

96. $\dfrac{15}{x} = 5$

Simplify by combining like terms and then find the solution.

97. $12n - n = 22$

98. $y + 3y + 2y = 12$

For the following questions:

(a) *Translate the English statement into an equation.*

(b) *Solve.*

99. What number subtracted from eighteen equals three?

100. What number increased by five equals eleven?

101. Triple what number is equal to twelve?

102. *Office Supply Purchases* Joseph must purchase supplies for his home office. The catalog from the supply warehouse lists the following prices for the items he plans to purchase. Round each amount to the nearest ten and then *estimate* the amount of money he will pay for these supplies.

Price List	
Statistical calculator	$25
Color inkjet print cartridge	26
Four-drawer metal file cabinet	87
Computer chair	156

You may want to use the Mathematics Blueprint for Problem Solving for exercises 103–105.

Mathematics Blueprint for Problem Solving

Gather the Facts	What Am I Asked to Do?	How Do I Proceed?	Key Points to Remember

103. *Payroll Deductions* A teacher's assistant receives a total salary per month of $3560. Deducted from his paycheck are taxes of $499, social security of $218, and retirement of $97. What is the total of his check after the deductions?

104. *Savings Account Balance* Jean's savings account had a balance of $5021. Over the course of a year, she made deposits of $759, $2534, and $532. She also made withdrawals of $799, $533, and $88.

(a) What was her ending balance?

(b) If Jean divides her savings equally into two money market accounts, how much money will there be in each account?

105. *Ceiling Molding Purchase* Ruth Ann must buy crown molding for the ceilings in her living and dining rooms. The living room measures 20 feet by 25 feet, and the dining room measures 15 feet by 18 feet. If crown molding costs $3 a linear foot, how much will it cost Ruth Ann to purchase the molding for both rooms?

How Am I Doing? Chapter 1 Test

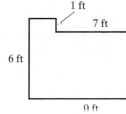

After you take this test read through the Math Coach on pages 101–102. Math Coach videos are available via MyMathLab and YouTube. Step-by-step test solutions on the Chapter Test Prep Videos are also available via MyMathLab and YouTube. (Search "BlairTobeyPrealgebra" and click on "Channels.")

1. Write 1525 in expanded notation.

2. Replace each question mark with the appropriate symbol $<$ or $>$.
   (a) $7 \, ? \, 2$ **(b)** $5 \, ? \, 0$

3. The total population of a small town is 2925. Round this population figure to
   (a) The nearest thousand   **(b)** The nearest hundred

4. Use the commutative and/or associative property of addition and then simplify.
   (a) $3 + (8 + x)$ **(b)** $5 + y + 2$ **(c)** $1 + (n + 2) + 4$

5. Add. $12{,}389 + 4 + 2302$ 6. Add. $244{,}869{,}201 + 19{,}077$

7. Subtract.
   (a) $613 - 75$ $^{\mathbb{M}}\mathbb{C}$ **(b)** $\begin{array}{r} 20{,}105 \\ - 7\,826 \end{array}$

▲ 8. Find the perimeter of the figure made of rectangles.

1 ft

7 ft

6 ft

0 ft

9. Multiply. $2(4)(y \cdot 2)$

10. Multiply.
    (a) $(432)(312)$ **(b)** $\begin{array}{r} 2031 \\ \times 129 \end{array}$

11. Divide.
    (a) $492 \div 12$ $^{\mathbb{M}}\mathbb{C}$ **(b)** $5523 \div 46$

12. Translate using numbers and symbols.
    (a) Seven subtracted from a number   **(b)** The product of ten and a number
    (c) y to the fourth power   **(d)** 7 cubed
    (e) Six times the sum of x and nine

13. Combine like terms.
    (a) $3xy + 2y + 4xy - 2$ **(b)** $2m + 5 + m + 6mn$

Use the distributive property to simplify.

14. $3(y + 4)$ 15. $8(x + 1) + 2$

16. Evaluate.
    $^{\mathbb{M}}\mathbb{C}$ **(a)** $2x - 3y$ if x is equal to 16 and y is equal to 4
         (b) $\dfrac{a^2 - 4}{b}$ if $a = 9$ and $b = 7$

17. Write in exponent form. $6 \cdot 6 \cdot 6 \cdot 6 \cdot 6 \cdot n \cdot n \cdot n$

1. _____ ☐
2. (a) _____ ☐
  (b) _____ ☐
3. (a) _____ ☐
  (b) _____ ☐
4. (a) _____ ☐
  (b) _____ ☐
  (c) _____ ☐
5. _____ ☐
6. _____ ☐
7. (a) _____ ☐
  (b) _____ ☐
8. _____ ☐
9. _____ ☐
10. (a) _____ ☐
  (b) _____ ☐
11. (a) _____ ☐
  (b) _____ ☐
12. (a) _____ ☐
  (b) _____ ☐
  (c) _____ ☐
  (d) _____ ☐
  (e) _____ ☐
13. (a) _____ ☐
  (b) _____ ☐
14. _____ ☐
15. _____ ☐
16. (a) _____ ☐
  (b) _____ ☐
17. _____ ☐

18. (a)

(b)

19.

20.

21.

22. (a)

(b)

(c)

(d)

(e)

23.

24. (a)

(b)

25. (a)

(b)

26. (a)

(b)

27.

28.

29.

30. (a)

(b)

31.

Total Correct:

18. Evaluate.

(a) 5^3 (b) 10^5

Evaluate.

19. $24 \div 4 - 2 \cdot 3$ **20.** $6^2 - 7 + 3 \cdot 4$ **21.** $3 \cdot 2 + 4(7 - 1)$

22. Solve each equation.

(a) $7 + x = 13$ (b) $\dfrac{x}{4} = 2$ (c) $x + 3x = 36$

(d) $5 + (b + 2) = 18$ (e) $9n - n = 32$

Translate into an equation.

23. Let B represent Fred's checking account balance. Fred's checking account balance decreased by \$155 equals \$275.

For problems 24 and 25:

(a) Translate into an equation.   *(b) Solve.*

24. What number divided by six equals two?

25. Three subtracted from what number equals one?

M_C **26.** Tickets to a play were \$25 for adults and \$18 for children. 412 adult tickets were sold, and 280 child tickets were sold.
(a) Find the total income from the sale of tickets.
(b) If the expenses for the play were \$7350, how much profit was made?

27. A restaurant sells 4 kinds of sandwiches: turkey, roast beef, veggie, and ham. Customers have a choice of 3 types of bread: wheat, white, or rye. How many different sandwiches are possible?

28. A store clerk receives a total salary per month of \$1540. Deducted from her paycheck are taxes of \$265, social security of \$78, and retirement of \$57. What is the total of her check after the deductions?

29. The rent on an apartment was \$525. To move in, Fred was required to pay the first and last months' rent, a security deposit of \$200, and a telephone installation fee of \$40. How much money did he need to move into the apartment?

30. Sylvia kept the following record of her living expenses for the month of February 2011:

Rent	\$790	Car payment	210
Phone and utilities	114	Gas and insurance	187
Food	318		

(a) Round each amount to the nearest hundred and then estimate Sylvia's monthly expenses for February.
(b) If her take-home (net) income for February was \$1921, estimate how much money was left after all the expenses were paid.

31. A frequent-flyer program offered by many major airlines to first-class passengers awards 3 frequent-flyer mileage points for every 2 miles flown. When customers accumulate a certain number of frequent-flyer points, they can cash in these points for free air travel, a ticket upgrade, or other awards. How many frequent-flyer points would Elizabeth accumulate if she flew 5000 miles in first-class?

MATH COACH

Mastering the skills you need to do well on the test.

Students often make the same types of errors when they do the Chapter 1 Test. Here are some helpful hints to keep you from making these common errors on test problems.

Subtract Whole Numbers with Borrowing—Problem 7(b)

Subtract.
$$\begin{array}{r} 20,105 \\ -7\,826 \\ \hline \end{array}$$

> **Helpful Hint** It is wise to show the borrowing steps. This will help you avoid a borrowing error.

Look at your work for Problem 7(b). Examine your steps. Do your borrowing steps match the solution below?

Yes ____ No ____

If you answered No, be sure to write your borrowing steps carefully and then check each subtraction step for errors. Stop now and rework the problem.

If you answered Yes, and still got an incorrect answer, check each subtraction step for errors.

Write out the borrowing steps:
$$\begin{array}{r} {\scriptstyle 1\ 9\ 10\,9\ 15} \\ 2\,0,1\,0\,5 \\ -\ 7\ 8\,2\,6 \\ \hline \end{array}$$

If you answered Problem 7(b) incorrectly, go back and rework the problem using these suggestions.

Performing Long Division with Whole Numbers—Problem 11(b)   Divide. $5523 \div 46$

> **Helpfule Hint** When performing long division, be sure to line up each column exactly as shown below. Accuracy in alignment will increase your chance of completing the problem correctly. Remember that a remainder results when there are no more numbers to bring down from the dividend.

Look at your work for Problem 11(b). Compare each line of your work with the problem below.

Did you align numbers correctly?

Yes ____ No ____

Did you obtain the correct value each time you subtracted?

Yes ____ No ____

If you answered No to either question, stop and rework the problem now.

$$\begin{array}{r} 46\overline{)5523} \\ \underline{46} \\ 92 \\ \underline{92} \\ 03 \end{array}$$

Did you get 12 R3 for your answer?

Yes ____ No ____

If you answered Yes, then you stopped dividing too soon. Because 46 cannot be divided into 3, we must write 0 in the quotient and continue to divide.

Now go back and rework the problem using these suggestions.

Need help? Watch the MATH COACH videos in MyMathLab® or on You.

101

Evaluate Algebraic Expressions—Problem 16(a)

Evaluate $2x - 3y$ if x is equal to 16 and y is equal to 4.

> **Helpful Hint** Take time to show the step of replacing the variable with the appropriate number before you do any calculations. To avoid errors, do not skip steps or complete calculations mentally.

As your first step, did you replace $2x - 3y$ with $2 \cdot 16 - 3 \cdot 4$?

Yes _____ No _____

If you answered No, stop and complete this step.

Did you multiply $2 \cdot 16$ for your next step?

Yes _____ No _____

If you answered No, then you *did not* follow the proper order of operations. Remember to complete all multiplications before you subtract.

If you answered Problem 11(b) incorrectly, go back and rework the problem using these suggestions.

Solving Applied Problems Using Several Operations—Problem 26(a) and (b)

Tickets to a play were $25 for adults and $18 for children. 412 adult tickets were sold, and 280 child tickets were sold.
  (a) Find the total income from the sale of tickets.
  (b) If the expenses for the play were $7350, how much profit was made?

> **Helpful Hint** Remember to use pictures, charts, or the Mathematics Blueprint for Problem Solving. First, try to **understand the problem** by rereading the problem carefully. What are the facts? What am I asked to do? What type of calculation(s) must I perform? Next, **calculate and state the answer**. If possible, take the time to **check** your answer.

Reread Problem 26 carefully, then fill in the information needed to solve the problem.

What are the facts? There are _____ adult tickets. They cost $_____ each. There are _____ child tickets.

They cost $_____ each.

What am I asked to do for Part (a)? You must find the total cost of tickets (income).

What type of calculation must I perform? Did you multiply 412 × $25 to find the income from adult tickets and 280 × $18 to find the income from child tickets? Then did you add these results?

Yes _____ No _____

If you answered No, stop now and use these hints to complete Part (a) correctly.

What am I asked to do for Part (b)? You must find the profit if the expenses were $7350.

What type of calculation must I perform? Did you calculate Income − Expenses = Profit?

Yes _____ No _____

If you answered No, stop now and use these hints to complete Part (b) correctly.

Try to write out all the facts of the situation to help you keep track of the details and determine what type of calculations are needed to solve the problem.

Now go back and rework the problem using these suggestions.

Need more help? Look for section examples marked with $^{\text{M}}\text{C}$ to review.

Whether you are planning to save for a vacation, new car, or other expenses, you must make decisions about how to spend money. Sometimes the costs are higher than you thought they would be, and you must adjust your spending habits to save money for these extra expenses. The mathematics you learn in this chapter will help you to perform the calculations necessary to make these decisions.

Integers

2.1 Understanding Integers

Student Learning Objectives

After studying this section, you will be able to:

① Use inequality symbols with integers.

② Find the opposite of numbers.

③ Find the absolute value of numbers.

④ Read line graphs.

A dive $20\frac{1}{4}$ feet below the surface can be written "$-20\frac{1}{4}$ feet."

① Using Inequality Symbols with Integers

In this section we enlarge the set of whole numbers to include numbers that are less than 0. These numbers are called **negative numbers.** We use the symbol "−" to indicate that the sign of a number is negative, and the symbol "+" to indicate that the sign is positive. Thus, we write the number *negative six* in symbols as -6. For **positive numbers** we usually do not write the plus sign. That is, we write $+6$ as 6.

We say	negative six	positive six
We write	-6	$+6$ or 6

We often encounter real-life applications that require us to consider numbers that are less than 0. For example, a weather report states that the Fahrenheit temperature is 20 degrees below 0. How can we write this temperature? We can use negative numbers. Thus the temperature reading of 20 degrees below 0 is written $-20°$. Here are some other examples of negative numbers.

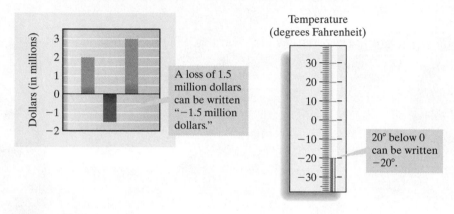

A loss of 1.5 million dollars can be written "-1.5 million dollars."

$20°$ below 0 can be written $-20°$.

We can also picture negative numbers using a **number line.** Positive numbers are to the right of 0 on the number line. Negative numbers are to the left of 0 on the number line. Note that the number 0 is neither positive nor negative.

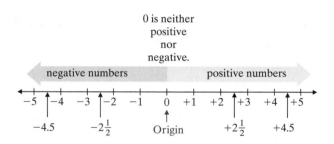

The positive numbers, negative numbers, and 0 are called **signed numbers.** In later chapters we will study signed numbers such as fractions and decimals. In this chapter we will study the set of signed numbers called **integers.** The integers are $\ldots, -3, -2, -1, 0, 1, 2, \ldots$.

Numbers decrease in value as we move from right to left on the number line. Therefore, 1 is less than 3 $(1 < 3)$ since 1 lies to the left of 3 on the number line, and -5 is less than -2 $(-5 < -2)$ since -5 lies to the left of -2 on the number line.

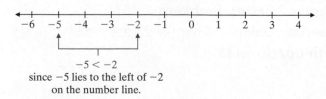

$-5 < -2$
since -5 lies to the left of -2
on the number line.

EXAMPLE 1 Graph −5, −3, 1, and 5 on a number line.

Solution We draw a dot in the correct location on the number line.

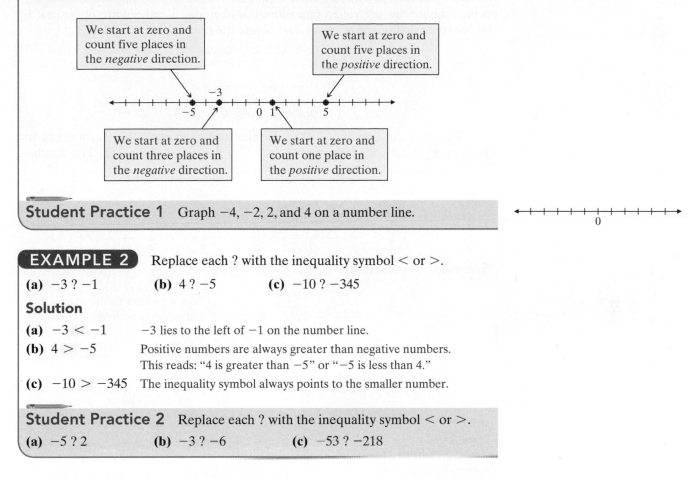

We start at zero and count five places in the *negative* direction.

We start at zero and count five places in the *positive* direction.

We start at zero and count three places in the *negative* direction.

We start at zero and count one place in the *positive* direction.

Student Practice 1 Graph −4, −2, 2, and 4 on a number line.

EXAMPLE 2 Replace each ? with the inequality symbol < or >.

(a) −3 ? −1   **(b)** 4 ? −5   **(c)** −10 ? −345

Solution

(a) −3 < −1   −3 lies to the left of −1 on the number line.

(b) 4 > −5   Positive numbers are always greater than negative numbers. This reads: "4 is greater than −5" or "−5 is less than 4."

(c) −10 > −345   The inequality symbol always points to the smaller number.

Student Practice 2 Replace each ? with the inequality symbol < or >.

(a) −5 ? 2   **(b)** −3 ? −6   **(c)** −53 ? −218

In everyday situations we use the concept of positive and negative numbers to represent many different things. We can use "+" to represent an increase, a rise, or whenever something goes up, and "−" to represent a decrease, a decline, or whenever something goes down. For example, when you use a checking account you can associate a deposit with "+" since your balance goes *up* and a check written with "−" since your balance goes *down*.

EXAMPLE 3 Fill in each blank with the appropriate symbol, + or −, to describe either an increase or a decrease.

(a) A discount of $5: _____ $5

(b) The temperature rises 10°F: _____ 10°F

Solution

(a) A discount of $5 results in the price decreasing: −$5

(b) The temperature rises 10°F: +10°F

Student Practice 3 Fill in each blank with the appropriate symbol, + or −, to describe either an increase or a decrease.

(a) A property tax increase of $130: _____ $130

(b) A dive of 7 ft below the surface of the sea: _____ 7 ft

NOTE TO STUDENT: Fully worked-out solutions to all of the Student Practice problems can be found at the back of the text starting at page SP-1.

② Finding the Opposite of Numbers

Numbers that are the same distance from zero but lie on the opposite sides of zero on the number line are called **opposites.** For example, 2 and −2 are opposites. By this we mean that the opposite of 2 is −2 and the opposite of −2 is 2.

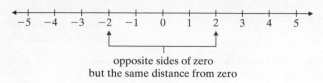

opposite sides of zero
but the same distance from zero

Now we can formally state the definition of an integer: Whole numbers and their opposites are called **integers:** $\{\ldots, -3, -2, -1, 0, 1, 2, 3, \ldots\}$. The numbers labeled on the number line above are integers.

EXAMPLE 4   Label −5 and the *opposite* of −5 on a number line.

Solution   We first label −5.

We start at −5.

Next, we locate the number that is the *same distance* from zero but lies on the opposite side of zero.

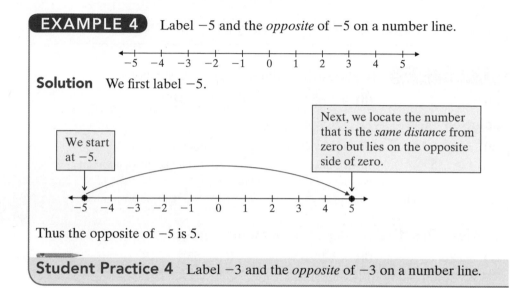

Thus the opposite of −5 is 5.

Student Practice 4   Label −3 and the *opposite* of −3 on a number line.

Another way to find the opposite of a number is to *change the sign* of the number. In Example 4 we can find the opposite of −5 by changing −5 to +5.

EXAMPLE 5   State the opposite of each number.

(a) 6   **(b)** −9

Solution   To find the opposite of a number, we change the sign of the number.

(a) The opposite of 6 is −6.   **(b)** The opposite of −9 is 9.

Student Practice 5   State the opposite of each number.

(a) −6   **(b)** −1   **(c)** 12   **(d)** 1

So far we have used the symbol "−" to indicate the operation *subtraction* and to write a *negative number.* We also use the "−" symbol to indicate "*the opposite of.*"

The opposite of	negative 1

$$-(-1)$$

The first − sign is read "the opposite of." The second − sign is read "negative one."

The interpretation of the symbol depends on the context in which it is used.

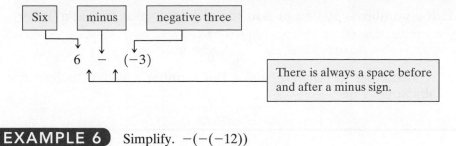

There is always a space before and after a minus sign.

EXAMPLE 6 Simplify. $-(-(-12))$

Solution $-(-(-12))$

$= -(12)$ The opposite of negative 12 is positive 12.

$= -12$ The opposite of positive 12 is negative 12.

Student Practice 6 Simplify. $-(-(-(-1)))$

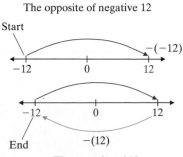

The opposite of negative 12

The opposite of 12

EXAMPLE 7 Evaluate $-(-x)$ for $x = -9$.

Solution To avoid errors involving negative signs, we can place parentheses around the variables and their replacements.

$$-(-x)$$
$$= -(-(x))$$ Place parentheses around x.
$$= -(-(-9))$$ Replace x with -9.
$$= -(9)$$ The opposite of -9 is 9: $-(-9) = 9$.
$$= -9$$ The opposite of 9 is -9.

Student Practice 7 Evaluate $-(-(-a))$ for $a = 6$.

③ Finding the Absolute Value of Numbers

Suppose that we want to find the distance from 0 to -4 and from 0 to $+4$. We can use the number line to measure this distance just as we use a ruler to measure feet or inches.

Distance from
0 to -4 is 4 units.

Distance from
0 to 4 is 4 units.

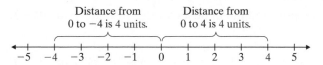

Both distances are the same; 4 and -4 are both 4 units from 0. As you can see, it is sometimes convenient to talk about just the numerical part of the number (the 4 of -4) and disregard the sign. The following definition gives us a way to do this.

> The **absolute value** of a number is the distance between that number and 0 on the number line.

We place the symbols "| |" around the number to indicate that we want the absolute value of that number. We write $|-4| = 4$ and $|4| = 4$. The absolute value is never negative because it is a distance, and distance is always measured in positive or zero units.

> **PROCEDURE TO FIND THE ABSOLUTE VALUE OF A NUMBER**
> 1. If a number is *positive* or *zero*, the absolute value is that number.
> $$|7| = 7$$
> $$|0| = 0$$
> 2. If a number is *negative*, make the number positive to find the absolute value.
> $$|-7| = 7$$

EXAMPLE 8 Simplify each absolute value expression.

(a) $|-9|$ **(b)** $|3|$

Solution

(a) $|-9| = 9$ The absolute value of a negative number is positive.

(b) $|3| = 3$ The absolute value of a positive number is positive.

Student Practice 8 Simplify each absolute value expression.

(a) $|-67|$ **(b)** $|8|$

EXAMPLE 9 Replace the ? with the symbol $<$, $>$, or $=$. $|-15|$? $|6|$

Solution $|-15|$? $|6|$
$$\downarrow \qquad \downarrow$$
$$15 \ ? \ 6 \quad \text{We find the absolute values.}$$
$$15 \ > \ 6 \quad \text{We write the appropriate inequality symbol, } >.$$
$$|-15| \ > \ |6| \quad -15 \text{ has a larger absolute value than 6.}$$

Note that when we say -15 has a larger absolute value, we mean that -15 is a greater distance from 0 than 6 is.

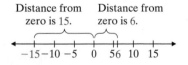

Distance from zero is 15.   Distance from zero is 6.

Student Practice 9 Replace the ? with the symbol $<$, $>$, or $=$. $|-12|$? $|2|$

EXAMPLE 10 Simplify. $-|-7|$

Solution We must find the opposite of the absolute value of -7.
$$-|-7|$$
$$= -(7) \quad \text{First we find the absolute value of } -7 \text{: } |-7| = 7.$$
$$= -7 \quad \text{Then we take the opposite of 7: } -(7) = -7.$$

Student Practice 10 Simplify. $-|-1|$

④ Reading Line Graphs

We can use a line graph to display information much the same way that we use a bar graph. On a line graph we use dots instead of bars to record data. Then we connect the dots with straight lines. The vertical number line on a graph is sometimes extended to include negative numbers.

EXAMPLE 11 The line graph below indicates the low temperatures for selected cities on a typical winter day.

(a) In which city was the temperature colder, Syracuse or Burlington?

(b) Which cities recorded a positive temperature for the day, and which cities recorded a negative temperature?

Solution On the graph negative numbers are located below zero and positive numbers above zero.

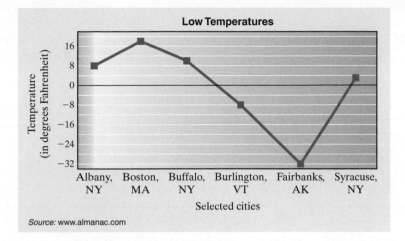

(a) The low temperature was 3°F in Syracuse and −8°F in Burlington. It was colder in Burlington.

Note that the dot representing Burlington's temperature is lower on the line graph than the dot representing Syracuse's temperature.

(b) The positive temperatures appear as dots above 0 and negative temperatures below 0.

Cities with positive temperatures: Albany, Boston, Buffalo, and Syracuse
Cities with negative temperatures: Burlington and Fairbanks

Student Practice 11 Use the line graph in Example 11 to answer the following.

(a) In which city was the temperature colder, Fairbanks or Buffalo?

(b) Name the city that recorded the highest temperature and the city that recorded the lowest temperature.

👣 STEPS TO SUCCESS Why Is the Math Coach Important?

To learn mathematics you must master each skill as you proceed through the course. Completing the Math Coach after you finish each How Am I Doing? Chapter Test will help you master each building block for the skills needed to succeed in mathematics.

How do I use the Math Coach? Take the Chapter Test without any assistance. Then place a ✓ beside each problem you **answered incorrectly.** Next, read the Math Coach to help you identify and fix your errors. If you missed any other problems, find an example similar to that problem and try to correct it yourself. Otherwise, contact your instructor or a tutor to see how to complete it correctly.

Making it personal: Try to correct all the problems you answered incorrectly on your in-class exam. Make it a priority to get help with the ones you cannot rework correctly. ▼

Watch the videos in MyMathLab

Download the MyDashBoard App

Verbal and Writing Skills, Exercises 1–4

Write in words.

1. $-(-1)$

2. $-|-4|$

Translate into numbers and symbols.

3. Negative four minus two

4. Five minus negative one

Fill in the blanks.

5. -9 is a _____ number, and 9 is a _____ number.

6. Numbers that are the same distance from zero but lie on opposite sides of zero on the number line are called _____.

Graph each number on the number line.

7. $-6, -4, 3,$ and 4

8. $-7, -2, 2,$ and 5

9. $-3, -1, 1,$ and 3

10. $-5, -4, 4,$ and 5

11. Which dot, A or B, represents a larger number on the following number line?

12. Which dot, X or Y, represents a larger number on the following number line?

Replace each ? with the inequality symbol $<$ or $>$.

13. $-9 ? 4$

14. $-9 ? 5$

15. $4 ? -3$

16. $2 ? -4$

17. $-5 ? 5$

18. $-3 ? 3$

19. $-9 ? -5$

20. $-10 ? -2$

21. $-8 ? -6$

22. $-51 ? -6$

23. $-298 ? -350$

24. $-765 ? -990$

Fill in the blanks with the appropriate symbol, $+$ or $-$, to describe either an increase or a decrease.

25. _____ A loss of $100

26. _____ A plane descends 900 ft

27. _____ A raise of $100

28. _____ The temperature rises 10°F

29. _____ A discount of $10

30. _____ A profit of $550

31. _____ A plane ascends 1000 ft

32. _____ A tax decrease of $250

33. Label -1 and the *opposite* of -1 on the number line.

34. Label 5 and the *opposite* of 5 on the number line.

Fill in the blanks.

35. The opposite of -3 is _____.

36. The opposite of -8 is _____.

37. The opposite of 16 is _____.

38. The opposite of 21 is _____.

Simplify.

39. $-(-4)$

40. $-(-7)$

41. $-(8)$

42. $-(1)$

Label the following on the number line.

43. $-(-(-7))$

44. $-(-(-8))$

<--+-->
                    0

<--+-->
                    0

Simplify.

45. $-(-(13))$

46. $-(-(30))$

47. $-(-(-(-1)))$

48. $-(-(-(-2)))$

Evaluate.

49. $-(-a)$ for $a = 6$

50. $-(-y)$ for $y = 13$

51. $-(-(-x))$ for $x = -1$

52. $-(-(-n))$ for $n = -6$

53. $-(-(-(-y)))$ for $y = -2$

54. $-(-(-(-x)))$ for $x = -5$

Simplify each absolute value expression.

55. $|8|$　　　　**56.** $|6|$　　　　**57.** $|-5|$　　　　**58.** $|-7|$

59. $|-16|$　　　**60.** $|-19|$　　　**61.** $|0|$　　　　**62.** $|42|$

Replace each ? with the symbol $<$, $>$, or $=$.

63. $|-3|$? $|1|$　　　**64.** $|-9|$? $|5|$　　　**65.** $|8|$? $|-8|$　　　**66.** $|6|$? $|-6|$

67. $|-9|$? $|16|$　　　**68.** $|19|$? $|-13|$　　　**69.** $|-35|$? $|-8|$　　　**70.** $|-71|$? $|-6|$

Simplify.

71. $-|-3|$　　　**72.** $-|-10|$　　　**73.** $|-14|$　　　**74.** $|-17|$

Applications

75. ***Charted Temperatures*** The line graph indicates the low temperature in selected cities during a week in January.

　　(a) In which city was the temperature colder, Fargo or Albany?

　　(b) Which cities recorded a positive temperature for the day, and which cities recorded a negative temperature?

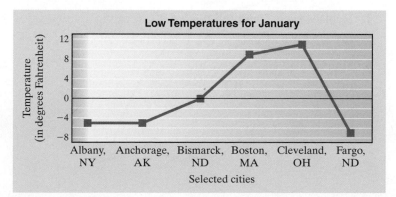

76. Use the line graph in exercise 75 to answer the following.

　　(a) In which city was the temperature colder, Anchorage or Bismarck?

　　(b) Name the cities that recorded the highest and lowest temperatures.

One Step Further *Simplify each statement, then replace the ? with the symbol <, >, or =.*

77. $-|-16|$? -16 **78.** $|-12|$? $-(-12)$ **79.** $|-9|$? $-|-9|$ **80.** $-(-(-1)$? $|-1|$

Simplify each statement, then perform the operation indicated.

81. $-(-6) + |-5|$ **82.** $-(-4) + |-9|$ **83.** $-(-|8|) - |-1|$ **84.** $-(-|5|) - |-2|$

To Think About

85. Which of the two numbers has the larger absolute value, -33 or 20?

86. Which of the two numbers has the larger absolute value, -43 or 40?

87. Which of the two numbers has the larger absolute value, 129 or -112?

88. Which of the two numbers has the larger absolute value, 231 or -98?

State whether each statement is true or false.

89. If $x > y$, then x must be a positive number.

90. The numbers $-2, -1, 0, 1,$ and 2 are called integers.

Fill in the blank.

91. If m is a negative number, then $-m$ is a _____ number.

92. There are two numbers that are 3 units from 1 on the number line. One of these numbers is 4 and the other number is _____ .

Cumulative Review *Perform the operation indicated.*

93. [**1.3.4**] $5009 - 258$

94. [**1.2.4**] $5699 + 351$

95. [**1.4.4**] $(256)(91)$

96. [**1.5.4**] $456 \div 3$

97. [**1.9.3**] *Budgeting* For a trip to Hawaii, Wanda has a vacation budget of $2600. The cost of airfare is $480, and lodging is $1200. If the travel agent estimates that Wanda will spend $350 on food, how much will Wanda have left in the budget to spend during the vacation?

98. [**1.9.3**] *Financing a Purchase* Tran Troung is buying new appliances for her home. She ordered a stove for $780, a washer for $520, a dryer for $450, and a refrigerator for $1150. The tax for the entire purchase is $203, and the delivery charge is $45. If Tran puts $800 down on the purchase, how much does she have to finance?

Quick Quiz 2.1

1. Replace each ? with the symbol <, >, or =.

    (a) -5 ? -6 **(b)** $|-11|$? $|13|$

2. Simplify. $-(-(-7))$

3. Simplify.

    (a) $|-3|$ **(b)** $-|-5|$

4. **Concept Check** Explain how to rearrange numbers in order from smallest to largest: $-1, -6, -4, -10, 0$

2.2 Adding Integers

① Adding Integers with the Same Sign

We often associate the + and − symbols with positive and negative situations. We can find the sum of integers by considering the outcome of these situations as illustrated below.

A price decrease of $10 followed by a price decrease of $20 results in a decrease of $30.

$$(-\$10) \quad + \quad (-\$20) \quad = \quad (-\$30)$$

In the same sense, an *increase* followed by an *increase* results in an *increase,* or a positive outcome.

Student Learning Objectives

After studying this section, you will be able to:

① Add integers with the same sign.

② Add integers with different signs.

③ Evaluate algebraic expressions involving addition of integers.

④ Solve applied problems involving addition of integers.

EXAMPLE 1

(a) Fill in the blank. A loss of $1000 followed by a loss of $1000 results in a _____.

(b) Write the math symbols that represent the situation described in (a).

Solution

(a) A loss of $1000 followed by a loss of $1000 results in a loss of $2000.

(b) A loss of $1000 followed by a loss of $1000 results in a loss of $2000.

$$-\$1000 \quad + \quad (-\$1000) \quad = \quad -\$2000$$

It is common to include parentheses around negative numbers when they appear after an operation symbol.

Student Practice 1

(a) Fill in the blank. A decrease in altitude of 100 feet followed by a decrease in altitude of 100 feet results in a _____.

(b) Write the math symbols that represent the situation described in (a).

NOTE TO STUDENT: Fully worked-out solutions to all of the Student Practice problems can be found at the back of the text starting at page SP-1.

Since integers are often used to indicate *direction* and *distance,* we can also use the number line to find the sum of numbers such as $-1 + (-2)$. A move to the *right* on the number line is a move in the *positive* direction, and a move to the *left* on the number line is a move in the *negative* direction. The direction we move on the number line is indicated by the sign of the number.

We start at 0 and move 1 unit in the negative direction, followed by another 2 units in the negative direction.

We see that $-1 + (-2) = -3$.

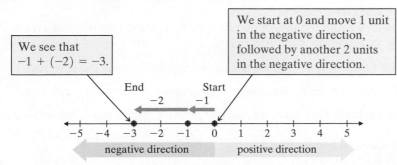

EXAMPLE 2

(a) Begin at 0 on the number line and move 3 units to the left followed by another 2 units to the left.

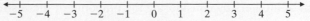

(b) Is the end result in the positive or the negative region?

(c) Write the math symbols that represent the situation.

(d) Use the number line to find the sum.

Solution

(a)

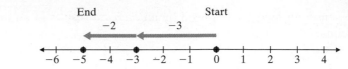

(b) From the illustration we see that the end result is in the negative region since we began at 0 and moved 3 units in the negative direction (left), followed by another 2 units in the negative direction.

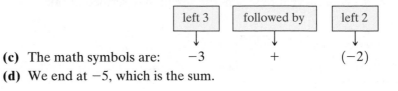

(c) The math symbols are: −3   +   (−2)

(d) We end at −5, which is the sum.

$$-3 + (-2) = -5$$

Student Practice 2

(a) Begin at 0 on the number line and move 4 units to the left followed by another 1 unit to the left.

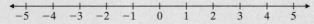

(b) Is the end result in the positive or the negative region?

(c) Write the math symbols that represent the situation.

(d) Use the number line to find the sum.

Example 2 shows that if we make a move in the negative direction followed by another move in the negative direction, the result is in the negative region. When we add negative numbers, we are repeatedly moving in the negative direction and thus the sum is a negative number.

How do we find the sum of numbers with the same sign without a number line? Let's look at the results from Example 2.

$-3 + (-2) = -$ We are adding two negative numbers, and the sum is negative.

$-3 + (-2) = -5$ We must add the absolute values to get 5: $2 + 3 = 5$.

Of course, we know that when we add two positive numbers, the answer is a positive number.

We state the formal rule.

RULE FOR ADDING TWO OR MORE NUMBERS WITH THE SAME SIGN

To add numbers with the *same* sign:

1. Use the common sign in the answer.
2. Add the absolute values of the numbers.

If all the numbers are positive, the sum is positive.
If all the numbers are negative, the sum is negative.

EXAMPLE 3   Add. $-1 + (-3)$

Solution We are adding two numbers with the same sign, so we keep the common sign (negative sign) and add the absolute values.

$-1 + (-3) = -$ The answer is *negative* since the common sign is negative.

$-1 + (-3) = -4$ Add. $1 + 3 = 4$

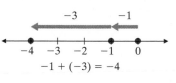

$-1 + (-3) = -4$

Student Practice 3   Add. $-2 + (-4)$

② Adding Integers with Different Signs

So far we have seen how to add numbers with the *same sign*. We use a similar approach to see how we add numbers that have *different signs*. Addition of numbers with *different signs* often involves situations such as a decrease followed by an increase or a quantity that rises and then falls.

If we wish, we can also use a vertical number line instead of a horizontal number line to illustrate these types of situations. If we move *up*, we are moving in the positive direction. If we move *down*, we are moving in the negative direction.

EXAMPLE 4   One night the temperature on Long Island, New York, dropped to $-25°F$. At dawn the temperature had risen $10°F$.

(a) Write the math symbols that represent the situation.

(b) Use the thermometer to determine if the temperature at dawn was positive or negative.

(c) Find the sum.

Solution

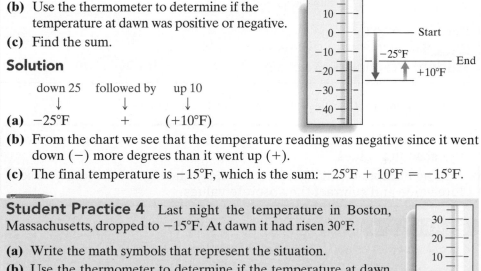

down 25   followed by   up 10
↓   ↓   ↓
(a) $-25°F$ $+$ $(+10°F)$

(b) From the chart we see that the temperature reading was negative since it went down $(-)$ more degrees than it went up $(+)$.

(c) The final temperature is $-15°F$, which is the sum: $-25°F + 10°F = -15°F$.

Student Practice 4 Last night the temperature in Boston, Massachusetts, dropped to $-15°F$. At dawn it had risen $30°F$.

(a) Write the math symbols that represent the situation.

(b) Use the thermometer to determine if the temperature at dawn was positive or negative.

(c) Find the sum.

Example 4 involves addition of integers with *different signs*. How do we perform this addition without using a chart? Let's look at the results from Example 4.

$$-25 + (+10) = -$$ The sign of the sum is *negative* since we move a larger distance in the *negative* direction.

$$-25 + (+10) = -15$$ We must subtract $25 - 10$ to get 15.

We see that to find the sum, we actually find the *difference* between 25 and 10. Also notice that if we do not account for the *sign*, the larger number is 25. The sign of 25 is *negative*, and the answer is also *negative*. This suggests the addition rule for two numbers with different signs.

RULE FOR ADDING TWO NUMBERS WITH DIFFERENT SIGNS

To add two numbers with *different* signs:

1. Use the *sign* of the number with the larger absolute value in the answer.

2. Find the difference between the larger absolute value and the smaller absolute value.

In other words, we keep the sign of the larger absolute value and subtract.

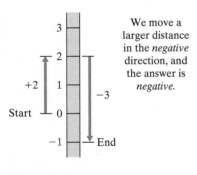

We move a larger distance in the *negative* direction, and the answer is *negative.*

EXAMPLE 5 Add.

(a) $2 + (-3)$ **(b)** $-2 + 3$

Solution We are adding numbers with *different* signs, so we keep the sign of the larger absolute value and subtract.

(a) $2 + (-3) = -$ The answer is *negative* since -3 is negative and has the larger absolute value.

$2 + (-3) = -1$ Subtract: $3 - 2 = 1$.

Thus, $2 + (-3) = -1$

(b) $-2 + 3 = +$ The answer is *positive* since 3 is positive and has the larger absolute value.

$-2 + 3 = +1$ Subtract: $3 - 2 = 1$.

Thus, $-2 + 3 = 1$

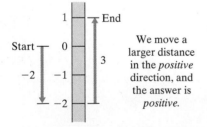

We move a larger distance in the *positive* direction, and the answer is *positive.*

Student Practice 5 Add. **(a)** $-4 + 7$ **(b)** $4 + (-7)$

We summarize the rules for adding numbers as follows.

Adding numbers with the *same sign*: Keep the common sign and add the absolute values.

Adding numbers with *different signs*: Keep the sign of the larger absolute value and subtract the absolute values.

EXAMPLE 6 Add.

(a) $8 + (-5)$ **(b)** $-8 + (-5)$

Solution

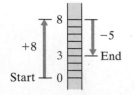

(a) $8 + (-5)$ We have *different* signs.

$8 + (-5) = +$ 8 is larger than 5, so the answer is positive.

$8 + (-5) = +3$ Subtract: $8 - 5 = 3$.

Thus, $8 + (-5) = 3$

(b) $-8 + (-5)$ We have the *same* sign.

$-8 + (-5) = -$ Keep the common sign.

$-8 + (-5) = -13$ Add: $8 + 5 = 13$.

Thus, $-8 + (-5) = -13$

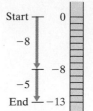

Start 0
-8
-8
-5
End -13

Student Practice 6 Add.

(a) $-3 + 8$ **(b)** $-3 + (-8)$

Recall from Section 2.1 that the numbers 3 and -3 are called opposites. Let's look at the number line in the margin and see what happens when we add the opposites $3 + (-3)$. When we add opposites the sum is 0 because we move the same distance in the positive direction as we move in the negative direction. This fact is referred to as the **additive inverse property,** and thus 3 and -3 are also called **additive inverses.**

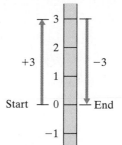

3

2

$+3$ -3

1

Start 0   End

-1

ADDITIVE INVERSE PROPERTY

For any number a,

$$a + (-a) = 0 \quad \text{and} \quad (-a) + a = 0$$

The sum of any number and its opposite is 0.

EXAMPLE 7

(a) Add. $1298 + (-1298)$ **(b)** Find x. $x + 21 = 0$

Solution

(a) Since 1298 and -1298 are additive inverses, their sum is 0.

$1298 + (-1298) = 0$

(b) The sum of additive inverses is 0. Thus if $x + 21 = 0$, then $x = -21$ since $-21 + 21 = 0$.

Student Practice 7

(a) Add. $3544 + (-3544)$ **(b)** Find y. $-13 + y = 0$

If there are three or more numbers to add, it may be easier to add positive numbers and negative numbers separately and then combine the results. We can do this because, just like with whole numbers, addition of integers is commutative and associative.

EXAMPLE 8   Add. $-3 + 9 + (-4) + 12$

Solution

$-3 + 9 + (-4) + 12 = [(-3) + (-4)] + (9 + 12)$ Change the order of addition and regroup.

$= -7 + (9 + 12)$ Add the negative numbers: $-3 + (-4) = -7$.

$= -7 + 21$ Add the positive numbers: $9 + 12 = 21$.

$= 14$ Add the result: $-7 + 21 = 14$.

Student Practice 8 Add. $-8 + 6 + (-2) + 5$

③ Evaluating Algebraic Expressions Involving Addition of Integers

We evaluate expressions involving integers just as we did expressions involving whole numbers. We replace the variables with the given numbers and perform the operations indicated.

EXAMPLE 9   Evaluate.

(a) $-7 + a + b$ for $a = -3$ and $b = 9$
(b) $-x + y + (-13)$ for $x = -2$ and $y = -6$

Solution   We place parentheses around the variables, and then replace each variable with the appropriate values.

(a) $-7 + (a) + (b)$

$= \underbrace{-7 + (-3)}_{} + (9)$

$= \quad -10 \quad + 9$

$= \quad -1$

(b) $-(x) + (y) + (-13)$

$= \underbrace{-(-2)}_{} + \underbrace{[(-6) + (-13)]}_{}$

$= \quad 2 \quad + \quad (-19)$

$= -17$

Student Practice 9   Evaluate.

(a) $-3 + x + y$ for $x = -5$ and $y = -11$
(b) $-x + 8 + y$ for $x = -2$ and $y = -15$

④ Solving Applied Problems Involving Addition of Integers

EXAMPLE 10   Information on Micro Firm Computer Sales' profit and loss situation is given on the graph. What was the company's overall profit or loss at the end of the third quarter?

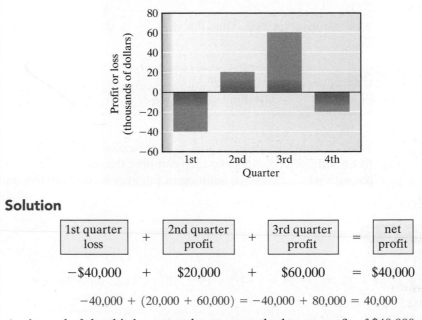

Solution

1st quarter loss	+	2nd quarter profit	+	3rd quarter profit	=	net profit

$\quad -\$40,000 \quad + \quad \$20,000 \quad + \quad \$60,000 \quad = \$40,000$

$-40,000 + (20,000 + 60,000) = -40,000 + 80,000 = 40,000$

At the end of the third quarter the company had a net profit of \$40,000.

Student Practice 10   What was Micro Firm Computer Sales' overall profit or loss at the end of the second quarter?

Verbal and Writing Skills, Exercises 1–6

1. Explain in your own words why a negative number added to a negative number is a negative number.

2. Write an application about temperatures to describe the addition of two negative numbers.

Fill in the blanks.

3. To add two numbers with different signs, we keep the sign of the _____ _____ _____ and _____.

4. The sum of two positive numbers is a _____ number. The sum of two negative numbers is a _____ number.

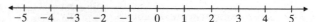

Fill in the box with + or −, then write the rule for adding integers that you used.

5. (a) $-2 + (-3) = \boxed{}\ 5$ Rule:

(b) $2 + (-3) = \boxed{}\ 1$ Rule:

(c) $-2 + 3 = \boxed{}\ 1$ Rule:

6. (a) $-4 + (-6) = \boxed{}\ 10$ Rule:

(b) $4 + (-6) = \boxed{}\ 2$ Rule:

(c) $-4 + 6 = \boxed{}\ 2$ Rule:

7. (a) Begin at 0 on the number line and move 2 units to the left followed by another 2 units to the left.

```
 +----+----+----+----+----+----+----+----+----+----+
-5   -4   -3   -2   -1    0    1    2    3    4    5
```

(b) Is the end result in the positive or negative region?

(c) Write the math symbols that represent the situation.

(d) Use the number line to find the sum.

8. (a) Begin at 0 on the number line and move 1 unit to the left followed by another 2 units to the left.

```
 +----+----+----+----+----+----+----+----+----+----+
-5   -4   -3   -2   -1    0    1    2    3    4    5
```

(b) Is the end result in the positive or negative region?

(c) Write the math symbols that represent the situation.

(d) Use the number line to find the sum.

9. (a) Begin at 0 on the number line and move 3 units to the right followed by another 2 units to the right.

```
 +----+----+----+----+----+----+----+----+----+----+
-5   -4   -3   -2   -1    0    1    2    3    4    5
```

(b) Is the end result in the positive or negative region?

(c) Write the math symbols that represent the situation.

(d) Use the number line to find the sum.

10. (a) Begin at 0 on the number line and move 1 unit to the right followed by another 4 units to the right.

```
 +----+----+----+----+----+----+----+----+----+----+
-5   -4   -3   -2   -1    0    1    2    3    4    5
```

(b) Is the end result in the positive or negative region?

(c) Write the math symbols that represent the situation.

(d) Use the number line to find the sum.

11. Fill in the blanks.

    (a) A decrease of 10°F followed by a decrease of 5°F results in a _____ .

    (b) Write the math symbols that represent the situation described in **(a)**. _____

12. Fill in the blanks.

    (a) A discount of $5 followed by a discount of $10 results in _____ .

    (b) Write the math symbols that represent the situation described in **(a)**. _____

13. Fill in the blanks.

    (a) A profit of $100 followed by a profit of $50 results in a _____ .

    (b) Write the math symbols that represent the situation described in **(a)**. _____

14. Fill in the blanks.

    (a) An increase of 150 units followed by an increase of 30 units results in a(n) _____ .

    (b) Write the math symbols that represent the situation described in **(a)**. _____

Add by using the rules for addition of integers.

15. **(a)** $-11 + (-13)$

    (b) $11 + 13$

16. **(a)** $-18 + (-14)$

    (b) $18 + 14$

17. **(a)** $-29 + (-39)$

    (b) $29 + 39$

18. **(a)** $-20 + (-30)$

    (b) $20 + 30$

19. **(a)** $-53 + (-18)$

    (b) $53 + 18$

20. **(a)** $-40 + (-10)$

    (b) $40 + 10$

Applications

21. ***Temperature Fluctuations*** During the early morning hours on a ski slope in Colorado, the temperature dropped to $-3°F$. At dawn it had risen 10°F.

    (a) Write the math symbols that represent the situation.

    (b) Use the chart to determine if the temperature at dawn was positive or negative.

    (c) Find the sum.

22. ***Temperature Fluctuations*** At midnight in Trenton, New Jersey, the temperature dropped to $-4°F$. At dawn it had risen 2°F.

    (a) Write the math symbols that represent the situation.

    (b) Use the chart to determine if the temperature at dawn was positive or negative.

    (c) Find the sum.

23. **(a)** Write the math symbols that represent the situation: move a marker up 2 units followed by a move down 4 units.

    (b) Use the chart to determine if the marker is in the positive or negative region.

    (c) Use the chart to find the sum.

24. **(a)** Write the math symbols that represent the situation: move a marker up 3 units followed by a move down 6 units.

    (b) Use the chart to determine if the marker is in the positive or negative region.

    (c) Use the chart to find the sum.

Express the outcome of each situation as an integer.

25. A 300-foot increase in altitude followed by a 400-foot decrease in altitude

26. Diving 10 feet downward followed by rising 2 feet

27. A loss of $400 followed by a profit of $500

28. An increase of 10 pounds in weight followed by a decrease of 5 pounds

Add by using the rules for addition of integers.

29. (a) $6 + (-8)$
   (b) $-6 + 8$

30. (a) $4 + (-9)$
   (b) $-4 + 9$

31. (a) $5 + (-1)$
   (b) $-5 + 1$

32. (a) $7 + (-3)$
   (b) $-7 + 3$

33. (a) $22 + (-16)$
   (b) $-22 + 16$

34. (a) $15 + (-24)$
   (b) $-15 + 24$

35. (a) $3 + (-1)$
   (b) $-3 + (-1)$
   (c) $-3 + 1$

36. (a) $5 + (-8)$
   (b) $-5 + (-8)$
   (c) $-5 + 8$

37. (a) $-9 + (-11)$
   (b) $9 + (-11)$
   (c) $-9 + 11$

38. (a) $-12 + (-3)$
   (b) $12 + (-3)$
   (c) $-12 + 3$

39. $2 + (-2)$

40. $10 + (-10)$

41. $-9 + 9$

42. $-5 + 5$

43. $-360 + 360$

44. $-500 + 500$

45. $452 + (-452)$

46. $786 + (-786)$

Find the value of x.

47. $x + 19 = 0$

48. $x + 35 = 0$

49. $-12 + x = 0$

50. $-42 + x = 0$

Mixed Practice, Exercises 51–70 *Add by using the rules for addition of integers.*

51. $12 + (-11)$

52. $14 + (-13)$

53. $-10 + 4$

54. $-6 + 3$

55. $-7 + 14$

56. $-4 + 9$

57. $22 + (-10)$

58. $34 + (-14)$

59. $-33 + (-5)$

60. $-42 + (-12)$

61. $-27 + (-12)$

62. $-43 + (-23)$

63. $-15 + 15$

64. $-92 + 92$

65. $-12 + 16$

66. $-11 + 15$

67. $6 + (-8)$

68. $5 + (-7)$

69. $15 + (-15)$

70. $13 + (-13)$

Add by using the rules for addition of integers.

71. $6 + (-9) + 1 + (-3)$

72. $4 + (-7) + 2 + (-5)$

73. $-21 + 16 + (-33)$

74. $-31 + 19 + (-25)$

75. $57 + (-29) + (-34) + 23$

76. $25 + (-17) + (-28) + 64$

77. $-15 + 7 + (-10) + 3$

78. $-12 + 4 + (-8) + 5$

79. Evaluate $y + 5$ for each of the following.
   (a) $y = -2$ (b) $y = -7$

80. Evaluate $x + 5$ for each of the following.
   (a) $x = -1$ (b) $x = -8$

81. Evaluate $a + (-8)$ for each of the following.
    (a) $a = 4$ **(b)** $a = -5$

82. Evaluate $x + (-6)$ for each of the following.
    (a) $x = 3$ **(b)** $x = -9$

83. Evaluate $-2 + x + y$ for each of the following.
    (a) $x = -6$ and $y = 4$
    (b) $x = 5$ and $y = -9$

84. Evaluate $-9 + a + b$ for each of the following.
    (a) $a = 7$ and $b = -3$
    (b) $a = -1$ and $b = 4$

85. Evaluate $-x + y + 6$ for $x = -3$ and $y = -1$.

86. Evaluate $-x + y + 4$ for $x = -2$ and $y = -5$.

87. Evaluate $-a + b + (-1)$ for $a = -3$ and $b = -5$.

88. Evaluate $-a + b + (-6)$ for $a = -5$ and $b = -1$.

Applications

Profit and Loss *The quarterly profit or loss in 2009 for Citron Foods is indicated on the graph. Use the graph to answer exercises 89 and 90.*

89. What was the company's overall profit or loss at the end of the second quarter?

90. What was the company's overall profit or loss at the end of the fourth quarter?

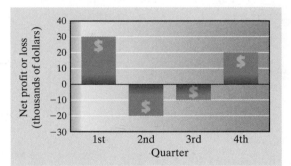

Solve each applied problem.

91. ***Checking Account*** While balancing her checkbook, Jessica discovered that her balance was $-\$121$. She hurried to the bank, hoping to prevent a bounced check, and deposited $200. What was the balance in her checking account after the deposit? (Assume there was no penalty from the bank.)

92. ***Checking Account*** Allison found that she had made an arithmetic mistake in her checkbook and the actual balance was $-\$97$. She quickly went to the bank and made a deposit of $150. What was the balance in her checking account after the deposit? (Assume there was no penalty from the bank.)

93. ***Submarine*** A submarine is 75 feet below sea level. It dives to a point 150 feet lower. Represent this distance below sea level as an integer.

94. ***Temperature*** The temperature at 3 P.M. was 2 degrees below zero. By midnight the temperature dropped another 8 degrees. Represent this temperature as an integer.

One Step Further *Evaluate.*

95. $3 + x + y + (-1) + z$ for $x = -1, y = 9$, and $z = -5$

96. $-2 + a + b + 6 + c$ for $a = 8, b = -4$, and $c = 4$

Add.

97. $-33 + 24 + (-38) + 19 + (-3)$

98. $9 + (-42) + (-88) + 10 + (-13)$

99. $12 + (-45) + (-9) + 5 + (-19)$

100. $2 + (-72) + (-41) + 11 + (-33)$

To Think About *Place the number in the box that makes each statement true.*

101. $-2 +$ ☐ $= -5$

102. $-6 +$ ☐ $= -7$

103. $3 +$ ☐ $= -1$

104. $4 +$ ☐ $= -3$

105. What number can be added to -11 to obtain -10?

106. What number can be added to -33 to obtain -30?

107. If $x + y + 30 = 0$, what does the sum $x + y$ equal? Find possible values of x and y for this equation.

108. If $22 + x + y = 0$, what does the sum $x + y$ equal? Find possible values of x and y for this equation.

Triangle Puzzle *To play the game, follow the steps listed below.*

1. *Start at any number on the bottom row of the triangle and move to any adjoining square, right, left, or up, and then add the two numbers.*

2. *Move to another adjoining square, adding the number in that square to your total.*

3. *Continue in the same manner, each time adding the number in the square you select to your total.*

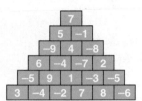

The goal is to reach the square at the top with the fewest number of squares selected while keeping the sum within the following guidelines.

109. At **no** time may the sum of the numbers be less than -6 or greater than or equal to 6.

110. At **no** time may the sum of the numbers be less than -5 or greater than or equal to 5.

Cumulative Review *Simplify.*

111. **[1.8.1]** $4x + 6x$

112. **[1.4.3]** $2(3x)$

113. **[1.8.1]** $8x - 3x$

114. **[1.7.3]** $3(x - 4)$

115. **[1.9.3]** *Distance Traveled* Vu Nguyen drives his car 150 miles from Phoenix to Tucson, Arizona, to visit a friend. Vu then proceeds from Tucson 110 miles to Sierra Vista for a one-week stay. He returns home to Phoenix from Sierra Vista driving the same route through Tucson. Vu had 23,566 miles on his odometer before he began his trip. If he did not drive any additional miles during his stay, what is the reading on the odometer at the end of the trip?

116. **[2.2.2]** *Boarding and Exiting a Bus* At the beginning of the day, 12 people boarded an empty bus on Route 33A. At the first stop, 4 people exited the bus and 8 people boarded. At the second stop, no one exited and 11 people boarded. At the third stop, 7 people exited and 15 boarded. Represent the number of people exiting with negative numbers and the number of people boarding with positive numbers. Then determine how many people were on the bus after the third stop.

Quick Quiz 2.2

1. Perform the operation indicated.

  (a) $-6 + (-4)$

  (b) $-6 + 4$

  (c) $-9 + 5 + (-3) + 6$

2. Evaluate $-a + b + 3$ for $a = -1$ and $b = -7$.

3. The profit or loss statement for Raskin Consulting, Inc. in 2006 showed a loss the first quarter of $3000, a loss the second quarter of $2500, a profit the third quarter of $8000, and a loss the fourth quarter of $3500. What was the company's overall profit or loss at the end of the fourth quarter?

4. **Concept Check** Without completing the calculations, explain how you can determine whether the answer is a positive or negative number.

Evaluate $132 + x + y + z$ for $x = -1, y = -3$, and $z = -2$.

👣👞 STEPS TO SUCCESS When to Use a Calculator

A calculator is an important tool and therefore we benefit by learning how to use it. You may be thinking, "Why learn math if I can use a calculator?" Well, often it is not practical or convenient to use a calculator. Many times we must perform calculations unexpectedly and do not have a calculator available, as in the following situations.

- You receive change for a purchase made. Is the change correct?

- You are shopping and must determine if you have enough money to buy an item marked "30% off the original price." Even if you have a calculator, you must know what operation is needed to find 30% of a number.

The calculator does only what you tell it to do; it does not plan the approach! You must determine if you should add, subtract, multiply, or divide.

Making it personal: Do the following exercises without and then with a calculator. Which way is faster, with or without a calculator?

1. Add. $-2 + (-3) + (-1)$

2. Combine like terms. $4xy + 2x + 3xy$

3. If you didn't know the rules for combining like terms, could you have done exercise 2 with a calculator?

2.3 Subtracting Integers

① Subtracting Integers

How do we subtract integers? What is the value of $-4 - 2$? Before we define a rule for subtracting integers, let's look at a few subtraction problems that we can do mentally.

Suppose that you have $20 in the bank and you write a check for $30. The bank will not be able to pay the $30 because you are *short $10*. If the bank cashed the check, your balance would represent a debt of $10, or $-$10. Thus we can see that $20 - $30 = -$10.

Now, what do you think the value of $6 - 7$ is? We can think of this as a situation in which we have 6 items and want to take away 7 items. We are *short 1 item*, or -1. Thus $6 - 7 = -1$.

Student Learning Objectives

After studying this section, you will be able to:

① Subtract integers.

② Perform several integer operations.

③ Solve applied problems involving subtraction of integers.

EXAMPLE 1   Subtract.

(a) $15 - $20

(b) $3 - 4$

Solution

(a) $15 - $20 = -$5$ If we have $15 and want to spend $20, we are short $5, or $-$5.

(b) $3 - 4 = -1$ If we have 3 items and try to take away 4 items, we are short 1 item, or -1.

Student Practice 1   Subtract.

(a) $10 - $20

(b) $5 - 7$

NOTE TO STUDENT: Fully worked-out solutions to all of the Student Practice problems can be found at the back of the text starting at page SP-1.

It is not always possible to subtract mentally, so we must find an efficient way to do more complicated subtraction problems. *We can rewrite a subtraction problem as a related addition problem and then use the rules for adding integers we learned in Section 2.2.* To illustrate, we write an addition problem that gives the same result as the subtraction problem in Example 1a. Look for a pattern.

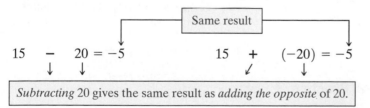

We see that $15 - 20$ is equivalent to $15 + (-20)$. They give the same result, -5. That is, subtracting 20 gives the same result as adding the opposite of 20. We see it is reasonable to generalize that subtracting is equivalent to adding the opposite.

EXAMPLE 2   Rewrite each subtraction as addition of the opposite.

(a) $40 - 10 = 30$ **(b)** $6 - 2 = 4$ **(c)** $25 - 5 = 20$

Solution

Subtraction	*Addition of the Opposite*
(a) $40 - 10 = 30$	$40 + (-10) = 30$
(b) $6 - 2 = 4$	$6 + (-2) = 4$
(c) $25 - 5 = 20$	$25 + (-5) = 20$

Student Practice 2   Rewrite each subtraction as addition of the opposite.

(a) $20 - 10 = 10$ **(b)** $5 - 2 = 3$ **(c)** $20 - 5 = 15$

We now state the rule for subtraction.

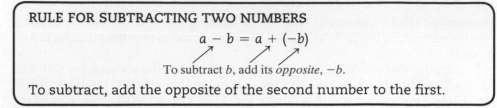

RULE FOR SUBTRACTING TWO NUMBERS

$$a - b = a + (-b)$$

To subtract b, add its *opposite*, $-b$.

To subtract, add the opposite of the second number to the first.

CAUTION: The *first* number does not change. Be sure to only change the *second* number.

EXAMPLE 3   Subtract.

(a) $-8 - 3$

(b) $-6 - (-4)$

Solution   We replace the *second* number by its opposite and then add using the rules for addition.

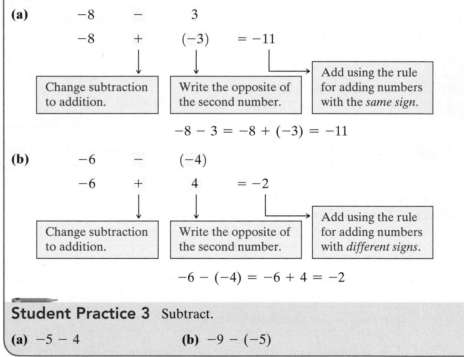

(a)

-8	$-$	3	
-8	$+$	(-3)	$= -11$

Change subtraction to addition.	Write the opposite of the second number.	Add using the rule for adding numbers with the *same sign*.

$$-8 - 3 = -8 + (-3) = -11$$

(b)

-6	$-$	(-4)	
-6	$+$	4	$= -2$

Change subtraction to addition.	Write the opposite of the second number.	Add using the rule for adding numbers with *different signs*.

$$-6 - (-4) = -6 + 4 = -2$$

Student Practice 3   Subtract.

(a) $-5 - 4$

(b) $-9 - (-5)$

At this point you should be able to do simple subtraction problems quickly.

Remember that in performing subtraction of two numbers:

1. The first number does not change.
2. The subtraction sign is changed to addition.
3. We write the opposite of the second number.
4. We find the result of this addition problem.

When you see $7 - 10$, you should think: "$7 + (-10)$." Try to think of each subtraction problem as a problem of *adding the opposite*.

If you see $-3 - 19$, think $-3 + (-19)$.

If you see $8 - (-2)$, think $8 + 2$.

EXAMPLE 4   Subtract.

(a) $8 - 9$ (b) $-3 - 16$
(c) $5 - (-4)$ (d) $-4 - (-4)$

Solution

(a) $8 - 9$ (b) $-3 - 16$
   $= 8 + (-9) = -1$ $= -3 + (-16) = -19$
(c) $5 - (-4)$ (d) $-4 - (-4)$
   $= 5 + 4 = 9$ $= -4 + 4 = 0$

Student Practice 4   Subtract.

(a) $7 - 10$ (b) $-4 - 15$ (c) $8 - (-3)$ (d) $-5 - (-1)$

② Performing Several Integer Operations

Subtraction of integers is not commutative or associative, but addition is. Thus if we first rewrite *all* subtraction as addition of the opposite, we can perform the addition in any order.

EXAMPLE 5   Perform the necessary operations. $4 - 7 - 5 - 3$

Solution

$\quad\quad 4 - 7 - 5 - 3$
$= 4 + (-7) + (-5) + (-3)$ First, write all subtraction as addition
$= 4 + [(-7) + (-5) + (-3)]$ of the opposite, and then group like signs.
$= 4 + (-15)$ Then add all *like signs:*
                             $(-7) + (-5) + (-3) = -15$.
$= -11$ Next, add *unlike signs:* $4 + (-15) = -11$.

Student Practice 5   Perform the necessary operations. $6 \quad 9 - 2 - 8$

▬ Understanding the Concept

Another Approach to Subtracting Several Integers   In Example 5, would you obtain the same answer if you first added $4 + (-7) = -3$ and then added the remaining numbers working from left to right? Let's try this approach.

$$4 - 7 - 5 - 3 = 4 + (-7) - 5 - 3$$
$$= -3 - 5 - 3$$
$$= -3 + (-5) - 3$$
$$= -8 - 3$$
$$= -8 + (-3)$$
$$= -11$$

Although this approach requires more calculations, it yields the same answer. Thus, when a problem contains only addition and subtraction, we see from Example 5 that we can use the associative and commutative properties of addition to simplify the process.

Exercise

1. Subtract. $2 - 6 - 8 - 11$
   (a) Change all subtraction to addition of the opposite, then perform the operations.

   (b) Complete the subtraction problem working from left to right.

   (c) Which method do you prefer?

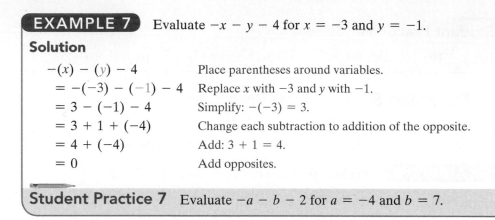

Mc **EXAMPLE 6**   Perform the necessary operations.
$$-9 - (-3) + (-4)$$

Solution

$$-9 - (-3) + (-4) = -9 + 3 + (-4)$$ Write subtraction as addition of the opposite.

$$= -13 + 3$$ Add like signs: $-9 + (-4) = -13$.

$$= -10$$ Add unlike signs: $-13 + 3 = -10$.

Student Practice 6   Perform the necessary operations.

$$-3 - (-5) + (-11)$$

EXAMPLE 7   Evaluate $-x - y - 4$ for $x = -3$ and $y = -1$.

Solution

$$-(x) - (y) - 4$$ Place parentheses around variables.

$$= -(-3) - (-1) - 4$$ Replace x with -3 and y with -1.

$$= 3 - (-1) - 4$$ Simplify: $-(-3) = 3$.

$$= 3 + 1 + (-4)$$ Change each subtraction to addition of the opposite.

$$= 4 + (-4)$$ Add: $3 + 1 = 4$.

$$= 0$$ Add opposites.

Student Practice 7   Evaluate $-a - b - 2$ for $a = -4$ and $b = 7$.

③ **Solving Applied Problems Involving Subtraction of Integers**

When we subtract $3000 - (-50)$, we obtain a result of 3050, which is larger than 3000. Why is the result larger than 3000 if we are subtracting a number from 3000? Because we are subtracting a negative number. We illustrate this idea next.

Suppose that we want to find the difference in altitude between the two mountains shown below. We subtract the *lower* altitude from the *higher* altitude. The difference in altitude between the two mountains is 3000 feet − 1000 feet = 2000 feet.

$3000 \text{ ft} - 1000 \text{ ft} = 2000 \text{ ft}$

Subtract a positive number, and the result is *less than* 3000.

The difference in altitude is the distance between the highest points.

Land that is below sea level is considered to have a negative altitude. A valley that is 50 feet below sea level is said to have an altitude of −50 feet. The difference in altitude between the mountain and the valley is found by subtracting 3000 feet − (−50) feet.

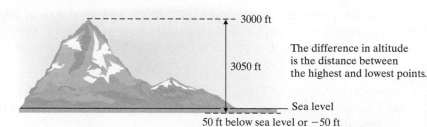

$3000 \text{ ft} - (-50 \text{ ft}) = 3050 \text{ ft}$

Subtract a negative number, and the result is *more than* 3000.

The difference in altitude is the distance between the highest and lowest points.

EXAMPLE 8   A portion of the Dead Sea is 1286 feet below sea level. What is the difference in altitude between Mount Carmel in Israel, which has an altitude of 1791 feet, and the Dead Sea?

Solution

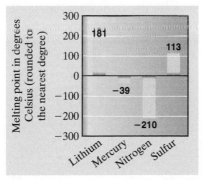

Mount Carmel ─── 1791 ft

Sea level ─ ─ ─ ─

Dead Sea

Difference in altitude is 3077 ft.

1286 ft below sea level or −1286 ft

We want to find the difference, so we must subtract.

higher altitude	minus	lower altitude	
↓	↓	↓	
1791 ft	−	(−1286 ft)	
= 1791 ft	+	1286 ft	= 3077 ft

The difference in altitude is 3077 ft.

Student Practice 8   Find the difference in altitude between a mountain 3800 feet high and a desert valley 895 feet below sea level.

EXAMPLE 9   The melting points in degrees Celsius of four chemical elements are listed on the bar graph. What is the difference between the melting point of mercury and the melting point of nitrogen?

Solution   We subtract the lower melting point from the higher melting point.

melting point of mercury	minus	melting point of nitrogen	
−39	−	(−210)	
−39	+	210	= 171

The difference is 171°C.

Student Practice 9   Refer to the bar graph in Example 9 to answer the following. What is the difference between the melting point of lithium and the melting point of nitrogen?

2.3 Exercises   MyMathLab®

Verbal and Writing Skills, Exercises 1–4

1. To subtract two numbers, we change the subtraction sign to _____ and take the _____ of the second number. Then we _____.

2. Explain what is wrong with the following problem.
$2 - 7 - 9 - 1 = 7 - 2 - 9 - 1$

3. When we subtract $25 - (-10)$, we get a number that is larger than 25. Explain why. You can use an illustration in your explanation.

4. Write an application about money where you subtract two integers and get a negative answer.

Fill in each box with the correct number.

5. To subtract -3, we add ▭.

6. To subtract -9, we add ▭.

7. To subtract 5, we add ▭.

8. To subtract 7, we add ▭.

9. $-7 - 6$
$= -7 + \boxed{}$
$= \boxed{}$

10. $-5 - 2$
$= -5 + \boxed{}$
$= \boxed{}$

11. $4 - 9$
$= 4 + \boxed{}$
$= \boxed{}$

12. $3 - 8$
$= 3 + \boxed{}$
$= \boxed{}$

13. $6 - (-3)$
$= 6 \,\boxed{}\, 3$
$= \boxed{}$

14. $8 - (-2)$
$= 8 \,\boxed{}\, 2$
$= \boxed{}$

15. $9 - (-5)$
$= 9 \,\boxed{}\, 5$
$= \boxed{}$

16. $7 - (-2)$
$= 7 \,\boxed{}\, 2$
$= \boxed{}$

Rewrite each subtraction as addition of the opposite.

17.

Subtraction	Addition of the Opposite
(a) $7 - 4 = 3$	
(b) $15 - 7 = 8$	
(c) $10 - 8 = 2$	

18.

Subtraction	Addition of the Opposite
(a) $5 - 3 = 2$	
(b) $12 - 6 = 6$	
(c) $7 - 1 = 6$	

Subtract.

19. $\$20 - \$35$

20. $\$3 - \$5$

21. $\$6 - \$7$

22. $\$5 - \$4$

23. $-6 - 4$

24. $-8 - 3$

25. $-5 - 4$

26. $-7 - 3$

27. $5 - (-2)$

28. $7 - (-4)$

29. $5 - (-9)$

30. $3 - (-7)$

31. $-8 - (-6)$

32. $-8 - (-3)$

33. $-8 - (-8)$

34. $-5 - (-5)$

35. $2 - 7$

36. $6 - 10$

37. $3 - 7$

38. $5 - 8$

39. $50 - 70$

40. $80 - 90$

41. $-85 - (-20)$

42. $-77 - (-11)$

Perform the necessary operations.

43. $12 - 9 - 5 - 8$

44. $5 - 2 - 6 - 10$

45. $2 - 1 - 9 - 7$

46. $9 - 3 - 7 - 25$

47. $9 - 10 - 2 + 3$

48. $8 - 11 - 4 + 7$

49. $-8 - (-3) + (-10)$

50. $-5 - (-2) + (-7)$

51. $-7 - (-2) - (-5)$

52. $-5 - (-9) - (-4)$

53. $-3 - (-8) + (-6)$

54. $-7 - (-2) + (-9)$

Mixed Practice, Exercises 55–68 *Perform the necessary operations.*

55. $7 - 21$

56. $9 - 13$

57. $-9 - (-9)$

58. $-11 - (-11)$

59. $-13 - 18$

60. $-18 - 56$

61. $40 - (-1)$

62. $39 - (-1)$

63. $8 - 1 - 9 - 5$

64. $3 - 7 - 5 - 16$

65. $7 + 8 - 6 - 11$

66. $6 + 4 - 8 - 22$

67. $9 - 10 - 2 + 3$

68. $-6 - 3 + (-7) - 2$

Evaluate.

69. $a - 9$ for $a = -8$

70. $x - 12$ for $x = -9$

71. $x - 11$ for $x = -3$

72. $x - 10$ for $x = -2$

73. $14 - m$ for $m = -5$

74. $19 - y$ for $y = -6$

75. $21 - y + x$ for $y = -1$ and $x = 2$

76. $14 - y + x$ for $y = -2$ and $x = 3$

77. $-8 - x - y$ for $x = -4$ and $y = 2$

78. $-7 - x - y$ for $x = -3$ and $y = 4$

79. $-1 - x + y$ for $x = -6$ and $y = -5$

80. $-2 - x + y$ for $x = -5$ and $y = -3$

Applications

High and Low Temperatures *The following chart displays the hottest and coldest spots for selected days in a recent winter. Use this chart to answer exercises 85 and 86.*

81. **(a)** Where was the temperature the highest during the five days listed on the chart?

(b) What was the difference in temperature between the record high and the record low on day 3?

82. **(a)** Where was the temperature the lowest during the five days listed on the chart?

(b) What was the difference in temperature between the record high and the record low on day 5?

	Record High		Record Low	
Day 1	Gila Bend, Arizona	72°F	Presque Isle, Maine	−18°F
2	Lajitas, Texas	79°F	Ely, Minnesota	−8°F
3	Indio, California	77°F	Devil's Lake, North Dakota	−9°F
4	Brownsville, Texas	88°F	Bodie State Park, California	−13°F
5	Del Rio, Texas	84°F	Presque Isle, Maine	−9°F

83. ***Calculating Distance*** How far above the floor of the basement is the roof of the office building shown in the figure?

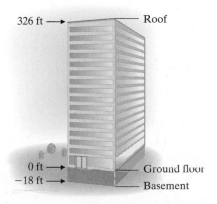

326 ft → Roof

0 ft → Ground floor

−18 ft → Basement

84. ***Temperature Readings*** On a particular day the temperature in a city was recorded.

(a) What was the difference between the temperatures at 10 A.M. and 10 P.M.?

(b) What was the difference between the temperatures at 4 P.M. and 4 A.M.?

85. *Altitude* Find the difference in altitude between a mountain that has an altitude of 3556 feet and a desert valley that is 150 feet below sea level.

86. *Altitude* Find the difference in altitude between a mountain that has an altitude of 5889 feet and a desert valley that is 175 feet below sea level.

Golf Scores *The following double bar graph indicates the scores for Sandra and Tran at selected holes at a National Pro-Am Tournament.*

87. What was the point difference between Sandra's and Tran's scores after the fourteenth hole?

88. What was the point difference between Sandra's and Tran's scores after the fifteenth hole?

89. If the lowest score after the eighteenth hole determines the winner, who won the National Pro-Am Tournament?

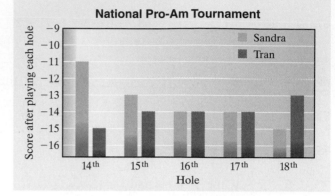

One Step Further *Perform the operations indicated.*

90. $-22 + 18 - 34 - 11 + (-16) - 2$

91. $-31 + 24 - 13 - 12 + (-14) - 3$

Evaluate.

92. $9 - x - y + z + 4$ for $x = -11, y = -2, z = -8$

93. $7 - a - b + c + 2$ for $a = -13, b = -4, c = -9$

Calculate.

94. $-345 - 768$

95. $-3009 - 893$

96. $632 - (-1346)$

97. $-2001 - (-987)$

To Think About *For exercises 98 and 99, (a) translate, and (b) solve.*

98. The sum of negative three and six is equal to what number?

99. Eight subtracted from negative one is equal to what number?

100. Is -1 a solution to $x - 4 = 6$?

101. Is -4 a solution to $x - 3 = -7$?

Fill in each box with the correct number.

102. $-1 - \square = -2$

103. $-2 - \square = -4$

For each of the following, find the next number in the sequence by identifying the pattern.

104. $-8, -3, 2, 7, 12, \ldots$

105. $-6, -3, 0, 3, 6, \ldots$

106. $2, -8, -18, -28, -38, \ldots$

107. $4, -1, -6, -11, -16, \ldots$

Cumulative Review *Simplify.*

108. **[1.6.4]** $2 + 3(5)$

109. **[1.6.4]** $12 - 3(4 - 1)$

110. [1.6.4] $3^2 + 4(2) - 5$

111. [1.6.4] $3 + [3 + 2(8 - 6)]$

112. [1.5.4] *Educational Supplies* L. R. William Elementary School must order 550 pencils for students to use on their standardized tests. If the pencils are packaged in boxes of 12, how many boxes should the school order?

113. [1.5.4] *Word Problems* A word processor can type 85 words per minute. At this rate, how long will it take to type 8670 words?

Quick Quiz 2.3

1. Perform the indicated operation.

   (a) $-9 - 15$ **(b)** $-6 - (-8)$

3. Find the difference in altitude between a mountain 6300 feet high and a desert valley that is 419 feet below sea level.

2. Perform the necessary operations.
$$-8 + 6 - (-5) - 10$$

4. **Concept Check** Is the following problem completed correctly? Why or why not?
$$-6 - (-3) + (-7)$$
$$= -6 - (-10)$$
$$= -6 + 10$$
$$= 4$$

👣 STEPS TO SUCCESS Why Is Homework Necessary?

You learn mathematics by practicing, not by watching.

Your instructor may make solving a mathematics problem look easy, but to learn the necessary skills you must practice them over and over again, just as your instructor once had to do. There is no other way. Learning mathematics is like learning how to play a musical instrument or to play a sport. *You must practice, not just observe, to do well.* Homework provides this practice. The amount of practice varies for each person. The more problems you do, the better you get.

Many students underestimate the amount of time that is required to learn math.

In general, two to three hours per week per unit or credit is a good rule of thumb. This means that for a three-unit class you should spend six to nine hours a week studying math. Spread this time throughout the week, not just in a few sittings. Your brain gets overworked just as your muscles do!

Making it personal:

1. Make a log of the time that you spend studying math. If your performance is not up to your expectations, increase your study time.

2. Review your time management plan to be sure you included sufficient hours per week to study mathematics.

   Number of hours per week studying math: ____

   Current grade in class: ____

   Should I increase study time? ____ Yes ____ No

Did You Know ...

That It May Be Better to Buy a Car than Lease It?

DECIDING TO BUY OR LEASE A CAR

Understanding the Problem:

Louvy has his eye on a brand new car. He thinks he should lease the car because his best friend Tranh has a car lease and says he can get the same deal for Louvy. On the other hand, Louvy's girlfriend Allie says it is always better to buy the car and finance it by taking out a loan.

Louvy does some research and finds that it is not at all simple. While a lease offers lower monthly payments, at the end of the lease period you are left with nothing.

	Lease	Purchase
Automobile price	$23,000	$23,000
Interest rate	6%	6%
Length of loan/lease	36 months	36 months
Down payment	$1000	$1000
Residual value (the value of the car you are turning in, or the price you would pay if you want to buy it)	$11,000	Not applicable
Monthly payment	$388.06	$669.28

Task 5: *How much will he save in payments each month if he leases the car?*

Task 6: *Which option should Louvy choose if he wants the best overall price? If he is concerned about his monthly payments?*

Making a Plan:

Step 1: Louvy needs to compare the total amounts he would pay for the entire loan.

Task 1: *Determine how much Louvy would pay to lease the car for three years.*

Task 2: *Determine how much Louvy would pay to buy the car with a three-year loan.*

Step 2: If Louvy wanted to buy the car at the end of the lease, he would have to pay an additional $11,000.

Task 3: *Determine the total cost if Louvy wants to buy the car at the end of the lease.*

Task 4: *Determine the total overall savings if Louvy buys the car instead of leasing it.*

Making a Decision:

Step 3: Louvy is unsure what to do. He likes the fact that buying the car would give him an overall savings, but a lower monthly payment is also important to him.

Applying the Situation to Your Life:

Task 7: *How much can you afford to pay for a car payment each month?*

Task 8: *How much would you pay in insurance, gas, and taxes?*

Task 9: *Would you rather buy or lease a car?*

Facts You Should Know:

You may wish to lease a car if

- you don't drive more than the specified number of miles in the lease on a yearly basis.
- you want a new vehicle every two to three years.

You may wish to buy a car if

- you intend to keep it a long time.
- you want to be debt-free after a time.

How Am I Doing?   Sections 2.1–2.3

How are you doing with your homework assignments in Sections 2.1 to 2.3? Do you feel you have mastered the material so far? Do you understand the concepts you have covered? Before you go further in the textbook, take some time to do each of the following problems.

2.1

Replace each ? with the symbol $<$, $>$, or $=$.

1. $-12 \ ? \ -7$

2. $|-11| \ ? \ |8|$

Simplify.

3. $-|-8|$

4. $-(-(-(3)))$

5. Evaluate $-(-x)$ for $x = -6$.

2.2

Add by using the rules for addition of integers.

6. $-2 + (-14)$

7. $-8 + 3 + (-1) + 4$

8. Evaluate $a + b + 12$ for $a = -9$ and $b = -5$.

9. Evaluate $-x + y + 7$ for $x = -8$ and $y = -11$.

10. The quarterly profit or loss in 2008 for Allied Moving and Storing is indicated on the graph. What was the company's overall profit or loss at the end of the fourth quarter?

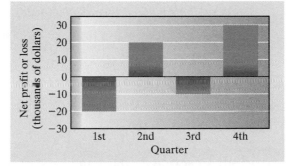

2.3

Perform the necessary operations.

11. $7 - 19$

12. $-3 - (-5)$

13. $-8 - (-2) - (-1)$

14. $-5 - 6 + (-1) - (-7)$

15. Evaluate $-5 - x - y$ for $x = -1$ and $y = -2$.

16. Find the difference in altitude between a mountain that has an altitude of 7622 feet and a valley that is 161 feet below sea level.

Now turn to page SA-4 for the answers to each of these problems. Each answer also includes a reference to the objective in which the problem is first taught. If you missed any of these problems, you should stop and review the Examples and Student Practice problems in the referenced objective. A little review now will help you master the material in the upcoming sections of the text.

1. _____

2. _____

3. _____

4. _____

5. _____

6. _____

7. _____

8. _____

9. _____

10. _____

11. _____

12. _____

13. _____

14. _____

15. _____

16. _____

2.4 Multiplying and Dividing Integers

Student Learning Objectives

After studying this section, you will be able to:

① Multiply two integers.

② Multiply two or more integers.

③ Use exponents with integers as bases.

④ Divide integers.

We are familiar with multiplying and dividing whole numbers. In this section we learn how to multiply and divide integers.

Before we begin our discussion, we must introduce some new vocabulary words. The set of whole numbers is made up of **odd numbers** and **even numbers.** The number 0 is the first even number. To find each consecutive even number, we add 2 to the previous number. Thus, the first four even numbers are 0, 2, 4, and 6. Whole numbers that are not even are odd. Thus, 1, 3, 5, and 7 are the first four odd numbers.

① Multiplying Two Integers

Recall the different ways we can indicate multiplication. You should be able to identify and use all of them.

$$-3 \times 3 \qquad -3 \cdot 3 \qquad 3(-3) \qquad (-3)(-3)$$

How do we determine whether a product is positive or negative? We follow a set of rules for multiplying integers. Before we state these rules, let's look at some situations involving multiplication of integers.

Since multiplication represents repeated addition, we can express some situations involving addition as multiplication problems.

Situation 1: Your business has a profit of $1000 a month for 3 months.
Math symbols that represent Situation 1:

$$\$1000 + \$1000 + \$1000 = \$3000 \quad \text{or} \quad (3)(\$1000) = \$3000$$

Situation 2: Your business suffers a loss of $1000 a month for 3 months.
Math symbols that represent Situation 2:

$$-\$1000 + (-\$1000) + (-\$1000) = -\$3000 \quad \text{or} \quad (3)(-\$1000) = -\$3000$$

EXAMPLE 1 Find the product by writing as repeated addition. $2(-3)$

Solution $2(-3) = -3 + (-3) = -6$

Therefore, $2(-3) = -6$.

Student Practice 1 Find the product by writing as repeated addition. $3(-1)$

NOTE TO STUDENT: Fully worked-out solutions to all of the Student Practice problems can be found at the back of the text starting at page SP-1.

Since it is not always practical to use repeated addition, we need a rule for multiplying integers. Let's look at the following pattern to help us understand the rule.

This column decreases by 1. ⟶ ⟵ This column decreases by 2.

$$
\begin{array}{l}
3 \cdot 2 = 6 \\
2 \cdot 2 = 4 \\
1 \cdot 2 = 2
\end{array} \Big\} \quad [+] \cdot [+] = [+] \quad \text{Positive product}
$$

$$0 \cdot 2 = 0$$

$$
\begin{array}{l}
-1 \cdot 2 = -2 \\
-2 \cdot 2 = -4 \\
-3 \cdot 2 = -6
\end{array} \Big\} \quad [-] \cdot [+] = [-] \quad \text{Negative product}
$$

Since multiplication is commutative, we know that $-3 \cdot 2 = 2 \cdot (-3)$. Thus

Negative product

$$2 \cdot (-3) = -6 \qquad [+] \cdot [-] = [-]$$

We see that whenever one number is positive and the other number is negative, the product is negative.

Now how do we multiply two *negative* numbers? Consider the following pattern.

This column decreases by 1.——⌐↓   ⌐——This column increases by 5.↓

$$3(-5) = -15$$
$$2(-5) = -10$$
$$1(-5) = -5$$
$$0(-5) = 0$$
$$-1(-5) = 5$$
$$-2(-5) = 10$$
$$-3(-5) = 15 \qquad [-] \cdot [-] = [+]$$

This seems to suggest that a *negative* number times a *negative* number gives a *positive* result, and this is the case. Let's summarize the results and look for a pattern.

The number of negative signs, 0, is **even.** → $[+] \cdot [+] = [+]$
The product is **positive.** ↗

The number of negative signs, 1, is **odd.** → $[-] \cdot [+] = [-]$
The product is **negative.** ↗

The number of negative signs, 1, is **odd.** → $[+] \cdot [-] = [-]$
The product is **negative.** ↗

The number of negative signs, 2, is **even.** → $[-] \cdot [-] = [+]$
The product is **positive.** ↗

We notice that when there is an *even* number of negative signs, the answer is *positive* and when there is an *odd* number of negative signs, the answer is *negative*. Why is this true? Because every pair of negative signs yields a positive result.

$[-] \cdot [-]$	$=$	$[+]$	Even number
one pair of	$=$	positive	of negative signs.
negative numbers		result	

$[-] \cdot [-]$	$\cdot$	$[-] \cdot [-]$	$\cdot$	$[-]$	$=$	$[-]$	Odd number
positive	$\times$	positive	$\times$	negative	$=$	negative	of negative
result		result		number		result	signs.

The pattern is summarized in the following procedure.

PROCEDURE TO DETERMINE THE SIGN OF A PRODUCT

For all nonzero numbers:

The product will be *positive* if there is an *even* number of negative signs.
The product will be *negative* if there is an *odd* number of negative signs.

When we multiply integers, we first determine the sign of the product and then multiply absolute values.

EXAMPLE 2 Multiply.

(a) $6(7)$ **(b)** $6(-7)$ **(c)** $-6(7)$ **(d)** $-6(-7)$

Solution

(a) $6(7) = +42$ The number of negative signs, 0, is *even* so the answer is *positive*.
We multiply absolute values: $6 \cdot 7 = 42$.

(b) $6(-7) = -$ The number of negative signs, 1, is *odd* so the answer is *negative*.
$\quad = -42$ We multiply absolute values: $6 \cdot 7 = 42$.

Continued on next page

(c) $-6(7) = -$ The number of negative signs, 1, is *odd* so the answer is *negative*.

     $= -42$ We multiply absolute values: $6 \cdot 7 = 42$.

(d) $-6(-7) = +$ The number of negative signs, 2, is *even* so the answer is *positive*.

     $= +42$ We multiply absolute values: $6 \cdot 7 = 42$.

Student Practice 2   Multiply.

(a) $3(8)$ **(b)** $3(-8)$ **(c)** $-3(8)$ **(d)** $-3(-8)$

② Multiplying Two or More Integers

Recall that multiplication is commutative and associative, so we can either multiply integers in the order they appear, or we can multiply any pair of numbers first and then multiply the result by another number. Then we continue until all of the factors have been multiplied.

EXAMPLE 3   Multiply. $(-3)(-1)(-2)$

Solution

$$(-3)(-1)(-2) = 3(-2) \quad \text{First we multiply } (-3)(-1) = 3.$$
$$= -6 \qquad \text{Then we multiply } 3(-2) = -6.$$

Student Practice 3   Multiply. $(-2)(-1)(-4)$

We can simplify the multiplication process if *before* we multiply, we count the *total* number of negative signs to determine the sign of the product. Then we multiply absolute values last.

$$(-2)(-1) = +2 \qquad \text{We multiply two negative numbers and the answer is positive.}$$

$$(-2)(-1)(-1) = -2 \qquad \text{We multiply three negative numbers and the answer is negative.}$$

$$(-2)(-1)(-1)(-1) = +2 \quad \text{We multiply four negative numbers and the answer is positive.}$$

EXAMPLE 4   Multiply. $(-3)(2)(-1)(4)(-3)$

Solution

$(-3)(2)(-1)(4)(-3)$ The answer is *negative* since there are 3 negative signs and 3 is an *odd* number.

$= -[3(2)(1)(4)(3)]$ Multiply the absolute values.

$= -72$

Student Practice 4   Multiply.
$(-3)(2)(-1)(-4)(-3)$

③ Using Exponents with Integers as Bases

In Chapter 1 we saw that we can use exponents to abbreviate repeated multiplication.

$$\underset{\substack{\text{repeated} \\ \text{multiplication}}}{(-3) \cdot (-3) \cdot (-3)} = \underset{\substack{\text{exponent} \\ \text{form}}}{(-3)^3}$$

Therefore, we can find the sign of a negative number raised to a power in the same way that we find the sign of a product.

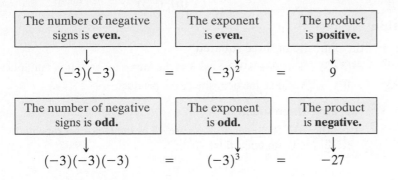

We see that the sign of the result depends on whether the exponent is odd or even. This can be generalized as follows.

> Suppose that a number is written in exponent form and the *base* is *negative*. The result is *positive* if the exponent is *even*. The result is *negative* if the exponent is *odd*.

EXAMPLE 5   Evaluate. $(-4)^3$

Solution

$(-4)^3 = (-4)(-4)(-4) = -64$ The answer is negative since the exponent 3 is odd.

Student Practice 5   Evaluate. $(-4)^4$

EXAMPLE 6   Evaluate.

(a) $(-1)^2$ **(b)** $(-1)^3$ **(c)** $(-1)^{31}$

Solution

(a) $(-1)^2 = 1$ The answer is positive since the exponent 2 is even.
(b) $(-1)^3 = -1$ The answer is negative since the exponent 3 is odd.
(c) $(-1)^{31} = -1$ The answer is negative since the exponent 31 is odd.

Student Practice 6   Evaluate.

(a) $(-2)^3$ **(b)** $(-2)^6$ **(c)** $(-2)^7$

We must be sure to use parentheses around the base when the base is a negative number. For example, $(-2)^4$ is *not* the same as -2^4. The base of a number in exponent form does not include the negative sign unless we use parentheses.

$(-2)^4$ means "-2 raised to the fourth power," since -2 is the base.
-2^4 means "the opposite of 2 raised to the fourth power," since 2 is the base.

$$(-2)^4 \;\; = \;\; (-2)(-2)(-2)(-2) \;\; = 16$$

The base is -2.   We use the -2 as the factor for repeated multiplication.

$$-2^4 \;\; = \;\; -(2\cdot2\cdot2\cdot2) = -(16) = -16$$

The base is 2.   We use 2 as the factor for repeated multiplication and take the opposite of the product.

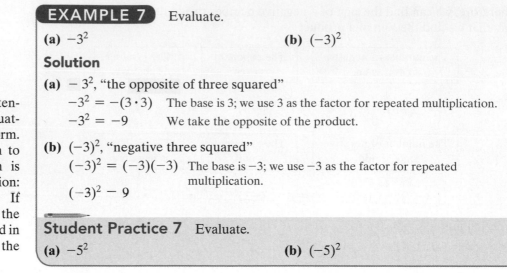

CAUTION: Pay special attention to parentheses when evaluating a number in exponent form. If the exponent is *hanging* on to the *parentheses* $(\ )^2$, the sign is included in the multiplication: $(-10)^2 = (-10)(-10) = 100$. If the exponent is *hanging* on to the *base* $-b^2$, the sign is not included in the multiplication: -4^2 means the opposite of $4^2 = -16$.

EXAMPLE 7 Evaluate.

(a) -3^2 (b) $(-3)^2$

Solution

(a) -3^2, "the opposite of three squared"

$\quad -3^2 = -(3 \cdot 3)$ The base is 3; we use 3 as the factor for repeated multiplication.

$\quad -3^2 = -9$ We take the opposite of the product.

(b) $(-3)^2$, "negative three squared"

$\quad (-3)^2 = (-3)(-3)$ The base is -3; we use -3 as the factor for repeated multiplication.

$\quad (-3)^2 = 9$

Student Practice 7   Evaluate.

(a) -5^2 (b) $(-5)^2$

④ Dividing Integers

What about division? Any division statement can be rewritten as a related multiplication statement. Therefore, the sign rules for division are very much like those for multiplication.

Division problem	$-20 \div (-4) = n$	
Related multiplication problem	$-20 = n(-4)$	Since $5(-4) = -20$, n must be positive.
Therefore,	$-20 \div (-4) = 5$	The number of negative signs is even and the result is positive.
Division problem	$-20 \div 4 = n$	
Related multiplication problem	$-20 = n(4)$	Since $(-5)(4) = -20$, n must be negative.
Therefore,	$-20 \div 4 = -5.$	The number of negative signs is odd and the result is negative.

Similarly, $20 \div (-4) = -5$ because $20 = (-5)(-4)$. As we can see, the sign rules for division are the same as those for multiplication. We will state them together.

> **RULE FOR MULTIPLYING OR DIVIDING NUMBERS**
>
> To multiply or divide nonzero numbers: First determine the sign of the answer as follows.
>
> The answer will be *positive* if the problem has an *even* number of negative signs.
>
> The answer will be *negative* if the problem has an *odd* number of negative signs.
>
> Next multiply or divide the absolute values of the numbers.

EXAMPLE 8   Divide.

(a) $36 \div 6$ (b) $36 \div (-6)$ (c) $-36 \div 6$ (d) $-36 \div (-6)$

Solution

(a) $36 \div 6 = 6$ (b) $36 \div (-6) = -6$

(c) $-36 \div 6 = -6$ (d) $-36 \div (-6) = 6$

Student Practice 8   Divide.

(a) $42 \div 7$ (b) $42 \div (-7)$ (c) $-42 \div 7$ (d) $-42 \div (-7)$

EXAMPLE 9   Perform each operation indicated.

(a) $56 \div (-8)$ **(b)** $9(-5)$ **(c)** $-20(-3)$ **(d)** $\dfrac{-72}{-8}$

Solution

(a) $56 \div (-8) = -7$ **(b)** $9(-5) = -45$

(c) $-20(-3) = 60$ **(d)** $\dfrac{-72}{-8} = -72 \div (-8) = 9$

Student Practice 9   Perform each operation indicated.

(a) $49 \div (-7)$ **(b)** $4(-9)$ **(c)** $-30(-4)$ **(d)** $\dfrac{-54}{-9}$

EXAMPLE 10   Evaluate.

(a) $\dfrac{m}{-n}$ for $m = -16$ and $n = -2$ **(b)** x^4 for $x = -2$

Solution   We place parentheses around the variables and then we replace each variable with the given value.

(a) $\dfrac{(m)}{-(n)} = \dfrac{(-16)}{-(-2)}$ We replace m with -16 and n with -2.

$\qquad = \dfrac{-16}{2}$ $-(-2) = 2$: The opposite of negative 2 is 2.

$\qquad = -8$ $-16 \div 2 = -8$

(b) $(x)^4 = (-2)^4$ We replace x with -2. The answer is positive since the

$\qquad = 16$ exponent is even. $2 \cdot 2 \cdot 2 \cdot 2 = 16$

Student Practice 10   Evaluate.

(a) $\dfrac{-x}{y}$ for $x = -22$ and $y = 11$ **(b)** a^3 for $a = -3$

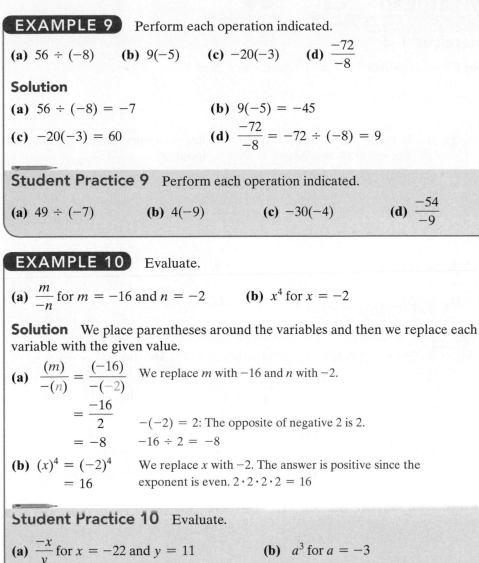

👣 STEPS TO SUCCESS Preparing for an Exam

If you have not read the Steps to Success, Reviewing for an Exam, in Section 1.7, it is a good idea to do so now so that you can learn how to prepare for an exam. Knowing *how to learn* and *how to study* are skills that we must acquire. Mastering these skills will help reduce anxiety and promote learning.

Making it personal: Place a ✓ in the blank as you complete each of the following.

1. Re-read the Steps to Success in Section 1.7 and follow the strategy recommended for Chapter 2 and all future chapter exams.

2. Watch the Chapter Test Prep Video.

3. Make a list of topics you do not understand, then meet with your instructor or tutor to help you with them.

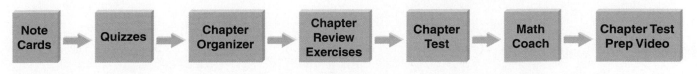

Note Cards → Quizzes → Chapter Organizer → Chapter Review Exercises → Chapter Test → Math Coach → Chapter Test Prep Video

Verbal and Writing Skills, Exercises 1–6

1. Why is the following statement false? Two negatives always gives us a positive result.

2. Explain why $-3^2 \neq 9$.

3. If you multiply 4 negative numbers, the product will be a _____ number.

4. If you multiply 7 negative numbers the product will be a _____ number.

5. The quotient of a positive number and a _____ number is negative.

6. The quotient of a negative number and a _____ number is positive.

Find the product by writing as repeated addition.

7. $3(-4)$

8. $4(-1)$

9. $4(-6)$

10. $2(-5)$

11. $2(-3)$

12. $3(-2)$

Fill in each box with the correct number.

13. (a) $3 \cdot \boxed{} = 9$
(b) $3 \cdot \boxed{} = -9$
(c) $-3 \cdot \boxed{} = -9$
(d) $-3 \cdot \boxed{} = 9$

14. (a) $5 \cdot \boxed{} = 25$
(b) $5 \cdot \boxed{} = -25$
(c) $-5 \cdot \boxed{} = -25$
(d) $-5 \cdot \boxed{} = 25$

15. (a) $\dfrac{12}{\boxed{}} = 3$
(b) $\dfrac{-12}{\boxed{}} = 3$
(c) $\dfrac{12}{\boxed{}} = -3$
(d) $\dfrac{-12}{\boxed{}} = -3$

16. (a) $\dfrac{18}{\boxed{}} = 9$
(b) $\dfrac{-18}{\boxed{}} = 9$
(c) $\dfrac{18}{\boxed{}} = -9$
(d) $\dfrac{-18}{\boxed{}} = -9$

Multiply.

17. (a) $9(2)$
(b) $9(-2)$
(c) $-9(2)$
(d) $-9(-2)$

18. (a) $11(7)$
(b) $11(-7)$
(c) $-11(7)$
(d) $-11(-7)$

19. (a) $5(2)$
(b) $-5(-2)$
(c) $-5(2)$
(d) $5(-2)$

20. (a) $1(8)$
(b) $-1(-8)$
(c) $-1(8)$
(d) $1(-8)$

21. $-2(-9)$

22. $-5(-5)$

23. $-1(-6)$

24. $-7(-3)$

25. $8(-7)$

26. $3(-11)$

27. $-5(9)$

28. $-6(3)$

To Think About, Exercises 29–32 *Determine the sign of each product without multiplying the integers.*

29. Is the product of $(-1)(3)(-236)(42)(-16)(-90)$ a positive or negative number?

30. Is the product of $(-2)(-96)(-69)(-72)(-6)(68)$ a positive or negative number?

31. Is the product of $(-943)(-721)(-816)(-96)(-51)$ a positive or negative number?

32. Is the product of $(-66)(-918)(-818)(-22)$ a positive or negative number?

Multiply.

33. $4(-5)(-2)$

34. $2(-4)(-5)$

35. $(-3)(-2)(-3)(-4)$

36. $(-4)(-3)(-2)(-2)$

37. $2(-1)(5)(-7)$

38. $8(-1)(2)(-3)$

39. $(-2)(-1)(4)(-5)$

40. $(-1)(-3)(2)(-3)$

41. $(-5)(4)(-3)(2)(-1)$

42. $(-4)(5)(-2)(1)(-3)$

To Think About, Exercises 43–48 *Determine the sign of each of the following without multiplying the integers.*

43. Is the value of $(-2)^{13}$ a positive or negative number?

44. Is the value of $(-8)^{12}$ a positive or negative number?

45. Is the value of $(-96)^{52}$ a positive or negative number?

46. Is the value of $(-81)^{51}$ a positive or negative number?

47. Is the value of -96^{52} a positive or negative number?

48. Is the value of -81^{51} a positive or negative number?

Evaluate.

49. $(-10)^2$

50. $(-7)^2$

51. $(-5)^3$

52. $(-7)^3$

53. (a) $(-4)^2$
   (b) $(-4)^3$

54. (a) $(-2)^2$
   (b) $(-2)^3$

55. (a) $(-1)^{13}$
   (b) $(-1)^{24}$

56. (a) $(-1)^{29}$
   (b) $(-1)^{16}$

57. (a) -4^2
   (b) $(-4)^2$

58. (a) -6^2
   (b) $(-6)^2$

59. (a) -2^3
   (b) $(-2)^3$

60. (a) -5^3
   (b) $(-5)^3$

61. (a) $(-4)^3$
   (b) -4^3

62. (a) $(-8)^2$
   (b) -8^2

63. (a) $(-9)^2$
   (b) -9^2

64. (a) $(-1)^{11}$
   (b) -1^{11}

Divide.

65. (a) $35 \div 7$
   (b) $35 \div (-7)$
   (c) $-35 \div 7$
   (d) $-35 \div (-7)$

66. (a) $50 \div 5$
   (b) $50 \div (-5)$
   (c) $-50 \div 5$
   (d) $-50 \div (-5)$

67. (a) $40 \div 8$
   (b) $40 \div (-8)$
   (c) $-40 \div 8$
   (d) $-40 \div (-8)$

68. (a) $20 \div 4$
   (b) $20 \div (-4)$
   (c) $-20 \div 4$
   (d) $-20 \div (-4)$

69. $30 \div (-5)$

70. $15 \div (-3)$

71. $\dfrac{-45}{5}$

72. $\dfrac{-24}{6}$

73. $-16 \div (-2)$

74. $-12 \div (-2)$

75. $\dfrac{-49}{-7}$

76. $\dfrac{-70}{-10}$

Mixed Practice *Perform each operation indicated.*

77. (a) $22 \div (-2)$
   (b) $22(-2)$

78. (a) $18 \div (-3)$
   (b) $18(-3)$

79. (a) $-4 \div (-2)$
   (b) $-4(-2)$

80. (a) $-8 \div (-4)$
   (b) $-8(-4)$

81. (a) $-15 \div 3$
   (b) $-15(3)$

82. (a) $-12 \div 3$
   (b) $-12(3)$

83. (a) $14 \div (-7)$
   (b) $-14(7)$

84. (a) $9 \div (-3)$
   (b) $-9(3)$

85. Evaluate x^2 for $x = -1$.

86. Evaluate x^3 for $x = -2$.

87. Evaluate $\dfrac{-x}{y}$ for $x = -42$ and $y = -7$.

88. Evaluate $\dfrac{-a}{b}$ for $a = -12$ and $b = -4$.

89. Evaluate $\dfrac{-m}{-n}$ for $m = -20$ and $n = 2$.

90. Evaluate $\dfrac{-x}{-y}$ for $x = -15$ and $y = 5$.

One Step Further

91. Evaluate.
   (a) $-y^3$ for $y = -2$
   (b) $-y^4$ for $y = -2$

92. Evaluate.
   (a) $-a^7$ for $a = -1$
   (b) $-a^{10}$ for $a = -1$

Applications *Distance Traveled* *The following formula is used to calculate the distance an object has traveled at a given rate and time. Use this formula to answer exercises 93 and 94.*

$$\text{distance} = \text{rate} \times \text{time}$$

93. The velocity (rate) of a projectile is -30 meters per second on a number line. The negative sign on the velocity indicates that it is moving to the left. At time $t = 0$ the projectile is at the zero mark on the number line. Find where it will be on the number line at $t = 3$ seconds.

$$\begin{array}{ccccccccccc} \hline -150 & -120 & -90 & -60 & -30 & 0 & 30 & 60 & 90 & 120 & 150 \end{array}$$

94. Find where the projectile in exercise 93 will be on the number line at $t = 4$ seconds.

$$\begin{array}{ccccccccccc} \hline -150 & -120 & -90 & -60 & -30 & 0 & 30 & 60 & 90 & 120 & 150 \end{array}$$

95. *Discounted Sporting Goods* Baker Sporting Goods marked $2 off the price of each baseball glove in stock. If there are 350 gloves in stock, write the total reduction in price of all gloves as an integer.

96. *Temperature Fluctuations* As a cold front passed through Minnesota, the temperature dropped 3°F each hour for 4 hours. Express the total drop in temperature as an integer.

To Think About

97. Is 8 a solution to the equation $\dfrac{x}{2} = -12$?

98. Is -10 a solution to the equation $\dfrac{x}{-5} = 2$?

Determine the value of x.

99. $\dfrac{x}{-3} = 8$

100. $\dfrac{x}{2} = -10$

Cumulative Review *Simplify.*

101. **[1.6.4]** $2^2 + 3(5) - 1$ **102.** **[1.6.4]** $8 + 2(9 \div 3)$ **103.** **[1.6.4]** $2^3 + (4 \div 2 + 6)$ **104.** **[1.6.4]** $3^2 + (6 \div 2 + 8)$

105. **[1.5.5]** *Speed of Sound* Kristina heard a train 3261 feet away approaching the station. How long did it take the sound of the train to reach Kristina's ear, if sound travels at a speed of approximately 1087 feet per second?

106. **[2.1.3]** Replace the ? with the symbol $<$, $>$, or $=$.
$|-1|$? $|-20|$

Quick Quiz 2.4

1. Perform the indicated operations.

    (a) $4(-3)$ **(b)** $-45 \div 5$ **(c)** $\dfrac{-48}{-8}$

2. Multiply. $(-2)(-6)(-1)(3)$

3. Evaluate.

    (a) x^5 for $x = -1$

    (b) $\dfrac{-a}{b}$ for $a = 10$ and $b = -2$

4. **Concept Check** When we evaluate $(-x)(y)$ for $x = -6$ and $y = 2$, we obtain a positive number. Is this true? Why or why not?

2.5 The Order of Operations and Applications Involving Integers

When there is more than one operation in a problem, we must follow the order of operations presented in Chapter 1. We use additional grouping symbols when more than one set are required in a problem. We start with parentheses, then use brackets [ ], then braces { }.

Do first.	**1.**	Perform operations inside grouping symbols such as parentheses and brackets.
↓	**2.**	Simplify any expressions with exponents.
	3.	Multiply or divide from left to right.
Do last.	**4.**	Add or subtract from left to right.

We perform one operation at a time and follow the sequence *identify, calculate, replace*. We identify the highest priority, do this calculation, and then replace the operation with the calculated amount.

We must be careful when working with integers, paying special attention to the signs of the numbers.

1 Following the Order of Operations with Integers

EXAMPLE 1 Simplify. $12 - 30 \div 5(-3)^2 - 2$

Solution

$12 - 30 \div 5\,(-3)^2 - 2$ *Identify:* The highest priority is exponents.
               Calculate: $(-3)^2 = 9$. *Replace:* $(-3)^2$ with 9.

$= 12 - 30 \div 5(\,9\,) - 2$ *Identify:* The highest priority is division.
               Calculate: $30 \div 5 = 6$. *Replace:* $30 \div 5$ with 6.

$= 12 - 6(9) - 2$ *Identify:* The highest priority is multiplication.
               Calculate: $6 \cdot 9 = 54$. *Replace:* $6 \cdot 9$ with 54.

$= 12 - 54 - 2$ We subtract last, changing all subtraction to addition of the opposite.

$= 12 + (-54) + (-2)$ We add: $12 + (-54) + (-2) = -44$.

$= -44$

Student Practice 1 Simplify. $-6 + 20 \div 2(-2)^2 - 5$

NOTE TO STUDENT: *Fully worked-out solutions to all of the Student Practice problems can be found at the back of the text starting at page SP-1.*

EXAMPLE 2 Simplify. $\dfrac{[-15 + 5(-3)]}{(13 - 18)}$

Solution We perform operations inside parentheses and brackets first.

$\dfrac{[-15 + 5(-3)]}{(13 - 18)} = \dfrac{[-15 + (-15)]}{(13 - 18)}$ We multiply: $5(-3) = -15$.

$= \dfrac{-30}{(13 - 18)}$ We add: $-15 + (-15) = -30$.

$= \dfrac{-30}{-5}$ We subtract: $13 - 18 = -5$.

$= 6$ We divide last: $-30 \div (-5) = 6$.

Student Practice 2 Simplify.

$$\frac{[-10 + 4(-2)]}{(11 - 20)}$$

EXAMPLE 3 Simplify. $-24 \div \{-3 \cdot [4 \div (-2)]\}$

Solution We perform operations within the innermost grouping symbols first.

$-24 \div \{-3[4 \div (-2)]\}$ We divide: $4 \div (-2) = -2$.

$= -24 \div \{-3(-2)\}$ We complete operations inside the brackets: $-3(-2) = 6$.

$= -24 \div 6$ Now we divide: $-24 \div 6 = -4$.

$= -4$

Student Practice 3 Simplify. $-18 \div \{3[12 \div (-2)]\}$

② Solving Applied Problems Involving More Than One Operation

Since real-life applications often require that we perform more than one operation, we must take care to follow the order of operations when solving these problems.

EXAMPLE 4 Ions are atoms or groups of atoms with positive or negative electrical charges. An oxide ion has an electrical charge of -2, while a magnesium ion has a charge of $+2$. Find the total charge of 8 oxide and 3 magnesium ions.

Solution We summarize the information.

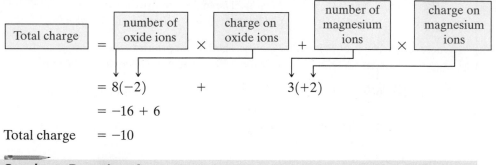

$= 8(-2) \quad + \quad 3(+2)$

$= -16 + 6$

Total charge $= -10$

Student Practice 4 Using the information from Example 4, find the total charge of 9 oxide and 4 magnesium ions.

👣 STEPS TO SUCCESS Re-evaluate Your Time Management Plan

It is a good idea to periodically re-evaluate your Time Management Plan to determine if adjustments are needed. If you did not read the Steps to Success, Time Management, in Section 1.5, it is a good idea to do so before you complete the following. This information will help you learn how to make a Time Management Plan.

Making it personal:

1. Record your instructor's office hours. Days: _____
   Times: _____.

2. Record the days and times the Math Tutoring Lab is open. Days: _____ Times: _____.

3. Record your grade in the class as of today.
   Grade: _____.

4. Record the number of hours a week you spend studying math. _____.

5. If your grade in the class is lower than you would like, increase your study time on your Time Management Plan.

6. Check the open days and times on your Time Management Plan to be sure you are free to meet with your instructor during office hours and to study in the Math Tutoring Lab. Make adjustments if necessary.

Verbal and Writing Skills, Exercises 1–4

1. Is $2 + 3(-1) = 5(-1) = -5$? Why or why not?

2. Is $3 + 4(-3) = 7(-3) = -21$?

3. Is $-2^2 + 8 = -4 + 8 = 4$?

4. Is $-2^4 - 4 = 16 - 4 = 12$?

Simplify.

5. $-2 + 3 \cdot 4$

6. $-5 + (10)2$

7. $1 + 7(2 - 6)$

8. $4 + 3(2 - 7)$

9. $-3 + 6(8 - 5)$

10. $-2 + 4(7 - 3)$

11. $12 - 5(2 - 6)$

12. $9 - 3(4 - 6)$

13. $5(-3)(4 - 7) + 9$

14. $5(-2)(3 - 9) + 1$

15. $-3(6 \div 3) + 7$

16. $-6(8 \div 2) + 2$

17. $3(-2)(9 - 5) - 10$

18. $5(-3)(5 - 2) - 3$

19. $-24 \div 12 - 8$

20. $-36 \div 12 - 5$

21. $(-3)^2 + 5(-9)$

22. $(-2)^2 + 4(-7)$

23. $(-3)^3 - 7(8)$

24. $(-2)^3 - 6(2)$

25. $(-2)^3 + 2(-8)$

26. $(-3)^3 + 6(-4)$

27. $36 \div (-6) + (-6)$

28. $16 \div (-4) + (-4)$

29. $12 - 20 \div 4(-4)^2 + 9$

30. $-15 - 50 \div 10(-3)^2 + 2$

31. $8 - 2(5 - 2^2) + 6$

32. $7 - 3(11 - 3^2) + 1$

Simplify.

33. $\dfrac{(-50 \div 2 + 3)}{(20 - 9)}$

34. $\dfrac{(-45 \div 5 + 1)}{[2 - (-2)]}$

35. $\dfrac{[3^2 + 4(-6)]}{[-3 + (-2)]}$

36. $\dfrac{[2^2 + 6(-3)]}{[-2 + (-5)]}$

37. $\dfrac{[-12 - 3(-2)]}{(15 - 17)}$

38. $\dfrac{[-10 - 4(-1)]}{(13 - 19)}$

39. $-16 \div \left\{ -4 \cdot [8 \div (-2)] \right\}$

40. $20 \div \left\{ 4 \cdot [15 \div (-3)] \right\}$

41. $-60 \div \left\{ 5 \cdot [-2 \cdot (-12 \div 4)] \right\}$

42. $-36 \div \left\{ 2 \cdot [-3 \cdot (-9 \div 3)] \right\}$

Applications

43. *Altitude of a Plane* A plane flying at an altitude of 35,000 feet descends 2000 feet three times before ascending 1000 feet. What is the current altitude of the plane?

44. *Temperature Fluctuations* During a storm in Anchorage, Alaska, the temperature was 8°F at noon. Then it dropped 2°F each hour for the next 4 hours, followed by an additional drop of 5°F the fifth hour. What was the temperature at 5 P.M.?

Ion Charges Ions are atoms or groups of atoms with positive or negative electrical charges. The charges of some ions are given below. Use these values to find the total charge in exercises 45–48.

Aluminum +3	Chloride −1
Phosphate −3	Silver +1

45. 14 phosphate and 9 silver

46. 11 chloride and 2 aluminum

47. 7 aluminum, 5 chloride, and 4 silver

48. 15 silver, 9 phosphate, and 8 chloride

Baseball Some baseball fans play a game called Fantasy Baseball. A fan can create a team made up of real baseball players and receive points based on their players' statistics (stats) for that day. Fans' fantasy teams compete against each other, and the team that accumulates the most points at the end of the season wins. The total points for each player are calculated based on the point value indicated in the chart. **Note that a strikeout receives points as an out and as a strikeout.**

Stat	Point Value	Stat	Point Value
Single	5	Walk	3
Double	10	Strikeout	−1
Triple	15	Out	−1
Home run	20	Stolen base	5
RBI*	5	Caught stealing	−5

*RBI: Runs Batted In

49. Derek Jeter is a player on Megan's team and had the following stats in a game: 1 double, 1 RBI, 1 walk, 1 strikeout, and 2 outs. Calculate the total points Derek received for Megan's team that night.

50. How many combined points does Megan's team receive for her first and second basemen's stats shown below?
First baseman: 1 single, 2 triples, 2 RBIs, 2 strikeouts, 2 outs, and caught stealing once.
Second baseman: 4 outs, 1 walk, 1 stolen base, and 2 strikeouts.

51. Vladimir Guerrero is a player on Ian's team and had the following stats in a game: 1 home run, 1 RBI, 2 walks, 2 strikeouts, and 2 outs. Calculate the total points Vladimir Guerrero received for Ian's team that night.

52. How many combined points does Ian's team receive for his catcher and third baseman's stats shown below?
Catcher: 1 home run, 2 doubles , 1 walk, 2 strikeouts, 3 RBIs, 2 outs, and caught stealing once.
Third baseman: 5 outs and 2 strikeouts.

One Step Further

53. $\dfrac{[(30 - 15 \div 3) + (-5)]}{(5 - 10)}$

54. $\dfrac{[(32 - 16 \div 4) + (-6)]}{(7 - 9)}$

55. $[(3 + 24) \div (-3)] \cdot [2 + (-3)^2]$

56. $[(-2 + 14) \div (-6)] \cdot [3 + (-2)^3]$

To Think About *Find the value of x.*

57. $3 + x - 2(-4) = 7 - (-13)$

58. $-2 + x + 3(-4) = -6 + (-4)$

Cumulative Review *Simplify.*

59. **[1.7.3]** $2(x + 3)$

60. **[1.7.3]** $3(a + 2)$

61. **[1.7.3]** $4(x - 2)$

62. **[1.7.3]** $7(x - 1)$

Quick Quiz 2.5

1. Perform the indicated operations.
$15 - 20 \div 5(-2)^2 + 3$

2. Perform the indicated operations. $\dfrac{(16 \div 8 - 4)}{(3 - 5)}$

3. Last winter in the northeast the temperature was 4 degrees Fahrenheit at midnight. It dropped 5 degrees every hour for 5 hours and then rose 8 degrees every hour after that. What was the temperature at noon?

4. Concept Check Explain in what order to do the operations to obtain the answer to the problem. $3^2 + 5(2 - 4)$

2.6 Simplifying and Evaluating Algebraic Expressions

Student Learning Objectives

After studying this section, you will be able to:

① Combine like terms with integer coefficients.

② Evaluate algebraic expressions with integers.

③ Use the distributive property with integers.

④ Solve applied problems with integers.

① Combining Like Terms with Integer Coefficients

Simplifying algebraic expressions with integers differs from doing so with whole numbers only in that we must consider the sign of the number when simplifying.

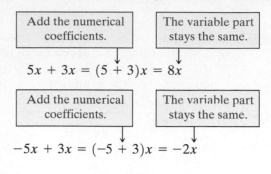

Add the numerical coefficients.	The variable part stays the same.

$$5x + 3x = (5 + 3)x = 8x$$

Add the numerical coefficients.	The variable part stays the same.

$$-5x + 3x = (-5 + 3)x = -2x$$

EXAMPLE 1 Simplify by combining like terms. $-4x + 7y + 2x$

Solution

$$
\begin{aligned}
-4x + 7y + 2x &= -4x + 2x + 7y &&\text{Rearrange terms.} \\
&= (-4 + 2)x + 7y &&\text{Add numerical coefficients of like terms.} \\
&= -2x + 7y &&\text{$-2x$ and $7y$ are not like terms.}
\end{aligned}
$$

Student Practice 1 Simplify by combining like terms. $-6y + 8x + 4y$

NOTE TO STUDENT: *Fully worked-out solutions to all of the Student Practice problems can be found at the back of the text starting at page SP-1.*

In Example 1 we were able to rearrange terms because addition is commutative. Since subtraction is *not commutative*, we must first change all subtraction statements to additions of the opposite and then rearrange the terms.

EXAMPLE 2 Simplify. $3x + 5y - x$

Solution

$$
\begin{aligned}
3x + 5y - x &= 3x + 5y + (-x) &&\text{First we change subtraction to addition of the opposite.} \\
&= 3x + (-1x) + 5y &&\text{Now we rearrange terms. Note that $-x = -1x$.} \\
&= 2x + 5y &&\text{We add like terms: $[3 + (-1)]x = 2x$.}
\end{aligned}
$$

Student Practice 2 Simplify. $7a + 4b - a$

The next example illustrates the similarity between the process of combining like terms and performing operations with integers.

EXAMPLE 3 Perform each operation indicated.

(a) $2 - 3 + 6$

(b) $2x - 3x + 6x$

Solution

(a)
$$
\begin{aligned}
2 - 3 + 6 &= 2 + (-3) + 6 \\
&= -1 + 6 \\
&= 5
\end{aligned}
$$

(b)
$$
\begin{aligned}
2x - 3x + 6x &= 2x + (-3x) + 6x \\
&= -1x + 6x \\
&= 5x
\end{aligned}
$$

Student Practice 3 Perform each operation indicated.
(a) $4 - 6 + 8$ **(b)** $4x - 6x + 8x$

In Example 3b we rewrote subtraction as the addition of the opposite. That is, we wrote

$$2x - 3x \quad \text{as} \quad 2x + (-3x).$$

When we write our final answer, it is sometimes necessary to reverse this process, and *write addition of the opposite as subtraction*. For example, consider the expression $-4b + 6a$. This expression can be written as follows.

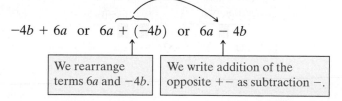

$$-4b + 6a \quad \text{or} \quad 6a + (-4b) \quad \text{or} \quad 6a - 4b$$

| We rearrange terms $6a$ and $-4b$. | We write addition of the opposite $+-$ as subtraction $-$. |

The expression $6a + (-4b)$ is not considered simplified since it can be written using fewer symbols. Therefore, we usually simplify: $6a + (-4b)$ to $6a - 4b$.

EXAMPLE 4 Simplify. $3a + 8b - 9a + 3ab - 10b$

Solution

$3a + 8b - 9a + 3ab - 10b$

First we change subtraction to addition of the opposite.

$= 3a + 8b + (-9a) + 3ab + (-10b)$

Next, we rearrange terms to group like terms.

$= 3a + (-9a) + 8b + (-10b) + 3ab$

Then we add coefficients of like terms.

$= [3 + (-9)]a + [8 + (-10)]b + 3ab$

$= -6a + (-2b) + 3ab$

Finally, we simplify by rewriting addition of the opposite as subtraction.

$= -6a - 2b + 3ab$

Student Practice 4 Simplify.
$2x + 8y - 5x + 4xy - 12y$

② Evaluating Algebraic Expressions with Integers

To evaluate expressions we replace the variables with the given numbers and then perform the indicated operations following the order of operations.

EXAMPLE 5 Evaluate. $\dfrac{(x^2 - y)}{4}$ for $x = -1$ and $y = -3$

Solution

$\dfrac{[(x)^2 - (y)]}{4} = \dfrac{[(-1)^2 - (-3)]}{4}$ Place parentheses around each variable. Replace x with -1 and y with -3.

$= \dfrac{[1 - (-3)]}{4}$ Calculate $(-1)^2 = 1$.

$= \dfrac{(1 + 3)}{4}$ Write subtraction as addition of the opposite.

$= \dfrac{4}{4} = 1$ Simplify.

Continued on next page

Student Practice 5   Evaluate.
$$\frac{(a^3 + b)}{2} \text{ for } a = -2 \text{ and } b = -4$$

③ Using the Distributive Property with Integers

Recall that to multiply $3(x + 1)$ we use the distributive property to "distribute" the 3. That is, we multiply every number or term inside the parentheses by 3, and then we simplify the result. We proceed the same way when working with integers.

We distribute 3.   We distribute -3.

$$3(x + 1) = 3x + 3 \qquad -3(x + 1) = -3x + (-3)$$
$$= -3x - 3$$

EXAMPLE 6   Simplify. $-2(y - 4)$

Solution

We distribute the -2 over subtraction.

$$-2(y - 4) = -2y - (-2)(4)$$

Next, we multiply: $(-2)(4)$.

$$= -2y - (-8)$$

We simplify by writing subtraction as addition of the opposite.

$$-2(y - 4) = -2y + 8$$

Student Practice 6   Simplify.
$-8(m - 1)$

④ Solving Applied Problems with Integers

EXAMPLE 7   To find the speed of a free-falling skydiver, we use the formula given below.

speed of skydiver		initial velocity		time since start of free fall
↓		↓		↓
s	$=$	v	$-$	$32t$

Find the speed of a skydiver at time $t = 5$ seconds if her initial downward velocity (v) is -7 feet per second.

Solution   We evaluate the formula for the values given: $v = -7$ and $t = 5$.

$$s = v - 32t$$
$$= -7 - 32(5)$$
$$= -7 - 160$$
$$= -167$$

A negative speed means that the object is moving in a downward direction. Therefore, the skydiver is falling 167 feet per second.

Student Practice 7   Use the formula in Example 7 to find the speed of the skydiver at $t = 4$ seconds if her initial downward velocity (v) is -5 feet per second.

Verbal and Writing Skills, Exercises 1–2

1. Is $-2x + 5x = -10x^2$? Why or why not?

2. Is $7x + 6y = 13xy$? Why or why not?

Fill in each box.

3. $-6x + \left(-3\boxed{}\right) = -9x$

4. $-4x + \left(-2\boxed{}\right) = -6x$

5. $5y + \boxed{}xy - 2y + 7xy = 3y + 10xy$

6. $8a + \boxed{}ab - 4a + 7ab = 4a + 11ab$

7. To simplify $9x + (-3y)$, we write $9x \boxed{} 3y$.

8. To simplify $4x + (-7y)$, we write $4x \boxed{} 7y$.

9. $-6(y - 1) = -6 \cdot \boxed{} - (-6) \cdot \boxed{} = -6y \boxed{} 6$

10. $-3(x - 2) = -3 \cdot \boxed{} - (-3) \cdot \boxed{} = -3x \boxed{} 6$

Simplify by combining like terms.

11. $-8x + 3x$

12. $-6x + 3x$

13. $4x + (-3x)$

14. $3y + (-2y)$

15. $-5x - 7x$

16. $-2a - 4a$

17. $-7a - (-2a)$

18. $-9b - (-3b)$

19. $14y + (-7y)$

20. $4x + (-2x)$

21. $-7x + (-6x)$

22. $-5y + (-7y)$

Write in simplest form.

23. $7a + (-9b)$

24. $5x + (-6y)$

25. $-5m + (-8n)$

26. $-4a + (-9b)$

27. $2x + (-y)$

28. $11x + (-y)$

29. $-2a - (-3b)$

30. $-12m - (-6n)$

Perform the operations indicated.

31. **(a)** $2 - 7 + 3$
    (b) $2x - 7x + 3x$

32. **(a)** $4 - 9 + 2$
    (b) $4x - 9x + 2x$

33. **(a)** $3 - 8 + 4$
    (b) $3x - 8x + 4x$

34. **(a)** $6 - 10 + 3$
    (b) $6x - 10x + 3x$

35. **(a)** $2 - 6 + 1$
    (b) $2x - 6x + 1x$

36. **(a)** $3 - 8 + 1$
    (b) $3x - 8x + 1x$

Simplify by combining like terms.

37. $-8y + 4x + 2y$

38. $-7x + 3y + 4x$

39. $6x + 4y + (-8x)$

40. $7a + 5b + (-11a)$

41. $9x + 3y + (-5x)$

42. $10y + 2x + (-4y)$

43. $-8x - 4x - y$

44. $-7a - 2a + b$

45. $3x + 8y - 10x - 2y$

46. $6x + 3y - 9x - 2y$

47. $4x + 2y - 6x - 7$

48. $8x + 3y - 11x - 6$

49. $4 + 3ab - 2 - 9ab$

50. $7 + 5xy + 9 - 8xy$

51. $5x + 7xy - 9x - xy$

52. $5y + 8xy - 11y - xy$

53. $7a - 2ab - 2 - 7ab + 3a$

54. $3x - 4xy - 7 - 9xy + 2x$

55. $3a + 2x - 5a + 7ax - x$

56. $5y + 4x - 8y + 3xy - x$

57. $6a + 7b - 9a + 5ab - 11b$

58. $-2y + 5x - 4y + 9xy + 3x$

59. $4x + 8y - 7x + 6xy - 10y$

60. $2a + 7b - 4a + 3ab - 12b$

Evaluate.

61. $x + 3y$ for $x = -3$ and $y = -2$

62. $a + 3b$ for $a = -1$ and $b = -3$

63. $m - 6n$ for $m = 6$ and $n = -3$

64. $x - 3y$ for $x = 4$ and $y = -1$

65. $a \cdot b - 6$ for $a = -1$ and $b = 5$

66. $m \cdot n - 8$ for $m = -5$ and $n = 2$

67. $\dfrac{(x + y)}{5}$ for $x = -9$ and $y = 4$

68. $\dfrac{(m + n)}{3}$ for $m = -10$ and $n = 7$

69. $9t^2$ for $t = -3$

70. $7a^2$ for $a = -2$

71. $8x - x^2$ for $x = -5$

72. $9m - m^2$ for $m = -3$

73. $\dfrac{(x^2 - x)}{2}$ for $x = -4$

74. $\dfrac{(t^2 - t)}{3}$ for $t = -3$

75. $\dfrac{(a - b^2)}{-3}$ for $a = 13$ and $b = 2$

76. $\dfrac{(x - y^2)}{-5}$ for $x = 30$ and $y = 5$

77. $\dfrac{(m^2 + 2n)}{-8}$ for $m = 6$ and $n = -2$

78. $\dfrac{(a^2 + 4b)}{-3}$ for $a = 5$ and $b = -4$

Simplify.

79. $-3(y + 1)$

80. $-5(x + 1)$

81. $-9(y - 1)$

82. $-8(a - 1)$

83. $-2(m - 3)$

84. $-3(x - 9)$

85. $-1(x + 5)$

86. $-1(a + 4)$

87. $6(-2 + y)$

88. $5(-3 + x)$

89. $2(-4 + a)$

90. $7(-1 + x)$

Applications

Skydiving *To find the speed of a free-falling skydiver, we use the formula $s = v - 32t$, where **s** is the speed of the skydiver, **v** is the intial velocity, and **t** is the time since the start of the free fall.*

91. Find the speed of a skydiver at time $t = 4$ seconds if his initial downward velocity (v) is -8 feet per second.

92. Find the speed of the skydiver at time $t = 3$ seconds if his initial downward velocity (v) is -7 feet per second.

Objects Rising and Descending A projectile is fired straight up with an initial velocity of 72 feet per second. It is known that the subsequent velocity of the projectile is given by the formula $v = 72 - 32t$, where t represents time in seconds. If $v > 0$, the object is rising, and if $v < 0$, it is descending.

93. At which of the following times is the object descending: $t = 1, 2$, and/or 3 seconds?

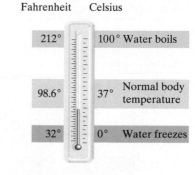

Velocity = $72 - 32t$

Velocity > 0 ↑ Object is rising

Velocity < 0 ↓ Object is descending

Temperature Conversion In the metric system temperature is measured on the Celsius scale. To convert Fahrenheit temperature to Celsius, we can use the formula

$$C = \frac{(5F - 160)}{9}$$

where F is the number of Fahrenheit degrees and C is the number of Celsius degrees.

94. When the temperature is 5°F, what is the Celsius reading?

95. When the temperature is 14°F, what is the Celsius reading?

Fahrenheit	Celsius	
212°	100°	Water boils
98.6°	37°	Normal body temperature
32°	0°	Water freezes

To Think About *Answer yes or no.*

96. Is 0 a solution to $\dfrac{x^3}{2} = 0$?

97. Is 7 a solution to $\dfrac{x^2}{7} = 13$?

Cumulative Review

▲ **98.** [1.2.5] *Geometry* Find the perimeter of a rectangle with a length of 6 feet and a width of 3 feet.

▲ **99.** [1.2.5] *Geometry* Find the perimeter of a square with a side of 7 inches.

100. [1.9.3] *Speed of Light* Light travels at a speed of 5,580,000 miles in 30 seconds. How far does it travel in 1 second? 1 minute? (60 seconds = 1 minute)

101. [1.9.3] *Heartbeat* The average heart beats 73 times per minute. How many times will the heart beat in 1 hour? 1 day? (60 minutes = 1 hour)

Quick Quiz 2.6

1. Simplify by combining like terms.
$-5x + 9y + 2 - 2y - 6x$

2. Simplify.

(a) $-3(a - 1)$

(b) $-8(x + 7)$

3. Use the formula $s = v - 32t$ to find the speed (s) of a free-falling skydiver at $t = 3$ seconds if his initial downward velocity (v) is −10 feet per second.

4. Concept Check State two other ways we can write $-3b + 7$.

▲ represents geometry-related content.

Chapter 2 Organizer

Topic and Procedure	Examples	✏ You Try It
Inequality, p. 104 Numbers decrease in value as we move from right to left on the number line. $\xleftarrow{\quad}$ -3 -2 -1 0 1 2 3 $\xrightarrow{\quad}$	Replace the ? with the inequality symbol $<$ or $>$. $-3 \; ? \; -1$ $-3 < -1$	1. Replace the ? with the inequality symbol $<$ or $>$. $-4 \; ? \; -9$
Opposites of numbers, p. 106 Numbers that are the same distance from zero but lie on opposite sides of zero on the number line are called *opposites*. We use "−" to indicate the opposite of a number.	Give the opposites of the following numbers: 1, −3. The opposite of 1 is −1. The opposite of −3 is 3.	2. Give the opposites of the following numbers: −16, 12
Absolute value, p. 107 The *absolute value* of a number a is the number of units between 0 and a on the number line.	Evaluate. **(a)** $\lvert -6 \rvert$ **(b)** $\lvert 8 \rvert$ $\lvert -6 \rvert = 6$ $\lvert 8 \rvert = 8$	3. Evaluate. **(a)** $\lvert 14 \rvert$ **(b)** $\lvert -9 \rvert$
Reading graphs involving signed numbers, p. 108 	1. Where was the temperature the coldest? The dot representing Bangor's temperature is lower on the line graph than Boston or Boise, indicating that it was coldest in Bangor. 2. Where was the temperature a positive number? Boston	4. **(a)** Where was the temperature the warmest? **(b)** Where was the temperature a negative number?
Adding two numbers with the same sign, p. 113 To add two or more numbers with the *same sign*: **1.** Use the common sign in the answer. **2.** Add the absolute values of the numbers.	Add. $-4 + (-6)$ $-4 + (-6) = -10$	5. Add. $-8 + (-5)$
Adding two numbers with different signs, p. 115 To add two numbers with *different signs*: **1.** Use the sign of the number with the larger absolute value in the answer. **2.** Subtract the absolute values of the numbers.	Add. **(a):** $9 + (-5)$ **(b):** $-8 + 6$ $9 + (-5) = 4$ $-8 + 6 = -2$	6. Add. **(a)** $-7 + 3$ **(b)** $9 + (-2)$
Subtracting integers, p. 125 To subtract two numbers, add the opposite of the second number to the first.	Subtract. **(a)** $-7 - 3 = -7 + (-3) = -10$ **(b)** $4 - (-8) = 4 + 8 = 12$	7. Subtract. **(a)** $-11 - 4$ **(b)** $5 - (-7)$
Multiplying integers, p. 136 To find the product of nonzero numbers, we first determine the *sign* of the product. The product will be positive if there is an even number of negative signs and negative if there is an odd number of negative signs. Then we multiply absolute values.	Multiply. **(a)** $(-2)(-5) = 10$ **(b)** $8(-3)(-1)(-2) = -48$	8. Multiply. **(a)** $(-2)(-3)(-5)(2)$ **(b)** $(-6)(-2)(3)$
Using exponents with integers, p. 138 If a number is written in exponent form and the base is negative: **1.** The result is positive if the exponent is even. **2.** The result is negative if the exponent is odd.	Evaluate. **(a)** $(-3)^3 = -27$ **(b)** $(-5)^2 = 25$ **(c)** $-2^2 = -(2 \cdot 2) = -4$ Note that the base in -2^2 is 2, so we square 2, then take the opposite.	9. Evaluate. **(a)** $(-6)^2$ **(b)** -6^2 **(c)** $(-2)^3$
Dividing integers, p. 140 To find the quotient of nonzero numbers, we determine the sign of the quotient as follows. The quotient will be positive if there is an even number of negative signs, and negative if there is an odd number of negative signs.	Divide. **(a)** $42 \div (-6) = -7$ **(b)** $-54 \div (-9) = 6$	10. Divide. **(a)** $(-18) \div (-2)$ **(b)** $-24 \div 6$

Topic and Procedure	Examples	✏ You Try It
Order of operations with integers, p. 145 We follow the same order of operations as presented in Chapter 1. **1.** Perform operations inside grouping symbols. **2.** Simplify any expression with exponents. **3.** Multiply and divide from left to right. **4.** Add and subtract from left to right.	Simplify. $\dfrac{[-12 + 3(-2)]}{(3 - 9)}$ We perform operations within the grouping symbols, then we divide. $\dfrac{[-12 + 3(-2)]}{(3 - 9)} = \dfrac{[-12 + (-6)]}{(3 - 9)}$ $= \dfrac{-18}{-6} = 3$	**11.** Simplify. $12 - 20 \div 2(-3)^2 - 6$
Combining like terms, p. 150 To combine like terms, we add or subtract the numerical coefficients of the like terms. The variable part stays the same.	Simplify. $4x + 5y - 7x$ We change subtraction to addition of the opposite and then simplify. $4x + 5y - 7x = 4x + 5y + (-7x)$ $= 4x + (-7x) + 5y$ $= -3x + 5y$	**12.** Simplify. $3a + 5b - 6a$
Evaluating expressions with integers, p. 151 To evaluate an expression, we replace the variable with the given number and simplify.	Evaluate. **(a)** $-3 + x$ for $x = -4$ **(b)** $-2 - ab^2$ for $a = -1$ and $b = -3$ **(a)** $-3 + (x) = -3 + (-4) = -7$ **(b)** $-2 - (-1)(-3)^2 = -2 + 9 = 7$	**13.** Evaluate. $-5 - xy^2$ for $x = -2$ and $y = 5$
Using the distributive property, p. 152 To multiply $a(b + c)$ and $a(b - c)$ we distribute the a by multiplying every number or variable inside the parentheses by a, and then simplify.	Simplify $-2(x + 1)$. $-2(x + 1) = -2(x + 1)$ $= -2x + (-2) \cdot 1$ $= -2x - 2$	**14.** Simplify. $-3(x - 2)$

Chapter 2 Review Problems

Vocabulary

Fill in the blank with the definition for each word.

1. (2.1) Negative numbers: _____

2. (2.1) Opposites: _____

3. (2.1) Integers: _____

4. (2.1) Absolute value: _____

Section 2.1

Replace each ? with the symbol $<$, $>$, or $=$.

5. -3 ? -1 **6.** $|5|$? $|-13|$ **7.** -9 ? -11

Fill in the blank with the appropriate symbol $+$ or $-$ to describe an increase or a decrease.

8. _____ A profit of $200   **9.** _____ A drop in temperature of $18°$

10. Fill in the blank: The opposite of 12 is _____.   **11.** Simplify. $-(-(-6))$

12. Simplify. $-|-11|$ **13.** Which of the two numbers has the larger absolute value, -23 or 12?

Profit and Loss *Justin invested in various stocks from January through May. At the end of each month he calculated the net profit or loss on all his stock transactions and then recorded this information on the following line graph. Use the graph to answer exercises 14 and 15.*

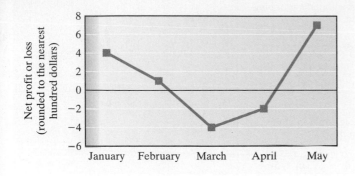

14. (a) In which month did Justin make the most money on his investments?

  (b) In which month did Justin lose the most money on his investments?

15. (a) List the months in which Justin had a net profit.

  (b) List the months in which Justin had a net loss.

Section 2.2

Evaluate.

16. (a) $-43 + (-16)$

  (b) $43 + 16$

17. (a) $-27 + (-39)$

  (b) $27 + 39$

18. *Profit and Loss* A company lost $25,000 in May and had a net gain (profit) of $15,000 in June. At the end of these 2 months, was there a net profit or a net loss?

19. *Profit and Loss* Terry lost $14 in the slot machines on Thursday. Later that evening he won $25 playing roulette. At the end of the evening, was there a net profit or a net loss?

20. *Temperature Fluctuations* Yesterday the temperature in Yosemite dropped to $-10°$F. Today it has risen $20°$F.

  (a) Write the math symbols that represent the situation.

  (b) Today, was the temperature a positive or a negative value?

  (c) Use the thermometer to find the sum.

Add by using the rules for addition of integers.

21. (a) $2 + (-8)$

  (b) $-2 + 8$

  (c) $-2 + (-8)$

22. (a) $27 + (-18)$

  (b) $-27 + 18$

  (c) $-27 + (-18)$

23. $3 + (-5) + 8 + (-2)$

24. $24 + (-52) + (-12) + (-56)$

25. Evaluate $x + 6$ for $x = -1$.

26. Evaluate $-x + y + 2$ for $x = -3$ and $y = -11$.

27. *Altitude of a Plane* To avoid the turbulence caused by a storm, the pilot adjusts the altitude of his plane as shown in the table. How many feet above or below the initial elevation of 35,000 feet is the plane at 5 P.M.? Express your answer as an integer.

Flight Recordings of Altitude			
4:15 P.M.	240-foot descent	4:45 P.M.	400-foot ascent
4:30 P.M.	350-foot ascent	5:00 P.M.	800-foot descent

Section 2.3

Perform the operations indicated.

28. $-7 - 5$

29. $-9 - (-4)$

30. $-4 - 4$

31. $-6 - (-6)$

32. $-6 - 9 + 4$

33. $6 - (-4) + (-5)$

34. $-4 - (-2)$

35. $6 - 9 - 2 - 8$

36. $-6 - (-9) + (-1)$

37. Evaluate $y - 15$ for $y = -2$.

38. Evaluate $-1 - x + y$ for $x = -4$ and $y = -2$.

Profit and Loss Use the graph to answer exercises 39 and 40.

39. What is the difference between the net profit in the fourth quarter and the net loss in the third quarter?

40. What is the difference between the net profit in the first quarter and the net loss in the second quarter?

41. *Altitude of a Plane* A jet is flying at an altitude of 2300 feet over the Atlantic Ocean where a submarine is submerged at a depth of 1312 feet below sea level. What is the difference in altitude between the jet and the submarine?

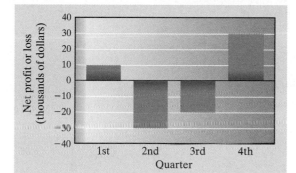

Section 2.4

Multiply.

42. **(a)** $6(3)$
(b) $6(-3)$
(c) $-6(3)$
(d) $-6(-3)$

43. **(a)** $5(2)$
(b) $5(-2)$
(c) $-5(2)$
(d) $-5(-2)$

44. $-7(-2)$

45. $-2(5)$

46. $3(-4)$

47. $-4(-1)$

48. $(-2)(-5)(-9)$

49. $(-2)(-8)(-1)(-4)$

50. $(-5)(1)(-2)(4)(-6)$

Evaluate.

51. $(-7)^2$

52. -9^2

53. $(-6)^3$

Divide.

54. **(a)** $49 \div 7$
(b) $49 \div (-7)$

55. **(a)** $-30 \div 5$
(b) $-30 \div (-5)$

Perform the operations indicated.

56. (a) $-44 \div (-4)$
 (b) $9(-5)$
 (c) $-11(-3)$
 (d) $\dfrac{25}{-5}$

57. (a) $12 \div (-4)$
 (b) $5(-8)$
 (c) $-12(-2)$
 (d) $\dfrac{36}{-9}$

58. Evaluate y^4 for $y = -1$.

59. Evaluate x^3 for $x = -3$.

60. Evaluate $\dfrac{-a}{b}$ for $a = -20$ and $b = 5$.

61. Evaluate $\dfrac{-m}{-n}$ for $m = 6$ and $n = -2$.

Section 2.5

Simplify.

62. $4 - 1(6 - 9)$

63. $3(-5)(2 - 6) + 8$

64. $-2^2 + 3(-4)$

65. $\dfrac{(-32 \div 8 + 4)}{(7 - 9)}$

66. ***Temperature Fluctuations*** The temperature of a small lake in Michigan was 12°F at 8 P.M. If the temperature of the lake dropped 5°F every hour for the next 3 hours and then dropped another 2°F the fourth hour, what was the temperature of the lake at midnight?

Section 2.6

Simplify by combining like terms.

67. $-4y + 3x + 9y$

68. $-6a - a$

69. $7x + 9y - 6x - 11y$

70. $3 + 5z - 7 + 2yz - 8z$

Evaluate.

71. $a + 3b$ for $a = 8$ and $b = -4$

72. $2x - y$ for $x = -2$ and $y = -1$

73. $\dfrac{(x^2 - y)}{4}$ for $x = -1$ and $y = -7$

74. $a^2 - b$ for $a = -3$ and $b = 9$

Use the formula $C = \dfrac{(5F - 160)}{9}$ to convert Fahrenheit temperature (F) to Celsius (C).

75. When the temperature is 41°F, what is the Celsius reading?

76. When the temperature is -4°F, what is the Celsius reading?

Simplify.

77. $-6(x + 1)$

78. $-2(a - 1)$

79. $4(-2 + x)$

How Am I Doing?   Chapter 2 Test

Replace each ? with the symbol $<$, $>$, *or* $=$.

1. -234 ? -5

2. $|4|$? $|-18|$

3. Replace the ? with the appropriate symbol, $+$ or $-$, to describe an increase or a decrease. The Dow Jones Industrial Average falls 14 points. ___?___

Simplify.

4. $-(-(-2))$

5. The opposite of 10 is _____.

6. Simplify.
   (a) $|12|$ (b) $-|-3|$

7. Last night the temperature dropped to $-10°F$. At dawn the temperature had risen $15°F$.
   (a) Write the math symbols that represent the situation.
   (b) At dawn, what was the temperature?

Perform the operations indicated.

8. (a) $-6 + 8$ (b) $6 + (-8)$

9. $-6 + (-4)$

10. $-20 + 5 + (-1) + (-3)$

11. $12 - 18$

12. (a) $-1 - 11$ (b) $-1 - (-11)$

13. $3 - (-10)$

ᴹᶜ 14. $-14 - 3 + (-6) - 1$

15. $(7)(-3)$

16. $(-8)(-4)$

ᴹᶜ 17. $(-5)(-2)(-1)(3)$

1.	☐
2.	☐
3.	☐
4.	☐
5.	☐
6. (a)	☐
(b)	☐
7. (a)	☐
(b)	☐
8. (a)	☐
(b)	☐
9.	☐
10.	☐
11.	☐
12. (a)	☐
(b)	☐
13.	☐
14.	☐
15.	☐
16.	☐
17.	☐

18. (a) _____ ☐

(b) _____ ☐

(c) _____ ☐

19. (a) _____ ☐

(b) _____ ☐

20. _____ ☐

21. _____ ☐

22. _____ ☐

23. (a) _____ ☐

(b) _____ ☐

24. _____ ☐

25. (a) _____ ☐

(b) _____ ☐

26. _____ ☐

27. _____ ☐

28. _____ ☐

29. _____ ☐

30. _____ ☐

31. _____ ☐

32. _____ ☐

33. _____ ☐

Total Correct: ☐

162

Evaluate.

18. (a) $(-5)^2$ (b) $(-5)^3$ (c) -5^2

Divide.

19. (a) $-8 \div 2$ (b) $-8 \div (-2)$

20. $\dfrac{-22}{11}$

21. $2 - 35 \div 5(-3)^2 - 6$

22. $\dfrac{[-8 + 2(-3)]}{(14 - 21)}$

23. Evaluate $-7 - x + y$ for each of the following.
(a) $x = -6$ and $y = -3$
(b) $x = -7$ and $y = 6$

24. Evaluate. $\dfrac{(2x - y^2)}{-9}$ for $x = -1$ and $y = -4$

25. Evaluate.
(a) x^4 for $x = -1$
(b) a^3 for $a = -2$

26. Evaluate. $\dfrac{-x}{y}$ for $x = -6$ and $y = -2$

Simplify.

27. $5x + 2y - 8x - 6y$

ℳ🄲 **28.** $-3x + 7xy + 8y - 12x - 11y$

ℳ🄲 **29.** $-6(a + 7)$

30. $-2(x - 1)$

Solve.

31. For the first quarter of the year, Earth Systems had a $20,000 profit. For the second quarter, the company had a $5000 loss. What was the company's overall profit or loss at the end of the second quarter?

32. Find the difference in altitude between a mountain 3700 feet high and a desert valley 592 feet below sea level.

33. Use the formula $s = v - 32t$ to find the speed (s) of a free-falling skydiver at $t = 5$ seconds if his initial downward velocity (v) is -7 feet per second.

MATH COACH

Mastering the skills you need to do well on the test.

Students often make the same types of errors when they do the Chapter 2 Test. Here are some helpful hints to keep you from making these common errors on test problems.

Adding and Subtracting Integers—Problem 14

$-14 - 3 + (-6) - 1$

> **Helpful Hint** You may find it easier to change all subtraction to addition of the opposite as your first step. This allows you to add numbers in any order. Write out your steps carefully to avoid careless errors.

Look at your work for Problem 14. Examine your steps.

As your first step, did you change $-14 - 3$ to $14 + (-3)$?

Yes _____ No _____

As your second step, did you rewrite $(-6) - 1$ as a related addition problem?

Yes _____ No _____

If you answered No to either question, stop now and complete these steps.

Did you get $-14 + (-3) + (-6) + (-1)$ as your next step?

Yes _____ No _____

If you answered No, think about how to rewrite subtraction as addition of the opposite. Recall that the subtraction sign changes to addition, and the number following the subtraction sign is changed to its opposite.

If you answered Problem 14 incorrectly, go back and rework the problem using these suggestions.

Multiplying Two or More Integers—Problem 17 $(-5)(-2)(-1)(3)$

> **Helpful Hint** As your first step, write down the sign of the product or quotient. Do this *before* you complete any calculations to be sure that you do not forget the sign of the answer.

As your first step, did you identify that there are three negative signs in the problem?

Yes _____ No _____

If you answered No, go back and count the number of negative signs again.

Next, did you determine that the product will be negative?

Yes _____ No _____

If you answered No, recall that an odd number of negative signs indicates a negative answer, while an even number of negative signs indicates a positive answer.

As your final step, did you multiply the absolute values of the numbers and add the negative sign to your answer?

Yes _____ No _____

If you answered No, stop and complete this step again.

Now go back and rework the problem using these suggestions.

Need help? Watch the MATH COACH videos in MyMathLab® or on YouTube™.

163

Combining Like Terms with Integer Coefficients—Problem 28
Simplify. $-3x + 7xy + 8y - 12x - 11y$

> **Helpful Hint** Make sure you *do not* combine two terms with different variables or two terms with a different number of variables. You may only combine like terms. Be careful to avoid sign errors.

Look at your work for Problem 28. Examine your steps.

Did you combine $7xy$ with any other terms?

Yes _____ No _____

If you answered Yes, stop and look at each term carefully. Notice that the only term with xy as the variable part is $7xy$, so we *cannot* combine it with any other terms.

Did you combine $-3x$ and $-12x$ and combine $8y$ and $-11y$?

Yes _____ No _____

If you answered No, go back and check your steps. Make sure you did not combine either of the two x-terms with either of the y-terms.

If you answered Yes, make sure you did not make a *sign error*.

If you answered Problem 28 incorrectly, go back and rework the problem using these suggestions.

Using the Distributive Property with Integers—Problem 29   Simplify. $-6(a + 7)$

> **Helpful Hint** Be sure to multiply every number or term inside the parentheses by the number outside the parentheses. Then simplify the result. Be careful to avoid sign errors.

Did you multiply both a and 7 by -6?

Yes _____ No _____

If you answered No, then it may help with accuracy to write arrows as follows: $-6(a + 7)$.

Are both of your terms in the final answer negative?

Yes _____ No _____

If you answered No, then you made at least one sign error. Try this step again.

Now go back and rework the problem using these suggestions.

Need more help? Look for section examples marked with ^{M}c to review.

164

Suppose that you decide to redecorate or remodel your home. Do you know how to determine how much the entire project will cost? After you have studied the topics in this chapter, you will have the knowledge necessary to determine the total cost for redecorating your home or other similar projects.

Introduction to Equations and Algebraic Expressions

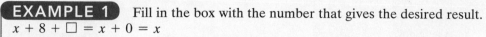

3.1 Solving Equations of the Form $x + a = c$ and $x - a = c$

Student Learning Objectives

After studying this section, you will be able to:

① Use the additive inverse property.

② Solve equations using the addition principle of equality.

③ Solve applied problems involving angles.

NOTE TO STUDENT: *Fully worked-out solutions to all of the Student Practice problems can be found at the back of the text starting at page SP-1.*

---

① Using the Additive Inverse Property

Recall from Chapter 2 that the numbers 2 and -2 are called **opposites** and their sum is equal to zero. This fact is often referred to as the **additive inverse property:** $a + (-a) = 0$ and $-a + a = 0$. We will use this property to help us solve equations involving addition and subtraction.

EXAMPLE 1 Fill in the box with the number that gives the desired result.
$x + 8 + \square = x + 0 = x$

Solution

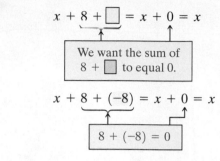

Thus, $x + 8 + (-8) = x + 0 = x$.

Student Practice 1 Fill in the box with the number that gives the desired result. $y - 6 + \square = y + 0 = y$

---

② Solving Equations Using the Addition Principle of Equality

We observe in our everyday world that if we add the same amount to two equal values, the results are equal. We illustrate this below.

We place 5 pounds on both sides of a seesaw

We add a 2-pound weight to the center of each 5-pound weight simultaneously.

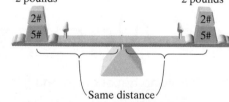

We would expect the seesaw to balance.   The seesaw should still balance.

In mathematics, a similar principle is observed. That is, we can add the same number to both sides of an equation without changing the solution. We state this principle of equality.

ADDITION PRINCIPLE OF EQUALITY

If the same number is added to both sides of an equation, the results on both sides are equal in value.

We can restate this principle in symbols this way. For any numbers a, b, and c,

$$\text{If } a = b, \text{ then } a + c = b + c.$$

To illustrate the addition principle, let's see what happens when we solve the equation $x - 2 = 7$ using basic arithmetic facts (as we did in Chapter 1). Then, let's solve the same equation by the addition principle of equality.

Solve using basic arithmetic facts.

$$x - 2 = 7 \quad \text{What number minus 2 equals 7?}$$
$$9 - 2 = 7 \quad \text{Nine minus 2 equals 7.}$$

Thus $x = 9$.

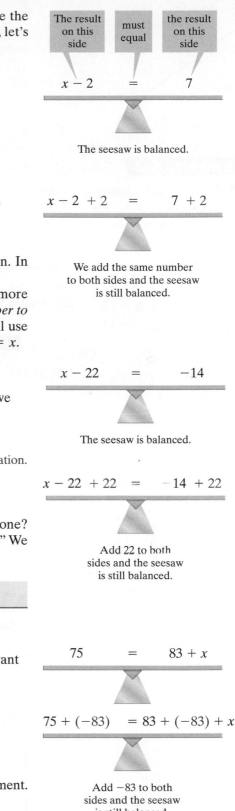

Solve using the addition principle.

$$x - 2 = 7$$
$$x - 2 + 2 = 7 + 2 \quad \text{We add the opposite of } -2 \text{ to both sides of the equation.}$$
$$x + 0 = 9 \quad \text{We simplify: } -2 + 2 = 0.$$
$$x = 9 \quad \text{The solution is the same.}$$

We see that adding 2 to both sides of the equation did not change the solution. In other words, using the addition principle of equality gives the same result.

You will find it easier to use the addition principle of equality to solve more complex equations. The goal when using this principle is to *add the same number to both sides of the equation so that when we simplify, x is left alone*. That is, we'll use it to get a simpler equation of the form $x = some\ number$ or $some\ number = x$.

EXAMPLE 2 Solve. $x - 22 = -14$

Solution We want an equation of the form $x = some\ number$. Therefore, we want to get x alone on one side of the equation.

$$x - 22 = -14$$
$$x - 22 + 22 = -14 + 22 \quad \text{Add the opposite of } -22 \text{ to both sides of the equation.}$$
$$x + 0 = 8 \quad\quad x - 22 + 22 = x + 0 \text{ and } -14 + 22 = 8$$
$$x = 8 \quad\quad \text{The solution is 8.}$$

How do we know what number to add to both sides of the equation to get x alone? We think, "What can we add to both sides so that $x - 22$ becomes simply x?" We add the *opposite* of -22, or $+22$, to both sides.

Student Practice 2 Solve. $x - 19 = -31$

EXAMPLE 3 Solve and check your solution. $75 = 83 + x$

Solution The variable x is on the right side of the equation; therefore we want an equation of the form $some\ number = x$.

$$75 = 83 + x$$
$$75 + (-83) = 83 + (-83) + x \quad \text{Add the opposite of 83 to both sides}$$
$$-8 = 0 + x \quad\quad\quad \text{since } 83 + (-83) = 0.$$
$$-8 = x$$

Check: To check our answer, we replace x with -8 and verify we get a true statement.

$$75 = 83 + x$$
$$75 \stackrel{?}{=} 83 + (-8)$$
$$75 = 75 \ \checkmark$$

Student Practice 3 Solve and check your solution. $92 = 46 + x$

Sometimes problems require that we write many steps. Writing too many steps can make the problem seem more complicated than it really is. We can eliminate some steps if, when adding a number to both sides of the equation, we place the addition *below* the terms rather than *beside* the terms. We can rewrite Example 3 using this format.

$$75 = \quad 83 + x$$
$$\underline{+ -83 \quad -83} \qquad \text{Add } -83 \text{ below the terms on both sides.}$$
$$-8 = \quad 0 + x$$

Try this with a few problems. You may find it easier and decide to use this format.

Mc **EXAMPLE 4**   Solve and check your solution. $3x - 2x - 4 = 9$

Solution   The variable x appears *more than once* on the left side of the equation. This means we'll need to complete an extra step, combining like terms, so that x appears only once in the equation.

$$3x - 2x - 4 = 9$$
$$\quad x - 4 = 9 \qquad \text{We combine like terms: } 3x - 2x = 1x \text{ or } x.$$
$$\underline{+ \qquad 4 \quad 4} \qquad \text{Think: "Add the opposite of } -4 \text{ to both sides}$$
$$\quad x + 0 = 13 \qquad \text{of the equation."}$$
$$\quad x = 13$$

Check: $\qquad 3x - 2x - 4 = 9$
$$3(13) - 2(13) - 4 \overset{?}{=} 9$$
$$39 - 26 - 4 \overset{?}{=} 9$$
$$9 = 9 \quad ✓$$

Student Practice 4   Solve and check your solution. $5x - 4x - 3 = 11$

There are three facts to remember when solving the equations we've studied so far.

1. First, if necessary, we must simplify each side of the equation by combining like terms or completing any addition and subtraction.
2. Next, we use the addition principle of equality to get the variable alone on one side of the equation, i.e., in the form $x = $ *some number* or *some number* $= x$.
3. An equation is like a balanced scale. Whatever we do to one side of the equation, we must do to the other side of the equation to maintain the balance.

EXAMPLE 5   Solve and check your solution. $2 - 6 = y - 7 + 12$

Solution   First, we must simplify each side of the equation separately by completing the addition and subtraction.

$$2 - 6 = y - 7 + 12$$
$$-4 = y + 5 \qquad \text{Simplify: } 2 - 6 = -4; -7 + 12 = 5.$$

Then we add -5 to both sides of the equation to get y alone on the right side: *some number* $= y$.

$$-4 = y \ + 5 \quad \text{Think: "Add the opposite of 5 to both sides of the equation."}$$
$$\underline{+ -5 \qquad -5} \quad \text{Add } -5 \text{ to both sides.}$$
$$-9 = y \qquad \text{We usually do not write the step } -9 = y + 0.$$

Check: $2 - 6 = y - 7 + 12$
$$2 - 6 \overset{?}{=} -9 - 7 + 12$$
$$-4 = -4 \quad ✓$$

Student Practice 5   Solve. $5 - 8 = y - 2 + 19$

③ Solving Applied Problems Involving Angles

In this section we will find the measure of an angle by finding the solution to an equation. Before we begin our discussion on this topic, we introduce terms and definitions.

In geometry a **line** $\longleftrightarrow$ extends indefinitely, but a portion of a line, called a **line segment** $\cdot\!\!-\!\!-\!\!\cdot$, has a beginning and an end. A **ray** $\cdot\!\!-\!\!\longrightarrow$ starts at a point and extends indefinitely in one direction. An **angle** is formed whenever two rays meet at the same endpoint. The point at which they meet is called the **vertex** of the angle.

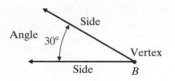

One way to name the angle in the figure above is $\angle B$.

The *amount of opening* of an angle can be measured. Angles are commonly measured in degrees. In the sketch above, the angle measures 30 degrees or 30°. The symbol ° indicates degrees. If you fix one side of an angle and keep moving the other side, the angle measure will get larger and larger until eventually you have gone around in one complete revolution.

One complete revolution is 360°.

One-half of a revolution is 180°.

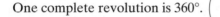

Two angles that have a sum of 180° are called **supplementary angles.** Two angles that share a common side are called **adjacent angles.** Adjacent angles formed by two intersecting lines are supplementary (they add up to 180°).

When a problem involves geometric shapes, we can draw pictures to help us visualize the situation and thus better *understand the problem*. Next, we *gather the facts*. In algebra, we often use these facts to find an equation that represents the situation. The following process will help you solve these types of problems.

1. Use the facts stated to *draw a picture*.
2. Write an *equation*.
3. *Replace variables* in the equation with the appropriate known values.
4. *Solve* the equation and answer the question.

We use this process to work Examples 6 and 7.

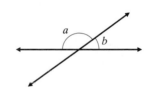

$\angle a + \angle b = 180°$

▲ **EXAMPLE 6** $\angle a$ and $\angle b$ in the margin are supplementary angles. Find the measure of $\angle a$ if the measure of $\angle b$ is 36°.

Solution   We *draw a picture*.

Since $\angle a$ and $\angle b$ are supplementary angles, we know that their sum is 180°. We use this information to *write an equation*.

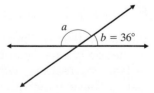

$\angle a + \angle b = 180°$	Write an equation.
$\angle a + 36° = 180°$	Replace $\angle b$ with 36°.
$+ \quad\; -36° \quad -36°$	Solve the equation.
$\angle a = 144°$	The measure of $\angle a$ is 144°.

Student Practice 6

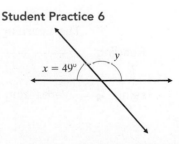

▲ **Student Practice 6** $\angle x$ and $\angle y$ are supplementary angles. Find the measure of $\angle y$ if the measure of $\angle x$ is 49°.

▲ **EXAMPLE 7**

(a) Translate into symbols. Angle y measures 40° more than angle x.

(b) Find the measure of $\angle x$ if the measure of $\angle y$ is 95°.

Solution (a) Angle y measures 40° more than angle x.

$$\angle y \quad = \quad 40° \quad + \quad \angle x$$

(b) $\angle y = 40° + \angle x$ Write an equation.

$95° = 40° + \angle x$ Replace $\angle y$ with 95°.

$\underline{+ -40° \qquad -40°}$ Solve the equation.

$55° = \angle x$ The measure of $\angle x$ is 55°.

▲ **Student Practice 7**

(a) Translate into symbols. Angle a measures 20° less than angle b.

(b) Find the measure of $\angle b$ if the measure of $\angle a$ is 80°.

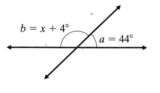

$b = x + 4°$

$a = 44°$

EXAMPLE 8   Find x and the measure of $\angle b$ for the pair of supplementary angles in the margin.

Solution   We know that $\angle a$ and $\angle b$ are supplementary angles and therefore their sum is 180°. We write the equation as follows.

$$\angle a + \angle b = 180°$$ Write an equation.

$44° + (x + 4°) = 180°$ Replace $\angle a$ with 44° and $\angle b$ with $x + 4°$.

$48° + x = 180°$ Simplify: $44° + 4° + x = 48° + x$.

$\underline{+ \qquad -48° \qquad\qquad -48°}$ Solve the equation.

$x = 132°$

Since $\angle b = x + 4°$ we must substitute 132° for x to find the measure of $\angle b$.

$$\angle b = x + 4°$$

$$\angle b = 132° + 4° = 136°$$

Therefore $x = 132°$ and $\angle b = 136°$.

Student Practice 8   Refer to Example 8 and find x and the measure of $\angle b$ if the measure of $\angle a$ is 55°.

👣 STEPS TO SUCCESS Getting the Most from Your Homework

Did you know that an important part of the learning process happens when you complete your homework? People learn by *reading, writing, listening, verbalizing,* and *seeing.* These activities are part of the learning cycle and always occur when you do your homework.

Making it personal:

Increase your learning while you do your homework. Follow the strategies described in Steps to Success in Section 1.6 (page 59) and learn more while you study less. Use the strategies described in Section 1.6 for this section. Which did you find most helpful? Write down other strategies you think may help you. ▼

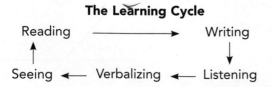

The Learning Cycle

Reading ⟶ Writing

Seeing ← Verbalizing ← Listening

Verbal and Writing Skills, Exercises 1–4 *Fill in the blanks.*

1. The sum of two opposite numbers is equal to _____.

2. The sum of two supplementary angles is equal to _____.

3. To solve $x - 6 = 3$, we _____ 6 to both sides of the equation.

4. To solve $x + 6 = 2$, we _____ 6 to both sides of the equation.

Fill in each ☐ with the number that gives the desired result.

5. $3 + \boxed{} = 0$

6. $9 + \boxed{} = 0$

7. $-9 + \boxed{} = 0$

8. $-1 + \boxed{} = 0$

9. $17 + \boxed{} = 0$

10. $58 + \boxed{} = 0$

11. $-28 + \boxed{} = 0$

12. $-13 + \boxed{} = 0$

13. $x + 5 + \boxed{} = x$

14. $y + 7 + \boxed{} = y$

15. $m - 2 + \boxed{} = m$

16. $a - 6 + \boxed{} = a$

Fill in each ☐ with the correct number to solve the equation.

17.
$$x + 12 = 16$$
$$+ \boxed{} \quad \boxed{}$$
$$x + \boxed{} = \boxed{}$$
$$x = \boxed{}$$

18.
$$x + 9 = 21$$
$$+ \boxed{} \quad \boxed{}$$
$$x + \boxed{} = \boxed{}$$
$$x = \boxed{}$$

19.
$$y - 16 = 32$$
$$+ \boxed{} \quad \boxed{}$$
$$y + \boxed{} = \boxed{}$$
$$y = \boxed{}$$

20.
$$y - 18 = 25$$
$$+ \boxed{} \quad \boxed{}$$
$$y + \boxed{} = \boxed{}$$
$$y = \boxed{}$$

Solve and check your solution.

21. (a) $x - 8 = 22$
   (b) $x + 8 = 22$

22. (a) $a - 5 = 29$
   (b) $a + 5 = 29$

23. (a) $x + 2 = -11$
   (b) $x - 2 = -11$

24. (a) $y + 4 = -10$
   (b) $y - 4 = -10$

25. (a) $-18 = x + 2$
   (b) $-18 = x - 2$

26. (a) $-19 = y - 1$
   (b) $-19 = y + 1$

Solve and check your solution.

27. $y - 10 = 5$

28. $m - 16 = 8$

29. $n - 43 = -74$

30. $y - 81 = -12$

31. $y + 20 = -35$

32. $x + 1 = -15$

33. $38 + x = 4$

34. $15 + x = 21$

35. $1 = x - 13$

36. $5 = x - 7$

37. $20 = y + 11$

38. $46 = a + 14$

39. $-13 = x + 1$

40. $-5 = x + 7$

Simplify and then solve. Check your solution.

41. $4x - 3x - 3 = 8$

42. $3x - 2x - 5 = 7$

43. $5y - 4y + 1 = -5$

44. $3y - 2y + 3 = 1$

45. $5 = 2y - y + 1$

46. $6 = 2y - y + 4$

Simplify each side of the equation, then solve the equation. Check your solution.

47. $-23 + 8 + x = -2 + 13$

48. $-33 + 9 + x = -1 + 7$

49. $4 - 9 = a - 1 + 14$

50. $3 - 8 = m - 4 + 7$

51. $-45 + 9 + m = -6 + 18$

52. $-27 + 7 + n = -5 + 8$

53. $-1 + 11 + x = -5 + 9$

54. $-3 + 9 + a = -4 + 7$

55. $3(7 - 11) = y - 5$

56. $5(8 - 13) = x - 1$

Find the measure of the unknown angle for each pair of supplementary angles.

▲ 57. Find the measure of $\angle a$ if the measure of $\angle b = 86°$.

▲ 58. Find the measure of $\angle x$ if the measure of $\angle y = 22°$.

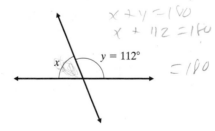

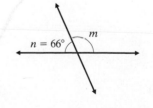

▲ 59. Find the measure of $\angle x$ if the measure of $\angle y = 112°$.

▲ 60. Find the measure of $\angle a$ if the measure of $\angle b = 101°$.

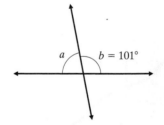

▲ 61. Find the measure of $\angle x$ if the measure of $\angle y$ is $43°$.

▲ 62. Find the measure of $\angle m$ if the measure of $\angle n$ is $66°$.

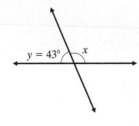

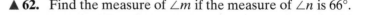

Translate each of the following statements into symbols. Then find the unknown angle.

▲ 63. **(a)** Translate into symbols. Angle x measures $70°$ more than angle y.

(b) Find the measure of $\angle y$ if $\angle x = 125°$.

▲ 64. **(a)** Translate into symbols. Angle a measures $50°$ more than angle b.

(b) Find the measure of $\angle b$ if $\angle a = 115°$.

▲ **65.** **(a)** Translate into symbols. Angle a measures 40° less than angle b.

  (b) Find the measure of $\angle b$ if the measure of $\angle a$ is 50°.

▲ **66.** **(a)** Translate into symbols. Angle x measures 80° less than angle y.

  (b) Find the measure of $\angle y$ if the measure of $\angle x$ is 40°.

Find x and the measure of $\angle b$ for the following pairs of supplementary angles.

▲ **67.**

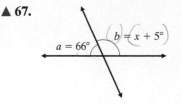

▲ **68.**

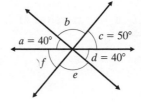

69. Refer to exercise 67. Find x and the measure of $\angle b$ if the measure of $\angle a$ is 52°.

70. Refer to exercise 68. Find x and the measure of $\angle b$ if the measure of $\angle a$ is 65°.

One Step Further *Simplify each side of each equation, then solve the equation.*

71. $2^2 + (5 - 9) = x + 3^3$

72. $4^2 + (3 - 7) = y + 2^3$

73. $5x + 1 - 2x = 4x - 2$

74. $8x + 6 - 5x = 2x - 3$

To Think About

▲ **75.** Find the measures of $\angle c$, $\angle d$, and $\angle f$.

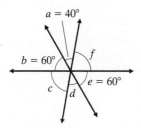

▲ **76.** Find the measure of $\angle b$, $\angle e$, and $\angle f$.

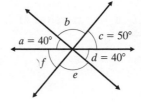

Cumulative Review *Translate into symbols.*

77. **[1.4.2]** Seven times a number.

78. **[1.4.2]** The product of three and a number.

79. **[1.8.4]** Eight times what number equals forty?

80. **[1.8.4]** Double what number equals thirty?

81. **[2.2.2]** *Temperature Fluctuations* If the low temperature on Tuesday was −8°F and the temperature rose 21°F during the day, what was the high temperature?

82. **[2.1.3]** Replace the ? with the symbol $<$, $>$, or $=$. $|-16|$? $|-8|$

Fantasy Basketball *Fantasy basketball is a game that many sports fans participate in. A fan can create a team made up of real basketball players and receive points based on their players' statistics (stats) for that day. Fans compete against other fantasy teams, and the team that accumulates the most points at the end of the season wins. The total points for each player are calculated based on the point values in the following chart.*

Statistic	Fantasy Point Value	Statistic	Fantasy Point Value
Field goal	2	Steal	2
Rebound	1	Block	2
Missed shot	−1	Turnover	−1

83. **[2.6.4]** How many fantasy points did a player on Delroy's fantasy team, the Lasers, earn based on the following stats: 15 field goals, 8 rebounds, 6 missed shots, 5 steals, 5 blocks, and 2 turnovers?

84. **[2.6.4]** How many fantasy points did a player on Joe's fantasy team, River Run, earn based on the following stats: 12 field goals, 14 rebounds, 8 missed shots, 3 steals, 7 blocks, and 4 turnovers?

Quick Quiz 3.1

1. Solve each equation and check your solution.
  (a) $x + 12 = -15$
  (b) $y - 7 = -20$

2. Solve each equation and check your solution.
  (a) $6x - 5x + 2 = 10 - 15$
  (b) $-8 + 10 = 6 - 9 + y$

3. **(a)** Translate into symbols. The measure of $\angle a$ is 35° less than the measure of $\angle b$.
  (b) Find the measure of $\angle b$ if the measure of $\angle a$ is 75°.

4. **Concept Check** To solve the equation $-9 + x = -15$, Damien subtracted 9 from both sides of the equation. Is this correct? Why or why not?

3.2 Solving Equations of the Form $ax = c$

① Solving Equations Using the Division Principle of Equality

Recall from Chapter 1 that any number divided by itself is equal to 1. For example, $2 \div 2 = \frac{2}{2} = 1$ and $-3 \div (-3) = \frac{-3}{-3} = 1$. We will use this division fact when we solve equations.

Student Learning Objectives

After studying this section, you will be able to:

① Solve equations using the division principle of equality.

② Translate English statements into equations.

③ Solve applied problems involving numbers and finance.

EXAMPLE 1 Fill in the box with the number that gives the desired result.

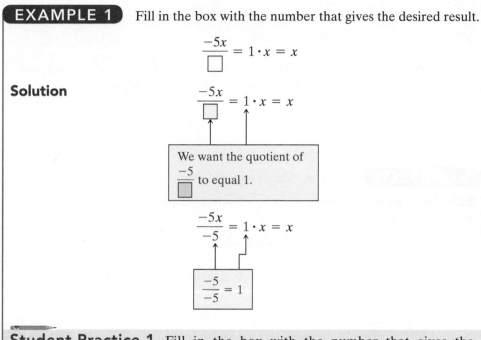

$$\frac{-5x}{\square} = 1 \cdot x = x$$

Solution

$$\frac{-5x}{\square} = 1 \cdot x = x$$

We want the quotient of $\dfrac{-5}{\square}$ to equal 1.

$$\frac{-5x}{-5} = 1 \cdot x = x$$

$$\frac{-5}{-5} = 1$$

Student Practice 1 Fill in the box with the number that gives the desired result.

$$\frac{-9x}{\square} = 1 \cdot x = x$$

NOTE TO STUDENT: Fully worked-out solutions to all of the Student Practice problems can be found at the back of the text starting at page SP-1.

Now let's see how we can use this division fact to help us solve equations. To solve an equation like $2x = 30$, we can think: "What can we do to the left side of the equation so that $2x$ becomes simply x?" We can divide by 2 since *dividing* $2 \cdot x$ by 2 *undoes* the *multiplication* by 2. Remember, whatever we do to one side of the equation, we must do to the other side of the equation.

$$2x = 30$$

$$\frac{2x}{2} = \frac{30}{2} \qquad \text{We divide both sides by 2.}$$

$$1 \cdot x = 15 \qquad 2 \div 2 = 1; \quad 30 \div 2 = 15$$

$$x = 15$$

To see why we must divide by 2 on *both sides* of the equation, let's return to our example of a balanced seesaw. If we cut the weight on each side in half (i.e., divide each side by 2), the seesaw should still balance.

In mathematics, a similar principle is observed. We now state this principle of equality.

DIVISION PRINCIPLE OF EQUALITY

If both sides of an equation are divided by the same nonzero number, the results on both sides are equal in value.

We can restate it in symbols this way. For any numbers a, b, c, with c not equal to 0,

$$\text{If } a = b, \text{ then } \frac{a}{c} = \frac{b}{c}.$$

It should be noted that, as stated in Chapter 1, division is defined in terms of multiplication. For example $\frac{10}{2} = 5$ because $10 = 5 \cdot 2$. In a similar manner, the division principle of equality is a form of the multiplication principle that we will learn about in Chapter 5 when we study fractions.

EXAMPLE 2 Solve and check your solution. $7x = -147$

Solution We want to make $7x = -147$ into a simpler equation, $x = $ *some number*.

$7x = -147$ The variable, x, is *multiplied* by 7.

$\dfrac{7x}{7} = \dfrac{-147}{7}$ Dividing both sides by 7 undoes the *multiplication* by 7.

$x = -21$ $\dfrac{7x}{7} = 1 \cdot x = x$ and $-147 \div 7 = -21$

Check: $7x = -147$

$7(-21) \overset{?}{=} -147; \quad -147 = -147$ ✓

Student Practice 2 Solve and check your solution. $5m = -155$.

Remember, it is best to first simplify each side of the equation, if necessary, before any other steps are taken.

Ⓜ **EXAMPLE 3** Solve. $3(x \cdot 5) = \dfrac{450}{5}$

Solution First, we simplify the equation.

$3(x \cdot 5) = \dfrac{450}{5}$

$3(x \cdot 5) = 90$ Divide: $450 \div 5 = 90$.

$3(5x) = 90$ Change the order of the factors.

$(3 \cdot 5)x = 90$ Regroup the factors.

$15x = 90$ Simplify.

$\dfrac{15x}{15} = \dfrac{90}{15}$ Dividing both sides by 15 undoes the multiplication by 15.

$x = 6$

We leave the check to the student.

Student Practice 3 Solve.

$$8(n \cdot 5) = \dfrac{320}{2}$$

EXAMPLE 4   Solve. $-50 + 10 = 6y - 4y$

Solution   We always simplify each side of the equation first.

$$-50 + 10 = 6y - 4y$$

$-40 = 6y - 4y$ Simplify the left side of the equation: $-50 + 10 = -40$.

$-40 = 2y$ Simplify the right side of the equation: $6y - 4y = 2y$.

$\dfrac{-40}{2} = \dfrac{2y}{2}$ Divide both sides of the equation by 2.

$-20 = y$ $-40 \div 2 = -20; 2 \div 2 = 1; 1 \cdot y = y$

We leave the check to the student.

Student Practice 4   Solve. $30 + (-12) = 7y - 5y$

② Translating English Statements into Equations

When we translate a statement into an equation, we must first *define the variables* that we are going to use. This helps us distinguish between the different things that are being compared. For example, consider the statement

> Kathy makes three times as many sales as Mark does.

We are comparing Kathy's sales to Mark's, so we can let the variable K represent the number of sales Kathy makes and M represent the number of sales Mark makes. Since Kathy makes three times as many sales as Mark, we have $K = 3M$. Keep in mind that we can use any letters; it is usually easiest to use the first letter of the word we are considering.

▲ **EXAMPLE 5**   Translate the statement into an equation. The measure of angle s (S) is two times the measure of angle y (Y).

Solution   We are representing the measure of angle s with the letter S and the measure of angle y with the letter Y. Now we translate.

The measure of angle s is two times the measure of angle y.

$$S = 2 \cdot Y$$

The equation that represents the statement is $S = 2Y$.

▲ **Student Practice 5**   Translate the statement into an equation. The measure of angle a (A) is four times the measure of angle b (B).

EXAMPLE 6 Translate the statement into an equation. There are five times as many dimes (D) as pennies (P) in a coin collection.

Solution This statement is phrased a little differently than others we have translated. Therefore, it is helpful to write a sentence that *compares* the two quantities.

There are more dimes (D) than pennies (P).

Think of a simple comparison of pennies and dimes, such as the case when there is only 1 penny.

If there is 1 penny in the collection, then there are 5 dimes.

five times as many dimes

Now we can rephrase the statement and translate into an equation.

The number of dimes is five times the number of pennies.

$$D = 5 \cdot P$$

The equation that represents the statement is $D = 5P$.

There are 5 times as many dimes as pennies.

Student Practice 6 Translate the statement into an equation. Sara (S) ran twice as many laps as Dave (D).

③ Solving Applied Problems Involving Numbers and Finance

EXAMPLE 7 The number of peanuts (P) is triple the number of cashews (C).

(a) Translate the statement into an equation.

(b) Find the number of cashews if there are 27 peanuts.

Solution P represents the number of peanuts, and C represents the number of cashews.

(a) The number of peanuts (P) is triple the number of cashews (C).

The word *triple* means *three times*.

The number of peanuts is triple the number of cashews.

$$P = 3 \cdot C$$

(b) Find the number of cashews if there are 27 peanuts.

$P = 3C$ We use the equation from part (a).

$27 = 3C$ We replace P with 27.

$\dfrac{27}{3} = \dfrac{3C}{3}$ We divide both sides by 3.

$9 = C$ There are 9 cashews.

Student Practice 7 The number of cars (C) is twice the number of trucks (T).

(a) Translate the statement into an equation.

(b) Find the number of trucks if there are 150 cars on the road.

EXAMPLE 8   Lena purchased x shares of stock at $35 per share. She sold all the stock for $56 per share and made a profit of $546. How many shares of stock did Lena purchase?

Solution   *Understand the problem.* We use a Mathematics Blueprint for Problem Solving to organize the information.

Mathematics Blueprint for Problem Solving

Gather the Facts	What Am I Asked to Do?	How Do I Proceed?	Key Points to Remember
Let x = the number of shares of stock purchased. Lena paid $35 for each share. She sold each share for $56.	Find the number of shares of stock purchased.	Let $35x$ = the purchase price and $56x$ = the sale price. Find the profit: profit = sale price − purchase price	Profit is how much money is made.

Solve and state the answer.

profit = sale price − purchase price

$546 = 56x - 35x$ We must simplify the equation.

$546 = 21x$ We combine like terms: $56x - 35x = 21x$.

$\dfrac{546}{21} = \dfrac{21x}{21}$ We divide both sides by 21.

$26 = x$ Lena purchased 26 shares of stock.

Check: We can estimate to see if our answer is reasonable. We round so that each number has 1 nonzero digit: $546 \to 500; \quad 56 \to 60; \quad 35 \to 40$

Now we estimate the value of x.

$$500 = 60x - 40x$$
$$500 = 20x$$
$$\frac{500}{20} = x$$
$$25 = x \qquad \text{Our answer is reasonable.} \quad ✓$$

Student Practice 8   Ian purchased x shares of stock at $25 per share. He sold all the stock for $45 per share and made a profit of $200. How many shares of stock did Ian purchase?

👣 STEPS TO SUCCESS Seeking Assistance

Getting the right kind of help at the right time can be a key factor to being successful in mathematics. When you have attended class on a regular basis, taken careful notes, methodically read your textbook, and diligently done your homework—in other words, when you have made every effort possible to learn the mathematics—you may still find that you are having difficulty. If this is the case, then you need to seek help.

Making it personal:

1. Make an appointment with your instructor to find out what you can do to be successful.

Appointment date and time: ⬚

Suggestions from instructor: ⬚

2. List the resources you will use: tutoring services, a mathematics lab, online videos, workshops, or computer software. ▼

Verbal and Writing Skills, Exercises 1–2

Fill in the blanks.

1. To solve the equation $-22x = 66$, we undo the multiplication by _____ both sides of the equation by _____.

2. To solve the equation $-16x = 64$, we undo the multiplication by _____ both sides of the equation by _____.

Fill in each □ with the number that gives the desired result.

3. $\dfrac{5x}{\square} = x$

4. $\dfrac{8x}{\square} = x$

5. $\dfrac{-2x}{\square} = x$

6. $\dfrac{-3x}{\square} = x$

7. $\dfrac{6 \cdot x}{\square} = x$

8. $\dfrac{2 \cdot x}{\square} = x$

9. $\dfrac{-1 \cdot x}{\square} = x$

10. $\dfrac{-9 \cdot x}{\square} = x$

Solve and check your solution.

11. $3x = 36$

12. $6x = 66$

13. $10x = 40$

14. $8x = 88$

15. $6y = -18$

16. $4a = -48$

17. $5m = -35$

18. $7y = -42$

19. $-3y = 15$

20. $-4x = 12$

21. $-7a = 49$

22. $-6y = 36$

23. $48 = 6x$

24. $30 = 3y$

25. $72 = 9x$

26. $56 = 14y$

27. $8x = 104$

28. $9y = 135$

29. $-19x = -76$

30. $-22y = -132$

31. $2(3x) = 54$

32. $5(2x) = 80$

33. $5(4x) = 40$

34. $3(2x) = 30$

35. $5(x \cdot 2) = \dfrac{40}{2}$

36. $3(x \cdot 5) = \dfrac{60}{2}$

37. $4(x \cdot 2) = \dfrac{96}{3}$

38. $6(x \cdot 5) = \dfrac{120}{2}$

Simplify each side of the equation, then solve. Check your solution.

39. $-26 - 18 = 11a$

40. $-44 - 16 = 6y$

41. $-4 - 4 = 8y$

42. $-3 - 3 = 6x$

43. $5x - 2x = 24$

44. $5x - 3x = 30$

45. $65 = 15x - 10x$

46. $81 = 15x - 6x$

Mixed Practice *Solve and check your solution.*

47. $-15y = 165$

48. $-14x = 168$

49. $-4x = 3 - 23$

50. $-3a = 4 - 13$

51. $12x - 4x = 56$

52. $15x - 10x = 95$

53. $55 = 5a$

54. $44 = 4y$

55. $(3x) \cdot 2 = \dfrac{36}{3}$

56. $(8x) \cdot 3 = \dfrac{48}{2}$

57. $9x + 3x = -120$

58. $7x + 3x = -90$

Applications

▲ **59.** *Dimensions of a Building* The length (L) of a building is three times the width (W). Translate the statement into an equation.

60. *Comparing Weights* The weight of a steel beam (B) is three times the weight of a steel pole (P). Translate the statement into an equation.

61. *Discount Price* The original price (R) of a ring is double the sale price (S). Translate the statement into an equation.

▲ **62.** The width (W) of a yard is triple the length (L). Translate the statement into an equation.

63. *Cost of Race Boat* The cost of a new race boat (R) is double the cost of an older model boat (B).

(a) Translate the statement into an equation.

(b) Find the cost of the older model if the new race boat costs $124,000.

▲ **64.** *Dimensions of Wood* The length of the larger piece of wood (L) is double the length of the smaller piece of wood (S).

(a) Translate the statement into an equation.

(b) Find the length of the smaller piece of wood if the larger piece of wood is 86 inches.

65. *Tickets Sold* There were three times as many child tickets (C) sold as adult tickets (A).

(a) Translate the statement into an equation.

(b) Find the number of adult tickets sold if there were 300 child tickets sold.

66. *Red vs. White Roses* There are four times as many red roses (R) in the garden as white roses (W).

(a) Translate the statement into an equation.

(b) Find the number of white roses if there are 64 red roses.

67. *Soccer Goals* The total number of soccer goals attempted (A) by a team is double the number of goals scored (S).

(a) Translate the statement into an equation.

(b) If the team attempted to score 42 times, find the number of goals scored.

68. *Coin Collection* There are four times as many nickels (N) in a coin collection as quarters (Q).

(a) Translate the statement into an equation.

(b) Find the number of quarters if there are 48 nickels in the collection.

69. *Stock Investment* Vu Nguyen purchased x shares of Baron Electric stock at $30 per share. He sold all the stock for $42 per share and made a profit of $360. How many shares of stock did Vu purchase?

70. *Stock Investment* Leslie purchased x shares of stock at $75 per share. She sold all the stock for $85 per share and made a profit of $800. How many shares of stock did Leslie purchase?

One Step Further

71. *Distance Traveled* On Monday Leah drove x miles on her road trip to visit her parents. On Tuesday she drove twice the number of miles she drove on Monday. If Leah drove a total of 360 miles on her two-day trip, how many miles did she drive on Monday?

72. *Hourly Rate* Mercelita charges $x per hour as a consultant. She worked on a consulting job for 8 hours on Monday, 6 hours on Tuesday, and 9 hours on Wednesday. If her total earnings for the three days were $1035, what is the hourly rate she charges?

To Think About *Solve.*

73. (a) $13x = 26$

(b) $-13x = 26$

(c) $x + 13 = 26$

(d) $x - 13 = 26$

74. (a) $9y = 72$

(b) $-9y = 72$

(c) $y + 9 = 72$

(d) $y - 9 = 72$

75. (a) $13x = -26$

(b) $-13x = -26$

(c) $x + 13 = -26$

(d) $x - 13 = -26$

76. (a) $9y = -72$

(b) $-9y = -72$

(c) $y + 9 = -72$

(d) $y - 9 = -72$

Find the value of x for each pair of supplementary angles. Then find the degree measure of each angle.

▲ **77.**

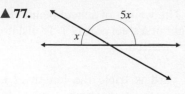

▲ **78.**

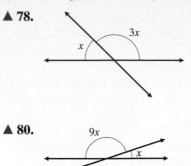

▲ **79.**

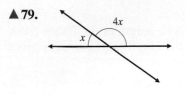

▲ **80.**

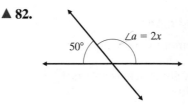

Find the value of x for each pair of supplementary angles.

▲ **81.**

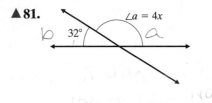

▲ **82.**

Cumulative Review

▲ **83.** **[1.2.5]** Find the perimeter of a rectangle with a length of 6 ft and a width of 4 ft.

▲ **84.** **[1.2.5]** Find the perimeter of a square with a side of 3 in.

85. **[2.6.2]** Evaluate. $2xy$ for $x = -7$ and $y = -9$

▲ **86.** **[1.7.2]** Evaluate. $L \cdot W \cdot H$ for $L = 2$, $W = 3$, and $H = 5$

Quick Quiz 3.2

1. Solve each equation and check your solution.

  (a) $7x - 10x = 18 + 45$

  (b) $-8 + 20 = 3(2x)$

2. Translate this statement into an equation. Mindy (M) has five times as many quarters as Sara (S).

3. The number of blue marbles (B) is twice the number of red marbles (R).

  (a) Translate the statement into an equation.

  (b) Find the number of red marbles if there are 10 blue marbles.

4. **Concept Check** Explain in words the steps that are needed to solve $4x + 3(2x) = -20$.

How Am I Doing? Sections 3.1–3.2

How are you doing with your homework assignments in Sections 3.1 and 3.2? Do you feel you have mastered the material so far? Do you understand the concepts you have covered? Before you go further in the textbook, take some time to do each of the following problems.

3.1

Solve.

1. $y - 15 = -26$

2. $4x - 3x - 2 = 9$

3. $2 - 8 = a - 1 + 11$

4. $\angle x$ and $\angle y$ are supplementary angles. Find the measure of $\angle x$ if the measure of $\angle y = 115°$.

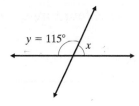

5. **(a)** Translate into symbols. Angle x measures 30° more than angle y.
   (b) Find the measure of $\angle y$ if the measure of $\angle x = 90°$.

3.2

Solve.

6. $-16 - 12 = 14a$

7. $2(3x) = -18$

8. $-48 - 10 = 8x - 6x$

9. The length (L) of a building is four times the width (W). Translate the statement into an equation.

10. The number of child tickets (C) sold is triple the number of adult tickets (A) sold.

    (a) Translate the statement into an equation.

    (b) Find the number of adult tickets sold if there were 150 child tickets sold.

11. Linda purchased x shares of stock at $50 per share. She sold all the stock for $65 per share and made a profit of $795. How many shares of stock did Linda purchase?

Now turn to page SA-6 for the answers to each of these problems. Each answer also includes a reference to the objective in which the problem is first taught. If you missed any of these problems, you should stop and review the Examples and Student Practice problems in the referenced objective. A little review now will help you master the material in the upcoming sections of the text.

1. _____

2. _____

3. _____

4. _____

5. (a) _____

   (b) _____

6. _____

7. _____

8. _____

9. _____

10. (a) _____

    (b) _____

11. _____

3.3 Equations and Geometric Formulas

Student Learning Objectives

After studying this section, you will be able to:

① Solve equations involving perimeter.

② Solve equations involving areas of rectangles and parallelograms.

③ Solve equations involving volume.

④ Solve applied geometry problems.

In this section we will work with the perimeter, area, and volume of various shapes. You will need to be familiar with the abbreviations used for units of measure in order to understand the material. For your review, Appendix B includes this information.

① Solving Equations Involving Perimeter

Earlier we learned that to find the perimeter of a rectangle, we find the sum of the lengths of all four sides: $P = L + L + W + W$. We can rewrite this formula as $P = 2L + 2W$. A square is the special case of a rectangle where all sides are equal. We find the perimeter using the formula $P = s + s + s + s$, or $P = 4s$.

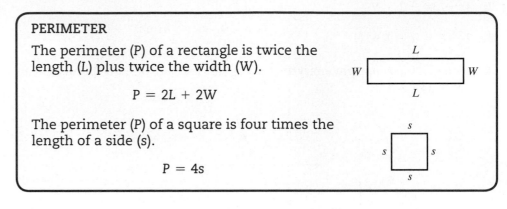

PERIMETER

The perimeter (P) of a rectangle is twice the length (L) plus twice the width (W).

$$P = 2L + 2W$$

The perimeter (P) of a square is four times the length of a side (s).

$$P = 4s$$

Recall that we use a four-step process to solve problems involving geometry.

1. Use the facts stated to *draw a picture*.
2. Write an *equation* or *formula*.
3. *Replace variables* with the appropriate known values.
4. *Solve* the equation and answer the question.

▲ **EXAMPLE 1** Find the perimeter of a rectangle with $L = 8$ feet and $W = 6$ feet.

Solution We use the four-step process.

$L = 8$ ft

$W = 6$ ft

Step 1: Draw a picture.

$P = 2\,L + 2\,W$ *Step 2:* Write the formula.

$= 2(\,8\,\text{ft}\,) + 2(\,6\,\text{ft}\,)$ *Step 3:* Replace L with 8 ft and W with 6 ft.

$= 16\,\text{ft} + 12\,\text{ft}$ *Step 4:* Simplify.

$P = 28\,\text{ft}$ The perimeter is 28 feet.

▲ **Student Practice 1** Find the perimeter of a square with sides 11 yards in length.

NOTE TO STUDENT: Fully worked-out solutions to all of the Student Practice problems can be found at the back of the text starting at page SP-1.

It is important that you understand that formulas are the beginning equations we need in working these problems. Once we write the formula, we replace the variables with known values and solve.

Sometimes we must solve an equation to find the unknown side of a geometric shape.

▲ **EXAMPLE 2**   The length of a rectangle is three times the width. If the perimeter of the rectangle is 24 feet, find the width.

$L = 3W$

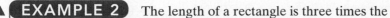

W

Solution   Since we are given the picture, we start with step 2 of the four-step process.

$P = 2L + 2W$	First, we write the formula for perimeter.
$24 = 2(3W) + 2W$	Next, we replace P with 24 and L with the given value $3W$.
$24 = 6W + 2W$	Then, we multiply: $2(3W) = 6W$.
$24 = 8W$	We combine like terms.
$\dfrac{24}{8} = \dfrac{8W}{8}$	Now we can divide each side by 8.
$3 = W$	The width of the rectangle is 3 feet.

▲ **Student Practice 2**   The length of a rectangle is twice the width. If the perimeter of the rectangle is 30 feet, find the width.

$L = 2W$

W

---

② **Solving Equations Involving Areas of Rectangles and Parallelograms**

Arrays can be used to illustrate the area of a rectangular region. In Chapter 1 we learned that to find the number of items in an array, we multiply the number of rows times the number of columns. Thus, the number of items in a 3 by 5 array is 3×5, or 15.

5

★ ★ ★ ★ ★

3 ★ ★ ★ ★ ★ $3 \times 5 = 15$

★ ★ ★ ★ ★

If each object in a 3×5 array is a square with sides of length 1 foot, then the number of squares needed to fill the region is 3 times 5, or 15. Thus the **area** of the rectangular array is 15 square feet. We can check this by counting the number of squares in the array

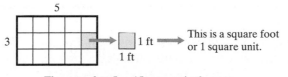

There are 3×5 or 15 squares in the array.
Area = 15 square feet

▲ **EXAMPLE 3**   What is the area of the rug pictured in the margin?

Solution   Think of an array with 3 rows and 9 columns.

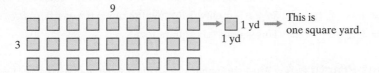

Just as we multiplied the number of rows times the number of columns to find the number of items in an array, we *multiply* the *length* times the *width* to find the area of the rug. The area is 3 yards $\times$ 9 yards $= 27$ square yards. Note, this means that there are 27 squares in the above illustration.

▲ **Student Practice 3**   What is the area of the flower garden pictured in the margin?

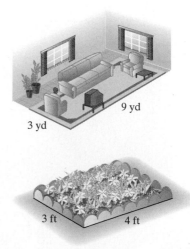

3 ft   4 ft

As we saw in Example 3, the area of a rectangle is the product of the length and the width, $A = L \cdot W$, where A represents the value of the area, L the value of the length, and W the value of the width.

Since the length and width of a square are equal, we can use the formula $A = s \cdot s$ or $A = s^2$ to find the area of a square.

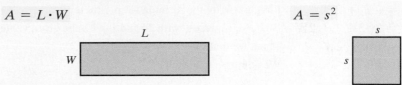

$$A = L \cdot W \qquad\qquad\qquad A = s^2$$

Parallelograms are figures that are related to rectangles. Actually, they are in the same "family," the **quadrilaterals** (four-sided figures).

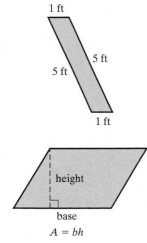

> A **parallelogram** is a four-sided figure in which both pairs of opposite sides are parallel.

Parallel lines are straight lines that are always the same distance apart. The opposite sides of a parallelogram are equal in length. The figures in the margin are parallelograms. Notice that adjoining sides need not be perpendicular.

To find the area of a parallelogram, we multiply the base times the height. Any side can be considered the base. The height is the length of a line segment perpendicular to the base. When we write the formula for area, we denote the length of the **base** as **b** and the **height** as **h**.

The formulas for area are summarized as follows.

> The area of a *rectangle* is the length times the width. $\quad A = LW$
> The area of a *square* is the length of one side squared. $\quad A = s^2$
> The area of a *parallelogram* is the base times the height. $\quad A = bh$

How do we write the *units* when we multiply to find area? We perform calculations with units in the same way as with variables.

▲ **EXAMPLE 4** Find the area of a rectangle with a length of 3 feet and a width of 2 feet.

Solution Follow the four-step process. *Step 1:* Draw a picture (shown at right). Now we complete steps 2–4.

$$
\begin{aligned}
A &= L \cdot W && \text{\textit{Step 2:} Write the formula.}\\
&= 3 \text{ ft} \cdot 2 \text{ ft} && \text{\textit{Step 3:} Replace } L \text{ and } W \text{ with the given values.}\\
&= (3 \cdot 2)(\text{ft} \cdot \text{ft}) && \text{\textit{Step 4:} Simplify. Multiply the units: ft \textit{times} ft = ft}^2.\\
&= 6 \text{ ft}^2 && \text{This is read "six square feet."}
\end{aligned}
$$

▲ **Student Practice 4** Find the area of a square with a side of 6 feet.

The units for area can be expressed two ways:

1. Using exponents: $\qquad\qquad$ ft^2, $\qquad\qquad$ yd^2, $\qquad\qquad$ in.^{2}, and so on.

$\qquad\qquad\qquad\qquad\qquad\qquad\quad$ ↓ $\qquad\qquad$ ↓ $\qquad\qquad$ ↓

2. Using abbreviations: $\qquad$ sq ft, $\qquad$ sq yd, $\qquad$ sq in.

$\qquad\qquad\qquad\qquad$ (square feet) (square yards) (square inches)

▲ **EXAMPLE 5**   Find the height of a parallelogram with base = 8 meters and area = 24 m².

Solution   First we draw a picture (see margin).

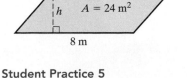

$A = bh$ Next, we write the formula.

$24 = 8h$ Then we replace A and b with the values given.

$\dfrac{24}{8} = \dfrac{8h}{8}$ Now, we solve the equation for h by dividing by 8 on both sides of the equation.

$3 = h$ The height is 3 meters.

Student Practice 5

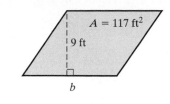

▲ **Student Practice 5**   Find the base of a parallelogram with area = 117 ft² and height = 9 ft.

When we calculate area or perimeter or solve any problem that deals with geometric figures and formulas, it is important that we know the type of figure and the correct formula associated with the problem.

▲ **EXAMPLE 6**   Find the area of the following shape made of rectangles.

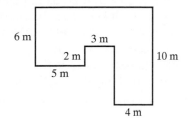

Solution   Divide the figure into three rectangles and then find the area of each rectangle separately. Next, add these three areas together to find the area of the figure.

Area of rectangle #1:

$A = L \cdot W$

$= 5\,\text{m} \cdot 6\,\text{m}$

$= 30\,\text{m}^2$ The area of rectangle #1 is 30 m².

To find the area of rectangle #2 we must find the *width* of the rectangle. We indicate this width with the variable x in the figure in the margin. Since the width of rectangle #1 is 6 m, we know that $x + 2 = 6$. Solving this equation we have $x = 4$, and thus the width of rectangle #2 equals 4 m.

Area of rectangle #2:

$A = L \cdot W$

$= 3\,\text{m} \cdot 4\,\text{m}$

$= 12\,\text{m}^2$ The area of rectangle #2 is 12 m².

Next we find the area of rectangle #3.

Area of rectangle #3:

$A = L \cdot W$

$= 4\,\text{m} \cdot 10\,\text{m}$

$= 40\,\text{m}^2$ The area of rectangle #3 is 40 m².

We add the three areas to find the area of the figure.

$30\,\text{m}^2 + 12\,\text{m}^2 + 40\,\text{m}^2 = 82\,\text{m}^2$ The area of the shape is 82 m².

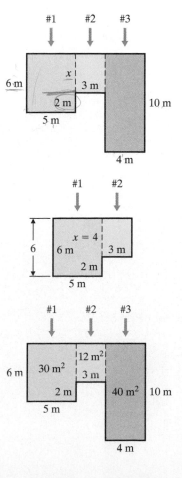

Continued on next page

Student Practice 6

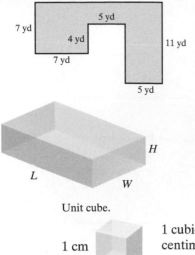

▲ **Student Practice 6** Find the area of the shape made of rectangles and squares.

③ Solving Equations Involving Volume

How much water can a pool hold? How much air is inside a box? These are questions of *volume*. We use volume to measure the space enclosed by a geometric figure that has three dimensions. We call the three dimensions length (*L*), width (*W*), and height (*H*).

Recall that the area of a rectangular region is the number of unit squares needed to fill the rectangular region. The **volume** (*V*) of a rectangular solid is the number of unit cubes needed to fill the figure completely. A unit cube is a rectangular solid with all the edges 1 unit in length.

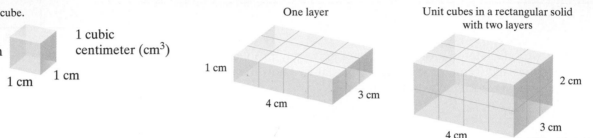

Unit cube.

The one-layer rectangular solid can be viewed as a 3 by 4 array (3 rows of cubes with 4 cubes in each row). There are 3(4) or 12 cubes in the layer. In the two-layer solid, there are 3(4)(2) cubes or 24 cubes. Therefore, the volume of the third figure is 3 cm · 4 cm · 2 cm or 24 cubic centimeters (cm³).

As you can see, to find the number of cubes in the solid figure, we can either count the cubes needed to fill the figure or multiply the length times the width times the height.

$V = LWH$

> **VOLUME**
>
> The volume of a rectangular solid is the product of the length times the width times the height.
>
> $$V = LWH$$

▲ **EXAMPLE 7** Find the unknown side of the rectangular solid.

6 m 4 m *L* $V = 216 \text{ m}^3$

Solution We complete the four-step process. The picture is drawn for us, so we first write the formula. Then we replace the appropriate variables with the values given. Finally, we solve the equation to find the unknown.

$V = L \cdot W \cdot H$ Write the formula.
$216 = L \cdot 4 \cdot 6$ Replace the variables with the values given.
$216 = 24L$ Simplify.
$\dfrac{216}{24} = \dfrac{24L}{24}$ Solve for *L*.
$9 = L$ The length of the box is 9 m.

Student Practice 7

8 m 6 m *W* $V = 192 \text{ m}^3$

▲ **Student Practice 7** Find the unknown side of the rectangular solid.

We summarize all the formulas in this section for your reference.

TABLE OF FORMULAS

Perimeter

The perimeter of a rectangle is twice the length plus twice the width. $P = 2L + 2W$

The perimeter of a square is four times the length of a side. $P = 4s$

Area

The area of a rectangle is the length times the width. $A = LW$

The area of a square is the length of one side squared. $A = s^2$

The area of a parallelogram is the base times the height. $A = bh$ $h = \text{height}$ $b = \text{base}$

Volume

The volume of a rectangular solid is the length times the width times the height. $V = LWH$

---

④ **Solving Applied Geometry Problems**

▲ **EXAMPLE 8**   The blueprints for an office building show that the length of the entryway is 9 feet and the width is 6 feet. The tile for the entryway costs $16 per square yard.

(a) How many square yards of tile must be purchased for the entryway?

(b) How much will the tile for the entry-way cost?

RECEPTION
← 9 ft →
ENTRY   6 ft
OFFICE 1
OFFICE 5

Solution   *Understand the problem.* We organize the information in a Mathematics Blueprint for Problem Solving.

Mathematics Blueprint for Problem Solving

Gather the Facts	What Am I Asked to Do?	How Do I Proceed?	Key Points to Remember
The dimensions of the entryway are: $L = 9$ ft $W = 6$ ft The tile costs $16 per square yard.	**(a)** Find the area to determine the number of square yards of tile that must be purchased. **(b)** Find the cost of the tile for the entryway.	**1.** Draw a 6-foot by 9-foot rectangle. **2.** Convert feet to yards. **3.** Find the area of the entryway. **4.** Multiply the area times $16 to find the cost of the tile.	Change feet to yards since we are asked to find square yards. There are 3 feet in 1 yard. Area = Length × Width

Continued on next page

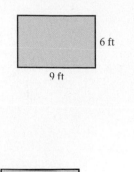

6 ft

9 ft

6 ft = 2 yd

9 ft = 3 yd

Solve and state the answer.

(a) To find the number of square yards of tile needed, we proceed as follows.
  1. We draw a rectangle and label the sides.
  2. We convert feet to yards, since the tile is sold in square yards.

$$9 \text{ ft} = 3 \text{ ft} + 3 \text{ ft} + 3 \text{ ft} \qquad\qquad 6 \text{ ft} = 3 \text{ ft} + 3 \text{ ft}$$
$$= 3 \times 3 \text{ ft} \qquad\qquad\qquad\qquad = 2 \times 3 \text{ ft}$$
$$= 3 \times 1 \text{ yd} \qquad\qquad\qquad\qquad = 2 \times 1 \text{ yd}$$
$$= 3 \text{ yd} \qquad\qquad\qquad\qquad\quad = 2 \text{ yd}$$

  3. We relabel the figure and find the area in square yards.

$$A = L \times W$$
$$= 3 \text{ yd} \times 2 \text{ yd}$$
$$= 3 \times 2 \times \text{yd} \times \text{yd}$$
$$= 6 \text{ yd}^2$$

  6 yd^2 of tile must be purchased.

(b) The tile sells for $16 per square yard and 6 square yards must be purchased: $16 \times 6 = \$96$. The tile will cost $96.

Check: Use your calculator to verify that these calculations are correct.

▲ **Student Practice 8**   Jesse is purchasing carpet for his living room. The room is 12 feet long and 15 feet wide and the carpet costs $11 per square yard.

(a) How many square yards of carpet must be purchased for the living room?

(b) How much will the carpet cost for the living room?

👣 STEPS TO SUCCESS Reading the Text

Homework time each day should begin with a careful reading of the assigned section(s) in your textbook.

Reading a mathematics textbook is unlike reading the books you use in literature and history. Mathematics texts are technical books that provide you with exercises to practice on. Using a mathematics text successfully requires that you read each word slowly and carefully. Completing the following will help you use a mathematics text more effectively.

Making it personal: Place a ✔ in the blank as you complete each of the following.

_____ **1.** Read your textbook with paper and a pencil in hand.

_____ **2.** Underline new definitions or concepts and write them in your notebook on a separate sheet labeled "Important Facts."

_____ **3.** Whenever you encounter unfamiliar terms, look them up and note their definitions on your "Important Facts" notebook pages.

_____ **4.** When you come to an example, work through it step by step. Be sure to read each word and to follow directions carefully.

_____ **5.** Be sure that you understand what you are reading. Make a note of any things that you do not understand and ask your instructor about them.

Do not hurry through the material. Learning mathematics takes time.

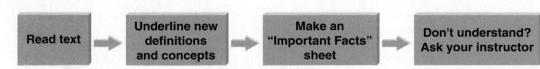

Read text → Underline new definitions and concepts → Make an "Important Facts" sheet → Don't understand? Ask your instructor

Verbal and Writing Skills, Exercises 1–4

1. To find out how much water is needed to fill a rectangular pool, do you need to find the perimeter, the area, or the volume? Why?

2. To determine how much brick you need to trim the edge of the pool, do you need to find the perimeter, the area, or the volume? Why?

3. Fill in each blank with the correct formula.

   (a) Perimeter of a rectangle: _____.

   (b) Perimeter of a square: _____.

   (c) Volume of a rectangular solid: _____.

   (d) Area of a rectangle: _____.

   (e) Area of a square: _____.

   (f) Area of a parallelogram: _____.

4. Write the four-step process for solving problems involving geometry.

   Step 1. _____.

   Step 2. _____.

   Step 3. _____.

   Step 4. _____.

▲ *For each of the following:* **(a)** *State the appropriate formula for perimeter.* **(b)** *Find the perimeter.*

5. A rectangle with $L = 2$ feet and $W = 7$ feet

   (a) _____   (b) _____

6. A rectangle with $L = 16$ feet and $W = 20$ feet

   (a) _____   (b) _____

7. A square with sides of length 11 feet

   (a) _____   (b) _____

8. A square with sides of length 25 inches

   (a) _____   (b) _____

9. A square with sides of length 54 yards

   (a) _____   (b) _____

10. A square with sides of length 30 centimeters

    (a) _____   (b) _____

▲ *For each of the following:* **(a)** *State the appropriate formula for perimeter.* **(b)** *Find the unknown side.*

11. A square has a perimeter of 40 feet. Find the length of each side of the square.

    (a) _____   (b) _____

12. A square has a perimeter of 80 inches. Find the length of each side of the square.

    (a) _____   (b) _____

13. The length of a rectangle is four times the width. If the perimeter of the rectangle is 30 feet, find the width.

    (a) _____   (b) _____

14. The length of a rectangle is twice the width. If the perimeter of the rectangle is 48 yards, find the width.

    (a) _____   (b) _____

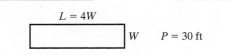

$L = 4W$ | W $P = 30$ ft

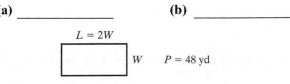

$L = 2W$ | W $P = 48$ yd

▲ *Find the length and width of each rectangle.*

15. Perimeter $= 66$ ft. The length is ten times the width.

W
$L = 10W$

16. Perimeter $= 36$ in. The length is eight times the width.

W
$L = 8W$

17. **(a)** How many squares with 1-inch sides can be placed in a space that is 50 square inches?
(b) What is the area of the rectangle?

1 in.

1 in.

18. **(a)** How many squares with 1-foot sides can be placed in a space that is 22 square feet?
(b) What is the area of the rectangle?

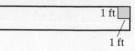

1 ft

1 ft

19. ***Floor Tiles*** Square tiles with sides 1 foot in length are placed in a space that has an area of 50 square feet. How many of the 1-foot squares are needed to fill the space?

20. ***Linoleum Floor Tiles*** How many squares of linoleum floor tiles with sides 1 foot in length would be needed to cover a floor space that has an area of 60 square feet?

For each of the following: **(a)** *State the appropriate formula for area.* **(b)** *Find the area.*

21. A driveway

18 ft   22 ft

(a) _____   **(b)** _____

22. An Oriental rug

3 yd

5 yd

(a) _____   **(b)** _____

23. A square with sides of 10 inches

(a) _____   **(b)** _____

24. A square with sides of 9 meters

(a) _____   **(b)** _____

25. A parallelogram with a base of 12 feet and a height of 9 feet

(a) _____   **(b)** _____

26. A parallelogram with a base of 50 inches and a height of 40 inches

(a) _____   **(b)** _____

▲ *Find the unknown side of each rectangle.*

27.

10 ft

x $A = 60\ \text{ft}^2$

28.

x

7 yd $A = 56\ \text{yd}^2$

▲ *Find the unknown measure for each parallelogram.*

29.

8 m

h $A = 88\ \text{m}^2$

30.

b

12 in. $A = 132\ \text{in.}^2$

▲ Find the area of each shape made of rectangles.

31.

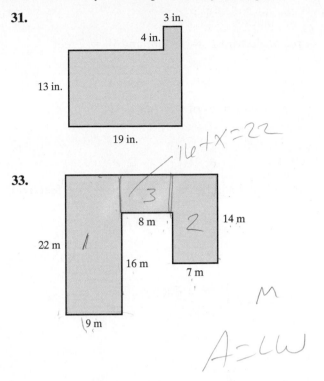

3 in.
4 in.
13 in.
19 in.

16 + x = 22

32.

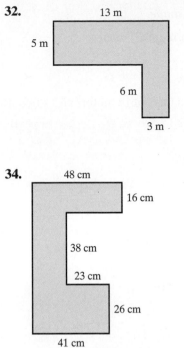

13 m
5 m
6 m
3 m

33.

22 m
8 m
3
2
14 m
16 m
7 m
9 m

A = LW

M

34.

48 cm
16 cm
38 cm
23 cm
26 cm
41 cm

▲ Solve each applied problem.

35. **(a)** How many cubes with 1-inch sides can be placed in a rectangular solid of length = 4 inches, width = 5 inches, and height = 2 inches?
(b) What is the volume of the rectangular solid?

36. **(a)** How many cubes with 1-centimeter sides can be placed in a rectangular solid of length = 5 centimeters, width = 2 centimeters, and height = 3 centimeters?
(b) What is the volume of the rectangular solid?

37. **Water in a Fish Tank** A fish tank is 3 feet wide, 4 feet long, and 3 feet high. How much water can be placed in the tank?

38. **Airspace in a Room** The ceiling height of a room is 8 feet. The width of the floor is 10 feet, and the length is 15 feet. How much airspace is in the room?

▲ For each of the following rectangular solids: **(a)** State the appropriate formula for volume. **(b)** Find the volume.

39. Length = 11 inches, width = 5 inches, height = 15 inches

(a) _____　　**(b)** _____ in

40. Length = 13 feet, width = 7 feet, height = 10 feet

(a) _____　　**(b)** _____

41. Length = 27 yards, width = 10 yards, height = 16 yards

(a) _____　　**(b)** _____

42. Length = 70 meters, width = 50 meters, height = 30 meters

(a) _____　　**(b)** _____

▲ Find the unknown side of each rectangular solid.

43.

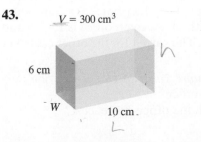

$V = 300 \text{ cm}^3$

6 cm
W
10 cm
L

h
300 = 6(5)

44.

$V = 315 \text{ m}^3$

H
7 m
9 m

Mixed Practice

Find the perimeter of each shape.

45. A square with sides of length 7 inches

46. A rectangle with $L = 12$ feet and $W = 7$ feet

Find the area of each shape.

47. A parallelogram with a base of 17 meters and a height of 8 meters

48. A square with sides of 9 inches

Find the volume of each solid.

49. A rectangular solid with $L = 15$ yards, $W = 7$ yards, and $H = 2$ yards

50. A rectangular solid with $L = 8$ feet, $W = 5$ feet, and $H = 6$ feet

Find the unknown side of each shape.

51.
6 ft

x [   ] $A = 30$ ft^2

52.
x

5 in. [   ] $A = 40$ in.^{2}

53. $V = 200$ m^3

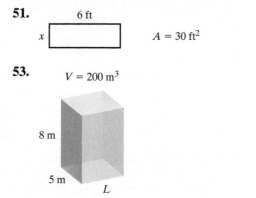

8 m

5 m

L

54. $V = 64$ ft^3

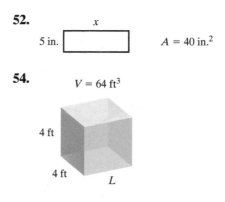

4 ft

4 ft L

Applications

55. *A Piece of Marble* Damien has a 2-foot-square piece of marble.

  (a) State the dimensions of the marble on the figure below in inches (1 ft $= 12$ in.).

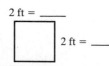

2 ft = _____

2 ft = _____

  (b) State the area of the marble in square inches.

56. *A Card Table* Marcy has a 3-foot-square card table.

  (a) State the dimensions of the table on the figure below in inches (1 ft $= 12$ in.).

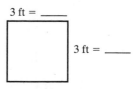

3 ft = _____

3 ft = _____

  (b) State the area of the table in square inches.

57. *A Vegetable Garden* A rectangular vegetable garden has a length of 12 feet and a width of 9 feet.

  (a) State the dimensions of the garden in yards (3 feet $= 1$ yard).

  (b) State the area of the garden in square yards.

58. *An Entryway* A rectangular entryway has a length of 12 feet and a width of 6 feet.

  (a) State the dimensions of the entryway in yards (3 feet $= 1$ yard).

  (b) State the area of the entryway in square yards.

59. *Carpeting a Patio* Carmen is purchasing outdoor carpet for her patio enclosure, which is 12 *feet* long and 9 *feet* wide. The carpet she plans to purchase is priced at $8 per *square yard*.

  (a) How many *square yards* of carpet must she purchase?

  (b) How much will the outdoor carpet for Carmen's patio cost?

60. *Carpeting an Office* The blueprints for a warehouse show that the length of the main office is 15 *feet* and the width is 12 *feet*. The carpet for the office costs $11 per *square yard*.

  (a) How many *square yards* of carpet must be purchased for the office?

  (b) How much will the office carpet cost?

61. *Linoleum Flooring* Shannon has a one-story house with a family room that has a length of 21 *feet* and width of 15 *feet*. The linoleum floor she will purchase for the family room costs $16 per *square yard*.

(a) How many *square yards* of linoleum must Shannon purchase for her family room?

(b) How much will the family room linoleum cost?

62. *Ceramic Tile Picture* Jamel is making a picture out of ceramic tile for a wall in the Teen Center. The Teen Center wants the picture to fit exactly in a space with $L = 24$ *inches* and $W = 36$ *inches*. Jamel charges $30 per *square foot* to make ceramic designs.

(a) How many *square feet* will the picture be? (12 in. = 1 ft)

(b) How much will it cost the Teen Center to have Jamel make the picture?

One Step Further

▲ **63.** *Fertilizer Cost* A 1-pound container of rose fertilizer sells for $3. Each 1-pound container will fertilize 100 *square feet*. If the length of the rose garden is 5 *yards* and the width is 4 *yards*, how much will it cost to fertilize the entire rose garden?

▲ **64.** *Chest Lining* One roll of felt material at the fabric store sells for $12 and can cover 10 *square feet* of area. Daisy wishes to place felt on the base of the interior of a large wooden chest. The base measures 2 *yards* by 1 *yard*. How much will it cost Daisy to purchase the felt for the chest?

To Think About

Painting a Room The walls of Anita's family room are illustrated below. All of the walls are 8 feet high, and the rear and front walls are each 22 feet long. Each of the side walls is 16 feet long. The French door is 3 feet wide and 7 feet high, while the sliding door is 6 feet wide and 7 feet high. The window on the first side wall is 4 feet by 3 feet; the window on the second side wall is 2 feet by 4 feet.

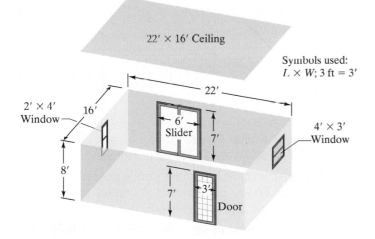

Paint Store Price List

Type	Price	Coverage
One-coat	$18 per gal	1 gal per 400 ft²
	$5 per qt	1 qt per 100 ft²
Primer	$12 per gal	1 gal per 300 ft²
	$3 per qt	1 qt per 75 ft²

▲ *Use the figure and table to answer exercises 65 and 66.*

65. Anita's landlord will paint her family room. He has agreed to let Anita purchase the paint so that she can choose the brand and color she wants. The landlord will reimburse Anita only for the cost of the paint used. Anita must buy only as much paint as she needs since the extra paint cannot be returned. Describe the quantity of paint that Anita should purchase if the landlord will put 1 coat of paint on all of the walls and the ceiling of the family room.

66. Refer to exercise 65. If the landlord will put 1 coat of primer on the walls and ceiling of the family room, describe the quantity of primer that Anita must purchase.

Find the area of the shaded region.

67.

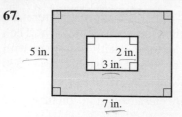

5 in. 2 in. 3 in. 7 in.

68. Find the area of the region that is *not* shaded.

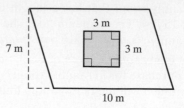

3 m 7 m 3 m 10 m

69. Find the next figure that would appear in the sequence.

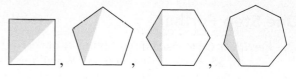

70. Find the next two figures that would appear in the sequence.

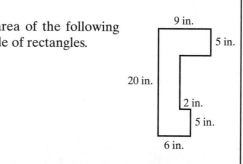

Cumulative Review

71. **[2.6.3]** Simplify. $-7(x - 2)$

72. **[2.6.1]** Combine like terms. $-7x + 3x - 2y$

73. **[1.4.3]** Simplify. $(2)(3x)(5)$

74. **[2.3.2]** Perform the indicated operations. $-8 - 3 - 1 - 4$

Quick Quiz 3.3

1. The width of a rectangle is double the length. If the perimeter of the rectangle is 72 feet, find the length.

2. Find the area of the following shape made of rectangles.

9 in. 5 in. 20 in. 2 in. 5 in. 6 in.

3. A storage facility claims the volume of its smallest rental unit is 200 cubic feet. If its length and width both measure 5 feet, what is the height of the unit?

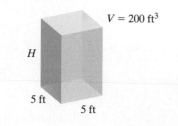

$V = 200 \text{ ft}^3$ H 5 ft 5 ft

4. **Concept Check** Hanna purchased a redwood box at the garden store and filled it with sand.

(a) To determine how much sand is needed to fill the box, do you find the area, perimeter, or volume?

(b) State the formula you must use.

(c) The volume of the box is 200 ft³, the height is 5 in., and the length of the box is double the height. Explain in words how you would find the width of the box.

Did You Know...
That Driving a Long Distance to Save Money May End Up Costing You More?

GAS MONEY AND DRIVING TIME

Understanding the Problem:

Mary needs to get gas before she goes shopping for sneakers. She has two cost-saving decisions to make: (1) she wonders if it makes sense to drive across town for gas that costs 15 cents less per gallon, and (2) she wonders if it is worth driving 18 miles away to save 20% at a sneaker sale.

Making a Plan:

Mary needs to first calculate what it is going to cost her to drive across town to get gas. Then she can compare this to what she saves by buying the cheaper gas. Secondly, she needs to calculate what it costs to drive 18 miles to the mall where the sneakers are on sale. Then she can compare this cost to what she would save at the 20%-off sale.

Step 1: It takes Mary 20 extra minutes to drive across town to the cheaper-priced gas station. Mary feels her time is worth $9 an hour. So 20 minutes at $9 an hour costs her $3. Mary also calculates that it costs her $1 in gas to drive to the farther gas station.

Task 1: What is the total cost to go to the farther gas station?

Step 2: Mary usually buys 20 gallons of gas, so next she calculates what that will cost. Her neighborhood gas station has gas for $3.70 a gallon, which will cost $74. Across town it is selling for $3.55 a gallon, which will cost $71.

Task 2: What is the savings from buying the gas at the farther station?

Finding a Solution:

Step 3: Now that Mary has the total cost and the amount of the savings, she can see if it is worth it to go to the farther gas station.

Task 3: Calculate the difference between what Mary will save and what it will cost to drive across town to get gas.

Task 4: Does it make sense to drive farther to get gas? Why or why not?

Step 4: Now Mary wants to decide if it is worth it to drive to the mall for the sneaker sale. It takes her an extra 40 minutes to drive to the mall instead of going to a local store. She calculates the cost of her time at $6 and the cost of gas at $2.

Task 5: What is the total cost to drive to the mall?

Step 5: The sneakers she wants are $95 at her local store but are on sale at the mall for $76, which is 20% off. How much will she save by driving for a deal?

Task 6: How much would she save by buying the sneakers on sale?

Finding a Solution:

Step 6: Now that Mary has the total cost and the amount of the savings, she can see if it is worth it to drive to the mall.

Task 7: Calculate the difference between what Mary will save and what it will cost to drive to the mall.

Task 8: Is it worth driving 18 miles away for the sale? Why or why not?

Applying the Situation to Your Life:

Have you ever driven out of your way to save money? Now you know you'll need to take into consideration the expenses incurred to determine if you are really saving money. If you enjoy the trip and have a lot of free time, you may not want to include the value of your time in the calculations. You could do the calculations for Mary again and see if that changes the decision. Can you think of any ways to reduce expenses? One way would be by making multiple purchases in the same trip.

3.4 Performing Operations with Exponents

Student Learning Objectives

After studying this section, you will be able to:

① Use the product rule for exponents.

② Multiply algebraic expressions with exponents.

③ Use the distributive property to multiply a monomial and a binomial.

④ Simplify geometric formulas involving monomials and binomials.

NOTE TO STUDENT: Fully worked-out solutions to all of the Student Practice problems can be found at the back of the text starting at page SP-1.

① Using the Product Rule for Exponents

Recall from Chapter 1 that $3 \cdot 3$ written in exponent form is 3^2, where 3 is the base and 2 is the exponent. The exponent 2 tells us how many times the base appears as a factor.

Now, how do we multiply $3^2 \cdot 3^3$? One way is to write the expressions as repeated multiplication and then rewrite it with one exponent.

$$\underbrace{3 \cdot 3}_{3^2} \cdot \underbrace{3 \cdot 3 \cdot 3}_{3^3} = 3^5 \qquad \text{Three appears as a factor 5 times; the exponent is 5.}$$

Thus $3^2 \cdot 3^3 = 3^5$.

EXAMPLE 1 Write $5^4 \cdot 5^3$ as repeated multiplication and then rewrite it with one exponent.

Solution

$$\underbrace{5 \cdot 5 \cdot 5 \cdot 5}_{5^4} \cdot \underbrace{5 \cdot 5 \cdot 5}_{5^3} = 5^7 \qquad \text{Since five appears as a factor 7 times, the exponent is 7.}$$

$$5^4 \cdot 5^3 = 5^7$$

Student Practice 1 Write $4^2 \cdot 4^4$ as repeated multiplication and then rewrite it with one exponent.

We can shorten this process by using rules that tell us how to multiply numbers that have the same base and are written in exponent form. Try to notice a pattern in the following products.

$$4^3 \cdot 4^4 = (\overbrace{4 \cdot 4 \cdot 4}^{3 \text{ factors}})(\overbrace{4 \cdot 4 \cdot 4 \cdot 4}^{4 \text{ factors}}) = 4^{\overset{7 \text{ factors}}{7}} \qquad \text{The exponent is 7 since there are 7 factors.}$$

$$x^2 \cdot x^3 = (\overbrace{x \cdot x}^{2 \text{ factors}})(\overbrace{x \cdot x \cdot x}^{3 \text{ factors}}) = x^{\overset{5 \text{ factors}}{5}} \qquad \text{The exponent is 5 since there are 5 factors.}$$

Let's summarize these products: $(4^3)(4^4) = 4^7$; $(x^2)(x^3) = x^5$. Do you see the pattern?

We state the rule as follows.

PRODUCT RULE FOR EXPONENTS

To multiply constants or variables in exponent form that have the *same base*, add the exponents but keep the base unchanged.

If a and b are positive integers, then

$$x^a \cdot x^b = x^{a+b}.$$

There are three things we must remember when we use the rule for multiplying expressions with exponents.

1. We can use the product rule *only* if the bases are the same.
2. We *add* exponents. We do *not* multiply exponents.
3. We do *not* multiply the bases.

EXAMPLE 2   Multiply and write the product in exponent form.

(a) $x^3 \cdot x^6$ **(b)** $x^7 \cdot x$ **(c)** $4^5 \cdot 3^4$ **(d)** $2^2 \cdot 2^4$

Solution

(a) $x^3 \cdot x^6 = x^{3+6} = x^9$

(b) $x^7 \cdot x = x^7 \cdot x^1 = x^{7+1} = x^8$

       Note that any number or variable that does not have a written exponent is understood to have an exponent of 1.

(c) $4^5 \cdot 3^4 = 4^5 \cdot 3^4$ The rule for multiplying numbers with the *same base* does not apply since the bases are *different*.

(d) $2^2 \cdot 2^4 = 2^{2+4} = 2^6$

CAUTION: In part (d), the base does not change; only the exponents change.

This is correct: $2^2 \cdot 2^4 = 2 \cdot 2 \cdot 2 \cdot 2 \cdot 2 \cdot 2 = 2^6$ There are 6 factors of 2 in the product.

This is wrong! $2^2 \cdot 2^4 = 2 \cdot 2 \cdot 2 \cdot 2 \cdot 2 \cdot 2 \neq 4^6$ There are *not* 6 factors of 4 in the product.

Student Practice 2   Multiply.

(a) $y^5 \cdot y$ **(b)** $a^4 \cdot a^5$ **(c)** $5^3 \cdot 3^4$ **(d)** $6^5 \cdot 6^6$

② Multiplying Algebraic Expressions with Exponents

For the algebraic expression $3x^4$, the number 3 is called the **numerical coefficient** or just the **coefficient.** A numerical coefficient is a number that is multiplied by a variable.

$$3x^4$$
$$\downarrow$$

3 is a numerical coefficient.

When we multiply two expressions, we first multiply the numerical coefficients and then we multiply the variable expressions.

$(3x^4)(4x^7) = 3 \cdot x^4 \cdot 4 \cdot x^7$
$\qquad\qquad = (3 \cdot 4) \cdot (x^4 \cdot x^7)$ Change the order of multiplication and regroup factors.

$(3x^4)(4x^7) = \quad 12 \quad \cdot \quad x^{11}$ Multiply the numerical coefficients. Multiply variables separately by adding their exponents:

$$x^4 \cdot x^7 = x^{4+7} = x^{11}.$$

PROCEDURE FOR MULTIPLYING ALGEBRAIC EXPRESSIONS WITH EXPONENTS

1. Multiply the numerical coefficients.
2. Use the product rule for exponents.

$$(5a^2)(3a^4) = 5 \cdot 3 \cdot a^2 \cdot a^4 = 15a^6$$

Not only should you know how to work with algebraic expressions, but it is critical that you know clearly which parts are the *numerical coefficients*, which parts are the *bases*, and which parts are the *exponents*. These words are used extensively in algebra.

As stated earlier, any variable that does not have a *visible exponent* is understood to have an *exponent of 1*. For example $x = x^1$ and $y = y^1$.

Likewise, any variable that does not have a *visible numerical coefficient* is understood to have a numerical *coefficient of 1*. For example $y^4 = 1y^4$ and $a^6 = 1a^6$.

EXAMPLE 3   Multiply. $(7a)(a^3)(4a^6)$

Solution

$$(7a)(a^3)(4a^6) = (7a^1)(1a^3)(4a^6)$$
$$= (7 \cdot 1 \cdot 4)(a^1 \cdot a^3 \cdot a^6) \quad \text{Change the order of the factors and regroup them.}$$
$$= 28 \cdot a^{10} \quad \text{Multiply the numerical coefficients and variables separately: } 7 \cdot 1 \cdot 4 = 28 \text{ and } a^1 \cdot a^3 \cdot a^6 = a^{1+3+6} = a^{10}.$$
$$(7a)(a^3)(4a^6) = 28a^{10}$$

Student Practice 3   Multiply. $(4y)(5y^2)(y^5)$

We must sometimes multiply algebraic expressions that involve more than one variable.

EXAMPLE 4   Multiply. $(5x^6)(7y^4)(2x^4)$

Solution $(5x^6)(7y^4)(2x^4) = (5 \cdot 7 \cdot 2)(x^6 \cdot y^4 \cdot x^4) = 70x^{10}y^4$

$$(5x^6)(7y^4)(2x^4) = 70x^{10}y^4$$

CAUTION: We cannot use the product rule for exponents to simplify $x^{10}y^4$ because the bases, x and y, are not the same.

Student Practice 4   Multiply. $(4y^3)(3x^2)(5y^2)$

Just as with whole numbers, we use the associative and commutative properties of multiplication to simplify algebraic expressions with integers.

$$4(2x) = (4 \cdot 2)x = 8x \qquad -4(2x) = (-4 \cdot 2)x = -8x$$

EXAMPLE 5   Multiply.

(a) $(5x)(-8x)$ **(b)** $(-7x^2)(-4y^4)$

Solution

(a) $(5x)(-8x) = (5)(-8)(x \cdot x)$ $x \cdot x = x^1 \cdot x^1 = x^2$
$$= -40x^2$$

(b) $(-7x^2)(-4y^4) = (-7)(-4)(x^2 \cdot y^4)$ x^2 and y^4 do not have the same base. Thus, we cannot add exponents.
$$= 28x^2y^4$$

Student Practice 5   Multiply.

(a) $(-6a)(-8a)$

(b) $(5x^3)(-2y^5)$

● **Understanding the Concept**

Do I Add or Multiply Coefficients?   When we combine the like terms $4n + 3n$, do we get the same answer as when we multiply $(4n)(3n)$? Let's simplify both and see what happens.

$$4n + 3n \qquad\qquad\qquad (4n)(3n)$$
$$(n + n + n + n) + (n + n + n) = 7n \quad 4 \cdot n \cdot 3 \cdot n = 4 \cdot 3 \cdot n \cdot n = 12n^2$$
$$4n + 3n = 7n \qquad\qquad (4n)(3n) = 12n^2$$

We *add* the numerical coefficients.   We *multiply* the numerical coefficients.
The variable stays the same.   Then we add the exponents of the variable.

We can see that the answers are not the same. When we simplify we must remember the difference between adding and multiplying algebraic expressions.

Exercise

1. Simplify.

   (a) $3x + 5x$ **(b)** $(3x)(5x)$

   (c) $7xy^2 - 5xy^2$ **(d)** $(7xy^2)(5xy^2)$

---

③ **Using the Distributive Property to Multiply a Monomial and a Binomial**

One type of expression we deal with in mathematics is called a **polynomial.** Polynomials are expressions that contain terms with variable parts that have only whole number exponents. Polynomials do not have variables in the denominator.

       Polynomials   *Not a Polynominal*

    $3xy + 1 \quad 2a^3 - 3$ $2x + \dfrac{1}{x}$ ←   Variable in the denominator

There are special names for polynomials with one, two, and three terms.

      A **monomial** has *one* term. $4a$

      A **binomial** has *two* terms. $5a^3 + 3b$

      A **trinomial** has *three* terms. $3x^2 + 7x - 4$

EXAMPLE 6   Identify each polynomial as a monomial, binomial, or trinomial.

(a) $2x^2 + 5$ **(b)** $8x$ **(c)** $5x^3 + 8x - 1$

Solution

(a) $2x^2 + 5$ is a binomial because there are two terms.

(b) $8x$ is a monomial because there is one term.

(c) $5x^3 + 8x - 1$ is a trinomial because there are three terms.

Student Practice 6   Identify each polynomial as a monomial, binomial, or trinomial.

   (a) $4x^2$ **(b)** $8x^2 - 9x + 1$ **(c)** $5x^3 + 8x$

Recall that to multiply $5 \cdot (x + 1)$ we use the distributive property: we multiply $5 \cdot x$ and then $5 \cdot 1$ to obtain $5x + 5$. We follow the same process when we multiply a monomial times a binomial, as illustrated in the next example.

$\mathbb{M}_{\mathbb{C}}$ **EXAMPLE 7**   Use the distributive property and then simplify.

$3x^2(x^5 + 8x) + 2x^7$

Solution

$$3x^2(x^5 + 8x) + 2x^7$$

$3x^2 \cdot x^5$

$3x^2(x^5 + 8x) + 2x^7 = 3x^2 \cdot x^5 + 3x^2 \cdot 8x + 2x^7$ Use the distributive property to multiply $3x^2(x^5 + 8x)$.

$3x^2 \cdot 8x$

$= 3x^2 \cdot 1x^5 + 3x^2 \cdot 8x^1 + 2x^7$ Write x^5 as $1x^5$ and x as x^1.

$= (3 \cdot 1) \cdot (x^2 \cdot x^5) + (3 \cdot 8) \cdot (x \cdot x) + 2x^7$ Multiply numerical coefficients.

$= 3x^{2+5} + 24x + 2x^7$ Multiply variables by adding their exponents.

$= 3x^7 + 24x^3 + 2x^7$

$3x^2(x^5 + 8x) + 2x^7 = 5x^7 + 24x^3$ Combine like terms: $3x^7 + 2x^7 = 5x^7$.

Student Practice 7   Use the distributive property to simplify.

$$4x^4(x^3 - 6x) + 3x^5$$

CAUTION: The solution, $5x^7 + 24x^3$, in Example 7 is simplified. We *cannot* combine like terms because $5x^7$ and $24x^3$ are *not* like terms. Variable parts of like terms must be identical, which means that both the variable and the exponent must be the same. In the expression $5x^7 + 24x^3$, the variables x^7 and x^3 do not have the same exponent, therefore $5x^7$ and $24x^3$ are not like terms.

④ Simplifying Geometric Formulas Involving Monomials and Binomials

Recall from Section 3.3 that we use the following formulas to find area.

Area of a rectangle: $A = LW$

Area of a parallelogram: $A = bh$

When we work with geometric shapes in algebra, we often use *expressions* for the length and width instead of *numerical* measurements, and thus our answer will be an *expression* containing a variable. We cannot get a numerical answer until we have a value for the variable.

EXAMPLE 8 Write the area of the rectangle below as an algebraic expression and then simplify.

x^2

$3x^3 - 2$

Solution First we write the formula for area and then we replace L and W with the given expressions.

$A = L\,W$

$= (3x^3 - 2)(x^2)$ $L = 3x^3 - 2$ and $W = x^2$

$= (3x^3 - 2)(x^2)$ We distribute the x^2.

$= (3x^3)(x^2) - (2)(x^2)$

$A = 3x^5 - 2x^2$ We simplify.

Note that the monomial x^2 was on the *right side* of the binomial. When this is the case we proceed the *same way* except that we distribute the x^2 from the *right side*.

CAUTION: We cannot combine the terms $3x^5$ and $2x^2$ in Example 8; they are not like terms. Thus our answer is the binomial expression $3x^5 - 2x^2$.

Student Practice 8 Write the area of the parallelogram below as an algebraic expression and then simplify.

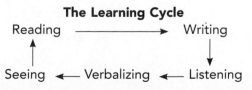

$h = x^3$

$7x^5 - 2x$

👣 STEPS TO SUCCESS Class Attendance and the Learning Cycle

Did you know that an important part of the learning process happens in the classroom? People learn by *reading, writing, listening, verbalizing,* and *seeing.* These activities are all part of the **learning cycle** and always occur in class.

- *Listening and seeing*: hearing and watching the instructor's lecture
- *Reading*: reading the information on the board and in handouts
- *Verbalizing*: asking questions and participating in class discussions
- *Writing*: taking notes and working problems assigned in class

The Learning Cycle

Reading ⟶ Writing

Seeing ← Verbalizing ← Listening

Attendance in class completes the entire learning cycle once. Completing the following steps activates the entire learning cycle one more time.

Making it personal: *Place a* ✔ *in the blank as you complete each of the following.*

Notebook for Class Notes

____ 1. Shortly after class, read your class notes and the text for the section covered in class. Add any facts that you recall from class that you may have missed, or other facts from the text.

____ 2. Rewrite your notes if they are too messy to understand.

____ 3. Place a ? beside any notes from class that you don't understand, then ask a classmate or the instructor for clarification.

____ 4. Place these notes in a notebook so you can refer to them when needed.

Keep in mind that you must pay attention and participate to learn. Just being there is not enough.

Verbal and Writing Skills, Exercises 1–6

1. Is $2^4 \cdot 2^2 = 4^6$? Why or why not?

2. Is $3^5 \cdot 4^3 = 12^8$? Why or why not?

3. **(a)** Is $2(3x^2) = 2 \cdot 3 + 2 \cdot x^2 = 6 + 2x^2$? Why or why not?

4. Why can you determine that the products in part **(a)** and part **(b)** are *equal* without completing the multiplication?
(a) $x(2x^4 - 3x^2)$
(b) $(2x^4 - 3x^2)x$

(b) Is $2(3x^2) = 2 \cdot 3 \cdot x^2 = 6x^2$? Why or why not?

(c) Is $2(3 + x^2) = 2 \cdot 3 + 2 \cdot x^2 = 6 + 2x^2 = 8x^2$? Why or why not?

Fill in the blank.

5. In the algebraic expression $4x^2$, the number 4 is called the _____.

6. A polynomial with one term is called a _____.
A polynomial with two terms is called a _____.
A polynomial with three terms is called a _____.

Multiply and write the product in exponent form.

7. **(a)** $(z \cdot z \cdot z) \cdot (z \cdot z)$
(b) $z^3 \cdot z^2$

8. **(a)** $(y \cdot y \cdot y \cdot y) \cdot (y \cdot y \cdot y)$
(b) $y^4 \cdot y^3$

9. **(a)** $(x \cdot x) \cdot (x \cdot x \cdot x \cdot x)$
(b) $x^2 \cdot x^4$

10. **(a)** $(b \cdot b \cdot b) \cdot (b \cdot b \cdot b)$
(b) $b^3 \cdot b^3$

11. **(a)** $m \cdot m \cdot m$
(b) $m^2 \cdot m$

12. **(a)** $(x \cdot x \cdot x) \cdot x$
(b) $x^3 \cdot x$

Write each expression as repeated multiplication and then rewrite it with one exponent.

13. $2^4 \cdot 2^2$

14. $7^3 \cdot 7^3$

15. $3^5 \cdot 3^3$

16. $5^3 \cdot 5^5$

Multiply and write the product in exponent form.

17. $x^5 \cdot x^2$

18. $a^7 \cdot a^7$

19. $a \cdot a$

20. $y^5 \cdot y$

21. $3^2 \cdot 3^3$

22. $7^3 \cdot 7^5$

23. $4 \cdot 4^5$

24. $6 \cdot 6^3$

25. $8^2 \cdot 7^5$

26. $3^9 \cdot 4^6$

27. $x^5 \cdot y^3$

28. $x^3 \cdot y^5$

29. $x^5 \cdot x^2 \cdot x^6$

30. $y^4 \cdot y^3 \cdot y^5$

31. $y^3 \cdot y^6 \cdot y^5$

32. $x^4 \cdot x^3 \cdot x^5$

33. $3^3 \cdot 3^2 \cdot 3^5$

34. $4^7 \cdot 4^4 \cdot 4^2$

35. $2^5 \cdot 3^2 \cdot 4^7$

36. $4^7 \cdot 5^3 \cdot 2^4$

Simplify.

37. $(4y^5)(6y^7)$

38. $(5y^3)(3y^3)$

39. $(6a^6)(9a^8)$

40. $(6a^4)(2a^4)$

41. $(-5x)(3x^2)$

42. $(-6x)(3x)$

43. $(-4y)(3y)$

44. $(5y)(-4y)$

45. $(5a)(7a^3)(2a^6)$ **46.** $(3x)(2x^4)(4x^5)$ **47.** $(x)(4x^6)(7x^5)$ **48.** $(4x)(x^7)(9x^6)$

49. $(5x)(3y)(-2x)$ **50.** $(-4x)(2y)(7x)$ **51.** $(6y)(-3x)(-3y)$ **52.** $(3a)(-2b)(-4a)$

Identify each polynomial as a monomial, binomial, or trinomial.

53. (a) $5z^3 + 4$
(b) 9
(c) $2x^7 - 3x^4 - 3$

54. (a) $6a^4$
(b) $5y^2 - 9y - 3$
(c) $7z - 6$

Simplify.

55. $2x(x^2 + 5)$ **56.** $5y(y^2 + 6)$ **57.** $6x^2(3x^3 - 1)$ **58.** $7x^3(2x^2 - 2)$

59. $5y^3(y^4 - 2y) + 3y^7$ **60.** $7x^2(x^3 - 6x) + 2x^5$ **61.** $-2x^3(x^2 + 4x) + 5x^5$ **62.** $-3x^2(x^3 + 3x) + 6x^5$

63. $(2y - 5)(-6y^2)$ **64.** $(5x - 1)(-3x^2)$ **65.** $(2x^3 - 4x)(3x)$ **66.** $(3y^2 - 2y)(9y^2)$

▲ *Write the area of each rectangle as an algebraic expression and then simplify.*

67.
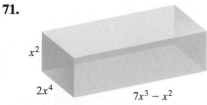
x^3
$2x^4 - 5$

68.
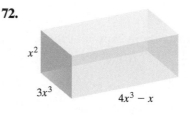
x^2
$4x^3 - 3$

▲ *Write the area of each parallelogram as an algebraic expression and then simplify.*

69.

x^2
$7x^3 - 4x$

70.
x^4
$5x^5 - 4$

One Step Further

▲ *Write the volume of each rectangular solid as an algebraic expression and then simplify.*

71.

x^2
$2x^4$
$7x^3 - x^2$

72.
x^2
$3x^3$
$4x^3 - x$

▲ *Write the perimeter of each rectangle as an algebraic expression and then simplify.*

73.
$3x^2$
$4x^3$

74.
x^3
$6x^4$

75.
$5x^3 + 3x^2$
$2x^2 - x$

76.
$4x^4 + 5x^3$
$2x^4 - x^3$

▲ *For each of the following rectangles:* **(a)** *Write the area as an algebraic expression and then simplify.* **(b)** *Write the perimeter as an algebraic expression and then simplify.*

77.
$3x^2 + 5$
$2x^2$

78.
$5a^3$
$2a^3 - 3$

To Think About   *Perform the operation indicated.*

79. (a) $4cd + 9cd$

  (b) $(9cd)(4cd)$

  (c) $4(cd + 9)$

80. (a) $5xz + 2xz$

  (b) $(5xz)(2xz)$

  (c) $5(xz + 2)$

81. (a) $9ab^5 - 7ab^5$

  (b) $(9ab^5)(7ab^5)$

  (c) $9(ab^5 + 7)$

82. (a) $6y^7z - 3y^7z$

  (b) $(6y^7z)(3y^7z)$

  (c) $6(y^7z + 3)$

Write the width of each rectangle as an algebraic expression.

83. The width of a rectangle is four more than the square of the length.

84. The width of a rectangle is two less than the square of the length.

Cumulative Review

85. **[2.4.4]** Divide. $\dfrac{-35}{7}$

86. **[2.4.3]** Evaluate. $(-3)^3$

87. **[1.5.4]** Divide. $20{,}566 \div 312$

88. **[1.4.4]** Multiply. $31{,}423 \times 28$

Golf Scores *In golf, par is the number of strokes that is standard for that particular hole. Scores under par (less than par) are recorded as negative numbers, and scores over par (more than par) are recorded as positive numbers. Zero is used for a score of par, and the lowest total score for an entire course determines the winner.*

89. **[2.3.2]** Determine Allen's total score after the third hole if his scores on the first three holes were as follows: first hole, 1 stroke under par; second hole, 2 strokes over par; third hole, par.

90. **[2.3.2]** Determine Laura's total score after the fourth hole if her scores on the first four holes were as follows: first hole, par; second hole, 1 stroke over par; third hole, 1 stroke under par; fourth hole, 1 stroke under par.

Quick Quiz 3.4

1. Multiply. Leave your answer in exponent form.

  (a) $x^2 \cdot x$ (b) $4^5 \cdot 4^2$

  (c) $-3y^3(2y)(7y^4)$

2. Simplify.

  (a) $6a(a^3 + 8a)$

  (b) $(x^5 - 3)(2x^2)$

3. Write the area of the following rectangle as an algebraic expression and then simplify.

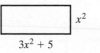
x^2

$3x^2 + 5$

4. **Concept Check** Explain in words the steps to simplify each of the following.

  (a) $2x^4(x^2 + y)$ (b) $(-3x^3)(x^3)$

  (c) Using the results for (a) and (b) explain the extra steps needed to complete the problem $2x^4(x^2 + y) + (-3x^3)(x^3)$, and then state the answer.

Chapter 3 Organizer

Topic and Procedure	Examples	✏️ You Try It
Solving equations using the addition principle, p. 166 1. Use the commutative and associative properties to simplify each side of the equation. 2. Add the appropriate value to both sides of the equation so that the variable is alone on one side and a number is on the other side of the equals sign: $x = $ *some number*, or *some number* $= x$ 3. Check by substituting your answer back into the original equation.	Solve. $6x - 5x - 1 = 10$ $\begin{aligned} x - 1 &= 10 \\ +\quad 1 &\quad 1 \\ \hline x + 0 &= 11 \\ x &= 11 \end{aligned}$ *Check:* $6x - 5x - 1 = 10$ $6(11) - 5(11) - 1 \overset{?}{=} 10$ $66 - 55 - 1 \overset{?}{=} 10$ $11 - 1 \overset{?}{=} 10$ $10 = 10$ ✓	1. Solve. $5x - 4x - 2 = 8$
Solving applied problems involving angles, p. 169 Two angles that share a common side are called adjacent angles. Adjacent angles formed by two intersecting lines are supplementary. That is, they add up to 180°. 	If $\angle a$ and $\angle b$ are supplementary angles, find the measure of $\angle a$ if the measure of $\angle b = 46°$. $\begin{aligned} \angle a + \angle b &= 180° \\ \angle a + 46° &= 180° \\ +\quad -46° &\quad -46° \\ \hline \angle a &= 134° \end{aligned}$	2. If $\angle a$ and $\angle b$ are supplementary angles, find the measure of $\angle b$ if the measure of $\angle a = 55°$.
Solving equations using the division principle, p. 175 1. Use the commutative and associative properties to simplify each side of the equation. 2. If the variable is *multiplied* by a number, undo the multiplication by *dividing* both sides of the equation by this number so that the variable is alone on one side and a number is on the other side of the equals sign: $x = $ *some number*, or *some number* $= x$ 3. Check by substituting your answer back into the original equation.	Solve. $-36 = 3(4x)$ $-36 = 12x$ Simplify. $\dfrac{-36}{12} = \dfrac{12x}{12}$ The variable is multiplied by 12. We divide both sides by 12. $-3 = x$ *Check:* $-36 = 3(4x)$ $-36 \overset{?}{=} 3(4 \cdot (-3))$ $-36 = -36$ ✓	3. Solve. $-48 = 3(4x)$
Translating English statements into equations, p. 177 1. Translate the statement into an equation. 2. Use the addition or the division principles of equality to solve the equation.	There are three times as many dimes (D) as nickels (N) in a coin collection. **(a)** Translate the statement into an equation. **(b)** Find the number of nickels if there are 42 dimes. **(a)** $D = 3N$ Translate. **(b)** $42 = 3N$ Solve. $\dfrac{42}{3} = \dfrac{3N}{3}$ $14 = N$ There are 14 nickels.	4. There are two times as many roses (R) as carnations (C) in the bouquet of flowers. **(a)** Translate the statement into an equation. **(b)** Find the number of carnations if there are 12 roses.

Topic and Procedure	Examples	— You Try It
Solving equations involving perimeter, p. 184 $$P = 2L + 2W$$ W ⎤ L	The length of a rectangle is double the width. If the perimeter is 36 meters, find the width. $L = 2W$ W $\begin{aligned} P &= 2L + 2W &&\text{Write the formula.} \\ 36 &= 2(2W) + 2W &&\text{Replace } P \text{ with 36 and } L \text{ with } 2W. \\ 36 &= 4W + 2W &&\text{Simplify.} \\ 36 &= 6W &&\text{Combine like terms.} \\ \frac{36}{6} &= \frac{6W}{6} &&\text{Divide each side by 6.} \\ 6 &= W &&\text{The width is 6 meters.} \end{aligned}$	**5.** The length of a rectangle is four times the width. If the perimeter is 70 feet, find the width. $L = 4W$ W
Solving equations involving the area of a rectangle, pp. 185–186 $$A = LW$$ W L	Find the width of a rectangle with length = 5 m and $A = 45\text{ m}^2$. $\begin{aligned} A &= LW &&\text{Write the formula.} \\ 45 &= 5W &&\text{Replace } A \text{ with 45 and } L \text{ with 5.} \\ \frac{45}{5} &= \frac{5W}{5} &&\text{Solve for } W. \\ 9 &= W &&\text{The width is 9 meters.} \end{aligned}$	**6.** Find the length of a rectangle with width 6 in. and $A = 42\text{ in}^2$.
Solving equations involving the area of a parallelogram, pp. 186–187 $A = bh$ b = length of base h = height	Find the height of a parallelogram with base = 12 m and area = 108 m^2. $\begin{aligned} A &= bh \\ 108 &= 12h \\ \frac{108}{12} &= \frac{12h}{12} \\ 9 &= h &&\text{The height is 9 meters.} \end{aligned}$	**7.** Find the base of a parallelogram with height = 15 m and area = 180 m^2.
Solving equations involving the volume of a rectangular solid (box), p. 188 $$V = LWH$$ H W L	Find the width of a box with dimensions $L = 8$ in., $H = 4$ in., and volume = 96 in.^3. $\begin{aligned} V &= L \cdot W \cdot H \\ 96 &= 8 \cdot W \cdot 4 \\ 96 &= 32W \\ \frac{96}{32} &= \frac{32W}{32} \\ 3 &= W &&\text{The width is 3 inches.} \end{aligned}$	**8.** Find the length of a box with dimensions $W = 5$ cm, $H = 7$ cm, and volume = 280 cm^3.
Product rule for exponents, p. 198 If bases are the same, we add exponents but keep the base unchanged. $$x^a \cdot x^b = x^{a+b}$$	Multiply and write the product in exponent form. **(a)** $x^3 \cdot x^6 = x^{3+6} = x^9$ **(b)** $3 \cdot 3^4 = 3^1 \cdot 3^4 = 3^{1+4} = 3^5$ **(c)** $5^8 \cdot 2^2$ The product rule does not apply—the bases are not the same.	**9.** Multiply and write the product in exponent form. **(a)** $y^4 \cdot y^5$ **(b)** $6^3 \cdot 6$ **(c)** $4^2 \cdot 5^3$
Multiplying algebraic expressions with exponents, p. 199 To multiply expressions: **1.** Multiply the numerical coefficients. **2.** Use the product rule for exponents.	Multiply: **(a)** $(-8x^4)(5x^3) = (-8 \cdot 5)(x^4)(x^3) = -40x^7$ **(b)** $(4x^4)(3x^2) = (4 \cdot 3) \cdot (x^4 \cdot x^2) = 12x^6$	**10.** Multiply. $(7x^3)(-4x^2)$
The distributive property and exponent form, p. 201 We can use the distributive property to multiply algebraic expressions with exponents.	Multiply. $x^4(x^2 + 3)$ $$x^4 \cdot x^2 + x^4 \cdot 3 = x^6 + 3x^4$$	**11.** Multiply. $x^5(x^3 + 5)$

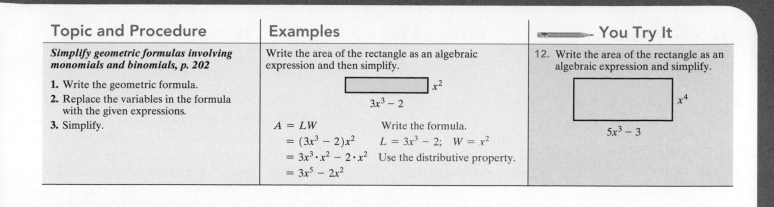

Topic and Procedure	Examples	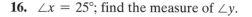 You Try It
Simplify geometric formulas involving monomials and binomials, p. 202 1. Write the geometric formula. 2. Replace the variables in the formula with the given expressions. 3. Simplify.	Write the area of the rectangle as an algebraic expression and then simplify. x^2 $3x^3 - 2$ $A = LW$ Write the formula. $= (3x^3 - 2)x^2$ $L = 3x^3 - 2$; $W = x^2$ $= 3x^3 \cdot x^2 - 2 \cdot x^2$ Use the distributive property. $= 3x^5 - 2x^2$	12. Write the area of the rectangle as an algebraic expression and simplify. x^4 $5x^3 - 3$

Chapter 3 Review Problems

Vocabulary

Write the definition for each word.

1. (3.1) Adjacent angles: _____
2. (3.1) Supplementary angles: _____
3. (3.3) Parallel lines: _____
4. (3.4) Polynomials: _____
5. (3.4) Numerical coefficient: _____
6. (3.4) Monomial: _____
7. (3.4) Binomial: _____
8. (3.4) Trinomial: _____

Section 3.1

Solve and check your solution.

9. $x - 15 = 12$

10. $x + 3 = -7$

11. $-8 + 1 = y + 4 - 16$

12. $6 - 13 + y = -4 + 1$

13. $4(2 - 6) = x - 2$

14. $4x - 3x - 6 = 6$

▲ *Find the measure of the unknown angle for each pair of supplementary angles.*

15. $\angle b = 81°$; find the measure of $\angle a$.

16. $\angle x = 25°$; find the measure of $\angle y$.

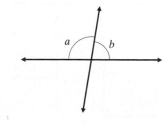

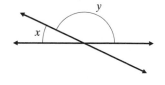

▲ 17. **(a)** Translate into symbols. Angle x measures 22° more than angle y.

     (b) Find the measure of $\angle y$ if the measure of $\angle x = 101°$.

Section 3.2

Solve and check your solution.

18. $4y = 48$

19. $7y = -63$

20. $6a = -42$

21. $7(3x) = 42$

22. $3(y \cdot 4) = 24$

23. $6(x \cdot 3) = \dfrac{-36}{2}$

24. $\dfrac{-48}{2} = 2(3 \cdot x)$ **25.** $2(5x) = 70$ **26.** $7(2y) = 28$

27. $6x - 4x = 20$ **28.** $5x + 2x = 42$

▲ **29.** *Dimensions of a Room* The length (L) of a room is three times the width (W).

(a) Translate the statement into an equation.

(b) Find the width of the room if the length is 36 feet.

31. *Hourly Rate* Sonia charges $\$x$ per hour as a consultant. She worked on a consulting job for 7 hours on Tuesday, 8 hours on Wednesday, and 8 hours on Thursday. If her total earnings for the three days were $1150, what is the hourly rate she charges?

30. *Blue vs. White Cars* There are twice as many white cars (W) on the road as blue cars (B).

(a) Translate the statement into an equation.

(b) Find the number of blue cars if there are 200 white cars on the road.

32. *Miles Jogged* On Friday Leo jogged x miles for his workout. On Saturday he jogged twice the number of miles he did on Friday. If Leo jogged a total of 12 miles during the two-day workout, how many miles did he jog on Friday?

Section 3.3

▲ **33.** The perimeter of a triangle is the sum of the lengths of the sides. Find the unknown side of the following triangle.

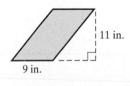

$P = 28$ in.

▲ **34.** The length of a rectangle is three times the width. If the perimeter of the rectangle is 32 feet, find the width.

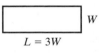

▲ **35.** Find the area of the square garden.

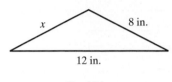

▲ **36.** Find the area of the rectangular tablecloth.

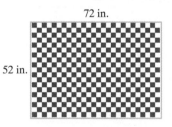

▲ **37.** Find the area of a parallelogram with base $= 9$ inches and height $= 11$ inches.

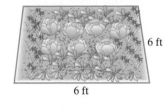

▲ **38.** Find the area of the shape made up of rectangles.

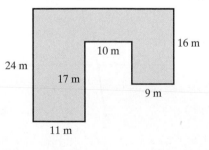

▲ **39.** Find the missing side of the rectangle.

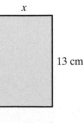

$A = 104$ cm^2

▲ **40.** Find the missing side of the parallelogram.

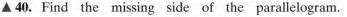
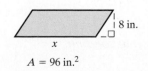

$A = 96$ in.^{2}

▲ *Find the unknown side of each rectangular solid.*

41.

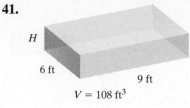

$V = 108 \text{ ft}^3$

42.

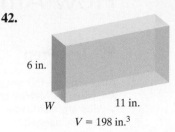

$V = 198 \text{ in.}^3$

▲ **43.** ***Water Needed to Fill Pool*** A rectangular children's wading pool has the following measurements: $L = 25$ feet, $W = 15$ feet, and $H = 2$ feet. How much water is needed to fill the pool to capacity?

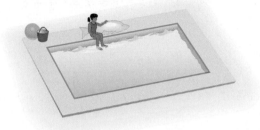

▲ **44.** ***Fluid Capacity of a Cube*** What is the maximum amount of fluid that a 5-inch cube can hold?

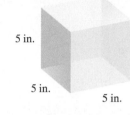

▲ **45.** ***Buying Weed Killer*** The McBeth's front yard measures 16 yards in length and 12 yards in width. A 1-pound container of weed killer sells for $4 and covers an area of 1000 square feet. How much will it cost the McBeths to purchase enough weed killer for their entire front yard?

▲ **46.** ***Carpeting an Office*** Douglas is purchasing carpet for his office, which measures 18 feet long and 12 feet wide. The carpet he plans to purchase is priced at $15 per square yard.
   (a) How many square yards of carpet must he purchase for the office?
   (b) How much will the carpet cost?

Section 3.4

Multiply and leave your answer in exponent form.

47. $2^3 \cdot 2^3$

48. $7^5 \cdot 7^4$

49. $a \cdot a^6$

50. $4^2 \cdot 3^3$

51. $(3y^4)(5y^4)$

52. $(4x^2)(3x^6)$

53. $(3a)(a^3)(7a^6)$

54. $(3z)(y^8)(4z^3)(2y^3)$

55. $(4x^5)(-3x^2)$

56. $(-7z^7)(5z^3)$

57. $(-3a^4)(9a^7)$

58. $(4y^5)(-5y^9)$

Identify each polynomial as a monomial, binomial, or trinomial.

59. $3x^2 - 1$

60. $4x^3$

61. $2xy^2 + 3x + 1$

Use the distributive property to simplify.

62. $x(x^2 + 3)$

63. $(x^3 - 4x)6x^2$

▲ **64.** Write the area of the rectangle as an algebraic expression and simplify.

$3x^2 + 6$, x^3

▲ **65.** Write the volume of the rectangular solid as an algebraic expression and simplify.

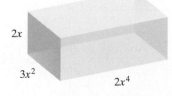

$2x$, $3x^2$, $2x^4$

How Am I Doing? Chapter 3 Test

After you take this test read through the Math Coach on pages 214–215. Math Coach videos are available via MyMathLab and YouTube. Step-by-step test solutions on the Chapter Test Prep Videos are also available via MyMathLab and YouTube. (Search "BlairTobeyPrealgebra" and click on "Channels.")

Solve each equation and check your solution.

1. $x - 2 = -8$ **2.** $y + 3 = 72$

3. $9 - 15 = a - 3$ **MC 4.** $3x - 2x - 7 = -1 + 6$

5. $12 = 5x - 2x$ **6.** $-3y = 42$

MC 7. $2(4x) = -72$ **8.** $\dfrac{-16}{2} = 2 + y$

Find the measure of the unknown angle for the supplementary angles.

9. $\angle x = 75°$; find the measure of $\angle y$.

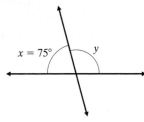

10. (a) Translate into symbols. Angle x measures $10°$ more than angle y.
  (b) Find the measure of $\angle y$ if the measure of $\angle x = 95°$.

11. The number of female students (F) in an English class is triple the number of male students (M) in the class.
  (a) Translate the statement into an equation.
  (b) If there are 21 female students in the class, find the number of male students in the class.

12. The length of a rectangle is triple the width. If the perimeter of the rectangle is 48 yards, find the width.

$L = 3W$, W

13. Find the area of a rectangle with length 5 ft and width 3 ft.

14. Find the area of a parallelogram with a base $= 6$ inches and height $= 12$ inches.

Answer column

1. _____
2. _____
3. _____
4. _____
5. _____
6. _____
7. _____
8. _____
9. _____
10. (a) _____
  (b) _____
11. (a) _____
  (b) _____
12. _____
13. _____
14. _____

15. Find the area of the following shape made of rectangles.

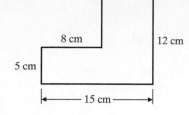

16. Find the unknown side.

$V = 288 \text{ in.}^3$

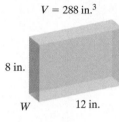

8 in.

W 12 in.

17. Write the area of the following rectangle as an algebraic expression and then simplify.

x^2

$4x^3 + x^2$

Solve.

18. The blueprints for an office building show that the length of the entryway is 9 feet and the width is 6 feet. How many square yards of tile must be purchased for the entryway?

19. A contractor's crew must dig up and haul away enough dirt to leave a hole 30 feet wide, 40 feet long, and 10 feet deep. How much dirt will they need to haul away?

Multiply. Leave your answer in exponent form.

20. $y^3 \cdot y^2$ **21.** $z \cdot z^3$ **22.** $2^3 \cdot 3^2$

23. $(5x)(x^3)(x^4)$ ℳℭ **24.** $(-8x^2)(-9x^4)$

Simplify.

25. $5y(y^4 + 8)$ ℳℭ **26.** $(3x^2 - 5x)6x^3$

15. _____ ☐

16. _____ ☐

17. _____ ☐

18. _____ ☐

19. _____ ☐

20. _____ ☐

21. _____ ☐

22. _____ ☐

23. _____ ☐

24. _____ ☐

25. _____ ☐

26. _____ ☐

Total Correct: ☐

MATH COACH

Mastering the skills you need to do well on the test.

Students often make the same types of errors when they do the Chapter 3 Test. Here are some helpful hints to keep you from making these common errors on test problems.

Solving Equations Using the Addition Principle of Equality—
Problem 4   Solve. $3x - 2x - 7 = -1 + 6$

> **Helpful Hint** When you solve equations, consider the following.
> - Did you simplify each side before you began the process to solve?
> - Did you use the correct principle?
> - Did you remember that whatever you do on one side of the equation, you must do on the other side?
> - Did you take the time to double-check your calculations and + and − signs?

Look at your work for Problem 4. Examine your steps.

Did you simplify each side of the equation as your **first step**?

Yes ____   No ____

If you answered No, stop and complete this step.

After simplifying, did you obtain the equation $x - 7 = 5$?

Yes ____   No ____

If you answered No, consider how to identify and combine like terms. Notice that there are like terms to combine on both sides of the equation. Be careful to avoid sign errors.

Did you use the correct principle and add 7 to both sides of the equation?

Yes ____   No ____

If you answered No, go back and complete this step.

If you answered Problem 4 incorrectly, go back and rework the problem using these suggestions.

Solving Equations Using the Division Principle of Equality—Problem 7   Solve. $2(4x) = -72$

> **Helpful Hint** When you solve equations, consider the following.
> - Did you simplify each side before you began the process to solve?
> - Did you use the correct principle?
> - Did you remember that whatever you do on one side of the equation, you must do on the other side?
> - Did you take the time to double-check your calculations and + and − signs?

Look at your work for Problem 7. Examine your steps.

Did you simplify the left side of equation as your **first step**?

Yes ____   No ____

If you answered No, stop and complete this step.

After simplifying, did you obtain the equation $8x = -72$?

Yes ____   No ____

If you answered No, go back and multiply $2(4x)$ again.

Did you use the correct principle and divide both sides of the equation by 8?

Yes ____   No ____

If you answered No, stop and complete this step. Be careful when working with + and − signs.

Now go back and rework the problem using these suggestions.

Need help? Watch the MATH COACH videos in MyMathLab® or on You Tube™.

214

Multiplying Algebraic Expressions with Exponents—Problem 24

Multiply. Leave your answer in exponent form. $(-8x^2)(-9x^4)$

> **Helpful Hint** First multiply the numerical coefficients. Then apply the product rule for exponents by adding the exponents of x.

Did you multiply the numerical coefficients -8 and -9?

Yes _____ No _____

If you answered No, go back and complete this step. Be careful to avoid sign errors.

Did you add the exponents $2 + 4$?

Yes _____ No _____

If you answered No, consider how the product rule for exponents works again. Notice that the two variables being multiplied have the same base but different exponents. Using the product rule, we can add the two exponents.

If you answered Problem 24 incorrectly, go back and rework the problem using these suggestions.

Using the Distributive Property to Multiply a Monomial and a Binomial—Problem 26

Simplify. $(3x^2 - 5x)6x^3$

> **Helpful Hint** To help with accuracy, you may find it easier to use the commutative property of multiplication to rewrite the problem with the monomial on the left side of the parentheses. Remember to consider the following:
> - If a variable does not have an exponent, then the exponent is understood to be 1.
> - Multiply both terms inside the parentheses by the monomial.

Look at your work for Problem 26. Examine your steps.

Did you remember to multiply both terms by $6x^3$?

Yes _____ No _____

If you answered No, then stop now and complete these calculations.

Did you multiply $6x^3$ times $3x^2$ to get $18x^5$?

Yes _____ No _____

If you answered No, remember to multiply the numerical coefficients 6 and 3 first and then use the product rule to multiply x^3 by x^2.

Did you multiply $6x^3$ times $-5x$ and get $-30x^3$?

If you answered Yes, you forgot that the x in $-5x$ has an exponent of 1. Stop now and apply the product rule again.

Now go back and rework the problem using these suggestions.

Need more help? Look for section examples marked with $\mathbb{M}\mathbb{C}$ to review.

215

Cumulative Test for Chapters 1–3

1. _____

2. _____

3. (a) _____

(b) _____

4. _____

5. _____

6. _____

7. _____

8. (a) _____

(b) _____

9. _____

10. _____

11. _____

12. _____

13. _____

14. _____

15. _____

16. _____

This test provides a comprehensive review of the key objectives for Chapters 1–3.

1. There are 5280 feet in a mile. Round 5280 to the nearest hundred.

2. Subtract. $18,700 - 896$

3. Video Time had income of $167,350 in 2009. The expenses for that year were $86,000.
   (a) How much profit did Video Time make?
   (b) If the 2 owners divided the profits equally, how much money did each owner receive?

Simplify.

4. $6(7x)(2)$

5. $3(y \cdot 8)$

Multiply.

6. $(209)(67)$

Divide.

7. $2844 \div 14$

Translate the English statement into an equation.

8. (a) Double some number equals twenty-eight.
   (b) The sum of a number and 9 equals 16.

Replace the each ? with the symbol <, >, or =.

9. $|-17| \ ? \ |-2|$

Perform the operations indicated.

10. $5 + (-6)$

11. $-10 - 8$

12. $-7 - 6 - 4 - 8$

13. $(-18) \div (-9)$

14. Evaluate. $(-2)^5$

15. Multiply. $(-2)(-1)(4)(3)(-2)$

16. Simplify. $-4 + 15 \div 5(-3)^2 - 1$

17. *Combine like terms.*

    (a) $3mn - 7mn + 4m$

    (b) $5 + 9y + 3 + y$

18. Evaluate. $x - 2y^3 + 1$ for $x = -1$ and $y = -3$

19. Simplify. $-2(x - 3)$

20. $-36 + y = -2$ **21.** $2x - x + 5 = -6$ **22.** $\dfrac{120}{10} = -3y$

23. Find the perimeter of a rectangle with $L = 3$ in. and $W = 2$ in.

24. Find the unknown side of the rectangular solid.

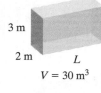

3 m

2 m L

$V = 30 \text{ m}^3$

25. $\angle a$ and $\angle b$ are supplementary angles. Find the measure of $\angle a$ if the measure of $\angle b = 105°$.

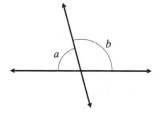

26. The number of children (C) at an amusement park is double the number of adults (A).

    (a) Translate the statement into an equation.

    (b) If there are 140 children in the park, find the number of adults.

Multiply.

27. $(-3x^2)(4x^6)$

28. $x^3(2x^2 + 5)$

17. (a) _____

(b) _____

18. _____

19. _____

20. _____

21. _____

22. _____

23. _____

24. _____

25. _____

26. (a) _____

(b) _____

27. _____

28. _____

CHAPTER 4

Fractions, Ratio, and Proportion

4.1 Factoring Whole Numbers

The set of whole numbers includes many special types of numbers, such as *prime numbers* and *composite numbers*. In this section you will learn the difference between prime and composite numbers and see how to write a whole number as a product of prime factors.

① Using the Divisibility Tests

A number x is said to be **divisible** by a number y if y divides x exactly, without a remainder (a remainder equal to 0). We say "25 is divisible by 5" because there is *no remainder* when we divide 25 by 5 ($25 \div 5 = 5$). We say that "26 is *not* divisible by 5" because there is a *remainder* when we divide 26 by 5: $26 \div 5 = 5$ R1.

Some students find the following rules helpful when deciding if a number is divisible by 2, 3, and/or 5.

DIVISIBILITY TESTS

1. A number is divisible by 2 if it is even. This means that the last digit is 0, 2, 4, 6, or 8.
2. A number is divisible by 3 if the sum of its digits is divisible by 3.
3. A number is divisible by 5 if its last digit is 0 or 5.

EXAMPLE 1 Determine if the number is divisible by 2, 3, and/or 5.

(a) 234   **(b)** 910   **(c)** 711   **(d)** 38,910

Solution

(a) 234	2 and 3	Divisible by 2 because 234 is even and by 3 since $2 + 3 + 4 = 9$ and 9 is divisible by 3.
(b) 910	5 and 2	Divisible by 5 because the last digit is 0 and by 2 since 910 is even.
(c) 711	3	Divisible by 3 because the sum of the digits is divisible by 3.
(d) 38,910	2, 3, 5	Divisible by 2 because 38,910 is even, by 3 since the sum of the digits is divisible by 3, and by 5 since the last digit is 0.

Student Practice 1 Determine if the number is divisible by 2, 3, and/or 5.

(a) 975   **(b)** 122   **(c)** 420   **(d)** 11,121

② Identifying Prime and Composite Numbers

A **prime number** is a whole number greater than 1 that is divisible only by itself and 1.

The number 5 is *prime* since it is divisible only by the numbers 5 and 1.
The number 6 is *not prime* since it is divisible by 2 and 3, in addition to 6 and 1.

A **composite number** is a whole number greater than 1 that can be divided by whole numbers other than itself and 1.

Student Learning Objectives

After studying this section, you will be able to:

① Use the divisibility tests.

② Identify prime and composite numbers.

③ Find the prime factors of whole numbers.

NOTE TO STUDENT: Fully worked-out solutions to all of the Student Practice problems can be found at the back of the text starting at page SP-1.

Any whole number (except 0 and 1) that is not prime is composite. The numbers 0 and 1 are neither prime nor composite numbers.

The number 10 is *composite* since it is divisible by 2 and 5 as well as by 10 and 1.

Composite numbers
↓   ↓
0, 1 2, 3 4 5 6
↑   ↑ ↖ ↑   ↑
Neither   Prime numbers
prime nor composite

EXAMPLE 2 State whether each number is prime, composite, or neither.

1, 4, 7, 11, 14, 15, 17, 22, 27, 31, 120

Solution 1 is neither prime nor composite.
4, 14, 15, 22, 27, and 120 are composite.
7, 11, 17, and 31 are prime.

Student Practice 2 State whether each number is prime, composite, or neither. 0, 3, 9, 13, 16, 19, 23, 32, 37, 41, 50

The first few prime numbers are 2, 3, 5, 7, 11, 13, 17, 19, 23, 29, ...

---

③ Finding the Prime Factors of Whole Numbers

Factors are numbers that are multiplied together. In the product $4 \cdot 6 = 24$, 4 and 6 are called factors of 24.

$24 = 4 \cdot 6$ These factors are *not* prime numbers.
↑ ↑
factor factor

Prime factors are factors that are prime. To write a number as a product of prime factors, we must break the multiplication down until each factor is prime. Thus, 24 written as a product of prime factors is

$$24 = 4 \cdot 6$$
$$24 = 2 \cdot 2 \cdot 2 \cdot 3 \quad \text{These factors are prime numbers.}$$

EXAMPLE 3 Express as a product of prime factors.

(a) 9   **(b)** 20

Solution

(a) $9 = 3 \cdot 3$ or 3^2

(b) $20 = 2 \cdot 2 \cdot 5$ or $2^2 \cdot 5$ $20 = 4 \cdot 5$ is *not correct* because 4 is not a prime number.

Note that when a number has duplicate factors such as $20 = 2 \cdot 2 \cdot 5$, we often express the factors in terms of *powers of prime factors:* $2^2 \cdot 5$.

Student Practice 3 Express as a product of prime factors.

(a) 14   **(b)** 27

When you divide two whole numbers and get a remainder of 0, both the divisor and the quotient are factors. Thus *we can divide to find prime factors*.

We use the division $15 \div 3 = 5$ to find the prime factors of 15.

Division Problem	*Related Multiplication*

$$\begin{array}{r} 5 \rightarrow \text{quotient} \\ 3\overline{)15} \\ \uparrow \\ \text{divisor} \end{array}$$

$$3 \cdot 5 = 15$$
$$\uparrow \quad \uparrow$$
Divisor and quotient are factors.

The following process uses repeated division to find prime factors. We often refer to this method as using a **division ladder** to find prime factors.

PROCEDURE TO FIND PRIME FACTORS USING A DIVISION LADDER

1. Determine if the original number is divisible by a *prime number*. If so, divide and find the quotient.
2. Divide the quotient by *prime numbers* until the final quotient is a prime number.
3. Write the divisors and the *final quotient* as a product of prime factors.

EXAMPLE 4   Express 28 as a product of prime factors.

Solution   Since 28 is even, it is divisible by the prime number 2. We start the division ladder by dividing 28 by 2.

Step 1 $2\overline{)28}$ with quotient 14 The quotient 14 is *not* a prime number.

We must continue to divide until the quotient is a prime number.

Step 2 $2\overline{)14}$ with quotient 7 The quotient 7 is a prime number.

This quotient is a prime number. Thus all the factors are prime. We are finished dividing. This process is simplified if we write the divisions as follows, placing step 1 on the bottom and moving up the ladder as we divide.

$$\begin{array}{r} 7 \\ \textbf{Step 2 } 2\overline{)14} \\ \uparrow \textbf{Step 1 } 2\overline{)28} \end{array}$$

Now we write all the divisors and the quotient as a product of prime factors.

$$28 = 2 \cdot 2 \cdot 7 \quad \text{or} \quad 2^2 \cdot 7$$

Student Practice 4   Express 50 as a product of prime factors.

CAUTION: It is important to note that all the divisors and the final quotient must be prime numbers to ensure that all factors are prime.

Mᴄ **EXAMPLE 5** Express 60 as a product of prime factors and check your answer.

Solution We must divide by prime numbers to ensure that all factors are prime. Since 60 is even, we know that we can divide 60 by 2. (*Note:* 0 is an even number.)

$$ \underline{5} \text{5 is prime, so we are finished dividing.}$$

Step 3 $3\overline{)15}$

Step 2 $2\overline{)30}$

Step 1 $2\overline{)60}$ We start by dividing by 2.

$$60 = 2 \cdot 2 \cdot 3 \cdot 5 \quad \text{or} \quad 2^2 \cdot 3 \cdot 5$$

We can check our answer by multiplying the prime factors.
    Check:

$$60 \stackrel{?}{=} 2^2 \cdot 3 \cdot 5$$
$$60 \stackrel{?}{=} 4 \cdot 15$$
$$60 = 60 \;\checkmark \quad \text{The answer checks.}$$

Student Practice 5 Express 96 as a product of prime factors and check your answer.

Understanding the Concept

The Various Division Ladders In Example 5, if we had started the division process with either 3 or 5 instead of 2, the result would have been the same. In fact, it does not matter what prime number is used to start the division—the results will be equivalent. Why? To illustrate, let's compare the following results.

	$\underline{5}$	$\underline{2}$	$\underline{2}$
Step 3 $3\overline{)15}$	$5\overline{)10}$	$2\overline{)4}$
Step 2 $2\overline{)30}$	$2\overline{)20}$	$3\overline{)12}$
Step 1 $2\overline{)60}$	$3\overline{)60}$	$5\overline{)60}$

$$60 = 2 \cdot 2 \cdot 3 \cdot 5 \quad 60 = 3 \cdot 2 \cdot 5 \cdot 2 \quad 60 = 5 \cdot 3 \cdot 2 \cdot 2$$
$$60 = 2^2 \cdot 3 \cdot 5$$

EXAMPLE 6 Express 210 as a product of prime factors.

Solution From the divisibility rules we know that 210 is divisible by 2, 3, and 5. We can start with 5.

$$ \underline{7} \text{7 is prime, so we are finished dividing.}$$

Step 3 $2\overline{)14}$

Step 2 $3\overline{)42}$

Step 1 $5\overline{)210}$

$$210 = 5 \cdot 3 \cdot 2 \cdot 7 = 2 \cdot 3 \cdot 5 \cdot 7$$

Note that we wrote all factors in ascending order since this is standard notation.

Student Practice 6 Express 315 as a product of prime factors.

We can also use a **factor tree** to find prime factors. This method uses the related multiplication instead of division to find the factors. Let's see how we would have factored the number 210 from Example 6 using a factor tree.

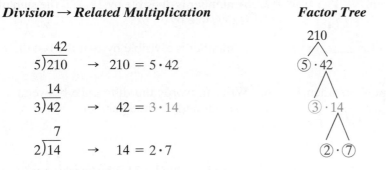

Division → Related Multiplication   *Factor Tree*

We write 210 as a product of circled numbers: $210 = 2 \cdot 3 \cdot 5 \cdot 7$

PROCEDURE TO BUILD A FACTOR TREE TO FIND PRIME FACTORS

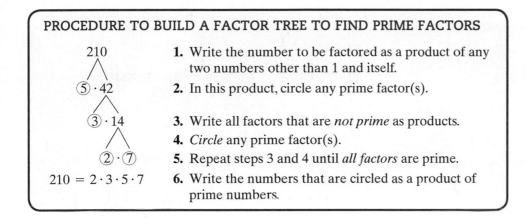

1. Write the number to be factored as a product of any two numbers other than 1 and itself.

2. In this product, circle any prime factor(s).

3. Write all factors that are *not prime* as products.
4. *Circle* any prime factor(s).
5. Repeat steps 3 and 4 until *all factors* are prime.
6. Write the numbers that are circled as a product of prime numbers.

Note that when we use the factor tree we can break the number into a product of any two numbers. These numbers do not need to be prime. However, we must continue to factor each number in the tree until we get all prime numbers.

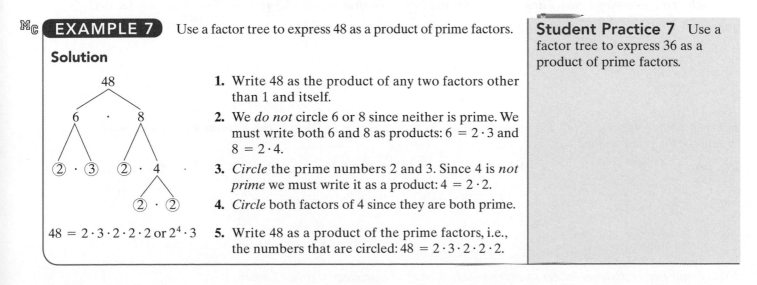

EXAMPLE 7   Use a factor tree to express 48 as a product of prime factors.

Solution

1. Write 48 as the product of any two factors other than 1 and itself.

2. We *do not* circle 6 or 8 since neither is prime. We must write both 6 and 8 as products: $6 = 2 \cdot 3$ and $8 = 2 \cdot 4$.

3. *Circle* the prime numbers 2 and 3. Since 4 is *not prime* we must write it as a product: $4 = 2 \cdot 2$.

4. *Circle* both factors of 4 since they are both prime.

$48 = 2 \cdot 3 \cdot 2 \cdot 2 \cdot 2$ or $2^4 \cdot 3$ 5. Write 48 as a product of the prime factors, i.e., the numbers that are circled: $48 = 2 \cdot 3 \cdot 2 \cdot 2 \cdot 2$.

Student Practice 7   Use a factor tree to express 36 as a product of prime factors.

Verbal and Writing Skills, Exercises 1–10

Fill in the blanks.

1. A number is divisible by _____ if it is even.

2. A number is divisible by _____ if the sum of the digits is divisible by 3.

3. A number is divisible by 5 if the last digit is _____ or _____.

4. A number is divisible by 10 if the last digit is _____.

5. Explain why you must divide by prime numbers when you use a division ladder to find prime factors.

6. Write in words the difference between a composite number and a prime number.

7. Is 165 divisible by 2? Why or why not?

8. Is 185 divisible by 2? Why or why not?

9. Is 232 divisible by 3? Why or why not?

10. Is 276 divisible by 3? Why or why not?

Determine if each number is divisible by 2, 3, and/or 5.

11. 102

12. 822

13. 705

14. 645

15. 330

16. 270

17. 22,971

18. 822,440

State whether each number is prime, composite, or neither.

19. 0, 9, 1, 17, 40, 8, 15, 22

20. 1, 42, 7, 12, 30, 6, 13, 31

21. Write 8 as a product of the following.
   (a) Any two factors
   (b) Prime factors

22. Write 20 as a product of the following.
   (a) Any two factors
   (b) Prime factors

Fill in the missing factors so that the number is expressed as a product of prime factors.

23. $28 = 2 \cdot \square \cdot 7 = 2^{\square} \cdot 7$

24. $27 = 3 \cdot 3 \cdot \square = 3^{\square}$

25. $75 = 3 \cdot \square \cdot 5 = 3 \cdot 5^{\square}$

26. $45 = 3 \cdot \square \cdot 5 = 3^{\square} \cdot 5$

For each of the following: **(a)** *Fill in the division ladder to complete each problem.* **(b)** *State the prime factors.*

27. (a) $\square$
   $2\overline{)10}$
   $\square\overline{)30}$
   $5\overline{)150}$
   (b) _____

28. (a) $\square$
   $3\overline{)15}$
   $\square\overline{)30}$
   $3\overline{)90}$
   (b) _____

For each of the following: **(a)** *Fill in the factor tree to complete each problem.* **(b)** *Express the original number as a product of the prime factors.*

29. (a) 220

(b) _____

30. (a) 140

(b) _____

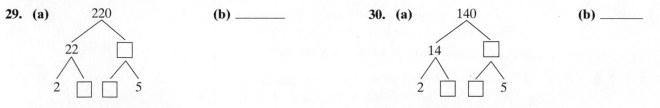

Use a factor tree or division ladder to express each number as a product of prime factors.

31. 15

32. 4

33. 20

34. 18

35. 24

36. 32

37. 70   **38.** 42   **39.** 64

40. 56   **41.** 80   **42.** 36

43. 75   **44.** 81   **45.** 45

46. 55   **47.** 99   **48.** 63

49. 300   **50.** 200   **51.** 110

52. 155   **53.** 136   **54.** 126

55. 90   **56.** 175   **57.** 225

One Step Further   *The calculator can be a useful tool for finding prime factors when 2, 3, and/or 5 are not factors. You can quickly check for factors by dividing by the prime numbers 7, 11, 13, 17, 19 or higher if needed. Use your calculator to help you express each number as a product of prime factors.*

58. 91   **59.** 1309   **60.** 561   **61.** 2737

62. We can also use a calculator to check that the prime factors in an answer are correct. Use your calculator to multiply the prime factors in exercises 58–61, and verify that your answers are correct.

To Think About

63. Write a five-digit number that is divisible by the number 3 and the number 5.

64. Write a six-digit number that is divisible by the number 2 and the number 3.

When we square a whole number or fraction, the number we obtain is called a **perfect square.** For example, 9 is a perfect square since when we square the number 3 we obtain 9. In exercises 65 and 66 we will look at a relationship that exists between the sequence of perfect squares and the sequence of positive odd numbers. This relationship was investigated in the thirteenth century by an Italian mathematician named Leonardo of Pisa, also known as Fibonacci.

65. **(a)** List the first 6 odd numbers.

(b) Complete the green table as follows. In the third row, write the first 3 odd numbers, in the fourth row, write the first 4 odd numbers, and so on.

(c) In the blue boxes, write the sum of the odd numbers.

(d) In the orange boxes, write each sum in exponent form.

(e) Describe the pattern observed with the set of numbers in the blue boxes.

(f) Describe the pattern observed with the set of numbers in the orange boxes.

(g) Based on the observations made in exercises 65(e) and 65(f), fill in the last four black boxes.

(h) Observe the pattern above and complete the following.

The sum of the first 12 positive odd numbers equals: $\boxed{}^2$

The sum of the first 20 positive odd numbers equals: $\boxed{}^2$

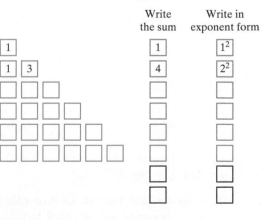

Complete exercise 65, then use those results to complete the following.

66. **(a)** Write each of the following sums in exponent form.

The sum of the first 2 positive odd numbers equals: $\boxed{}^2$

The sum of the first 3 positive odd numbers equals: $\boxed{}^2$

The sum of the first 4 positive odd numbers equals: $\boxed{}^2$

The sum of the first 5 positive odd numbers equals: $\boxed{}^2$

The sum of the first 6 positive odd numbers equals: $\boxed{}^2$

(b) Without actually adding the first 30 positive odd numbers, describe how you could find the sum.

Cumulative Review *Simplify.*

67. **[3.4.2]** $(2x^2)(5x^3y)$

68. **[3.4.3]** $7y^2 + 2y^2$

69. **[2.6.1]** $5x + 3x + 2$

70. **[3.4.2]** $(5x)(3x)(2)$

Quick Quiz 4.1

1. Determine if 504 is divisible by 2, 3, and/or 5.

2. Is the number 57 prime, composite, or neither?

3. Express 315 as the product of prime factors.

4. **Concept Check** Delroy is having a hard time factoring the number 318.

   (a) Explain in words how Delroy can determine one factor of 318.

   (b) State this factor.

👣🖊 STEPS TO SUCCESS Taking Notes in Class

During a lecture you are *listening, seeing, reading,* and *writing* all at the same time. Although an important part of mathematics studying is taking notes, you must also focus on what the instructor is saying so you can follow the logic.

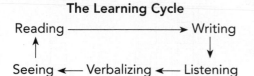

The Learning Cycle

Reading ⟶ Writing

Seeing ⟵ Verbalizing ⟵ Listening

Making it personal: *Place a* ✓ *in the blank as you complete each of the following.*

___ **1. Preview the lesson before class.** The important concepts will be more familiar and you will have a better idea of what to write down.

___ **2.** Ask the teacher if you can **record the lecture.**

___ **3. Write down only the important ideas and examples.** Make sure that you're also listening to the lecture so you can follow the reasoning.

___ **4. Include helpful hints and references to the text.** You will be amazed how easily you forget these if you don't write them down.

___ **5. Review your notes, clarifying whatever appears vague.** Do this on the *same day* sometime after class, so that you will be able to recall material from the lecture that you did not include in your notes.

You may find that you learn more by *seeing* and *listening* than by the other modes of learning. You may prefer to take fewer notes, focusing more on what the instructor is saying. This is fine as long as you write a brief outline during class of the instructor's lecture. Then immediately after class, you should add the details.

4.2 Understanding Fractions

① Understanding Fractions

Whole numbers are used to describe whole objects or entire quantities. However, often we have to represent parts of whole quantities. In mathematics, **fractions** are a set of numbers used to describe parts of whole quantities. The *whole* can be an object (a pizza), or, just as often, the whole can be a set of things (pieces of pizza) that we choose to consider as a whole unit.

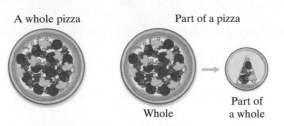

A whole pizza   Part of a pizza

Whole   Part of a whole

The vertical bar in the margin is divided into eight equal parts. The five shaded parts represent *part* of the whole bar and are represented by the fraction 5/8 or $\frac{5}{8}$.

In the fraction $\frac{5}{8}$ the number 5 is called the **numerator** and the number 8 is called the **denominator.**

$\frac{5}{8}$ $\rightarrow$ The *numerator* specifies how many of these parts are being considered.
    $\rightarrow$ The *denominator* specifies the total number of equal parts.

> The *denominator* of a fraction shows the number of equal parts in the whole.
> The *numerator* shows the number of parts being talked about or being used.

When you say "$\frac{3}{4}$ of a pizza has been eaten," what you are indicating is that if the pizza was cut into four equal pieces, then three of the four pieces have been eaten.

Remember that the *numerator* is always the *top number* and the *denominator* is always the *bottom number*.

EXAMPLE 1   Use a fraction to represent the shaded part of each object.

(a)   (b)   (c)

Solution

(a) One out of four parts is shaded, or $\frac{1}{4}$.

(b) Seven out of nine parts are shaded, or $\frac{7}{9}$.

(c) Three out of three parts are shaded, or $\frac{3}{3} = 1$.

Student Practice 1   Use a fraction to represent the shaded part of each object.

(a)   (b)   (c)

Student Learning Objectives

After studying this section, you will be able to:

① Understand fractions.

② Identify proper fractions, improper fractions, and mixed numbers.

③ Change improper fractions to mixed numbers.

④ Change mixed numbers to improper fractions.

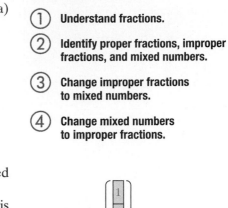

5 shaded parts or $\frac{5}{8}$

8 equal parts

3 pieces eaten

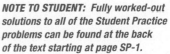

4 pieces in one pizza

NOTE TO STUDENT: Fully worked-out solutions to all of the Student Practice problems can be found at the back of the text starting at page SP-1.

Note that in Example 1c when 3 out of 3 parts are shaded, we have 1 whole amount. This illustrates a division fact, which states that any nonzero number divided by itself is 1.

What does it mean when a fraction has a 0 in the numerator, or when there is a 0 in the denominator? Let's look at the following situations to see how to answer these questions.

If $20 is divided equally among 5 people, each person receives $4.

$$20 \div 5 = 4 \quad \text{or} \quad \frac{20}{5} = 4$$

If $0 is divided equally among 5 people, each person receives $0.

$$0 \div 5 = 0 \quad \text{or} \quad \frac{0}{5} = 0$$

If $20 is divided among 0 people?—This cannot be done.

$$20 \div 0 \quad \text{and} \quad \frac{20}{0} \text{ are undefined.}$$

We summarize as follows.

DIVISION PROBLEMS INVOLVING THE NUMBERS ONE AND ZERO

1. Any nonzero number divided by itself is 1.

$$3 \div 3 = \frac{3}{3} = 1$$

2. Zero can never be the divisor in a division problem.

$$5 \div 0 = \frac{5}{0} \quad \text{We say division by 0 is } \textbf{undefined.}$$

3. Zero may be divided by any number except zero; the result is always zero.

$$0 \div 4 = \frac{0}{4} = 0 \quad \text{In other words, } \textit{any fraction with 0 in the numerator} \\ \textit{and a nonzero denominator equals 0.}$$

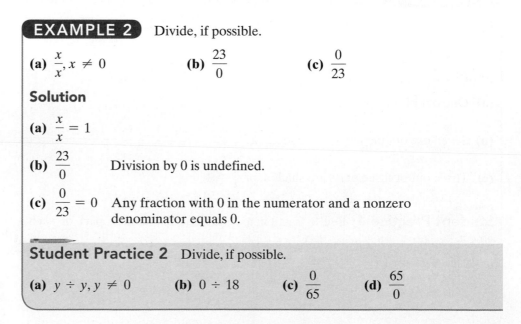

EXAMPLE 2   Divide, if possible.

(a) $\dfrac{x}{x}, x \neq 0$ **(b)** $\dfrac{23}{0}$ **(c)** $\dfrac{0}{23}$

Solution

(a) $\dfrac{x}{x} = 1$

(b) $\dfrac{23}{0}$ Division by 0 is undefined.

(c) $\dfrac{0}{23} = 0$ Any fraction with 0 in the numerator and a nonzero denominator equals 0.

Student Practice 2   Divide, if possible.

(a) $y \div y, y \neq 0$ **(b)** $0 \div 18$ **(c)** $\dfrac{0}{65}$ **(d)** $\dfrac{65}{0}$

Circle graphs are especially helpful for showing the relationship of parts to a whole. The circle represents the whole, and the pie-shaped pieces represent parts of the whole.

EXAMPLE 3   The approximate number of inches of rain that falls during selected periods of one year in Seattle, Washington, is shown by the circle graph.

(a) What fractional part of the total yearly rainfall occurs from October to December?

(b) What fractional part of the total yearly rainfall does *not* occur from July to September?

Solution   First we must find the total rainfall for 1 year.

$$13 \text{ in.} + 15 \text{ in.} + 4 \text{ in.} + 5 \text{ in.} = 37 \text{ in.}$$

(a) From October to December there were 15 inches of rain out of a total of 37 inches.

$$\frac{15}{37} \quad \text{Fractional part of rainfall that does occur.}$$

(b) From July to September there were 4 inches of rain out of a total of 37 inches.

$$37 \text{ in.} - 4 \text{ in.} = 33 \text{ in.} \quad \text{Rainfall that does } \textit{not} \text{ occur from July to September.}$$

$$\frac{33}{37} \quad \text{Fractional part of rainfall that does } \textit{not} \text{ occur.}$$

Student Practice 3   Use the circle graph in Example 3 to answer the following.

(a) What fractional part of the total yearly rainfall occurs from January to March?

(b) What fractional part of the total yearly rainfall does *not* occur from April to June?

② Identifying Proper Fractions, Improper Fractions, and Mixed Numbers

We have names for different kinds of fractions. A **proper fraction** is used to describe a quantity *less than 1*. If the numerator is *less* than the denominator, the fraction is a proper fraction. The fraction $\frac{3}{4}$ is a proper fraction.

An **improper fraction** is used to describe a quantity *greater than or equal to 1*. If the numerator is *greater than or equal to* the denominator, the fraction is an improper fraction. The fraction $\frac{7}{4}$ is an improper fraction because the numerator is larger than the denominator. Since $\frac{4}{4}$ describes a quantity equal to 1, it is also an improper fraction.

A **mixed number** is the sum of a whole number greater than zero and a proper fraction, and is used to describe a quantity greater than 1. An improper fraction can also be written as a mixed number.

The following chart will help you visualize the different fractions and their names.

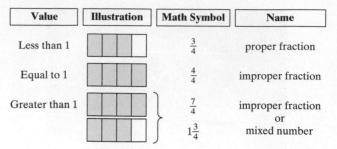

Value	Illustration	Math Symbol	Name
Less than 1		$\frac{3}{4}$	proper fraction
Equal to 1		$\frac{4}{4}$	improper fraction
Greater than 1		$\frac{7}{4}$ $1\frac{3}{4}$	improper fraction or mixed number

The last figure can be represented by 1 whole added to $\frac{3}{4}$ of a whole, or $1 + \frac{3}{4}$. This is written $1\frac{3}{4}$ (we do not write the addition symbol) and is a mixed number. Thus the improper fraction $\frac{7}{4}$ is equivalent to the mixed number $1\frac{3}{4}$.

EXAMPLE 4   Identify each as a proper fraction, an improper fraction, or a mixed number.

(a) $\dfrac{9}{8}$ **(b)** $\dfrac{8}{9}$ **(c)** $7\dfrac{3}{4}$ **(d)** $\dfrac{3}{3}$

Solution

(a) $\dfrac{9}{8}$ Improper fraction   The numerator is larger than the denominator.

(b) $\dfrac{8}{9}$ Proper fraction   The numerator is less than the denominator.

(c) $7\dfrac{3}{4}$ Mixed number   A whole number is added to a proper fraction.

(d) $\dfrac{3}{3}$ Improper fraction   The numerator is equal to the denominator.

Student Practice 4   Identify each as a proper fraction, an improper fraction, or a mixed number.

(a) $\dfrac{6}{5}$ **(b)** $\dfrac{x}{x}, x \neq 0$ **(c)** $6\dfrac{2}{9}$ **(d)** $\dfrac{1}{2}$

③ Changing Improper Fractions to Mixed Numbers

We can see the relationship between an improper fraction and a mixed number by drawing pictures. For example, the fraction $\frac{13}{5}$ is represented by the shaded boxes below. Note that each box is divided into 5 pieces. Thus 13 pieces, each of which is $\frac{1}{5}$ of a box, are shaded. We see that $\frac{13}{5} = 2\frac{3}{5}$ since 2 whole boxes and $\frac{3}{5}$ of another box are shaded.

$$\frac{13}{5} = 2\frac{3}{5}$$

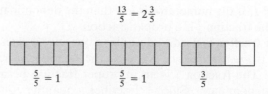

$$\frac{5}{5} = 1 \qquad \frac{5}{5} = 1 \qquad \frac{3}{5}$$

Since $\frac{13}{5} = 13 \div 5$, we can divide to change an improper fraction to a mixed number.

$$\frac{13}{5} = 13 \div 5 = 5 \overline{)13} \begin{array}{c} 2\,R3 \\ \hline \underline{10} \\ 3 \end{array} = 2\frac{3}{5}$$

We summarize this procedure as follows.

PROCEDURE TO CHANGE AN IMPROPER FRACTION TO A MIXED NUMBER

1. Divide the numerator by the denominator.
2. The quotient is the whole number part of the mixed number.
3. The remainder from the division will be the numerator of the fraction. The denominator of the fraction remains unchanged.

A mixed number is in the following form: quotient $\dfrac{\text{remainder}}{\text{denominator}}$.

EXAMPLE 5 Write $\dfrac{19}{7}$ as a mixed number.

Solution The answer is in the form quotient $\dfrac{\text{remainder}}{\text{denominator}}$.

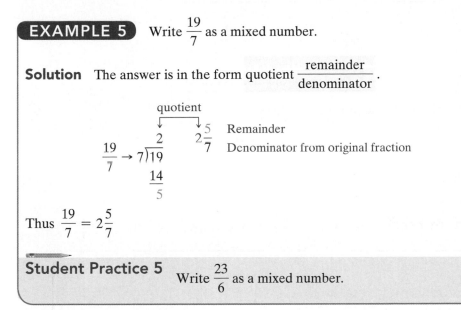

Thus $\dfrac{19}{7} = 2\dfrac{5}{7}$

Student Practice 5 Write $\dfrac{23}{6}$ as a mixed number.

---

④ Changing Mixed Numbers to Improper Fractions

We often change mixed numbers to improper fractions since improper fractions are usually easier to work with. To illustrate the method used to change a mixed number to an improper fraction, we can draw a picture. For example, suppose that we want to write $2\frac{1}{4}$ as an improper fraction. We can illustrate the quantity $2\frac{1}{4}$ as two whole boxes that are divided into fourths plus $\frac{1}{4}$ of a third box. Now if we count the shaded pieces, we see that we have $\frac{9}{4}$.

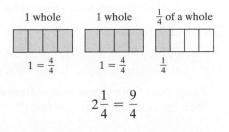

$$2\frac{1}{4} = \frac{9}{4}$$

We can perform the same procedure using the following method.

> **PROCEDURE TO CHANGE A MIXED NUMBER TO AN IMPROPER FRACTION**
> 1. Multiply the whole number by the denominator of the fraction.
> 2. Add this product to the numerator. The result is the numerator of the improper fraction. The denominator does not change.
>
> Improper fraction: $\dfrac{(\text{denominator} \cdot \text{whole number}) + \text{numerator}}{\text{denominator}}$

EXAMPLE 6   Change $6\frac{1}{2}$ to an improper fraction.

Solution

Improper fraction: $\dfrac{(\text{denominator} \cdot \text{whole number}) + \text{numerator}}{\text{denominator}}$

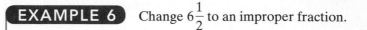

| Multiply the whole number by the denominator. | Add the numerator to the product. |

$$6\frac{1}{2} = \frac{(2 \cdot 6) + 1}{2} = \frac{12 + 1}{2} = \frac{13}{2}$$

The denominator does not change.

We can also write the process as follows.

$$6\frac{1}{2} = \frac{2 \text{ times } 6 \text{ plus } 1}{2} = \frac{13}{2}$$

Student Practice 6   Change $8\frac{2}{3}$ to an improper fraction.

STEPS TO SUCCESS Problems with Accuracy

Strive for accuracy. Mistakes are often made because of accidental errors rather than a lack of understanding. Such mistakes are frustrating. A simple arithmetic or sign error can lead to an incorrect answer.

Complete the following steps and discover how you can reduce careless errors!

Making it personal: *Place a* ✔ *in the blank as you complete each of the following.*

_____ **1.** Work carefully and take your time.

_____ **2.** Concentrate on the problem. Sometimes problems become mechanical, and your mind begins to wander. You become careless and make a mistake.

_____ **3.** Check your problem. Be sure that you copied it correctly from the book.

_____ **4.** Check each step of the problem for sign errors as well as computation errors. Does your answer make sense?

_____ **5.** Becoming aware of where you make errors will help you avoid making them again. Make a note of the types of errors you make most often.

List of errors:

Keep practicing new skills. Remember the old saying that "practice makes perfect."

Verbal and Writing Skills, Exercises 1–4 *Fill in the blanks and boxes.*

1. Fractions are a set of numbers used to describe _____ of a whole quantity. In the fraction $\frac{2}{3}$, the 2 is called the _____ and the 3 is called the _____.

2. When you say $\frac{3}{7}$ of a cake has been eaten, you are indicating that if the cake were cut into _____ equal pieces, then _____ of these pieces were eaten.

3. To change the mixed number $6\frac{2}{3}$ to an improper fraction, we multiply _____ × _____, and then add _____ to the product. We write the improper fraction as $\frac{\square}{\square}$.

4. To change the improper fraction $\frac{31}{4}$ to a mixed number, we divide _____ by _____ and get a quotient of _____. We write the mixed number as _____ $\frac{\square}{\square}$.

Use a fraction to represent the shaded area in each figure.

5.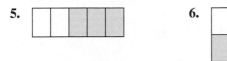

6.

7.

8.

Divide, if possible. Assume all variables in the denominators are nonzero.

9. $\frac{7}{0}$ 10. $\frac{4}{0}$ 11. $\frac{0}{z}$ 12. $\frac{0}{x}$

13. $\frac{44}{44}$ 14. $\frac{a}{a}$ 15. $\frac{0}{9}$ 16. $\frac{0}{4}$

17. $\frac{8}{0}$ 18. $\frac{6}{0}$ 19. $\frac{y}{y}$ 20. $\frac{66}{66}$

Applications, Exercises 21–32

21. **Baseball** A baseball player had 5 base hits in 12 times at bat. Write the fractional part of the times at bat that describes the number of times the player had a base hit.

22. **Archery** An archer hit the target 7 times out of 19 shots. Write the fractional part of the total shots that describes the number of times the archer hit the target.

23. **Men vs. Women** There are 57 women and 37 men in the hospital cafeteria. What fractional part of the customers in the hospital cafeteria consists of men?

24. **Male vs. Female** There are 83 men and 67 women working for a small corporation. What fractional part of the employees consists of men?

25. **Dance Production Class** There are 26 dancers in the dance production class at a high school. Nine of the dancers are juniors. Write the fraction that describes the dancers who are *not* juniors.

26. **Salad Bar Choices** At a salad bar, there are 29 different items to choose from. Eleven of the choices contain pasta. Write the fraction that describes the choices that do *not* contain pasta.

27. **Baseball** The city baseball team won 21 of the 32 games they played. What fractional part of the games did the team lose?

28. **Manufacturing Defects** Lois inspected 137 computer chips and found that 11 were defective. What fractional part of the chips were *not* defective?

Payroll Deductions *The deductions from Arnold's paycheck are shown on the circle graph. Use this circle graph to answer exercises 29–32.*

29. What fractional part of the deductions is for state or federal income tax?

30. What fractional part of the deductions is for insurance or social security?

233

31. What fractional part of the deductions is *not* for insurance?

32. What fractional part of the deductions is *not* for state tax?

Identify each as a proper fraction, an improper fraction, or a mixed number.

33. $\dfrac{11}{9}$

34. $4\dfrac{1}{2}$

35. $\dfrac{8}{8}$

36. $\dfrac{11}{4}$

37. $\dfrac{5}{6}$

38. $\dfrac{z}{z}, z \neq 0$

39. $7\dfrac{1}{9}$

40. $\dfrac{3}{7}$

Change each improper fraction to a mixed number or a whole number.

41. $\dfrac{15}{8}$

42. $\dfrac{19}{4}$

43. $\dfrac{48}{5}$

44. $\dfrac{94}{7}$

45. $\dfrac{41}{2}$

46. $\dfrac{25}{3}$

47. $\dfrac{32}{5}$

48. $\dfrac{79}{7}$

49. $\dfrac{47}{5}$

50. $\dfrac{54}{7}$

51. $\dfrac{33}{33}$

52. $\dfrac{44}{44}$

Change each mixed number to an improper fraction.

53. $8\dfrac{3}{7}$

54. $7\dfrac{2}{3}$

55. $24\dfrac{1}{4}$

56. $10\dfrac{1}{9}$

57. $15\dfrac{2}{3}$

58. $13\dfrac{3}{5}$

59. $33\dfrac{1}{3}$

60. $41\dfrac{1}{2}$

61. $8\dfrac{9}{10}$

62. $3\dfrac{1}{50}$

63. $8\dfrac{7}{15}$

64. $5\dfrac{19}{20}$

Cumulative Review *Simplify.*

65. **[2.4.1]** $(-5)(-8)$

66. **[2.4.1]** $(-7)(9)$

67. **[4.1.3]** Express 63 as a product of prime factors.

68. **[4.1.3]** Express 54 as a product of prime factors.

Quick Quiz 4.2

1. Divide, if possible. $\dfrac{0}{9}$

2. Change $4\dfrac{3}{5}$ to an improper fraction.

3. The local humane society has 8 cats, 5 dogs, and 2 rabbits that need homes. What fractional part of these pets are cats?

4. **Concept Check** Explain in words how you change the mixed number $2\dfrac{3}{4}$ to an improper fraction.

① Finding Equivalent Fractions

There are many fractions that name the same quantity. For example, in the following illustration the two pieces of wood are the same length.

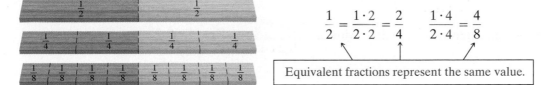

One piece of wood cut into
2 equal parts
One piece of wood cut into
4 equal parts

As you can see, 1 of 2 pieces of wood represents the same quantity as 2 of 4 pieces of wood. We say that $\frac{1}{2} = \frac{2}{4}$ and call these fractions **equivalent fractions.**

There are many other fractions that represent the same value as $\frac{1}{2}$ and thus are equivalent to $\frac{1}{2}$.

$$\frac{1}{2} = \frac{2}{4} = \frac{4}{8} = \frac{8}{16} \text{ are equivalent fractions}$$

Equivalent fractions *look different* but have the *same value* because they represent the same quantity. Now, how can we find an equivalent fraction without using a picture or diagram? Observe the following pattern.

$$\frac{1}{2} = \frac{1 \cdot 2}{2 \cdot 2} = \frac{2}{4} \qquad \frac{1 \cdot 4}{2 \cdot 4} = \frac{4}{8}$$

> Equivalent fractions represent the same value.

We see that when we multiply the numerator and denominator of $\frac{1}{2}$ by the same nonzero number, we get an equivalent fraction. Why is this true? Recall that the identity property of 1 states that if we multiply a number by 1, the value of that number does not change. Since $1 = \frac{1}{1} = \frac{2}{2} = \frac{3}{3}$, we have the following.

> We factor the
> the fraction.

$$\frac{1}{2} = \frac{1 \cdot 2}{2 \cdot 2} = \frac{1}{2} \cdot \frac{2}{2} = \frac{1}{2} \cdot 1 = \frac{1}{2}$$

> Multiplying the numerator and denominator by
> the same number is the same as multiplying by 1.

We see that when we multiply the numerator and denominator of a fraction by the same nonzero number, we are actually *multiplying by a form of 1*, and therefore we do not change the value of the fraction. Now we can state a procedure to find an equivalent fraction.

PROCEDURE TO FIND EQUIVALENT FRACTIONS

To find an equivalent fraction, we multiply *both* the numerator and denominator by the *same* nonzero number.

$$\frac{a}{b} = \frac{a \cdot c}{b \cdot c} \quad \text{where } b \text{ and } c \text{ are not } 0$$

Student Learning Objectives

After studying this section, you will be able to:

① Find equivalent fractions.

② Reduce fractions to lowest terms.

③ Simplify fractions containing variables.

④ Solve applied problems involving fractions.

When there is a variable in the denominator, we will assume that the variable does not equal zero, since division by zero is not defined.

EXAMPLE 1

(a) Multiply the numerator and denominator of $\frac{3}{4}$ by 2 to find an equivalent fraction.

(b) Multiply the numerator and denominator of $\frac{3}{4}$ by $3x$ to find an equivalent fraction.

Solution

(a) $\frac{3}{4} = \frac{3 \cdot 2}{4 \cdot 2} = \frac{6}{8}$

(b) $\frac{3}{4} = \frac{3 \cdot 3x}{4 \cdot 3x} = \frac{9x}{12x}$

NOTE TO STUDENT: Fully worked-out solutions to all of the Student Practice problems can be found at the back of the text starting at page SP-1.

Student Practice 1

(a) Multiply the numerator and denominator of $\frac{4}{7}$ by 3 to find an equivalent fraction.

(b) Multiply the numerator and denominator of $\frac{4}{7}$ by $5x$ to find an equivalent fraction.

---

^{Mc} **EXAMPLE 2** Write $\frac{3}{4}$ as an equivalent fraction with a denominator of $16x$.

Solution

$$\frac{3}{4} = \frac{\square}{16x}$$

$$\frac{3 \cdot ?}{4 \cdot ?} = \frac{\square}{16x} \quad \text{4 times what number equals } 16x? \quad 4x$$

Since we must multiply the denominator by $4x$ to obtain $16x$, we must also multiply the numerator by $4x$.

$$\frac{3 \cdot 4x}{4 \cdot 4x} \quad \frac{12x}{16x}$$

Student Practice 2 Write $\frac{2}{9}$ as an equivalent fraction with a denominator of $36x$.

---

② Reducing Fractions to Lowest Terms

A fraction is considered to be **reduced to lowest terms** (or written in simplest form) if the numerator and denominator have no common factors other than 1.

reduced   not reduced
↓   ↓

$$\frac{2}{3} = \frac{2 \cdot 1}{3 \cdot 1} \qquad \frac{4}{6} = \frac{2 \cdot 2}{3 \cdot 2}$$

no common factor   common factor
other than 1   of 2

As we saw earlier, when we multiply a fraction by any form of 1 such as $\frac{2}{2}$, or $\frac{3}{3}$, or $\frac{4}{4}$, and so on, we obtain an equivalent fraction. However, this equivalent fraction is *not reduced*. We reverse this process to reduce a fraction.

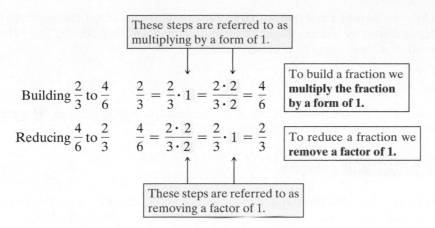

Building $\dfrac{2}{3}$ to $\dfrac{4}{6}$ $\dfrac{2}{3} = \dfrac{2}{3} \cdot 1 = \dfrac{2 \cdot 2}{3 \cdot 2} = \dfrac{4}{6}$

Reducing $\dfrac{4}{6}$ to $\dfrac{2}{3}$ $\dfrac{4}{6} = \dfrac{2 \cdot 2}{3 \cdot 2} = \dfrac{2}{3} \cdot 1 = \dfrac{2}{3}$

These steps are referred to as multiplying by a form of 1.

To build a fraction we **multiply the fraction by a form of 1.**

To reduce a fraction we **remove a factor of 1.**

These steps are referred to as removing a factor of 1.

As you can see, reducing fractions requires that you recognize common factors of the numerator and denominator. One way to do this is to write the numerator and denominator each as a product of prime factors. Then use slashes / to indicate that you are rewriting common factors as the equivalent fraction $\frac{1}{1}$ so that you can remove a factor of 1.

$$\dfrac{4}{6} = \dfrac{2 \cdot \overset{1}{\cancel{2}}}{3 \cdot \underset{1}{\cancel{2}}} = \dfrac{2}{3}$$ Slashes indicate that we are rewriting $\dfrac{2}{2}$ as $\dfrac{1}{1}$.

Since $\dfrac{1}{1} = 1$, we have $\dfrac{2}{3} \cdot 1 = \dfrac{2}{3}$.

PROCEDURE TO REDUCE A FRACTION TO LOWEST TERMS

1. Write the numerator and denominator of the fraction each as a product of prime factors. For example, $\dfrac{4}{6} = \dfrac{2 \cdot 2}{3 \cdot 2}$.

2. Any factor that appears in both the numerator and denominator is a common factor. Rewrite the common factors as the equivalent fraction $\dfrac{1}{1}$ and multiply. $\dfrac{2 \cdot 2}{3 \cdot 2} = \dfrac{2 \cdot \overset{1}{\cancel{2}}}{3 \cdot \underset{1}{\cancel{2}}} = \dfrac{2}{3}$

Note that you could reduce $\frac{4}{6}$ by dividing the numerator and denominator by 2. This method *does not work* as well as the one presented when variables are involved.

It is important that you reduce fractions using the method of removing factors of 1 in order to prepare for algebra.

The direction *simplify* or *reduce* means to reduce to lowest terms.

EXAMPLE 3   Simplify. $\dfrac{15}{35}$

Solution   Since the last digit of both 15 and 35 is 5, we know that both numbers are divisible by 5. $\dfrac{15}{35} = \dfrac{3 \cdot 5}{7 \cdot 5}$ First, write the numerator and denominator as products of prime numbers.

$$= \dfrac{3 \cdot \overset{1}{\cancel{5}}}{7 \cdot \underset{1}{\cancel{5}}}$$ Then, rewrite $\dfrac{5}{5}$ as the equivalent fraction $\dfrac{1}{1}$.

$$= \dfrac{3 \cdot 1}{7 \cdot 1} = \dfrac{3}{7}$$

Student Practice 3   Simplify. $\dfrac{18}{54}$

We can reduce a fraction with a negative number in either the numerator or the denominator by writing the negative sign in *front* of the fraction. The value of the fraction will not change, as illustrated below.

$$\frac{-15}{5} = -15 \div 5 = -3 \qquad \frac{15}{-5} = 15 \div (-5) = -3 \qquad -\frac{15}{5} = -(15 \div 5) = -(3) = -3$$

When we reduce a fraction, we write the negative sign in front of the fraction. We do not include it as part of the prime factors.

EXAMPLE 4 Simplify. $\dfrac{-72}{48}$

Solution If we recognize that the numerator and denominator have common factors that are *not* prime, we can use these factors to reduce the fraction.

$$\frac{-72}{48} = -\frac{72}{48} \qquad \text{Write the negative sign in front of the fraction.}$$

$$= -\frac{\overset{1}{\cancel{8}} \cdot 9}{\underset{1}{\cancel{8}} \cdot 6} \qquad \text{8 is a common factor of 72 and 48.}$$

$$= -\frac{3 \cdot \overset{1}{\cancel{3}}}{2 \cdot \underset{1}{\cancel{3}}} \qquad \text{Write the remaining factors as products of primes.}$$

$$= -\frac{3}{2} \qquad \text{Simplify.}$$

CAUTION: In Example 4 we used common factors (not prime factors) when we reduced the fraction: $\frac{-72}{48} = -\frac{8 \cdot 9}{8 \cdot 6}$. When we use common factors to reduce fractions, it is important that we rewrite any remaining factors as products of primes: $-\frac{9}{6} = -\frac{3 \cdot 3}{3 \cdot 2}$. This will ensure that the fraction is reduced to lowest terms.

Student Practice 4 Simplify. $\dfrac{60}{-36}$

③ Simplifying Fractions Containing Variables

We use the same process to simplify fractions containing variables. That is, we rewrite common factors in the numerator and denominator as the equivalent fraction $\frac{1}{1}$ and simplify.

EXAMPLE 5 Simplify. $\dfrac{150n^2}{200n}$

Solution Since 25 is a common factor of the numerator and denominator, we can write each as a product of 25 times some prime numbers.

$$\frac{150n^2}{200n} = \frac{25 \cdot 3 \cdot 2 \cdot n \cdot n}{25 \cdot 2 \cdot 2 \cdot 2 \cdot n} \qquad \begin{array}{l}\text{Write all other factors as products of prime numbers to}\\ \text{ensure that the fraction is reduced to lowest terms.}\end{array}$$

$$= \frac{\overset{1}{\cancel{25}} \cdot 3 \cdot \overset{1}{\cancel{2}} \cdot n \cdot \overset{1}{\cancel{n}}}{\underset{1}{\cancel{25}} \cdot 2 \cdot \underset{1}{\cancel{2}} \cdot 2 \cdot \underset{1}{\cancel{n}}}$$

$$= \frac{3 \cdot n}{2 \cdot 2}$$

$$\frac{150n^2}{200n} = \frac{3n}{4}$$

Note that we could have also used $50n$ as a common factor in both the numerator and denominator of $\dfrac{150n^2}{200n}$.

Student Practice 5 Simplify. $\dfrac{80x^2}{140x}$

CAUTION: Students sometimes apply slashes incorrectly as follows.

$$\frac{\cancel{3} + 4}{\cancel{3}} = 4 \quad \text{THIS IS WRONG!}$$

$$\frac{3 + 4}{3} = \frac{7}{3} \quad \text{THIS IS RIGHT!} \qquad \frac{\cancel{3} \cdot 4}{\cancel{3}} = \frac{4}{1} = 4 \quad \text{THIS IS RIGHT!}$$

We may not use slashes with addition or subtraction signs. We may use slashes only if we are multiplying factors.

④ Solving Applied Problems Involving Fractions

EXAMPLE 6 The yearly sales report in the margin shows the number and type of real estate sales made by Tri-Star Realty.

(a) What fractional part of the total sales were single-family homes?

(b) What fractional part of the total sales were *not* condominiums?

End of the Year Sales Report

Type of Sale	Total Sales
Condominium	14
Town home	21
Single-family home	45

Solution

(a) *Understand the problem.* We use the information on the chart to find the total sales. Then we write the fraction and simplify.

Solve and state the answer.

Condominium sales	+	Town home sales	+	Single-family home sales	=	Total sales
14	+	21	+	45	=	80

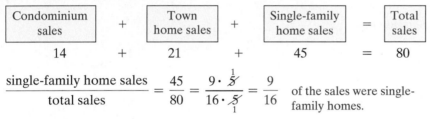

of the sales were single-family homes.

Note that we should always reduce fractions to lowest terms.

(b) *Understand the problem.* We refer to the chart and determine how many sales were *not* condominiums. Then we write the fraction and simplify.

Solve and state the answer.

Type of Sale	Total Sales
~~Condominium~~	~~14~~
Town home	21
Single-family home	45

We want the total sales that were *not* condominiums.

We find the sum: $21 + 45 = 66$.

$$\frac{\text{sales that were not condominiums}}{\text{total sales}} = \frac{66}{80} = \frac{33 \cdot \overset{1}{\cancel{2}}}{40 \cdot \underset{1}{\cancel{2}}} = \frac{33}{40}$$

of the sales were *not* condominiums.

Check:

(a) The fraction $\frac{9}{16}$ is a little more than $\frac{8}{16}$, which equals $\frac{1}{2}$. This means that approximately one-half of the total yearly sales were single-family homes. Since 45 sales is a little more than one-half of the total sales of 80, our answer is reasonable.

(b) The fraction $\frac{33}{40}$ represents a number close to the whole amount $\frac{40}{40}$. This means that most of the sales were *not* condominiums. Since 66 of the 80 sales were *not* condominiums, it is true that most of the sales were *not* condominiums, and our answer is reasonable.

Student Practice 6 Refer to Example 6 to answer the following. Write each fraction in simplest form.

(a) What fractional part of the sales were town homes?

(b) What fractional part of the sales were not single-family homes?

Verbal and Writing Skills, Exercises 1–2 *Fill in the blanks.*

1. To build an equivalent fraction, we _____ the numerator and denominator by the _____ number.

2. State three ways that we can write the fraction *negative two-thirds*.

$$\underline{\quad} \ \underline{\quad} \ \underline{\quad}$$

Fill in each box so that the fractions are equivalent.

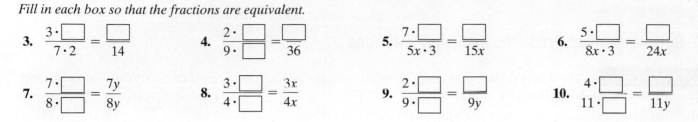

3. $\dfrac{3 \cdot \boxed{}}{7 \cdot 2} = \dfrac{\boxed{}}{14}$

4. $\dfrac{2 \cdot \boxed{}}{9 \cdot \boxed{}} = \dfrac{\boxed{}}{36}$

5. $\dfrac{7 \cdot \boxed{}}{5x \cdot 3} = \dfrac{\boxed{}}{15x}$

6. $\dfrac{5 \cdot \boxed{}}{8x \cdot 3} = \dfrac{\boxed{}}{24x}$

7. $\dfrac{7 \cdot \boxed{}}{8 \cdot \boxed{}} = \dfrac{7y}{8y}$

8. $\dfrac{3 \cdot \boxed{}}{4 \cdot \boxed{}} = \dfrac{3x}{4x}$

9. $\dfrac{2 \cdot \boxed{}}{9 \cdot \boxed{}} = \dfrac{\boxed{}}{9y}$

10. $\dfrac{4 \cdot \boxed{}}{11 \cdot \boxed{}} = \dfrac{\boxed{}}{11y}$

Multiply the numerator and denominator of each given fraction by the following numbers to find two different equivalent fractions: (a) 4 and (b) 5x. Assume $x \neq 0$.

11. $\dfrac{7}{9}$

12. $\dfrac{6}{7}$

13. $\dfrac{4}{11}$

14. $\dfrac{9}{13}$

Find an equivalent fraction with the given denominator.

15. $\dfrac{3}{8} = \dfrac{?}{32}$

16. $\dfrac{11}{15} = \dfrac{?}{60}$

17. $\dfrac{5}{6} = \dfrac{?}{30}$

18. $\dfrac{7}{8} = \dfrac{?}{40}$

19. $\dfrac{9}{13} = \dfrac{?}{39}$

20. $\dfrac{8}{11} = \dfrac{?}{44}$

21. $\dfrac{35}{40} = \dfrac{?}{80}$

22. $\dfrac{20}{25} = \dfrac{?}{50}$

23. $\dfrac{8}{9} = \dfrac{?}{9y}$

24. $\dfrac{5}{14} = \dfrac{?}{14n}$

25. $\dfrac{3}{7} = \dfrac{?}{28y}$

26. $\dfrac{3}{12} = \dfrac{?}{60y}$

27. $\dfrac{3}{6} = \dfrac{?}{18a}$

28. $\dfrac{7}{9} = \dfrac{?}{81x}$

29. $\dfrac{5}{7} = \dfrac{?}{21x}$

30. $\dfrac{6}{11} = \dfrac{?}{22x}$

Simplify.

31. $\dfrac{20}{25}$

32. $\dfrac{21}{28}$

33. $\dfrac{12}{16}$

34. $\dfrac{24}{30}$

35. $\dfrac{30}{36}$

36. $\dfrac{12}{32}$

37. $\dfrac{16}{28}$

38. $\dfrac{18}{27}$

39. $\dfrac{24}{36}$

40. $\dfrac{32}{64}$

41. $\dfrac{30}{85}$

42. $\dfrac{33}{55}$

43. $\dfrac{48}{56}$

44. $\dfrac{63}{81}$

45. $\dfrac{36}{72}$

46. $\dfrac{23}{46}$

47. $\dfrac{49}{35}$

48. $\dfrac{81}{72}$

49. $\dfrac{75}{60}$

50. $\dfrac{62}{54}$

Simplify.

51. (a) $\dfrac{-12}{18}$ (b) $\dfrac{12}{-18}$ (c) $-\dfrac{12}{18}$

52. (a) $\dfrac{-15}{25}$ (b) $\dfrac{15}{-25}$ (c) $-\dfrac{15}{25}$

53. $\dfrac{-15}{30}$

54. $\dfrac{-35}{40}$

55. $\dfrac{-42}{48}$

56. $\dfrac{-60}{70}$

57. $\dfrac{30}{-42}$

58. $\dfrac{25}{-60}$

59. $-\dfrac{16}{18}$

60. $-\dfrac{14}{18}$

Simplify. Since division by zero is undefined, assume that any variable in a denominator is nonzero.

61. $\dfrac{21a}{24a}$

62. $\dfrac{20n}{35n}$

63. $\dfrac{20y}{35y}$

64. $\dfrac{14x}{21x}$

65. $\dfrac{24xy}{42x}$

66. $\dfrac{20nx}{45n}$

67. $\dfrac{14y}{28xy}$

68. $\dfrac{12x}{36nx}$

69. $\dfrac{27x^2}{45x}$

70. $\dfrac{28x^2}{49x}$

71. $\dfrac{20y}{24y^2}$

72. $\dfrac{21y}{24y^2}$

73. $\dfrac{36n^2}{-42n}$

74. $\dfrac{64y^2}{-72y}$

75. $\dfrac{-35x}{45x^2}$

76. $\dfrac{-20y}{30y^2}$

Applications *Solve. Write each fraction in simplest form.*

77. *Incorrect Test Answers* Shawn answered 22 questions correctly on a test of 36 questions. What fractional part of the questions did he answer *incorrectly*?

78. *Floor Tiles* During an earthquake 18 of the 81 floor tiles in Hamza's family room were cracked. What fractional part of the tiles in the family room were *not* cracked?

Correct/Incorrect Test Answers *The chart shows the number of questions Alexsandra answered correctly on a three-part English test. Use this chart to answer exercises 79 and 80.*

79. What fractional part of the questions did Alexsandra answer correctly?

80. What fractional part of the questions did Alexsandra answer incorrectly?

English Test Results

	Number of Correct Answers	Number of Questions
Part 1	22	36
Part 2	15	20
Part 3	20	25

Shark Attacks *Use the bar graph to answer exercises 81–83. Use the divisibility tests to write each fraction in simplest form.*

The highest number of recorded unprovoked shark attacks on humans worldwide occurred in the year 2000. Over half of the attacks happened on offshore reefs or banks.

81. What fractional part of the total number of shark attacks happened in 2006?

82. What fractional part of the total number of shark attacks happened in 2000?

83. What fractional part of the total number of shark attacks happened in 2008 or 2009?

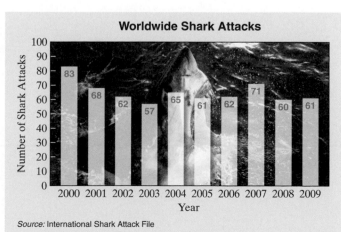

Source: International Shark Attack File

One Step Further *Simplify. Assume that any variable in a denominator is nonzero.*

84. $\dfrac{25x^2y^2z^4}{135x^3y}$

85. $\dfrac{40a^2b^2c^4}{88ab}$

86. $\dfrac{156ab^3}{144bc^4}$

87. $\dfrac{256xy^3}{300yz^5}$

To Think About

▲ **88.** *Garden Dimensions* Sean's rectangular garden is 84 inches by 48 inches. If he increases the length from 84 inches to 168 inches, what will the width have to be in order for the garden to have the same area?

▲ **89.** *Garden Enlargement* A rectangular garden has a length of 90 inches and a width of 80 inches. If the length is increased from 90 inches to 120 inches, what will the width have to be in order for the garden to have the same area?

Cumulative Review *Simplify.*

90. **[3.4.1]** $x^4 \cdot x^3 \cdot x^3$

91. **[3.4.1]** $2^4 \cdot 2^6$

92. **[3.4.2]** $(-3a)(2a^4)$

93. **[3.2.3]** *Peanuts and Cashews* The number of peanuts (P) in a mix is double the number of cashews (C).

   (a) Translate the statement into an equation.

   (b) If there are 34 peanuts in the mix, how many cashews are in the mix?

▲ **94.** **[3.3.4]**   *Comparing Room Sizes* Which has the greater area, a rectangular room that is 13 feet by 16 feet or a square room that is 15 feet on each side? Which has the greater perimeter?

Quick Quiz 4.3

1. Find an equivalent fraction with the given denominator. $\dfrac{3}{7} = \dfrac{?}{56}$

2. Simplify. $\dfrac{-24}{36}$

3. Simplify. $\dfrac{33a^2}{44a}$

4. **Concept Check** Can the fraction $\frac{105}{231}$ be reduced? Why or why not?

Did You Know...

That Different Investments Grow at Different Rates So You Should Rebalance Your Investments Annually?

INVESTING

Understanding the Problem:

Jason is just learning how to manage his investments. A year ago he opened up an investment account. Through his account, he can choose from a variety of mutual funds that represent different segments of the market. When he first opened his account, he met with a financial advisor to get help deciding which funds to invest in. Looking at Jason's long-term goals, his risk tolerance, and other factors, they came up with a plan for how to divide up his money in different funds. The amount of money he invested in the different mutual fund categories is displayed in the pie chart below.

Original Investments

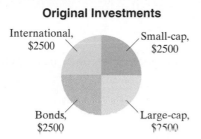

After a year, due to differences in performance, the value of some categories grew faster than others. The current distribution of his money is displayed in the pie chart below.

Current Investments

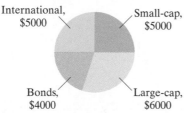

Making a Plan:

Jason's advisor told him that different funds may grow at different rates. Therefore, to keep his level of risk the same, Jason should rebalance his money annually to return to the original fractions they came up with. This is because some funds are riskier than others, and if a risky fund grows the most, then a greater fraction of his money will be at risk. By keeping the original fraction of his money invested in each category, he keeps the risk level he is comfortable with. Jason will need to sell shares in some categories and buy shares in others so that the fraction invested in each category is equal to what he started with originally.

Step 1: Jason must determine which categories no longer match his original investment plan.

Task 1: *What fractional part of the total original investments was the Large-cap? Small-cap? Bonds? International?*

Task 2: *What fractional part of the current investments is the Large-cap? Small-cap? Bonds? International?*

Task 3: *Which categories have changed?*

Step 2: For the categories that need to be adjusted, Jason must calculate the amount he should change, and whether he will be buying or selling for each particular fund.

Task 4: *What amount should Jason have in each category in order to maintain the original fractions?*

Task 5: *Complete step 2 for Jason.*

Finding a Solution:

Jason now knows what he needs to do to get back to his original investment plan. He can now contact his investment company and tell them which funds he wants to sell shares in and which funds he wants to buy shares in. Then he'll be back on track to reach his investment goals.

Applying the Situation to Your Life:

In your own investments, you will see how some investments perform better than others. One reason to rebalance your investments is to control risk. Some funds are riskier than others. That is, there is a greater chance of losing money. But when they are not losing money, riskier funds can make you a lot of money. They may grow to be an overly large fraction of your investments. Then, if the risky funds have a bad year, you could lose a lot of money. By selling shares in a risky fund when it is making money, you get two benefits: One, you are making a profit by selling high, and two, you are limiting the amount of money that is at risk if that fund has a bad year.

How Am I Doing? Sections 4.1–4.3

How are you doing with your homework assignments in Sections 4.1 to 4.3? Do you feel you have mastered the material so far? Do you understand the concepts you have covered? Before you go further in the textbook, take some time to do each of the following problems.

4.1

1. Determine if 312 is divisible by 2, 3, and/or 5.

2. Express 120 as a product of prime factors.

4.2

3. Divide, if possible. Assume all variables in the denominator are nonzero.

   (a) $\dfrac{0}{3}$ (b) $\dfrac{3}{0}$ (c) $\dfrac{a}{a}$ (d) $\dfrac{22}{22}$

4. Change $\frac{51}{7}$ to a mixed number.

5. Change to $5\frac{3}{4}$ an improper fraction.

6. There were 17 women and 19 men in a popular aerobics class. What fractional part of the class were men?

4.3

Find an equivalent fraction with the given denominator.

7. $\dfrac{3}{7} = \dfrac{?}{35}$

8. $\dfrac{2}{9} = \dfrac{?}{27y}$

Simplify.

9. $\dfrac{20}{-42}$

10. $\dfrac{25y}{45y^2}$

11. During a hurricane 7 out of 42 windows in a hotel were shattered. What fractional part of the windows were *not* shattered by the hurricane?

Now turn to page SA-7 for the answers to each of these problems. Each answer also includes a reference to the objective in which the problem is first taught. If you missed any of these problems, you should stop and review the Examples and Student Practice problems in the referenced objective. A little review now will help you master the material in the upcoming sections of the text.

1. _____

2. _____

3. (a) _____

  (b) _____

  (c) _____

  (d) _____

4. _____

5. _____

6. _____

7. _____

8. _____

9. _____

10. _____

11. _____

4.4 Simplifying Fractional Expressions with Exponents

① Using the Quotient Rule for Exponents

Frequently, we must divide variable expressions such as in $x^6 \div x^4$. We can rewrite the expression as the fraction $\dfrac{x^6}{x^4}$ and simplify using repeated multiplication.

$$\frac{x^6}{x^4} = \frac{x \cdot x \cdot x \cdot x \cdot x \cdot x}{x \cdot x \cdot x \cdot x} = \frac{\overset{1}{\cancel{x}} \cdot \overset{1}{\cancel{x}} \cdot \overset{1}{\cancel{x}} \cdot \overset{1}{\cancel{x}} \cdot x \cdot x}{\underset{1}{\cancel{x}} \cdot \underset{1}{\cancel{x}} \cdot \underset{1}{\cancel{x}} \cdot \underset{1}{\cancel{x}}} = \frac{x^2}{1} = x^2$$

When exponents are large, this process can be time consuming. Let's examine some divisions and look for a pattern to discover a division rule. Notice in the previous division that there are 6 factors in the numerator and 4 factors in the denominator. After we simplify, we have $6 - 4 = 2$ factors left in the numerator. Thus we can write

$$\frac{x^6}{x^4} = \frac{\overbrace{\cancel{x} \cdot \cancel{x} \cdot \cancel{x} \cdot \cancel{x} \cdot x \cdot x}^{6 \text{ factors}}}{\underbrace{\cancel{x} \cdot \cancel{x} \cdot \cancel{x} \cdot \cancel{x}}_{4 \text{ factors}}} = x^2 \quad 6 - 4 = 2 \text{ factors left in the } \textit{numerator.}$$

Let's consider another division problem.

$$\frac{2^3}{2^4} = \frac{2 \cdot 2 \cdot 2}{2 \cdot 2 \cdot 2 \cdot 2} = \frac{\overbrace{\cancel{2} \cdot \cancel{2} \cdot \cancel{2}}^{3 \text{ factors}}}{\underbrace{\cancel{2} \cdot \cancel{2} \cdot \cancel{2} \cdot 2}_{4 \text{ factors}}} = \frac{1}{2^1} \quad 4 - 3 = 1 \text{ factor left in the } \textit{denominator.}$$

We see that we *subtract exponents* to divide these expressions.
Let's summarize the results.

$$\frac{x^6}{x^4} = x^2 \qquad \text{Since there are more factors in the } \textit{numerator,} \text{ after we simplify we have } 6 - 4 = 2 \text{ factors left in the } \textit{numerator.}$$

$$\frac{2^3}{2^4} = \frac{1}{2} \qquad \text{Since there are more factors in the } \textit{denominator,} \text{ after we simplify we have } 4 - 3 = 1 \text{ factor left in the } \textit{denominator.}$$

> ### THE QUOTIENT RULE
>
> If the bases in the numerator and denominator of a fractional expression are the same and a and b are positive integers, then
>
> $$\frac{x^a}{x^b} = x^{a-b} \quad \text{Use this form if the } \textit{larger exponent} \text{ is in the } \textit{numerator} \text{ and } x \neq 0.$$
>
> $$\frac{x^a}{x^b} = \frac{1}{x^{b-a}} \quad \text{Use this form if the } \textit{larger exponent} \text{ is in the } \textit{denominator} \text{ and } x \neq 0.$$

Since division by zero is undefined, in all problems in this book we assume that the denominator of any variable expression is not zero.

EXAMPLE 1 Simplify. Leave your answer in exponent form.

(a) $\dfrac{n^9}{n^6}$

(b) $\dfrac{5^8}{5^9}$

(c) $\dfrac{2^7}{3^4}$

Solution

(a) $\dfrac{n^9}{n^6} = n^{9-6}$ There are more factors in the numerator.
The leftover factors are in the numerator.

$\dfrac{n^9}{n^6} = \dfrac{n^3}{1} = n^3 \longleftarrow$

(b) $\dfrac{5^8}{5^9} = \dfrac{1}{5^{9-8}}$ There are more factors in the denominator.

$\dfrac{5^8}{5^9} = \dfrac{1}{5^1}$ or $\dfrac{1}{5}$ The leftover factor is in the denominator.

(c) $\dfrac{2^7}{3^4}$ We cannot divide using the rule for exponents. The bases are not the same.

Student Practice 1 Simplify. Leave your answer in exponent form.

(a) $\dfrac{4^{11}}{4^7}$

(b) $\dfrac{6^9}{8^{14}}$

(c) $\dfrac{y^5}{y^9}$

NOTE TO STUDENT: *Fully worked-out solutions to all of the Student Practice problems can be found at the back of the text starting at page SP-1.*

TO THINK ABOUT: When Can We Use the Quotient Rule? Write the numerator and denominator of $\dfrac{2^7}{3^4}$ using repeated multiplication. Do you see why we cannot use the rule for exponents to simplify?

Let's see what happens when the expressions in the numerator and denominator are equal. For example, consider $\dfrac{5^2}{5^2}$. Since any number divided by itself is equal to 1, we know that we can rewrite 1 as $\dfrac{5^2}{5^2}$. Now, using the rules of exponents we have

$$1 = \frac{5^2}{5^2} = 5^0 \quad 5^{2-2} = 5^0 \quad \text{Thus, } 1 = 5^0$$

We can generalize that any number (except 0) can be raised to the zero power. The result is 1.

> For any nonzero number a, $a^0 = 1$. The expression 0^0 is not defined.

EXAMPLE 2 Simplify. $\dfrac{16x^6y^0}{20x^8}$

Solution

$\dfrac{16x^6y^0}{20x^8} = \dfrac{\overset{1}{\cancel{4}} \cdot 2 \cdot 2 \cdot x^6 \; y^0}{\underset{1}{\cancel{4}} \cdot 5 \cdot x^8}$ We factor 16 and 20.

$= \dfrac{2 \cdot 2 \cdot 1}{5 \cdot x^{8-6}}$ $\dfrac{x^6}{x^8} = \dfrac{1}{x^{8-6}}; \; y^0 = 1$

$= \dfrac{4}{5x^2}$ The leftover x factors are in the denominator.

Student Practice 2 Simplify.

$$\dfrac{25y^5x^0}{45y^8}$$

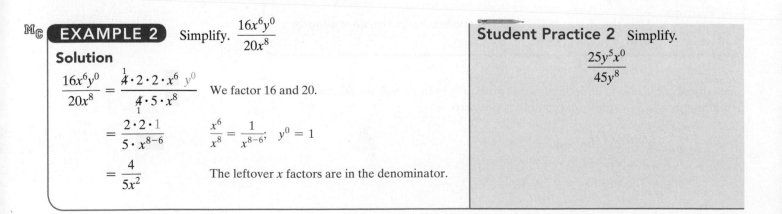

⬤ **Understanding the Concept**

Do I Add, Subtract, Multiply, or Divide Coefficients and Exponents?

What can we do if we forget algebraic rules? Do we add, subtract, or multiply exponents? If we start with a simple problem and then *think* about what it is we are actually trying to do, we can often determine the rules by observing our calculations.

1. $6x^5 + 3x^5 = ?$

We have six x^5's and add three more x^5's; we have nine x^5's.

$6x^5 + 3x^5 = 9x^5$

In this case, we *add* coefficients and the variable stays the same.

2. $(6x^2)(3x^4) = ?$

$$6 \cdot x \cdot x \cdot 3 \cdot x \cdot x \cdot x \cdot x = 6 \cdot 3 \cdot x \cdot x \cdot x \cdot x \cdot x \cdot x$$
$$= 18 \cdot x^6$$
$$(6x^2)(3x^4) = 18x^6$$

In this case, we *multiply* coefficients and then *add* exponents of like bases.

3. $\dfrac{6x^2}{3x} = ?$　　$\dfrac{6x^2}{3x} = \dfrac{2 \cdot 3 \cdot x \cdot x}{3 \cdot x} = 2x$

In this case, we *divide* coefficients and then *subtract* exponents of like bases.

② Raising a Power to a Power

If we have $(x^3)^4$, we say that we are *raising a power to a power*. A problem such as $(x^3)^4$ can done by first writing $(x^3)^4$ as a product and then simplifying.

$(x^3)^4$ means $x^3 \cdot x^3 \cdot x^3 \cdot x^3$　　By definition of raising a value to the fourth power

$\qquad = x^{3+3+3+3}$　　Use the product rule of exponents

$\qquad = x^{12}$　　Add exponents $3 + 3 + 3 + 3$ to simplify.

EXAMPLE 3　Write $(2^4)^2$ as a product and then simplify. Leave your answer in exponent form.

Solution　$(2^4)^2$ means $2^4 \cdot 2^4 = 2^{4+4} = 2^8$

Student Practice 3　Write $(4^2)^3$ as a product and then simplify. Leave your answer in exponent form.

Since repeated addition can be written as multiplication, we can simplify the calculations in Example 3 as follows.

$(2^4)^2 = 2^{4 \cdot 2} = 2^8$　　We multiply exponents since $4 + 4 = 4 \cdot 2$.

This leads to the following rule.

RAISING A POWER TO A POWER OR A PRODUCT TO A POWER

To raise a power to a power, keep the same base and multiply the exponents.

$$(x^a)^b = x^{ab}$$

To raise a product to a power, raise each factor to that power.

$$(xy)^b = x^b y^b$$

EXAMPLE 4   Use the rules for raising a power to a power or a product to a power to simplify. Leave your answer in exponent form.

(a) $(3^3)^3$ **(b)** $(x^2)^0$ **(c)** $(4x^4)^5$

Solution

(a) $(3^3)^3 = 3^{(3)(3)} = 3^9$ We multiply exponents.

The base does not change when raising a power to a power.

(b) $(x^2)^0 = x^{(2)(0)} = x^0 = 1$

(c) $(4x^4)^5 = (4^1 \cdot x^4)^5$ We write 4 as 4^1.

$\qquad\quad = 4^{(1)(5)} \cdot x^{(4)(5)}$ We raise each factor to the power 5.

$\qquad\quad = 4^5 \cdot x^{20}$ We multiply exponents.

$(4x^4)^5 = 4^5 x^{20}$

Student Practice 4   Use the rules for raising a power to a power or a product to a power to simplify. Leave your answer in exponent form.

(a) $(3^3)^4$ **(b)** $(n^0)^7$ **(c)** $(3y^4)^6$

Now we introduce a similar rule involving quotients that is very useful.

ADDITIONAL POWER RULE

If a fraction in parentheses is raised to a power, the parentheses indicate that the numerator and denominator are *each* raised to that power.

$$\left(\frac{x}{y}\right)^a = \frac{x^a}{y^a} \quad \text{if } y \neq 0$$

EXAMPLE 5   Simplify. $\left(\dfrac{2}{x}\right)^3$

Solution   We must remember to raise both the numerator and the denominator to the power.

$$\left(\frac{2}{x}\right)^3 = \left(\frac{2^1}{x^1}\right)^3 = \frac{2^{(1)(3)}}{x^{(1)(3)}} = \frac{2^3}{x^3} = \frac{8}{x^3}$$

Student Practice 5   Simplify. $\left(\dfrac{x}{3}\right)^3$

👣 STEPS TO SUCCESS Success in Mathematics

The Learning Cycle

Reading $\longrightarrow$ Writing

Seeing $\longleftarrow$ Verbalizing $\longleftarrow$ Listening

Mathematics is a building process, mastered one step at a time. The foundation of this process is composed of a few basic requirements. Those who are successful in mathematics realize the absolute necessity of building a study of mathematics on the firm foundation of the six minimum requirements in the following.

Making it personal: *Place a* ✓ *in blank as you complete each of the following.*

____ **1.** Attend class every day.

____ **2.** Read the textbook.

____ **3.** Take notes in class.

____ **4.** Do assigned homework every day.

____ **5.** Get help immediately when needed.

____ **6.** Review regularly.

Verbal and Writing Skills, Exercises 1–2

State the rule for simplifying each of the following, then simplify.

1. (a) $15x^3 + 5x^3$

(b) $(15x^3)(5x^3)$

(c) $\dfrac{15x^3}{5x}$

(d) $(x^3)^2$

2. (a) $14x^2 + 6x^2$

(b) $(14x^2)(6x^2)$

(c) $\dfrac{14x^2}{6x}$

(d) $(x^2)^2$

Simplify. In this exercise set, assume that all variables in any denominator are nonzero. Leave your answers in exponent form.

3. $\dfrac{7^4}{7^3}$

4. $\dfrac{4^9}{4^8}$

5. $\dfrac{a^8}{a^3}$

6. $\dfrac{x^7}{x^4}$

7. $\dfrac{5^8}{5^9}$

8. $\dfrac{2^6}{2^7}$

9. $\dfrac{3}{3^6}$

10. $\dfrac{8}{8^4}$

11. $\dfrac{z^4}{y^8}$

12. $\dfrac{z^6}{x^8}$

13. $\dfrac{9^3}{8^8}$

14. $\dfrac{6^6}{7^8}$

15. $\dfrac{z^8}{z^8}$

16. $\dfrac{a^5}{a^5}$

17. $\dfrac{y^3 z^4}{y^5 z^7}$

18. $\dfrac{a^3 b^5}{a^7 b^7}$

19. $\dfrac{m^9 3^6}{m^7 3^7}$

20. $\dfrac{5^4 r^7}{5^5 r^2}$

21. $\dfrac{a^5 7^4}{a^3 7^7}$

22. $\dfrac{4^3 z^9}{4^9 z^2}$

23. $\dfrac{b^9 9^9}{b^7 9^{11} 3^0}$

24. $\dfrac{7^7 r^2 s^0}{7^2 r^6}$

25. $\dfrac{4^6 a b^0}{4^9 a^4 b}$

26. $\dfrac{6^3 x y^0}{6^9 x^6 y}$

Simplify.

27. $\dfrac{20y^4}{35y}$

28. $\dfrac{24x^5}{36x}$

29. $\dfrac{9a^4}{27a^3}$

30. $\dfrac{7m^5}{21m^4}$

31. $\dfrac{56x^9 y^0}{64x^3}$

32. $\dfrac{32y^7}{48y^5 z^0}$

33. $\dfrac{12x^4 y^2}{15xy^3}$

34. $\dfrac{20a^5 b^3}{30ab^5}$

To Think About, Exercises 35–38 *Multiply and write in exponent form.*

35. (a) $z^2 \cdot z^2 \cdot z^2$
    (b) $(z^2)^3$

36. (a) $a^3 \cdot a^3 \cdot a^3$
    (b) $(a^3)^3$

37. (a) $x^4 \cdot x^4$
    (b) $(x^4)^2$

38. (a) $y^6 \cdot y^6$
    (b) $(y^6)^2$

Simplify. Leave your answer in exponent form.

39. $(z^2)^4$

40. $(x^7)^4$

41. $(3^5)^4$

42. $(2^3)^2$

43. $(b^1)^6$

44. $(x^3)^1$

45. $(x^0)^4$

46. $(y^5)^0$

47. $(y^3)^3$

48. $(6^2)^3$

49. $(2^4)^5$

50. $(x^3)^9$

51. $(x^2)^0$

52. $(y^0)^4$

53. $(6^3)^9$

54. $(8^2)^3$

55. $(x^2)^2$

56. $(b^3)^3$

57. $(3y^2)^6$

58. $(5x^3)^4$

59. $(4x^2)^3$

60. $(5b^4)^6$

61. $(3a^4)^8$

62. $(3y^3)^7$

63. $(2^2x^5)^3$

64. $(3^4y^6)^2$

65. $(8^3n^4)^6$

66. $(7^4b^5)^5$

Simplify.

67. $\left(\dfrac{4}{x}\right)^2$

68. $\left(\dfrac{3}{y}\right)^3$

69. $\left(\dfrac{a}{b}\right)^7$

70. $\left(\dfrac{b}{a}\right)^7$

71. $\left(\dfrac{3}{x}\right)^3$

72. $\left(\dfrac{4}{y}\right)^2$

73. $\left(\dfrac{m}{n}\right)^4$

74. $\left(\dfrac{x}{y}\right)^7$

75. $\left(\dfrac{x}{6}\right)^2$

76. $\left(\dfrac{y}{5}\right)^3$

77. $\left(\dfrac{3}{7}\right)^2$

78. $\left(\dfrac{1}{2}\right)^3$

One Step Further *Simplify.*

79. $\dfrac{25x^2y^3z^4}{135x^7y}$

80. $\dfrac{40a^9b^2c^4}{88a^3b}$

81. $\dfrac{156a^0b^8}{144b^6c^9}$

82. $\dfrac{256x^0y^{15}}{300y^9z^8}$

83. $\left(\dfrac{2y}{5x}\right)^2$

84. $\left(\dfrac{5x}{3y}\right)^2$

85. $\left(\dfrac{3a^2}{2b^3}\right)^3$

86. $\left(\dfrac{4a^3}{3b^5}\right)^3$

Mixed Problems *Simplify.*

87. (a) $15x^3 + 5x^3$
(b) $(15x^3)(5x^3)$
(c) $(x^3)^3$
(d) $\dfrac{15x^3}{5x^5}$

88. (a) $24x^5 + 6x^5$
(b) $(24x^5)(6x^5)$
(c) $(x^5)^5$
(d) $\dfrac{24x^5}{6x^3}$

89. (a) $3x^3 + 9x^3$
(b) $(3x^3)(9x^3)$
(c) $(3x^3)^3$
(d) $\dfrac{3x^3}{9x^4}$

90. (a) $7x^6 + 14x^6$
(b) $(7x^6)(14x^6)$
(c) $(7x^6)^2$
(d) $\dfrac{7x^6}{14x^2}$

91. (a) $12x^4 + 3x^4$
(b) $(12x^4)(3x^4)$
(c) $(2x^4)^4$
(d) $\dfrac{2x^4}{3x}$

92. (a) $14x^3 + 3x^3$
(b) $(14x^3)(3x^3)$
(c) $(4x^3)^3$
(d) $\dfrac{4x}{3x^4}$

93. (a) $5y^2 + 15y^2$
(b) $(5y^2)(15y^2)$
(c) $(5y^2)^2$
(d) $\dfrac{5y^2}{15y^7}$

94. (a) $8a^5 + 16a^5$
(b) $(8a^5)(16a^5)$
(c) $(8a^5)^5$
(d) $\dfrac{8a^5}{16a}$

To Think About

95. Find the next row in the triangular pattern below.

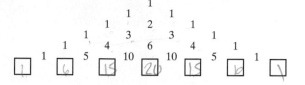
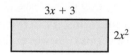

Cumulative Review *Solve.*

96. [3.2.1] $3x = 42$

97. [3.2.1] $48 = 16x$

98. [3.2.1] $(18 - 4) = 7x$

99. [3.2.1] $13x = 130$

▲ **100. [3.4.4]** Write the area of the rectangle as an algebraic expression and then simplify. ($A = LW$)

$3x + 3$

$2x^2$

101. [1.9.3] *College Expenses* Kristina has a \$45,000 trust fund for her college expenses. She plans to attend college for 9 months a year, for 4 years, to earn a B.A. During her first two years the on-campus room and board will be \$620 per month. For her last two years Kristina will live off campus, and she estimates that her rent will be \$350 and food expenses \$250 per month for the entire 18 months. Tuition and books will cost \$7500 per year. How much should Kristina plan to borrow on a student loan to cover the college expenses that will exceed the amount in her trust fund?

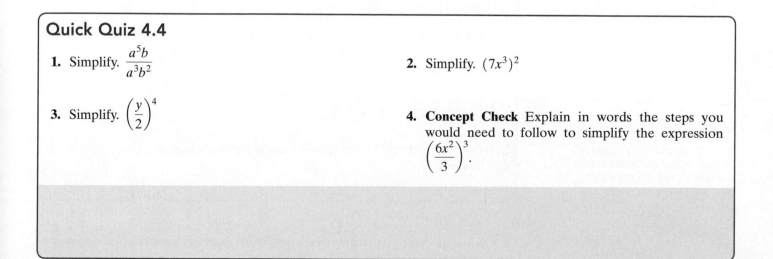

Quick Quiz 4.4

1. Simplify. $\dfrac{a^5 b}{a^3 b^2}$

2. Simplify. $(7x^3)^2$

3. Simplify. $\left(\dfrac{y}{2}\right)^4$

4. Concept Check Explain in words the steps you would need to follow to simplify the expression $\left(\dfrac{6x^2}{3}\right)^3$.

4.5 Ratios and Rates

① Writing Two Quantities with the Same Units as a Ratio

A **ratio** is a comparison of two quantities that have the same units. For example, if we compare the 5-foot width of a garden to the 22-foot width of a backyard, the ratio of the lengths would be 5 to 22.

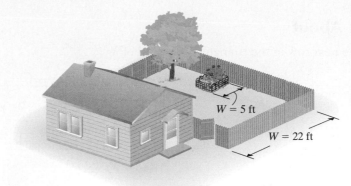

We can express a ratio in three ways.

In words:	the ratio of 5 to 22
Using a colon:	$5:22$
Using a fraction:	$\dfrac{5}{22}$

Each of the ways of expressing a ratio is read "5 to 22."

Although a ratio can be written in different forms, it is a fraction and therefore should always be simplified (reduced to lowest terms). However, improper fractions *are not* changed to mixed numbers.

EXAMPLE 1 Write each ratio in simplest form. Express your answer as a fraction.

(a) The ratio of 20 dollars to 35 dollars   **(b)** $14:21$

Solution

(a) 20 dollars to 35 dollars $= \dfrac{20 \text{ dollars}}{35 \text{ dollars}} = \dfrac{\cancel{5} \cdot 4}{\cancel{5} \cdot 7} = \dfrac{4}{7}$

We treat units in the same way we do numbers and variables, for example, $\dfrac{3}{3} = 1, \dfrac{a}{a} = 1,$ and $\dfrac{\text{dollars}}{\text{dollars}} = 1.$

(b) $14:21 = \dfrac{14}{21} = \dfrac{\cancel{7} \cdot 2}{\cancel{7} \cdot 3} = \dfrac{2}{3}$

Student Practice 1 Write each ratio in simplest form. Express your answer as a fraction.

(a) The ratio of 28 feet to 49 feet   **(b)** $27:81$

It is important that you read ratio problems carefully since the *order of quantities* is important, as shown in the next example.

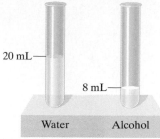

EXAMPLE 2   A mixture consists of 20 milliliters (mL) of water and 8 milliliters of alcohol. Write each ratio as a fraction and reduce to lowest terms.

(a) The ratio of alcohol to water   **(b)** The ratio of water to alcohol

Solution

(a) $\dfrac{\text{alcohol}}{\text{water}} \Rightarrow \dfrac{8 \text{ mL}}{20 \text{ mL}} = \dfrac{4 \cdot 2}{4 \cdot 5} = \dfrac{2}{5}$ **(b)** $\dfrac{\text{water}}{\text{alcohol}} \Rightarrow \dfrac{20 \text{ mL}}{8 \text{ mL}} = \dfrac{4 \cdot 5}{4 \cdot 2} = \dfrac{5}{2}$

Student Practice 2   15 women and 21 men are enrolled in a physical science class. Write each ratio as a fraction and reduce to lowest terms.

(a) The ratio of men to women   **(b)** The ratio of women to men

EXAMPLE 3   Some regions of Alaska receive less annual snowfall than mainland areas of the United States. For example, Barrow, Alaska, located near the Arctic Ocean, receives only 29 inches annually on average, while Chicago's Midway Airport receives an annual average of 46 inches. What is the ratio of the annual average snowfall in Barrow, Alaska, to the annual average snowfall at Midway Airport in Chicago?

Solution

$$\frac{\text{Barrow}}{\text{Midway Airport}} \Rightarrow \frac{29 \text{ inches}}{46 \text{ inches}} = \frac{29}{46}$$

Barrow 29″

Fairbanks 31″

Anchorage 43″

Juneau 40″

Student Practice 3   Refer to Example 3 to complete the following problem. Write the ratio of the annual average snowfall at Midway Airport, Chicago, to the annual average snowfall in Barrow, Alaska.

② Writing Two Quantities with Different Units as a Unit Rate

A **rate** is a comparison of two quantities with *different* units. Usually, to avoid misunderstanding, we include the units when we write a rate. For example, we write $44 earned in 4 hours as the following rate.

$44 earned in 4 hours: $\dfrac{\$44}{4 \text{ hours}} = \dfrac{\$11}{1 \text{ hour}}$ We include units in a rate.

Since the denominator is 1, we can write our rate as "*$11 for each hour*" or "*$11 per hour.*" When the denominator is 1, we have the rate for a single unit, which is the **unit rate.** Since the key words *per* and *for each* mean that we divide, to find a unit rate we divide.

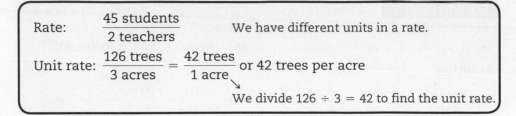

Rate: $\dfrac{45 \text{ students}}{2 \text{ teachers}}$ We have different units in a rate.

Unit rate: $\dfrac{126 \text{ trees}}{3 \text{ acres}} = \dfrac{42 \text{ trees}}{1 \text{ acre}}$ or 42 trees per acre

We divide $126 \div 3 = 42$ to find the unit rate.

EXAMPLE 4   Bertha drove her car 416 miles in 8 hours. Find the unit rate in miles per hour.

Solution

$$\dfrac{416 \text{ miles}}{8 \text{ hours}} \qquad \text{We divide: } 8\overline{)416}. \;\; \substack{52}$$

$$\dfrac{416 \text{ miles}}{8 \text{ hours}} = \dfrac{52 \text{ miles}}{1 \text{ hour}} \quad \text{or} \quad 52 \text{ miles per hour}$$

Note that miles per hour is sometimes written *mph*.

Student Practice 4   Iris travels 90 miles on 5 gallons of gas. Find the unit rate in miles per gallon.

EXAMPLE 5   The calories and fat content for 1 medium bag of french fries are given in the chart. What is the unit rate in calories per gram of fat in 1 medium bag of french fries at each restaurant?

(a) Burger King   **(b)** McDonald's

Solution

(a) Burger King: We divide $370 \div 17$ to find the unit rate.

$$\dfrac{370 \text{ calories}}{17 \text{ grams of fat}} = 21\dfrac{13}{17} \text{ calories per gram of fat}$$

	Calories	Fat
Burger King	370	17 g
McDonald's	440	22 g

(b) McDonald's: We divide $440 \div 22$ to find the unit rate.

$$\dfrac{440 \text{ calories}}{22 \text{ grams of fat}} = 20 \text{ calories per gram of fat}$$

Student Practice 5   C & R construction can dispose of 400 pounds of trash for $26 at the local dump. What is the unit rate in pounds per dollar?

③ Solving Applied Problems Involving Rates

We work with **unit rates** in many areas, such as sports, business, budgeting, and science. Sometimes we want to find the number of boards needed to panel each wall of a room, or we may need to find the number of people needed to complete a task. Often we want to find the best buy for the dollar. These are all applications of unit rates.

EXAMPLE 6 University of Chicago tornado researcher Tetsuya Theodore Fujita cataloged 31,054 tornados in the United States during the 70 years 1916–1985 and found that $\frac{7}{10}$ of the tornados occurred in the spring and early summer.

Write as a unit rate: the average number of tornados per year that occurred in the month of May. Round your answer to the nearest whole number.

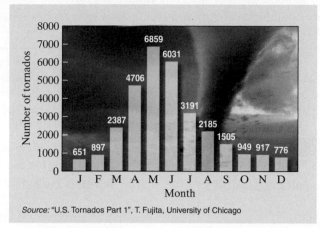

Source: "U.S. Tornados Part 1", T. Fujita, University of Chicago

Solution $\dfrac{6859 \text{ tornados in May}}{70 \text{ years}}$

We divide to find the unit rate.

$$70\overline{)6859} = 97\tfrac{69}{70} \quad \text{or approximately 98 tornados per year in May}$$

Note that we rounded $97\frac{69}{70}$ to 98 because the fraction $\frac{69}{70}$ is close to 1.

Student Practice 6 Refer to Example 6. Write as a unit rate: the average number of tornados per year that occur in the month of June. Round your answer to the nearest whole number.

EXAMPLE 7 Sunshine Preschool has a staffing policy requiring that for every 60 children, there are 3 preschool teachers, and for every 24 children, there are 2 aides.

(a) How many children per teacher does the preschool have?

(b) How many children per aide does the preschool have?

(c) If there are 60 students at the preschool, how many aides must there be to satisfy the staffing policy?

Solution

(a) Children per teacher:

$$\frac{\text{children}}{\text{teacher}} \Rightarrow \frac{60 \text{ children}}{3 \text{ teachers}} = \frac{20 \text{ children}}{1 \text{ teacher}} \text{ or 20 children per teacher}$$

(b) Children per aide:

$$\frac{\text{children}}{\text{aide}} \Rightarrow \frac{24 \text{ children}}{2 \text{ aides}} = \frac{12 \text{ children}}{1 \text{ aide}} \text{ or 12 children per aide}$$

(c) Since every 12 children require 1 aide, we divide $60 \div 12$ to find how many aides are needed for 60 children.

$$60 \div 12 = 5 \text{ aides for 60 children}$$

Continued on next page

Student Practice 7 Autumn Home, a private nursing home, has a medical staffing policy requiring that for every 40 patients there are 2 registered nurses (RNs), and for every 30 patients there are 2 nurse's aides.

(a) How many patients per RN does Autumn Home have?

(b) How many patients per aide does Autumn Home have?

(c) If there are 60 patients at Autumn Home, how many aides must there be to satisfy the staffing policy?

We often ask ourselves questions like, "Which package is the better buy, the pack of 3 or the pack of 7?" "Is it cheaper to buy the 12-ounce box or the 16-ounce box?" We find the *unit price* (price per item) to answer these types of questions.

EXAMPLE 8 The Tech Store has black print cartridges on sale. A package of 6 sells for $96, and the same brand in a package of 8 sells for $136.

(a) Find each unit price. **(b)** Which is the better buy?

Solution

(a) $\dfrac{\$96}{6} = \$16$ per cartridge; $\quad \dfrac{\$136}{8} = \$17$ per cartridge

(b) The package of 6 cartridges is the better buy.

Student Practice 8 The Linen Factory is having a sale on their designer hand towels. The Hazelette Collection is on sale at 6 for $78, and the Springview Collection is on sale at 9 for $108.

(a) Find each unit price. **(b)** Which is the better buy?

STEPS TO SUCCESS Re-evaluate Your Test-Taking Skills

It is time to look at your test results to see what you can do to improve your scores. Did you know that there are key strategies you can use when taking a test that can help improve your score? Complete the following to see one of the ways you can improve your test scores.

Making it personal:

1. List all your test scores for this class: _____

2. If you already read the Steps to Success in Section 1.8 it is time to reread it. If you did not complete that study skill, then you should do it now. Then use the strategy presented in that study skill on the next test.

3. Record your score for the next test. _____

If your test score did not improve, make an appointment with your instructor to discuss what else you can do to improve your mastery. Then make a list of what you will do to improve your test scores. ▼

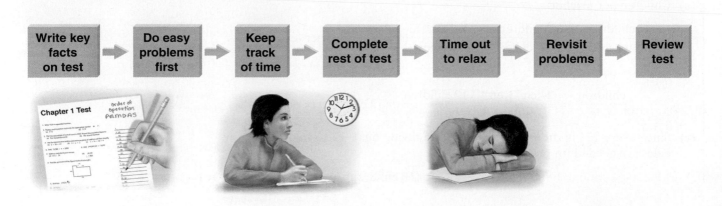

4.5 Exercises

MyMathLab®

Verbal and Writing Skills, Exercises 1–2

1. Explain the difference between a ratio and a rate.

2. Explain the difference between a rate and a unit rate.

Write each ratio as a fraction in simplest form.

3. 25 to 45

4. 10 to 22

5. 35 : 10

6. 46 : 14

7. 54 : 70

8. 20 : 35

9. 34 minutes to 12 minutes

10. 25 dollars to 15 dollars

11. 14 gallons to 35 gallons

12. 24 yards to 16 yards

13. 17 hours to 41 hours

14. 3 hours to 11 hours

15. $121 to $423

16. $85 to $151

Applications, Exercises 17–24 *Write each ratio as a fraction and simplify.*

17. *Solution Mixture* A mixture contains 35 milliliters of water and 15 milliliters of chlorine.

(a) State the ratio of chlorine to water.

(b) State the ratio of water to chlorine.

18. *Field Trip* The marine science field trip consisted of 40 juniors and 22 seniors.

(a) State the ratio of seniors to juniors.

(b) State the ratio of juniors to seniors.

19. *Basketball* The Willow Brook recreational basketball team had a season record of 29 wins and 13 losses.

(a) State the ratio of wins to losses.

(b) State the ratio of losses to wins.

20. *Men vs. Women* A choir consists of 17 men and 11 women.

(a) State the ratio of men to women.

(b) State the ratio of women to men.

Fatty Acids *Write as a ratio and simplify.*

21. The number of grams of trans-fatty acids in a serving of corn flakes cereal to the number of grams in a serving of margarine

22. The number of grams of trans-fatty acids in a serving of white bread to the number of grams in a serving of potato chips

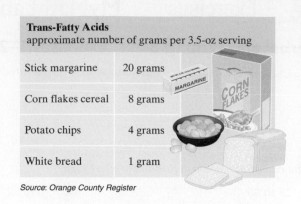

Trans-Fatty Acids
approximate number of grams per 3.5-oz serving

Stick margarine	20 grams
Corn flakes cereal	8 grams
Potato chips	4 grams
White bread	1 gram

Source: Orange County Register

Suspension Bridge *Use the following information to answer exercises 23 and 24. The world's longest suspension bridge is Japan's Akashi Kaikyo Bridge. The bridge is 6532 feet long and links Kobe and Awaji Island.*

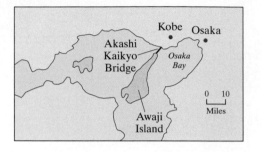

11 Longest Suspension Bridges

Akashi Kaikyo, Japan	6532 ft
Xihoumen, China	5412 ft
Great Belt East, Denmark	5328 ft
Runyang South, China	4888 ft
Humber River, England	4625 ft
Jiangyin, China	4543 ft
Tsing Ma, China	4517 ft
Verrazano Narrows, New York City	4258 ft
Golden Gate, San Francisco	4200 ft
Yangluo, China	4200 ft
Hoga Kusten, Sweden	3970 ft

Source: www.tkk.fi

Write as a ratio and simplify.

23. The length of the shortest of the 11 suspension bridges to the length of the longest suspension bridge

24. The length of the Golden Gate Bridge to the length of the Great Belt East Bridge

What is the unit rate in calories per gram of fat?

25. 410 calories for 19 grams of fat

26. 247 calories for 6 grams of fat

What is the unit rate in miles per gallon?

27. Traveling 300 miles on 15 gallons of gas

28. Traveling 405 miles on 18 gallons of gas

What is the unit rate in dollars per hour?

29. Earning $304 in 38 hours

30. Earning $455 in 35 hours

What is the unit rate in miles per hour (mph)?

31. Traveling 320 miles in 6 hours

32. Traveling 410 miles in 7 hours

Applications

33. *Gas Mileage* A car travels 616 miles on 28 gallons of gas. Find how many miles the car can be driven on one gallon of gas.

34. *Walking Speed* Michelle and Debi walk for 75 minutes. They average 6525 steps in that time. How many steps per minute do they take on average?

35. *Book Club* Marci joined a book club that charges $108 for 9 books. What is the cost per book?

36. *CD Club* Delroy joined a CD club that charges $189 for 21 CDs. How much per CD did Delroy pay?

37. *Student-Teacher Ratio* The J. D. Robertson Academy of Arts school has a staffing policy requiring that for every 90 students there are 5 instructors, and for every 30 students there are 2 tutors.

(a) How many students per instructor does the academy have?

(b) How many students per tutor does the academy have?

(c) How many tutors are needed to satisfy the staffing policy if there are 90 students in the academy?

38. *Agent-Client Ratio* The All Point Insurance Group requires each office to have 4 insurance agents for every 620 clients and 5 clerical staff members for every 310 clients.

(a) How many clients per agent does the group have?

(b) How many clients per clerical staff member does the group have?

(c) How many clerical staff members would be required for an office that has 930 clients?

39. *Comparison Shopping* The Crystal Shop has their Gold Lace crystal wine glasses on sale. A box of 8 glasses is $96, and a box of 6 is $78.

(a) Find each unit price.

(b) Which is the better buy?

40. *Comparison Shopping* The tanning salon has a special on their tanning sessions: 12 sessions for $96 or 15 sessions for $135.

(a) Find each unit price.

(b) Which is the better deal?

41. *Comparison Shopping* The music shop sells used CDs 4 for $32 or 6 for $48.

(a) Find each unit price.

(b) Which is the better deal?

42. *Comparison Shopping* Computer World is having a sale on printer paper: 4 reams of paper for $12 or 7 reams of paper for $21.

(a) Find each unit price.

(b) Which is the better deal?

Sales Statistics Johnson and Brothers Suits recorded the sales shown in the bar graph for the third quarter of the year. Use the graph to answer exercises 43 and 44.

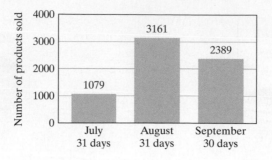

43. Write as a unit rate: the average number of sales per day in the month of August. Round your answer to the nearest whole number.

44. Write as a unit rate: the average number of sales per day in the month of July. Round your answer to the nearest whole number.

Cumulative Review *Solve and check your answer.*

45. [3.1.2] $x - 12 = 25$

46. [3.1.2] $x + 15 = 40$

47. [3.1.2] $5x - 4x + 6 = 14$

48. [3.1.2] $-2 - 5 + a = 15$

49. [2.3.3] *High/Low Temperature* The record high temperature for a city in the Midwest was 101°F, and the record low recorded was −8°F. Find the difference in temperature between the record high and low.

50. [2.2.4] *Football* On the first play of a college football game, the home team lost 4 yards. On the second play, a pass resulted in a gain of 8 yards. If there was a penalty against the home team on the third play resulting in a loss of 5 yards, determine the net gain (or loss) of the home team after the third play.

Quick Quiz 4.5

1. Sara and Mark ran for president of their elementary school. Mark received 45 of the votes while Sara received 75 of the votes. Write the ratio and simplify: the number of votes Mark received to the number of votes Sara received.

2. Reza cooked 30 hot dogs in 12 minutes. How many hot dogs did he cook per minute?

3. Chocolate Delite is having a sale on its sugar free chocolate bars: $72 for 36 bars or $60 for 18 bars. Which is the better buy?

4. **Concept Check** A large furniture store determined they needed to have 8 salespeople in the store for every 160 customers. Explain how to determine how many salespeople per customer the store has.

4.6 Proportions and Applications

① Writing Proportions

Recall that the fractions $\frac{4}{8}$ and $\frac{1}{2}$ are equivalent fractions: $\frac{4}{8} = \frac{1}{2}$. If these fractions represent ratios or rates, we say that they are *proportional*. In other words, a **proportion** states that two ratios or two rates are equal. For example, $\frac{4}{8} = \frac{1}{2}$ is a proportion and $\frac{2\text{ trees}}{7\text{ feet}} = \frac{4\text{ trees}}{14\text{ feet}}$ is also a proportion. The proportion $\frac{4}{8} = \frac{1}{2}$ is read "four *is to* eight as one *is to* two."

Student Learning Objectives

After studying this section, you will be able to:

① Write proportions.

② Determine if statements are proportions.

③ Find the missing number in proportions.

④ Solve applied problems involving proportions.

> A **proportion** states that two ratios or two rates are equal.
>
> If $\dfrac{a}{b}$ and $\dfrac{c}{d}$ are two equal ratios, then $\dfrac{a}{b} = \dfrac{c}{d}$ is a proportion.

When we write a proportion we must be sure that the units are in the appropriate position. One way to write a proportion is for the numerators to have the same units and the denominators to have the same units. In other words, we write each fraction as we would a rate or a ratio. In this book we write proportions in this manner.

EXAMPLE 1 Translate the statement into a proportion. 2 waiters is to 6 tables as 8 waiters is to 24 tables.

Solution   2 waiters is to 6 tables as 8 waiters is to 24 tables

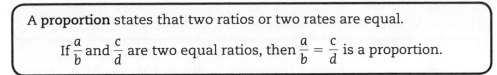

$$\frac{2\text{ waiters}}{6\text{ tables}} = \frac{8\text{ waiters}}{24\text{ tables}}$$

Student Practice 1 Translate the statement into a proportion. 3 nails is to 6 feet as 9 nails is to 18 feet.

NOTE TO STUDENT: Fully worked-out solutions to all of the Student Practice problems can be found at the back of the text starting at page SP-1.

EXAMPLE 2 Translate the statement into a proportion. If 6 pounds of flour cost $2, then 18 pounds will cost $6.

Solution   We can restate as follows: 6 pounds is to $2 as 18 pounds is to $6.

$$\frac{6\text{ pounds}}{2\text{ dollars}} = \frac{18\text{ pounds}}{6\text{ dollars}}$$ We write pounds in the numerator.
We write dollars in the denominator.

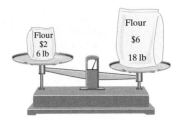

Student Practice 2 Translate the statement into a proportion. If it takes 4 hours to drive 144 miles, it will take 6 hours to drive 216 miles.

② Determining If Statements Are Proportions

A proportion states that two ratios or two rates are equal. Since ratios and rates are fractions, a proportion is a statement that 2 fractions are equal. Therefore, to determine if a statement is a proportion, we must verify that the fractions in the proportion are equal.

How can we *check* to see if two fractions are equal? We use the **equality test for fractions,** which states that if two fractions are equal, their cross products are equal. By **cross product** we mean the denominator of one fraction times the numerator of the other fraction.

EQUALITY TEST FOR FRACTIONS

For any two fractions where $b \neq 0, d \neq 0$,

$$\frac{a}{b} = \frac{c}{d} \text{ if and only if } d \cdot a = b \cdot c$$

In other words, two fractions are equal if the cross products are equal.

If two fractions are unequal (we use the symbol $\neq$), their cross products are unequal.

EXAMPLE 3   Use the equality test for fractions to see if the fractions are equal.

(a) $\dfrac{2}{11} \overset{?}{=} \dfrac{18}{99}$

(b) $\dfrac{3}{16} \overset{?}{=} \dfrac{12}{62}$

Solution   We form the cross products to determine if the fractions are equal.

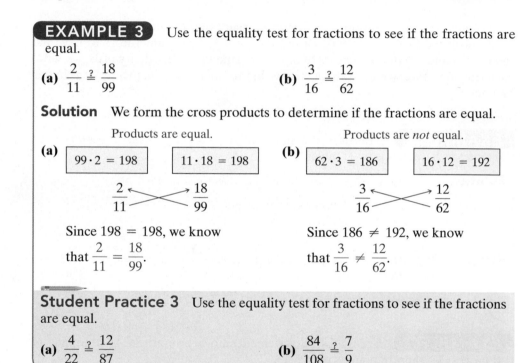

Products are equal.

(a)

| $99 \cdot 2 = 198$ | | $11 \cdot 18 = 198$ |

$$\frac{2}{11} \diagtimes \frac{18}{99}$$

Since $198 = 198$, we know that $\dfrac{2}{11} = \dfrac{18}{99}$.

Products are *not* equal.

(b)

| $62 \cdot 3 = 186$ | | $16 \cdot 12 = 192$ |

$$\frac{3}{16} \diagtimes \frac{12}{62}$$

Since $186 \neq 192$, we know that $\dfrac{3}{16} \neq \dfrac{12}{62}$.

Student Practice 3   Use the equality test for fractions to see if the fractions are equal.

(a) $\dfrac{4}{22} \overset{?}{=} \dfrac{12}{87}$

(b) $\dfrac{84}{108} \overset{?}{=} \dfrac{7}{9}$

Since a proportion is just a statement that two fractions are equal, we can use the equality test for fractions to determine if a statement is a proportion.

DETERMINING IF A STATEMENT IS A PROPORTION

To determine if a statement is a proportion, we use the equality test for fractions, which states that two fractions are equal if their cross products are equal.

EXAMPLE 4   Determine if the statement is a proportion.

(a) $\dfrac{16 \text{ points}}{35 \text{ games}} \overset{?}{=} \dfrac{48 \text{ points}}{125 \text{ games}}$

(b) $\dfrac{22}{29} \overset{?}{=} \dfrac{88}{116}$

Solution   We check $\dfrac{16}{35} \overset{?}{=} \dfrac{48}{125}$ by forming the two cross products.

(a) $\boxed{125 \cdot 16 = 2000}$ $\dfrac{16}{35} \overset{?}{=} \dfrac{48}{125}$ $\boxed{35 \cdot 48 = 1680}$

The two cross products are *not* equal.

Thus $\dfrac{16 \text{ points}}{35 \text{ games}} \neq \dfrac{48 \text{ points}}{125 \text{ games}}$.   This is not a proportion.

(b) We form the two cross products.

$\boxed{116 \cdot 22 = 2552}$ $\dfrac{22}{29} \overset{?}{=} \dfrac{88}{116}$ $\boxed{29 \cdot 88 = 2552}$

The two cross products are equal. Thus $\dfrac{22}{29} = \dfrac{88}{116}$.   This is a proportion.

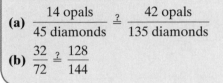

Student Practice 4   Determine if the statement is a proportion.

(a) $\dfrac{14 \text{ opals}}{45 \text{ diamonds}} \overset{?}{=} \dfrac{42 \text{ opals}}{135 \text{ diamonds}}$

(b) $\dfrac{32}{72} \overset{?}{=} \dfrac{128}{144}$

③ Finding the Missing Number in Proportions

Sometimes one of the quantities in a proportion is unknown. We can find this unknown quantity by finding the cross products and solving the resulting equation.

> **TO SOLVE FOR A MISSING NUMBER IN A PROPORTION**
>
> 1. Find the cross products and form an equation.
> $$\dfrac{x}{21} = \dfrac{2}{7}$$
> $$7 \cdot x = 21 \cdot 2$$
>
> 2. Solve the equation by dividing on both sides so that the variable stands alone.
> $$\dfrac{7x}{7} = \dfrac{42}{7}$$
>
> 3. Simplify the result.
> $$x = 6$$
>
> 4. Check your answer.
> $$\dfrac{6}{21} \overset{?}{=} \dfrac{2}{7}$$
> $$7 \cdot 6 \overset{?}{=} 21 \cdot 2$$
> $$42 = 42 \quad ✓$$

EXAMPLE 5   Find the value of n in $\dfrac{n}{24} = \dfrac{15}{60}$.

Solution

$$\dfrac{n}{24} = \dfrac{15}{60}$$

$60 \cdot n = 24 \cdot 15$ Find the cross products and form an equation.

$60n = 360$ Simplify.

$\dfrac{60 \cdot n}{60} = \dfrac{360}{60}$ Divide by 60 on both sides of the equation.

$n = 6$ Divide: $60 \div 60 = 1$; $360 \div 60 = 6$

Continued on next page

Check whether the proportion is true.

$$\frac{6}{24} \overset{?}{=} \frac{15}{60}$$

$$60 \cdot 6 \overset{?}{=} 24 \cdot 15 \quad \text{We check cross products.}$$

$$360 = 360 \quad \checkmark$$

Student Practice 5   Find the value of n in

$$\frac{n}{18} = \frac{28}{72}.$$

Understanding the Concept

Reducing a Proportion   Can we reduce a proportion before we solve for the missing number in the proportion? Yes. If you can see that the ratio or rate without the variable can be reduced, you may reduce it, and the answer will still be correct. Let's look at the proportion in Example 5 and observe what happens when we reduce the ratio $\frac{15}{60}$.

$$\frac{n}{24} = \frac{15}{60} \qquad \text{We see that we can reduce } \frac{15}{60}.$$

$$\frac{n}{24} = \frac{1}{4} \qquad \text{Reduce: } \frac{15}{60} = \frac{\cancel{3} \cdot \cancel{5}}{\cancel{3} \cdot 4 \cdot \cancel{5}} = \frac{1}{4}.$$

$$4 \cdot n = 24 \cdot 1 \qquad \text{Cross products.}$$

$$4n = 24 \qquad \text{This step is easier when we reduce. Do you see why?}$$

$$\frac{4n}{4} = \frac{24}{4} \qquad \text{Divide by 4 on both sides.}$$

$$n = 6 \qquad \text{We see that the answer is the same.}$$

Exercise

1. Why can we reduce a ratio or rate and still get the correct answer when we solve a proportion?

④ Solving Applied Problems Involving Proportions

When a situation involves a ratio or rate, we can use a proportion to find the solution. Let's examine a variety of applied problems that can be solved with proportions.

EXAMPLE 6   Johanna owns two rectangular plots of land that have the same dimensions. She subdivides one into 5 equal parcels and the other into 7 equal parcels. For the same price you can buy either 3 of the 5 parcels or 4 of the 7 parcels. Do the two parcels for sale yield the same amount of land?

Solution   We must determine if 3 out of 5 parcels is equivalent to 4 out of 7 parcels. We have 3 of the 5 parcels $\rightarrow \frac{3}{5}$; 4 of the 7 parcels $\rightarrow \frac{4}{7}$.

We use the equality test for fractions to determine if we have a proportion.

$$\frac{3}{5} \overset{?}{=} \frac{4}{7}$$

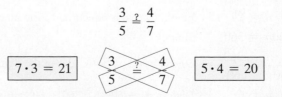

$$\boxed{7 \cdot 3 = 21} \qquad \boxed{5 \cdot 4 = 20}$$

The cross products are *not* equal. Thus $\frac{3}{5} \neq \frac{4}{7}$ and this is *not* a proportion. The two parcels do not yield the same amount of land.

Student Practice 6   Brett owns two square plots of land that have the same dimensions. He subdivides one into 8 equal parcels and the other into 11 equal parcels. For the same price you can buy either 7 of the 8 parcels or 10 of the 11 parcels. Do the two parcels yield the same amount of land?

You can solve many problems using proportions. To solve the problem in Example 6 we had to determine if the statement was a proportion. Often we can write a proportion that has a missing number to solve a problem, as you will see in the next example.

EXAMPLE 7   Kim Nguyen is planning a company party for 120 people. The delicatessen told her that 3 quarts of potato salad will serve 24 people. How many quarts of potato salad should Kim order for the party?

Solution   We write the statement in words that represents the proportion, and then set up the proportion. We let n represent the number of quarts of potato salad Kim must order.

3 quarts of potato salad is to 24 people as n quarts of potato salad is to 120 people.

$$\frac{3 \text{ quarts}}{24 \text{ people}} = \frac{n \text{ quarts}}{120 \text{ people}}$$

We must solve for n.

$$\frac{3}{24} = \frac{n}{120}$$

$\frac{1}{8} = \frac{n}{120}$ We see that the fraction $\frac{3}{24}$ can be reduced to $\frac{1}{8}$.

$120 \cdot 1 = 8n$ We form the cross product.

$\frac{120 \cdot 1}{8} = \frac{8n}{8}$ We divide both sides by 8 to solve for n.

$15 = n$ We simplify.

Kim must order 15 quarts of potato salad for the party.

Student Practice 7   Mary Lou's Catering has a policy that when planning a buffet, there should be 18 desserts for every 15 people who will be attending the buffet. How many desserts should the catering company plan to serve at a buffet if 180 people are expected to attend?

EXAMPLE 8   Estelle has a fence in her yard around her vegetable garden. The garden is 6 feet wide and 7 feet long. The yard's dimensions are proportional to the garden's. What is the length of the yard if the width is 25 feet?

Solution   Set up the proportion. Let the letter x represent the length of the yard.

$$\frac{\text{width of garden}}{\text{length of garden}} = \frac{\textbf{width of yard}}{\text{length of yard}} \longrightarrow \frac{6 \text{ ft}}{7 \text{ ft}} = \frac{\textbf{25 ft}}{x \text{ ft}}$$

Now we solve for x. $\frac{6}{7} = \frac{25}{x}$

$6x = 7 \times 25$ We form the cross product.

$6x = 175$ We simplify.

$\frac{6x}{6} = \frac{175}{6}$

$x = 29\frac{1}{6}$ We change $\frac{175}{6}$ to the mixed number $29\frac{1}{6}$.

The length of the yard is $29\frac{1}{6}$ feet.

Continued on next page

▲ **Student Practice 8**   Refer to Example 8 to answer the following. If the width of the yard is 20 feet, what must the length of the yard be for the dimensions of the yard to be proportional to those of the garden?

EXAMPLE 9   Two partners, Cleo and Julie, invest money in their small business at the ratio 3 to 5, with Cleo investing the smaller amount. If Cleo invested $6000, how much did Julie invest?

Solution   The ratio *3 to 5* represents Cleo's investment *to* Julie's investment.

$$\frac{3}{5} = \frac{\text{Cleo's investment of \$6000}}{\text{Julie's investment of \$}x}$$

$$\frac{3}{5} = \frac{6000}{x}$$

$$3x = 30{,}000$$

$$\frac{3x}{3} = \frac{30{,}000}{3}$$

$$x = 10{,}000$$

Julie invested $10,000 in their business.

Student Practice 9   Refer to Example 9 to answer the following. Cleo and Julie also split the profits from the partnership in the same ratio, 3 to 5. If Cleo receives $2400 for her share of the profit, how much does Julie receive in profits?

4.6 Exercises

 MyMathLab®

Watch the videos in MyMathLab

Download the MyDashBoard App

Verbal and Writing Skills, Exercises 1–2

1. Why should we be able to see that $\frac{55}{55} = \frac{621}{621}$ without doing any calculations?

2. Why should we be able to see that $\frac{1}{8} < \frac{8}{9}$ without doing any calculations?

Translate each statement into a proportion.

3. 2 is to 7 as 24 is to 84.

4. 5 is to 13 as 30 is to 78.

5. 12 goals is to 7 games as 24 goals is to 14 games.

6. 16 doctors is to 5 nurses as 48 doctors is to 15 nurses.

7. 3 is to 8 as 18 is to 48.

8. 3 is to 14 as 15 is to 70.

Translate each statement into a proportion.

9. *Fat Content* If 14 crackers contain 6 grams of fat, then 70 crackers contain 30 grams of fat.

10. *Reading Maps* If 3 inches on a map represent 270 miles, 6 inches represent 540 miles.

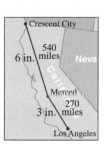

11. *Pulley Rotations* If a pulley can complete $3\frac{1}{2}$ rotations in 2 minutes, it should complete 14 rotations in 8 minutes.

12. *Fabric Needed* If it takes $2\frac{1}{4}$ yards of material to make 1 skirt, it will take 9 yards to make 4 skirts.

13. *Basketball* If Matt averages 4 baskets out of 7 free throws attempted in a basketball game, he should make 12 out of 21 free throws.

14. *Basketball* If Sal averages 6 baskets out of 11 free throws attempted in a basketball game, he should make 18 out of 33 free throws.

Use the equality test for fractions to determine if the fractions are equal.

15. $\frac{5}{8} \stackrel{?}{=} \frac{30}{45}$

16. $\frac{4}{7} \stackrel{?}{=} \frac{28}{49}$

17. $\frac{6}{11} \stackrel{?}{=} \frac{42}{77}$

18. $\frac{10}{17} \stackrel{?}{=} \frac{20}{27}$

267

Determine if each statement is a proportion.

19. $\dfrac{2}{7} \overset{?}{=} \dfrac{8}{28}$

20. $\dfrac{8}{9} \overset{?}{=} \dfrac{40}{45}$

21. $\dfrac{14}{19} \overset{?}{=} \dfrac{26}{29}$

22. $\dfrac{13}{37} \overset{?}{=} \dfrac{15}{39}$

23. $\dfrac{2 \text{ American dollars}}{11 \text{ Euros}} \overset{?}{=} \dfrac{65 \text{ American dollars}}{135 \text{ Euros}}$

24. $\dfrac{6 \text{ defective parts}}{109 \text{ parts produced}} \overset{?}{=} \dfrac{20 \text{ defective parts}}{401 \text{ parts produced}}$

Find the value of x in each proportion. Check your answer.

25. $\dfrac{x}{8} = \dfrac{5}{2}$

26. $\dfrac{x}{10} = \dfrac{6}{5}$

27. $\dfrac{12}{x} = \dfrac{2}{5}$

28. $\dfrac{6}{x} = \dfrac{2}{7}$

29. $\dfrac{12}{18} = \dfrac{x}{21}$

30. $\dfrac{15}{21} = \dfrac{x}{14}$

31. $\dfrac{15}{6} = \dfrac{10}{x}$

32. $\dfrac{25}{10} = \dfrac{20}{x}$

Find the value of n.

33. $\dfrac{80 \text{ gallons}}{24 \text{ acres}} = \dfrac{20 \text{ gallons}}{n \text{ acres}}$

34. $\dfrac{70 \text{ pints}}{25 \text{ ft}^2} = \dfrac{14 \text{ pints}}{n \text{ ft}^2}$

35. $\dfrac{n \text{ grams}}{15 \text{ liters}} = \dfrac{12 \text{ grams}}{45 \text{ liters}}$

36. $\dfrac{n \text{ miles}}{15 \text{ gallons}} = \dfrac{16 \text{ miles}}{3 \text{ gallons}}$

Applications

37. *Typing Speed* Mark can type 400 words in 5 minutes, and John can type 675 words in 9 minutes. Do they type at the same rate?

38. *Soccer Goals* Amy scored 4 goals in 7 soccer games, and Sara scored 6 goals in 9 soccer games. Do they score at the same rate?

39. *Size Comparison* Two cakes for a banquet are the same size. One cake is cut into 30 pieces and the other is cut into 25 pieces. Is 18 out of 30 slices of cake the same amount as 15 out of 25 slices of cake?

▲ **40.** *Land Parcels* Lester owns two large farms with square plots of the same dimensions. He subdivides one farm into 3 equal parcels and the other into 4 equal parcels. For the same price you can buy either 2 of 3 parcels or 3 of 4 parcels. Do the parcels yield the same amount of land?

▲ **41.** *Parcels of Land* Jason owns two square plots of land that have the same dimensions. He subdivides one into 4 equal parcels and the other into 5 equal parcels. For the same price you can buy either 3 of the 4 parcels or 4 of the 5 parcels. Do the two parcels yield the same amount of land?

42. *Pizza Slices* Two pizzas are the same size. One pizza is cut into 8 slices, while the other pizza is cut into 12 slices. Is 6 out of 8 slices of pizza the same amount as 9 out of 12 slices?

43. *Calories* If 2 servings of cereal contain 126 calories, then how many calories are there in 5 servings of cereal?

44. *Medicine Dosage* If a 200-pound man can have 1000 milligrams of a medicine a day, how much can a 120-pound woman have?

45. *Calories* A 1-ounce serving of Deluxe Mixed Nuts contains 170 calories. How many calories are there in a 40-ounce jar?

46. *Baseball* A baseball player gets 20 hits out of 50 times at bat. How many hits must she get in her next 150 times at bat to keep her batting average the same?

47. Snickers Candy In 1999 the Snickers candy bar was the most popular candy bar, with sales reaching $120 million. In one day, 16 million are made. At this rate how many Snickers bars are made in 7 days? (*Source: Orange County Register*)

48. M&M Candy Over 300 million M&Ms were produced each day in the United States in 1999. The process to make one M&M takes 4 hours from the initial mix to printing an M on every shell. If 1 bag of M&Ms contains approximately 60 M&Ms, how many processing hours does it take for 1 bag of M&Ms? (*Source: Orange County Register*)

49. Scale Drawing In a scale drawing, a 210-foot-tall building is drawn 3 inches high. If another building is drawn 5 inches high, how tall is that building?

50. Ice Cream If 100 grams of ice cream contain 15 grams of fat, how much fat is in 260 grams of ice cream?

51. Weight on Pluto If a 120-pound person weighs approximately 8 pounds on Pluto, how much does a 150-pound person weigh on Pluto?

52. Reading Maps On a tour guide map of Ohio, 2 inches on the map represent 260 miles. How many miles do 3 inches represent?

53. Shares of Stock In a stock split, each person received 8 shares for each 5 shares that he or she held. If a person had 850 shares of stock in the company, how many shares did she receive in the stock split?

54. Bicycle Speed If Wendy pedals her bicycle at 84 revolutions per minute, she travels at 14 miles per hour. How fast does she go if she pedals at 96 revolutions per minute?

55. Stereo Speaker A 100-watt stereo system needs copper speaker wire that is 30 millimeters thick to handle the output of sound clearly. How thick would the speaker wire need to be if you had a 140-watt stereo and you wanted the same ratio of watts to millimeters?

56. Weed Killer A bottle of spurge and oxalis killer for your lawn states that you need to use 2 tablespoons to treat 300 square feet of lawn. How many tablespoons will you need to use to treat 1500 square feet of lawn?

▲ **57. Pool Fence Dimension** Julio wants to put a fence around his rectangular pool, which is 12 feet wide and 18 feet long. If the size of the yard will only allow for a fence that is 30 feet long and Julio wants the dimensions of the enclosed area to be proportional to those of the pool, how wide should the fence be?

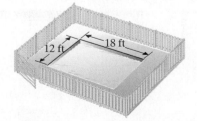

▲ **58. Patio Enlargement** Devon has a small concrete patio 5 feet wide and 7 feet long in his yard. He wants to enlarge the patio, keeping the dimensions of the new patio proportional to those of the old patio. If he has room to increase the length to 21 feet, how wide should the patio be?

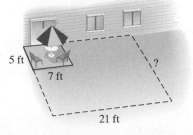

Profit and Loss *Use the following information to answer exercises 59 and 60.*

Two partners, John Ling and Kelvey Marks, each invest money in their business at a ratio of 6 to 7, with Kelvey investing the larger amount.

59. If John invested $2400, how much did Kelvey invest?

60. If the profits from the partnership are distributed to John and Kelvey based on the ratio of their investment, how much profit will John receive if Kelvey receives $798 in profits?

One Step Further

Write as a proportion.

61. $\frac{1}{3}$ is to $\frac{1}{8}$ as $\frac{1}{4}$ is to $\frac{3}{32}$.

62. $\frac{1}{5}$ is to $\frac{1}{9}$ as $\frac{1}{6}$ is to $\frac{5}{54}$.

Payroll Deductions *Renée Sharp received a promotion from file clerk to receptionist at Elen Insurance Group. She earns a monthly salary of $1950 in her new position as a receptionist instead of a weekly salary of $325 as a file clerk. Assume that Renée's deductions as a receptionist remain proportional to her deductions as a file clerk as you answer exercises 63–68.*

Weekly paycheck

Employee	Position	*ELEN*
Renee Sharp	File Clerk	ELEN INSURANCE GROUP

Total Gross Pay	Federal Withholding	State Withholding	Retirement	Insurance	Net Pay
$ 325	$ 40	$ 22	$ 32	$ 16	$ 215

Monthly paycheck

Employee	Position	*ELEN*
Renée Sharp	Receptionist	ELEN INSURANCE GROUP

Total Gross Pay	Federal Withholding	State Withholding	Retirement	Insurance	Net Pay
$ 1950					

Determine the following information about her new position as a receptionist.

63. Find the federal withholding.

64. Find the state withholding.

65. Find the retirement deduction.

66. Find the insurance deduction.

67. Find Renée's take-home pay.

68. When Renée worked as a file clerk, she placed $20 a week in her savings account. How much should Renée place in her savings account each month so that her monthly savings contribution is proportional to the amount she saved as a clerk?

To Think About

▲ **69.** ***Box Dimensions*** A box has dimensions of $L = 2$ inches, $W = 3$ inches, and $H = 5$ inches. If you increase the length of this box to 6 inches, what do the width and height have to be so that the dimensions of the new box are proportional to the dimensions of the original box?

5 in.

3 in.   2 in.

▲ **70.** ***Picture Frames*** Helena is making three frames for her living room wall. She wants three different-sized frames with dimensions that are proportional. If the smallest frame is 5 inches wide by 7 inches high, and the largest frame is 21 inches high, what must the width of the largest frame be? What must the dimensions of the medium-sized frame be? Assume the dimensions of all the frames are whole numbers.

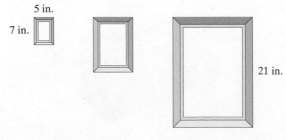

5 in.

7 in.

21 in.

71. (a) Fill in the boxes with the **sum** of the two numbers that are marked in the following sequence.

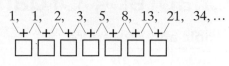

1, 1, 2, 3, 5, 8, 13, 21, 34, …

(b) Observe the pattern in 73(a), then explain how to find the next number in the sequence.

(c) Find the next 3 numbers in the sequence.

Cumulative Review

Exam Grading Scale To discourage guessing on the multiple-choice final exam, a history professor chose the following grading scale.

72. Find the final exam score for Victor if he left 5 questions blank, got 50 correct, and got 10 incorrect.

73. What score would Victor have earned if he had guessed incorrectly instead of leaving the 5 questions blank?

Response	Point Value
Blank	-1
Correct	$+5$
Incorrect	-3

Quick Quiz 4.6

1. Determine if $\dfrac{18}{30} = \dfrac{34}{56}$ is a proportion.

2. Find the value of n. $\dfrac{12}{15} = \dfrac{n}{90}$

3. On a map of New England, 2 inches represent 72 miles. How many miles do 9 inches represent?

4. Concept Check Justin and Sara share the profits from their business based on the ratio of their investment. The ratio of the investment is 5 to 7, with Sara investing the larger amount. Explain how you would determine how much profit Justin will receive if Sara gets $840.

Chapter 4 Organizer

Topic and Procedure	Examples	✏ You Try It
Divisibility tests, p. 219 1. A number is divisible by 2 if it is even. 2. A number is divisible by 3 if the sum of the digits is divisible by 3. 3. A number is divisible by 5 if the last digit is 0 or 5.	Determine if 6740 is divisible by 2, 3, or 5. 6740 is even, so it is divisible by 2. The sum of the digits is 17, which is not divisible by 3. 6740 is not divisible by 3. Since the last digit of 6740 is 0, it is divisible by 5.	1. Determine if 4731 is divisible by 2, 3, or 5.
Prime factors, p. 220 We write a number as a product of prime factors using a division ladder or a factor tree.	Express 75 as a product of prime factors. Division ladder: $$\begin{array}{r} 3 \\ 5\overline{)15} \\ 5\overline{)75} \end{array} \quad \begin{array}{l} 75 = 5 \cdot 5 \cdot 3 \\ \text{or } 3 \cdot 5^2 \end{array}$$ Factor tree: $\quad 75 \quad 75 = 3 \cdot 5^2$	2. Express 98 as a product of prime factors.
Changing an improper fraction to a mixed number, p. 230 1. Divide the numerator by the denominator. 2. The whole number part of the mixed number is the quotient. 3. The fraction is the remainder over the divisor.	Change to a mixed number: $\dfrac{37}{7}$ $$\begin{array}{r} 5 \\ 7\overline{)37} \\ \underline{35} \\ 2 \end{array} = 5\frac{2}{7}$$	3. Change to a mixed number. $\dfrac{45}{7}$
Changing a mixed number to an improper fraction, p. 231 1. Multiply the whole number by the denominator. 2. Add the product to the numerator. 3. Place the sum over the denominator.	Write as an improper fraction. $4\frac{2}{3}$ $$4\frac{2}{3} = \frac{(3 \times 4) + 2}{3} = \frac{12 + 2}{3} = \frac{14}{3}$$	4. Write as an improper fraction. $5\frac{1}{5}$
Finding equivalent fractions, p. 235 To find an equivalent fraction with a given denominator: 1. Find the number or expression by which you must multiply the original denominator to get the new denominator. 2. Multiply both the numerator and denominator of the original fraction by this number or expression.	Write $\frac{4}{7}$ as an equivalent fraction with a denominator of $21x$. $$\frac{4}{7} = \frac{?}{21x}$$ 7 times what expression equals $21x$? $3x$. $$\frac{4 \times 3x}{7 \times 3x} = \frac{12x}{21x}$$	5. Write an equivalent fraction with a denominator of $24x$. $\dfrac{7}{8} = \dfrac{?}{24x}$
Reducing fractions, p. 236 To reduce a fraction to lowest terms: 1. Write the numerator and denominator of the fraction each as a product of prime numbers. 2. Rewrite each factor that appears in both the numerator and denominator as $\frac{1}{1}$. 3. Multiply.	Simplify. $\dfrac{30xy}{42x}$ $$\frac{30xy}{42x} = \frac{\overset{1}{2} \cdot \overset{1}{3} \cdot 5 \cdot \overset{1}{x} \cdot y}{\underset{1}{2} \cdot \underset{1}{3} \cdot 7 \cdot \underset{1}{x}} = \frac{5y}{7}$$	6. Simplify. $\dfrac{27xy}{36y}$
Dividing expressions in exponent form, p. 245 If the bases are the same and $x \neq 0$, then $$\frac{x^a}{x^b} = \begin{cases} x^{a-b} & \text{if the larger exponent is in the numerator} \\ \dfrac{1}{x^{b-a}} & \text{if the larger exponent is in the denominator} \end{cases}$$	Simplify. (a) $\dfrac{x^5}{x^3}$ (b) $\dfrac{3^2}{3^4}$ (a) $\dfrac{x^5}{x^3} = x^2$ (b) $\dfrac{3^2}{3^4} = \dfrac{1}{3^2} = \dfrac{1}{9}$	7. Simplify. (a) $\dfrac{x^7}{x^2}$ (b) $\dfrac{4^3}{4^6}$

Topic and Procedure	Examples	✏️ You Try It
Raising a power to a power, pp. 247–248 We keep the same base and multiply exponents. $(x^a)^b = x^{ab}$ The parentheses indicate that *each factor* within the parentheses must be raised to a power. $(xy)^b = x^b y^b$ $\left(\dfrac{x}{y}\right)^a = \dfrac{x^a}{y^a}$, if $y \neq 0$	Simplify. **(a)** $(4x^4)^6$ **(b)** $\left(\dfrac{x}{2}\right)^3$ **(a)** $(4x^4)^6 = 4^6 x^{24}$ **(b)** $\left(\dfrac{x}{2}\right)^3 = \dfrac{x^3}{8}$	8. Simplify. **(a)** $(3x^5)^6$ **(b)** $\left(\dfrac{x}{3}\right)^2$
Forming a ratio, p. 252 A *ratio* is a comparison of two quantities that have the same units. A ratio can be written with the word "to," with a *colon*, or as a *fraction*.	We write the ratio "12 dollars to 41 dollars" in three ways: the ratio of 12 to 41; 12 : 41; $\dfrac{12}{41}$	9. Write the ratio "15 dollars to 26 dollars" in three ways.
Forming a rate, p. 253 A *rate* is a comparison of two quantities that have different units. A rate is usually expressed as a fraction in reduced form.	A college dormitory has 10 washing machines for every 85 students living in the dorm. What is the rate of washing machines to students? $\dfrac{10 \text{ machines}}{85 \text{ students}} = \dfrac{2 \text{ machines}}{17 \text{ students}}$	10. A day care center has a ratio of 16 staff for every 120 children. What is the rate of staff to children?
Forming a unit rate, p. 253 A *unit rate* is a rate with a denominator of 1. To find a unit rate, divide the denominator into the numerator.	Leslie drove her car 357 miles in 7 hours. Find the unit rate. $\dfrac{357 \text{ miles}}{7 \text{ hours}} = 51$ miles per hour	11. A machine can seal 427 boxes in 7 hours. Find the unit rate.
Writing proportions, p. 261 A *proportion* is a statement that two rates or ratios are equal. The proportion statement "*a* is to *b* as *c* is to *d*" can be written $\dfrac{a}{b} = \dfrac{c}{d}$	Write the proportion 33 is to 44 as 15 is to 20. $\dfrac{33}{44} = \dfrac{15}{20}$	12. Write the proportion 35 is to 55 as 14 is to 22.
Using the equality test for fractions, p. 262 For any two fractions where $b \neq 0, d \neq 0, \frac{a}{b} = \frac{c}{d}$ if and only if $d \cdot a = b \cdot c$. In other words, two fractions are equal if the cross products are equal.	Use the equality test for fractions to determine if the fractions are equal. $\dfrac{3}{7} \overset{?}{=} \dfrac{2}{9}$ We form the cross products. $9 \cdot 3 \overset{?}{=} 7 \cdot 2$ $27 \neq 14$; therefore $\dfrac{3}{7} \neq \dfrac{2}{9}$	13. Use the equality test for fractions to see if the fractions are equal. $\dfrac{14}{9} = \dfrac{42}{27}$
Determining if a statement is a proportion, p. 262 To determine if a statement is a proportion, we use the equality test for fractions. The cross products *must be equal* for the statement to be a proportion.	Is $\dfrac{9}{31} = \dfrac{7}{28}$ a proportion? $28 \times 9 \overset{?}{=} 31 \times 7$ $252 \neq 217$ This is *not* a proportion.	14. Is $\dfrac{6}{17} = \dfrac{13}{27}$ a proportion?
Finding the missing number in a proportion, p. 263 To solve a proportion for the value of the variable: **1.** Form the cross product. **2.** Simplify the result. **3.** Solve the equation by dividing on both sides so that the variable stands alone.	Solve for n. $\dfrac{13}{n} = \dfrac{52}{8}$ $8 \cdot 13 = n \cdot 52$ Form the cross product. $104 = n \cdot 52$ $\dfrac{104}{52} = n$ Divide by 52. $2 = n$	15. Solve for n. $\dfrac{7}{n} = \dfrac{35}{115}$
Solving applied problems, pp. 264–265 **1.** Write a proportion using a variable to represent the unknown value. **2.** Solve the proportion.	A hockey player can score 3 goals in every 7 games. At this rate, how many goals should this hockey player score in 35 games? $\dfrac{3 \text{ goals}}{7 \text{ games}} = \dfrac{n \text{ goals}}{35 \text{ games}}$ Write the proportion. $35 \times 3 = 7 \times n$ $105 = 7 \times n$ $\dfrac{105}{7} = n$ $15 = n$ He should score 15 goals.	16. A telephone solicitor usually sells 4 products for every 15 calls made. At this rate, how many products should the soliticitor sell after making 60 calls?

Chapter 4 Review Problems

Vocabulary

Fill in the blank with the definition for each word.

1. (4.1) Prime number: _____

2. (4.1) Composite number: _____

3. (4.2) Proper fraction: _____

4. (4.2) Improper fraction: _____

5. (4.2) Mixed number: _____

6. (4.3) Equivalent fractions: _____

7. (4.5) Ratio: _____

8. (4.5) Rate: _____

9. (4.6) Proportion: _____

Section 4.1

Determine if each number is divisible by 2, 3, and/or 5.

10. 588,640

11. 41,595

State whether each number is prime, composite, or neither.

12. 0, 7, 21, 50, 11, 25, 51

13. 1, 32, 7, 12, 50, 6, 13, 41

Express as a product of prime factors.

14. 36

15. 56

16. 425

17. 90

Section 4.2

Evaluate, if possible. Assume that all variables in the denominator are nonzero.

18. $\dfrac{3}{0}$

19. $\dfrac{0}{3}$

20. $\dfrac{y}{y}$

Solve each applied problem.

21. **Males vs. Females** In a college history class, 20 of the 69 students are men. Write the fraction that describes the part of the class that are men.

22. **Dinner Desserts** At a potluck dinner, 8 of the 25 dishes are desserts. Write the fraction that describes the part of the dishes that are *not* desserts.

Write each improper fraction as a mixed number or a whole number.

23. $\dfrac{43}{5}$

24. $\dfrac{55}{6}$

25. $\dfrac{56}{8}$

Change each mixed number to an improper fraction.

26. $2\dfrac{1}{3}$

27. $6\dfrac{3}{5}$

28. $10\dfrac{2}{5}$

Section 4.3

Find an equivalent fraction with the given denominator.

29. $\dfrac{2}{9} = \dfrac{?}{18}$

30. $\dfrac{3}{4} = \dfrac{?}{36}$

31. $\dfrac{4}{5} = \dfrac{?}{35x}$

32. $\dfrac{6}{11} = \dfrac{?}{33y}$

Simplify. Assume that all variables in the denominator are nonzero.

33. $\dfrac{55}{75}$

34. $\dfrac{48}{54}$

35. $\dfrac{108}{36}$

36. $\dfrac{175}{75}$

37. $\dfrac{25x}{60x}$

38. $\dfrac{84x}{105xy}$

39. $\dfrac{-16}{18}$

40. $\dfrac{24}{-36}$

Section 4.4

Simplify.

41. $\dfrac{y^5}{y^3}$

42. $\dfrac{3^2}{3^3}$

43. $\dfrac{8^2}{3^4}$

44. $\dfrac{x^5 y^3}{x^2 y^9}$

45. $\dfrac{2^3 x^0}{2^6 x^9}$

46. $\dfrac{3^2 y^0}{3^3 y^6}$

47. $\dfrac{20x^5}{35x^9}$

48. $\dfrac{18y^6}{6y^4}$

49. $(3y^2)^3$

50. $(2^4 x)^2$

51. $\left(\dfrac{3}{y}\right)^2$

52. $\left(\dfrac{x}{2}\right)^3$

Section 4.5

Write each ratio as a fraction in simplest form.

53. 30 to 46

54. $15 : 35$

55. Write the ratio as a fraction and simplify. A mixture contains 20 milliliters of acid and 55 milliliters of water.

(a) State the ratio of acid to water. (b) State the ratio of water to acid.

Find the rate.

56. What is the unit rate in dollars per washcloth? $35 for 7 washcloths

57. What is the unit rate in inches per hour? 112 inches in 28 hours

58. *Gas Mileage* Juan drove his new car 286 miles on 11 gallons of gas. What is the unit rate in miles per gallon for his car?

59. *Typing Speed* It took Tan 30 minutes to type his 3090-word essay. How many words per minute can he type?

60. *Law Firm Staffing* The Johnson and Associates law firm has 32 legal secretaries for every 16 lawyers and 12 paralegals for every 4 lawyers.

(a) How many legal secretaries per lawyer does the law firm have?

(b) How many paralegals per lawyer does the law firm have?

(c) How many paralegals would be required if the law firm had 60 lawyers?

61. *Comparison Shopping* The KB Music Store is having a special on their '60s CDs: 6 for $72 or 8 for $96.

(a) Find each unit price.

(b) Which is the better buy?

62. *Profit/Loss* Jenny's Clothing Store recorded the profits shown in the chart in the second quarter of the year.

(a) Write as a ratio. The profit in April to the profit in May

(b) Write as a unit rate. The average profit per day in the month of June

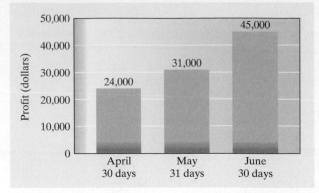

Section 4.6

Write each proportion.

63. 3 is to 7 as 21 is to 49.

64. 2 teachers is to 50 students as 6 teachers is to 150 students.

Write the proportion to express each statement.

65. *Reading Maps* If 2 inches on a map represent 190 miles, 6 inches represent 570 miles.

66. *Gas Mileage* If Tuan can drive 234 miles in his Honda Accord on 9 gallons of gas, he should be able to drive 468 miles on 18 gallons of gas.

Use the equality test for fractions to determine if the fractions are equal.

67. $\dfrac{3}{4} \overset{?}{=} \dfrac{50}{70}$

68. $\dfrac{13}{91} \overset{?}{=} \dfrac{12}{84}$

Determine if each statement is a proportion.

69. $\dfrac{3}{7} \overset{?}{=} \dfrac{12}{28}$

70. $\dfrac{4 \text{ goals}}{7 \text{ attempts}} \overset{?}{=} \dfrac{20 \text{ goals}}{46 \text{ attempts}}$

Find the value of the variable in each proportion. Check your answer.

71. $\dfrac{x}{30} = \dfrac{2}{15}$

72. $\dfrac{6}{5} = \dfrac{54}{x}$

73. $\dfrac{17 \text{ quarts}}{47 \text{ square feet}} = \dfrac{n \text{ quarts}}{94 \text{ square feet}}$

74. *Reading Maps* On a map of California, 1 inch on the map represents 120 miles. How many miles do 3 inches represent?

75. *Bicycle Speed* If Dale pedals his bicycle at 75 revolutions per minute, he travels 12 miles per hour. How fast does he go if he pedals at 100 revolutions per minute?

▲ 76. *Patio Cover Enlargement* Lacey has a small rectangular patio cover 4 feet wide and 7 feet long in his yard. He wants to enlarge the cover, keeping the dimensions of the new patio cover proportional to the old one. If he has room to increase the length to 14 feet, how wide should the patio cover be?

77. *Hourly Wages* Leslie and Gloria work at the same hospital but earn different salaries. Assume that Gloria's deductions for taxes are proportional to Leslie's deductions, and answer the following.
(a) Find Gloria's federal withholding.
(b) Find Gloria's state withholding.

Weekly paycheck			MIRA HOSPITAL
Employee			
Leslie Brook			
Total Gross Pay	Federal Withholding	State Withholding	
$400	$60	$20	

Weekly paycheck			MIRA HOSPITAL
Employee			
Gloria Smart			
Total Gross Pay	Federal Withholding	State Withholding	
$340			

How Am I Doing? Chapter 4 Test

CHAPTER Test Prep VIDEOS   MATH COACH   MyMathLab®   You Tube™

After you take this test read through the Math Coach pages 279–280. Math Coach videos are available via MyMathLab and YouTube. Step-by-step test solutions on the Chapter Test Prep Videos are also available via MyMathLab and YouTube. (Search "BlairTobeyPrealgebra" and click on "Channels.")

1. Determine if 230 is divisible by 2, 3, and/or 5.

2. Identify each number as prime, composite, or neither.
   (a) 27   (b) 1   (c) 19

Express as a product of prime factors.

Mc **3.** 84   **4.** 120

Divide, if possible. Assume all variables in the denominator are nonzero.

5. (a) $\dfrac{0}{4}$ (b) $\dfrac{t}{t}$ (c) $\dfrac{12}{0}$

Solve.

6. There are 36 students in a prealgebra class. Seventeen of these students are men. What fraction of the students are men?

7. The local parks and recreation department offers 12 classes for children, 7 classes for teens, and 16 classes for adults.
   (a) What fraction of the classes are for adults?
   (b) What fraction of the classes are *not* for children?

8. Change to a mixed number or whole number. $\dfrac{8}{5}$

9. Change $7\dfrac{1}{6}$ to an improper fraction.

Find an equivalent fraction with the given denominator.

10. $\dfrac{7}{8} = \dfrac{?}{40}$ Mc **11.** $\dfrac{4}{9} = \dfrac{?}{27y}$

Simplify.

12. $\dfrac{-18}{56}$ **13.** $\dfrac{16x}{32x^2y}$ Mc **14.** $\dfrac{y^3z^4}{y^7z}$ **15.** $\dfrac{8^2}{7^3}$

1.	☐
2. (a)	☐
(b)	☐
(c)	☐
3.	☐
4.	☐
5. (a)	☐
(b)	☐
(c)	☐
6.	☐
7. (a)	☐
(b)	☐
8.	☐
9.	☐
10.	☐
11.	☐
12.	☐
13.	☐
14.	☐
15.	☐

16. $\dfrac{42x^7y^6}{36x^0y^9}$ **17.** $(2y^4)^3$ **18.** $(x^3)^5$ **19.** $\left(\dfrac{x}{3}\right)^2$

Solve.

20. Write the ratio as a fraction and simplify. A soccer team had a season record of 18 wins and 4 losses.

   (a) State the ratio of wins to losses.

   (b) State the ratio of losses to wins.

21. Joe played tennis for 20 minutes and burned 150 calories. How many calories did he burn per minute?

22. A car travels 525 miles on 25 gallons of gas. Find how many miles the car can be driven on one gallon of gas.

23. Micro Computers is having a sale on copy paper: $36 for 12 reams of paper or $96 for 48 reams of paper.

   (a) Find each unit price.

   (b) Which is the better deal?

24. *Translate the statement into a proportion.*
If 2 inches on a map represents 225 miles, 6 inches represent 675 miles.

25. Determine if $\dfrac{20}{52} \overset{?}{=} \dfrac{5}{13}$ is a proportion.

26. Find the value of x. $\dfrac{4}{6} = \dfrac{20}{x}$

27. Two large birthday cakes are the same size. One cake is cut into 16 slices, while the other cake is cut into 24 slices. Is 4 out of 16 slices of cake the same amount as 6 out of 24 slices?

ᴹꞔ **28.** A bottle of fertilizer for your lawn states that you need 2 tablespoons to fertilize 400 square feet of lawn. How many tablespoons will you need to fertilize 1600 square feet of lawn?

17.

18.

19.

20. (a)

(b)

21.

22.

23. (a)

(b)

24.

25.

26.

27.

28.

Total Correct:

MATH COACH

Mastering the skills you need to do well on the test.

Students often make the same types of errors when they do the Chapter 4 Test. Here are some helpful hints to keep you from making these common errors on test problems.

Finding the Prime Factors of Whole Numbers—Problem 3
Express as a product of prime factors. 84

> **Helpful Hint** Choose the factoring method (division ladder or factor tree) that you prefer to solve this problem. Make sure all factors in your answer are *prime numbers*. Check your answer by multiplying all factors to be sure this product equals the number you are factoring.

Look at your work for Problem 3. Examine your steps.

Division ladder: Did you divide 84 by 4 or 6?

Yes _____ No _____

If you answered Yes, you forgot to use *only prime numbers* as divisors. Stop now and make this correction.

Did you include the final quotient as a factor?

Yes _____ No _____

If you answered No, stop and complete this step.

Factor Tree: Are all the factors in your answer prime numbers?

Yes _____ No _____

Did you forget to include a prime number listed in the tree as part of your product?

Yes _____ No _____

If you answered No to either of these questions, rework the problem and factor each number until all factors are prime. Circle all prime numbers as you factor to be sure you can easily see which factors to include in your answer.

Remember to check your answer by multiplying all the prime factors to be sure that the product is equal to the original number.

If you answered Problem 3 incorrectly, go back and rework the problem using these suggestions.

Finding Equivalent Fractions—Problem 11
Find an equivalent fraction with the given denominator. $\dfrac{4}{9} = \dfrac{?}{27y}$

> **Helpful Hint** Remember to multiply *both* the numerator and denominator by the same nonzero number or expression. The equivalent fraction should have the given denominator of 27y.

Did you choose 3 as the nonzero number or expression?

Yes _____ No _____

If you answered Yes, your resulting denominator of the equivalent fraction does not equal 27y. Consider what expression, when multiplied by 9, equals 27y.

Did you remember to multiply both the numerator and denominator by the same nonzero number or expression?

Yes _____ No _____

If you answered No, stop and make this correction.

Always double-check the product of the original denominator and your nonzero number or expression in the denominator. The result should be equal to the given denominator.

Now go back and rework the problem using these suggestions.

Need help? Watch the MATH COACH videos in MyMathLab® or on You Tube.

Using the Quotient Rule for Exponents—Problem 14 Simplify. $\dfrac{y^3 z^4}{y^7 z}$

> **Helpful Hint** For this problem, you must apply the quotient rule twice: once for the y variables and once for the z variables. Recall that when an exponent is not written, it is understood to be 1.

Did you notice that the exponent for y is larger in the denominator than in the numerator, and that the exponent for z is larger in the numerator than in the denominator?

Yes ____ No ____

If you answered No, go back and look carefully at the problem again.

Is y^4 in the denominator of your answer?

Yes ____ No ____

If you answered No, remember to subtract the exponents of y and to place the resulting y expression in the denominator since the *original exponent* for y is *larger* in the *denominator*.

Did you remember to subtract exponents for z and obtain the expression z^3?

Yes ____ No ____

If you answered No, make sure that you write a 1 as the exponent for z in the denominator and complete this step again.

If you answered Problem 14 incorrectly, go back and rework the problem using these suggestions.

Solving Applied Problems Involving Proportions—Problem 28 A bottle of fertilizer for your lawn states that you need 2 tablespoons to fertilize 400 square feet of lawn. How many tablespoons will you need to fertilize 1600 square feet of lawn?

> **Helpful Hint** When writing and setting up your proportion, remember to write the *unit names* in your proportion to avoid setting up the proportion incorrectly. Make sure that the same unit name appears in the numerator of both fractions and the same unit name appears in the denominator of both fractions.

Did you write $\dfrac{2 \text{ tablespoons}}{400 \text{ square feet}}$ as your first fraction and $\dfrac{x \text{ tablespoons}}{1600 \text{ square feet}}$ as your second fraction?

Yes ____ No ____

If you answered No, stop and reread the problem to see how these fractions were obtained.

Did you simplify $\dfrac{2}{400}$ before you solved the proportion?

Yes ____ No ____

If you answered No, go back and complete this step again. Simplifying first will make the calculations easier to perform and will help with accuracy.

Remember to include the units for x in your final answer.

If you answered Problem 28 incorrectly, go back and rework the problem using these suggestions.

Need more help? Look for section examples marked with $\mathbb{MC}$ to review.

280

We deal with fractions often in our everyday lives. Many situations call for precise measurements, and this requires us to use operations with fractions instead of decimals. Whether you are preparing a recipe; planning a landscaping, remodeling, or sewing project; or determining your gas mileage, you encounter operations with fractions. As you master the topics in this chapter, you will gain the basic skills necessary for many situations in your everyday life.

Operations on Fractional Expressions

5.1 Multiplying and Dividing Fractional Expressions

Student Learning Objectives

After studying this section, you will be able to:

① **Multiply fractions.**

② **Divide fractions.**

③ **Solve applied problems involving fractions.**

① Multiplying Fractions

Let's look at multiplication with fractions. We begin with an illustration representing the fraction $\frac{1}{3}$.

One whole divided into 3 parts with each part equal to the fraction $\frac{1}{3}$ (1 out of 3 parts)

Let's see what it means to have one-half of $\frac{1}{3}$. Imagine cutting each of the 3 pieces in the figure above in half. Instead of 3 parts, we would then have 6 parts.

One whole divided into 6 parts with each part equal to the fraction $\frac{1}{6}$ (1 out of 6 parts)

This is one-half of $\frac{1}{3}$.

From the illustration above, we see that one-half of $\frac{1}{3}$ is $\frac{1}{6}$. To find *one-half* of $\frac{1}{3}$, *we multiply* $\frac{1}{2} \cdot \frac{1}{3} = \frac{1}{6}$. We state the rule for multiplying fractions.

RULE FOR MULTIPLYING FRACTIONS

To multiply fractions, we multiply numerator times numerator and denominator times denominator.

In general, for all numbers a, b, c, and d, with $b \neq 0$, and $d \neq 0$,

$$\frac{a}{b} \cdot \frac{c}{d} = \frac{a \cdot c}{b \cdot d} \qquad \frac{2}{3} \cdot \frac{1}{7} = \frac{2 \cdot 1}{3 \cdot 7} = \frac{2}{21}$$

We often use multiplication of fractions to describe *taking a fractional part of something*. The word *of* indicates that we perform the operation multiplication.

EXAMPLE 1   Find $\frac{3}{7}$ of $\frac{2}{9}$.

Solution

$$\frac{3}{7} \text{ of } \frac{2}{9} = \frac{3}{7} \cdot \frac{2}{9} = \frac{3 \cdot 2}{7 \cdot 9} = \frac{3 \cdot 2}{7 \cdot 3 \cdot 3} = \frac{\overset{1}{\cancel{3}} \cdot 2}{7 \cdot \underset{1}{\cancel{3}} \cdot 3} = \frac{2}{21}$$

Student Practice 1   Find $\frac{5}{8}$ of $\frac{2}{3}$.

NOTE TO STUDENT: *Fully worked-out solutions to all of the Student Practice problems can be found at the back of the text starting at page SP-1.*

In Example 1 we removed the common factor 3. This process is often referred to as removing a factor of 1, or **factoring out common factors.**

Factoring out common factors.

$$\frac{3 \cdot 2}{7 \cdot 3 \cdot 3} = \frac{\overset{1}{\cancel{3}}}{\underset{1}{\cancel{3}}} \cdot \frac{2}{7 \cdot 3} \qquad \text{We factor the fraction and write } \frac{3}{3} \text{ as } \frac{1}{1}.$$

We often skip this step and *factor out the common factor* by placing slashes on common factors as we did in Example 1.

We should always factor out common factors before we multiply the numerators and denominators; otherwise, we must simplify the product. This is a lot of extra work! Let's see what would happen if in Example 1 we multiplied the numbers in the numerator and denominator before we factored out common factors.

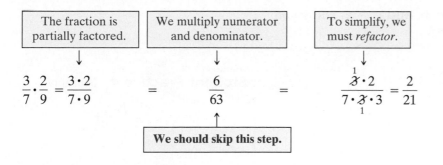

When multiplying fractions, always try to simplify before performing the multiplication in the numerator and the denominator of the fraction. This will ensure that the product will be in lowest terms.

EXAMPLE 2   Multiply. $\dfrac{9}{20} \cdot \dfrac{5}{21}$

Solution

$$\frac{9}{20} \cdot \frac{5}{21} = \frac{9 \cdot 5}{20 \cdot 21} = \frac{3 \cdot 3 \cdot 5}{2 \cdot 2 \cdot 5 \cdot 3 \cdot 7} = \frac{3 \cdot \overset{1}{\cancel{3}} \cdot \overset{1}{\cancel{5}}}{2 \cdot 2 \cdot \underset{1}{\cancel{3}} \cdot \underset{1}{\cancel{5}} \cdot 7} = \frac{3}{28}$$

Student Practice 2   Multiply.

$$\frac{8}{15} \cdot \frac{12}{14}$$

EXAMPLE 3   Multiply. $\dfrac{-2}{18} \cdot \dfrac{9}{11}$

Solution   When multiplying positive and negative fractions, we determine the sign of the product and then multiply and simplify.

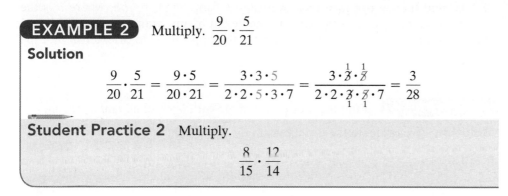

$\dfrac{-2}{18} \cdot \dfrac{9}{11} = (-)$ The product is negative because there is 1 negative sign and 1 is an odd number.

$= -\dfrac{2 \cdot 9}{18 \cdot 11}$

$= -\dfrac{2 \cdot 9}{2 \cdot 9 \cdot 11}$ Factor.

$= -\dfrac{\overset{1}{\cancel{2}} \cdot \overset{1}{\cancel{9}}}{\underset{1}{\cancel{2}} \cdot \underset{1}{\cancel{9}} \cdot 11}$ Factor out common factors and simplify.

$= -\dfrac{1}{11}$

Student Practice 3   Multiply.

$$\frac{-12}{24} \cdot \frac{-2}{13}$$

CAUTION: If you do not write the steps showing the 1 above the slashes, you *must remember* that the factor 1 is part of the multiplication. For example, in Example 3

$$-\frac{\cancel{2} \cdot \cancel{9}}{\cancel{2} \cdot \cancel{9} \cdot 11} = -\frac{1}{11} \text{ not } -\frac{0}{11}.$$

When multiplying a fraction by a whole number expression, it is helpful to write the expression with a denominator of 1. We can do this since $\frac{6}{1} = 6$, and $\frac{x}{1} = x$, and so on.

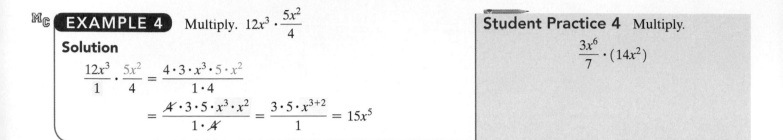

EXAMPLE 4   Multiply. $12x^3 \cdot \dfrac{5x^2}{4}$

Solution

$$\frac{12x^3}{1} \cdot \frac{5x^2}{4} = \frac{4 \cdot 3 \cdot x^3 \cdot 5 \cdot x^2}{1 \cdot 4}$$

$$= \frac{\cancel{4} \cdot 3 \cdot 5 \cdot x^3 \cdot x^2}{1 \cdot \cancel{4}} = \frac{3 \cdot 5 \cdot x^{3+2}}{1} = 15x^5$$

Student Practice 4   Multiply.

$$\frac{3x^6}{7} \cdot (14x^2)$$

 height
base

To find the area of a triangle we use the formula $A = \frac{1}{2}bh$, where b is the base of the triangle and h is the height. The height of any triangle is the distance of a line drawn from a vertex perpendicular to the opposite side or extension of the opposite side.

▲ **EXAMPLE 5**   Find the area of a triangle with $b = 12$ in. and $h = 7$ in.

Solution   We evaluate the formula with the given values.

$$A = \frac{1}{2}bh$$

$$= \frac{1}{2} \cdot 12 \text{ in.} \cdot 7 \text{ in.} \qquad \text{We replace } b \text{ with 12 in. and } h \text{ with 7 in.}$$

$$= \frac{1 \cdot 12 \text{ in.} \cdot 7 \text{ in.}}{2 \cdot 1 \cdot 1} \qquad \text{We multiply: } \frac{1}{2} \cdot \frac{12 \text{ in.}}{1} \cdot \frac{7 \text{ in.}}{1}.$$

$$= \frac{1 \cdot 2 \cdot 6 \cdot 7 \text{ in.} \cdot \text{in.}}{2} \qquad \text{We factor: } 12 = 2 \cdot 6. \text{ Then we simplify.}$$

$$A = 42 \text{ in.}^2 \qquad\qquad \frac{\cancel{2} \cdot 6 \cdot 7}{\cancel{2}} = 6 \cdot 7 = 42; \text{ in.} \cdot \text{in.} = \text{in.}^2$$

▲ **Student Practice 5**   Find the area of a triangle with $b = 20$ cm and $h = 9$ cm.

② Dividing Fractions

Before we discuss division with fractions, we introduce *reciprocal fractions*. The fractions $\frac{2}{3}$ and $\frac{3}{2}$ are called *reciprocals*, and $\frac{x}{a}$ and $\frac{a}{x}$ are also reciprocals. Let's see what happens when we multiply reciprocal fractions.

$$\frac{2}{3} \cdot \frac{3}{2} = \frac{2 \cdot 3}{3 \cdot 2} = \frac{\cancel{2} \cdot \cancel{3}}{\cancel{2} \cdot \cancel{3}} = 1 \qquad\qquad \frac{x}{a} \cdot \frac{a}{x} = \frac{x \cdot a}{a \cdot x} = \frac{\cancel{x} \cdot \cancel{a}}{\cancel{x} \cdot \cancel{a}} = 1$$

Notice that both products are equal to 1. If the product of two numbers is 1, we say that these two numbers are **reciprocals** of each other.

To find the reciprocal of a nonzero fraction, we *interchange* the numerator and denominator. This is often referred to as **inverting the fraction**.

EXAMPLE 6   Find the reciprocal.

(a) $\dfrac{-7}{8}$

(b) 6

Solution   To find the reciprocal, we invert the fraction.

(a) $\underbrace{\dfrac{-7}{8} \rightarrow \dfrac{8}{-7}}_{\text{invert}} = -\dfrac{8}{7}$

(b) $6 = \underbrace{\dfrac{6}{1} \rightarrow \dfrac{1}{6}}_{\text{invert}} = \dfrac{1}{6}$

Student Practice 6   Find the reciprocal.

(a) $\dfrac{a}{-y}$

(b) 4

Now let's see how we use reciprocals when we divide fractions. Consider this problem.

A total of $\frac{3}{4}$ pound of peanuts is to be placed in $\frac{1}{4}$-pound bags. How many $\frac{1}{4}$-pound bags will there be?

We must find how many $\frac{1}{4}$'s are in $\frac{3}{4}$. This is the division situation $\frac{3}{4} \div \frac{1}{4}$. To illustrate, we draw a picture.

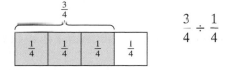

$$\frac{3}{4} \div \frac{1}{4}$$

How many bags?

Notice that there are three $\frac{1}{4}$'s in $\frac{3}{4}$. Thus, $\frac{3}{4} \div \frac{1}{4} = 3$. Therefore, we can fill three $\frac{1}{4}$-pound bags.

Now, how can we find $\frac{3}{4} \div \frac{1}{4} = 3$ without drawing a diagram? Notice that if we *find the reciprocal of (invert) the second fraction* and then *multiply,* we get 3.

Invert the second
fraction and multiply.
↓

$$\frac{3}{4} \div \frac{1}{4} = \frac{3}{4} \cdot \frac{4}{1} = \frac{3 \cdot \cancel{4}}{\cancel{4} \cdot 1} = \frac{3}{1} = 3$$

Thus dividing $\frac{3}{4}$ by $\frac{1}{4}$ is the same as multiplying $\frac{3}{4}$ by the reciprocal of $\frac{1}{4}$.

RULE FOR DIVIDING FRACTIONS

To divide two fractions, we find the reciprocal of (invert) the second fraction and *multiply.*

$$\frac{a}{b} \div \frac{c}{d} = \frac{a}{b} \cdot \frac{d}{c} \quad \text{(when } b, c, \text{ and } d \text{ are not 0)}$$

EXAMPLE 7 Divide. $\dfrac{-4}{11} \div \left(\dfrac{-3}{5} \right)$

Solution There are 2 negative signs in the division. The number 2 is even, so the answer is positive.

$$\dfrac{-4}{11} \div \left(\dfrac{-3}{5} \right) = \dfrac{-4}{11} \cdot \left(\dfrac{5}{-3} \right) \qquad \text{Invert the second fraction and multiply.}$$

$$= \dfrac{4 \cdot 5}{11 \cdot 3} = \dfrac{20}{33} \qquad \text{The product of two negative numbers is positive.}$$

Student Practice 7 Divide.

$$\dfrac{-7}{8} \div \dfrac{5}{13}$$

Understanding the Concept

Reciprocals and Division   Why do we invert the second fraction and multiply when we divide? Let's write the fraction $\frac{8}{2}$ as both a division problem and a multiplication problem and see what happens.

Division problem: $\dfrac{8}{2} = 8 \div 2 = 4$ We divide by 2.

Multiplication problem: $\dfrac{8}{2} = 8 \cdot \dfrac{1}{2} = 4$ We invert 2 and multiply.

Thus we see that whether we *divide* 8 by 2 or *multiply* 8 by $\frac{1}{2}$ (the reciprocal of 2), we get the same result.

Since we cannot divide by zero, we will assume that all variables in denominators are nonzero.

EXAMPLE 8 Divide. $\dfrac{7x^4}{20} \div \left(\dfrac{-14x^2}{45} \right)$

Solution

$$\dfrac{7x^4}{20} \div \left(\dfrac{-14x^2}{45} \right) = \dfrac{7x^4}{20} \cdot \left(\dfrac{45}{-14x^2} \right) \qquad \text{Invert the second fraction and multiply.}$$

$$= - \qquad \begin{array}{l} \text{The product is negative since there is 1} \\ \text{negative sign and 1 is an odd number.} \end{array}$$

$$= - \dfrac{\cancel{7} \cdot \cancel{5} \cdot 3 \cdot 3 \cdot x^4}{2 \cdot 2 \cdot \cancel{5} \cdot 2 \cdot \cancel{7} \cdot x^2} \qquad \text{Factor and simplify.}$$

$$= - \dfrac{9x^2}{8} \qquad\qquad \dfrac{x^4}{x^2} = x^{4-2} = x^2.$$

Student Practice 8 Divide.

$$\dfrac{9x^6}{21} \div \left(\dfrac{-42x^4}{18} \right)$$

CAUTION: To obtain the right answer, it is important that you *invert the second fraction,* not the first.

EXAMPLE 9 Divide. $16x^5 \div \dfrac{8x^2}{11}$

Solution

$$16x^5 \div \frac{8x^2}{11} = \frac{16x^5}{1} \cdot \frac{11}{8x^2} = \frac{2 \cdot \cancel{8} \cdot 11 \cdot x^5}{1 \cdot \cancel{8} \cdot x^2}$$

$$= 22x^3 \quad \text{Simplify: } \frac{2 \cdot 11}{1} = 22 \text{ and } \frac{x^5}{x^2} = x^{5-2} = x^3.$$

Student Practice 9 Divide.

$$28x^5 \div \frac{4x}{19}$$

③ Solving Applied Problems Involving Fractions

One of the most important steps in solving a real-life application is determining which operation to use. We often ask ourselves, "Should I multiply or divide?" The multiplication and division situations for fractions are similar to those for whole numbers and are stated below for your review. Drawing pictures and making charts can also help you determine whether to multiply or divide.

> **MULTIPLICATION SITUATIONS**
>
> We multiply in situations that require repeated addition or taking a fractional part *of* something.

1. *Repeated addition.* A recipe requires $\frac{1}{4}$ cup of flour for each serving. How many cups of flour are needed to make 3 servings?

$$
\begin{array}{ccccccc}
1 \text{ serving} & + & 1 \text{ serving} & + & 1 \text{ serving} & = & 3 \text{ servings} \\
\downarrow & & \downarrow & & \downarrow & & \\
\dfrac{1}{4} & + & \dfrac{1}{4} & + & \dfrac{1}{4} & = & 3 \cdot \dfrac{1}{4} = \dfrac{3}{4} \text{ cup of flour}
\end{array}
$$

We add $\frac{1}{4}$ cup 3 times, or $3 \cdot \frac{1}{4}$.

2. *Taking a fractional part of something.* A recipe requires $\frac{3}{4}$ cup of flour. How much flour is needed to make $\frac{1}{2}$ of the recipe?

We want to find $\dfrac{1}{2}$ of $\dfrac{3}{4}$ or $\dfrac{1}{2} \cdot \dfrac{3}{4} = \dfrac{3}{8}$ cup of flour.

> **DIVISION SITUATIONS**
>
> We divide when we want to split an amount into a certain number of equal parts or to find how many groups of a certain size are in a given amount.

1. *Split an amount into equal parts.* A pipe $\frac{3}{5}$ foot long must be cut into 2 equal parts. How long is each part?

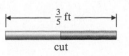

$$\frac{3}{5} \div 2 = \frac{3}{5} \cdot \frac{1}{2} = \frac{3}{10} \quad \text{Each part is } \frac{3}{10} \text{ foot long.}$$

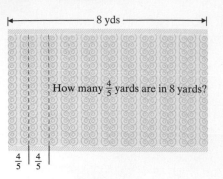

How many $\frac{4}{5}$ yards are in 8 yards?

$\frac{4}{5}$ $\frac{4}{5}$

2. *Find how many groups of a certain size are in a given amount.* A scarf requires $\frac{4}{5}$ yard of material. How many scarves can be made from 8 yards of material? We must find how many $\frac{4}{5}$-yard segments are in 8 yards.

$$8 \div \frac{4}{5} = 8 \cdot \frac{5}{4} = \frac{2 \cdot 4 \cdot 5}{4} = 10 \qquad \text{10 scarves can be made.}$$

When we solve applied problems, we must decide whether we have a division situation or a multiplication situation so we know which operation to use.

EXAMPLE 10   Samuel Jensen has $\frac{9}{40}$ of his income withheld for taxes and retirement. What amount is withheld each week if he earns $1440 per week?

Solution   The key phrase is "$\frac{9}{40}$ *of* his income." The word *of* often indicates multiplication.

$$\frac{9}{40} \text{ of } income \text{ is withheld for taxes and retirement}$$

$$\frac{9}{40} \cdot \$1440 = \frac{9}{40} \cdot \frac{\$1440}{1} = \$324$$

$324 is withheld for taxes and retirement each week.

Student Practice 10   Nancy Levine places $\frac{2}{13}$ of her income in a savings account each week. How much money does she place in her savings account if her income is $1703 per week?

▲ **EXAMPLE 11**   Harry must install 44 feet of baseboard along the edge of the floor of a library. After placing a nail at the corner, he must place a nail in the baseboard every $\frac{2}{3}$ foot. How many more nails will he need?

Solution   We draw a picture.

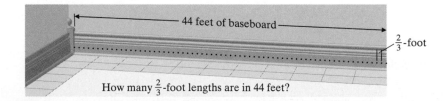

44 feet of baseboard

$\frac{2}{3}$-foot

How many $\frac{2}{3}$-foot lengths are in 44 feet?

Since we need to find out how many $\frac{2}{3}$-foot lengths are in 44 feet, we divide.

$$44 \div \frac{2}{3} = 44 \cdot \frac{3}{2} = \frac{44 \cdot 3}{2} = \frac{2 \cdot 22 \cdot 3}{2} = 66$$

He will need 66 more nails.

Student Practice 11   Alice must place $\frac{3}{4}$ pound of sugar in 2 equal-size containers. How much sugar should Alice place in each container?

In this exercise set, assume that all variables in any denominator are nonzero.

Verbal and Writing Skills, Exercises 1–6

1. Explain how to multiply fractions.

2. Explain how to divide fractions.

3. To split $\frac{1}{4}$ into 6 equal parts, do we multiply or divide? Why?

4. To find out how much money we have if we inherit $\frac{1}{4}$ of $6000, do we multiply or divide? Why?

5. Write an application that requires multiplication of $\frac{1}{3}$ and 90.

6. Write an application that requires division of 27 and $\frac{3}{4}$.

Fill in each box to complete each problem.

7. $\dfrac{1}{2} \cdot \dfrac{3}{7} = \dfrac{3}{\boxed{}}$

8. $\dfrac{7}{11} \cdot \dfrac{2}{3} = \dfrac{14}{\boxed{}}$

9. $\dfrac{1}{4} \div \dfrac{3}{7} = \dfrac{1}{4} \cdot \dfrac{\boxed{}}{\boxed{}} = \dfrac{7}{12}$

10. $\dfrac{1}{8} \div \dfrac{5}{3} = \dfrac{1}{8} \cdot \dfrac{\boxed{}}{\boxed{}} = \dfrac{3}{40}$

11. $\dfrac{1}{2} \cdot \dfrac{\boxed{}}{\boxed{}} = \dfrac{5}{12}$

12. $\dfrac{3}{7} \cdot \dfrac{\boxed{}}{\boxed{}} = \dfrac{9}{14}$

13. $\dfrac{5}{7} \div \dfrac{4}{3} = \dfrac{5}{7} \boxed{} \dfrac{\boxed{}}{\boxed{}} = \dfrac{15}{28}$

14. $\dfrac{4}{9} \div \dfrac{3}{7} = \dfrac{4}{9} \boxed{} \dfrac{\boxed{}}{\boxed{}} = \dfrac{28}{27}$

Find the following.

15. $\dfrac{1}{4}$ of $\dfrac{1}{3}$

16. $\dfrac{1}{8}$ of $\dfrac{1}{2}$

17. $\dfrac{5}{21}$ of $\dfrac{7}{8}$

18. $\dfrac{3}{21}$ of $\dfrac{7}{8}$

Multiply. Be sure your answer is simplified.

19. $\dfrac{7}{12} \cdot \dfrac{8}{28}$

20. $\dfrac{6}{21} \cdot \dfrac{9}{18}$

21. $\dfrac{3}{20} \cdot \dfrac{8}{9}$

22. $\dfrac{4}{35} \cdot \dfrac{5}{24}$

23. $\dfrac{-3}{8} \cdot \left(\dfrac{14}{-6}\right)$

24. $\dfrac{-5}{7} \cdot \left(\dfrac{21}{-25}\right)$

25. $\dfrac{16}{11} \cdot \left(\dfrac{-18}{36}\right)$

26. $\dfrac{4}{3} \cdot \left(\dfrac{-45}{18}\right)$

27. $\dfrac{-2}{21} \cdot \left(\dfrac{-14}{18}\right)$

28. $\dfrac{-8}{20} \cdot \left(\dfrac{-25}{32}\right)$

29. $-14 \cdot \dfrac{1}{28}$

30. $-11 \cdot \dfrac{1}{22}$

31. $\dfrac{6}{35} \cdot 5$

32. $\dfrac{3}{25} \cdot 15$

33. $\dfrac{2x}{3} \cdot \dfrac{3x}{5}$

34. $\dfrac{4x}{5} \cdot \dfrac{5x}{3}$

35. $\dfrac{6x^4}{7} \cdot 28x$

36. $\dfrac{6x^5}{15} \cdot 30x^7$

37. $8x^2 \cdot \dfrac{3x^3}{2}$

38. $9x^3 \cdot \dfrac{2x^4}{3}$

Mixed Practice, Exercises 39–46 *Multiply. Be sure your answer is simplified.*

39. $\dfrac{2}{10} \cdot \dfrac{6}{8}$

40. $\dfrac{4}{25} \cdot \dfrac{5}{8}$

41. $\dfrac{6x}{25} \cdot \dfrac{15}{12x^2}$

42. $\dfrac{4x}{35} \cdot \dfrac{7}{6x^2}$

43. $\dfrac{-3y^3}{20} \cdot \dfrac{12}{21y^2}$

44. $\dfrac{15y^3}{26} \cdot \left(\dfrac{-13}{10y}\right)$

45. $\dfrac{3x^2}{15} \cdot \dfrac{18x^3}{20}$

46. $\dfrac{5x^4}{6} \cdot \dfrac{2x^2}{25}$

▲ *Find the area of each triangle with the given base and height.*

47. $b = 12$ m and $h = 8$ m

48. $b = 19$ in. and $h = 26$ in.

49. $b = 21$ in. and $h = 40$ in.

50. $b = 8$ cm and $h = 11$ cm

Find the reciprocal.

51. $\dfrac{1}{3}$

52. $\dfrac{1}{7}$

53. 5

54. 7

55. $\dfrac{2}{-5}$

56. $\dfrac{3}{-4}$

57. $\dfrac{-x}{y}$

58. $\dfrac{-a}{b}$

Divide. Be sure your answer is simplified.

59. $\dfrac{6}{14} \div \dfrac{3}{8}$

60. $\dfrac{8}{12} \div \dfrac{5}{6}$

61. $\dfrac{7}{24} \div \dfrac{9}{16}$

62. $\dfrac{9}{28} \div \dfrac{4}{7}$

63. $\dfrac{-1}{12} \div \dfrac{3}{4}$

64. $\dfrac{-1}{15} \div \dfrac{2}{3}$

65. $\dfrac{-7}{24} \div \left(\dfrac{7}{-8}\right)$

66. $\dfrac{-9}{28} \div \left(\dfrac{4}{-7}\right)$

67. $\dfrac{8x^6}{15} \div \dfrac{16x^2}{5}$

68. $\dfrac{6y^4}{20} \div \dfrac{36y^2}{10}$

69. $\dfrac{7x^4}{12} \div \dfrac{-28}{36x^2}$

70. $\dfrac{3x^4}{45} \div \dfrac{-27}{45x^5}$

71. $14 \div \dfrac{2}{7}$

72. $18 \div \dfrac{2}{3}$

73. $\dfrac{7}{22} \div 14$

74. $\dfrac{8}{26} \div 16$

75. $21x^4 \div \dfrac{7x}{3}$

76. $15x^3 \div \dfrac{5x^2}{8}$

77. $22x^3 \div \dfrac{11}{6x^5}$

78. $18x^4 \div \dfrac{9}{5x^6}$

Mixed Practice *Perform the operation indicated.*

79. (a) $\dfrac{1}{15} \cdot \dfrac{25}{21}$

(b) $\dfrac{1}{15} \div \dfrac{25}{21}$

80. (a) $\dfrac{1}{6} \cdot \dfrac{24}{15}$

(b) $\dfrac{1}{6} \div \dfrac{24}{15}$

81. (a) $\dfrac{2x^2}{3} \div \dfrac{12}{21x^5}$

(b) $\dfrac{2x^2}{3} \cdot \dfrac{12}{21x^5}$

82. (a) $\dfrac{3x^3}{7} \div \dfrac{21}{25x^4}$

(b) $\dfrac{3x^3}{7} \cdot \dfrac{21}{25x^4}$

83. $\dfrac{5x^7}{-27} \cdot \dfrac{-9}{20x^4}$

84. $\dfrac{7x^4}{-6} \cdot \dfrac{-30}{21x^2}$

85. $\dfrac{12x^6}{35} \div \dfrac{-16}{25x^2}$

86. $\dfrac{32x^3}{-15} \div \dfrac{28}{35x^4}$

Applications

87. ***Payroll Deductions*** Lilly Smith has $\frac{2}{15}$ of her weekly income withheld for taxes. What amount is withheld each week if she earns $1350 per week?

88. ***Savings Plan*** Elliott has $\frac{3}{16}$ of his weekly income placed in a savings account. What amount is placed in his savings account if he earns $1600 per week?

▲ 89. ***Pipe Length*** Babette must cut pipes into lengths of $\frac{3}{4}$ foot. How many pipes can she make from a pipe that is 12 feet long?

90. ***Satin Material*** Beth has a piece of satin 12 feet long. How many $\frac{2}{3}$-foot pieces of satin can she cut from the material?

91. ***Miles Run*** Julie runs 32 laps for her daily workout. If each lap is $\frac{1}{4}$ mile, how many miles does Julie run in her 32-lap workout?

92. ***Propeller Revolutions*** The propeller on the Ipswich River Cruise Boat turns 320 revolutions per minute. How fast would it turn at $\frac{3}{4}$ of that speed?

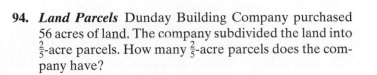

93. ***Factory Vats*** In the Westerfield Factory, products are made in vats that have the capacity to hold 120 quarts. If each bottle of a product contains $\frac{3}{4}$ quart, how many bottles can be made from each vat?

94. ***Land Parcels*** Dunday Building Company purchased 56 acres of land. The company subdivided the land into $\frac{2}{5}$-acre parcels. How many $\frac{2}{5}$-acre parcels does the company have?

One Step Further

95. $\dfrac{5}{14} \div \dfrac{2}{21} \div \left(\dfrac{15}{-3}\right)$

96. $\dfrac{8}{21} \div \left(\dfrac{4}{-7}\right) \div \dfrac{4}{3}$

97. After building a house the homeowners discover the land is sinking $\frac{2}{3}$ inch every year. How many years will it take the house to sink 4 inches?

98. On a recent history test, you answered $\frac{3}{5}$ of the questions and left the rest blank. Of the ones you answered, you got $\frac{2}{3}$ right. What fractional part of the questions did you answer correctly on the test?

99. *Pizza Party* James is planning a party at which he intends to serve pizza. If James estimates that each guest will eat $\frac{3}{8}$ of a pizza, how many pizzas should he order if 17 people will attend the party?

100. *Students' Home States* The records at a private college with 9600 students indicate that $\frac{1}{6}$ of these students are from California and $\frac{1}{5}$ are from Texas. How many students attending the college are not from Texas or California?

To Think About *Find the value of the variable(s) in each of the following.*

101. $\dfrac{3}{4} \cdot \dfrac{x}{27} = \dfrac{4}{9}$

102. $\dfrac{1}{2} \div \dfrac{3}{x} = \dfrac{8}{3}$

103. $\dfrac{2}{7}, \dfrac{4}{7}, \dfrac{x}{7}, \dfrac{8}{7}, \dfrac{10}{7}, \dfrac{y}{7}, \ldots$

104. $\dfrac{1}{5}, \dfrac{2}{5}, \dfrac{4}{5}, \dfrac{x}{5}, \dfrac{16}{5}, \dfrac{32}{5}, \dfrac{y}{5}, \ldots$

105. $\dfrac{1}{2}, \dfrac{2}{6}, \dfrac{4}{18}, \dfrac{x}{54}, \dfrac{16}{y}, \dfrac{32}{486}, \ldots$

106. $\dfrac{1}{3}, \dfrac{3}{6}, \dfrac{9}{x}, \dfrac{27}{24}, \dfrac{y}{48}, \dfrac{243}{96}, \ldots$

Cumulative Review *Find an equivalent fraction with the given denominator.*

107. [4.3.1] $\dfrac{2}{3} = \dfrac{?}{15}$

108. [4.3.1] $\dfrac{3}{4} = \dfrac{?}{20}$

Express as a product of prime factors.

109. [4.1.3] 120

110. [4.1.3] 145

Quick Quiz 5.1 *Perform the operation indicated.*

1. Find $\dfrac{3}{14}$ of $\dfrac{14}{27}$.

2. $\dfrac{-2x^2}{5} \cdot \dfrac{15x^4}{8}$

3. $\dfrac{5x^5}{8} \div \dfrac{x^3}{20}$

4. **Concept Check** Explain how you would divide $\frac{-16x^2}{3}$ by $8x$.

5.2 Multiples and Least Common Multiples of Algebraic Expressions

① Finding Multiples of Algebraic Expressions

To generate a list of **multiples** of a number, we multiply that number by 1, and then by 2, and then by 3, and so on. For example, we can list some multiples of 4 by multiplying 4 by 1, 2, 3, . . . , 8.

$$4 \cdot 1 \quad 4 \cdot 2 \quad 4 \cdot 3 \quad 4 \cdot 4 \quad 4 \cdot 5 \quad 4 \cdot 6 \quad 4 \cdot 7 \quad 4 \cdot 8 \ldots$$

Multiples of 4:   4,   8,   12,   16,   20,   24,   28,   32, . . .

EXAMPLE 1

(a) List the first six multiples of $8x$ and the first six multiples of $12x$.

(b) Which of these multiples are common to both lists?

Solution

(a)

$$8x \cdot 1 \quad 8x \cdot 2 \quad 8x \cdot 3 \quad 8x \cdot 4 \quad 8x \cdot 5 \quad 8x \cdot 6$$

Multiples of $8x$: $8x$, $16x$, $24x$, $32x$, $40x$, $48x$

$$12x \cdot 1 \quad 12x \cdot 2 \quad 12x \cdot 3 \quad 12x \cdot 4 \quad 12x \cdot 5 \quad 12x \cdot 6$$

Multiples of $12x$: $12x$, $24x$, $36x$, $48x$, $60x$, $72x$

(b) The multiples common to both lists are $24x$ and $48x$.

Student Practice 1

(a) List the first five multiples of $12x$ and the first five multiples of $20x$.

(b) Which of these multiples are common to both lists?

NOTE TO STUDENT: *Fully worked-out solutions to all of the Student Practice problems can be found at the back of the text starting at page SP-1.*

② Finding the Least Common Multiple of Numerical or Algebraic Expressions

An expression that is a multiple of two different expressions is called a *common multiple* of those two expressions. Therefore, in Example 1, we call $24x$ and $48x$ common multiples of $8x$ and $12x$. The number $24x$ is the *smaller of these common multiples* and is called the **least common multiple,** or **LCM,** of $8x$ and $12x$.

EXAMPLE 2   Find the LCM of 10 and 15.

Solution   First, we list some multiples of 10: 10, 20, 30, 40, 50, 60.

Next, we list some multiples of 15: 15, 30, 45, 60.

We see that both 30 and 60 are common multiples. Since 30 is the smaller of these common multiples, we call 30 the least common multiple (LCM).

Student Practice 2   Find the LCM of 4 and 5.

Suppose that we must find the LCM of 14, 32, 78, and 210. Listing the multiples of each number to find the LCM would be time consuming. A quicker, more efficient way to find LCMs uses prime factorizations. With this method we **build the LCM**

Student Learning Objectives

After studying this section, you will be able to:

① Find multiples of algebraic expressions.

② Find the least common multiple of numerical or algebraic expressions.

③ Solve applied problems involving the least common multiple (LCM).

using the prime factors of each number. In Example 2 we listed some multiples for 10 and 15 to find the LCM, 30. Now let's see how to find the LCM of 10 and 15 using prime factors.

$$
\begin{array}{c}
\overset{\text{factors of } 10}{\overset{\downarrow\ \ \downarrow}{}} \\
10 = 2 \cdot 5 \quad\quad \text{LCM} \rightarrow 30 = 2 \cdot 3 \cdot 5 \\
15 = 3 \cdot 5 \quad\quad\quad\quad\quad\quad\quad \underset{\text{factors of } 15}{\overset{\uparrow\ \uparrow}{}}
\end{array}
$$

Notice that the LCM has all the factors of 10 and all the factors of 15 in its prime factorization. Thus the LCM of 10 and 15 must satisfy at least two requirements.

> First, the LCM must have all factors of 10 in its prime factorization: a 2 and a 5.
>
> Second, the LCM must have all factors of 15 in its prime factorization: a 3 and a 5.

> To build the LCM, we write a prime factorization that satisfies the first requirement and then *add to this factorization* to satisfy the second requirement.

We build the LCM of 10 and 15 as follows.

1. The LCM must have a 2 and a 5 as factors. $\boxed{\text{LCM} = 2 \cdot 5 \cdot ?}$

2. The LCM must have a 3 and a 5 as factors. $\boxed{\text{LCM} = 2 \cdot 5 \cdot 3}$

5 is already a factor, so we just multiply by a 3.

Now, we multiply the factors to find the LCM, 30: $2 \cdot 5 \cdot 3 = 30$.

Notice that when we built the LCM, we were also constrained by a third requirement.

3. The LCM must contain the *minimum* number of factors necessary to satisfy the first two requirements.

Without this third requirement, we cannot create the *smallest* common multiple (LCM). For example, in the above illustration, to satisfy the second requirement—the LCM must have a 3 and a 5 in its prime factorization—we only inserted a 3 to the existing prime factorization, not a 3 and a 5. There was a 5 in the factorization already, so we did not need to insert another one. *If we insert an extra 5, we build a multiple of 10 and 15 that is not the smallest common multiple.*

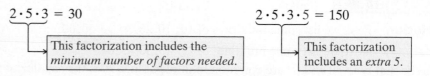

$2 \cdot 5 \cdot 3 = 30$	$2 \cdot 5 \cdot 3 \cdot 5 = 150$
This factorization includes the *minimum number of factors needed.*	This factorization includes an *extra 5.*

Although both 30 and 150 are common multiples of 10 and 15, the smallest common multiple, or LCM, is 30.

> **PROCEDURE TO FIND THE LCM**
>
> **1.** Factor each number into a product of prime factors.
> **2.** List the requirements for the factorization of the LCM.
> **3.** Build the LCM using the minimum number of factors.

EXAMPLE 3 Find the LCM of 18, 42, and 45.

Solution

Factor each → number	List requirements for → factorization of LCM	Build the LCM

$18 = 2 \cdot 3 \cdot 3 \;\rightarrow$ must have a 2 and a pair of 3's $\rightarrow$ $\boxed{\text{LCM} = 2 \cdot 3 \cdot 3 \cdot ?}$

$42 = 2 \cdot 3 \cdot 7 \;\rightarrow$ must have a 2, a 3, and a 7 $\rightarrow$ $\boxed{\text{LCM} = 2 \cdot 3 \cdot 3 \cdot 7 \cdot ?}$

         2 and 3 are already factors, so we just multiply by a 7.

$45 = 3 \cdot 3 \cdot 5 \;\rightarrow$ must have a pair of 3's and a 5 $\rightarrow$ $\boxed{\text{LCM} = 2 \cdot 3 \cdot 3 \cdot 7 \cdot 5}$

         A pair of 3's already exists, so we just multiply by a 5.

The LCM of 18, 42, and 45 is $2 \cdot 3 \cdot 3 \cdot 7 \cdot 5 = 630$.

Student Practice 3   Find the LCM of 28, 36, and 70.

EXAMPLE 4 Find the LCM of $2x$, x^2, and $6x$.

Solution

Factor each → expression	List requirements for → factorization of LCM	Build the LCM

$2x = 2 \cdot x \quad\rightarrow$ must have a 2 and an x $\rightarrow$ $\boxed{\text{LCM} = 2 \cdot x \cdot ?}$

$x^2 = x \cdot x \quad\rightarrow$ must have a pair of x's $\rightarrow$ $\boxed{\text{LCM} = 2 \cdot x \cdot x \cdot ?}$

         One x is already a factor, so we just multiply by another x.

$6x = 2 \cdot 3 \cdot x \quad\rightarrow$ must have a 2, a 3, and an x $\rightarrow$ $\boxed{\text{LCM} = 2 \cdot x \cdot x \cdot 3}$

         A 2 and an x are already factors, so we just multiply by a 3.

The LCM of $2x$, x^2, and $6x$ is $2 \cdot x \cdot x \cdot 3 = 6x^2$.

Student Practice 4   Find the LCM of $4x$, x^2, and $10x$.

---

③ **Solving Applied Problems Involving the Least Common Multiple (LCM)**

EXAMPLE 5   Shannon and Marsha are swimming laps from the dock on one side of a lake to a marker and back. Shannon can swim a lap in 12 minutes and Marsha can swim the same distance in 10 minutes. If Shannon and Marsha start swimming at the same time, in how many minutes will they meet to begin their next lap together?

Solution   *Understand the problem.* We organize the facts on a chart and then fill in the Mathematics Blueprint for Problem Solving.

Time	End of 1st Lap	End of 2nd Lap	...	
Shannon	12 min	24 min	...	← These are multiples of 12.
Marsha	10 min	20 min	...	← These are multiples of 10.

Continued on next page

Mathematics Blueprint for Problem Solving

Gather the Facts	What Am I Asked to Do?	How Do I Proceed?	Key Points to Remember
See the chart.	Find how many minutes it takes for them to begin another lap together.	Find the LCM of 12 and 10, which is how long before they begin a new lap together.	The LCM is the least (smallest) common multiple.

Solve and state the answer.
We factor 12 and 10 into a product of prime factors and then find the LCM.

$$12 = 2 \cdot 2 \cdot 3 \quad \text{LCM} = 2 \cdot 2 \cdot 3 \cdot 5 = 60$$
$$10 = 2 \cdot 5$$

Time	End of 1st Lap	End of 2nd Lap	...
Shannon	12 min	24 min	...60 min
Marsha	10 min	20 min	...60 min

Shannon and Marsha will begin another lap together 60 minutes after they start swimming.

Student Practice 5 Refer to Example 5 to complete this problem. Shannon can swim a lap in 15 minutes and Marsha can swim a lap in 18 minutes. In how many minutes will they meet to begin their next lap together?

EXAMPLE 6 Sonia and Leo are tour guides at a castle. Sonia gives a 40-minute tour of the interior of the castle, and Leo gives a 30-minute tour of the castle grounds. There is a 10-minute break after each tour. If tours start at 8 A.M., what is the next time that both tours will start at the same time?

Solution *Understand the problem.* We make a chart to help us develop a plan to solve the problem.

Tours	Number of Minutes after 8 A.M. Tours Start
Interior tours start every 50 minutes (40 min + 10-min break).	50, 100, ...
Grounds tours start every 40 minutes (30 min + 10-min break).	40, 80, ...

Mathematics Blueprint for Problem Solving

Gather the Facts	What Am I Asked to Do?	How Do I Proceed?	Key Points to Remember
See the chart.	Determine the next time when both tours start at the same time.	To find the *next* time that both tours start at the same time, we must find the LCM of 50 and 40.	60 min = 1 hr. Once we have the LCM in minutes, we change the LCM to hours and minutes to find the common start time.

Solve and state the answer.
First we factor 40 and 50 into a product of prime factors and then find the LCM.

$$40 = 2 \cdot 2 \cdot 2 \cdot 5 \quad \text{LCM} = 2 \cdot 2 \cdot 2 \cdot 5 \cdot 5 = 200$$
$$50 = 2 \cdot 5 \cdot 5$$

Tours	Number of Minutes after 8 A.M. Tours Start
Interior tours start every 50 minutes	50, 100, . . . , 200
Grounds tours start every 40 minutes	40, 80, . . . , 200

Both tours will start at the same time 200 minutes after 8 A.M.

Next, we change minutes to hours and minutes. Since we need to know how many 60's are in 200, we divide.

$$200 \text{ minutes} \div 60 \text{ minutes per hour} = 3 \text{ hours and } 20 \text{ minutes after 8 A.M.}$$
$$8 \text{ A.M.} + 3 \text{ hours and } 20 \text{ minutes} = 11{:}20 \text{ A.M.}$$

At 11:20 A.M. both tours will start at the same time.

Student Practice 6   Refer to Example 6 to complete this problem. Leo's tour of the grounds is reduced to 25 minutes. Determine the next time that both tours will start at the same time.

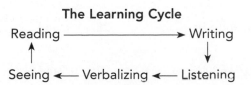 **STEPS TO SUCCESS** Improving Your Success in Class

In Sections 1.3 and 3.4 we discussed why an important part of the learning process happens in the classroom.

If you want to improve your grade and *work less while you learn more*, you must attend class.

Missing a class means you must work that much harder to learn the material taught in class. **Make learning easier by attending class!**

People learn by *reading, writing, listening, verbalizing*, and *seeing*. These activities are all part of the **learning cycle** and always occur in class.

- *Listening and seeing*: hearing and watching the instructor's lecture
- *Reading*: reading the information on the board and in handouts
- *Verbalizing*: asking questions and participating in class discussions
- *Writing*: taking notes and working problems assigned in class

The Learning Cycle

Reading ──────────→ Writing

Seeing ←── Verbalizing ←── Listening

Making it personal: *Place a* ✓ *in the blank as you complete each of the following.*

_____ 1. Meet with your instructor to find the following information.

Number of classes missed: _____

Number of assignments missed: _____

Grade average: _____

_____ 2. Discuss with your instructor what you can do to improve your grade. Instructor suggestions:

_____

_____

_____

_____

_____ 3. Read the Steps to Success on page 203 and complete the exercises on classroom attendance.

5.2 Exercises

MyMathLab®

Watch the videos
in MyMathLab

Download the
MyDashBoard App

Verbal and Writing Skills, Exercises 1–8

1. Explain why 12 is a multiple of 3 and not a multiple of 5.

2. Explain the relationship between a multiple and an LCM.

3. (a) List the first four multiples of 6 and the first four multiples of 8.
   (b) Which of these multiples are common to both lists?

4. (a) List the first five multiples of 4 and the first five multiples of 5.
   (b) Which of these multiples are common to both lists?

5. (a) List the first five multiples of 2 and the first five multiples of 5.
   (b) Which of these multiples are common to both lists?

6. (a) List the first six multiples of 4 and the first six multiples of 6.
   (b) Which of these multiples are common to both lists?

7. (a) List the first four multiples of $12x$ and the first four multiples of $18x$.
   (b) Which of these multiples are common to both lists?

8. (a) List the first four multiples of $15x$ and the first four multiples of $5x$.
   (b) Which of these multiples are common to both lists?

There are two factors missing in the LCM for the numbers listed. State the two missing factors.

9. $525 = 3 \cdot 5 \cdot 5 \cdot 7$
   $90 = 2 \cdot 3 \cdot 3 \cdot 5$
   $28 = 2 \cdot 2 \cdot 7$
   $\boxed{LCM = 2 \cdot 3 \cdot 5 \cdot 5 \cdot 7 \cdot ? \cdot ?}$

10. $220 = 2 \cdot 2 \cdot 5 \cdot 11$
    $189 = 3 \cdot 3 \cdot 3 \cdot 7$
    $385 = 5 \cdot 7 \cdot 11$
    $\boxed{LCM = 2 \cdot 2 \cdot 3 \cdot 3 \cdot 7 \cdot 11 \cdot ? \cdot ?}$

11. $10x^2 = 2 \cdot 5 \cdot x \cdot x$
    $18x = 2 \cdot 3 \cdot 3 \cdot x$
    $49x = 7 \cdot 7 \cdot x$
    $\boxed{LCM = 2 \cdot 3 \cdot 3 \cdot 5 \cdot 7 \cdot x \cdot ? \cdot ?}$

12. $15y^2 = 3 \cdot 5 \cdot y \cdot y$
    $25y = 5 \cdot 5 \cdot y$
    $9y^2 = 3 \cdot 3 \cdot y \cdot y$
    $\boxed{LCM = 3 \cdot 3 \cdot 5 \cdot y \cdot ? \cdot ?}$

Find the LCM of each group of numbers or expressions.

13. 5 and 15

14. 6 and 12

15. 8 and 28

16. 6 and 28

17. 15 and 20

18. 18 and 10

19. 40 and 60

20. 30 and 40

21. 5, 8, and 12

22. 6, 14, and 26

23. 7, 14, and 20

24. 8, 12, and 42

25. $4x$ and $18x$

26. $20x$ and $25x$

27. $21a$ and $81a$

28. $15a$ and $35a$

29. $18x$ and $45x^2$

30. $15x$ and $63x^2$

31. $22x^2$ and $4x^3$

32. $9x^2$ and $3x^3$

33. $12x^2, 5x, 3x^3$

34. $15x, 3x^3, 10x^2$

35. $12x, 14,$ and $4x^2$

36. $14x, 36,$ and $7x^2$

Applications

37. *Running Laps* Jessica and Luis are running laps around a field. Luis runs 1 lap every 4 minutes, while Jessica runs 1 lap every 6 minutes. If Luis and Jessica begin their run at the same time and location on the field, in how many minutes will they meet to begin their next lap together?

38. *Power Walking* Olga will power walk around the track field, and Annette will jog around the same field. Annette can complete 1 lap every 8 minutes, while Olga takes 12 minutes to complete a lap. If Annette and Olga begin their workout at the same time and location on the track, in how many minutes will they meet to begin their next lap together?

39. *Bottle Labeling* One machine takes 6 minutes to place labels on each bottle of juice in a carton. A second machine takes 8 minutes to place labels on each bottle of juice in a carton. If both machines start labeling at the same time, in how many minutes will each machine begin labeling a carton of juice at the same time again?

40. *Bike Assembly* Tuan and Sal work at a bike store. Sal puts together the 21-speed bikes and Tuan the dirt bikes. Sal takes 80 minutes to put together a 21-speed bike, and Tuan takes 60 minutes to put together a dirt bike. If they both start putting bikes together at 1 P.M., at what time will they both start on a new bike at the same time again?

41. *Track and Field Event* An elementary school track and field competition is held on two different fields. Each event on the first field is 20 minutes long, while each event on the second field is 30 minutes long. There are 15-minute breaks between events on both of the fields. If the competition on both fields starts at 8 A.M., determine the next time that both fields will start their track events at the same time.

42. *History Museum Shows* A history museum offers two shows, a 30-minute presentation that overviews all the displays in the museum and a 20-minute film on dinosaurs. There is a 10-minute break after each show. If the shows start at 10 A.M., what is the next time that both shows will start at the same time?

One Step Further *Find the LCM of each group of expressions.*

43. $2x^3, 8xy^2,$ and $10x^2y$

44. $4y^2, 2xy,$ and $9x^3y$

45. $2z^2, 5xyz,$ and $15xy$

46. $3x, 9xy^2,$ and $18xyz$

Cumulative Review

47. [4.2.3] Change $\dfrac{19}{2}$ to a mixed number.

48. [4.2.4] Change $4\dfrac{2}{5}$ to an improper fraction.

49. [2.5.1] Evaluate. $2 + 6(-1) \div 3$

50. [2.5.1] Evaluate. $12 - 5 \cdot 2^2 \div 4$

Quick Quiz 5.2 *Find the least common multiple (LCM) of each group of numbers or expressions.*

1. $9, 15$

2. $12x, 4x^3, 9x^2$

3. $3, 18, 30$

4. Concept Check Is $= 3 \times 5 \times 5 \times 7 \times x \times x$ the correct factorization for the LCM of $63x^2$ and $75x^3$? Why or why not?

5.3 Adding and Subtracting Fractional Expressions

Student Learning Objectives

After studying this section, you will be able to:

① Add and subtract fractional expressions with a common denominator.

② Add and subtract fractional expressions with different denominators.

① Adding and Subtracting Fractional Expressions with a Common Denominator

What does it mean to add or subtract fractions? Well, the idea is similar to counting like items or adding and subtracting like terms. For example, if we have 3 CDs and we get 2 more, we end up with 5 CDs: $3 + 2 = 5$. Similarly, we can add $3x + 2x = 5x$ because $3x$ and $2x$ are like terms. In the case of fractions, instead of adding like items or terms, we add *like parts* of a whole. By this we mean fractions with the same denominators: $\frac{3}{6} + \frac{2}{6} = \frac{5}{6}$. This is because when the denominators of fractions are the same, we are comparing *like parts* of a whole. The following situation and illustration can help us understand this idea.

Suppose that a large piece of wood is cut into 6 equal-size parts and we have 3 of these parts and then acquire 2 more. We add to find the total number of parts of wood we have.

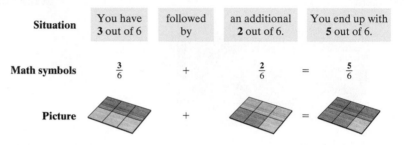

Situation	You have **3** out of 6	followed by	an additional **2** out of 6.	You end up with **5** out of 6.
Math symbols	$\frac{3}{6}$	$+$	$\frac{2}{6}$	$= \quad \frac{5}{6}$
Picture		$+$		$=$

When fractions have the *same denominator,* we say that these fractions have a **common denominator.** From the previous illustration, we observe that to add fractions that have a common denominator, we add the numerators and write the sum over the common denominator. A similar rule is followed for subtraction, except that the numerators are subtracted.

> **PROCEDURE TO ADD OR SUBTRACT FRACTIONAL EXPRESSIONS WITH COMMON DENOMINATORS**
>
> 1. The fractions added or subtracted must have a common denominator (denominators that are the same).
> 2. Add or subtract the numerators only.
> 3. The denominator stays the same.

EXAMPLE 1 Subtract. $\dfrac{7}{15} - \dfrac{3}{15}$

Solution

$$\frac{7}{15} - \frac{3}{15} = \frac{7 - 3}{15} \qquad \text{Subtract numerators.}$$
$$\text{The denominator stays the same.}$$
$$= \frac{4}{15}$$

Student Practice 1 Add.

$$\frac{3}{13} + \frac{7}{13}$$

NOTE TO STUDENT: Fully worked-out solutions to all of the Student Practice problems can be found at the back of the text starting at page SP-1.

EXAMPLE 2 Add. $\dfrac{-11}{20} + \left(\dfrac{-13}{20}\right)$

Solution

$$\dfrac{-11}{20} + \left(\dfrac{-13}{20}\right) = \dfrac{-11 + (-13)}{20} \quad \text{Add numerators.}$$
$$\text{The denominator stays the same.}$$

$$= \dfrac{-24}{20}$$

$$= \dfrac{\cancel{4}(-6)}{\cancel{4}(5)} \quad \text{Factor and simplify.}$$

$$= -\dfrac{6}{5}$$

Although the answer may also be written as the mixed number $-1\frac{1}{5}$, we generally leave it as an improper fraction. In either case, the answer must be reduced to lowest terms.

Student Practice 2 Add.

$$\dfrac{-7}{6} + \left(\dfrac{-21}{6}\right)$$

EXAMPLE 3 Perform the operation indicated.

(a) $\dfrac{6}{y} - \dfrac{2}{y}$ **(b)** $\dfrac{x}{5} + \dfrac{4}{5}$

Solution

(a) $\dfrac{6}{y} - \dfrac{2}{y} = \dfrac{6-2}{y} = \dfrac{4}{y}$ **(b)** $\dfrac{x}{5} + \dfrac{4}{5} = \dfrac{x+4}{5}$

The answer to part **(b)**, $\dfrac{x+4}{5}$, is simplified. We cannot add x and 4. They are not like terms.

Student Practice 3 Perform the operation indicated.

(a) $\dfrac{8}{x} - \dfrac{3}{x}$ **(b)** $\dfrac{y}{9} + \dfrac{5}{9}$

CAUTION: It is important to remember that to add or subtract fractions, the *denominators* must *be* the *same* (common denominators) and *stay* the *same*. We do not add or subtract the denominators— only the numerators.

② Adding and Subtracting Fractional Expressions with Different Denominators

As we stated earlier, we can add and subtract fractions if the fractions have common denominators (denominators that are the same). Now, what do we do when the denominators are not the same? The following example will help us determine the answer to this question.

Suppose that we have 2 equal-size blocks of molding clay. One block of clay is cut into 2 equal-size parts and another is cut into 3 equal-size parts. If we take 1 of the 2 parts, then 1 of the 3 parts, we must add to find the total amount of clay we have.

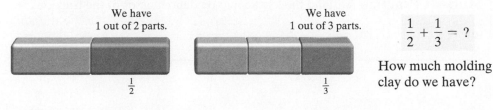

We have
1 out of 2 parts.

We have
1 out of 3 parts.

$\dfrac{1}{2} + \dfrac{1}{3} = ?$

How much molding clay do we have?

$\dfrac{1}{2}$ $\dfrac{1}{3}$

Since we are adding pieces of molding clay that came from blocks that were cut into a different number of parts, **we are not adding like parts of a whole.** The total amount of clay could be determined more easily if each block of clay had been cut into 6 equal-size parts.

$\frac{1}{2}$ or $\frac{3}{6}$ $\frac{1}{3}$ or $\frac{2}{6}$

$$\frac{1}{2} + \frac{1}{3} = ?$$
$$\downarrow \quad \downarrow \quad \downarrow$$
$$\frac{3}{6} + \frac{2}{6} = \frac{5}{6}$$

As we can see, cutting the blocks of clay into the same number of equally-sized pieces makes it possible to work with fractions that have a common denominator. We know how to add and subtract these fractions. *We use a similar idea to add and subtract fractions with different denominators.*

PROCEDURE TO ADD OR SUBTRACT FRACTIONAL EXPRESSIONS WITH DIFFERENT DENOMINATORS

1. Find the least common denominator (LCD).
2. Write equivalent fractions that have the LCD as the denominator.
3. Add or subtract the numerators of the fractions with common denominators.
4. Simplify the answer if necessary.

How do we find a *least common denominator?* Let's look at the fractions $\frac{1}{2}$ and $\frac{1}{3}$ from the earlier illustration. Notice that the least common denominator, 6, is also the least common multiple of 2 and 3. In fact, the **least common denominator** of two fractions is the least common multiple (LCM) of the two denominators. Since we are working with denominators, we call the LCM the least common denominator, or LCD.

EXAMPLE 4   Find the least common denominator (LCD) of the fractions.

(a) $\frac{1}{5}, \frac{1}{3}$ **(b)** $\frac{1}{12}, \frac{5}{18}$

Solution

(a) $\frac{1}{5}, \frac{1}{3}$

The LCD of $\frac{1}{5}$ and $\frac{1}{3}$ is 15.

(b) We find the LCD of $\frac{1}{12}$ and $\frac{5}{18}$.

$$12 = 2 \cdot 2 \cdot 3$$
$$18 = 2 \cdot 3 \cdot 3$$
$$LCD = 2 \cdot 2 \cdot 3 \cdot 3 = 36$$

The LCD of $\frac{1}{12}$ and $\frac{5}{18}$ is 36.

Student Practice 4   Find the least common denominator of the fractions.

(a) $\frac{1}{4}, \frac{1}{5}$ **(b)** $\frac{1}{12}, \frac{3}{28}$

How do we write equivalent fractions that have the LCD as the denominator? In Chapter 4 we learned that to find equivalent fractions, we multiply the numerator and denominator by the same nonzero number. Therefore, we must determine the value of the nonzero number that when multiplied by the denominator yields the LCD.

EXAMPLE 5 Write the equivalent fraction for $\dfrac{1}{5}$ that has 40 as the denominator.

$$\frac{1}{5} = \frac{?}{40}$$

Solution

$$\frac{1}{5} = \frac{?}{40} \quad \text{What number multiplied by the denominator, 5, yields 40?}$$
It is 8, since $5(8) = 40$.

$$\frac{1 \cdot 8}{5 \cdot 8} = \frac{8}{40} \quad \text{We multiply the numerator and denominator of } \frac{1}{5} \text{ by 8.}$$

$$\frac{1}{5} = \frac{8}{40}$$

Student Practice 5 Write the equivalent fractions that have 10 as the denominator.

(a) $\dfrac{3}{5} = \dfrac{?}{10}$ **(b)** $\dfrac{1}{2} = \dfrac{?}{10}$

Once we find the LCD and write equivalent fractions with the LCD as the denominator, we simply add or subtract the numerators of the fractions with the common denominators. We summarize the process.

> 1. Find the LCD.
> 2. Write equivalent fractions.
> 3. Add or subtract fractions.
> 4. Simplify if necessary.

EXAMPLE 6 Perform the operation indicated.

(a) $\dfrac{-5}{7} + \dfrac{3}{4}$ **(b)** $\dfrac{11}{12} - \dfrac{3}{20}$

Solution

(a) $\dfrac{-5}{7} + \dfrac{3}{4}$ **Step 1** Find the LCD of $\dfrac{-5}{7}$ and $\dfrac{3}{4}$.

$$\boxed{\text{LCD} = 28}$$

Step 2 Write equivalent fractions.

$$\frac{-5 \cdot 4}{7 \cdot 4} = \boxed{\frac{-20}{28}} \qquad \frac{3 \cdot 7}{4 \cdot 7} = \boxed{\frac{21}{28}}$$

Step 3 Add the numerators of the fractions with common denominators.

$$\frac{-5}{7} + \frac{3}{4} = \boxed{\frac{-20}{28} + \frac{21}{28}} = \frac{1}{28}$$

Student Practice 6 Perform the operation indicated.

(a) $\dfrac{-3}{8} + \dfrac{7}{9}$

Continued on next page

(b) $\dfrac{11}{12} - \dfrac{3}{20}$

Step 1 Find the LCD of $\dfrac{11}{12}$ and $\dfrac{3}{20}$.

$$LCD = \overbrace{2 \cdot 2 \cdot 3 \cdot \underbrace{5}} = 60$$
$$\longrightarrow 20 \longleftarrow$$

$$\boxed{LCD = 60}$$

Step 2 Write equivalent fractions.

$$\dfrac{11}{12} = \dfrac{11 \cdot 5}{12 \cdot 5} = \boxed{\dfrac{55}{60}} \qquad \dfrac{3}{20} = \dfrac{3 \cdot 3}{20 \cdot 3} = \boxed{\dfrac{9}{60}}$$

Step 3 Subtract the numerators of the fractions with common denominators.

$$\dfrac{11}{12} - \dfrac{3}{20} = \boxed{\dfrac{55}{60} - \dfrac{9}{60}} = \dfrac{46}{60}$$

Step 4 Simplify.

$$\dfrac{46}{60} = \dfrac{\cancel{2} \cdot 23}{\cancel{2} \cdot 2 \cdot 3 \cdot 5} = \dfrac{23}{30}$$

$$\dfrac{11}{12} - \dfrac{3}{20} = \dfrac{23}{30}$$

(b) $\dfrac{13}{30} - \dfrac{2}{15}$

EXAMPLE 7 Perform the operation indicated.

(a) $\dfrac{6}{x} + \dfrac{5}{3x}$

(b) $\dfrac{4}{y} - \dfrac{2}{x}$

Solution

(a) $\dfrac{6}{x} + \dfrac{5}{3x}$

Step 1 Find the LCD of $\dfrac{6}{x}$ and $\dfrac{5}{3x}$. $\boxed{LCD = 3x}$

Step 2 Write equivalent fractions.

$$\dfrac{6}{x} = \dfrac{6 \cdot 3}{x \cdot 3} = \boxed{\dfrac{18}{3x}} \qquad \dfrac{5}{3x} = \boxed{\dfrac{5}{3x}}$$

Step 3 Add fractions with common denominators.

$$\dfrac{6}{x} + \dfrac{5}{3x} = \boxed{\dfrac{18}{3x} + \dfrac{5}{3x}} = \dfrac{18 + 5}{3x} = \dfrac{23}{3x}$$

We did not need to write $\frac{5}{3x}$ as an equivalent fraction with the common denominator since its denominator, $3x$, is the LCD.

(b) $\dfrac{4}{y} - \dfrac{2}{x}$

Step 1 Find the LCD of $\dfrac{4}{y}$ and $\dfrac{2}{x}$. $\boxed{LCD = xy}$

Step 2 Write equivalent fractions.

$$\dfrac{4}{y} = \dfrac{4 \cdot x}{y \cdot x} = \boxed{\dfrac{4x}{xy}} \qquad \dfrac{2}{x} = \dfrac{2 \cdot y}{x \cdot y} = \boxed{\dfrac{2y}{xy}}$$

Step 3 Subtract fractions with common denominators.

$$\dfrac{4}{y} - \dfrac{2}{x} = \boxed{\dfrac{4x}{xy} - \dfrac{2y}{xy}} = \dfrac{4x - 2y}{xy}$$

We cannot subtract $4x - 2y$ since $4x$ and $2y$ are not like terms. Therefore, we leave the numerator as the expression $4x - 2y$.

CAUTION: We may not use slashes to divide out *part* of an addition or subtraction problem. We may use slashes only if we are multiplying factors.

$$\frac{4\cancel{x} - 2\cancel{y}}{\cancel{x}\,\cancel{y}} = 4 - 2 = 2 \quad \text{This is wrong!}$$

$$\frac{4x - 2y}{xy} \qquad \begin{array}{l}\text{This expression \textit{cannot} be simplified any}\\ \text{further since we are \textit{subtracting} } 4x \text{ and } 2y.\end{array}$$

$$\frac{(4\cancel{x})(2\cancel{y})}{\cancel{x}\,\cancel{y}} = 8 \qquad \text{This is correct.}$$

$$\frac{(4x)(2y)}{xy} = 8 \qquad \begin{array}{l}\text{This expression \textit{can} be simplified since we are}\\ \textit{multiplying } 4x \text{ and } 2y.\end{array}$$

Student Practice 7   Perform the operation indicated.

(a) $\dfrac{8}{x} + \dfrac{2}{5x}$ \qquad\qquad **(b)** $\dfrac{7}{y} - \dfrac{4}{x}$

EXAMPLE 8   Add. $\dfrac{7x}{16} + \dfrac{3x}{32}$

Solution

$\dfrac{7x}{16} + \dfrac{3x}{32}$ \qquad **Step 1** Find the LCD of $\dfrac{7x}{16}$ and $\dfrac{3x}{32}$. $\boxed{\text{LCD} = 32}$

Step 2 Write equivalent fractions.

$$\frac{7x \cdot 2}{16 \cdot 2} = \boxed{\frac{14x}{32}} \qquad \frac{3x}{32} = \boxed{\frac{3x}{32}}$$

Step 3 Add fractions with common denominators.

$$\frac{7x}{16} + \frac{3x}{32} = \boxed{\frac{14x}{32} + \frac{3x}{32}} = \frac{14x + 3x}{32} = \frac{17x}{32}$$

Student Practice 8   Add. $\dfrac{8x}{15} + \dfrac{9x}{24}$

When solving applied problems, we must determine what operation to use. As we saw in Chapter 1, problems that involve subtraction often use the phrases "how much more," or "how much is left." Those that require addition often state "how many," or "find the total." Identifying these key phrases in an applied problem will help you solve the problem.

EXAMPLE 9   Leila finished $\frac{1}{8}$ of her English term paper before spring break and $\frac{1}{2}$ of the paper during spring break. How much more did she complete during the break than before the break?

Solution   *Understand the problem.* The phrase "how much more" indicates that we subtract.

Solve and state the answer.

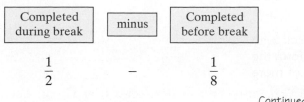

Completed during break	minus	Completed before break
$\dfrac{1}{2}$	$-$	$\dfrac{1}{8}$

Continued on next page

Step 1 The LCD is 8.

Step 2 Write equivalent fractions. $\dfrac{1}{2} = \dfrac{1 \cdot 4}{2 \cdot 4} = \boxed{\dfrac{4}{8}}$ $\quad \dfrac{1}{8} = \boxed{\dfrac{1}{8}}$

Step 3 Subtract. $\dfrac{1}{2} - \dfrac{1}{8} = \boxed{\dfrac{4}{8} - \dfrac{1}{8}} = \dfrac{3}{8}$

Leila finished $\frac{3}{8}$ more of the term paper *during* the break than *before* the break.

Check. Double-check your calculations on a separate piece of paper.

Student Practice 9 Jesse painted $\frac{1}{4}$ of his home on Monday and $\frac{1}{6}$ of his home on Tuesday. How much more did he complete on Monday than on Tuesday?

EXAMPLE 10 Ranak bought $\frac{3}{4}$ pound of sliced turkey and $\frac{5}{8}$ pound of sliced ham at the delicatessen. How many pounds of deli meat did she buy?

Solution *Understand the problem.* Since we want to find *how many* pounds she purchased, we must add $\frac{3}{4}$ and $\frac{5}{8}$.

To add $\frac{3}{4} + \frac{5}{8}$ we must first find the least common denominator, which is 8.

$$\frac{3}{4} = \frac{3 \cdot 2}{4 \cdot 2} = \frac{6}{8} \qquad \text{We write } \frac{3}{4} \text{ as the equivalent fraction } \frac{6}{8}.$$

$$\frac{6}{8} + \frac{5}{8} = \frac{11}{8} \text{ or } 1\frac{3}{8} \text{ pounds} \quad \text{We add and then change } \frac{11}{8} \text{ to the mixed number } 1\frac{3}{8}.$$

Ranak bought $1\dfrac{3}{8}$ pounds of deli meat.

Student Practice 10 Lester jogged $\frac{2}{3}$ mile on Monday and $\frac{4}{5}$ mile on Tuesday. How many miles did he jog in the two-day period?

Notice in Example 10 we changed the improper fraction to a mixed number. As answers to applications, mixed numbers are generally easier to understand.

👣 STEPS TO SUCCESS A Positive Attitude Toward Fractions

Often, students panic when they begin to study fractions in their math classes. They may think, "I have never understood fractions, and I never will!" Not true. You experience situations involving fractions every day. You may not actually perform mathematical calculations with fractions, but you find the results. Consider these two situations.

Situation: You know that if you slice an apple into quarters, you can take 1 of the 4 pieces or $\frac{1}{4}$ of the apple.

Calculation: $1 \div 4 = \frac{1}{4}$

Situation: If you take $\frac{1}{2}$ of $6, you know you have $3.

Calculation: $\frac{1}{2} \cdot 6 = 3$

As you can see, you already know how to work with fractions. Now you are ready to see how to perform the calculations so that you can find the outcome of more complex situations.

Making it personal:

1. Describe one situation similar to the ones just discussed where you use simple fraction calculations in your everyday life. Share this with your classmates. ▼

Verbal and Writing Skills, Exercises 1–4

Fill in the blanks.

1. When we add two fractions with the same denominator, we add the _____, and the _____ stays the same.

2. When we add two fractions with different denominators, we must first find the _____.

3. Is $\dfrac{4}{5} + \dfrac{5}{9} = \dfrac{4+5}{5+9} = \dfrac{9}{14}$? Why or why not?

4. Is $\dfrac{5}{9} - \dfrac{4}{5} = \dfrac{5-4}{9-5} = \dfrac{1}{4}$? Why or why not?

Fill in each box with the correct value.

5. $\dfrac{\square}{7} + \dfrac{3}{7} = \dfrac{5}{7}$

6. $\dfrac{\square}{5} + \dfrac{2}{5} = \dfrac{4}{5}$

7. $\dfrac{\square}{4} - \dfrac{1}{4} = \dfrac{1}{4}$

8. $\dfrac{\square}{9} - \dfrac{3}{9} = \dfrac{5}{9}$

Perform the operation indicated. Be sure to simplify your answer.

9. $\dfrac{6}{17} + \dfrac{9}{17}$

10. $\dfrac{4}{13} + \dfrac{2}{13}$

11. $\dfrac{6}{23} - \dfrac{5}{23}$

12. $\dfrac{7}{41} - \dfrac{6}{41}$

13. $\dfrac{-13}{28} + \left(\dfrac{-11}{28}\right)$

14. $\dfrac{-17}{74} + \left(\dfrac{-41}{74}\right)$

15. $\dfrac{-31}{51} + \dfrac{11}{51}$

16. $\dfrac{-27}{43} + \dfrac{15}{43}$

17. $\dfrac{9}{y} - \dfrac{8}{y}$

18. $\dfrac{7}{x} - \dfrac{3}{x}$

19. $\dfrac{31}{a} + \dfrac{8}{a}$

20. $\dfrac{12}{y} + \dfrac{21}{y}$

21. $\dfrac{x}{7} - \dfrac{5}{7}$

22. $\dfrac{x}{5} - \dfrac{4}{5}$

23. $\dfrac{y}{3} + \dfrac{14}{3}$

24. $\dfrac{y}{11} + \dfrac{25}{11}$

Find the least common denominator (LCD) of the fractions.

25. $\dfrac{1}{5}, \dfrac{1}{6}$

26. $\dfrac{1}{3}, \dfrac{1}{8}$

27. $\dfrac{1}{9}, \dfrac{1}{15}$

28. $\dfrac{1}{21}, \dfrac{1}{14}$

Write equivalent fractions that have 60 as the denominator.

29. $\dfrac{1}{5} = \dfrac{?}{60}$

30. $\dfrac{2}{15} = \dfrac{?}{60}$

31. $\dfrac{5}{6} = \dfrac{?}{60}$

32. $\dfrac{3}{10} = \dfrac{?}{60}$

Perform the operation indicated. Be sure to simplify your answer.

33. $\dfrac{11}{15} - \dfrac{31}{45}$

34. $\dfrac{29}{12} - \dfrac{23}{24}$

35. $\dfrac{17}{24} - \dfrac{1}{6}$

36. $\dfrac{11}{28} - \dfrac{1}{7}$

37. $\dfrac{3}{8} + \dfrac{4}{7}$

38. $\dfrac{7}{4} + \dfrac{5}{9}$

39. $\dfrac{-3}{4} + \dfrac{1}{10}$

40. $\dfrac{-5}{6} + \dfrac{3}{4}$

41. $\dfrac{-2}{13} + \dfrac{7}{26}$

42. $\dfrac{-4}{15} + \dfrac{11}{30}$

43. $\dfrac{-3}{14} + \left(\dfrac{-1}{10}\right)$

44. $\dfrac{-4}{15} + \left(\dfrac{-1}{6}\right)$

45. $\dfrac{7}{10} - \dfrac{5}{14}$

46. $\dfrac{7}{20} - \dfrac{3}{16}$

47. $\dfrac{11}{15} - \dfrac{5}{12}$

48. $\dfrac{4}{15} - \dfrac{2}{25}$

Perform the operation indicated. Be sure to simplify your answer.

49. $\dfrac{5}{2x} + \dfrac{8}{x}$

50. $\dfrac{7}{5x} + \dfrac{3}{x}$

51. $\dfrac{2}{7x} + \dfrac{3}{x}$

52. $\dfrac{2}{5x} + \dfrac{5}{x}$

53. $\dfrac{3}{2x} + \dfrac{5}{6x}$

54. $\dfrac{2}{4x} + \dfrac{5}{8x}$

55. $\dfrac{3}{x} + \dfrac{4}{y}$

56. $\dfrac{5}{x} + \dfrac{3}{y}$

57. $\dfrac{4}{a} - \dfrac{9}{b}$

58. $\dfrac{6}{x} - \dfrac{4}{y}$

59. $\dfrac{2x}{15} + \dfrac{3x}{5}$

60. $\dfrac{7x}{12} + \dfrac{5x}{6}$

61. $\dfrac{-3x}{10} - \dfrac{7x}{20}$

62. $\dfrac{-11x}{14} - \dfrac{3x}{28}$

63. $\dfrac{x}{3} + \left(\dfrac{-11x}{12}\right)$

64. $\dfrac{x}{4} + \left(\dfrac{-7x}{16}\right)$

Mixed Practice

65. $\dfrac{5}{6} + \dfrac{3}{10}$

66. $\dfrac{3}{7} + \dfrac{7}{2}$

67. $\dfrac{3}{16} + \left(\dfrac{-9}{20}\right)$

68. $\dfrac{7}{18} + \left(\dfrac{-5}{27}\right)$

69. $\dfrac{9}{y} + \dfrac{1}{x}$

70. $\dfrac{5}{x} + \dfrac{1}{y}$

71. $\dfrac{2x}{15} + \dfrac{3x}{20}$

72. $\dfrac{7x}{10} - \dfrac{2x}{15}$

Applications

73. *Sweets Purchased* Pat bought $\frac{3}{4}$ pound of peanut butter fudge and $\frac{7}{8}$ pound of fudge with nuts. How many pounds of fudge did Pat buy?

74. *Miles Walked* Leon walked $\frac{3}{8}$ mile to his friend's house and then $\frac{2}{3}$ mile to the park. How far did Leon walk?

75. *Fruit Bar Recipe* A chewy fruit bar recipe calls for $\frac{3}{4}$ cup of brown sugar and $\frac{1}{2}$ cup of granulated sugar.

(a) How many cups of sugar are in the recipe?

(b) How many more cups of brown sugar than granulated sugar are in the recipe?

76. *Ham Recipe* A ham glaze recipe calls for $\frac{1}{2}$ teaspoon of dry mustard, $\frac{1}{4}$ teaspoon of ground ginger, and $\frac{1}{4}$ teaspoon of salt.

(a) What is the quantity of ingredients in the recipe?

(b) How many more teaspoons of dry mustard than salt are in the recipe?

77. *Painting a Home* A mother and her teenage son are painting their home. In one day, the mother completes $\frac{1}{3}$ of the job and the son completes $\frac{1}{4}$ of the job. How much more of the job did the mother complete than the son?

78. *Homework Completed* Mary Ann finished $\frac{3}{4}$ of her math homework before dinner and $\frac{1}{5}$ after dinner. How much more homework did Mary Ann complete before dinner than after dinner?

79. *Inheritance* Eric inherited $\frac{1}{8}$ of his grandfather's estate, and his cousin inherited $\frac{1}{12}$ of the estate.

   (a) What part of the estate did Eric and his cousin inherit?

   (b) How much more of the estate did Eric inherit than his cousin?

80. *Family Business* Thu Tran owns $\frac{2}{9}$ of the family business, and her brother Bao owns $\frac{4}{15}$ of the business.

   (a) What part of the family business do Thu and Bao own together?

   (b) How much more of the family business does Bao own than Thu?

One Step Further *Perform the operations indicated.*

81. $\dfrac{7}{30} + \dfrac{3}{40} + \dfrac{1}{8}$

82. $\dfrac{1}{12} + \dfrac{3}{14} + \dfrac{4}{21}$

83. $\dfrac{1}{3} + \dfrac{1}{12} - \dfrac{1}{6}$

84. $\dfrac{1}{5} + \dfrac{2}{3} - \dfrac{11}{15}$

Cumulative Review *Evaluate each of the following for the given values.*

85. **[2.4.4]** $\dfrac{-a}{b}$ for $a = -24$ and $b = 6$

86. **[2.4.4]** $\dfrac{-x}{-y}$ for $x = -10$ and $y = -2$

87. **[2.6.2]** $\dfrac{(a^2 - b)}{-4}$ for $a = 3$ and $b = 1$

88. **[2.6.2]** $9x - x^2$ for $x = -2$

Solve the applications in exercises 89 and 90.

89. **[1.9.3]** *Household Budget* A family's monthly budget for household expenses is $3033. After $1295 is spent for the house payment, $469 for food, $387 for clothing, and $287 for entertainment, how much is left in the budget?

90. **[1.9.3]** *College Living Expense* Joan has a $500 monthly expense allowance from her parents while she is living in the dorm at college. Her expenses this month were: food $190, gas $43, telephone $42, school supplies $96, and entertainment $55. How much money did she have left after expenses?

Quick Quiz 5.3 *Perform the operations indicated.*

1. $\dfrac{1}{y} - \dfrac{2}{9y}$

3. $\dfrac{2x}{15} + \dfrac{5x}{6}$

2. (a) $\dfrac{5}{12} + \dfrac{3}{20}$ **(b)** $\dfrac{3}{14} - \dfrac{1}{21}$

4. Concept Check

   (a) What is a common denominator for the fractions $\frac{3x}{20}$ and $\frac{5x}{6}$?

   (b) Explain how you would add the fractions.

5.4 Operations with Mixed Numbers

Student Learning Objectives

After studying this section, you will be able to:

① Add and subtract mixed numbers.

② Multiply and divide mixed numbers.

③ Solve applied problems involving mixed numbers.

① Adding and Subtracting Mixed Numbers

We add and subtract mixed numbers in a manner similar to the one used for proper fractions. The only difference is that we work with the whole number and the fractional parts separately.

> **ADDING AND SUBTRACTING MIXED NUMBERS**
> We add or subtract the fractions first and then the whole numbers.

EXAMPLE 1 Add. $4\frac{1}{8} + 3\frac{3}{8}$

Solution

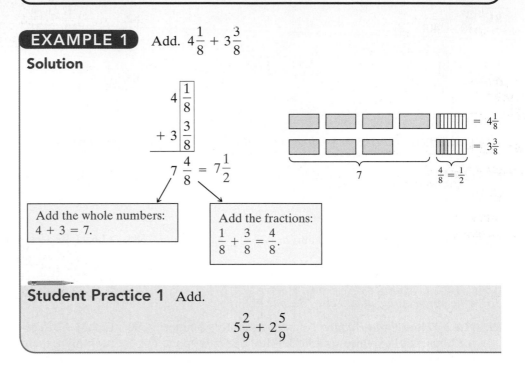

Add the whole numbers: $4 + 3 = 7$.

Add the fractions: $\frac{1}{8} + \frac{3}{8} = \frac{4}{8}$.

NOTE TO STUDENT: *Fully worked-out solutions to all of the Student Practice problems can be found at the back of the text starting at page SP-1.*

Student Practice 1 Add.

$$5\frac{2}{9} + 2\frac{5}{9}$$

If the fractional parts of the mixed numbers do not have common denominators, we find the LCD and build equivalent fractions to obtain common denominators before adding.

EXAMPLE 2 Add. $4\frac{2}{3} + 2\frac{1}{4}$

Solution The LCD of $\frac{2}{3}$ and $\frac{1}{4}$ is 12, so we build equivalent fractions with this denominator.

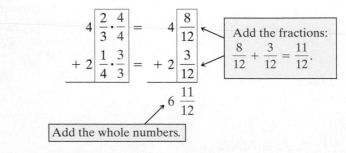

Add the fractions: $\frac{8}{12} + \frac{3}{12} = \frac{11}{12}$.

Add the whole numbers.

Student Practice 2 Add.

$$5\frac{1}{3} + 6\frac{3}{5}$$

EXAMPLE 3 Add. $2\frac{5}{7} + 6\frac{2}{3}$

Solution The LCD of $\frac{5}{7}$ and $\frac{2}{3}$ is 21.

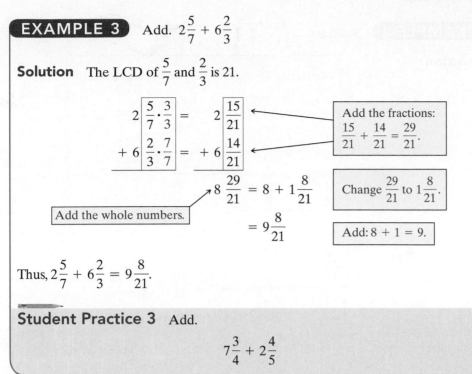

Add the fractions:
$\frac{15}{21} + \frac{14}{21} = \frac{29}{21}$.

$8\frac{29}{21} = 8 + 1\frac{8}{21}$ Change $\frac{29}{21}$ to $1\frac{8}{21}$.

Add the whole numbers.

$= 9\frac{8}{21}$ Add: $8 + 1 = 9$.

Thus, $2\frac{5}{7} + 6\frac{2}{3} = 9\frac{8}{21}$.

Student Practice 3 Add.

$$7\frac{3}{4} + 2\frac{4}{5}$$

CARRYING

In Example 3 we simplified $8\frac{29}{21}$ to $9\frac{8}{21}$ because the fractional part of a mixed number must be a proper fraction. We used a process similar to *carrying* with whole numbers.

Change to a mixed number

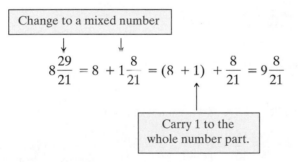

$$8\frac{29}{21} = 8 + 1\frac{8}{21} = (8 + 1) + \frac{8}{21} = 9\frac{8}{21}$$

Carry 1 to the whole number part.

BORROWING

When subtracting mixed numbers, it is sometimes necessary to *borrow*. The process used to borrow with mixed numbers is the opposite of carrying with mixed numbers.

Write 9 as the sum $8 + 1$.

Borrow 1 from the whole number part and add to the fraction part.

$$9\frac{8}{21} = (8 + 1) + \frac{8}{21} = 8 + 1\frac{8}{21} = 8\frac{29}{21}$$

Change $1\frac{8}{21}$ to the improper fraction $\frac{29}{21}$.

Thus $9\frac{8}{21}$ is written as $8\frac{29}{21}$ when borrowing is necessary.

EXAMPLE 4 Subtract. $7\frac{4}{15} - 2\frac{7}{15}$

Solution

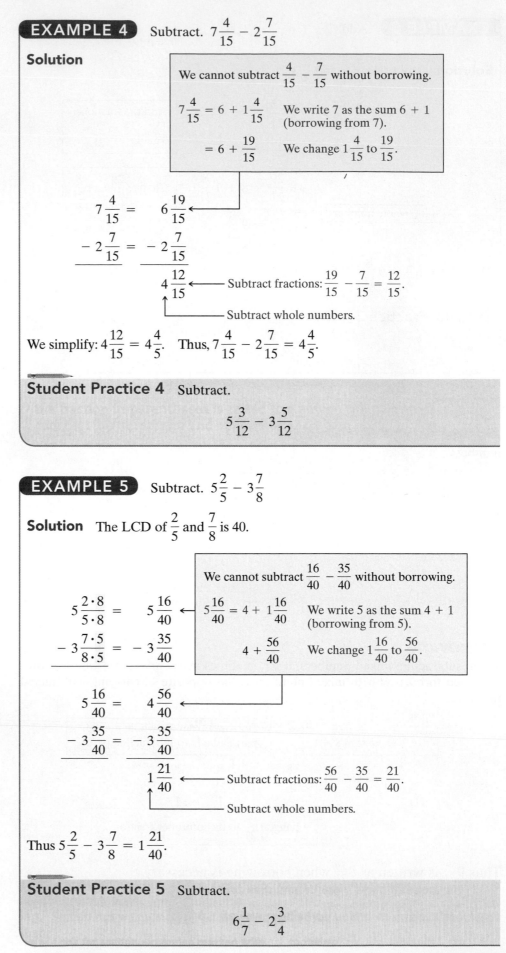

We cannot subtract $\frac{4}{15} - \frac{7}{15}$ without borrowing.

$7\frac{4}{15} = 6 + 1\frac{4}{15}$ We write 7 as the sum $6 + 1$ (borrowing from 7).

$= 6 + \frac{19}{15}$ We change $1\frac{4}{15}$ to $\frac{19}{15}$.

$$7\frac{4}{15} = 6\frac{19}{15}$$
$$-2\frac{7}{15} = -2\frac{7}{15}$$
$$4\frac{12}{15}$$

Subtract fractions: $\frac{19}{15} - \frac{7}{15} = \frac{12}{15}$.

Subtract whole numbers.

We simplify: $4\frac{12}{15} = 4\frac{4}{5}$. Thus, $7\frac{4}{15} - 2\frac{7}{15} = 4\frac{4}{5}$.

Student Practice 4 Subtract.

$$5\frac{3}{12} - 3\frac{5}{12}$$

EXAMPLE 5 Subtract. $5\frac{2}{5} - 3\frac{7}{8}$

Solution The LCD of $\frac{2}{5}$ and $\frac{7}{8}$ is 40.

$$5\frac{2 \cdot 8}{5 \cdot 8} = 5\frac{16}{40}$$
$$-3\frac{7 \cdot 5}{8 \cdot 5} = -3\frac{35}{40}$$

We cannot subtract $\frac{16}{40} - \frac{35}{40}$ without borrowing.

$5\frac{16}{40} = 4 + 1\frac{16}{40}$ We write 5 as the sum $4 + 1$ (borrowing from 5).

$4 + \frac{56}{40}$ We change $1\frac{16}{40}$ to $\frac{56}{40}$.

$$5\frac{16}{40} = 4\frac{56}{40}$$
$$-3\frac{35}{40} = -3\frac{35}{40}$$
$$1\frac{21}{40}$$

Subtract fractions: $\frac{56}{40} - \frac{35}{40} = \frac{21}{40}$.

Subtract whole numbers.

Thus $5\frac{2}{5} - 3\frac{7}{8} = 1\frac{21}{40}$.

Student Practice 5 Subtract.

$$6\frac{1}{7} - 2\frac{3}{4}$$

● **Understanding the Concept**

Should We Change to an Improper Fraction?   When combining mixed numbers, we can change the numbers to improper fractions, and then add or subtract. To illustrate, we work Example 5 this way.

$$\begin{array}{ccc} & \text{Change to} & \text{Build} \\ & \text{improper} & \text{equivalent} \\ & \text{fractions.} & \text{fractions.} \\ & \downarrow & \downarrow \end{array}$$

$$\begin{array}{ccc} 5\dfrac{2}{5} & = & \dfrac{27}{5} & = & \dfrac{216}{40} \\[2mm] -3\dfrac{7}{8} & = & -\dfrac{31}{8} & = & -\dfrac{155}{40} \\[2mm] \hline & & & & \dfrac{61}{40} \text{ or } 1\dfrac{21}{40} \end{array}$$ Subtract improper fractions.

As you can see, the result is the same as the result obtained in Example 5. Which method should you use? Well, there are advantages to both methods. We do not have to carry or borrow when we change to improper fractions first. But if the numbers are large, changing to improper fractions can be more difficult.

Exercise

1. Add the following numbers using both methods. Which way is easier? Why?

$$25\dfrac{3}{5} + 32\dfrac{5}{8}$$

EXAMPLE 6   Subtract. $8 - 3\dfrac{1}{4}$

Solution

$$\begin{array}{rl} 8 = & 7\dfrac{4}{4} \\ -3\dfrac{1}{4} = & -3\dfrac{1}{4} \\ \hline & 4\dfrac{3}{4} \end{array}$$ $8 = (7 + 1) = 7 + \dfrac{4}{4}$ or $7\dfrac{4}{4}$

When we borrowed 1 from 8 we changed the 1 to $\frac{4}{4}$ because we wanted a fraction that had the same denominator as $\frac{1}{4}$.

Student Practice 6   Subtract. $9 - 4\dfrac{1}{3}$

② Multiplying and Dividing Mixed Numbers

In Chapter 4 we learned that the mixed number $2\frac{1}{4}$ means $2 + \frac{1}{4}$. To multiply $3 \cdot 2\frac{1}{4}$ we have $3\left(2 + \frac{1}{4}\right)$, which requires use of the distributive property.

$$3 \cdot 2\dfrac{1}{4} = 3\left(2 + \dfrac{1}{4}\right) = 3 \cdot \left(2 + \dfrac{1}{4}\right) = 3 \cdot 2 + 3 \cdot \dfrac{1}{4} = 6 + \dfrac{3}{4} = 6\dfrac{3}{4}$$

Changing the mixed number to an improper fraction simplifies the calculations.

$$3 \cdot 2\dfrac{1}{4} = \dfrac{3}{1} \cdot \dfrac{9}{4} = \dfrac{27}{4} = 6\dfrac{3}{4}$$

CAUTION: When *adding* and *subtracting* mixed numbers, we normally *don't* change to improper fractions before we add or subtract. However, when *multiplying* and *dividing* mixed numbers, we must change to improper fractions *before* we multiply or divide to avoid using the distributive property.

> Change mixed numbers to improper fractions before multiplying or dividing.

EXAMPLE 7 Multiply. $5\frac{5}{12} \cdot 3\frac{11}{15}$

Solution We change the mixed numbers to improper fractions and then multiply.

$$5\frac{5}{12} \cdot 3\frac{11}{15} = \frac{65}{12} \cdot \frac{56}{15} = \frac{\cancel{5} \cdot 13 \cdot \cancel{4} \cdot 14}{3 \cdot \cancel{4} \cdot \cancel{5} \cdot 3} = \frac{182}{9} \text{ or } 20\frac{2}{9}$$

Student Practice 7 Multiply.

$$6\frac{3}{7} \cdot 1\frac{13}{15}$$

EXAMPLE 8 Divide. $2\frac{1}{4} \div (-5)$

Solution Recall that to divide we invert the second fraction and multiply.

$$2\frac{1}{4} \div (-5) = \frac{9}{4} \div \frac{(-5)}{1} \quad \text{Change to an improper fraction.}$$

$$= \frac{9}{4} \cdot \left(-\frac{1}{5}\right) \quad \text{Find the reciprocal of (invert) } \frac{-5}{1}.$$

$$= -\frac{9}{20} \quad \text{Multiply.}$$

Student Practice 8 Divide.

$$1\frac{1}{4} \div (-2)$$

③ Solving Applied Problems Involving Mixed Numbers

When solving applied problems, it is important to identify which operation to use. In Sections 5.1 and 5.3 we reviewed the addition, subtraction, multiplication, and division situations we learned in Chapter 1. We summarize these, as well as other key words and phrases, below.

> **Multiplication Situations and Key Words**
> - Situations that require repeated addition or taking a fraction *of* something
> - Key words: *double, triple,* and *times*
>
> **Division Situations and Key Words**
> - Situations that require splitting an amount into a certain number of equal parts or finding how many times one number goes into another number
> - Key words: *quotient, divided by*
>
> **Addition Key Phrases**
> - "How many," "find the total"
>
> **Subtraction Key Phrases**
> - "How much more," "how much is left"

When solving applications involving fractions, it is helpful to draw pictures or diagrams.

▲ **EXAMPLE 9**   Ester uses a small piece of painted wood as the base for each centerpiece she makes for banquet tables. She has a long piece of wood that measures $13\frac{1}{2}$ feet. She needs to cut it into pieces that are $\frac{1}{2}$ foot long for the centerpiece bases. How many centerpiece bases will she be able to cut from the long piece of wood?

Solution   *Understand the problem.* We draw a picture.

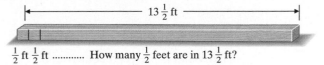

$\frac{1}{2}$ ft $\frac{1}{2}$ ft .... How many $\frac{1}{2}$ feet are in $13\frac{1}{2}$ ft?

Solve and state the answer. This is a division situation since we want to know how many $\frac{1}{2}$s are in $13\frac{1}{2}$. Therefore we must divide $13\frac{1}{2} \div \frac{1}{2}$.

$$13\frac{1}{2} \div \frac{1}{2} = \frac{27}{2} \div \frac{1}{2} \quad \text{We change } 13\frac{1}{2} \text{ to an improper fraction.}$$

$$= \frac{27}{2} \cdot \frac{2}{1} \quad \text{We invert } \frac{1}{2} \text{ and change the division to multiplication.}$$

$$= 27 \quad \text{We simplify.}$$

Ester can make 27 centerpiece bases from the piece of wood.

Student Practice 9   A recipe uses $4\frac{1}{2}$ tablespoons of brown sugar. If Sara only has a $\frac{1}{2}$-tablespoon measuring utensil, how many times must she fill this utensil to get the desired amount of sugar?

▲ **EXAMPLE 10**   A plumber has a pipe $5\frac{3}{4}$ feet long. He needs $\frac{1}{3}$ of the pipe for a repair job. What length must he cut off the pipe to get the desired size?

Solution   *Understand the problem.* We draw a picture.

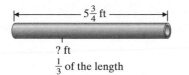

? ft
$\frac{1}{3}$ of the length

Solve and state the answer. The key word of ($\frac{1}{3}$ of the pipe) indicates that we must multiply to find the length the plumber must cut off.

$$\frac{1}{3} \cdot 5\frac{3}{4} = \frac{1}{3} \cdot \frac{23}{4} = \frac{23}{12} \text{ or } 1\frac{11}{12}$$

The plumber must cut off $1\frac{11}{12}$ feet of pipe.

▲ **Student Practice 10**   Jerome has $32\frac{1}{2}$ feet of fencing material. He needs $\frac{1}{10}$ of this material to repair a portion of the fence on his land. How much fencing material does Jerome need to repair his fence?

Verbal and Writing Skills, Exercises 1–2

1. Marcy multiplied two mixed numbers and got the following results.

$$2\frac{2}{3} \cdot 3\frac{4}{5} \rightarrow 2 \cdot 3 = 6 \quad \text{and} \quad \frac{2}{3} \cdot \frac{4}{5} = \frac{8}{15}$$

$$2\frac{2}{3} \cdot 3\frac{4}{5} = 6\frac{8}{15} \quad \text{This answer is wrong.}$$

(a) What did Marcy do wrong?

(b) What is the correct answer?

2. Lester divided two mixed numbers and got the following results.

$$8\frac{1}{2} \div 4\frac{4}{7} \rightarrow 8 \div 4 = 2 \quad \text{and} \quad \frac{1}{2} \cdot \frac{4}{7} = \frac{2}{7}$$

$$8\frac{1}{2} \div 4\frac{4}{7} = 2\frac{2}{7} \quad \text{This answer is wrong.}$$

(a) What did Lester do wrong?

(b) What is the correct answer?

Add or subtract. Simplify all answers. Express as a mixed number.

3. $10\frac{4}{9} + 11\frac{1}{9}$

4. $8\frac{1}{5} + 3\frac{3}{5}$

5. $5\frac{5}{8} + 11\frac{1}{8}$

6. $4\frac{1}{10} + 2\frac{3}{10}$

7. $5\frac{2}{3} + 8\frac{1}{4}$

8. $22\frac{3}{5} + 16\frac{1}{10}$

9. $14\frac{1}{4} + 6\frac{1}{3}$

10. $13\frac{1}{2} + 7\frac{4}{5}$

11. $7\frac{5}{6} + 4\frac{3}{8}$

12. $5\frac{14}{15} + 10\frac{3}{10}$

13. $7\frac{4}{5} - 2\frac{1}{5}$

14. $6\frac{3}{8} - 2\frac{1}{8}$

15. $9\frac{2}{3} - 6\frac{1}{6}$

16. $15\frac{3}{4} - 13\frac{1}{6}$

17. $11\frac{1}{5} - 6\frac{3}{5}$

18. $25\frac{2}{7} - 16\frac{5}{7}$

19. $10\frac{5}{12} - 3\frac{9}{10}$

20. $12\frac{4}{9} - 7\frac{5}{6}$

21. $9 - 2\frac{1}{4}$

22. $4 - 2\frac{2}{7}$

Mixed Practice, Exercises 23–30 *Add or subtract. Simplify all answers. Express as a mixed number.*

23. $8\frac{2}{5} - 6\frac{1}{7}$

24. $45\frac{3}{8} - 26\frac{1}{10}$

25. $1\frac{1}{6} + \frac{3}{8}$

26. $1\frac{2}{3} + \frac{5}{18}$

27. $8\frac{1}{4} + 3\frac{5}{6}$

28. $7\frac{3}{4} + 6\frac{2}{5}$

29. $32 - 1\frac{2}{9}$

30. $24 - 3\frac{4}{11}$

Multiply or divide and simplify your answer.

31. $(-2) \cdot 3\frac{1}{5}$

32. $(-3) \cdot 2\frac{1}{4}$

33. $4\frac{1}{3} \cdot 2\frac{1}{4}$

34. $6\frac{1}{3} \cdot 2\frac{1}{4}$

35. $-\frac{3}{4} \cdot 3\frac{5}{7}$

36. $-\frac{8}{11} \cdot 4\frac{3}{4}$

37. $2\frac{1}{4} \div (-4)$

38. $2\frac{1}{2} \div (-3)$

39. $4\frac{1}{2} \div 2\frac{1}{4}$

40. $8\frac{1}{4} \div 2\frac{3}{4}$

41. $3\frac{1}{4} \div \frac{3}{8}$

42. $2\frac{5}{8} \div \frac{1}{2}$

43. $-6 \div \frac{1}{4}$

44. $-8 \div \frac{1}{4}$

Mixed Practice, Exercises 45–50 *Multiply or divide and simplify your answer.*

45. $1\frac{1}{4} \cdot 3\frac{2}{3}$

46. $2\frac{3}{5} \cdot 1\frac{4}{7}$

47. $4\frac{1}{2} \div \left(-\frac{6}{7}\right)$

48. $1\frac{7}{8} \div \left(-\frac{1}{3}\right)$

49. $7\frac{1}{2} \div (-8)$

50. $8\frac{2}{9} \div (-9)$

Applications

51. *Camping Supplies* To put up the tents for a camping trip, Andy needs several pieces of rope each $6\frac{1}{2}$ feet long. If Andy has a rope that is 26 feet long, how many pieces can he cut for his tents?

52. *Tie Pattern* Quynh has $4\frac{1}{5}$ yards of silk. It takes $\frac{3}{5}$ of a yard to make a tie. How many ties can she make from this silk?

▲ **53.** *Oak Wood* A carpenter has a piece of oak wood $7\frac{1}{5}$ feet long. He needs $\frac{1}{3}$ of the length for a shelf. What length must he cut off the wood to get the desired size?

54. *Project* Ella has $7\frac{1}{2}$ weeks to finish a project for her employer. She must plan to allow $\frac{1}{5}$ of the time to develop a formal presentation. How many weeks should Ella allow to develop the presentation?

Recipes Use the given recipe to answer exercises 55–58

55. To double the recipe, how much flour do you need?

56. To triple the recipe, how much brown sugar do you need?

57. To make four times the recipe, how many cups of chocolate chips do you need?

58. To make $\frac{1}{2}$ of the recipe, how much granulated sugar do you need?

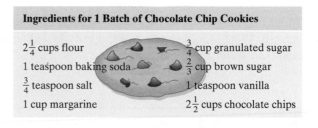

Ingredients for 1 Batch of Chocolate Chip Cookies

$2\frac{1}{4}$ cups flour

$\frac{3}{4}$ cup granulated sugar

1 teaspoon baking soda

$\frac{2}{3}$ cup brown sugar

$\frac{3}{4}$ teaspoon salt

1 teaspoon vanilla

1 cup margarine

$2\frac{1}{2}$ cups chocolate chips

59. *20-mile Run* As part of a 3-day workout plan to prepare for a 20-mile endurance run, Jeff ran $9\frac{1}{8}$ miles Monday, $12\frac{1}{3}$ miles Tuesday, and $17\frac{1}{6}$ miles Wednesday. How many miles did Jeff run in the 3-day period?

60. *Purchase* Will bought $\frac{5}{8}$ pound of M&M's and $1\frac{7}{16}$ pounds of jellybeans. How many pounds of candy did Will buy?

To Think About   *Find the values for each of the variables.*

61. $\dfrac{1}{4}, \dfrac{1}{2}, \dfrac{3}{4}, 1, \dfrac{a}{4}, \dfrac{3}{2}, \dfrac{7}{b}, 2, \dfrac{9}{4}, \dfrac{5}{2}, \dfrac{11}{4}, c, \ldots$

62. $\dfrac{1}{8}, \dfrac{1}{4}, \dfrac{3}{8}, \dfrac{1}{2}, \dfrac{a}{8}, \dfrac{3}{4}, \dfrac{7}{b}, 1, \dfrac{9}{8}, \dfrac{5}{4}, \dfrac{11}{8}, \dfrac{c}{2}, \ldots$

Cumulative Review   *Perform the operations indicated.*

63. [1.6.4] $2 + 9 \cdot 8$

64. [1.6.4] $12 - 4 \div 2$

65. [1.6.4] $\dfrac{(5 + 7)}{(2 \cdot 3)}$

66. [1.6.4] $\dfrac{(11 - 5)}{2}$

67. [1.9.3] *Water Dispenser* Aqua Water Company charges a $7 per month rental fee for each dispenser and $2 for each gallon of water delivered. The company offers a $3 per month discount on the total rental fee for every 8 dispensers rented and a $5 per month discount for every 15 gallons of water purchased within a 1-month period. From the invoice below, determine how much the large advertising firm *Sell It Now* paid Aqua Water Company for the rental of dispensers and the water delivered over a period of 3 months.

Sell It Now	Aqua Water Company Invoice		
	January	February	March
Number of units rented	19	26	26
Gallons of water delivered	28	36	31
Monthly charge	_____	_____	_____

68. [1.9.3] Refer to exercise 67 to answer the following. A water dispenser that has hot and cold water rents for $8 per month. What would the cost of the rentals and water be for the 3-month period if the advertising firm rented all hot and cold water dispensers?

Quick Quiz 5.4   *Perform the operation indicated.*

1. (a) $1\dfrac{7}{9} + 3\dfrac{5}{12}$ **(b)** $10\dfrac{1}{3} - 5\dfrac{5}{12}$

2. $(-9) \cdot 2\dfrac{1}{3}$

3. $4\dfrac{2}{5} \div \left(-1\dfrac{1}{10}\right)$

4. Concept Check Explain how you would multiply $2\frac{1}{2} \times 3\frac{2}{3}$.

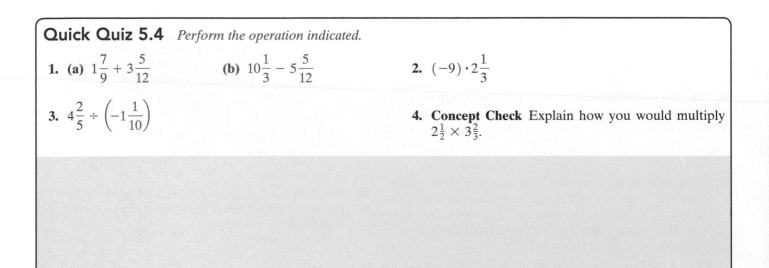

Did You Know...
That Saving a Few Dollars Every Day Can Add Up to Over a Thousand Dollars a Year?

EXPENSES THAT COST MORE THAN YOU THINK

Understanding the Problem:
Tracy and Max would like to go on an exciting vacation. They need to start saving so that they will be able to afford the trip, but they don't have a lot of extra cash left over at the end of the month. They wonder how much they could save if they stopped going out for lunch and instead made lunch themselves.

Making a Plan:
They decide to figure out how much they are spending now on lunches and what it will cost to make their own lunches. Then they can calculate how long it will take to save up enough money if they stop going out for lunch.

Step 1: They reviewed their weekly expenses, and they determined how much they are spending on average Monday through Thursday. Tracy goes to a sub sandwich shop and spends $8 a day. On Fridays she goes out with friends to a restaurant and spends $13 on lunch. Max orders out Monday through Friday and spends $10 a day. They total up their weekly spending.

Task 1: *Add up what Tracy and Max spend every week on lunch.*

Step 2: They calculate that it will cost $3 a day per person if they make their own lunches.

Task 2: *What is the weekly cost of bringing their own lunches?*

Step 3: Next, they figure out how much they will save per week by not going out for lunch.

Task 3: *How much will they save every week?*

Finding a Solution:
Step 4: Now that they know how much they can save per week, they want to know how long it will take to save up $2275, the cost of the trip.

Task 4: *After how many weeks of saving will they have enough money?*

Applying the Situation to Your Life:
There is probably money you spend on a daily basis that you could save by making changes. You could cut down on eating out by making more of your own meals. Instead of buying coffee, you could brew your own coffee. See where you could cut spending and tally up the weekly savings. Is there something you've thought of buying, but you haven't had the money? Use your predicted weekly savings to figure out how many weeks it will take you to save up enough money. Having a goal will help you stick to a savings plan.

How Am I Doing? Sections 5.1–5.4

1. _____

2. _____

3. _____

4. _____

5. _____

6. _____

7. _____

8. _____

9. _____

10. _____

11. (a) _____

(b) _____

12. _____

13. _____

14. _____

15. _____

How are you doing with your homework assignments in Sections 5.1 to 5.4? Do you feel you have mastered the material so far? Do you understand the concepts you have covered? Before you go further in the textbook, take some time to do each of the following problems.

5.1

Perform the operation indicated. Simplify your answer.

1. $\dfrac{16}{9} \cdot \left(\dfrac{-18}{36} \right)$

2. $\dfrac{-3y^3}{20} \cdot \dfrac{-12y^2}{21}$

3. $25 \div \dfrac{5}{7}$

4. $\dfrac{3y^4}{20} \div \dfrac{12y^2}{5}$

5. Mary Beth owns 63 acres of land. She wishes to subdivide the land into $\frac{3}{4}$-acre parcels. How many $\frac{3}{4}$-acre parcels will Mary Beth have?

5.2

6. Find the LCM of 12 and 21.

7. Find the LCM of $7x$, 21, and $2x^2$.

5.3

Perform the operation indicated.

8. $\dfrac{-3}{7} + \dfrac{2}{21}$

9. $\dfrac{11}{16} - \dfrac{5}{20}$

10. $\dfrac{5x}{8} - \dfrac{3x}{14}$

11. Abbas is painting his house. The first day he completed $\frac{1}{4}$ of the job and the second day he completed $\frac{1}{5}$ of the job.

(a) How much more did he complete the first day than the second day?

(b) What part of the total job did Abbas finish the first two days?

5.4

Perform the operation indicated.

12. $2\dfrac{2}{5} + 4\dfrac{6}{7}$

13. $6\dfrac{1}{12} - 2\dfrac{7}{15}$

14. $2\dfrac{2}{3} \cdot 1\dfrac{5}{16}$

15. $10\dfrac{2}{9} \div 2\dfrac{1}{3}$

Now turn to page SA-10 for the answers to each of these problems. Each answer also includes a reference to the objective in which the problem is first taught. If you missed any of these problems, you should stop and review the examples and Student Practice problems in the referenced objective. A little review now will help you to master the material in the upcoming sections of the text.

5.5 Order of Operations and Complex Fractions

① Following the Order of Operations with Fractions

Recall that when we work a problem with more than one operation, we must follow the **order of operations.**

Student Learning Objectives

After studying this section, you will be able to:

① Follow the order of operations with fractions.

② Simplify complex fractions.

③ Solve applied problems involving complex fractions.

> **ORDER OF OPERATIONS**
>
> If a fraction has operations written in the numerator or in the denominator or both, these operations must be done first. We perform operations in the following order.
>
> 1. Perform operations inside grouping symbols.
> 2. Simplify exponents.
> 3. Do multiplication and division in order from left to right.
> 4. Do addition and subtraction last, working from left to right.

EXAMPLE 1 Simplify. $\left(\dfrac{2}{3}\right)^2 - \dfrac{2}{9} \cdot \dfrac{1}{3}$

Solution

$\left(\dfrac{2}{3}\right)^2 - \dfrac{2}{9} \cdot \dfrac{1}{3} = \dfrac{4}{9} - \dfrac{2}{9} \cdot \dfrac{1}{3}$ First we simplify exponents: $\left(\dfrac{2}{3}\right)^2 = \left(\dfrac{2}{3}\right)\left(\dfrac{2}{3}\right) = \dfrac{4}{9}.$

$= \dfrac{4}{9} - \dfrac{2}{27}$ Then we multiply: $\dfrac{2}{9} \cdot \dfrac{1}{3} = \dfrac{2}{27}.$

$= \dfrac{4 \cdot 3}{9 \cdot 3} - \dfrac{2}{27}$ Next we find the LCD, which is 27, and build equivalent fractions.

$= \dfrac{12}{27} - \dfrac{2}{27}$ Now we subtract.

$= \dfrac{10}{27}$ Last, we simplify.

CAUTION: Do not add or subtract before multiplying or dividing even though addition or subtraction comes first in the problem or appears to be easier. *Be careful not to make this error.*

Student Practice 1 Simplify.

$\left(\dfrac{5}{3}\right)^2 + \dfrac{20}{9} \cdot \dfrac{1}{5}$

② Simplifying Complex Fractions

When the numerator and/or the denominator is a fraction, we have a *complex fraction*.

> **COMPLEX FRACTION**
>
> A fraction that contains at least one fraction in the numerator or in the denominator is a **complex fraction.**

NOTE TO STUDENT: Fully worked-out solutions to all of the Student Practice problems can be found at the back of the text starting at page SP-1.

These three fractions are complex fractions.

$$\dfrac{\dfrac{2}{x}}{\dfrac{12}{x}}, \quad \dfrac{\dfrac{3}{7} + 2}{\dfrac{1}{8}}, \quad \dfrac{2 - \dfrac{1}{x}}{\dfrac{1}{6} + \dfrac{2}{9}} \leftarrow \text{main fraction bar}$$

Although we usually do not write grouping symbols (parentheses or brackets) around the numerator and denominator of a complex fraction, it is understood that *they exist*. Thus we must perform operations above and then below the main fraction bar before we divide.

EXAMPLE 2 Simplify. $\dfrac{(-2)^2 + 8}{\dfrac{2}{3}}$

Solution We must follow the order of operations.

$\dfrac{[(-2)^2 + 8]}{\left(\dfrac{2}{3}\right)}$ We write grouping symbols in the numerator and denominator, since it is understood they exist.

$= \dfrac{12}{\left(\dfrac{2}{3}\right)}$ Within the brackets we simplify exponents first, and then add: $(-2)^2 = 4$ and $4 + 8 = 12$.

$= 12 \div \dfrac{2}{3}$ The main fraction bar means divide.

$= 12 \cdot \dfrac{3}{2}$ We invert the second fraction and multiply.

$= \dfrac{\overset{1}{\cancel{2}} \cdot 2 \cdot 3 \cdot 3}{\underset{1}{\cancel{2}}} = 18$ We factor 12: $12 = 2 \cdot 2 \cdot 3$. Then we simplify.

Student Practice 2 Simplify.

$$\dfrac{\dfrac{4}{7}}{3^2 + (-5)}$$

EXAMPLE 3 Simplify. $\dfrac{\dfrac{x^2}{8}}{\dfrac{x}{4}}$

Solution Since the main fraction bar indicates division, we can divide the top fraction by the bottom fraction to simplify.

$$\dfrac{\dfrac{x^2}{8}}{\dfrac{x}{4}} = \dfrac{x^2}{8} \div \dfrac{x}{4} = \dfrac{x^2}{8} \cdot \dfrac{4}{x} = \dfrac{x^2 \cdot \cancel{4}}{2 \cdot \cancel{4} \cdot x} = \dfrac{\cancel{x} \cdot x}{2 \cdot \cancel{x}} = \dfrac{x}{2}$$

Student Practice 3 Simplify.

$$\dfrac{\dfrac{x^2}{5}}{\dfrac{x}{10}}$$

EXAMPLE 4   Simplify. $\dfrac{\dfrac{2}{3} + \dfrac{1}{6}}{\dfrac{3}{4} - \dfrac{1}{2}}$

Solution   We write parentheses in the numerator and denominator and follow the order of operations.

$$\dfrac{\left(\dfrac{2}{3} + \dfrac{1}{6}\right)}{\left(\dfrac{3}{4} - \dfrac{1}{2}\right)} = \dfrac{\left(\dfrac{2 \cdot 2}{3 \cdot 2} + \dfrac{1}{6}\right)}{\left(\dfrac{3}{4} - \dfrac{1 \cdot 2}{2 \cdot 2}\right)} = \dfrac{\left(\dfrac{4}{6} + \dfrac{1}{6}\right)}{\left(\dfrac{3}{4} - \dfrac{2}{4}\right)} = \dfrac{\dfrac{5}{6}}{\dfrac{1}{4}}$$ Add top fractions.

Subtract bottom fractions.

Now we divide the top fraction by the bottom fraction.

$$\dfrac{5}{6} \div \dfrac{1}{4} = \dfrac{5}{6} \cdot \dfrac{4}{1} = \dfrac{5 \cdot 2 \cdot 2}{3 \cdot 2} = \dfrac{10}{3}$$ Thus, $\dfrac{\dfrac{2}{3} + \dfrac{1}{6}}{\dfrac{3}{4} - \dfrac{1}{2}} = \dfrac{10}{3}$

Student Practice 4

Simplify. $\dfrac{\dfrac{3}{5} + \dfrac{1}{2}}{\dfrac{5}{6} - \dfrac{1}{3}}$

③ Solving Applied Problems Involving Complex Fractions

EXAMPLE 5   A recipe requires $3\frac{2}{3}$ cups of flour to make bread to feed 50 people. How much flour do we need to make bread to feed 120 people?

Solution   Since the problem concerns the rate of cups of flour per 50 people, we set up a proportion and solve for the missing number.

$$\dfrac{3\dfrac{2}{3} \text{ cups}}{50 \text{ people}} = \dfrac{x \text{ cups}}{120 \text{ people}}$$

$120 \cdot 3\dfrac{2}{3} = 50x$ We find the cross products.

$120 \cdot \dfrac{11}{3} = 50x$ We change $3\frac{2}{3}$ to an improper fraction.

$\dfrac{40 \cdot \overset{1}{3} \cdot 11}{\underset{1}{3}} = 50x$ Since the sum of the digits of 120 is divisible by 3, we factor $120 = 40 \cdot 3$ and simplify.

$40 \cdot 11 = 50x$ Now we solve the equation for x.

$\dfrac{40 \cdot 11}{50} = \dfrac{50x}{50}$ We divide by 50 on both sides.

$\dfrac{4 \cdot \overset{1}{10} \cdot 11}{5 \cdot \underset{1}{10}} = x$ We factor and simplify.

$\dfrac{44}{5} = x$ or $x = 8\dfrac{4}{5}$ We need $8\frac{4}{5}$ cups of flour.

▲ **Student Practice 5**   A carpenter needs $6\frac{4}{5}$ feet of oak wood to make shelves for 2 small bookcases. How many feet of oak wood will the carpenter need to make 15 bookcases?

MyMathLab®

Watch the videos
in MyMathLab

Download the
MyDashBoard App

Verbal and Writing Skills, Exercises 1–2

1. When we perform a series of calculations without grouping symbols, do we add first or multiply first?

2. When we perform a series of calculations without grouping symbols, do we divide first or subtract first?

Simplify.

3. $\dfrac{3}{5} - \dfrac{1}{3} \div \dfrac{5}{6}$

4. $\dfrac{1}{2} + \dfrac{3}{8} \div \dfrac{3}{4}$

5. $\dfrac{3}{4} + \dfrac{1}{4} \cdot \dfrac{3}{5}$

6. $\dfrac{4}{5} - \dfrac{1}{5} \cdot \dfrac{2}{3}$

7. $\dfrac{5}{7} \cdot \dfrac{1}{3} \div \dfrac{2}{7}$

8. $\dfrac{2}{7} \cdot \dfrac{3}{4} \div \dfrac{1}{2}$

9. $\left(\dfrac{3}{2}\right)^2 - \dfrac{1}{3} \div \dfrac{1}{2}$

10. $\left(\dfrac{5}{6}\right)^2 + \dfrac{1}{12} - \dfrac{1}{4}$

11. $\dfrac{5}{6} \cdot \dfrac{1}{2} + \dfrac{2}{3} \div \dfrac{4}{3}$

12. $\dfrac{3}{5} \cdot \dfrac{1}{2} + \dfrac{1}{5} \div \dfrac{2}{3}$

13. $\dfrac{2}{9} \cdot \dfrac{1}{4} + \left(\dfrac{2}{3} \div \dfrac{6}{7}\right)$

14. $\dfrac{3}{4} \cdot \left(\dfrac{1}{6} + \dfrac{1}{2}\right) \div \dfrac{4}{5}$

15. $\dfrac{3}{4} \cdot \dfrac{1}{4} + \left(\dfrac{3}{4}\right)^2$

16. $\left(\dfrac{2}{5}\right)^2 + \dfrac{3}{5} \cdot \dfrac{1}{5}$

17. $\left(-\dfrac{2}{5}\right) \cdot \left(\dfrac{1}{4}\right)^2$

18. $\left(\dfrac{4}{3}\right)^2 \cdot \left(-\dfrac{1}{2}\right)$

Simplify.

19. $\dfrac{7 + (-3)^2}{\dfrac{8}{9}}$

20. $\dfrac{6 + (-4)^2}{\dfrac{2}{3}}$

21. $\dfrac{\dfrac{4}{7}}{2^3 + 8}$

22. $\dfrac{\dfrac{3}{10}}{2^2 + 5}$

23. $\dfrac{2 \cdot 3 - 1}{\dfrac{5}{8}}$

24. $\dfrac{4 \cdot 3 + 2}{\dfrac{2}{7}}$

25. $\dfrac{\dfrac{6}{7}}{\dfrac{9}{14}}$

26. $\dfrac{\dfrac{7}{15}}{\dfrac{21}{10}}$

27. $\dfrac{\dfrac{x^2}{2}}{\dfrac{x}{4}}$

28. $\dfrac{\dfrac{x^2}{7}}{\dfrac{x}{14}}$

29. $\dfrac{\dfrac{x}{3}}{\dfrac{x^2}{9}}$

30. $\dfrac{\dfrac{x}{6}}{\dfrac{x^2}{12}}$

31. $\dfrac{\dfrac{1}{2} + \dfrac{3}{4}}{\dfrac{4}{5} + \dfrac{1}{10}}$

32. $\dfrac{\dfrac{3}{7} + \dfrac{1}{14}}{\dfrac{2}{3} + \dfrac{1}{6}}$

33. $\dfrac{\dfrac{4}{25} - \dfrac{3}{50}}{\dfrac{3}{10} + \dfrac{5}{20}}$

34. $\dfrac{\dfrac{5}{12} - \dfrac{7}{24}}{\dfrac{1}{2} + \dfrac{1}{8}}$

Mixed Practice

35. $\dfrac{\dfrac{x}{10}}{\dfrac{x^2}{20}}$

36. $\dfrac{\dfrac{x}{8}}{\dfrac{x^2}{32}}$

37. $\left(\dfrac{1}{2}\right)^2 + \dfrac{2}{3} \cdot \dfrac{6}{7}$

38. $\left(\dfrac{2}{3}\right)^2 - \dfrac{1}{3} \cdot \dfrac{3}{4}$

39. $\dfrac{\dfrac{3}{8} + \dfrac{1}{4}}{\dfrac{4}{5} + \dfrac{3}{4}}$

40. $\dfrac{\dfrac{5}{6} + \dfrac{1}{3}}{\dfrac{3}{8} + \dfrac{1}{2}}$

41. $\dfrac{11 + (3)^2}{\dfrac{2}{3}}$

42. $\dfrac{\dfrac{6}{7}}{(-2)^3 + 14}$

Applications

43. *Dessert Recipe* A recipe requires $2\frac{3}{4}$ cups of sugar to make a dessert dish for 20 people. How much sugar is needed to make the recipe for 36 people?

44. *Punch Recipe* Monica is using a punch recipe that requires $5\frac{1}{2}$ cups of concentrated punch for every 4 cups of water. How many cups of concentrated punch should she use with 16 cups of water?

45. *Lawn Fertilizer* A gardener needs about $3\frac{3}{4}$ bags of fertilizer for every 2 lawns he feeds. Approximately how many bags of fertilizer will he need to feed 12 lawns?

46. *Casserole Recipe* A casserole recipe requires $2\frac{1}{2}$ teaspoons of salt to prepare a serving for 8 people. How much salt should be added to prepare a serving for 12 people?

47. *Acres of Land* A developer needs $2\frac{1}{4}$ acres of land for every 3 houses she builds. How many acres does she need to build 28 homes?

48. *Inches and Centimeters* There are approximately $2\frac{1}{2}$ centimeters in 1 inch. Approximately how many centimeters are in 12 inches?

One Step Further *Simplify.*

49. $\dfrac{\dfrac{25xy^2}{49}}{\dfrac{15x^2y}{14}}$

50. $\dfrac{\dfrac{36x^2y^2}{45}}{\dfrac{12xy}{30}}$

51. $\dfrac{\dfrac{1}{x} - \dfrac{1}{y}}{\dfrac{1}{x} + \dfrac{1}{y}}$

52. $\dfrac{\dfrac{1}{a} + \dfrac{1}{b}}{\dfrac{1}{a} - \dfrac{1}{b}}$

To Think About *Evaluate each expression for the value given.*

53. $x - \dfrac{2}{5} \div \dfrac{4}{15}$ for $x = \dfrac{7}{2}$

54. $x - \dfrac{5}{6} \div \dfrac{25}{12}$ for $x = \dfrac{7}{5}$

55. $-\dfrac{3}{8} \cdot \dfrac{16}{21} + x$ for $x = -\dfrac{4}{7}$

56. $-\dfrac{4}{9} \cdot \dfrac{18}{24} + x$ for $r = -\dfrac{1}{3}$

Cumulative Review *Perform the operations indicated.*

57. [2.4.2] $(-2)(3)(-1)(-5)$

58. [2.4.2] $(-1)(-3)(-1)(-5)$

59. [2.4.4] $-50 \div 5$

60. [2.4.4] $-36 \div (-6)$

Quick Quiz 5.5 *Perform the operations indicated.*

1. $\left(\dfrac{1}{3}\right)^2 + \dfrac{1}{3} \div \dfrac{5}{6}$

2. $\dfrac{\dfrac{a}{15}}{\dfrac{a}{10}}$

3. $\dfrac{\dfrac{8}{15} - \dfrac{1}{5}}{\dfrac{1}{2} + \dfrac{1}{3}}$

4. Concept Check Explain how you would simplify $\dfrac{1 + 2 \times 3}{\dfrac{1}{2}}$.

5.6 Solving Applied Problems Involving Fractions

① Solving Applied Problems Involving Fractions

For applied problems involving fractions, we may need to draw a picture to help us determine which operation to use.

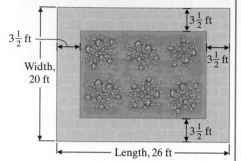

▲ **EXAMPLE 1** Jason planted a rectangular rose garden in the center of his 26-foot by 20-foot backyard. Around the garden there is a sidewalk that is $3\frac{1}{2}$ feet wide. The garden and sidewalk take up the entire 26-foot by 20-foot yard.

(a) What are the dimensions of the rose garden?

(b) How much will it cost to put a fence around the rose garden if the fencing costs $2\frac{1}{2}$ per linear foot?

Solution *Understand the problem.* We draw a picture of the situation to help us develop a plan to solve the problem.

Mathematics Blueprint for Problem Solving

Gather the Facts	What Am I Asked to Do?	How Do I Proceed?	Key Points to Remember
Refer to the picture for dimensions.	**(a)** Find the dimensions of the rose garden.	1. Find the length and width of the garden.	Use the perimeter formula $P = 2L + 2W$.
Fencing: costs $2\frac{1}{2}$ per foot	**(b)** Calculate the cost of a fence around the rose garden.	2. Use the length and width to find the perimeter. 3. Multiply the perimeter times $2\frac{1}{2}$ to find the cost.	The garden is in the center of the yard.

Solve and state the answer.

1. We find the length and width of the rose garden. From the picture, we see the following.

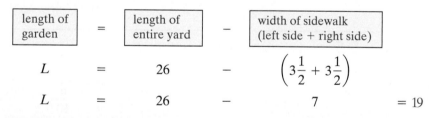

We find the width of the rose garden.

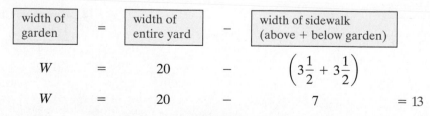

(a) The dimensions of the garden are 19 feet by 13 feet.

2. Now we find the perimeter of the garden.

$$P = 2L + 2W = 2(19) + 2(13)$$
$$= 38 + 26 = 64 \quad \text{The perimeter is 64 ft}$$

3. We multiply $2\frac{1}{2}$ times 64 feet to find the total cost of fencing.

$$2\frac{1}{2} \cdot 64 = \frac{5}{2} \cdot \frac{64}{1} = \frac{5 \cdot \cancel{2} \cdot 32}{\cancel{2} \cdot 1} = 160$$

(b) It will cost \$160 to put a fence around the rose garden.

Check. We can place the answers on the diagram to check.

Width: $3\frac{1}{2} + 13 + 3\frac{1}{2} = 20$ ft   ✓

Length: $3\frac{1}{2} + 19 + 3\frac{1}{2} = 26$ ft   ✓

Round the price: $2\frac{1}{2} \rightarrow \$3$.

Round the perimeter: $64 \rightarrow 60$.

Then multiply to estimate the cost of the fence: $\$3 \times 60 = \$180$, which is close to our answer. ✓

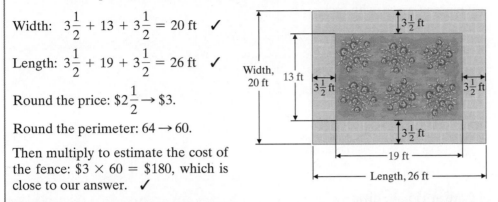

NOTE TO STUDENT: *Fully worked-out solutions to all of the Student Practice problems can be found at the back of the text starting at page SP-1.*

▲ **Student Practice 1**   Larry wants to place a rug in the center of his living room floor, which measures 22 feet by 15 feet. He wants to center the rug so that there is $2\frac{1}{2}$ feet of wood flooring showing on each side of the rug.

(a) What are the dimensions of the rug Larry must buy?

(b) Larry wants to place binding around the outer edges of the carpet. If the cost to bind the carpet is \$2 per linear foot, how much will Larry pay for the binding?

▲ **EXAMPLE 2**   Marian is planning to build a fence on her farm. She determines that she must make 115 wooden fence posts that are each $3\frac{3}{4}$ feet in length. The wood to make the fence posts is sold in 20-foot lengths. How many 20-foot pieces of wood must Marian purchase so that she can make 115 fence posts?

Solution   *Understand the problem.* We draw a picture to help us develop a plan.

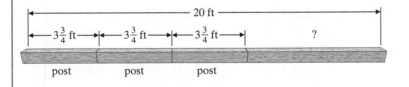

▲ **Student Practice 2**   Nancy wishes to make two bookcases, each with 4 shelves. Each shelf is $3\frac{1}{8}$ feet long. The wood for the shelves is sold as 10-foot boards. How many boards does Nancy need to buy for the shelves?

Mathematics Blueprint for Problem Solving

Gather the Facts	What Am I Asked to Do?	How Do I Proceed?	Key Points to Remember
Refer to the picture.	Find the number of 20-foot lengths of wood needed to make 115 posts.	**1.** Determine the number of posts that can be made from one 20-foot piece of wood. **2.** Use the information in step 1 to find how many 20-foot pieces are needed to make 115 posts.	We must change mixed numbers to improper fractions before we perform division.

Continued on next page

Solve and state the answer.

1. From the picture, we see that we must divide to find how many $3\frac{3}{4}$-foot sections are in 20 feet.

$$20 \div 3\frac{3}{4} = 20 \div \frac{15}{4} = 20 \cdot \frac{4}{15} = \frac{4 \cdot \cancel{5} \cdot 4}{\cancel{5} \cdot 3} = \frac{16}{3} \quad \text{or} \quad 5\frac{1}{3}$$

5 posts can be cut from each 20-foot piece of wood, with some wood left over.

2. Now we must find how many of the 20-foot pieces are needed.

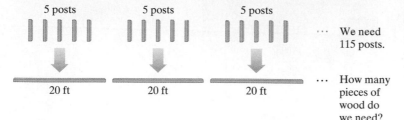

We need 115 posts.

How many pieces of wood do we need?

We must find how many groups of 5 are in 115. We divide $115 \div 5$.

$$115 \div 5 = \frac{115}{5} = \frac{\cancel{5} \cdot 23}{\cancel{5}} = 23$$

Marian must purchase 23 pieces of wood.

Check. We can estimate our answer by rounding the fraction $3\frac{3}{4}$ to 4 and reworking the problem.

20-foot length ÷ 4 feet per post = 5 posts per length of wood

115 posts ÷ 5 posts per length of wood = 23 lengths of wood ✓

👣 STEPS TO SUCCESS   Why Is Homework Necessary?

You learn mathematics by practicing, not by watching. Your instructor may make solving a mathematics problem look easy, but to learn the necessary skills, you must practice them over and over again, just as your instructor once had to do. There is no other way. Learning mathematics is like learning how to play a musical instrument or to play a sport. *You must practice, not just observe, to do well.* Homework provides this practice. The amount of practice varies for each person. The more problems you do, the better you get.

Many students underestimate the amount of time each week that is required to learn math. In general, two to three hours per week per unit or credit is a good rule of thumb.

Making it personal:

1. Make a log of the time that you spend studying math. If your performance is not up to your expectations, increase your study time.

2. Review your time management plan to be sure you included sufficient hours per week to study mathematics.

   Number of hours per week studying math: _____

   Current grade in class: _____

   I must increase study time: _____ Yes _____ No

Applications

When solving the following application problems, you may want to use a Mathematics Blueprint for Problem Solving to help organize your work.

1. **Workout Program** Joan planned a workout program that includes increasing the amount of time she exercises by $\frac{1}{4}$ hour each week for the first 4 weeks. If Joan starts week 1 with a $\frac{1}{2}$-hour workout, how long will her workout program be at the beginning of week 4?

▲ 2. **Bolt Length** In assembling her bookshelves, Keri finds that she needs a bolt that will reach through the $\frac{5}{8}$-inch wood, a $\frac{1}{8}$-inch-thick washer, and a $\frac{1}{4}$-inch-thick nut. How long must the bolt be?

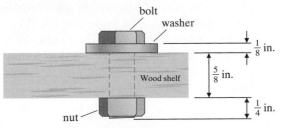

3. **Average Speed** Andrew must drive 305 miles to a sales conference. If it takes him $4\frac{1}{2}$ hours to drive to the sales conference, what is his average speed in miles per hour?

4. **Metal Cutter** A machine that cuts metal is calibrated to cut strips $1\frac{3}{8}$ inches long from a piece of metal that is 22 inches long. How many strips will the machine cut from this one piece?

Recipe vs. Servings Use the given recipe to answer exercises 5 and 6.

5. To make 6 servings, how much of each ingredient would you need?

6. To make 12 servings, how much of each ingredient would you need?

Cereal Preparation Directions

Ingredient	Servings	
	1	2
Water	$1\frac{1}{4}$ cups	$2\frac{1}{2}$ cups
Salt	$\frac{1}{8}$ tsp.	$\frac{1}{4}$ tsp.
Cereal	$\frac{1}{4}$ cup	$\frac{1}{2}$ cup

Use the given chart to answer exercises 7 and 8.

7. **Fabric Needed** To make 5 long skirts in size 8, how much 45-inch-wide material would you need?

8. To make a short skirt and a bodice in size 14, how much material would it take if the material were 60 inches wide?

Pattern Directions

Skirt Pattern		4	6	8	10	12	14
Sizes		4	6	8	10	12	14
Sizes–European		30	32	34	36	38	40
Long skirt	45"	$2\frac{3}{4}$	$2\frac{3}{4}$	$2\frac{3}{4}$	$2\frac{3}{4}$	$2\frac{3}{4}$	$2\frac{3}{4}$ yd
	60"	$2\frac{1}{4}$	$2\frac{1}{4}$	$2\frac{3}{8}$	$2\frac{1}{2}$	$2\frac{1}{2}$	$2\frac{3}{4}$ yd
Short skirt	45"	$1\frac{3}{4}$	$1\frac{3}{4}$	$1\frac{3}{4}$	$1\frac{3}{4}$	$1\frac{3}{4}$	$1\frac{3}{4}$ yd
	60"	1	1	1	1	1	$1\frac{1}{4}$ yd
Bodice A or B	45"	$\frac{1}{2}$	$\frac{1}{2}$	$\frac{1}{2}$	$\frac{5}{8}$	$\frac{5}{8}$	$\frac{3}{4}$ yd
	60"	$\frac{1}{2}$	$\frac{1}{2}$	$\frac{1}{2}$	$\frac{1}{2}$	$\frac{1}{2}$	$\frac{1}{2}$ yd

9. **Gas Mileage** John had $15\frac{2}{3}$ gallons of gasoline in his car before traveling to Los Angeles. When he arrived, he only had $9\frac{1}{2}$ gallons left.

   (a) How much gasoline did he use to travel to Los Angeles?

   (b) If his car gets 24 miles per gallon, how many miles did John travel to Los Angeles?

10. **Pep Club Banner** The Pep Club has $18\frac{2}{3}$ feet of paper on a roll used to make banners. The club must make 3 banners, each $3\frac{3}{4}$ feet in length. How much paper will be left on the roll after the club makes the 3 banners?

11. *Money and Savings* The Tran family is receiving a $2400 tax refund. They decide to spend $\frac{1}{4}$ of the money on a vacation, and use $\frac{1}{8}$ of the money to pay off bills. They will put the remainder of the tax refund money in a savings account. How much money will they put in the savings account?

▲ **12.** *Bathroom Tiles* Each tile that Amy wants covers $3\frac{1}{2}$ square inches of space. She wants to cover 245 square inches in her bathroom with the tile.
 (a) How many tiles will she need to purchase?

 (b) If each box of tiles contains 12 tiles, how many boxes must Amy purchase?

▲ **13.** *Flower Bed Dimensions* Monica planted a rectangular flower bed in the center of her 40-foot by 25-foot front lawn. There is a $4\frac{3}{4}$-foot-wide grass area around the entire flower bed. The grass and the flower bed take up the entire 40-foot by 25-foot area.
 (a) What are the dimensions of the flower bed?
 (b) How much will it cost to put a fence around the flower bed if the fencing costs $2\frac{1}{4}$ per linear foot?

▲ **14.** *Pool Dimensions* Howard put a rectangular pool in his yard, which is $45\frac{1}{2}$ feet by 20 feet. There is a grass area 5 feet wide around the entire perimeter of the pool. The pool and grass take up the entire $45\frac{1}{2}$-foot by 20-foot yard.
 (a) What are the dimensions of the pool?
 (b) How much will it cost to put a row of tile around the edge of his pool if the tile costs $3\frac{1}{2}$ per linear foot?

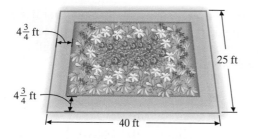

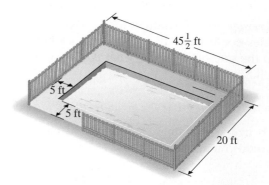

15. *Bookcase Material* Brenda wishes to build 3 bookcases, each with 4 shelves. Each shelf will be $3\frac{3}{4}$ feet long. The wood for the shelves is sold in 8-foot boards. How many boards does Brenda need to buy for the shelves?

16. *Bookcase Material* Julie wishes to build 2 bookcases, each with 5 shelves. Each shelf will be $2\frac{3}{4}$ feet long. The wood for the shelves is sold in 6-foot boards. How many boards does Julie need to buy for the shelves?

▲ **17.** *Pool Tile Cost* Mamadou put a rectangular pool in his yard, which is 50 feet by 30 feet. There is a slate tile area $5\frac{1}{2}$ feet wide around the entire perimeter of the pool. The pool and the tile area take up the entire 50-foot by 30-foot yard.
 (a) What are the dimensions of the pool?

 (b) How much will it cost to put a row of slate tile around the edge of the pool if the tile costs $2\frac{1}{2}$ per linear foot?

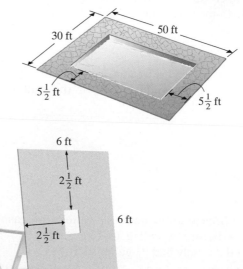

18. *Cost for Metal Strip* Lester wants to cut a square hole in the center of a square piece of wood that has a 6-foot side. He wants each edge of the square hole to be $2\frac{1}{2}$ feet from the edge of the wood.
 (a) What are the dimensions of the square hole?

 (b) How much will it cost for Lester to put a strip of metal around the hole, if the metal costs $1\frac{1}{2}$ per linear foot?

▲ **19.** *Yard Fence* Shannon is planning to build a picket fence around 3 sides of her front yard. She determines that she must make 56 wooden posts that are each $2\frac{3}{4}$ feet in length. The wood to make the posts comes in 10-foot lengths. How many 10-foot pieces of wood must Shannon purchase so that she can make 56 wooden posts?

20. *Wood Needed* Jesse is a carpenter for a new housing project. He must cut the wood for all the floor molding. Jesse determined that one set of molding will consist of 64 pieces of wood that are each $1\frac{3}{4}$ feet in length. The type of wood he must use is sold in 12-foot lengths. How many 12-foot pieces of wood must Jesse purchase so he can cut 64 pieces of molding?

One Step Further

21. *Distance Traveled* A space probe took $4\frac{1}{2}$ days to travel 240,000 miles.
   (a) How many miles per day did the spacecraft travel?
   (b) How many miles per hour did the spacecraft travel?

22. *Titanic Cruise Ship* The night of the *Titanic* cruise ship disaster, the captain decided to run his ship at $22\frac{1}{2}$ knots (nautical miles per hour). The *Titanic* traveled at that speed for $4\frac{3}{4}$ hours before it met its tragic demise. How far did the *Titanic* travel at this excessive speed before the disaster?

Cumulative Review *Solve and check your solution.*

23. **[3.2.1]** $3x = 12$ **24.** **[3.2.1]** $5x = 45$ **25.** **[3.1.2]** $x - 5 = 12$ **26.** **[3.1.2]** $x + 3 = -1$

Quick Quiz 5.6

1. Alex wishes to build 3 bookcases, each with 5 shelves. Each shelf is $3\frac{1}{3}$ feet long. The wood to make the shelves is sold in 12-foot lengths. How many 12-foot pieces of wood must Alex purchase so he can make 3 bookcases?

2. Ella weighed 134 pounds on August 1. She weighed herself weekly and found that the first week her weight went up $2\frac{1}{4}$ pounds, and the second week her weight went down $2\frac{3}{8}$ pounds. The third week it rose $1\frac{1}{2}$ pounds.

   (a) What was Ella's weight at the end of week 3?

   (b) How much more did she weigh in week 3 than on August 1?

3. Martha put a rectangular pavilion in the center of her $32\frac{1}{2}$-foot by 28-foot backyard. There is a grass area $5\frac{1}{2}$ feet wide around the entire perimeter of the pavilion. The pavilion and grass take up the entire $32\frac{1}{2}$-foot by 28-foot yard.

   (a) What are the dimensions of the pavilion?

   (b) How much will it cost to put a row of tile around the outside of the pavilion if the tile costs $2 per linear foot?

4. **Concept Check** Choose the correct operation you must use to answer each of the following questions: Add, Subtract, Multiply, or Divide. You do not need to calculate the answer.

   (a) Jason ran $2\frac{1}{3}$ miles, and Lester ran $2\frac{7}{8}$ miles. How much farther did Lester run than Jason?

   (b) Beatrice earns $780 per week and has $\frac{1}{13}$ of her paycheck placed in a savings account. How much money does she put in her savings each week?

   (c) Samuel has 14 pounds of candy and must place it in $\frac{2}{3}$-pound bags. How many bags can he fill?

5.7 Solving Equations of the Form $\frac{x}{a} = c$

① Solving Equations Using the Multiplication Principle of Equality

As we saw in Chapter 3, to solve the equation $5x = 30$, we can divide both sides of the equation by the same nonzero number.

$$5x = 30$$

$$\frac{5x}{5} = \frac{30}{5}$$

$$\frac{\overset{1}{\cancel{5}}x}{\underset{1}{\cancel{5}}} = \frac{30}{5}$$

$$x = 6$$

We *divided* by 5 to *undo* the *multiplication* by 5 so that we could get x alone on one side of the equation, that is, in the form $x = $ some number. Now, to solve $\frac{x}{2} = 8$, can we *multiply* by 2 on both sides of the equation to *undo* the *division* by 2 and get x alone?

Let's return to our example of a balanced seesaw to answer this question. If we doubled the number of weights on each side (we are multiplying each side by 2), the seesaw should still balance. In mathematics, a similar principle is observed. Thus we can multiply both sides of the equation by a number without changing the solution.

We can now state the multiplication principle.

MULTIPLICATION PRINCIPLE OF EQUALITY

If both sides of an equation are multiplied by the same nonzero number, the results on both sides are equal in value.

For any numbers a, b, and c, with c not equal to 0:

If $a = b$, then $ca = cb$.

To solve an equation like $\frac{x}{5} = 30$, we can simply think: "What can we do to the left side of the equation so that $\frac{x}{5}$ becomes simply x?" Since x is divided by 5, we can *undo* the *division* and obtain x alone by *multiplying* by 5. However, whatever we do to one side of the equation, we must do to the other side of the equation.

$$\frac{x}{5} = 30 \rightarrow \frac{5 \cdot x}{5} = 30 \cdot 5 \quad \text{We multiply both sides of the equation by 5.}$$

$$\frac{\overset{1}{\cancel{5}} \cdot x}{\underset{1}{\cancel{5}}} = 150$$

$$x = 150$$

EXAMPLE 1   Solve. $\dfrac{x}{-4} = 28$

Solution   Since we are dividing the variable x by -4, we can *undo* the division and get x alone by multiplying by -4.

$$\frac{x}{-4} = 28 \qquad \text{The variable } x \text{ is } divided \text{ by } -4.$$

$$\frac{-4 \cdot x}{-4} = 28 \cdot (-4) \quad \text{We undo the division by } multiplying \text{ both sides by } -4.$$

$$x = -112 \qquad \text{Simplify: } \frac{-4x}{-4} = x, \text{ and } 28 \cdot (-4) = -112.$$

Be sure that you *check* your solution.

Student Practice 1   Solve.

$$\frac{a}{-2} = 17$$

NOTE TO STUDENT: *Fully worked-out solutions to all of the Student Practice problems can be found at the back of the text starting at page SP-1.*

It is important that we remember to perform any necessary simplification of an equation before we find the solution.

EXAMPLE 2   Solve. $\dfrac{x}{2^3} = \dfrac{1}{2} + \dfrac{1}{4}$

Solution   We simplify each side of the equation first and then we find the solution.

$$\frac{x}{2^3} = \frac{1}{2} + \frac{1}{4}$$

$$\frac{x}{8} = \frac{1}{2} + \frac{1}{4} \quad \text{Simplify: } 2^3 = 8$$

$$\frac{x}{8} = \frac{3}{4} \qquad \text{Add: } \frac{1}{2} + \frac{1}{4} = \frac{2}{4} + \frac{1}{4} - \frac{3}{4}$$

$$\frac{8 \cdot x}{8} = \frac{3}{4} \cdot 8 \qquad \text{We } undo \text{ the division by multiplying both sides by 8.}$$

$$x = 6 \qquad \text{Multiply to find the solution: } 8 \cdot \frac{3}{4} = 6$$

We leave the check for the student.

Student Practice 2   Solve.

$$\frac{x}{3^2} = \frac{1}{3} + \frac{1}{9}$$

Sometimes the coefficient of the variable is a fraction such as in $\frac{2}{3}x = 5$. Think: "what can we do to the left side of the equation so x will stand alone?" Recall that when you multiply a fraction by its reciprocal, the product is 1. For example, $\frac{3}{2} \cdot \frac{2}{3} = 1$. We will use this idea to solve the equation $\frac{2}{3}x = 5$.

$$\frac{2}{3}x = 5$$

$$\frac{3}{2} \cdot \frac{2}{3}x = \frac{5}{1} \cdot \frac{3}{2}$$

$$1 \cdot x = \frac{15}{2}$$

$$x = \frac{15}{2}$$

EXAMPLE 3 Solve for the variable and check your solution. $-\dfrac{3}{4}x = 12$

Solution

$$-\frac{3}{4}x = 12$$

$$\left(-\frac{4}{3}\right)\left(-\frac{3}{4}\right)x = 12\left(-\frac{4}{3}\right)$$

Multiply both sides of the equation by $-\dfrac{4}{3}$ because $\left(-\dfrac{4}{3}\right)\left(-\dfrac{3}{4}\right) = 1$.

$$1\,x = -\frac{4 \cdot \cancel{3} \cdot 4}{\cancel{3}}$$

$$x = -16$$

Check. $\quad -\dfrac{3}{4}x = 12$

$$\left(-\frac{3}{4}\right)(-16) \overset{?}{=} 12 \quad \text{Replace } x \text{ with } -16.$$

$$12 = 12 \ \checkmark$$

Student Practice 3 Solve for the variable and check your solution.

$$-\frac{3}{8}x = 9$$

STEPS TO SUCCESS Keep Trying

Do you wish you could improve your math grade? Are you frustrated and starting to become discouraged? Don't give up! Take note of the following suggestions. They will help make a difference.

- *Be patient.* Would you expect to learn how to play the piano easily, without a lot of effort? Developing the skills to do math is like learning to play a musical instrument. *It takes time and effort.*

  Of course, those who have had experience with various instruments earlier in life might learn more easily and faster than someone who has not.

  The same is true with mathematics. But for many students this is catch-up time.

- *Increase your study time.* It is not unusual to study mathematics for 8 to 12 hours a week. Perfecting math skills requires the same intensity as preparing to play a sport. Baseball players practice many

hours each day to perfect their swing or curveball. *Increasing your study time will help you improve your understanding of math.* Yes, it can be slow moving at first, but eventually, as your skills develop, you will find that math will become easier and you will understand concepts more quickly.

Making it personal:

1. *Seek help.* Make an appointment with your instructor for help or use any tutorial services that are available.

   Date of appointment: _____.

   Day and time I will use tutorial services: _____.

2. *Be positive.* Don't let past frustrations stand in your way. Start fresh with a positive attitude.

   Find a study partner.

   Name of partner: _____.

   Day and time we study together: _____.

Verbal and Writing Skills, Exercises 1–4 *Fill in the blanks.*

1. When you multiply a nonzero fraction by its reciprocal, the product is _____ .

2. A nonzero number divided by itself is equal to _____ .

3. To solve $\dfrac{x}{6} = 2$, we _____ by 6 on both sides of the equation.

4. To solve $\frac{6}{7}x = 42$, we multiply by the _____ of $\frac{6}{7}$ on both sides of the equation.

Fill in each box to complete each solution.

5. $\dfrac{x}{12} = -3$

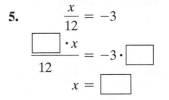

$\dfrac{\boxed{} \cdot x}{12} = -3 \cdot \boxed{}$

$x = \boxed{}$

6. $\dfrac{x}{7} = -8$

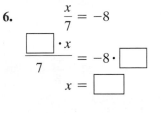

$\dfrac{\boxed{} \cdot x}{7} = -8 \cdot \boxed{}$

$x = \boxed{}$

7. $\dfrac{x}{-5} = 4$

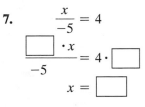

$\dfrac{\boxed{} \cdot x}{-5} = 4 \cdot \boxed{}$

$x = \boxed{}$

8. $\dfrac{x}{-3} = 6$

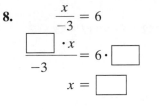

$\dfrac{\boxed{} \cdot x}{-3} = 6 \cdot \boxed{}$

$x = \boxed{}$

9. $\dfrac{2}{5}x = -8$

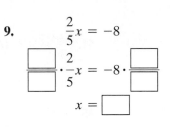

$\dfrac{\boxed{}}{\boxed{}} \cdot \dfrac{2}{5}x = -8 \cdot \dfrac{\boxed{}}{\boxed{}}$

$x = \boxed{}$

10. $\dfrac{3}{4}x = -6$

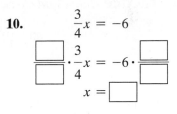

$\dfrac{\boxed{}}{\boxed{}} \cdot \dfrac{3}{4}x = -6 \cdot \dfrac{\boxed{}}{\boxed{}}$

$x = \boxed{}$

11. $\dfrac{-5}{7}x = 10$

$\dfrac{\boxed{}}{\boxed{}} \cdot \dfrac{(-5)}{7}x = 10 \cdot \dfrac{\boxed{}}{\boxed{}}$

$x = \boxed{}$

12. $\dfrac{-4}{9}x = 16$

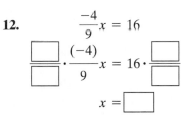

$\dfrac{\boxed{}}{\boxed{}} \cdot \dfrac{(-4)}{9}x = 16 \cdot \dfrac{\boxed{}}{\boxed{}}$

$x = \boxed{}$

Solve and check your solutions.

13. $\dfrac{y}{8} = 12$

14. $\dfrac{x}{9} = 16$

15. $\dfrac{x}{7} = 31$

16. $\dfrac{m}{6} = 10$

17. $\dfrac{m}{13} = -30$

18. $\dfrac{a}{6} = -13$

19. $\dfrac{x}{15} = -5$

20. $\dfrac{m}{17} = -4$

21. $-15 = \dfrac{a}{4}$

22. $-35 = \dfrac{y}{2}$

23. $-4 = \dfrac{a}{-20}$

24. $-5 = \dfrac{m}{-22}$

Simplify, then solve and check your solution.

25. $\dfrac{x}{3^2} = 2 + 6 \div 3$

26. $\dfrac{x}{2^2} = 6 + 8 \div 4$

27. $\dfrac{x}{8} = \dfrac{1}{4} + \dfrac{5}{8}$

28. $\dfrac{x}{10} = \dfrac{2}{5} + \dfrac{3}{10}$

29. $\dfrac{y}{2^3} = \dfrac{1}{8} + \dfrac{1}{2}$

30. $\dfrac{y}{3^3} = \dfrac{2}{27} + \dfrac{1}{3}$

Solve and check your solutions.

31. $\dfrac{3}{5}y = 12$

32. $\dfrac{6}{7}y = 30$

33. $\dfrac{4}{9}x = 12$

34. $\dfrac{2}{3}x = 14$

35. $\dfrac{1}{2}x = -15$

36. $\dfrac{1}{5}x = -12$

37. $\dfrac{-5}{8}x = 30$

38. $\dfrac{-3}{7}x = 21$

Mixed Practice *Solve and check your solution.*

39. $3 = \dfrac{a}{11}$

40. $10 = \dfrac{x}{7}$

41. $\dfrac{x}{3^2} = 4 + 6 \div 2$

42. $\dfrac{y}{2^2} = 12 - 3 \div 3$

43. $\dfrac{-3}{4}y = -12$

44. $\dfrac{-2}{9}y = -12$

45. $-1 = \dfrac{m}{30}$

46. $-9 = \dfrac{x}{15}$

To Think About *Solve and check your solution.*

47. **(a)** $4x = 52$

  (b) $\dfrac{x}{4} = 52$

48. **(a)** $3y = 39$

  (b) $\dfrac{y}{3} = 39$

49. **(a)** $4x = -52$

  (b) $\dfrac{x}{4} = -52$

50. **(a)** $3y = -39$

  (b) $\dfrac{y}{3} = -39$

51. **(a)** $x - 7 = 21$
  (b) $x + 12 = 33$
  (c) $5x = 3$
  (d) $\dfrac{x}{5} = 11$

52. **(a)** $x - 4 = 19$
  (b) $x + 6 = 23$
  (c) $5x = 4$
  (d) $\dfrac{x}{3} = 12$

53. **(a)** $x - 7 = -12$
  (b) $3x = -2$
  (c) $\dfrac{x}{6} = -9$
  (d) $x + 11 = -34$

54. **(a)** $x + 15 = -31$
  (b) $13x = -7$
  (c) $\dfrac{x}{2} = -8$
  (d) $x - 3 = -10$

One Step Further *Name the fraction $\frac{a}{b}$ in lowest terms that makes each statement true.*

55. $\frac{a}{b}x = 6$

$\frac{3}{2} \cdot \frac{a}{b}x = 6 \cdot \frac{3}{2}$

$x = 9$

$\frac{a}{b} = ?$

56. $\frac{a}{b}x = 10$

$\frac{3}{5} \cdot \frac{a}{b}x = 10 \cdot \frac{3}{5}$

$x = 6$

$\frac{a}{b} = ?$

57. $\frac{a}{b}x = 9$

$\frac{7}{3} \cdot \frac{a}{b}x = 9 \cdot \frac{7}{3}$

$x = 21$

$\frac{a}{b} = ?$

58. $\frac{a}{b}x = 8$

$\frac{5}{2} \cdot \frac{a}{b}x = 8 \cdot \frac{5}{2}$

$x = 20$

$\frac{a}{b} = ?$

Cumulative Review *Combine like terms.*

59. [2.6.1] $3a + 7b - 9a - 10ab + 2b$

60. [2.6.1] $-2x + 3xy - 4y - 6x - 5xy$

Simplify.

61. [3.4.3] $3x^2(x^4 + 8x)$

62. [3.4.3] $-2x^3(x^2 - 6)$

▲ **63.** [5.6.1] ***Stained Glass Window*** Find the area of the stained glass window with length $L = 4\frac{2}{3}$ feet and width $W = 3\frac{3}{8}$ feet.

64. [5.6.1] ***Bread Recipe*** Juan uses $2\frac{1}{2}$ cups of flour for one loaf of bread. How many cups of flour will he need for 4 loaves of bread?

Quick Quiz 5.7 *Solve.*

1. $\frac{x}{5} = 2 - 3^2$

2. $\frac{3}{5}a = 18$

3. $\frac{y}{8} = -9$

4. Concept Check To solve the equation $\frac{x}{-5} = 6$, Amy multiplied both sides of the equation by 5 to obtain $x = 30$. Is this correct? Why or why not?

Chapter 5 Organizer

Topic and Procedure	Examples	✏️ You Try It
Multiplying fractions, p. 282 **1.** Simplify by factoring out common factors whenever possible. **2.** Multiply numerators. **3.** Multiply denominators.	Multiply. $-\dfrac{15}{49}\cdot\dfrac{7}{12}$ $-\dfrac{15}{49}\cdot\dfrac{7}{12}=-\dfrac{\cancel{3}\cdot5\cdot\cancel{7}}{7\cdot\cancel{7}\cdot\cancel{3}\cdot4}=-\dfrac{5}{28}$	1. Multiply. $\dfrac{18}{50}\cdot\left(\dfrac{-20}{21}\right)$
Dividing fractions, p. 284 To divide two fractions, we find the reciprocal of (invert) the second fraction and multiply.	Divide. $12x^4\div\dfrac{8x^2}{5}$ $12x^4\div\dfrac{8x^2}{5}=\dfrac{12x^4}{1}\cdot\dfrac{5}{8x^2}$ $=\dfrac{3\cdot\cancel{4}\cdot5\cdot x^4}{1\cdot2\cdot\cancel{4}\cdot x^2}=\dfrac{15x^2}{2}$	2. Divide. $15x^3\div\dfrac{10x^4}{7}$
Finding the least common denominator (LCD), pp. 293–294 To find the LCD: **1.** Write each denominator as the product of prime factors. **2.** List the requirements for the factorization of the LCD. **3.** Build an LCD that has all the factors of each denominator, using a minimum number of factors.	Find the LCD. $\dfrac{2}{9},\dfrac{4}{15},\dfrac{1}{10}$ $9=3\cdot3\quad 15=3\cdot5\quad 10=2\cdot5$ $\text{LCD}=2\cdot3\cdot3\cdot5=90$	3. Find the LCD. $\dfrac{3}{20},\dfrac{7}{30},\dfrac{1}{8}$
Adding or subtracting fractions with a common denominator, p. 300 **1.** Add or subtract the numerators. **2.** Keep the common denominator. **3.** Simplify the answer if necessary.	Add. $\dfrac{2}{11}+\dfrac{3}{11}$ Subtract. $\dfrac{5}{x}-\dfrac{2}{x}$ $\dfrac{2}{11}+\dfrac{3}{11}=\dfrac{5}{11}$ $\dfrac{5}{x}-\dfrac{2}{x}=\dfrac{3}{x}$	4. **(a)** Subtract. $\dfrac{4}{13}-\dfrac{1}{13}$ **(b)** Add. $\dfrac{3}{y}+\dfrac{6}{y}$
Adding or subtracting fractions with different denominators, p. 302 **1.** Find the LCD of the fractions. **2.** Write equivalent fractions that have the LCD as the denominator. **3.** Follow the steps for adding and subtracting fractions with a common denominator.	Add. $\dfrac{2}{5}+\dfrac{4}{7}+\dfrac{3}{10}$ $\text{LCD}=70$ $\dfrac{2\cdot14}{5\cdot14}+\dfrac{4\cdot10}{7\cdot10}+\dfrac{3\cdot7}{10\cdot7}$ $=\dfrac{28}{70}+\dfrac{40}{70}+\dfrac{21}{70}=\dfrac{89}{70}\text{ or }1\dfrac{19}{70}$	5. Add. $\dfrac{5}{6}+\dfrac{1}{9}+\dfrac{7}{15}$
Adding mixed numbers, p. 310 **1.** Change fractional parts to equivalent fractions with the LCD as a denominator, if needed. **2.** Add whole numbers and fractions separately. **3.** If improper fractions occur, change to mixed numbers and simplify.	Add. $7\dfrac{1}{2}+2\dfrac{5}{6}$ $7\dfrac{1\cdot3}{2\cdot3}=\ 7\dfrac{3}{6}$ $+2\dfrac{5}{6}\ =+2\dfrac{5}{6}$ $9\dfrac{8}{6}=9\dfrac{4}{3}=9+1\dfrac{1}{3}=10\dfrac{1}{3}$	6. Add. $8\dfrac{1}{3}+3\dfrac{8}{9}$
Subtracting mixed numbers, p. 312 **1.** Change fractional parts to equivalent fractions with the LCD as a denominator, if needed. **2.** If necessary, borrow from the whole number to subtract fractions. **3.** Subtract whole numbers and fractions separately. **4.** Simplify the answer if necessary.	Subtract. $9\dfrac{1}{5}-4\dfrac{3}{4}$ $9\dfrac{1\cdot4}{5\cdot4}=\ 9\dfrac{4}{20}=\ 8\dfrac{24}{20}$ We must borrow. $-4\dfrac{3\cdot5}{4\cdot5}=-4\dfrac{15}{20}=-4\dfrac{15}{20}$ $9\dfrac{4}{20}=8+1\dfrac{4}{20}$ $4\dfrac{9}{20}$	7. Subtract. $9\dfrac{2}{7}-4\dfrac{3}{4}$

Topic and Procedure	Examples	You Try It
Multiplying and dividing mixed and/or whole numbers, pp. 313–314 1. Change any whole number to a fraction with a denominator of 1. 2. Change any mixed numbers to improper fractions. 3. Use the rule for multiplication or division of fractions.	Divide. $8\frac{1}{3} \div 5\frac{5}{9}$ $8\frac{1}{3} \div 5\frac{5}{9} = \frac{25}{3} \div \frac{50}{9}$ $= \frac{25}{3} \cdot \frac{9}{50}$ $= \frac{\cancel{5} \cdot \cancel{5} \cdot \cancel{3} \cdot 3}{\cancel{3} \cdot 2 \cdot \cancel{5} \cdot \cancel{5}}$ $= \frac{3}{2}$ or $1\frac{1}{2}$ Multiply. $4 \cdot 2\frac{1}{3}$ $4 \cdot 2\frac{1}{3} = \frac{4}{1} \cdot \frac{7}{3}$ $= \frac{28}{3}$ or $9\frac{1}{3}$	8. (a) Divide. $4\frac{2}{5} \div 2\frac{2}{15}$ (b) Multiply. $5 \cdot 3\frac{1}{3}$
Order of operations with fractions, p. 321 To simplify fractions, perform operations in the following order: 1. Perform operations inside parentheses. 2. Simplify exponents. 3. Multiply and divide, working left to right. 4. Add and subtract, working left to right.	Simplify. $\frac{5}{6} + \frac{1}{3} \cdot \left(\frac{3}{4} - \frac{1}{2} \right)$ $\frac{5}{6} + \frac{1}{3} \cdot \left(\frac{3}{4} - \frac{1}{2} \right) = \frac{5}{6} + \frac{1}{3} \cdot \frac{1}{4}$ $= \frac{5}{6} + \frac{1}{12}$ $= \frac{10}{12} + \frac{1}{12} = \frac{11}{12}$	9. Simplify. $\left(\frac{1}{3} + \frac{2}{7} \right) \cdot \frac{3}{4} - \frac{1}{4}$
Simplifying complex fractions, pp. 321–322 To simplify a complex fraction, simplify above and then below the main fraction bar. Perform operations in the following order: 1. Perform all operations in the numerator. 2. Perform all operations in the denominator. 3. Divide the top fraction by the bottom fraction by inverting the bottom fraction and multiplying it by the top fraction.	Simplify. $\dfrac{\frac{3}{4} + \frac{1}{2}}{\frac{1}{3} - \frac{1}{6}}$ $\dfrac{\frac{3}{4} + \frac{1}{2}}{\frac{1}{3} - \frac{1}{6}} = \dfrac{\frac{3}{4} + \frac{2}{4}}{\frac{2}{6} - \frac{1}{6}} = \dfrac{\frac{5}{4}}{\frac{1}{6}}$ $= \frac{5}{4} \div \frac{1}{6} = \frac{5}{4} \cdot \frac{6}{1}$ $= \frac{15}{2}$ or $7\frac{1}{2}$	10. Simplify. $\dfrac{\frac{1}{7} + \frac{9}{14}}{\frac{3}{4} - \frac{1}{2}}$
Solving equations using the multiplication principle, p. 332 1. If the variable is *divided* by a number, *undo* the *division* by *multiplying* both sides of the equation by this number. 2. If the variable is multiplied by a fraction, multiply both sides of the equation by the reciprocal of that fraction. 3. Check by substituting your answer back into the original equation.	Solve for x. $\frac{x}{-2} = 16$ $\frac{x}{-2} = 16$ The variable is divided by -2. $(-2) \cdot \frac{x}{-2} = 16 \cdot (-2)$ We multiply both sides by -2. $x = -32$ Solve for x. $\frac{-2}{3}x = 4$ $\left(-\frac{3}{2} \right)\left(-\frac{2}{3} \right)x = 4\left(-\frac{3}{2} \right)$ $x = -6$ We leave the check to the student.	11. Solve for x. $\frac{x}{-10} = \frac{1}{3} + \frac{1}{6}$

Chapter 5 Review Problems

Vocabulary

Fill in the blank with the definition for each word.

1. (5.2) LCM: _____

2. (5.3) LCD: _____

3. (5.5) Complex fraction: _____

In this exercise set assume that all variables in any denominator are nonzero.

Section 5.1

Find the reciprocal.

4. **(a)** $\dfrac{4}{9}$

   (b) -6

5. **(a)** 7

   (b) $\dfrac{1}{11}$

6. Find $\dfrac{3}{4}$ of $\dfrac{-8}{9}$.

Perform the operation indicated.

7. $\dfrac{5}{21} \cdot \dfrac{3}{15}$

8. $\dfrac{-4}{35} \cdot \dfrac{14}{18}$

9. $\dfrac{9x}{15} \cdot \dfrac{21}{18x^3}$

10. $\dfrac{8x^3}{25} \cdot \dfrac{-45}{18x}$

11. $\dfrac{-5x^4}{6} \cdot 12x^3$

12. $\dfrac{9}{14} \div \dfrac{45}{12}$

13. $\dfrac{7}{15} \div \dfrac{-35}{20}$

14. $\dfrac{8}{42} \div \dfrac{-22}{7}$

15. $\dfrac{11x^5}{25} \div \dfrac{3}{5x^2}$

16. $\dfrac{16x^2}{9} \div \dfrac{24x^4}{6}$

▲ *Find the area of each triangle with the given base and height.*

17. $b = 18$ m and $h = 7$ m

18. $b = 11$ in. and $h = 20$ in.

19. **Payroll Deductions** Michelle Smith has $\frac{2}{7}$ of her income withheld for taxes, dues, and medical coverage. What amount is withheld each month for taxes, dues, and medical coverage if her rate of pay is $3500 per month?

20. **Food Container** Les wants to store $\frac{1}{2}$ pound of flour in 3 equal-size containers. How much flour should Les place in each container?

Section 5.2

Find the LCM of the given expressions.

21. 7 and 21

22. 10 and 20

23. 18 and 30

24. 3, 9, and 12

25. $4x$, 8, and $16x$

26. $7x^2$, $14x^4$, and $20x$

27. $18x$ and $45x^2$

28. $20x$ and $25x^2$

Section 5.3

Perform the operation indicated. Be sure to simplify your answer.

29. $\dfrac{6}{17} - \dfrac{3}{17}$

30. $\dfrac{-23}{27} + \dfrac{-11}{27}$

31. $\dfrac{7}{x} - \dfrac{5}{x}$

32. $\dfrac{x}{7} - \dfrac{4}{7}$

33. $\dfrac{1}{6} + \dfrac{4}{9}$

34. $\dfrac{15}{32} - \dfrac{7}{28}$

35. $\dfrac{-3}{14} + \dfrac{8}{21}$

36. $\dfrac{5}{2x} + \dfrac{8}{3x}$

37. $\dfrac{4x}{15} + \dfrac{2x}{45}$

38. $\dfrac{3x}{14} - \dfrac{5x}{42}$

Section 5.4

Perform the operation indicated. Be sure to simplify your answer.

39. $10\dfrac{1}{3} + 2\dfrac{4}{5}$

40. $12\dfrac{7}{9} + 6\dfrac{2}{3}$

41. $11\dfrac{1}{5} - 6\dfrac{11}{25}$

42. $25 - 16\dfrac{5}{14}$

43. $-3 \cdot \left(4\dfrac{1}{3}\right)$

44. $2\dfrac{3}{4} \div 5$

45. $4\dfrac{2}{5} \div 8\dfrac{1}{3}$

46. $-12 \div \dfrac{2}{3}$

▲ **47.** *Yard Fencing* Shawnee purchased 75 feet of fencing for a rectangular dog run. How much fencing will she have left after she encloses the $20\frac{1}{2}$-foot by $10\frac{1}{4}$-foot dog run?

▲ **48.** *Piece of Wood* Mike has a piece of wood $7\frac{1}{2}$ feet long. If he needs $\frac{1}{4}$ of the length of the wood, what length of wood does he need?

Section 5.5

Simplify.

49. $\dfrac{3}{4} + \dfrac{1}{2} \cdot \dfrac{2}{5}$

50. $\left(\dfrac{3}{4}\right)^2 + \dfrac{1}{8} \div \dfrac{1}{2}$

51. $\dfrac{(-2)^2 + 12}{\dfrac{4}{7}}$

52. $\dfrac{\dfrac{4}{3}}{\dfrac{1}{9}}$

53. $\dfrac{\dfrac{x}{12}}{\dfrac{x^2}{20}}$

54. $\dfrac{\dfrac{1}{2} + \dfrac{1}{4}}{\dfrac{2}{3} - \dfrac{1}{9}}$

55. $\dfrac{\dfrac{1}{3} + \dfrac{2}{5}}{\dfrac{1}{5} + \dfrac{1}{10}}$

56. A developer needs $3\frac{1}{4}$ acres of land for every 4 houses he builds. How many acres does the developer need to build 32 homes?

Section 5.6

You may want to use a Mathematics Blueprint for Problem Solving for the following applied problems.

▲ **57.** *Pool Dimensions* Leslie put a rectangular pool in her yard, which is $38\frac{1}{4}$ feet by 25 feet. There is a $4\frac{1}{2}$-foot concrete path around the entire pool. The pool and the concrete area take up the entire $38\frac{1}{4}$-foot by 25-foot yard.

   (a) What are the dimensions of the pool?

   (b) How much will it cost to put a row of tile around the edge of the pool if the tile costs $3\frac{1}{4}$ per linear foot?

58. *Putting Green Grass* The Pleasantville Country Club maintains the putting greens with a grass height of $\frac{7}{8}$ inch. The grass on the fairways is maintained at $2\frac{1}{2}$ inches. How much must the blade be lowered by a person mowing the fairways if that person will be using the same mowing machine on the putting green?

Section 5.7

Solve and check your solution.

59. $\dfrac{x}{3} = 9$

60. $\dfrac{y}{2} = -6$

61. $\dfrac{x}{-6} = -4$

62. $\dfrac{2}{3}y = 22$

63. $\dfrac{-3}{4}x = 18$

64. $\dfrac{1}{3}y = 4$

65. $\dfrac{x}{2^3} = 2 \cdot 6 + 1$

66. $\dfrac{y}{3^2} = 1 + 8 \div 2$

How Am I Doing? Chapter 5 Test

After you take this test read through the Math Coach on pages 344–345. Math Coach videos are available via MyMathLab and YouTube. Step-by-step test solutions on the Chapter Test Prep Videos are also available via MyMathLab and YouTube. (Search "BlairTobeyPrealgebra" and click on "Channels.")

Perform the operation indicated.

1. Find $\dfrac{2}{3}$ of $\dfrac{9}{10}$.

2. $\dfrac{-1}{4} \cdot \dfrac{2}{5}$

3. $\dfrac{7x^2}{8} \cdot \dfrac{16x}{14}$

4. $\dfrac{-1}{2} \div \dfrac{1}{4}$

5. $\dfrac{2x^8}{5} \div \dfrac{22x^4}{15}$

ᴹᴄ 6. $\dfrac{8x}{15} \cdot \dfrac{25}{12x^3}$

7. List the first six multiples of 6 and 9, and then find the common multiples.

Find the least common multiple (LCM) of the given expressions.

8. 14, 21

9. $5a, 10a^4, 20a^2$

10. 5, 7, 10

11. Find the least common denominator (LCD) for $\dfrac{17}{30}$ and $\dfrac{1}{4}$.

Perform the operation indicated.

12. $\dfrac{12}{11} + \dfrac{3}{11}$

13. $\dfrac{1}{5a} + \dfrac{3}{a}$

14. $4\dfrac{5}{6} + 3\dfrac{1}{3}$

ᴹᴄ 15. $\dfrac{2}{21} + \dfrac{5}{9}$

16. $\dfrac{7}{12} - \dfrac{2}{15}$

17. $10\dfrac{1}{8} - 2\dfrac{2}{3}$

18. $\dfrac{7x}{15} - \dfrac{x}{20}$

19. $(-4) \cdot 5\dfrac{1}{3}$

1. _____

2. _____

3. _____

4. _____

5. _____

6. _____

7. _____

8. _____

9. _____

10. _____

11. _____

12. _____

13. _____

14. _____

15. _____

16. _____

17. _____

18. _____

19. _____

20. $1\dfrac{2}{3} \div 3$

21. $2\dfrac{1}{2} \div \left(-\dfrac{3}{5}\right)$

Simplify.

Ⓜ 22. $\left(\dfrac{2}{3}\right)^2 + \dfrac{1}{2} \cdot \dfrac{1}{4}$

23. $\dfrac{\dfrac{x}{2}}{\dfrac{x}{4}}$

24. $\dfrac{\dfrac{1}{3} + \dfrac{1}{2}}{\dfrac{5}{9} - \dfrac{1}{3}}$

Solve.

25. $\dfrac{x}{-3} = 6 + 2^2$

26. $\dfrac{4}{7}y = 28$

27. $\dfrac{x}{4} = -20$

Ⓜ 28. Anna wishes to build 2 bookcases, each with 5 shelves. Each shelf is $3\frac{1}{2}$ feet long. The wood for the shelves is sold in 10-foot boards. How many boards does Anna need to buy for the shelves?

29. Rudy's normal body temperature is approximately $98\frac{1}{2}°$F. Due to a cold, his temperature went up 3°F on Saturday. Then on Sunday it went down $1\frac{1}{2}°$F.

(a) What was Rudy's temperature on Sunday?

(b) How many degrees higher was his temperature on Sunday than his normal body temperature?

30. Jo Anne put a rectangular pool in her yard, which is $35\frac{1}{2}$ feet by 22 feet. There is a grass area 5 feet wide around the entire perimeter of the pool. The pool and grass take up the entire $35\frac{1}{2}$-foot by 22-foot yard.

(a) What are the dimensions of the pool?

(b) How much will it cost to put a row of tile around the edge of the pool if the tile costs $1\frac{1}{2}$ per linear foot?

20.	☐
21.	☐
22.	☐
23.	☐
24.	☐
25.	☐
26.	☐
27.	☐
28.	☐
29. (a)	☐
(b)	☐
30. (a)	☐
(b)	☐
Total Correct:	☐

MATH COACH

Mastering the skills you need to do well on the test.

Students often make the same types of errors when they do the Chapter 5 Test. Here are some helpful hints to keep you from making these common errors on test problems.

Multiplying Fractions—Problem 6 $\dfrac{8x}{15} \cdot \dfrac{25}{12x^3}$

> **Helpful Hint** First, factor out the common factors. Use the correct rule for exponents to simplify the factors x *and* x^3. Then multiply the remaining numerators and multiply the remaining denominators.

Look at your work for Problem 6. Examine your steps.

Did you simplify the fractions before multiplying?
Yes ____ No ____

If you answered No, stop and complete these calculations.

Did you divide 8 and 12 by 2, and 25 and 15 by 5?
Yes ____ No ____

If you answered Yes, make sure you did not stop with 20 in the numerator and 18 in the denominator. That fraction can be simplified further.

Did you use the quotient rule to simplify?
Yes ____ No ____

If you answered No, go back and perform this step.

If you answered Problem 6 incorrectly, go back and rework the problem using these suggestions.

Adding Fractions with Different Denominators—Problem 15 $\dfrac{2}{21} + \dfrac{5}{9}$

> **Helpful Hint** Make sure you understand how to find the LCD. Then rewrite all fractions with the LCD as the denominator. Remember, you do not add the denominators of fractions.

Did you factor 21 into $3 \cdot 7$, and 9 into $3 \cdot 3$?
Yes ____ No ____

If you answered No, stop and complete this step.

Did you get $3 \cdot 3 \cdot 7$ or 63 as the LCD?
Yes ____ No ____

If you answered No, review how to find the LCD of 21 and 9.

Did you multiply $\dfrac{2}{21}$ by $\dfrac{3}{3}$ and $\dfrac{5}{9}$ by $\dfrac{7}{7}$ to rewrite fractions?
Yes ____ No ____

If you answered No, review how to write equivalent fractions using 63 as the LCD.

Now go back and rework the problem using these suggestions.

Need help? Watch the MATH COACH videos in MyMathLab® or on You Tube™.

344

Following the Order of Operations with Fractions—Problem 22 $\left(\frac{2}{3}\right)^2 + \frac{1}{2} \cdot \frac{1}{4}$

> **Helpful Hint** Write out the rule for the order of operations and refer to it as you complete each step. Then be sure to write down each step. Skipping steps often leads to errors.

Did you square both 2 and 3 in $\left(\frac{2}{3}\right)^2$ as your first step?

Yes ____ No ____

Did you multiply $\frac{1}{2} \cdot \frac{1}{4}$ as your next step?

Yes ____ No ____

If you answered No to either question, go back and make these corrections.

Did you rewrite $\frac{4}{9}$ and $\frac{1}{8}$ with the LCD 72 as the denominator and then add?

Yes ____ No ____

If you answered No, stop and consider why this step must occur before adding the final two fractions.

If you answered Problem 22 incorrectly, go back and rework the problem using these suggestions.

Solving Applied Problems Involving Fractions—Problem 28

Anna wishes to build 2 bookcases, each with 5 shelves. Each shelf is $3\frac{1}{2}$ feet long. The wood for the shelves is sold in 10-foot boards. How many boards does Anna need to buy for the shelves?

> **Helpful Hint** Use the Mathematics Blueprint for Problem Solving to help organize your work. Be sure to change mixed numbers to improper fractions before doing any other steps. Then change any division to multiplication by inverting the second fraction.

Did you realize that the problem requires you to perform the calculation 10 divided by $3\frac{1}{2}$?

Yes ____ No ____

If you answered No, draw a diagram to help you better understand the problem.

Did you change $3\frac{1}{2}$ to $\frac{7}{2}$ and then rewrite the division as $10 \cdot \frac{2}{7}$?

Yes ____ No ____

If you answered No to either step, stop and perform these calculations.

Did you think that your calculations were now complete?

Yes ____ No ____

If you answered Yes, go back and read the problem again. Your answer refers to the number of shelves that can be made out of one 10-foot board. You still need to find out how many 10-foot boards are needed to make 10 shelves, each with a length of $3\frac{1}{2}$ feet. Make sure that you answer the question asked in the problem.

Now go back and rework the problem using these suggestions.

Need more help? Look for section examples marked with $^{M}\mathbb{C}$ to review.

345

As selected trees grow, the number of new branches forms a pattern of numbers known as the Fibonacci sequence. With this knowledge we can then determine how many new branches will form at each stage of their growth. Learning mathematics helps us develop the ability to recognize patterns and enables us to use this knowledge in many fields. Scientists, researchers, and doctors use the ability to understand patterns to learn about nature, health care, and medicine. Detectives and psychologists use this knowledge to make inferences within their fields. The mathematics covered in this chapter will help you develop the skills to recognize patterns.

Polynomials

6.1 Adding and Subtracting Polynomials

① Identifying the Terms of a Polynomial

Student Learning Objectives

After studying this section, you will be able to:

① Identify the terms of a polynomial.

② Add polynomials.

③ Find the opposite of a polynomial.

④ Subtract polynomials.

Recall from Chapter 3 that polynomials are expressions that contain terms with variable parts that have only whole number exponents. For example, $3x^2 + 4x + 5$ and $2y + 6$ are polynomials.

Operations involving polynomials often require that we identify the terms of a polynomial. In a variable expression such as a polynomial, the sign in front of the term is considered part of the term. Thus the terms of the polynomial $12ab^2 - 3b^2 + 2a - 5b$ are $+12ab^2$, $-3b^2$, $+2a$, and $-5b$.

$$\text{Polynomial:} \quad 12ab^2 - 3b^2 + 2a - 5b$$
$$\text{Terms:} \qquad\quad +12ab^2, -3b^2, +2a, -5b$$

EXAMPLE 1 Identify the terms of the polynomial. $xy^2 - 7y^2 - 2x + 5y$

Solution We include the sign in front of the term as part of the term.

$$\text{Polynomial:} \quad xy^2 - 7y^2 - 2x + 5y$$
$$\text{Terms:} \qquad\quad +xy^2, -7y^2, -2x, +5y$$

Student Practice 1 Identify the terms of the polynomial.

$$y^2 - 4x^2 + 5x - 9y$$

NOTE TO STUDENT: Fully worked-out solutions to all of the Student Practice problems can be found at the back of the text starting at page SP-1.

② Adding Polynomials

When we write operations with polynomials, we often place grouping symbols such as parentheses, brackets, and braces around each polynomial to distinguish between the polynomials. We write the sum of the polynomials $8x^2 + 2x - 4$ and $6x^2 - 2$ as follows: $(8x^2 + 2x - 4) + (6x^2 - 2)$. To find the sum of polynomials, we use the associative and commutative properties of addition to rearrange terms so that like terms are grouped together, and then we add like terms.

> To add polynomials, combine like terms.

EXAMPLE 2 Perform the operations indicated.
$$(-4x^2 + 5x - 2) + (3x^2 + 4)$$

Solution We must combine like terms.
$(-4x^2 + 5x - 2) + (3x^2 + 4)$

$= -4x^2 + 3x^2 + 5x - 2 + 4$ We rearrange terms so that like terms are grouped together.

$= -1x^2 + 5x + 2$ We add like terms:
$-4x^2 + 3x^2 = -1x^2$; $-2 + 4 = +2$.

$= -x^2 + 5x + 2$ $-1x^2 = -x^2$

Student Practice 2 Perform the operations indicated.
$$(6x^2 - 7x + 3) + (-3x^2 + 6x)$$

③ Finding the Opposite of a Polynomial

To simplify expressions such as $(-1)(2x + 6)$, we must remove parentheses. Of course we can't just take them away: we must use the distributive property to multiply each term inside the parentheses by -1. Therefore, for the remainder of this book we will say "remove parentheses" as a shorthand direction for "use the distributive property to multiply."

Multiplying using the distributive property removes the parentheses from the expression.

$$(-1)(2x + 6) = (-1)(2x) + (-1)(6) = -2x - 6$$

Notice that multiplying an expression by -1 has the same effect as taking the opposite of each term inside the parentheses. Therefore, if a grouping symbol has a negative sign in front, we can mentally multiply by -1 by changing the sign of each term inside the parentheses. Note that the expression *the opposite of* $(2x + 6)$ is written $-(2x + 6)$.

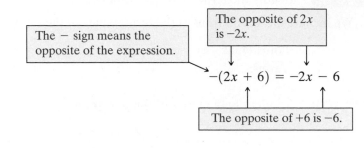

When a negative sign precedes parentheses, we find the *opposite* of the expression by changing the sign of each term inside the parentheses.

EXAMPLE 3 Simplify. $-(-2a + 5b - 7c)$

Solution

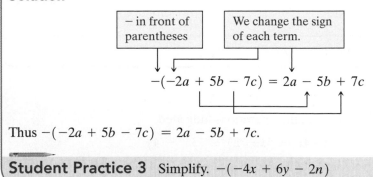

Thus $-(-2a + 5b - 7c) = 2a - 5b + 7c$.

Student Practice 3 Simplify. $-(-4x + 6y - 2n)$

④ Subtracting Polynomials

We write the difference of the polynomials $6x^2 + 2$ and $4x^2 - 1$ as follows: $(6x^2 + 2) - (4x^2 - 1)$. Now, subtracting polynomials is very much like subtracting numbers. Recall that to subtract a number, we add the opposite of the number: $2 - 5 = 2 + (-5) = -3$. Similarly, to subtract a polynomial, we *add the opposite of the polynomial.* Recall that to find the *opposite* of a *polynomial*, we change the sign of each term inside the parentheses.

$$(6x^2 + 2) - (4x^2 - 1)$$

$$= 6x^2 + 2 + (-4x^2) + 1 \quad \text{We change the sign of each term inside the parentheses}$$
$$\text{and add.}$$
$$= 2x^2 + 3 \quad\quad\quad\quad\quad\quad \text{We combine like terms: } 6x^2 + (-4x^2) = 2x^2; \;\; 2 + 1 = 3.$$

> To subtract two polynomials, change the sign of each term in the second polynomial and then add.

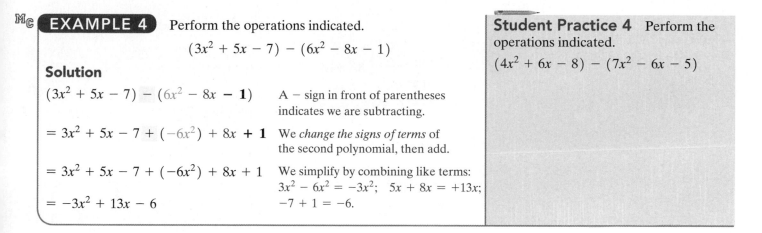

Mc **EXAMPLE 4** Perform the operations indicated.

$$(3x^2 + 5x - 7) - (6x^2 - 8x - 1)$$

Solution

$$(3x^2 + 5x - 7) - (6x^2 - 8x - 1) \quad \text{A} - \text{sign in front of parentheses}$$
$$\text{indicates we are subtracting.}$$

$$= 3x^2 + 5x - 7 + (-6x^2) + 8x + 1 \quad \text{We } \textit{change the signs of terms} \text{ of}$$
$$\text{the second polynomial, then add.}$$

$$= 3x^2 + 5x - 7 + (-6x^2) + 8x + 1 \quad \text{We simplify by combining like terms:}$$
$$3x^2 - 6x^2 = -3x^2; \;\; 5x + 8x = +13x;$$
$$= -3x^2 + 13x - 6 \quad\quad\quad\quad\quad\quad -7 + 1 = -6.$$

Student Practice 4 Perform the operations indicated.

$$(4x^2 + 6x - 8) - (7x^2 - 6x - 5)$$

Sometimes it is necessary to simplify before we add or subtract polynomials. Often simplifying involves removing parentheses using the distributive property. For example, $2(-3x + 1) = 2(-3x) + 2(1) = -6x + 2$.

EXAMPLE 5 Perform the operations indicated.

$$6x - 3(-4x^2 + 3) - (-2x^2 + x - 5)$$

Solution First we multiply (-3) times the binomial $(-4x^2 + 3)$.

$$6x - 3(-4x^2 + 3) - (-2x^2 + x - 5) \quad \text{We multiply: } -3(-4x^2 + 3);$$

$$= 6x + 12x^2 - 9 - (-2x^2 + x - 5) \quad\quad -3(-4x^2) = +12x^2; \;\; -3(+3) = -9.$$

$$= 6x + 12x^2 - 9 + 2x^2 - x + 5 \quad \text{We remove parentheses and change the}$$
$$\text{sign of each term inside the parentheses.}$$

$$= 5x + 14x^2 - 4 \quad\quad\quad \text{We combine like terms: } 6x - x = 5x;$$
$$12x^2 + 2x^2 = +14x^2; \;\; -9 + 5 = -4.$$

$$= 14x^2 + 5x - 4 \quad\quad\quad \text{We write the polynomial so that the powers of } x \text{ decrease}$$
$$\text{as we read from left to right.}$$

Student Practice 5 Perform the operations indicated.

$$4x - 2(3x^2 + 1) - (-3x^2 + x - 6)$$

Verbal and Writing Skills, Exercises 1–2

Fill in the blanks.

1. To subtract two polynomials, we add the _____ of the second polynomial to the _____ polynomial.

2. To add two polynomials, we _____ _____ terms.

Identify the terms of each polynomial.

3. $2z^2 + 4z - 2y^4 + 3$

4. $4x^5 + 2c^2 - 6c + 4$

5. $6x^6 - 3x^3 - 3y - 1$

6. $3a^7 + 4b^4 - 7a - 7$

Perform the operations indicated.

7. $(7y - 3) + (-4y + 9)$

8. $(4y - 3) + (-6y + 6)$

9. $(2a^2 - 3a + 6) + (4a - 2)$

10. $(3c^2 - 6c + 3) + (2c - 7)$

11. $(5y^2 + 2y - 5) + (-4y^2 - 8y + 2)$

12. $(-7z^2 + 9z - 3) + (8z^2 - 6z + 9)$

Simplify.

13. $-(5x + 2y)$

14. $-(8x + 5y)$

15. $-(-8x + 4)$

16. $-(-5a + 3)$

17. $-(-3x + 6z - 5y)$

18. $-(-3x + 4y - 8z)$

Perform the operations indicated.

19. $(10x + 7) - (3x + 5)$

20. $(8a + 7) - (3a + 2)$

21. $(7x - 3) - (-4x + 6)$

22. $(5y + 2) - (-7y - 8)$

23. $(-8a + 5) - (4a - 3)$

24. $(-5c + 2) - (3c - 6)$

25. $(3y^2 + 4y - 5) - (4y^2 - 6y - 8)$

26. $(5z^2 + 8z - 5) - (6z^2 - 3z - 9)$

27. $(2x^2 + 6x - 5) - (6x^2 - 4x - 8)$

28. $(3a^2 + 4a - 7) - (7a^2 - 2a - 5)$

29. $(-6z^2 + 9z - 1) - (3z^2 + 8z - 7)$

30. $(-9x^2 + 4x - 9) - (6x^2 + 2x - 8)$

31. $(2a^2 + 9a - 1) - (-5a^2 - 8a - 4)$

32. $(7c^2 - 3c + 6) - (-9c^2 + 2c - 8)$

33. $(-7x^2 - 6x - 1) - (2x^2 + 8x + 4)$

34. $(-5m^2 - 2m - 9) - (5m^2 + 2m + 7)$

35. $3x - 2(5x^2 - 6) - (-2x^2 + x - 1)$

36. $4x - 3(6x^2 + 2) - (-3x^2 - x + 1)$

37. $6x - (3x^2 + 8x + 2) + 2(-x^2 - 9)$

38. $9x - (5x^2 + 6x + 2) + 3(-x^2 - 5)$

Mixed Practice

Perform the operations indicated.

39. $(-4x^2 + 7x + 1) - (x^2 - 5)$

40. $(-3x^2 + 7x + 2) - (x^2 - 2)$

41. $(4x^2 - 5x - 7) + (3x^2 - 8x - 2)$

42. $(7x^2 - 5x - 6) + (6x^2 - 9x + 1)$

43. $(4x^2 + 6x - 2) - (3x^2 + 7x + 1) + (x^2 - 1)$

44. $(8x^2 - 3x + 4) + (-5x^2 + 6x + 3) - (6x^2 - 1)$

45. $4(-9x - 6) - (3x^2 - 7x + 1) + 4x$

46. $2(-4x + 2) - (2x^2 + 2x - 4) + 3x$

One Step Further

Perform the operations indicated.

47. $(-7x^2 + 3x - 4) - (5x^2 + 7x - 1) + (x^2 - 8)$

48. $(-3x^2 + 2x - 4) - (4x^2 + 6x - 5) + (x^2 - 7)$

49. $5x - 2(4x^2 + 3x - 2) - (2x^2 + 8x - 1)$

50. $2x - 3(5x^2 + 2x - 6) - (2x^2 + 8x - 1)$

51. $7x - 2^3(x + 3) - (-3)^2(x^2 - 2x - 1)$

52. $9x - 2^2(x + 1) - (-4)^2(x^2 - 3x - 1)$

To Think About

Determine the value of a if x ≠ 0.

53. $(ax + 3) + (2x^2 + 5x - 6) + (8x - 2) = 2x^2 - 10x - 5$

54. $(ax - 5) + (4x^2 + 6x + 9) + (3x - 1) = 4x^2 - 2x + 3$

Determine the values of a and b if x ≠ 0.

55. $(ax^2 - bx - 7) + (5x^2 - 2x + 3) = 9x^2 - 5x - 4$ **56.** $(ax^2 - bx + 2) + (3x^2 - 5x - 1) = 8x^2 - 7x + 1$

Cumulative Review   *Perform the operation indicated.*

57. **[4.4.1]** $\dfrac{-6x^8}{2x^2}$ **58.** **[4.4.1]** $\dfrac{-8x^6y^2}{2x^2y^7}$ **59.** **[3.4.2]** $(-4x)(2x^2)$ **60.** **[3.4.2]** $(3y)(-2y)(5y)$

61. **[5.6.1]** *Miles Walked* Maria walked $2\frac{2}{7}$ miles and Juan walked $3\frac{1}{2}$ miles in the Walk for the Orphans fundraiser. How many more miles did Juan walk than Maria?

62. **[5.6.1]** *Recipe Mixture* A cook mixed $\frac{3}{4}$ cup of brown sugar and $\frac{1}{2}$ cup of white sugar in a mixture. How much sugar was in the mixture?

Quick Quiz 6.1   *Perform the operations indicated.*

1. (a) $(-3x + 1) + (-5x - 2)$

(b) $(-6x^2 + 4x - 7) + (8x^2 - 9x + 2)$

2. (a) $(6a - 4) - (3a - 2)$

(b) $(9y^2 - 3y - 8) - (2y^2 - 4y + 8)$

3. (a) $-(-2x^2 - 3x) + (-5x^2 + 9x + 2) - (5x^2 - 10x + 3)$

(b) $9x - (6x^2 - 2x + 3) - 3(-4x^2 + 7x - 1)$

4. Concept Check Mitchell subtracted two polynomials as follows.

$$(-6x^2 + 3x - 1) - (4x^2 + 2x - 7)$$
$$= -6x^2 + 3x - 1 - 4x^2 + 2x - 7$$
$$= -10x^2 + 5x - 8$$

Did Mitchell complete the problem correctly? Why or why not?

6.2 Multiplying Polynomials

Recall from Chapter 3 that there are special names for polynomials: a monomial has one term, a binomial has two terms, and a trinomial has three terms. It is important for you to review the meanings of these names since we will use them in our discussion of polynomials.

① Multiplying a Monomial Times a Polynomial

We are familiar with using the distributive property to multiply a monomial times a binomial: $2(x + 3)$. We use the same process to multiply a monomial times a trinomial. That is, we multiply each term of the trinomial by the monomial.

When we use the distributive property, it is important to identify each term carefully so that we take the *sign of the term* into consideration when multiplying.

Student Learning Objectives

After studying this section, you will be able to:

① Multiply a monomial times a polynomial.

② Multiply a binomial times a trinomial.

③ Multiply binomials using FOIL.

EXAMPLE 1 Multiply. $-4x(2x - 6y - 7)$

Solution We multiply each term by $-4x$.

$$-4x(2x - 6y - 7)$$

$$= -4x(2x) - 4x(-6y) - 4x(-7)$$

$$= -8x^2 + 24xy + 28x$$

First term of the product:
$(-4x)(2x) = -8x^{1+1} = -8x^2$

Second term of the product:
$(-4x)(-6y) = +24xy$

Third term of the product:
$(-4x)(-7) = +28x$

Student Practice 1 Multiply. $-6x(3x - 8y - 2)$

NOTE TO STUDENT: Fully worked-out solutions to all of the Student Practice problems can be found at the back of the text starting at page SP-1.

Notice what happened in Example 1 when we multiplied by the negative monomial $-4x$.

A *positive* term changes to a *negative* term.

$$-4x(2x - 6y - 7) = -8x^2 + 24xy + 28x$$

Negative terms change to *positive* terms.

Therefore, when multiplying by a negative monomial, it is a good idea to check the product, verifying that the sign of each term changes.

The distributive property also works if the monomial is on the right. The process may be more familiar if we move the monomial to the left side of the polynomial before we multiply.

EXAMPLE 2 Multiply. $(3x^2 + x - 6)(-7x^3)$

Solution We move the monomial to the left side.

$$-7x^3(3x^2 + x - 6)$$

We multiply each term by $-7x^3$.

$$-7x^3(3x^2 + x - 6)$$

$$= -21x^5 - 7x^4 + 42x^3$$

$-7x^3(3x^2) = -21x^{3+2} = -21x^5$
$-7x^3(+x) = -7x^{3+1} = -7x^4$
$-7x^3(-6) = +42x^3$

Since we are multiplying by a negative monomial, we check the sign of each term in the product to be sure it changed.

Continued on next page

Positive terms change to *negative* terms.

$$-7x^3(3x^2 + x - 6) = -21x^5 - 7x^4 + 42x^3 \ \checkmark$$

A *negative* term changes to a *positive* term.

Student Practice 2   Multiply. $(2y^2 + y - 5)(-3y^4)$

② Multiplying a Binomial Times a Trinomial

When we multiply a binomial times a trinomial, we use the distributive property twice since we must multiply *each term* of the binomial times the trinomial. Then we simplify the expression by combining like terms.

EXAMPLE 3   Multiply. $(2x + 3)(3x^2 + 5x - 1)$

Solution   We multiply $2x$ times $3x^2 + 5x - 1$ and then $+3$ times $3x^2 + 5x - 1$.

$$(2x + 3)(3x^2 + 5x - 1)$$
$$= 2x(3x^2 + 5x - 1) + 3(3x^2 + 5x - 1)$$
$$= 2x \cdot 3x^2 + 2x \cdot 5x + 2x(-1) + 3 \cdot 3x^2 + 3 \cdot 5x + 3(-1)$$
$$= 6x^3 + 10x^2 - 2x + 9x^2 + 15x - 3 \quad \text{We multiply.}$$
$$= 6x^3 + 19x^2 + 13x - 3 \qquad\qquad \text{We combine like terms:}$$
$$\qquad\qquad\qquad\qquad\qquad\qquad 10x^2 + 9x^2 = 19x^2; \quad -2x + 15x = 13x.$$

Student Practice 3   Multiply. $(3y - 1)(4y^2 + 2y - 6)$

③ Multiplying Binomials Using FOIL

In this section we use a repeated application of the distributive property to multiply a binomial times a binomial.

EXAMPLE 4   Use the distributive property to multiply. $(x + 1)(x + 4)$

Solution   We multiply $x(x + 4)$ and $+1(x + 4)$.

$$x \cdot (x + 4)$$

$$(x + 1) \ \boxed{(x + 4)} = x(x + 4) + 1(x + 4) \qquad \text{We multiply each term of } (x + 1)$$
$$\qquad\qquad\qquad\qquad\qquad\qquad\qquad\qquad \text{times the binomial } (x + 4).$$

$$(+1) \cdot (x + 4)$$

$$(x + 1)(x + 4) = x(x + 4) + 1(x + 4) \qquad \text{Use the distributive property}$$
$$\qquad\qquad\qquad\qquad\qquad\qquad\qquad\qquad \text{again; multiply } x(x + 4) \text{ and}$$
$$\qquad\qquad\qquad\qquad\qquad\qquad\qquad\qquad (+1)(x + 4).$$

$$= x \cdot x + x \cdot 4 + 1 \cdot x + 1 \cdot 4$$
$$= x^2 + 4x + 1x + 4$$
$$(x + 1)(x + 4) = x^2 + 5x + 4 \qquad\qquad \text{Combine like terms.}$$

Student Practice 4   Use the distributive property to multiply.
$$(x + 3)(x + 5)$$

The distributive property shows us how the problem can be done and why it can be done. In actual practice there is a shorter way to obtain the answer. It is often referred to as the **FOIL method.** The letters FOIL stand for: First terms, Outer terms, Inner terms, and Last terms.

$$(x + 1)(x + 4) \qquad (x + 1)(x + 4) \qquad (x + 1)(x + 4) \qquad (x + 1)(x + 4)$$

First terms   Outer terms   Inner terms   Last terms

The FOIL letters are simply a way to remember how the four terms in the final product are obtained. Let's return to Example 4 and rewrite the steps we used to multiply the binomials using the distributive property and then compare these results with the FOIL method.

The distributive property:

$$(x + 1)(x + 4) = x(x + 4) + 1(x + 4)$$
$$= x \cdot x + x \cdot 4 + 1 \cdot x + 1 \cdot 4$$
$$= \boxed{x^2 + 4x + 1x + 4}$$
$$= x^2 + 5x + 4 \qquad\qquad \text{The result is the same.}$$

The FOIL method:

$(x + 1)(x + 4)$	F	Multiply *first* terms:	$x \cdot x = \boxed{x^2}$
$(x + 1)(x + 4)$	O	Multiply *outer* terms:	$(+4)x = \boxed{+4x}$
$(x + 1)(x + 4)$	I	Multiply *inner* terms:	$(+1)x = \boxed{+1x}$
$(x + 1)(x + 4)$	L	Multiply *last* terms:	$(+1)(+4) = \boxed{+4}$

$$(x + 1)(x + 4) = x^2 + 4x + 1x + 4$$
$$= x^2 + 5x + 4$$

Our result is the same as when we used the distributive property. Now let's study the use of the FOIL method in a few examples.

EXAMPLE 5 Multiply. $(x + 4)(x + 3)$

Solution

	First	Outer	Inner	Last
$(x + 4)(x + 3)$	F	O	I	L
	$x \cdot x$	$(+3)x$	$(+4)x$	$(+4)(+3)$
	x^2	$+3x$	$+4x$	$+12$

$$(x + 4)(x + 3) = x^2 + 7x + 12 \quad \text{Combine like terms.}$$

Student Practice 5 Multiply. $(x + 2)(x + 4)$

CAUTION: You can use FOIL only when you multiply a binomial times a binomial. You *cannot* use FOIL to multiply a binomial times any other type of polynomial.

EXAMPLE 6   Multiply. $(y - 1)(y - 7)$

Solution

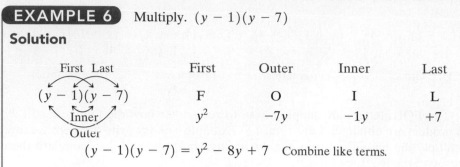

	First	Outer	Inner	Last
	F	O	I	L
	y^2	$-7y$	$-1y$	$+7$

$$(y - 1)(y - 7) = y^2 - 8y + 7 \quad \text{Combine like terms.}$$

In Example 6, the products of the outer terms and the inner terms are negative. We must be sure to pay special attention to the signs of the terms when we multiply.

Student Practice 6   Multiply. $(y - 5)(y - 3)$

EXAMPLE 7   Multiply. $(2x + 3)(x - 1)$

Solution

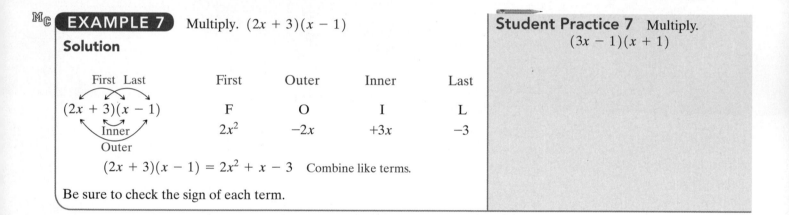

	First	Outer	Inner	Last
	F	O	I	L
	$2x^2$	$-2x$	$+3x$	-3

$$(2x + 3)(x - 1) = 2x^2 + x - 3 \quad \text{Combine like terms.}$$

Be sure to check the sign of each term.

Student Practice 7   Multiply.
$$(3x - 1)(x + 1)$$

👣 STEPS TO SUCCESS Positive Thinking

Some people convince themselves that they cannot learn mathematics. If you are concerned that you may have difficulty in this course, it is time to reprogram your thinking. Replace those negative thoughts with positive ones such as

"I can learn mathematics if I work at it. I will give this math class my best shot."

You will be pleasantly surprised at the difference this positive attitude makes!

Now, some might wonder how a change in attitude can make a difference. Well, the approach and features in this book can be the keys to your success.

• Reread the Steps to Success boxes, which offer tips on how to study. If you know how to study, learning will be far easier.

• Do your homework. Exercises develop the fundamentals on which the concepts are built. Many students

do not learn because they miss certain building blocks. For example, how can someone learn long division if he or she doesn't know multiplication facts?

If you take this advice to heart, you'll be off to a good start. Keep up the good work.

Making it personal:

1. Name a few things that you plan to do in this class to help you be successful.

2. Write in words two positive thoughts about mathematics and/or your ability to complete this course.

Verbal and Writing Skills, Exercises 1–4

1. Erin multiplied $(-4)(x^2 + 2x + 1)$ and obtained this result: $-4x^2 - 8x + 4$. What is wrong with this multiplication?

2. Write in words the multiplication that the word FOIL represents.

Fill in the blanks and boxes to complete each problem.

3. **(a)** A _____ is a polynomial with one term.

   (b) A _____ is a polynomial with two terms.

   (c) A _____ is a polynomial with three terms.

4. **(a)** The polynomial $x^2 + 2x + 1$ is called a _____.

   (b) The polynomial $3x^2$ is called a _____.

   (c) The polynomial $5x^3 + 9$ is called a _____.

5. Multiply. $-5(3y^2 - 2y + 6)$

   First term of the product is: ☐

   Second term of the product is: ☐

   Third term of the product is: ☐

   Therefore, $-5(3y^2 - 2y + 6) =$ ☐ .

6. Multiply. $-3(2y^2 - 4y + 1)$

   First term of the product is: ☐

   Second term of the product is: ☐

   Third term of the product is: ☐

   Therefore, $-3(2y^2 - 4y + 1) =$ ☐ .

7. To multiply $(x - 1)(x^2 + 3x + 1)$:

   We multiply x times the trinomial: x ☐(☐) = ☐ .

   Then we multiply -1 times the trinomial: -1 ☐(☐) = ☐ .

   Finally, we combine like terms: $(x - 1)(x^2 + 3x + 1) =$ ☐ .

8. To multiply $(y - 2)(y^2 + 4y + 3)$:

   We multiply the trinomial by the first term of the binomial: ☐$(y^2 + 4y + 3) =$ ☐ .

   Then we multiply the trinomial by the second term of the binomial: ☐$(y^2 + 4y + 3) =$ ☐ .

   Finally, we combine like terms: $(y - 2)(y^2 + 4y + 3) =$ ☐ .

9. $(x + 1)(x - 3)$ F = ☐

   O = ☐

   I = ☐

   L = ☐

   Therefore, $(x + 1)(x - 3) =$ ☐ .

10. $(x - 2)(x - 1)$ F = ☐

    O = ☐

    I = ☐

    L = ☐

    Therefore, $(x - 2)(x - 1) =$ ☐ .

Use the distributive property to multiply.

11. $8(2x^2 + 3x - 2)$

12. $4(3y^2 + 7y - 3)$

13. $-4y(2y^2 - 3y + 6)$

14. $-2y(5y^2 + 2y - 8)$

15. $-3x^2(x - 2)$

16. $-4x^3(x - 3)$

17. $(y^6 + 9)(-5y^2)$

18. $(x^3 + 3)(-2x^2)$

19. $(x^3 - 5x - 2)(-3x^4)$ **20.** $(4y^2 + 3y - 2)(-2y^4)$ **21.** $(x - 1)(2x^2 - 3x - 2)$ **22.** $(x - 5)(x^2 - 2x - 1)$

23. $(2y + 4)(3y^2 + y - 5)$ **24.** $(3x - 2)(x^2 + 3x + 1)$ **25.** $(3x - 1)(x^2 + 2x + 1)$ **26.** $(3x + 1)(2x^2 + x - 3)$

Use FOIL to multiply.

27. $(x + 6)(x + 7)$ **28.** $(a + 2)(a + 1)$ **29.** $(x + 3)(x + 9)$ **30.** $(y + 2)(y + 5)$

31. $(a + 6)(a + 2)$ **32.** $(x + 4)(x + 1)$ **33.** $(y + 4)(y - 8)$ **34.** $(a + 7)(a - 4)$

35. $(x + 2)(x - 4)$ **36.** $(x + 3)(x - 5)$ **37.** $(x - 4)(x + 2)$ **38.** $(m - 3)(m + 5)$

39. $(2x + 1)(x + 2)$ **40.** $(3x + 1)(x + 2)$ **41.** $(3x - 3)(x - 1)$ **42.** $(4x - 3)(x - 1)$

43. $(2y - 1)(y + 2)$ **44.** $(4y - 2)(y + 1)$ **45.** $(2y + 1)(y - 2)$ **46.** $(4y + 2)(y - 1)$

Mixed Practice

Multiply.

47. $-5a(2a - 4b - 6)$ **48.** $-4x(-3x + 5y - 7)$ **49.** $-7x^3(x - 3)$ **50.** $-8x^3(x - 5)$

51. $(x - 2)(x^2 - 3x + 1)$ **52.** $(x - 4)(x^2 + x - 2)$ **53.** $(z + 2)(z - 5)$ **54.** $(b + 1)(b - 3)$

55. $(2x + 1)(4x^2 + 2x - 8)$ **56.** $(3x + 1)(2x^2 + 3x - 2)$

57. $(y - 7)(y + 2)$

58. $(z - 9)(z + 5)$

59. (a) $(z + 5)(z + 1)$ **60. (a)** $(x + 4)(x + 1)$ **61. (a)** $(x + 3)(x - 1)$ **62. (a)** $(z - 2)(z + 4)$

(b) $(z - 5)(z - 1)$ **(b)** $(x - 4)(x - 1)$ **(b)** $(x - 3)(x + 1)$ **(b)** $(z + 2)(z - 4)$

One Step Further

Simplify.

63. $(x + 2)(x - 1) + 2(3x + 3)$

64. $(x - 3)(x + 1) + 4(2x + 1)$

65. $-2x(x^2 + 3x - 1) + (x - 2)(x - 3)$

66. $-3x(x^2 + x - 2) + (x - 1)(x - 2)$

To Think About

67. If $a(2x - 3) = -14x + 21$, what is the value of a?

68. If $b(-3x + 4) = 15x - 20$, what is the value of b?

Cumulative Review

Perform the operations indicated.

69. **[3.2.3]** *Coin Problem* The number of dimes (D) is three times the number of nickels (N).

(a) Write the statement as an equation.

(b) Find the number of nickels if there are 21 dimes.

70. **[3.2.3]** *Geometry* The length (L) of a piece of wood is double the width (W).

(a) Translate the statement into an equation.

(h) Find the width if the length is 40 feet.

71. **[4.6.3]** *Calories* If two Donna Deluxe cookies contain 250 calories, then how many calories are in five Donna Deluxe cookies?

72. **[4.5.3]** *Earnings* Joanna earned $64 last month babysitting a total of 16 hours on the weekends. How much did she earn in dollars per hour?

Quick Quiz 6.2 *Multiply.*

1. $(-3z - 1)(5z^2)$

2. $(x - 1)(4x^2 - 2x + 8)$

3. (a) $(y - 2)(y + 1)$

 (b) $(2a + 3)(a + 5)$

4. Concept Check Multiply each of the following.
 1. $(x + 1)(x + 2)$ **2.** $(x - 1)(x - 2)$

 (a) Explain why the middle terms in each product of 1 and 2 have opposite signs.

 (b) Explain why the last terms in each product of 1 and 2 have the same sign.

How Am I Doing? Sections 6.1–6.2

How are you doing with your homework assignments in Sections 6.1 and 6.2? Do you feel you have mastered the material so far? Do you understand the concepts you have covered? Before you go further in the textbook, take some time to do each of the following problems.

1. _____

2. _____

3. _____

4. _____

5. _____

6. _____

7. _____

8. _____

9. _____

10. _____

11. _____

12. _____

6.1

Perform the operations indicated.

1. $(3y^2 + 5y - 2) + (4y - 7)$

2. $(-7a + 5) - (2a - 3)$

3. $(-2x^2 + 4x - 7) - (5x^2 + 3x - 4)$

4. $2x - 3(5x^2 + 4) + (-3x^2 - x + 6)$

5. $(3x^2 - 6x - 8) - (2x^2 + 7x + 5) + (x^2 - 7)$

6.2

Multiply.

6. $-8(2a^2 - 3a + 1)$

7. $-2y(-6y + 4x - 5)$

8. $-4x^2(x^2 + 6)$

9. $(y + 2)(4y^2 + 3y - 2)$

10. $(y + 2)(y - 4)$

11. $(x - 4)(x - 1)$

12. $(2y + 3)(y + 4)$

Now turn to page SA-11 for the answers to each of these problems. Each answer also includes a reference to the objective in which the problem is first taught. If you missed any of these problems, you should stop and review the Examples and Student Practice problems in the referenced objective. A little review now will help you master the material in the upcoming sections of the text.

6.3 Translating from English to Algebra

In Chapter 3 we wrote variable expressions for several quantities using *different* variables. In this section we'll learn how to write variable expressions for several quantities using the *same* variable. In Chapter 7 we'll use this skill to write and solve equations.

Student Learning Objectives

After studying this section, you will be able to:

① Write variable expressions when comparing two or more quantities.

② Write variable expressions from written phrases.

① Writing Variable Expressions When Comparing Two or More Quantities

Often, real-life applications involve comparing two or more quantities. When this is the case, we describe one quantity *in terms of another*. It is helpful to let a variable represent the quantity *to which things are being compared*. For example, if John is 4 inches taller than Ed, and Chris is 2 inches shorter than Ed, we are *comparing* all the heights to *Ed's height*. Thus, we let the variable x represent Ed's height, and we describe John's and Chris's heights in terms of Ed's.

$x + 4$
John's height is 4 inches more than Ed's.

x
Ed's height

$x - 2$
Chris's height is 2 inches less than Ed's.

Let x = Ed's height, then $(x + 4)$ = John's height, and $(x - 2)$ = Chris's height. x, $(x + 4)$, and $(x - 2)$ are called *variable expressions*. When we write these expressions, we are *defining the variable expressions*. As we will see in Chapter 7, this is an important step; it will help us create the equation needed to find the solution to the problem.

▲ **EXAMPLE 1** The length of a garden is double the width. Define the variable expressions for the length and width of the garden.

Solution Since we are comparing the length *to the width*, we let the variable represent the *width* of the garden. We can choose any variable, so we choose W.

$$\text{Let the width} = W.$$

Now we write the expression for the length in terms of the width by translating the statement.

The length of a garden is double the width.

$$\text{length} = 2 \cdot W$$

We define our variable expressions as follows.

$$\text{width} = W$$
$$\text{length} = 2W$$

Continued on next page

361

NOTE TO STUDENT: Fully worked-out solutions to all of the Student Practice problems can be found at the back of the text starting at page SP-1.

▲ **Student Practice 1** The height of a pole is one-third the height of a building. Define the variable expressions for the height of the pole and the height of the building.

Remember that it is helpful to let the variable represent the quantity to which things are being compared. For example, if you are comparing one quantity to another, usually the *second quantity* is represented by the variable.

ᴹᴄ **EXAMPLE 2** The second side of a triangle is 2 inches longer than the first; the third side is 8 inches shorter than three times the length of the first side. Define the variable expression for the length of each side of the triangle.

Solution It helps to draw a picture. Since we are *comparing* all sides *to* the *first side,* we let the *variable represent* the length of the *first side.* We may choose any variable, so we use the letter *f.*

▲ **Student Practice 2** The second side of a triangle is 3 inches longer than the first; the third side is 12 inches shorter than three times the length of the first side. Define the variable expression for the length of each side of the triangle.

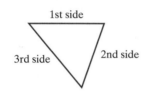

Let the length of the first side $= f$.

The second side is *2 inches longer than the first side.*

$$\text{second side} = f + 2$$

To find the expression for the third side, we can separate the sentence "The third side is 8 inches shorter than 3 times the first side" into two phrases and translate each phrase separately.

The third side is 8 inches shorter than

$$\text{third side} = \boxed{} - 8 \qquad \text{We are subtracting } 8 \text{ } from \text{ an amount}$$

3 times the first

$$\boxed{3 \cdot f} - 8 \qquad \text{We are subtracting } 8 \text{ from } 3f.$$

$$\text{third side} = 3f - 8$$

We define our variable expressions as follows.

length of the first side $= f$
length of the second side $= (f + 2)$
length of the third side $= (3f - 8)$

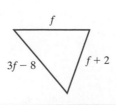

② Writing Variable Expressions from Written Phrases

EXAMPLE 3　A company's profit for the first quarter is three times the profit for the second quarter. The profit for the third quarter is $22,100 more than the profit for the second quarter.

(a) Write the variable expression for the profit for each quarter.

(b) Write the following phrase using math symbols: The profit for the first quarter minus the profit for the third quarter plus the profit for the second quarter.

(c) Simplify the expression from part **(b)**.

Solution

(a) We are comparing the first- and third-quarter profits *to the second-quarter profit,* so we let the variable represent the *second-quarter profit.*

$$\text{second-quarter profit} = x$$

Now we write the first-quarter profit in terms of the second.

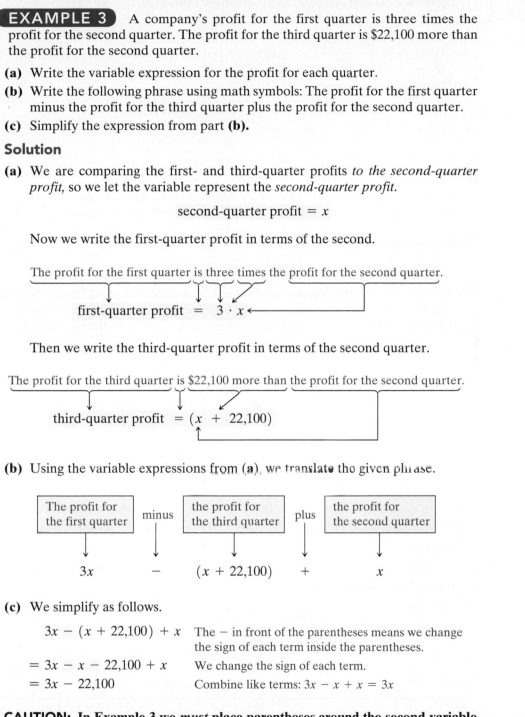

The profit for the first quarter is three times the profit for the second quarter.

$$\text{first-quarter profit} = 3 \cdot x$$

Then we write the third-quarter profit in terms of the second quarter.

The profit for the third quarter is $22,100 more than the profit for the second quarter.

$$\text{third-quarter profit} = (x + 22{,}100)$$

(b) Using the variable expressions from **(a)**, we translate the given phrase.

The profit for the first quarter	minus	the profit for the third quarter	plus	the profit for the second quarter
$3x$	$-$	$(x + 22{,}100)$	$+$	x

(c) We simplify as follows.

$$3x - (x + 22{,}100) + x \quad \text{The } - \text{ in front of the parentheses means we change the sign of each term inside the parentheses.}$$
$$= 3x - x - 22{,}100 + x \quad \text{We change the sign of each term.}$$
$$= 3x - 22{,}100 \quad \text{Combine like terms: } 3x - x + x = 3x$$

CAUTION: In Example 3 we *must* place parentheses around the second variable expression because it is preceded by a minus sign. The entire expression $x + 22{,}100$ is being subtracted, not just the first term, x. Therefore, we must change the sign of *each term* inside the parentheses. Without the parentheses we would *not* change $+22{,}100$ to $-22{,}100$ and the answer would be wrong!

Continued on next page

Student Practice 3   A company's profit for the second quarter is two times the profit for the first quarter. The profit for the third quarter is $30,000 more than the profit for the first quarter.

(a) Write the variable expression for the profit for each quarter.
(b) Write the following phrase using math symbols: The profit for the second quarter minus the profit for the third quarter plus the profit for the first quarter.
(c) Simplify the expression from part **(b).**

Understanding the Concept

Variable Expression or Equation?   Do you understand the difference between a variable expression and an equation? Consider the following.

Variable Expression	*Equation*
$2x + 5$	$2x + 5 = 40$

Notice that the **variable expression does not have an equals sign.** This means that we cannot use the principles of equality to solve the expression since these principles require that we add, multiply, or divide on both sides of the equals sign.

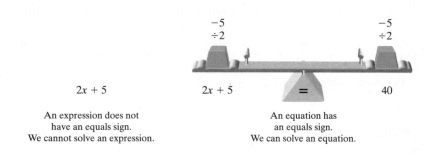

$$2x + 5 \qquad\qquad 2x + 5 \quad = \quad 40$$

An expression does not   An equation has
have an equals sign.   an equals sign.
We cannot solve an expression.   We can solve an equation.

Be sure to pay attention to whether you are working with an expression or an equation. **You can simplify an expression; you cannot solve an expression.**

Exercises

Fill in each blank with the appropriate phrase. Choose from the following: simplify and solve; or simplify.

1. We can _____ $3x + 1 + 4x$.

2. We can _____ $3x + 1 + 4x = 9$.

3. We can _____ $4 + 2x + 6x = 11$.

4. We can _____ $4 + 2x + 6x$.

Verbal and Writing Skills, Exercises 1–6 *Fill in the blanks.*

1. *Age Comparison* Juan is two times as old as Mark. We are comparing Juan's age to Mark's age; therefore we can let the variable x = _____ age and $2x$ = _____.

2. *Age Comparison* Rhonda is three times as old as Tara. We are comparing Rhonda's age to Tara's age; therefore we can let the variable x = _____ age and $3x$ = _____ age.

3. *Miles Run* Alice can run 1 mile in 20 seconds less time than Leslie. Shannon can run 1 mile in 50 seconds less time than Leslie. We are comparing Alice's and Shannon's 1-mile running records to Leslie's; therefore we can let x = _____ running record, $x - 20$ = _____ running record, and $x - 50$ = _____ running record.

4. *Home Runs* Last season Jose made 6 more home runs than Arnold. Ernie made 4 fewer home runs than Arnold. We are comparing Jose's and Ernie's home run records to Arnold's; therefore we can let x = _____ record, $x + 6$ = _____ record, and $x - 4$ = _____ record.

Write an applied problem using the following variable definitions.

5. *Miles Driven* Number of miles driven by Scott = x; number of miles driven by Mark = $x + 500$

6. *Wage Comparison* Jesse's weekly salary = x; Ian's weekly salary = $x + 50$

Applications

7. *Company Profit* A company's profit for April was $8000 less than the profit for March. Define the variable expressions for the company's profits in March and April.

8. *Price Comparison* The price of a used car is $19,500 less than the price of the new model. Define the variable expressions for the prices of the used and new cars.

9. *Company Profit* A company's profit for the fourth quarter is $31,100 more than the profit for the second quarter. Define variable expressions for the profits for the second and fourth quarters.

10. *Festival Profit* Last year's profit from the summer art festival was $2200 more than this year's profit. Define variable expressions for last year's and this year's profits.

11. *Height Comparison* The height of a pole is one-half the height of a tree. Define variable expressions for the height of the pole and the height of the tree.

12. *Entertainment Tickets* The number of children's tickets sold is one-third the number of adults' tickets sold. Define variable expressions for the number of adults' and children's tickets sold.

▲ **13.** *Geometry* The length of a rectangle is double the width. Define variable expressions for the length and width of the rectangle.

▲ **14.** *Geometry* The length of a rectangle is triple the width. Define variable expressions for the length and width of the rectangle.

▲ **15.** *Geometry* The width of a rectangle is 13 inches shorter than twice the length. Define variable expressions for the length and width of the rectangle.

▲ **16.** *Geometry* The width of a rectangle is 25 inches shorter than three times the length. Define variable expressions for the length and width of the rectangle.

17. *Music DVDs* The number of music DVDs that Carl has in his collection is three less than double the number of music DVDs that Toni has in her collection. Define variable expressions for the number of music DVDs that Carl and Toni have.

18. *Company Profit* A company's profit for the second quarter was $4000 less than twice the first quarter's profit. Define variable expressions for the first- and second-quarter profits.

▲ **19.** *Geometry* The length of a rectangular box is double the width. The height is three times the width. Define variable expressions for the width, length, and height of the rectangular box.

▲ **20.** *Geometry* The width of a rectangular box is double the length. The height is five times the length. Define variable expressions for the width, length, and height of the rectangular box.

21. *Model Car Collection* Jim has sixteen more blue cars in his model car collection than red cars. He has seven fewer white cars than red cars in the collection. Define variable expressions for the number of blue, red, and white cars Jim has in his collection.

22. *Height Comparison* Sion is 3 inches taller than Damien, and Brad is 4 inches shorter than Damien. Define variable expressions for the heights of Sion, Damien, and Brad.

▲ **23.** *Geometry* The second side of a triangle is 4 inches longer than the first. The third side is 10 inches shorter than two times the first. Define the variable expression for the length of each side of the triangle.

▲ **24.** *Geometry* The second side of a triangle is 3 inches longer than the first. The third side is 7 inches shorter than three times the first. Define the variable expression for the length of each side of the triangle.

▲ **25. *Height Comparison*** The height of a building is four times the height of a tree.

(a) Define variable expressions for the height of the building and the height of the tree.

(b) Write the following phrase using math symbols: The difference between the height of the building and the height of the tree.

(c) Simplify the expression from part **(b)**.

▲ **26. *Geometry*** The length of a yard is triple the length of the garden.

(a) Define variable expressions for the length of the yard and the length of the garden.

(b) Write the following phrase using math symbols: The difference between the length of the yard and the length of the garden.

(c) Simplify the expression from part **(b)**.

27. *School Election* In a school election for class president, Jeri received thirty more votes than Max. Lee received forty-five fewer votes than Max.

(a) Define variable expressions for the votes received by Jeri, Max, and Lee.

(b) Write the following phrase using math symbols: The number of votes received by Max minus the number of votes received by Jeri plus the number of votes received by Lee.

(c) Simplify the expression from part **(b)**.

28. *Election Results* Tina received 2000 fewer votes than Michael in an election. Samantha received 4200 more votes than Michael in the election.

(a) Define variable expressions for the numbers of votes received by Tina, Michael, and Samantha.

(b) Write the following phrase using math symbols: The number of votes received by Samantha plus the number of votes received by Michael minus the number of votes received by Tina.

(c) Simplify the expression from part **(b)**.

29. *Wage Comparison* Vu's salary is $125 more than Sam's salary. Evan's salary is $80 less than Sam's salary.

(a) Define variable expressions for the salaries for Vu, Sam, and Evan.

(b) Write the following phrase using math symbols: Vu's salary plus Sam's salary minus Evan's salary.

(c) Simplify the expression from part **(b)**.

30. *Price Comparison* A gold bracelet costs $130 less than a gold necklace. A gold locket costs $60 more than the gold necklace.

(a) Define variable expressions for the prices of the bracelet, necklace, and locket.

(b) Write the following phrase using math symbols: The price of the bracelet plus the price of the necklace minus the price of the locket.

(c) Simplify the expression from part **(b)**.

▲ **31. *Geometry*** The height of a box is 8 inches longer than the width. The length is 2 inches shorter than double the width.

(a) Define variable expressions for height, length, and width of the box.

(b) Write the following phrase using math symbols: The sum of the height, width, and length.

(c) Simplify the expression from part **(b)**.

32. *Geometry* The second angle of a triangle is triple the first. The third angle of the triangle is 35° larger than the first.

(a) Define the variable expressions for each angle of the triangle.

(b) Write the following phrase using math symbols: The sum of all three angles of the triangle.

(c) Simplify the expression from part **(b)**.

To Think About

Answer true or false.

33. We can solve $3x + 6$.

34. We can solve $3x + 6 = 12$.

Cumulative Review

Solve.

35. **[3.2.1]** $11x = 44$

36. **[3.1.2]** $y + 77 = -6$

37. **[5.7.1]** $\dfrac{m}{7} = -5$

38. **[3.1.2]** $4x - 3x + 8 = 62$

39. **[3.3.2]** Find the area of a rectangle with $L = 14$ inches and $W = 10$ inches.

40. **[3.3.2]** Find the volume of a rectangular solid with $L = 6$ feet, $W = 4$ feet, and $H = 5$ feet.

Quick Quiz 6.3

1. Tina's monthly salary is triple Mai's monthly salary. Define variable expressions for Tina's monthly salary and Mai's monthly salary.

2. Dixie is 4 years older than Sugar. Pumpkin is 3 years younger than twice Sugar's age. Define the variable expressions for the ages of the three cats.

3. Phoebe purchased a watch, ring, and bracelet at the jewelry store. The watch cost seventy-five dollars more than the ring. The bracelet was fifty dollars less than the cost of the ring.

  (a) Define variable expressions for the prices of the watch, ring, and bracelet.

  (b) Write the following phrase using math symbols: The price of the watch minus the price of the bracelet plus the price of the ring.

  (c) Simplify the expression from part **(b)**.

4. **Concept Check** The width of a box is triple the height. The length of the box is six inches shorter than twice the height.

  (a) Which side of the box will you choose to let the variable represent: length, width, or height? Explain why.

  (b) Define the variable expressions for each side of the box.

Did You Know...

That Starting to Save for Retirement a Few Years Sooner Can Earn You a Lot More Than You Think?

RETIREMENT

Understanding the Problem:
Sofia is thinking about saving for retirement. She has been told it is best to start as soon as possible, but she doesn't have a lot of extra money right now. Is waiting really going to make that much of a difference?

Making a Plan:
The local branch of Sofia's bank is offering a seminar on retirement. During the seminar, they will go over some retirement savings scenarios using different starting ages. This will give Sofia a chance to see what effect waiting can have on retirement planning. Once she knows more, she'll decide what she should do.

Step 1: At the seminar they consider three different people who are just starting to save for retirement. Below is a table showing at what age they started saving and how much they will have accumulated (assuming an 8% annual return) when they retire at 65.

Name	Age	Savings per year	Accumulated at 65
Doug	45	$3600	$178,000
Taylor	35	$3600	$440,000
Ben	25	$3600	$1,007,000

Task 1: How much additional money will Taylor have accumulated over Doug?

Task 2: How much additional money will Ben have accumulated over Taylor?

Step 2: Ben's ten years of extra savings will grow to more than twice that of Taylor's ten years of extra savings due to compound interest. Compound interest is the interest earned on savings that then gets added to the savings, which then earns more interest. With compound interest and a rate of 8%, money will double in about 9 years. Ben's extra savings will be compounding for 40 years while Taylor's extra savings will be compounding for only 30 years.

Task 3: Ben will be saving ten years longer than Taylor. What is the extra amount Ben will put in his retirement account over those ten years?

Task 4: Since Ben saves ten years longer than Taylor, how much of the additional money that Ben will accumulate by age 65 is due to compound interest?

Finding a Solution:
Step 3: Sofia was very surprised to learn all this. She had not realized that starting early could make such a huge difference in how much she would have at retirement. Sofia decides that starting early is so important that she is going to cut her spending so she can start to save $3000 a year. At the seminar they told her that for every $1000 she saves a year, she will have about $280,000 at retirement. Using that figure, she calculates how much she will have if she saves $3000 a year.

Task 5: How much will Sofia have at retirement if she saves $3000 a year?

Applying the Situation to Your Life:
How early you begin saving for retirement has a big effect on how much you will have to save each year to reach your goals. So you should start as soon as possible. Now that you know how big an impact starting early can make, do you think this will encourage you to start saving?

6.4 Factoring Using the Greatest Common Factor

Student Learning Objectives

After studying this section, you will be able to:

① Find the greatest common factor (GCF).

② Factor out the greatest common factor from a polynomial.

NOTE TO STUDENT: *Fully worked-out solutions to all of the Student Practice problems can be found at the back of the text starting at page SP-1.*

① Finding the Greatest Common Factor (GCF)

Recall from Chapter 4 that we factor a whole number by writing an equivalent product. For example, we can factor 6 as $6 \cdot 1$ or $3 \cdot 2$, and 12 as $12 \cdot 1$, $6 \cdot 2$, or $3 \cdot 4$. As you can see, factoring is the reverse of multiplying. When we look at the factors of 6 and 12, we notice that 1, 2, 3, and 6 are factors of both 6 and 12. We call 1, 2, 3, and 6 **common factors** of 6 and 12.

$$6 = 6 \cdot 1 \qquad 6 = 3 \cdot \mathbf{2}$$
$$12 = 12 \cdot 1 \qquad 12 = 3 \cdot 4 \qquad 12 = 6 \cdot \mathbf{2}$$

We see that 1, 2, 3, and 6 are common factors.

6 is the largest of the common factors because it is the largest number that divides into both 6 and 12 without a remainder. We call the largest common factor the **greatest common factor (GCF).**

EXAMPLE 1 Find the GCF of 8 and 16.

Solution Think of a factor, the *largest factor*, that will divide into *both* 8 and 16. The largest common factor is 8, therefore the GCF of 8 and 16 is 8.

Student Practice 1 Find the GCF of 7 and 14.

When numbers are large, it may be hard to find the GCF by inspection. When this is the case, we can write each number as a product of prime factors to find the GCF.

EXAMPLE 2 Find the GCF of 8, 20, and 28.

Solution First we write each number as a product of prime factors in exponent form. Then we identify the common factors and use the *smallest power* that appears on these factors to find our GCF.

$$8 = 2 \cdot 2 \cdot 2 \quad = 2^3 \qquad \text{We line up the common factors.}$$
$$20 = 2 \cdot 2 \cdot 5 \quad = 2^2 \cdot 5 \qquad 2 \text{ is the common prime factor.}$$
$$28 = 2 \cdot 2 \cdot 7 = 2^2 \cdot 7$$

Notice that $2 \cdot 2$ $2^2 = 4$ The smallest power that appears on
or 2^2 is common   the common factor is 2.
to 8, 20, 28.

The greatest common factor of 8, 20, and 28 is 4.

Student Practice 2 Find the GCF of 9, 21, and 42.

We find the GCF of the terms of a variable expression in a similar manner. We summarize the procedure below.

FINDING THE GREATEST COMMON FACTOR

1. Write each term of the expression as a product of prime factors in exponent form.
2. Identify common prime factors or variables that have the *smallest power*.
3. Multiply the common factors from step 2 to find the GCF.

If there are no common prime factors or variables, the GCF is 1.

EXAMPLE 3　Find the GCF.

(a) $15x^2y + 18x^3$　　　　**(b)** $a^4bc + a^2b^2$

Solution

(a) $15x^2y + 18x^3$

Rewrite each term using prime factors written in exponent form.

$$15x^2y = 3^1 \cdot 5 \cdot x^2 \cdot y$$
$$18x^3 = 3^2 \cdot 2 \cdot x^3$$
$$\downarrow \quad \downarrow$$

Identify the common prime factors and use the *smaller power* that appears on each of these factors.

$$3^1 \cdot x^2$$　The smaller power on 3 is 1 and on x is 2.

The GCF of $15x^2y + 18x^3$ is $3x^2$.

CAUTION: Note that 2, 5, and y are not part of the GCF because they do not appear as factors in both terms.

(b) $a^4bc + a^2b^2$

Identify the common prime factors and use the smaller power that appears on each of these factors.

$$a^4bc = a^4 \cdot b^1 \cdot c$$
$$a^2b^2 = a^2 \cdot b^2$$
$$\downarrow \quad \downarrow$$
$$a^2 \cdot b^1$$　The smaller power on a is 2 and on b is 1.

The GCF of $a^4bc + a^2b^2$ is a^2b.

CAUTION: Note that a^4b^2 is *not* the GCF of $a^4bc + a^2b^2$ because a^4 is *not* a factor of a^2b^2 and b^2 is *not* a factor of a^4bc. To avoid this error be sure to use the *smallest power* that appears on common prime factors to find the GCF.

Student Practice 3　Find the GCF.

(a) $16a^3b^4 + 12a^4b$　　　　**(b)** $x^2yz^3 + xz^2$

② Factoring Out the Greatest Common Factor from a Polynomial

Now let's see what it means to **factor an expression.** To factor $9x + 18$, we find an equivalent expression written as a product. We can think of it as reversing the distributive property.

$$9x + 18 = 9(x + 2)$$　We start with an expression and reverse the distributive property to write a product.

How did we come up with the product $9(x + 2)$? We illustrate the steps below.

1. Find the GCF of the terms in the expression $9x + 18$.　　9 is the GCF of $9x$ and 18.

2. Write each term of $9x + 18$ as a product.

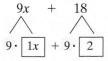

To complete the next step, we "factor out 9 from each term." In other words, we remove 9 as a factor from each term and write $9 \cdot$ (remaining factors separated by the $+$ sign).

3. Factor out 9 from each term.

$$9 \cdot 1x + 9 \cdot 2$$　Remove 9 from each term.

$9(1x + 2)$ or $9(x + 2)$　Write the remaining factors separated by the $+$ sign inside parentheses.

Now we have $9x + 18 = 9(x + 2)$ and we can *check our answer* by using the distributive property to multiply $9(x + 2)$: the product is $9x + 18$.

EXAMPLE 4   Factor $12x + 15$ and check your solution.

Solution   Step 1: 3 is the greatest common factor of $12x + 15$.

$$12x \; + \; 15$$

$$= 3 \cdot 4x + 3 \cdot 5 \quad \text{Step 2: Write each term as a product.}$$

$$12x + 15 = 3(4x + 5) \quad \text{Step 3: Factor out the GCF, 3.}$$

Check. We multiply using the distributive property.

$$3(4x + 5) \stackrel{?}{=} 12x + 15$$

$$3(4x + 5) = 3 \cdot 4x + 3 \cdot 5 = 12x + 15 \; \checkmark$$

Student Practice 4   Factor $8y + 12$ and check your solution.

EXAMPLE 5   Factor $8x - 12y + 16$ and check your solution.

Solution   Step 1: 4 is the greatest common factor of $8x - 12y + 16$.

$$8x \; - \; 12y \; + \; 16$$

$$= 4 \cdot 2x - 4 \cdot 3y + 4 \cdot 4 \quad \text{Step 2: Write each term as a product.}$$

$$8x - 12y + 16 = 4(2x - 3y + 4) \quad \text{Step 3: Factor out the GCF, 4.}$$

Check. We should always check our answer using the distributive property.

$$4(2x - 3y + 4) \stackrel{?}{=} 8x - 12y + 16$$

$$8x - 12y + 16 = 8x - 12y + 16 \; \checkmark$$

Student Practice 5   Factor $5a - 15b + 20$ and check your solution.

When we are asked to factor an expression, we must factor "completely." By this we mean that *none* of the terms in the product has any common factors. To illustrate, let's revisit Example 5. We can also factor $8x - 12y + 16$ as $2(4x - 6y + 8)$ since they are equivalent expressions. But the product $2(4x - 6y + 8)$ is *not* factored completely because $4x - 6y + 8$ has the common factor 2. This happens when we do not factor out the *greatest* common factor. To complete the process we need to factor out another 2.

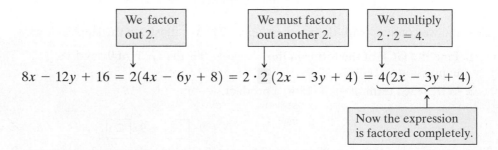

This extra step will not be necessary if we make sure to factor out the GCF.

If you are having difficulty factoring out the GCF, you may want to try an approach that uses multiplication and division. Consider what happens when we multiply and divide 10 by the same number.

$$\frac{2 \cdot 10}{2} = 10 \quad \text{We multiply and divide 10 by 2.}$$

As you can see, we do not change the value of a number when we multiply and divide that number by the same number. Thus, to factor out the GCF we can multiply the expression by the GCF and then divide the expression by the GCF. Let's factor $3x + 12$ using this approach.

$$3x + 12 = \text{GCF}\left(\frac{3x}{\text{GCF}} + \frac{12}{\text{GCF}}\right)$$

We multiply by the GCF and then divide *each term* by the GCF.

$$= 3\left(\frac{3x}{3} + \frac{12}{3}\right)$$

The GCF is 3.

$$3x + 12 = 3(x + 4)$$

$$\frac{3x}{3} = 1 \cdot x = x \text{ and } \frac{12}{3} = 4$$

Try this approach on a few problems and see which method you like best.

$\mathbb{M_C}$ **EXAMPLE 6**   Factor. $5xy^2 + 15xy^3$

Solution   We find the GCF of the expression.

$$5xy^2 = \quad\ 5 \cdot x \cdot y^2$$
$$15xy^3 = 3 \cdot 5 \cdot x \cdot y^3$$
$$\downarrow \ \ \downarrow \ \ \downarrow$$
$$\text{The GCF} = \quad 5 \cdot x \cdot y^2 = 5xy^2$$

We factor out the GCF from the expression.

$$\text{GCF}\left(\frac{5xy^2}{\text{GCF}} + \frac{15xy^3}{\text{GCF}}\right)$$

$$5xy^2 + 15xy^3 = 5xy^2\left(\frac{5xy^2}{5xy^2} + \frac{15xy^3}{5xy^2}\right)$$

$$5xy^2 + 15xy^3 = 5xy^2(1 + 3y) \qquad \frac{5xy^2}{5xy^2} = 1; \quad \frac{15xy^3}{5xy^2} = \frac{3 \cdot 5xy^{3-2}}{5x} = 3y$$

Check. We multiply $5xy^2(1 + 3y)$ using the distributive property.

$$5xy^2(1 + 3y) \overset{?}{=} 5xy^2 + 15xy^3$$
$$5xy^2(1 + 3y) = 5xy^2(1) + 5xy^2(3y) = 5xy^2 + 15xy^3 \ \checkmark$$

Student Practice 6   Factor.
$$6ab^3 - 18ab^4$$

👣 STEPS TO SUCCESS   Evaluating Your Study Habits

Take a close look at your study habits. You will be surprised how you can make a difference in your performance by **learning how to learn**. Practice the key techniques discussed in the Steps to Success and improve your performance in all your classes as well as in your place of employment.

Making it personal:

1. The chart on the inside front cover shows where you can find each Steps to Success section. Reread these techniques for success. Which of the suggestions do you think will help you the most? Make a list of these suggestions.

2. Based on this list, write out a study plan that you will follow for the remainder of the semester.

3. Discuss this plan with your instructor or counselor to see if they have any other suggestions.

Now follow this plan for the remainder of the semester! ▼

6.4 Exercises

MyMathLab®

Watch the videos
in MyMathLab

Download the
MyDashBoard App

Verbal and Writing Skills, Exercises 1–2

1. Jessie incorrectly factored $6x - 12$ as follows: $6(x + 2)$. What did Jessie do wrong?

2. Explain why the following polynomial is not factored completely. $10x^2y^3 + 20x^3y^2 = 5xy(2xy^2 + 4x^2y)$

3. For 8 and 12:

(a) State the common factors.

(b) State the GCF.

4. For 9 and 18:

(a) State the common factors.

(b) State the GCF.

Find the GCF for each set of numbers.

5. 4, 16

6. 6, 18

7. 18, 27

8. 12, 20

9. 6, 9, 15

10. 8, 10, 12

11. 10, 15, 20

12. 12, 18, 24

13. For the polynomial $a^3bc + a^6c$:

(a) What variables are common to the terms?

(b) State the smaller power on each of these variables.

(c) State the greatest common factor of the polynomial.

14. For the polynomial $x^4yz - x^2z$:

(a) What variables are common to the terms?

(b) State the smaller power on each of these variables.

(c) State the greatest common factor of the polynomial.

Find the GCF for each expression.

15. $xy^2 + xy^3$

16. $mn^3 + mn^4$

17. $a^2b^5 + a^3b^4$

18. $x^3y^4 + x^2y^5$

19. $a^3bc^2 + ac^3$

20. $x^2yz^3 + xz^2$

21. $x^3yz^3 + xy^4$

22. $a^2bc^3 + ab^3$

Fill in the missing numbers or variables needed to factor each expression.

23. $9x + 15 = \boxed{}(3x + 5)$

24. $4x + 16 = \boxed{}(x + 4)$

25. $10xy^2 - 15y = \boxed{}(2xy - 3)$

26. $6x^2y - 12x = \boxed{}(xy - 2)$

27. $2x + 4 = 2(\boxed{} + \boxed{})$

28. $3x + 9 = 3(\boxed{} + \boxed{})$

29. $6x^2 - 3x = 3x(\boxed{} - \boxed{})$

30. $14y^2 - 21y = 7y(\boxed{} - \boxed{})$

31. $18x^3 - 3x^2 - 9x = 3x(\boxed{} - \boxed{} - \boxed{})$

32. $10y^3 + 4y^2 - 8y = 2y(\boxed{} + \boxed{} - \boxed{})$

Fill in the missing + or − sign.

33. (a) $2x - 10 = 2(x \boxed{} 5)$

(b) $2x + 10 = 2(x \boxed{} 5)$

34. (a) $3y - 6 = 3(y \boxed{} 2)$

(b) $3y + 6 = 3(y \boxed{} 2)$

Factor. Check by multiplying.

35. $3a - 6$

36. $7x - 14$

37. $5y + 5$

38. $9x + 9$

39. $10a + 4b$

40. $6x + 10y$

41. $15m + 3n$

42. $5a + 25b$

43. $7x + 14y + 21$

44. $6a + 24b + 30$

45. $8a + 18b - 6$

46. $15x + 20y - 10$

47. $2a^2 - 4a$

48. $15y^2 - 3y$

49. $4ab - b^2$

50. $5xy - y^2$

51. $5x + 10xy$

52. $9x + 18xy$

53. $7x^2y - 14xy$

54. $8a^2b - 16ab$

55. $12a^2b - 6a^2$

56. $15ab^3 - 5b^3$

57. $3x^2 - 9x + 18$

58. $2x^2 - 8x + 12$

Mixed Practice

Factor and check your answer.

59. $4x^2 + 8x^3$

60. $3y^3 + 9y^2$

61. $2x^2y + 4xy$

62. $3a^2b + 6ab$

63. $4y + 2$

64. $10x + 5$

65. $15a - 20$

66. $9b - 15$

67. $5x - 10xy$

68. $9x - 18xy$

69. $9xy^3 - 3xy$

70. $4xy^2 - 2xy$

71. $6x - 3y + 12$

72. $5a + 15b - 10$

73. $4x^2 + 8x - 4$

74. $2x^2 + 6x - 2$

One Step Further

Factor and check your answer.

75. $2x^3y^3 - 8x^2y^2$

76. $5x^3y^3 - 10x^2y^2$

77. $6x^2y + 2xy + 4x$

78. $8xy^2 + 4xy + 2y$

To Think About

When factoring a polynomial whose first coefficient is negative, we can either factor out a negative or a positive common factor. The way we factor depends on the situation.

79. Multiply.

(a) $-2(x - 5y)$

(b) $2(-x + 5y)$

(c) What can you say about the results of (a) and (b)?

80. Factor $-3x + 9$:

(a) Using -3 as the GCF.

(b) Using 3 as the GCF.

Cumulative Review

Find the least common denominator of each set of fractions.

81. [5.3.2] $\dfrac{2}{3}, \dfrac{1}{4}, \dfrac{5}{2}$

82. [5.3.2] $\dfrac{3}{4}, \dfrac{4}{2}, \dfrac{1}{5}$

83. [5.3.2] $\dfrac{1}{2x}, \dfrac{7}{x}$

84. [5.3.2] $\dfrac{2}{x}, \dfrac{3}{5x}$

85. **[5.6.1]** *Rainfall Measured* A rain gauge collected $2\frac{2}{3}$ inches of rain in January and again in March and collected $3\frac{1}{4}$ inches in February. If the normal rainfall for the four months from January through April is 10 inches, how much rain must fall in April for this normal rainfall figure to be achieved?

86. **[4.6.4]** *Potato Salad Servings* Louise ordered 45 pounds of potato salad for her wedding. If one serving size is estimated to be $\frac{1}{3}$ pound, will Louise have enough potato salad for the 125 guests?

Quick Quiz 6.4

1. Find the GCF.

(a) $12, 20, 36$

(b) $x^2yz^2 - x^2y^2$

2. Factor. $4x^2 - 10y + 2$

3. Factor. $5ab^2 - 15ab$

4. **Concept Check** For the expression $12xy + 16x$

(a) Is xy part of the GCF? Why or why not?

(b) State the GCF.

(c) Factor $12xy + 16x$.

Chapter 6 Organizer

Topic and Procedure	Examples	✏ You Try It
Adding polynomials, p. 347 To add two polynomials, we add like terms.	Perform the operations indicated. $(-6x^2 + 9x - 7) + (2x^2 - 5x - 4)$ $= -6x^2 + 2x^2 + 9x - 5x - 7 - 4$ $= -4x^2 + 4x - 11$	1. Perform the operations indicated. $(-5x^2 + 8x - 3) + (3x^2 - 4x - 2)$
Finding the opposite of a polynomial, p. 348 When a negative sign precedes parentheses, we remove parentheses and change the sign of *each* term inside the parentheses.	Simplify. $-(3a - 5b + 8c)$ a − sign in front of parentheses $-(3a - 5b + 8c) = -3a + 5b - 8c$ We change the sign of each term.	2. Simplify. $-(6y - 4z + 7)$
Subtracting polynomials, pp. 348–349 To subtract two polynomials, change the signs of all terms in the second polynomial and then add.	Perform the operations indicated. $(5x^2 - 7x - 9) - (3x^2 - 4x - 2)$ $= 5x^2 - 7x - 9 - 3x^2 + 4x + 2$ $= 2x^2 - 3x - 7$	3. Perform the operations indicated. $(7x^2 - 4x - 3) - (4x^2 - 2x - 5)$
Multiply a monomial times a polynomial, p. 353 Use the distributive property and multiply each term of the polynomial by the monomial.	Multiply. $-3x(5x - 3y + 8)$ $-3x(5x - 3y + 8) = -15x^2 + 9xy - 24x$	4. Multiply. $-2a(4a - 5b + 6)$
Multiply a binomial times a trinomial, p. 354 When we multiply a binomial times a trinomial, we multiply each term of the binomial times the trinomial.	Multiply. $(3x + 2)(4x^2 - 2x + 5)$ We multiply $3x(4x^2 - 2x + 5)$. $= 3x \cdot 4x^2 + 3x \cdot (-2x) + 3x \cdot 5$ $= 12x^3 - 6x^2 + 15x$ We multiply $2(4x^2 - 2x + 5)$. $= 2 \cdot 4x^2 + 2 \cdot (-2x) + 2 \cdot 5$ $= 8x^2 - 4x + 10$ We add like terms. $12x^3 - 6x^2 + 15x + 8x^2 - 4x + 10$ $= 12x^3 + 2x^2 + 11x + 10$	5. Multiply. $(4x + 3)(2x^2 - x + 3)$
Multiplying binomials using FOIL, pp. 354–355 1. Multiply the *First* terms: F 2. Multiply the *Outer* terms: O 3. Multiply the *Inner* terms: I 4. Multiply the *Last* terms: L 5. Combine like terms.	Multiply using FOIL. $(x - 3)(x + 4)$ First Last F O I L $(x - 3)(x + 4) = x^2 + 4x - 3x - 12$ Inner Outer $= x^2 + x - 12$ Combine like terms.	6. Multiply using FOIL. $(x + 5)(x - 2)$
Factoring out the greatest common factor, p. 370 1. Write each term of the expression as a product of prime factors or variables in exponent form. 2. Identify common prime factors or variables that have the *smallest power*. 3. Multiply the common factors from step 2 to find the GCF. 4. Factor out the GCF.	Factor. $5x^2y - 10xy$ The greatest common factor is $5xy$. $5x^2y - 10xy = 5xy \cdot x - 5xy \cdot 2$ $= 5xy(x - 2)$ Check. $5xy(x - 2) \overset{?}{=} 5x^2y - 10xy$ $5xy(x - 2) = 5xy \cdot x - 5xy \cdot 2$ $= 5x^2y - 10xy$ ✓	7. Factor. $3ab - 9a^2b$

Topic and Procedure	Examples	You Try It
Write variable expressions when two or more quantities are being compared, p. 361 When two or more quantities are being compared, we let a variable represent the quantity *to which things are being compared*.	The first side of a triangle is 3 inches longer than the third; the second side is 9 inches shorter than triple the length of the third side. **(a)** Define the variable expression for the length of each side of the triangle. **(b)** Write the following phrase using math symbols and simplify: the length of the first side plus the length of the third side minus the length of the second side.	8. The height of a rectangular box is triple the width. The length is 3 inches shorter than double the width. **(a)** Define the variable expression for the length of each side of the rectangular box. **(b)** Write the following phrase using math symbols and simplify: the height plus the width minus the length.

(a) We are comparing the first and second side *to the third side*, so we let the variable represent the length of *the third side*.

$$\text{length of third side} = x \qquad \text{length of first side} = x + 3 \qquad \text{length of second side} = 3x - 9$$
$$\text{(3 inches longer than the third)} \qquad \text{(9 inches shorter than triple the third)}$$

(b) The first side plus the third side minus the second side.

$$(x + 3) + x - (3x - 9) \qquad \text{Translate.}$$
$$= 2x + 3 - (3x - 9) \qquad (x + 3) + x = x + x + 3 = 2x + 3$$
$$= 2x + 3 - 3x + 9 \qquad \text{Change the signs of all terms inside}$$
$$= -x + 12 \qquad \text{the parentheses and simplify.}$$

Chapter 6 Review Problems

Section 6.1

Identify the terms of each polynomial.

1. $2x^2 + 5x - 3z^3 + 4$

2. $a^4 - 2b^2 - 3b - 4$

Simplify.

3. $-(2a - 3)$

4. $-(-6x + 4y - 2)$

Perform the operations indicated.

5. $(-3x - 9) + (5x - 2)$

6. $(4x + 8) - (8x + 2)$

7. $(9a^2 - 3a + 5) - (-4a^2 - 6a - 1)$

8. $(-4x^2 - 3) - (3x^2 + 7x + 1) + (-x^2 - 4)$

9. $2(-2x^2 + 2) - (3x^2 + 5x - 6)$

10. $4x - (x^2 + 2x) + 3(3x^2 - 6x + 4)$

Section 6.2 *Multiply.*

11. $-4(6x^2 - 8x + 5)$

12. $-2y(y - 6)$

13. $3x(9x - 3y + 2)$

14. $-5n(-4n - 9m - 7)$

15. $4x^2(x^4 - 4)$

16. $x^4(x^5 - 2x - 3)$

17. $(z + 4)(5z)$

18. $(y + 10)(-6y)$

19. $(x^3 - 6x)(4x^2)$

20. $(x - 2)(2x^2 + 3x - 1)$

21. $(y + 5)(3y^2 - 2y + 3)$

22. $(y - 1)(-3y^2 + 4y + 5)$

23. $(2x + 3)(x^2 + 3x - 1)$

Use the FOIL method to multiply.

24. $(x + 2)(x + 4)$ **25.** $(y + 4)(y - 7)$ **26.** $(x - 2)(3x + 4)$ **27.** $(x - 3)(5x - 6)$

28. **Company Profit** A company's profit for the third quarter is $22,300 more than the profit for the first quarter. Define variable expressions for the profits for the first and third quarters.

▲ 29. **Geometry** The width of a field is 22 feet shorter than the length. Define variable expressions for the length and width of the field.

▲ 30. **Geometry** The measure of $\angle a$ is 30° more than the measure of $\angle b$. The measure of $\angle c$ is twice the measure of $\angle b$. Define variable expressions for the measures of $\angle a$, $\angle b$, and $\angle c$.

31. **Floral Bouquet** A floral shop puts three times as many carnations as roses in a popular friendship bouquet. They put five more lilies in the bouquet than roses. Define the variable expressions for the numbers of carnations, roses, and lilies used in the bouquet.

32. **Wage Comparison** Phoebe's salary is $145 more than Erin's salary. Kelly's salary is $60 less than Erin's salary.

   (a) Define variable expressions for the salaries for Phoebe, Erin, and Kelly.

   (b) Write the following phrase using math symbols: Erin's salary plus Phoebe's salary minus Kelly's salary.

   (c) Simplify the expression from part **(b)**.

33. **Eye Color** In a first-period history class at a local school, the number of children with brown eyes is seven more than the number of children with blue eyes. There are nine fewer children with green eyes than children with blue eyes.

   (a) Define the variable expressions for the numbers of children with brown, blue, and green eyes.

   (b) Write the following phrase using math symbols: The number of children with blue eyes plus the number of children with brown eyes minus the number of children with green eyes.

   (c) Simplify the expression from part **(b)**.

▲ 34. **Geometry** The length of the second side of a triangle is double the first. The third side of the triangle is 10 inches longer than the first.

   (a) Define the variable expression for each side of the triangle.

   (b) Write the following phrase using math symbols: The sum of all sides of the triangle.

   (c) Simplify the expression from part **(b)**.

▲ 35. **Geometry** The length of a box is 7 inches longer than the width. The height is 4 inches shorter than three times the width.

   (a) Define the variable expressions for the length, width, and height of the box.

   (b) Write the following phrase using math symbols: The sum of the height, width, and length.

   (c) Simplify the expression from part **(b)**.

Section 6.4

Find the GCF for each of the following.

36. 14, 21   **37.** 6, 21   **38.** 25, 45   **39.** 18, 36

40. 8, 14, 18   **41.** 12, 16, 20   **42.** $a^2bc + ab^3$ **43.** $xy^3z + x^2y^2$

Factor.

44. $6x - 14$ **45.** $5x + 15$ **46.** $4a + 12b$ **47.** $3y - 9z$

48. $6xy^2 - 12xy$ **49.** $8a^2b - 16ab$ **50.** $10x^3y + 5x^2y$

51. $4y^3 - 6y^2 + 2y$ **52.** $3a - 6b + 12$ **53.** $2x + 4y - 10$

How Am I Doing? Chapter 6 Test

Write the answers.

1. Identify terms of the polynomial. $x^2y - 2x^2 + 3y - 5x$

1. _____ ☐

2. Simplify. $-(4x - 2y - 6)$

2. _____ ☐

Perform the operations indicated.

3. $(-5x + 3) + (-2x + 4)$

3. _____ ☐

4. $(4y + 5) - (2y - 3)$

4. _____ ☐

$^{M}\!c$ **5.** $(-7p - 2) - (3p + 4)$

5. _____ ☐

6. $(4x^2 + 8x - 3) + (9x^2 - 10x + 1)$

6. _____ ☐

7. $(-6m^2 - 3m - 8) - (6m^2 + 3m - 4)$

7. _____ ☐

8. $(x^2 - x + 7) + (-2x^2 + 4x + 6) - (x^2 + 8)$

8. _____ ☐

9. $3x - 2(7x^2 + 2x - 1) - (3x^2 + 8x - 2)$

9. _____ ☐

Multiply.

10. $-7a(2a + 3b - 4)$

10. _____ ☐

11. $-2x^3(4x^2 - 3)$

11. _____ ☐

12. $(x + 5)(x + 9)$

12. _____ ☐

^{MC} **13.** $(x + 3)(x - 2)$

14. $(2x + 1)(x - 3)$

15. $(3x^3 - 1)(-4x^4)$

16. $(y - 3)(4y^2 + 2y - 6)$

^{MC} ▲ **17.** The width of a piece of wood is three inches shorter than the length. Define the variable expressions for the length and width of the piece of wood.

▲ **18.** The second side of a triangle is 6 inches longer than the first. The third side is 2 inches shorter than two times the first. Define the variable expression for the length of each side of the triangle.

19. Jason received 3000 fewer votes than Lena in an election. Nhan received 5100 more votes than Lena in the election.

  (a) Define the variable expressions for the numbers of votes received by Jason, Lena, and Nhan.

  (b) Write the following phrase using math symbols: The number of votes received by Nhan plus the number of votes received by Lena minus the number of votes received by Jason.

  (c) Simplify the expression from part **(b)**.

Find the GCF.

20. $9, 21$

21. $8, 16, 20$

22. $x^2yz + x^3z$

Factor.

23. $3x + 12$

24. $7x^2 - 14x + 21$

^{MC} **25.** $2x^2y - 6xy^2$

13.	☐
14.	☐
15.	☐
16.	☐
17.	☐
18.	☐
19. (a)	☐
(b)	☐
(c)	☐
20.	☐
21.	☐
22.	☐
23.	☐
24.	☐
25.	☐
Total Correct:	☐

MATH COACH

Mastering the skills you need to do well on the test.

Students often make the same types of errors when they do the Chapter 6 Test. Here are some helpful hints to keep you from making these common errors on test problems.

Subtracting Polynominals—Problem 5

Perform the operations indicated. $(-7p - 2) - (3p + 4)$

> **Helpful Hint** Be careful when subtracting expressions. Students often forget to change the sign of *each term* in the second polynomial. Take the extra time and check your work to be sure you did not make this error.

After removing parentheses, did you get $-3p + 4$ as the last two terms?

Yes ____ No ____

If you answered Yes, you forgot to change the sign of every term in the second polynomial. Stop now and make this correction.

Did you get $-7p - 3p = -10p$ and $-2 - 4 = -6$ after removing parentheses?

Yes ____ No ____

If you answered No, consider grouping like terms together before subtracting. Then go back and complete this step again.

If you answered any of Problem 5 incorrectly, go back and rework the problem using these suggestions.

Multiplying Binomials Using FOIL—Problem 13   Multiply. $(x + 3)(x - 2)$

> **Helpful Hint** To help with accuracy, try to draw arrows when following the FOIL method. Pay particular attention to the sign of each number to avoid errors.

Using FOIL, did you get $\mathbf{F} = x^2$, $\mathbf{O} = -2x$, $\mathbf{I} = +3x$, and $\mathbf{L} = -6$?

Yes ____ No ____

If you answered No, check to see if you made a sign error or used the FOIL method incorrectly.

Is the middle term of your answer equal to $+x$?

Yes ____ No ____

If you answered No, look at your work for adding the inner plus the outer terms. It may help with accuracy to write out the step $-2x + 3x = x$.

Now go back and rework the problem using these suggestions.

Need help? Watch the **MATH COACH** videos in MyMathLab® or on You Tube™.

Writing Variable Expressions When Comparing Two or More Quantities—Problem 17

The width of a piece of wood is 3 inches shorter than the length. Define the variable expressions for the length and width of the piece of wood.

Helpful Hint It is a good idea to let the variable represent the quantity to which things are being compared. When writing expressions, remember that math symbols are not always written in the same order as they are read in the statement.

Did you let the variable represent the length?

Yes ____ No ____

If you answered No, stop now and make this correction.

Did you write the variable expression: width $= 3 - L$?

Yes ____ No ____

If you answered Yes, go back and reread the problem carefully. Notice that the phrase "is shorter than the length" means 3 taken away from L.

Remember that the question asks for the variable expressions for both length and width, so you will need to write both of these answers.

If you answered any of Problem 17 incorrectly, go back and rework the problem using these suggestions.

Factoring Out the GCF from a Polynomial—Problem 25   Factor. $2x^2y - 6xy^2$

Helpful Hint
- Write each number as a product using *only* prime factors. Then align all common prime factors.
- When forming the GCF, choose factors that are common to every term and use the smallest power on these common factors.
- After factoring, double-check to see if the product is factored completely.

Did you factor each term as follows? $2x^2y = 2^1 \cdot \quad x^2 \cdot y^1$

$$6xy^2 = 2^1 \cdot 3^1 \cdot x^1 \cdot y^2$$

Notice that common factors are aligned. ↑   ↑ ↑

Yes ____ No ____

If you answered No, go back and perform this step again.

Did you factor $2x^2y - 6xy^2$ into $2(x^2y - 3xy^2)$?

Yes ____ No ____

If you answered Yes, the expression is *not factored completely*. To avoid this situation, you must find the correct GCF. Align common factors as shown in the first step. Then you will see that 2, x, and y are common to both terms and therefore part of the GCF.

Did you choose $2x^2y^2$ as the GCF?

Yes ____ No ____

If you answered Yes, check your work again. You *did not use* the *smallest* power on each common factor: $2^1, x^1$, and y^1.

Now go back and rework the problem using these suggestions.

Need more help? Look for section examples marked with $\mathbb{MC}$ to review.

Cumulative Test for Chapters 1–6

This test provides a comprehensive review of the key objectives for Chapters 1–6.

1. Sheila is preparing for her first semester of college and must determine if she has enough in her savings account to pay for all her college expenses. Since her scholarship pays for part of her tuition, she will pay only $560 for her first semester's tuition. College service fees are $35 and her books will cost $410. She must also buy a parking permit for $30 and lab materials for $22. She is on the track team and needs to buy a uniform for $120. Sheila has $1100 in her savings account. Does she have enough money to pay her college expenses? How much extra money will she have, or how much more money will she need for her college expenses?

2. An amusement park charges $10 for adults, $6 for children, and $7 for senior citizens. Alexandra and Stanley are sharing the cost to treat their relatives to the park. All together there will be five adults, seven children, and four senior citizens. How much will Stanley have to pay for his share of the cost?

Combine like terms.

3. $5x - 3x + x + 5$

4. $-8r + 3 - 5r - 8$

Perform the operations indicated.

5. $8^2 - 10 + 4$

6. $-4 + 2^3 - 9$

7. $7 - 24 \div 6(-2)^2 - 3$

8. $3 - 12 \div (-2) + 4^2$

9. $(-10x^2)(5x)$

10. $(3x^2)(x^3)(x)$

11. $5x(x^2 + 3)$

12. Simplify. $\dfrac{90n^2}{54n}$

13. Simplify. $\dfrac{8a^3}{32a^5}$

14. Simplify. $\left(\dfrac{x}{3}\right)^3$

Find the area of the following.

15. (a) A square with sides of 12 inches.
    (b) A parallelogram with a base of 10 feet and a height of 15 feet.

16. Find the volume of the rectangular solid with $L = 5$ yards, $W = 3$ yards, and $H = 4$ yards.

17. Write $\dfrac{37}{4}$ as a mixed number.

18. Write $\dfrac{40}{3}$ as a mixed number.

19. Express 550 as a product of prime factors.

20. Lester traveled 310 miles in 6 hours. What is the unit rate in miles per hour?

21. If a 4-ounce serving of yogurt contains 9 grams of carbohydrates, how many grams of carbohydrates does 12 ounces of yogurt contain?

22. Evaluate $\dfrac{n^2 - 6}{m}$ for $n = -4$ and $m = 2$.

23. Solve for x.
   (a) $x + 45 = -2$
   (b) $6x - 5x - 9 = 34$

24. Solve for n.
   (a) $\dfrac{n}{-8} = 6$
   (b) $-9n = 99$

25. Multiply and simplify. $\dfrac{2x^4}{5x} \cdot \dfrac{10x^2}{4}$

Divide.

26. $-\dfrac{1}{6} \div \left(-\dfrac{2}{3}\right)$

27. Find the least common denominator of the fractions. $\dfrac{1}{12}, \dfrac{1}{28}$

28. Frank had $12\frac{1}{2}$ gallons of gasoline in his car before leaving for work one morning. When he returned home that evening, he had $9\frac{1}{3}$ gallons of gas left. How many gallons of gas did Frank use that day?

29. Jaci bought $5\frac{1}{4}$ pounds of Halloween candy. If she brings $\frac{1}{3}$ of the candy to work, how much will she have left?

30. Perform the operations indicated. $(-8x^2 + 3x - 6) - (x^2 + 5x)$

Multiply.

31. $(x + 1)(3x^2 - 2x + 6)$

32. $(x + 2)(x + 6)$

33. $(2x + 7)(x - 3)$

34. A building is 4 feet higher than twice the height of a tree. Define variable expressions for the height of the building and the height of the tree.

Factor.

35. $9a + 18b + 9$

36. $12x^2y + 6x^2$

19.	_____
20.	_____
21.	_____
22.	_____
23. (a)	_____
(b)	_____
24 . (a)	_____
(b)	_____
25.	_____
26.	_____
27.	_____
28.	_____
29.	_____
30.	_____
31.	_____
32.	_____
33.	_____
34.	_____
35.	_____
36.	_____

Using the mathematics covered in this chapter enables business owners to make wise business decisions. Determining cost, profit, and the break-even point is an important part of operating a successful business. To find this information, we solve cost, profit, and revenue equations so we can maximize profit. The mathematics covered in this chapter will help you with many other everyday decisions. Can you compare and make decisions about salary versus commission income or alternative remodeling projects? After studying the mathematics in this chapter, you will have the knowledge to do this.

Solving Equations

7.1 Solving Equations Using One Principle of Equality

① Solving Equations Using the Addition Principle of Equality

In Chapter 3 we learned how to solve equations of the form $x - a = c$, $x + a = c$, and $ax = c$ using the addition and division principles of equality. In Chapter 5 we used the multiplication principle to solve equations of the form $\frac{x}{a} = c$. In this section we will study all four forms, $x - a = c$, $x + a = c$, $ax = c$, and $\frac{x}{a} = c$, within one section and focus on how to determine which principle to use. This skill will prepare you for the next section, which requires that you use more than one principle of equality to solve equations of the form $ax + b = c$.

Before we begin our discussion on solving equations, we restate in words the principle of equality presented in Chapter 3.

> **ADDITION PRINCIPLE OF EQUALITY**
>
> If the *same number* or *variable term* is added to both sides of the equation, the results on each side are equal in value.

We can restate it in symbols this way.

> For any numbers a, b, and c, if $a = b$, then $a + c = b + c$.

It is important to keep in mind that an equation is like a balanced seesaw. Whatever we do to one side of the equation, we must do to the other side to maintain the balance.

The goal when solving an equation is to get the variable alone on one side of the equation, that is, to get the equation in the form $x = some\ number$ or $some\ number = x$.

EXAMPLE 1 Solve $x - 2 = -42$ and check your solution.

Solution We want an equation of the form $x = some\ number$. To obtain this we think, "What can we add to each side of the equation so that $x - 2$ becomes simply x?" We add the opposite of -2, or $+2$, to both sides.

$$
\begin{array}{ll}
x - 2 = -42 & \text{We want to get } x \text{ alone.} \\
\underline{+\quad +2 \qquad +2} & \text{We add the opposite of } -2, \text{ or } +2, \text{ to both sides.} \\
x + 0 = -40 & -2 + 2 = 0; -42 + 2 = -40 \\
x = -40 & x + 0 = x
\end{array}
$$

$$
\begin{array}{ll}
Check. & x - 2 = -42 \\
& -40 - 2 \overset{?}{=} -42 \\
& \qquad -42 = -42 \; \checkmark
\end{array}
$$

Student Practice 1 Solve $x - 6 = -56$ and check your solution.

EXAMPLE 2 Solve. $-12 - 15 = 9x + 15 - 8x - 6$

Solution First we simplify each side of the equation, and then we solve for x.

$$
\begin{array}{ll}
-12 - 15 = 9x + 15 - 8x - 6 & \\
-27 = 9x + 15 - 8x - 6 & \text{We calculate: } -12 - 15 = -12 + (-15) = -27. \\
-27 = 9x + 15 - 8x - 6 & \\
-27 = x + 9 & \text{We simplify: } 9x - 8x = x, \text{ and } 15 - 6 = 9.
\end{array}
$$

Continued on next page

Student Learning Objectives

After studying this section, you will be able to:

① Solve equations using the addition principle of equality.

② Solve equations using the multiplication principle of equality.

③ Solve equations using the division principle of equality.

We have a balanced seesaw.

We add the same amount to both sides and the seesaw remains balanced.

NOTE TO STUDENT: *Fully worked-out solutions to all of the Student Practice problems can be found at the back of the text starting at page SP-1.*

Now we solve for x.

$$-27 = x + 9 \qquad \text{The opposite of } +9 \text{ is } -9.$$
$$\underline{+\qquad -9 \qquad -9} \qquad \text{We add } -9 \text{ to both sides of the equation.}$$
$$-36 = x$$

We leave the check to the student.

Student Practice 2 Solve. $-11 - 17 = 5y + 12 - 4y - 1$.

② Solving Equations Using the Multiplication Principle of Equality

We learned how to use the multiplication principle in Chapter 5. We restate this principle in words for your review.

> **MULTIPLICATION PRINCIPLE OF EQUALITY**
>
> If both sides of an equation are *multiplied* by the same nonzero number, the results on each side are equal in value.

We can restate it in symbols as follows.

> For any numbers a, b, and c with c not equal to 0, if $a = b$, then $a \cdot c = b \cdot c$.

Our goal when solving equations is to get the variable alone on one side of the equation. We must remember that whatever we do to one side of the equation, we must do to the other side of the equation.

EXAMPLE 3 Solve $\dfrac{y}{-5} = -10 + 2^3$ and check your solution.

Solution First we simplify each side of the equation, and then we solve for y.

$$\frac{y}{-5} = -10 + 2^3$$

$$\frac{y}{-5} = -2 \qquad \text{We simplify: } 2^3 = 8 \text{ and } -10 + 8 = -2.$$

To solve the equation $\dfrac{y}{-5} = -2$, we can think, "What can we do to the left side of the equation so that $\dfrac{y}{-5}$ becomes simply y?" Since y is *divided by* -5, we can *undo* the division and obtain y alone by *multiplying by* -5.

$$\frac{-5 \cdot y}{-5} = -2 \cdot (-5) \qquad \text{We undo the division by multiplying by } -5.$$

$$y = 10 \qquad \text{We simplify: } \frac{-5y}{-5} = y \text{ and } -2 \cdot (-5) = 10.$$

$$\text{Check.} \qquad \frac{y}{-5} = -10 + 2^3$$

$$\frac{10}{-5} \stackrel{?}{=} -10 + 2^3 \qquad \text{We replace } y \text{ with 10.}$$

$$-2 = -2 \ \checkmark$$

Student Practice 3 Solve $\dfrac{y}{-3} = 2^2 - 8$ and check your solution.

Recall from Chapter 5 that to solve an equation such as $\frac{2}{3}x = 5$, we multiply both sides of the equation by the reciprocal of $\frac{2}{3}$.

$$\frac{2}{3}x = 5$$

$$\frac{3}{2} \cdot \frac{2}{3}x = 5 \cdot \frac{3}{2} \qquad \text{We multiply both sides by } \frac{3}{2} \text{, the reciprocal of } \frac{2}{3}.$$

$$x = \frac{15}{2}$$

Be sure to simplify each side of the equation as your first step before solving for the variable.

EXAMPLE 4   Solve. $\frac{3}{4}a = 5^2 + 2$

Solution   First we simplify $5^2 + 2$, and then we solve for a.

$$\frac{3}{4}a = 5^2 + 2$$

$$\frac{3}{4}a = 27 \qquad \text{Simplify: } 5^2 + 2 = 25 + 2 = 27.$$

$$\frac{4}{3} \cdot \frac{3}{4}a = 27 \cdot \frac{4}{3} \qquad \text{Multiply both sides by } \frac{4}{3}.$$

$$a = 36 \qquad \frac{27 \cdot 4}{3} = \frac{\overset{1}{\cancel{3}} \cdot 9 \cdot 4}{\underset{1}{\cancel{3}}} = 36$$

Check. $$\frac{3}{4}a = 5^2 + 2$$

$$\frac{3}{4} \cdot 36 \overset{?}{=} 5^2 + 2 \qquad \text{Replace } a \text{ with 36.}$$

$$27 = 27 \ \checkmark \qquad \frac{3}{4} \cdot 36 = \frac{3 \cdot \overset{1}{\cancel{4}} \cdot 9}{\underset{1}{\cancel{4}}} = 27$$

Student Practice 4   Solve. $\frac{4}{5}a = 3^2 + 7$

③ Solving Equations Using the Division Principle of Equality

In Chapter 3 we saw that to solve an equation in the form $3x = -27$, we can use the division principle of equality, which states that we can divide both sides of the equation by 3. We restate this principle in words for your review.

DIVISION PRINCIPLE OF EQUALITY

If both sides of the equation are divided by the same nonzero number, the results on each side are equal in value.

The division principle of equality is an alternate form of the multiplication principle. This is because dividing by c is the same as multiplying by $\frac{1}{c}$. We will explore this in the To Think About on page 391.

Since division by zero is not defined, we put a restriction on the number by which we are dividing. We restate this principle in this way.

For any numbers a, b, and c, with c not equal to 0, if $a = b$, then $\dfrac{a}{c} = \dfrac{b}{c}$.

There are two things we must remember when we solve an equation.

1. First, we must be sure that each side of the equation is simplified.

2. Next, we use the principles of equality to get the variable alone on one side of the equation; that is, we find an equivalent equation of the form $x = some\ number$ or $some\ number = x$.

EXAMPLE 5 Solve $\dfrac{-20}{5} = -2(5x) + 7x$ and check your solution.

Solution We begin by simplifying each side of the equation.

$$\frac{-20}{5} = -2(5x) + 7x$$

$$-4 = -2(5x) + 7x \qquad \text{Simplify: } \frac{-20}{5} = -4.$$

$$-4 = -10x + 7x \qquad \text{Multiply: } -2(5x) = -10x.$$

$$-4 = -3x \qquad \text{Add: } -10x + 7x = -3x.$$

Now that both sides of the equation are simplified, we use the division principle of equality to transform the equation into the form $some\ number = x$.

$$\frac{-4}{-3} = \frac{-3x}{-3} \qquad \text{Dividing by } -3 \text{ on both sides } undoes \text{ the multiplication by } -3.$$

$$\frac{4}{3} = x \qquad \text{Simplify: } \frac{-4}{-3} = \frac{4}{3} \text{ and } \frac{-3x}{-3} = x.$$

If we leave the solution as an improper fraction, it will be easier to check it in the original equation.

$$Check. \qquad \frac{-20}{5} = -2(5\,x) + 7\,x$$

$$\frac{-20}{5} \stackrel{?}{=} -2\left(5 \cdot \frac{4}{3}\right) + 7 \cdot \frac{4}{3}$$

$$-4 \stackrel{?}{=} \frac{-40}{3} + \frac{28}{3} \qquad -2\left(5 \cdot \frac{4}{3}\right) = -2\left(\frac{20}{3}\right) = \frac{-40}{3};\ \ 7 \cdot \frac{4}{3} = \frac{28}{3}$$

$$-4 \stackrel{?}{=} \frac{-12}{3}$$

$$-4 = -4 \ \checkmark$$

Student Practice 5 Solve $\dfrac{16}{-4} = 4x + 3(-3x)$ and check your solution.

You may have to rewrite an equation so that a coefficient of 1 or -1 is obvious: $-x = -1x$ or $x = 1x$. With practice you will be able to recognize the coefficient without actually rewriting it.

EXAMPLE 6 Solve. $-x = 36$

Solution

$-1x = 36$ Rewrite the equation: $-x$ is the same as $-1x$.

$\dfrac{-1x}{-1} = \dfrac{36}{-1}$ Dividing by -1 on both sides *undoes* the multiplication by -1.

$x = -36$

Although checking the solution is not always illustrated in the examples, you should always do so.

Student Practice 6 Solve. $-x = 25$

TO THINK ABOUT: The Multiplication Principle of Equality Can you think of another way to solve the equation $-x = 36$ in Example 6? What other operation can we do to transform the equation to the form $x = some\ number$? Try multiplying both sides of the equation by -1 and see what happens. Do you see why we can solve the equation this way?

Let's summarize a few facts to help you solve equations using the principles of equality.

- Subtraction *undoes* addition.
- Addition *undoes* subtraction.
- Division *undoes* multiplication.
- Multiplication *undoes* division.

👣 STEPS TO SUCCESS Preparing for Your Career

Open the door to a better-paying job. Studying mathematics sharpens your mind and helps you sort out and solve real-life situations that do not require the use of mathematics, thus opening the door to many jobs.

The world of technology requires the use of mathematics daily. Many vocational and professional areas—such as the fields of business, statistics, economics, psychology, finance, computer science, chemistry, physics, medicine, engineering, electronics, nuclear energy, banking, quality control, and teaching—require a certain level of expertise in mathematics. Those who want to work in these fields must be able to function at a given mathematical level. Those who cannot will not be able to enter these job areas.

Master the basics of this course. It will very likely help you advance to the career of your choice.

Keep in mind that mathematical thinking does not always require performing math calculations. Organizing and planning skills are enhanced when you study mathematics!

Making it personal:

1. Make a list of the types of jobs you plan to seek when you finish your education.

2. Talk with your instructor, counselor, or job placement director about the level and type of mathematics, as well as the planning and organizational skills, that are required for these jobs. ▼

Verbal and Writing Skills, Exercises 1–12

Fill in the blanks.

1. To solve the equation $y + 8 = -17$, we add the opposite of $+8$, which is ___, to both sides of the equation.

2. To solve the equation $a - 1 = -7$, we add the opposite of -1, which is ___, to both sides of the equation.

3. To solve the equation $x - 15 = 82$, we add ___ to both sides of the equation.

4. To solve the equation $y + 15 = 22$, we add ___ to both sides of the equation.

5. Before we solve the equation $7x + 2 - 6x - 9 = 2 - 8$, we must first _____ each side of the equation.

6. Before we solve the equation $4x + 2 - 3x - 1 = 10 - 11$, we must first _____ each side of the equation.

Fill in each blank with the missing phrase.

7. To solve the equation $-9y = -36$, we _____ on both sides of the equation.

8. To solve the equation $-12 = -3x$, we _____ on both sides of the equation.

9. To solve the equation $\dfrac{y}{-9} = 2$, we _____ on both sides of the equation.

10. To solve the equation $-12 = \dfrac{x}{-3}$, we _____ on both sides of the equation.

11. We _____ on both sides of the equation when we solve $14 = x + 3$.

12. We _____ on both sides of the equation when we solve $12 = x - 3$.

Solve and check your answer.

13. $x - 7 = -20$

14. $y - 5 = -12$

15. $a + 9 = -1$

16. $x + 12 = -4$

17. $-5 = x + 5$

18. $-2 = y + 3$

19. $7 - 9 = y - 4$

20. $8 - 11 = x - 4$

21. $7 - 12 = x - 6 + 3^2$

22. $8 - 13 = x - 2 + 3^2$

23. $-8 + 4^2 = a + 6 - 4$

24. $-5 + 3^2 = a + 5 - 7$

25. $12x - 1 - 11x - 1 = -5$

26. $8x + 7 - 7x + 2 = -3$

27. $12 - 16 = 4y + 8 - 3y$

28. $8 - 9 = 3x - 4 - 2x$

29. $6x - 6 - 5x + 1 = -2 + 4$

30. $11x + 12 - 10x - 1 = -4 + 11$

Solve and check your solution.

31. $-16 = \dfrac{x}{3}$

32. $3 = \dfrac{x}{-2}$

33. $14 = \dfrac{y}{-2}$

34. $-4 = \dfrac{a}{6}$

35. $\dfrac{a}{-4} = -6 + 9$

36. $\dfrac{y}{-3} = 9 - 11$

37. $\dfrac{x}{4} = -8 + 6$

38. $\dfrac{x}{5} = -6 + 2$

39. $\dfrac{y}{-2} = 4 + 2^2$

40. $\dfrac{x}{-1} = 5 + 4^2$

41. $\dfrac{3}{5}x = 9$

42. $\dfrac{5}{7}x = 15$

43. $\dfrac{-2}{3}a = 18$

44. $\dfrac{-5}{8}a = 15$

45. $\dfrac{7}{6}y = 4^2 + 5$

46. $\dfrac{3}{7}y = 5^2 - 4$

Solve and check your solution.

47. $3(2x) = 24$

48. $4(6x) = 48$

49. $-6 = 2(-5x)$

50. $-2 = 6(-4x)$

51. $8x + 4(4x) = 48$

52. $6x + 3(3x) = 45$

53. $6(-2x) - 3x = -30$

54. $5(-5x) - 4x = -29$

55. $\dfrac{-10}{2} = 3(4x) + 2x$

56. $\dfrac{6}{-3} = 2(3x) + 3x$

57. $\dfrac{21}{3} = 5x + 3(-4x)$

58. $\dfrac{32}{8} = 8x + 2(-2x)$

59. $-x = 9$

60. $-x = 16$

61. $-x = -4$

62. $-x = -6$

Mixed Practice *Solve.*

63. **(a)** $2x = -12$

(b) $\dfrac{x}{2} = -12$

(c) $x - 2 = -12$

(d) $x + 2 = -12$

64. **(a)** $5y = -20$

(b) $\dfrac{y}{5} = -20$

(c) $y - 5 = -20$

(d) $y + 5 = -20$

65. **(a)** $y - 10 = 9$

(b) $y + 10 = 9$

(c) $\dfrac{y}{-10} = 9$

(d) $-10y = 9$

66. **(a)** $x - 6 = 11$

(b) $x + 6 = 11$

(c) $\dfrac{x}{-6} = 11$

(d) $-6x = 11$

67. $-15 = a + 5$

68. $-2 = y + 7$

69. $5x + 2 - 4x = 9$

70. $7x + 3 - 6x = 4$

71. $\dfrac{x}{7} = -2 + 3^2$

72. $\dfrac{y}{9} = -6 + 2^2$

73. $-x = 12$

74. $-y = 8$

75. $\dfrac{20}{5} = 3x + 2(-6x)$

76. $\dfrac{12}{-6} = 4(2x) + 3x$

77. $\dfrac{6}{7}x = 3^2 + 3$

78. $\dfrac{3}{4}y = 2^2 + 5$

To Think About *As stated earlier, the division principle is another form of the multiplication principle. To see why this is true, complete exercise 79.*

79. (a) Solve $-2x = 8$ using the division principle.

(b) Solve $-2x = 8$ using the multiplication principle. (*Hint:* Multiply each side of the equation by the reciprocal of -2.)

(c) What can you say about the answers?

(d) Why do you think this is true?

A magic square is an array of numbers with a special property. In this type of array the sum of the numbers in each column, each row, and in each of the two diagonals is equal to the same number.

80. The sum of each row, each column, and each diagonal in the magic square is 15.

8	z	6
3	5	7
x	9	y

(a) Form an equation using the 1st column by writing the sum of 8, 3, and x equal to 15. Solve the equation for x. Then in the array below, replace x with the value of x.

(b) Complete the same process using the 2nd and 3rd columns to find the values of y and z.

(c) Check your answer by verifying that the sum of each diagonal is 15.

8		6
3	5	7
	9	

81. The sum of each row, each column, and each diagonal in the magic square is 18.

9	z	y
x	6	10
7	8	3

(a) Form an equation using the 1st column by writing the sum of 9, x, and 7 equal to 18. Solve the equation for x. Then in the array below, replace x with the value of x.

(b) Complete the same process using the 2nd and 3rd columns to find the values of y and z.

(c) Check your answer by verifying that the sum of each diagonal is 18.

9		
	6	10
7	8	3

82. Use the properties of magic squares to find the values of a, b, c, d, and e.

5	a	18
b	15	c
d	e	25

5		18
	15	
		25

▲ *The sum of the interior angles of a triangle is 180°. A right triangle is a special kind of triangle in which one angle is 90°. This angle is identified by the symbol* ⌐. *Find the unknown angle for the following right triangles.*

Interior angle

83.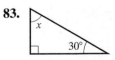
x 30°

84.
x 60°

▲ *Find the measure of each unknown angle.*

85.

86.

Cumulative Review *Express as a product of prime numbers.*

87. **[4.1.3]** 210

88. **[4.1.3]** 112

89. **[1.3.5]** *Earth and Sun* The distance between the sun and Earth varies because Earth travels around the sun in an orbit that has an elliptical (oval) shape. The shortest distance from Earth to the sun is approximately 91,400,000 miles, and the greatest distance is about 94,500,000 miles. What is the difference between the greatest and shortest distance?

Quick Quiz 7.1 *Solve and check your solution.*

1. $-3y + 4y + 8 = -9 + 5$

2. $\dfrac{x}{-5} - 3^2 = 7$

3. $2(-7b) + 8b = -24$

4. **Concept Check** Which of the following equations would you solve by dividing by 7 on both sides of the equation? Explain why this operation is used to solve for x.

(a) $x - 7 = -21$

(b) $\dfrac{x}{7} = -21$

(c) $7x = -21$

(d) $x + 7 = -21$

After studying this section, you will be able to:

① Solve equations using more than one principle of equality.

② Simplify and solve equations.

To solve equations such as $4x - 9 = 78$, we use more than one principle of equality. In this section we will see how to use the principles of equality to solve these more complex equations.

① Solving Equations Using More Than One Principle of Equality

When we solve an equation that requires using more than one principle of equality, we must be able to determine which principle to use first. We can use the following sequence of steps to solve an equation such as $3x + 6 = 9$.

> **PROCEDURE TO SOLVE AN EQUATION IN THE FORM**
> $ax + b =$ SOME NUMBER
> 1. First, use the addition principle to get the variable term ax alone on one side of the equation:
> $$ax = some\ number$$
> 2. Then apply the multiplication or division principle to get the variable x alone on one side of the equation:
> $$x = some\ number$$

Calculator

Checking Solutions to Equations

We can use a scientific calculator to check our solutions. To check that 5 is a solution to $3x - 1 = 14$, enter:

3 ⨯ 5 − 1 =

The display should read:

14

EXAMPLE 1 Solve $3x + 7 = 88$ and check your solution.

Solution First, we use the addition principle to get the variable term, $3x$, alone.

$$3x + 7 = 88$$
$$\underline{+\quad -7\quad -7}\quad \text{\textit{Add} } -7 \text{ to both sides of the equation.}$$
$$3x\quad = 81\quad \text{The variable term, } 3x, \text{ is alone.}$$

Then we apply the division principle to get the variable, x, alone.

$$3x = 81$$
$$\frac{3x}{3} = \frac{81}{3}\quad \text{\textit{Divide by} 3 on both sides of the equation.}$$
$$x = 27\quad \text{The variable, } x, \text{ is alone.}$$

Check.
$$3x + 7 = 88$$
$$3(27) + 7 \stackrel{?}{=} 88$$
$$81 + 7 \stackrel{?}{=} 88$$
$$88 = 88\ ✓$$

Student Practice 1 Solve $8x + 9 = 105$ and check your solution.

NOTE TO STUDENT: *Fully worked-out solutions to all of the Student Practice problems can be found at the back of the text starting at page SP-1.*

Remember that you should use the addition principle before you use the division or multiplication principle.

Mc **EXAMPLE 2**   Solve. $-6x - 4 = 74$

Solution   First, we must get the variable term, $-6x$, alone by using the addition principle.

$$
\begin{array}{rcl}
-6x - 4 &=& 74 \\
+\quad\;\; + 4 \quad +4 & & \quad\quad \textit{Add } 4 \text{ to both sides of the equation.} \\
\hline
-6x \quad\quad &=& 78 \quad\quad \textit{The variable term, } -6x, \text{ is alone.}
\end{array}
$$

Then we apply the division principle to get the variable, x, alone.

$$-6x = 78$$

$$\frac{-6x}{-6} = \frac{78}{-6} \quad \textit{Divide by } -6 \text{ on both sides of the equation.}$$

$$x = -13 \quad \text{The variable, } x, \text{ is alone.}$$

We leave the check to the student.

Student Practice 2   Solve.
$$-5m - 10 = 115$$

EXAMPLE 3   Solve $-3 = 6 - 4y$ and check your solution.

Solution   We use the addition principle to get the variable term, $-4y$, alone.

$$
\begin{array}{rcl}
-3 &=& 6 - 4y \\
+ \;\; -6 & & -6 \quad\quad \text{Add } -6 \text{ to both sides of the equation.} \\
\hline
-9 &=& \quad -4y
\end{array}
$$

Now we use the division principle to get y alone on the right side of the equation.

$$\frac{-9}{-4} = \frac{-4y}{-4} \quad \text{Divide by } -4 \text{ on both sides of the equation.}$$

$$\frac{9}{4} = y \quad\quad \frac{-9}{-4} = \frac{9}{4} \text{ and } \frac{-4y}{-4} = y.$$

Check.
$$
\begin{aligned}
-3 &= 6 - 4y \\
-3 &\overset{?}{=} 6 - 4\left(\frac{y}{4}\right) \\
-3 &\overset{?}{=} 6 - 9 \\
-3 &= -3 \;\checkmark
\end{aligned}
$$

Student Practice 3   Solve $-2 = 4 - 5x$ and check your solution.

② Simplifying and Solving Equations

As we have said, the process of solving an equation is easier if we simplify each side of the equation before we begin to solve it. Many students find that it is helpful to have a written procedure to follow when solving more-involved equations.

PROCEDURE TO SOLVE EQUATIONS

1. *Parentheses.* Remove any parentheses.
2. *Simplify each side of the equation.* Combine like terms and simplify numerical work.
3. *Isolate the ax term.* Use the addition principle to get *all* terms with the variable on *one side* of the equation and the numerical values on the other side.
4. *Isolate the x.* Use the multiplication or division principle to get the variable alone on one side of the equation.
5. *Check* your solution.

Sometimes there is a variable term on *both* sides of the equation. In this case, we must rewrite the equation so that all variable terms are on one side. Then we continue to work toward the form $x =$ *some number*.

EXAMPLE 4   Solve $5x - 3 = 6x + 2$ and check your solution.

Solution   First, we add $-6x$ to both sides of the equation so that all variable terms are on one side of the equation.

$$
\begin{array}{rcl}
5x - 3 &=& 6x + 2 \\
+\ -6x & & -6x \\
\hline
-x - 3 &=& 2
\end{array}
$$

Then, we solve the equation.

$$
\begin{array}{rcl}
-x - 3 &=& 2 \\
+\quad\ + 3 & & +3 \\
\hline
-x &=& 5
\end{array}
$$
Add 3 to both sides of the equation.

$$
\frac{-1x}{-1} = \frac{5}{-1}
$$
Write $-x$ as $-1x$, and divide by -1 on both sides.

$$
x = -5
$$

Another way to solve $-1x = 5$ is to multiply both sides of the equation by -1. $-1x = 5 \rightarrow -1(-1x) = -1(5) \rightarrow x = -5$. We leave the check to the student.

Student Practice 4   Solve $3x - 1 = 4x - 6$ and check your solution.

Understanding the Concept

Variables on Both Sides of an Equation   When a variable appears on both sides of an equation, does it matter which side is cleared of variable terms? No. To see why, let's look at the equation $9x + 1 = 7x - 4$.

Add $-7x$ to both sides:

$$
\begin{array}{rcl}
9x + 1 &=& 7x - 4 \\
+\ -7x & & -7x \\
\hline
2x + 1 &=& -4
\end{array}
$$

$$\boxed{2x + 1 = -4}$$

$$2x = -5$$ Add -1 to both sides.

$$x = -\frac{5}{2}$$ Divide both sides by 2.

Add $-9x$ to both sides:

$$
\begin{array}{rcl}
9x + 1 &=& 7x - 4 \\
+\ -9x & & -9x \\
\hline
1 &=& -2x - 4
\end{array}
$$

$$\boxed{1 = -2x - 4}$$

$$5 = -2x$$ Add 4 to both sides.

$$-\frac{5}{2} = x$$ Divide both sides by -2.

As we can see, the answers are the same—the only difference being that in one case we have an equation with the variable on the *left side* and in the other case we have an equation with the variable on the *right side*.

Some students prefer to bring the variable terms together on one side of the equation so that there is a positive coefficient on the remaining variable term. Others prefer always to have the variable on the left side. This is just a preference; either way is correct.

^{Mc} **EXAMPLE 5**   Solve $2y - 3 + 3y + 7 = 6y + 21$ and check your solution.

Student Practice 5   Solve $-2 + 6y + 4 = 8y + 15$ and check your solution.

Solution

$$
\begin{array}{rl}
2y - 3 + 3y + 7 = & 6y + 21 \\
5y - 3 \quad\quad + 7 = & 6y + 21 \qquad 2y + 3y = 5y \\
5y + 4 = & 6y + 21 \qquad \text{Combine: } -3 + 7 = 4. \\
\underline{+ \quad -5y \quad\quad\quad -5y} & \qquad\quad \text{Add } -5y \text{ to both sides of the equation.} \\
4 = & y + 21 \\
\underline{+ \; -21 \quad\quad\quad - 21} & \qquad\quad \text{Add } -21 \text{ to both sides of the equation.} \\
-17 = & y
\end{array}
$$

Check.

$$
\begin{array}{c}
2y - 3 + 3y + 7 = 6y + 21 \\
2(-17) - 3 + 3(-17) + 7 \overset{?}{=} 6(-17) + 21 \\
-34 - 3 - 51 + 7 \overset{?}{=} -102 + 21 \\
-81 = -81 \quad \checkmark
\end{array}
$$

👣 STEPS TO SUCCESS   Mastering Word Problems

Learning mathematics is like learning to play a sport. Learning mathematics without ever doing word problems is similar to learning all the skills of a sport without ever playing a game or learning all the notes on an instrument without ever playing a song.

   The key to success is practice. Make yourself do as many problems as you can. You may not be able to work them all correctly at first, but keep trying.

   Do not give up. Whenever you reach a difficult one, if you cannot solve it, just try another one. Then come back and try the problem again later.

   Ask for help from your teacher or the tutoring lab. Ask other classmates how they solved the problem.

   You will succeed. At first the problems may seem like each one is different, but as you practice more and more, you will begin to see the similarities, the different "types." You will see patterns in solving problems, which will enable you to solve problems of a given type more easily.

Making it personal:

1. Spend this week thinking about situations that you have encountered that require the use of mathematics. Write out the facts for one of these situations, then form an applied problem. Share this applied problem with other students in your class. ▼

Verbal and Writing Skills, Exercises 1–2

Write in words the question asked by each equation.

1. $1 + 2n = 5$

2. $3x - 2 = 7$

Fill in the boxes in each step with the values needed to solve each equation.

3.
$$3x + 6 = 5x + 9$$
$$+\boxed{}\quad\boxed{}$$
$$0 + 6 = 2x + 9$$
$$+\boxed{}\quad\boxed{}$$
$$-3 = 2x + 0$$
$$\frac{-3}{\boxed{}} = \frac{2x}{\boxed{}}$$
$$-\frac{3}{2} = x$$

4.
$$8 - 5x = 7x + 3$$
$$+\boxed{}\quad\boxed{}$$
$$8 + 0 = 12x + 3$$
$$+\boxed{}\quad\boxed{}$$
$$5 = 12x + 0$$
$$\frac{5}{\boxed{}} = \frac{12x}{\boxed{}}$$
$$\frac{5}{12} = x$$

5.
$$6 - 2x = 5 - 9x$$
$$+\boxed{}\quad\boxed{}$$
$$6 + 7x = 5 + 0$$
$$\boxed{}\quad\boxed{}$$
$$0 + 7x = -1$$
$$\frac{7x}{\boxed{}} = \frac{-1}{\boxed{}}$$
$$x = -\frac{1}{7}$$

6.
$$-3 + 8x = 9x - 4$$
$$+\boxed{}\ \boxed{}$$
$$-3 + 0 = x - 4$$
$$+\boxed{}\qquad\boxed{}$$
$$1 = x$$

7.
$$-4x + 2 + 3x = 13$$
$$\boxed{}x + 2 = 13$$
$$+\qquad\boxed{}\quad\boxed{}$$
$$\frac{-1x}{\boxed{}} + 0 = \frac{11}{\boxed{}}$$
$$x = -11$$

8.
$$-6x + 4 + 5x = 10$$
$$\boxed{}x + 4 = 10$$
$$+\qquad\boxed{}\quad\boxed{}$$
$$\frac{-1x}{\boxed{}} + 0 = \frac{6}{\boxed{}}$$
$$x = -6$$

Solve and check your solution.

9. $3x - 9 = 27$

10. $6x + 2 = 14$

11. $5x - 15 = 20$

12. $8x - 14 = 10$

13. $18 = 4x - 10$

14. $17 = 4x + 1$

15. $5x - 1 = 16$

16. $3x - 2 = 12$

17. $-4y + 9 = 65$

18. $-6w + 7 = 67$

19. $-6m - 10 = 88$

20. $-5y - 9 = 74$

21. $52 = -2x - 10$

22. $36 = -4x - 8$

23. $-1 = -7 - 5y$

24. $-2 = -9 - 4x$

25. $2 = 4 + 2x$

26. $10 = 4 - 6x$

27. $-6 = 6 - 3y$

28. $-3 = 9 - 2y$

Simplify the left side of the equation by combining like terms and then solve.

29. $8y + 6 - 2y = 18$

30. $7x + 3 - 4x = 12$

31. $9x - 2 + 2x = 6$

32. $3m - 1 + 4m = 5$

33. $7x + 8 - 8x = 11$

34. $4x + 7 - 5x = 12$

Solve and check your solution. You may collect the variable terms on the right or on the left.

35. $15x = 9x + 48$

36. $16x = 7x + 36$

37. $4x = -12x + 7$

38. $14x = -2x + 11$

39. $8x + 2 = 5x - 4$

40. $7x + 6 = 2x - 9$

41. $11x + 20 = 12x + 2$

42. $6y - 5 = 13y + 9$

43. $-2 + y + 5 = 3y + 9$

44. $-9 + 2y + 6 = y + 11$

45. $13y + 9 - 2y = 6y - 8$

46. $14y - 6 + 4y = 8y - 1$

47. $-9 - 3x + 8 = -6x + 3 - 3x$

48. $-1 - 6x + 2 = -4 + 3x - 5x$

Mixed Practice *Solve and check your solution.*

49. $-2y + 6 = 12$

50. $-8x + 4 = 20$

51. $3x - 4 = 11$

52. $7x - 2 = 19$

53. $5x + 5 - 2x = 15$

54. $4x + 3 - 2x = 12$

55. $13x = 8x + 20$

56. $15x = 6x + 30$

57. $4x - 24 = 6x - 8$

58. $9x - 4 = 3x - 10$

One Step Further *Solve.*

59. $2x - 6 + 2(4x) + 13 = -2 - 5x + (3^2 - 3)$

60. $9x - 8 - 2(3x) + (4^2 + 3) = -4x - 3 + 2x$

Cumulative Review *Multiply.*

61. **[2.6.3]** $-3(x - 4)$

62. **[2.6.3]** $2(-3 + y)$

Find the value of x in each proportion.

63. **[4.6.3]** $\dfrac{13}{21} = \dfrac{65}{x}$

64. **[4.6.3]** $\dfrac{15}{17} = \dfrac{x}{85}$

Quick Quiz 7.2 *Solve and check your solution.*

1. $-5a = 3a + 40$

2. $3y + 8 - 6y = -13$

3. $4x + 9 = 6x - 3$

4. **Concept Check** Explain the steps to solve the equation $8x + 3 - 4x = 15$.

Student Learning Objective

After studying this section, you will be able to:

① Solve equations with parentheses.

① Solving Equations with Parentheses

As we have seen, we must simplify each side of an equation before we can begin the process of solving the equation for a variable. If equations contain parentheses, we remove the parentheses first. Then when we combine like terms, the equations become just like those encountered previously. We review the procedure to solve equations presented in Section 7.2.

PROCEDURE TO SOLVE EQUATIONS

1. *Parentheses.* Remove any parentheses.
2. *Simplify each side of the equation.* Combine like terms and simplify numerical work.
3. *Isolate the ax term.* Use the addition principle to get all terms with the variable on one side of the equation and the numerical values on the other side.
4. *Isolate the x.* Use the multiplication or division principle to get the variable alone on one side of the equation.
5. *Check* your solution.

EXAMPLE 1 Solve $-3(2x + 1) + 4x = 27$ and check your solution.

Solution We use the distributive property to remove parentheses and then we simplify and solve.

$$-3(2x + 1) + 4x = 27$$
$$-6x - 3 + 4x = 27 \quad \text{Multiply: } (-3)(2x) = -6x; (-3)(+1) = -3.$$
$$-2x - 3 = 27 \quad \text{Combine like terms: } -6x + 4x = -2x.$$
$$\underline{ +3 \quad +3} \quad \text{Add 3 to both sides of the equation.}$$
$$-2x = 30$$
$$\frac{-2x}{-2} = \frac{30}{-2} \quad \text{Divide by } -2 \text{ on both sides of the equation.}$$
$$x = -15$$

Check.
$$-3(2x + 1) + 4x = 27$$
$$-3[2(-15) + 1] + 4(-15) \overset{?}{=} 27$$
$$-3(-30 + 1) + (-60) \overset{?}{=} 27$$
$$-3(-29) + (-60) \overset{?}{=} 27$$
$$27 = 27 \ ✓$$

Student Practice 1 Solve $-5(3x + 2) - 6x = 32$ and check your solution.

NOTE TO STUDENT: *Fully worked-out solutions to all of the Student Practice problems can be found at the back of the text starting at page SP-1.*

When simplifying and solving equations, pay special attention to the *sign* of each term.

ᴹ𝒸 **EXAMPLE 2** Solve $5(y + 8) = -6(y - 2) + 94$ and check your solution.

Solution

$$5(y + 8) = -6(y - 2) + 94$$

$$5 \cdot y + 5 \cdot 8 = -6 \cdot y - 6(-2) + 94 \quad \text{Remove parentheses.}$$

$$5y + 40 = -6y + 12 + 94$$

$$5y + 40 = -6y + 106$$

Simplify each side of the equation separately.

$$\begin{array}{rcl} 5y + 40 &=& -6y + 106 \\ + 6y && +6y \\ \hline 11y + 40 &=& 106 \end{array}$$

Add $6y$ to both sides so that all y terms are on one side of the equation.

$$\begin{array}{rcl} 11y + 40 &=& 106 \\ + \quad -40 && -40 \\ \hline 11y &=& 66 \end{array}$$

Continue to use the principles of equality to solve for y.

$$\frac{11y}{11} = \frac{66}{11}$$

$$y = 6$$

Check. $5(y + 8) = -6(y - 2) + 94$

$5(6 + 8) \overset{?}{=} -6(6 - 2) + 94$

$5(14) \overset{?}{=} -6(4) + 94$

$70 \overset{?}{=} -24 + 94$

$70 = 70$ ✓

Student Practice 2 Solve $2(x + 4) = -7(x - 3) + 2$ and check your solution.

Recall from Section 6.1 that when a negative sign is in front of parentheses, we find the opposite of the expression by changing the sign of each term inside the parentheses. We sometimes encounter this situation when solving equations.

EXAMPLE 3 Solve. $(2x^2 + 3x + 1) - (2x^2 + 6) = 4x + 1$

Solution   Since there is a minus sign in front of the parentheses, $-(2x^2 + 6)$, we change the sign of each term inside the parentheses.

$$(2x^2 + 3x + 1) - (2x^2 + 6) = 4x + 1$$

$$2x^2 + 3x + 1 - 2x^2 - 6 = 4x + 1 \quad \begin{array}{l}\text{Change the sign of each term} \\ \text{and remove parentheses.}\end{array}$$

$$3x - 5 = 4x + 1 \quad \begin{array}{l} 2x^2 - 2x^2 = 0; \quad 1 - 6 = -5. \\ \text{Solve for } x.\end{array}$$

$$\begin{array}{rcl} + \quad -3x && -3x \\ \hline -5 &=& x + 1 \\ + \quad -1 && - 1 \\ \hline -6 &=& x \end{array} \quad \text{or } x = -6$$

We leave the check for the student.

Student Practice 3 Solve. $(4x^2 + 6x + 3) - (4x^2 + 2) = 3x + 1$

Verbal and Writing Skills, Exercises 1–2

Fill in the blanks with the correct justification for each step of the solution to each equation.

1. $-5(3x + 2) + 2x = 16$
$-15x - 10 + 2x = 16$

_____

$-13x - 10 = 16$ _____
$-13x = 26$ _____

$x = -2$ _____

2. $-3(4x + 1) + 3x = 15$
$-12x - 3 + 3x = 15$

_____

$-9x - 3 = 15$ _____
$-9x = 18$ _____

$x = -2$ _____

Solve and check your solution.

3. $-3(2x + 1) = 15$

4. $-5(3x + 2) = 20$

5. $4(3x - 1) = 12$

6. $3(2x - 3) = 14$

7. $36 = -2(4y + 2)$

8. $42 = -6(2y - 3)$

9. $-5(2x + 1) + 3x = 37$

10. $-7(4y + 1) + 15y = 6$

11. $-2(5y - 1) + 4y = -4$

12. $-4(2y - 2) + 4y = -12$

13. $-4(2x + 1) = -5 - 3x$

14. $-2(3x + 2) = -9 - 2x$

15. $3(y - 4) + 6(y + 1) = 57$

16. $5(y - 2) + 2(y + 4) = 26$

17. $2(x - 1) + 4(x + 2) = 18$

18. $3(x - 4) + 6(x + 1) = 21$

19. $10 = -2(x - 2) + 6(x - 1)$

20. $16 = -3(x - 3) + 4(x - 1)$

21. $6(x - 4) = -9(x + 1) + 10$

22. $4(x - 2) = -2(x + 6) + 31$

23. $2(y + 3) = -6(y - 1) - 8$

24. $3(y + 1) = -2(y - 2) - 6$

25. $(6x^2 + 4x - 1) - (6x^2 + 9) = 14$

26. $(9y^2 + 8y - 2) - (9y^2 + 5) = 9$

27. $(5x^2 + x + 3) - (5x^2 + 9) = 4x + 1$

28. $(8x^2 + 8x - 2) - (8x^2 + 5) = 6x + 2$

Mixed Practice *Solve and check your solution.*

29. $6(x - 1) + 2(x + 1) = 10 - 5x$

30. $7(x - 1) + 3(x + 1) = 20 - 4x$

31. $4(-6x + 2) - 6x = 68$

32. $2(-7x + 3) - 2x = 38$

33. $(4x^2 + 3x + 1) - (4x^2 - 2) = 5x - 2$

34. $(3x^2 - 2x + 4) - (3x^2 + 1) = 6x + 2$

One Step Further *Solve and check your solution.*

35. $(2x^2 - 6x - 3) - (x^2 + 4) = x^2 + 6$

36. $3x^2 + 4x - 9 - (2x^2 - 1) = x^2 - 5$

37. $(x^2 + 3x - 1) - (2x + 1) = x^2 - 5$

38. $(y^2 + 2y - 5) - (3y + 2) = y^2 - 1$

To Think About

39. Is $x = 19$ a solution to $(2x + 9) + (3x - 2) - (5x + 1) = 2(x - 6) - (x - 1)$?

40. Is $a = 2$ a solution to $(9a + 2) + (4a - 1) - (6a + 3) = 4(a - 1) + 8$?

Cumulative Review *Find the least common denominator.*

41. [5.3.2] $\dfrac{2}{3}, \dfrac{1}{4}, \dfrac{5}{2}$

42. [5.3.2] $\dfrac{3}{4}, \dfrac{4}{2}, \dfrac{1}{5}$

43. [5.3.2] $\dfrac{1}{2x}, \dfrac{7}{x}$

44. [5.3.2] $\dfrac{2}{x}, \dfrac{3}{5x}$

45. [5.6.1] **Weighing Produce** The cashier at the produce store charged Israel for $\frac{4}{5}$ pound of grapes. When the cashier's scale was later tested for accuracy, it was determined that the scale was inaccurate. The actual weight of the grapes was $\frac{2}{3}$ pound. Was Israel overcharged or undercharged for his purchase?

46. [5.6.1] **Pressure on Diver** The formula for calculating the pressure P in pounds per square inch on an object submerged to a depth D in feet is given by $P = 15 + \frac{1}{2}D$. Find the pressure on a diver who is submerged underwater at a depth of $13\frac{1}{4}$ feet.

Quick Quiz 7.3 *Solve and check your solution.*

1. $-4(3x - 5) - 2x = 34$

2. $-8(x + 1) = 4(2x - 5) - 20$

3. $(4a^2 + 2a - 1) - (4a^2 - 3a + 4) = 3a + 1$

4. **Concept Check** **(a)** The first step we should perform to solve $2(y - 1) + 2 = -3(y - 2)$ is to simplify the equation using the _____ property.

(b) Complete this simplification, then explain the rest of the steps needed to solve the equation so that the variable is on the left side of the equation: $y = $ some number.

How Am I Doing?   Sections 7.1–7.3

How are you doing with your homework assignments in Sections 7.1 to 7.3? Do you feel you have mastered the material so far? Do you understand the concepts you have covered? Before you go further in the textbook, take some time to do each of the following problems.

7.1

Solve and check your solution.

1. $-12 = a + 3$

2. $\dfrac{x}{-2} = 4 + 5^2$

3. $-x = 4$

4. $\dfrac{21}{-7} = 9x + 3(-4x)$

7.2

Solve and check your solution.

5. $4x + 10 = 14$

6. $-5y - 9 = 24$

7. $-3 = 7 - 2y$

8. $6x - 4 = 14x - 9$

9. $15y - 6 + 5y = 8y - 1$

10. $7 - 8 = 2x - 5 + 5x$

7.3

Solve and check your solution.

11. $-3(2x + 1) + 4x = 11$

12. $2(x + 1) = -3(x + 2) + 10$

13. $(3y^2 + 8y - 1) - (3y^2 + 2) = 5$

Now turn to page SA-13 for the answers to each of these problems. Each answer also includes a reference to the objective in which the problem is first taught. If you missed any of these problems, you should stop and review the Examples and Student Practice problems in the referenced objective. A little review now will help you master the material in the upcoming sections of the text.

7.4 Solving Equations with Fractions

① Solving Equations Using the LCD Method

Solving equations such as $\frac{x}{2} + \frac{x}{7} = 11$ can be a lengthy process. To avoid unnecessary work, we can transform the given equation containing fractions to an equivalent equation that does not contain fractions. This process is often referred to as **clearing the fractions.** How do we *clear the fractions* from the equation? We multiply *all terms* on both sides of the equation by the least common denominator (LCD) of all the fractions contained in the equation.

EXAMPLE 1 Solve $\frac{x}{3} + \frac{x}{2} = 5$ and check your solution.

Solution First, we clear the fractions from the equation by multiplying each term by the LCD = 6; then we solve the equation.

$$6\left(\frac{x}{3}\right) + 6\left(\frac{x}{2}\right) = 6(5)$$ Multiply each term by the LCD to clear the fractions.

$$2x + 3x = 30$$ Simplify: $\frac{6x}{3} = 2x$ and $\frac{6x}{2} = 3x$. We cleared the fractions from the equation.

$$5x = 30$$ Combine like terms.

$$\frac{5x}{5} = \frac{30}{5}$$ Divide both sides by 5.

$$x = 6$$

Note that the equation $2x + 3x = 30$ is equivalent to $\frac{x}{3} + \frac{x}{2} = 5$ and much easier to work with.

Check. $\qquad \frac{x}{3} + \frac{x}{2} = 5$

$$\frac{6}{3} + \frac{6}{2} \overset{?}{=} 5$$

$$2 + 3 = 5 \quad \checkmark$$

CAUTION: It is important to multiply **every term on both sides** of the equation by the least common denominator. A common mistake made when solving $\frac{x}{3} + \frac{x}{2} = 5$ is to multiply the fractions, $\frac{x}{3}$ and $\frac{x}{2}$, by the LCD but not the 5.

Student Practice 1 Solve $\frac{x}{5} + \frac{x}{2} = 7$ and check your solution.

Let's now write down the steps we have used.

PROCEDURE TO SOLVE AN EQUATION CONTAINING FRACTIONS

1. Determine the LCD of all the denominators.
2. Multiply *every term* of the equation by the LCD (clear the fractions).
3. Solve the resulting equation.
4. Check your solution.

EXAMPLE 2 Solve. $-4x + \dfrac{3}{2} = \dfrac{2}{5}$

Solution The LCD is 10. We multiply each term by 10.

$$10\,(-4x) + 10\left(\dfrac{3}{2}\right) = 10\left(\dfrac{2}{5}\right) \quad \text{Multiply each term by the LCD, 10.}$$

$$-40x + 15 = 4 \qquad \text{Simplify: } 10\cdot\dfrac{3}{2} = 15 \text{ and } 10\cdot\dfrac{2}{5} = 4.$$

$$
\begin{array}{rcr}
-40x + 15 = & & 4 \qquad \text{Solve for } x. \\
\underline{+\quad\ -15} & & \underline{-15} \\
-40x\qquad\ \ = & & -11
\end{array}
$$

$$\dfrac{-40x}{-40} = \dfrac{-11}{-40}$$

$$x = \dfrac{11}{40}$$

We leave the check to the student.

Student Practice 2 Solve. $-5x + \dfrac{2}{7} = \dfrac{3}{2}$

NOTE TO STUDENT: Fully worked-out solutions to all of the Student Practice problems can be found at the back of the text starting at page SP-1.

EXAMPLE 3 Solve. $\dfrac{x}{7} + x = 8$

Solution There is only one denominator; thus the LCD is 7.

$$7\left(\dfrac{x}{7}\right) + 7\,(x) = 7\,(8) \quad \text{Multiply each term by the LCD, 7.}$$

$$1x + 7x = 56 \qquad \text{Simplify: } \dfrac{7x}{7} = 1x.$$

$$8x = 56 \qquad\ \ \text{Combine like terms.}$$

$$\dfrac{8x}{8} = \dfrac{56}{8} \qquad\ \ \text{Divide both sides by 8.}$$

$$x = 7$$

Student Practice 3 Solve. $\dfrac{x}{4} + x = 5$

Verbal and Writing Skills, Exercises 1–2

Fill in the blanks.

1. To solve $\frac{x}{3} + \frac{x}{5} = 10$, we multiply each term by ___, so that we clear the fractions.

2. To solve $\frac{x}{4} - \frac{x}{2} = -6$, we multiply each term by ___ , so that we clear the fractions.

Fill in each box with the correct number to complete each solution.

3. $\frac{x}{4} + \frac{x}{3} = 7$

$\square \cdot \frac{x}{4} + \square \cdot \frac{x}{3} = \square \cdot 7$

$\square x + \square x = 84$

$\square x = 84$

$x = \square$

4. $\frac{x}{5} + \frac{x}{3} = 8$

$\square \cdot \frac{x}{5} + \square \cdot \frac{x}{3} = \square \cdot 8$

$\square x + \square x = 120$

$\square x = 120$

$x = \square$

Solve and check your solution.

5. $\frac{x}{6} + \frac{x}{2} = 8$

6. $\frac{x}{5} + \frac{x}{10} = 3$

7. $\frac{x}{6} + \frac{x}{4} = 5$

8. $\frac{x}{8} + \frac{x}{12} = 5$

9. $\frac{x}{2} - \frac{x}{5} = 6$

10. $\frac{x}{2} - \frac{x}{9} = 7$

11. $3x + \frac{2}{3} = \frac{9}{2}$

12. $2x + \frac{1}{4} = \frac{2}{3}$

13. $5x + \frac{1}{8} = \frac{3}{4}$

14. $3x + \frac{2}{5} = \frac{1}{2}$

15. $5x - \frac{1}{2} = \frac{1}{8}$

16. $3x - \frac{1}{3} = \frac{1}{9}$

17. $-2x + \frac{1}{2} = \frac{3}{7}$

18. $-3x + \frac{1}{4} = \frac{1}{3}$

19. $-4x + \frac{2}{3} = \frac{1}{6}$

20. $-2x + \frac{1}{5} = \frac{1}{10}$

21. $\frac{x}{3} + x = 8$

22. $\frac{x}{4} + x = 5$

23. $\frac{x}{3} + x = 6$

24. $\frac{x}{6} + x = 7$

25. $\frac{x}{2} + x = 6$

26. $\frac{x}{5} - 4x = 4$

27. $\frac{x}{4} - 2x = 3$

28. $\frac{x}{3} - 4x = 5$

Mixed Practice   *Solve and check your solution.*

29. $2x + \frac{1}{3} = \frac{1}{6}$

30. $5x + \frac{1}{2} = \frac{1}{4}$

31. $\frac{x}{3} + x = 8$

32. $\frac{x}{5} + x = 6$

33. $\frac{x}{2} + \frac{x}{5} = 7$

34. $\frac{x}{3} + \frac{x}{2} = 5$

One Step Further *Solve and check your solution.*

35. $\dfrac{5}{2} + \dfrac{x}{3} = \dfrac{1}{6}$

36. $\dfrac{1}{9} + \dfrac{x}{6} = \dfrac{2}{3}$

37. $\dfrac{3}{2} + \dfrac{x}{10} = \dfrac{1}{5}$

38. $\dfrac{2}{7} + \dfrac{x}{14} = \dfrac{1}{2}$

39. $\dfrac{x}{2} - \dfrac{2}{6} = -\dfrac{5}{6}$

40. $\dfrac{x}{3} - \dfrac{4}{5} = -\dfrac{2}{5}$

Solve.

41. $x + \dfrac{2}{3} + 2 = \dfrac{3}{2} + \dfrac{1}{4}$

42. $2x + \dfrac{3}{4} + 3 = \dfrac{2}{3} + \dfrac{1}{6}$

43. $4 + \dfrac{6}{x} + \dfrac{2}{5} = \dfrac{3}{2x}$

44. $3 + \dfrac{2}{x} + \dfrac{5}{6} = \dfrac{5}{3x}$

45. $\dfrac{1}{3}\left(\dfrac{x}{2} + 3\right) + \dfrac{1}{4} = \dfrac{5}{6}$

46. $\dfrac{3}{4}(x - 3) + \dfrac{1}{2} = \dfrac{2}{3}$

Cumulative Review *Translate using symbols.*

47. **[1.7.1]** Six more than twice a number

48. **[1.3.2]** Twelve less than some number

49. **[1.2.1]** The sum of 4 and x

50. **[1.7.1]** Two times the sum of 5 and y

51. **[5.6.1]** *Partnership Investments* Three investors own a restaurant. One partner owns $\frac{1}{3}$ of the restaurant, while the second partner owns $\frac{1}{4}$. How much does the third partner own?

52. **[5.6.1]** *Weight Loss* A boxer must lose 11 pounds in 3 weeks to be eligible for his weight category in the next boxing match. If he loses $4\frac{1}{4}$ pounds the first week and $3\frac{1}{2}$ pounds the second week, how much does he need to lose the third week to achieve the 11-pound goal?

Quick Quiz 7.4 *Solve.*

1. $\dfrac{x}{3} + \dfrac{x}{2} = 20$

2. $3a + \dfrac{2}{3} = \dfrac{4}{5}$

3. $\dfrac{y}{4} + y = 25$

4. **Concept Check** Explain the steps to solve the equation $-2x + \dfrac{3}{4} = \dfrac{1}{2}$.

7.5 Using Equations to Solve Applied Problems

One of the first steps in solving many real-life applications is writing the equation that represents the situation. In this section we learn how to write the equations needed to solve an applied problem.

Student Learning Objectives

After studying this section, you will be able to:

① Solve applied problems involving geometric figures.

② Solve applied problems involving comparisons.

① Solving Applied Problems Involving Geometric Figures

Before we begin our discussion, let's review some of the formulas we learned in earlier chapters. To find the perimeter of a shape, we find the sum of all the sides. For a rectangle, we can use the formula $P = 2L + 2W$ to find the perimeter, and $A = LW$ to find the area.

▲ **EXAMPLE 1** Art has a rectangular planter box in his front yard that has width = 2 ft and length = 6 ft. Art plans to increase the length by x ft so that the new perimeter is 32 ft.

$$L = 6\text{ ft} + x$$
$$W = 2\text{ ft} \quad \boxed{} \quad P = 32\text{ ft}$$

(a) How much should the length be enlarged?

(b) What will the length of the enlarged planter box be?

Solution *Understand the problem.* We organize the information in a Mathematics Blueprint for Problem Solving.

Mathematics Blueprint for Problem Solving

Gather the Facts	What Am I Asked to Do?	How Do I Proceed?	Key Points to Remember
$L = x + 6$ $W = 2$ $P = 32$ ft	**(a)** *Find x:* the number of feet the length should be enlarged **(b)** *Find L:* the length of the enlarged planter box.	**(a)** To find x, start with $P = 2L + 2W$ and replace P, L, and W with the given values. Then solve for x. **(b)** To find L, use the equation $L = x + 6$.	Place the unit "ft" in the answer.

Solve and state the answer.

(a) How much should the length be enlarged? We must find x to answer this question.

$$P = 2L + 2W \qquad \text{Write the formula for the perimeter.}$$
$$32 = 2(x + 6) + 2(2) \quad \text{Replace } P, L, \text{ and } W \text{ with the values given.}$$

Now we simplify and solve for x.

$$32 = 2x + 12 + 4 \quad \text{Multiply.}$$
$$32 = 2x + 16 \qquad \text{Simplify.}$$
$$\underline{+\,-16 \qquad -\,16} \qquad \text{Solve the equation.}$$
$$16 = 2x$$
$$\frac{16}{2} = \frac{2x}{2}$$
$$8 = x$$

The length should be enlarged 8 ft.

Continued on next page

$L = x + 6$ or 14 ft

$W = 2$ ft

(b) What will the length of the enlarged planter box be?

To find L, we replace x with 8 in the expression that represents the length.

$$L = x + 6$$
$$L = 8 + 6 = 14$$

The length of the planter box will be 14 ft.

Check. We evaluate the formula for perimeter with our answer to check our calculations:

$$P = 2L + 2W$$
$$32 \stackrel{?}{=} 2(14) + 2(2)$$
$$32 = 32 \checkmark$$

NOTE TO STUDENT: *Fully worked-out solutions to all of the Student Practice problems can be found at the back of the text starting at page SP-1.*

▲ **Student Practice 1** Refer to Example 1 to answer the following. Art increases the length by x ft so that the new perimeter is 36 ft.

$L = 6$ ft $+ x$

$W = 2$ ft

$P = 36$ ft

(a) How much should the length be enlarged?

(b) What will the length of the enlarged planter box be?

② Solving Applied Problems Involving Comparisons

In Section 6.3 we saw how to write variable expressions. In this section we will review this topic and then see how to use these expressions to form an equation and solve applied problems.

When an applied problem involves comparing two or more quantities, we must write a variable expression that describes one quantity *in terms of another* as the first step to solving the problem. We usually let a variable represent the quantity *to which things are being compared.* For example, if Laura earns $3 per hour more than Jessica, and Wendy earns $2 per hour less than Jessica, we are *comparing all the hourly wages to Jessica's* hourly wage. Therefore, we use the variable x to represent Jessica's hourly wage, and then we describe Laura's and Wendy's wages in terms of Jessica's.

$$\text{Jessica's hourly wage} = x$$
$$\text{Laura's hourly wage} = x + 3$$
$$\text{Wendy's hourly wage} = x - 2$$

We include writing variable expressions as part of a six-step process that will help you organize a Mathematics Blueprint for Problem Solving.

PROCEDURE TO SOLVE APPLIED PROBLEMS

1. Read the problem carefully to get an overview.
2. Write down formulas or draw pictures if possible.
3. Define the variable expressions.
4. Write an equation using the variable expressions defined in step 3.
5. Solve the equation and determine the values asked for in the problem.
6. Check your answer. Ask yourself if the answers obtained are reasonable.

EXAMPLE 2   For the following applied problem:

(a) Define the variable expressions.

(b) Write an equation.

(c) Solve the equation and determine the values asked for.

(d) Check your answers.

Linda is a store manager. The assistant manager, Erin, earns $8500 less annually than Linda does. The sum of Linda's annual salary and Erin's annual salary is $72,000. How much does each earn annually?

Solution   *Understand the problem.* To help us understand the problem, we use a Mathematics Blueprint for Problem Solving.

Mathematics Blueprint for Problem Solving

Gather the Facts	What Am I Asked to Do?	How Do I Proceed?	Key Points to Remember
Erin earns $8500 less than Linda. The sum of the two salaries is $72,000.	Find Linda's and Erin's salaries.	**(a)** Let L represent Linda's salary and write an expression for Erin's salary. **(b)** Form an equation by setting the sum of the expressions equal to $72,000. **(c)** Solve the equation.	I must find both Linda's and Erin's salaries, and check my answers.

(a) We define the variable expressions. Since we are *comparing* Erin's salary *to Linda's,* we let the variable L represent Linda's salary.

$$\text{Linda's salary} = L$$

$$\text{Erin's salary} = (L - 8500) \quad \text{\$8500 less than Linda's salary}$$

(b) We write an equation.

$$\boxed{\begin{array}{c}\text{Linda's}\\\text{salary}\end{array}} \quad + \quad \boxed{\begin{array}{c}\text{Erin's}\\\text{salary}\end{array}} \quad = \quad \boxed{\begin{array}{c}\text{total annual salary}\\\text{for both people}\end{array}}$$

$$L \quad + \quad (L - 8500) \quad = \quad 72,000$$

The equation is $L + (L - 8500) = 72,000$.

Solve and state the answer.

(c) We solve the equation.

$$L + (L - 8500) = 72,000$$
$$2L - 8500 = 72,000$$
$$\underline{+ \qquad 8500 \quad 8,500}$$
$$2L = 80,500$$
$$\frac{2L}{2} = \frac{80,500}{2} \qquad L = 40,250$$

Linda earns $40,250 annually.

Now we find Erin's salary.

$$L - \$8500 = \text{Erin's salary (\$8500 less than Linda's salary)}$$
$$\$40,250 - \$8500 = \$31,750$$

Erin earns $31,750 annually.

Continued on next page

(d) *Check.* Is the sum of their salaries equal to $72,000?

$$\$31,750 + \$40,250 \overset{?}{=} \$72,000 \qquad \$72,000 = \$72,000 \;\checkmark$$

Student Practice 2   For the following applied problem:

(a) Define the variable expressions.

(b) Write an equation.

(c) Solve the equation and determine the values asked for.

(d) Check your answers.

Jason is a foreman for a construction company. The apprentice for the company earns $7400 less annually than Jason does. The sum of Jason's annual salary and the apprentice's annual salary is $83,000. How much does each earn annually?

● Understanding the Concept

Forming Equations   In this section we saw how to translate English expressions into mathematical symbols, and then form an equation. Sometimes we must write the English expression ourselves by observing a pattern in a set of numbers as illustrated below.

Suppose we must investigate the mathematical relationship between the following two sequences.

$$x \text{ is: } 1, 2, 3, 4, 5, \ldots$$
$$y \text{ is: } 1, 4, 9, 16, 25, \ldots$$

The mathematical operations that are commonly used in sequences are listed in the table below.

To Obtain Each Number in Sequence y:	
(a) Add the same number to each x.	**(d)** Multiply or divide each x by the same number.
(b) Subtract the same number from each x.	
(c) Add (or subtract) a sequence of numbers to (from) each x: 1, 5, 1, 5, $\ldots$	**(e)** Raise x to a power
	(f) Any combination of **(a)**–**(e)**.

Referring to **(e)** in the table we can state the relationship between x and y in words as follows.

We raise each value in x to the power of 2 to obtain each corresponding value in y.

Now we can translate this statement and write the equation or formula, $x^2 = y$. Then we can use this equation to find other numbers in the sequences. For example,

The next number in the sequence x is 6. $\quad x$ is: $1, 2, 3, 4, 5, 6, \ldots$
We evaluate $6^2 = 36$ to find the next $\quad y$ is: $1, 4, 9, 16, 25, 36, \ldots$
number in y.

Exercise

1. Find the next two numbers in the sequence x, and then find the corresponding values in y.

For exercises 1–6:

(a) *Write an equation.* **(b)** *Solve the equation.*

1. If three times a number is increased by nine, the result is fifteen. What is the number?

2. If four times a number is increased by six, the result is ten. What is the number?

3. If triple a number is decreased by four, the result is five. What is the number?

4. If double a number is decreased by eight, the result is six. What is the number?

5. If the sum of 4 and a number is multiplied by 2, the result is 12. What is the number?

6. If the sum of 8 and a number is multiplied by 5, the result is 45. What is the number?

Use the appropriate formula to solve exercises 7–12.

▲ **7.** Find the value of x in the following rectangle if the perimeter is 60 meters.

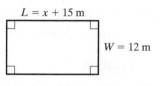

$L = x + 15$ m

$W = 12$ m

▲ **8.** Find the value of x in the following rectangle if the perimeter is 66 meters.

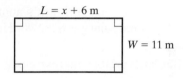

$L = x + 6$ m

$W = 11$ m

▲ **9.** Find the length of each side of the following triangle if the perimeter is 12 centimeters.

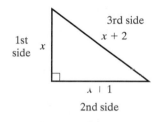

1st side

x

3rd side

$x + 2$

$x + 1$

2nd side

▲ **10.** Find the length of each side of the following triangle if the perimeter is 24 decimeters.

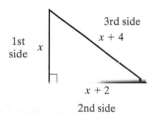

1st side

x

3rd side

$x + 4$

$x + 2$

2nd side

▲ **11.** Find the value of x in the following rectangle if the area is 90 square inches.

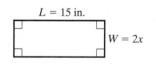

$L = 15$ in.

$W = 2x$

▲ **12.** Find the value of x in the following rectangle if the area is 180 square inches.

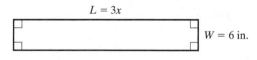

$L = 3x$

$W = 6$ in.

Applications

▲ **13.** *Geometry* Natalie's family room has width $= 12$ ft and length $= 15$ ft. She is enlarging the room by increasing the width x ft so that the new perimeter of the room is 70 ft.

(a) How much should the width be enlarged?

(b) What will the width of the enlarged family room be?

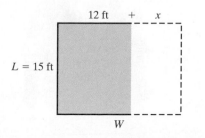

12 ft $+$ x

$L = 15$ ft

W

▲ **14.** *Geometry* Juan's concrete patio has width $= 5$ ft and length $= 12$ ft. He is enlarging the patio by increasing the width x ft so that the new perimeter of the patio is 44 ft.

(a) How much should the width be enlarged?

(b) What will the width of the enlarged concrete patio be?

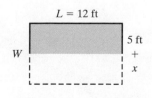

$L = 12$ ft

W

5 ft $+$ x

▲ **15.** *Geometry* Richard is planning to enlarge a rectangular space with length = 9 ft and width = 7 ft so that the new area is 105 ft².

    (a) How much should the length be enlarged? (Recall that $A = LW$.)

    (b) What will the length of the enlarged space be?

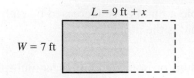

$L = 9 \text{ ft} + x$

$W = 7 \text{ ft}$

▲ **16.** *Geometry* Samantha is planning to enlarge a small rectangular space with length = 20 in. and width = 15 in. so that the new area is 450 in.².

    (a) How much should the length be enlarged? (Recall that $A = LW$.)

    (b) What will the length of the enlarged space be?

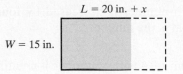

$L = 20 \text{ in.} + x$

$W = 15 \text{ in.}$

17. *Salary Comparison* Mark is a teaching assistant at a local community college. A tutor at the college works the same number of hours as Mark but earns $11,400 less annually. The sum of Mark's annual salary and the tutor's annual salary is $28,890. How much does each earn?

    (a) Define the variable expressions.

    (b) Write an equation.

    (c) Solve the equation and determine the values asked for.

    (d) Check your answer.

18. *Salary Comparison* Lena is a sales supervisor. The salesclerk earns $6200 less annually than Lena does. The sum of Lena's annual salary and the clerk's annual salary is $58,000. How much does each earn?

    (a) Define the variable expressions.

    (b) Write an equation.

    (c) Solve the equation and determine the values asked for.

    (d) Check your answer.

19. *Distance Traveled* Over two days Cal drove 825 miles to attend a wedding. He drove 165 more miles the second day than the first day. How far did he drive each day?

    (a) Define the variable expressions.

    (b) Write an equation.

    (c) Solve the equation and determine the values asked for.

    (d) Check your answer.

20. *Miles Walked* Andrew walked 4 miles less than Dave last week in a Boys and Girls Club "walk for the homeless" program. The two boys together walked 34 miles. How many miles did each boy walk?

    (a) Define the variable expressions.

    (b) Write an equation.

    (c) Solve the equation and determine the values asked for.

    (d) Check your answer.

WALK FOR
THE HOMELESS

21. *Flying Time* The total flying time for two flights is 15 hours. The flight time for the first flight is half of the second. How long is each flight?

(a) Define the variable expressions.

(b) Write an equation.

(c) Solve the equation and determine the values asked for.

(d) Check your answer.

▲ **22.** *Geometry* The perimeter of a rectangle is 48 feet. The length is 4 feet less than triple the width. What are the dimensions of the rectangle?

(a) Define the variable expressions.

(b) Write an equation.

(c) Solve the equation and determine the values asked for.

(d) Check your answer.

▲ **23.** *Geometry* A triangle has a perimeter of 120 meters. The length of the second side is double the first side. The length of the third side is 12 meters longer than the first side. Find the length of each side.

(a) Define the variable expressions.

(b) Write an equation.

(c) Solve the equation and determine the values asked for.

(d) Check your answer.

▲ **24.** *Geometry* A triangle has a perimeter of 176 feet. The second side is 25 feet longer than the first. The third side is 5 feet shorter than the first. Find the length of each side.

(a) Define the variable expressions.

(b) Write an equation.

(c) Solve the equation and determine the values asked for.

(d) Check your answer.

▲ **25.** *Geometry* The perimeter of a rectangle is 68 meters. The length is 2 meters less than triple the width. What are the dimensions of the rectangle?

(a) Define the variable expressions.

(b) Write the equation.

(c) Solve the equation and determine the values asked for.

(d) Check your answer.

▲ **26.** *Geometry* The perimeter of a rectangle is 74 feet. The length is 2 feet less than double the width. What are the dimensions of the rectangle?

(a) Define the variable expressions.

(b) Write the equation.

(c) Solve the equation and determine the values asked for.

(d) Check your answer.

27. *Enrollment Comparison* Last year a total of 395 students took English. 95 more students took it in the spring than in the fall. 75 fewer students took it in the summer than in the fall. How many students took it during each of the three semesters?

(a) Define the variable expressions.

(b) Write the equation.

(c) Solve the equation and determine the values asked for.

(d) Check your answer.

28. *Student Housing* A small community college has a total of 1704 students. 115 more students live on campus than live in nearby off-campus housing. 55 fewer students live at home and commute than live in nearby off-campus housing. How many students are there in each of the three categories?

(a) Define the variable expressions.

(b) Write the equation.

(c) Solve the equation and determine the values asked for.

(d) Check your answer.

One Step Further

▲ **29.** Find the value of x in the following pair of supplementary angles.

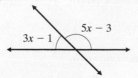

▲ **30.** Since the triangle below is an isosceles triangle, the measures of $\angle a$ and $\angle b$ are equal. Find x in the following isosceles triangle.

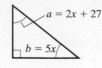

To Think About

▲ **31.** *Geometry* If the perimeter and area of a square are equal in value, what is the length of the side of this square?

▲ **32.** *Geometry* A triangle with all sides of equal length is called an *equilateral triangle*. If the perimeter of an equilateral triangle is equal to the length of a side squared, what is the length (in feet) of each side of the triangle?

The charts in exercises 33–36 list the numbers in two different but related sequences. The first sequence is labeled x, and the second y. Refer to the table below to determine the mathematical relationship between x and y.

> **To Obtain Each Number in Sequence y:**
> **(a)** Add the same number to each x.
> **(b)** Subtract the same number from each x.
> **(c)** Add (or subtract) a sequence of numbers to (from) each x: 1, 5, 1, 5, . . .
> **(d)** Multiply or divide each x by the same number.
> **(e)** Raise x to a power.
> **(f)** Any combination of **(a)–(e)**.

33. **(a)** Find the missing numbers for sequence x, then place these values in the appropriate place in the chart.

x	1	2	3	4	5	6			...	30
y	3	6	9	12	15				...	

(b) Determine the mathematical relationship between x and y, then write this relationship in your own words.

(c) Translate the statement in **(b)** into an equation.

(d) Use this equation to find the missing numbers for y and place these values in the chart.

34. **(a)** Find the missing numbers for sequence x, then place these values in the appropriate place in the chart.

x	25	20	15	10		0		...	−50
y	19	14	9	4		−6		...	

(b) Determine the mathematical relationship between x and y, then write this relationship in your own words.

(c) Translate the statement in **(b)** into an equation.

(d) Use this equation to find the missing numbers for y and place these values in the chart.

35. **(a)** Find the missing numbers for sequence x, then place these values in the appropriate place in the chart.

x	0	1	2	3		5		...	45
y	1	4	7	10		16		...	

(b) Determine the mathematical relationship between x and y, then write this relationship in your own words.

(c) Translate the statement in **(b)** into an equation.

(d) Use this equation to find the missing numbers for y and place these values in the chart.

36. **(a)** Find the missing numbers for sequence x, then place these values in the appropriate place in the chart.

x	2	3	4	5		7		...	20
y	3	8	15	24		48		...	

(b) Determine the mathematical relationship between x and y, then write this relationship in your own words.

(c) Translate the statement in **(b)** into an equation.

(d) Use this equation to find the missing numbers for y and place these values in the chart.

Cumulative Review

▲ **37.** **[5.6.1]** *Checker Board* The dimensions of an opened checker game board are $13\frac{1}{2}$ in. by $13\frac{1}{2}$ in. by $\frac{5}{8}$ in. What are the dimensions of the board when it is folded in half?

▲ **38.** **[5.6.1]** *Brick Steps*

(a) What are the length and the width of the step made of bricks if there is grout between the bricks and between the bricks and the wall?

(b) What is the perimeter of the step?

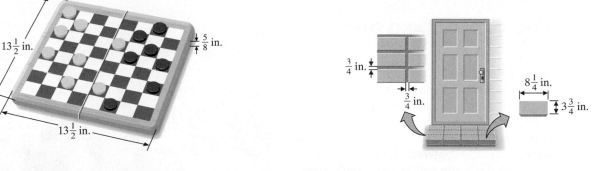

Quick Quiz 7.5 *Use the following information for problems 1–3.*

The perimeter of a triangle is 29 meters. The first side is 5 meters longer than the third side. The second side is twice the length of the third side. Find the length of each side of the triangle.

1. Define the variable expressions.

2. Write an equation.

3. Solve the equation and determine the values asked for.

4. **Concept Check** The high school marching band fundraiser for uniforms requires that students find donors to pledge money based on the number of laps they run on the school track field. This year Miguel ran 2 more laps than Sam, and Alicia ran 4 fewer laps than Sam. The total laps completed by all three was 29.

Eduardo wanted to calculate the number of laps Alicia completed. To find this information, he let S represent the number of laps that Sam ran, and then solved the equation $(S + 2) + (4 - S) = 29$ to find the number of laps completed by Alicia.

(a) Did Eduardo use the correct equation? Why or why not?

(b) How many laps did Miguel complete for the fundraiser?

Did You Know...
That Hidden Costs Can Make a Seemingly Affordable Purchase Cost More Than You Can Afford?

UNDERSTANDING TAXES AND OTHER HIDDEN COSTS

Understanding the Problem:

Arnold has been shopping for a new suit and has narrowed down his choices to one particular suit, which is sold at two different stores. The first store is a local department store which lists the price at $315. Because Arnold lives near the border between two states, he checked the price for the same suit at The Men's Store located in the neighboring state. The Men's Store is the same distance from his home as the department store. The same suit at The Men's Store is advertised for $335. Arnold was curious as to why there was such a price difference for the same suit, so he decided to look into other costs that may increase the final cost of the purchase, such as sales tax and tailoring charges.

Making a Plan:

Arnold can afford to set aside only a small amount of money in his overall budget for clothes, so he must buy the suit that will cost the least amount of money. Therefore, he must determine if there are any extra costs when he considers which suit to buy. First, Arnold checked the sales tax on purchases at each store. The local department store charges 7%, while The Men's Store in the neighboring state charges 4%. Next, he determined that the department store charges a $20 tailoring fee, while The Men's Store doesn't charge extra for tailoring suits. Because Arnold has short legs, he knows he will need some adjustments when he buys a suit.

Step 1: First, Arnold calculates the cost of the suit at the department store. The price tag says $315 and he calculates that the 7% sales tax is about $22.

Task 1: After adding the tax and tailoring fee to the price, how much will it cost Arnold to buy this suit?

Step 2: Arnold calculates the cost of the suit at The Men's Store. This suit has a price tag of $335 and he calculates that the 4% sales tax is about $13.

Task 2: After adding the tax to the price, how much will it cost Arnold to buy the this suit?

Finding a Solution:

Step 3: Now that he knows the total cost for each suit, Arnold can decide which suit to buy.

Task 3: Because Arnold must buy the suit that costs the least amount of money, which suit should Arnold buy?

Task 4: If the suit at the department store came with free tailoring, what would be the cost of that suit?

Task 5: Would this change Arnold's decision? If yes, which suit should Arnold buy?

Applying the Situation to Your Life:

The total cost of an item can be more than what is on the price tag. You need to take sales taxes into consideration when making purchases, or you might come up short and not be able to buy what you had planned. Also include necessary fees in your calculations. In this case, without tailoring, the suit wouldn't fit Arnold very well. As we saw with the suits, the item with the lowest price tag is not always the cheapest option. Remember to calculate the total cost when comparing items to purchase. An easy way to calculate the final cost is to use the calculator function that is found on most cell phones. This will help you pick the best deal and will prevent any surprises when you check out.

Chapter 7 Organizer

Topic and Procedure	Examples	✏ You Try It
Solving equations using the addition principle, p. 387 1. Simplify each side of the equation. 2. *Add* the appropriate value to both sides of the equation so that the variable is on one side and a number is on the other side of the equals sign: $\quad x = $ *some number* 3. Check by substituting your answer back into the original equation.	Solve. $-3 - 9 = 8x + 6 - 7x - 1$ $-3 - 9 = 8x + 6 - 7x - 1$ $-12 = 8x + 6 - 7x - 1 \quad -3 - 9 = -12.$ $-12 = x + 5 \qquad\qquad 8x - 7x = x; 6 - 1 = 5.$ $\begin{aligned} + \quad -5 \quad\quad -5 \\ \hline -17 = x \end{aligned}$ $\qquad$ Solve for x. *Check.* $\quad -3 - 9 = 8x + 6 - 7x - 1$ $\quad -3 - 9 \overset{?}{=} 8(-17) + 6 - 7(-17) - 1$ $\qquad\qquad -12 = -12 \; ✓$	1. Solve. $\quad -1 - 6 = 9x + 5 - 8x - 2$
Solving equations using the multiplication principle, p. 388 1. Simplify each side of the equation. 2. If the variable is *divided* by a number, we *undo* the *division* by *multiplying* both sides of the equation by this number so that the variable is on one side and a number is on the other side of the equals sign: $\quad x = $ *some number* or *some number* $= x$ 3. Check by substituting your answer back into the original equation.	Solve. $\dfrac{y}{-2} = 3^2 - 10$ $\dfrac{y}{-2} = 3^2 - 10$ $\dfrac{y}{-2} = -1 \qquad$ Simplify: $3^2 - 10 = -1.$ $\dfrac{-2\,y}{-2} = -1(-2) \quad$ Solve for y. $y = 2$	2. Solve. $\dfrac{y}{-3} = 2^2 - 8$
Solving equations using the division principle, pp. 389–390 1. Simplify each side of the equation. 2. If the variable is *multiplied* by a number, we *undo* the *multiplication* by *dividing* both sides of the equation by this number so that the variable is on one side and a number is on the other side of the equals sign: $\quad x = $ *some number* or *some number* $= x$ 3. Check by substituting your answer back into the original equation.	Solve. $\dfrac{-16}{4} = -3(2x) + 9x$ $\dfrac{-16}{4} = -3(2x) + 9x$ $-4 = -3(2x) + 9x \quad$ Simplify. $\dfrac{-16}{4} = -4.$ $-4 = -6x + 9x \quad$ Simplify: $-3(2x) = -6x.$ $-4 = 3x \qquad\quad$ Combine like terms. $\dfrac{-4}{3} = \dfrac{3x}{3} \qquad\quad$ Solve for x. $-\dfrac{4}{3} = x$	3. Solve. $-\dfrac{15}{5} = -2(4x) + 10x$
Solving equations using more than one principle of equality, p. 396 1. Remove any parentheses. 2. Combine like terms and simplify numerical work. 3. Use the addition principle to get all variable terms on one side of the equals sign and numerical values on the other side. 4. Use the multiplication or division principle to get the variable alone on one side of the equals sign. 5. Check your solution.	Solve. $4x - 7 = 2 + 6x + 9$ $4x - 7 = 2 + 6x + 9$ $4x - 7 = 11 + 6x$ $\begin{aligned} +\,-4x \qquad\qquad -4x \\ \hline -7 = 11 + 2x \end{aligned}$ $-7 = 11 + 2x$ $\begin{aligned} +\,-11 \quad -11 \\ \hline -18 = \qquad 2x \end{aligned}$ $\dfrac{-18}{2} = \dfrac{2x}{2}$ $-9 = x$ *Check.* $\quad 4x - 7 = 2 + 6x + 9$ $\quad 4(-9) - 7 \overset{?}{=} 2 + 6(-9) + 9$ $\quad -36 - 7 \overset{?}{=} 2 + (-54) + 9$ $\qquad -43 = -43 \; ✓$	4. Solve. $6x - 4 = 3 + 8x + 7$

Topic and Procedure	Examples	You Try It
Solving equations with parentheses, p. 402 Simplify the equation and then solve for the variable.	Solve. $-2(9x - 1) = 4(8x + 3) + 15$ $-18x + 2 = 32x + 12 + 15$ Remove parentheses. $\begin{aligned} -18x + 2 &= 32x + 27 \quad \text{Simplify.} \\ + \ +18x \quad\quad\ &\ +18x \\ \hline 2 &= 50x + 27 \quad \text{Solve for } x. \\ + -27 \quad\quad\ &\ -27 \\ \hline -25 &= 50x \\ \dfrac{-25}{50} &= \dfrac{50x}{50} \\ -\dfrac{1}{2} &= x \end{aligned}$	5. Solve. $-3(8x - 2) = 2(7x + 1) + 23$
Solving equations involving fractions, p. 407 **1.** Determine the LCD of all the denominators. **2.** Multiply every term of the equation by the LCD. **3.** Solve the resulting equation. **4.** Check your solution.	Solve. $4x + \dfrac{1}{2} = \dfrac{2}{5}$ $4x + \dfrac{1}{2} = \dfrac{2}{5}$ The LCD is 10. $(10)\, 4x + (10) \cdot \dfrac{1}{2} = (10) \cdot \dfrac{2}{5}$ Multiply each term by 10. $\begin{aligned} 40x + 5 &= 4 \\ + \quad -5 \quad &\ -5 \\ \hline 40x &= -1 \end{aligned}$ $x = -\dfrac{1}{40}$	6. Solve. $2x + \dfrac{1}{3} = \dfrac{2}{7}$

Using Equations to Solve Applied Problems, p. 411

Use the six-step process to organize a Mathematics Blueprint for Problem Solving.

> **PROCEDURE TO SOLVE APPLIED PROBLEMS**
> 1. Read the problem carefully to get an overview.
> 2. Write down formulas or draw pictures if possible.
> 3. Define the variable expressions.
> 4. Write an equation using the variable expressions defined in step 3.
> 5. Solve the equation and determine the values asked for in the problem.
> 6. Check your answer. Ask yourself if the answers obtained are reasonable.

EXAMPLE For the following problem:

(a) Define the variable expressions.

(b) Write an equation.

(c) Solve the equation and determine the values asked for.

(d) Check your answers.

The cost of two paintings sold at an art show totals $35,000. The cost of the painting *The Flower Garden* is $5000 more than double the cost of the *Sunset* painting. How much does each painting cost?

Understand the problem. Fill in the Mathematics Blueprint for Problem Solving.

Mathematics Blueprint for Problem Solving

Gather the Facts	What Am I Asked to Do?	How Do I Proceed?	Key Points to Remember
The Flower Garden costs $5000 more than double the cost of the *Sunset* painting. The total of the two paintings is $35,000.	Find the cost of each painting.	**(a)** Let the variable x represent the cost of the *Sunset* painting and write an expression for the cost of *The Flower Garden*. **(b)** Set the sum of the expressions equal to $35,000. **(c)** Solve the equation.	I must find the cost of both paintings, and check my answers.

Solve and state the answer.

(a) Define the variable expressions. We are comparing the cost of *The Flower Garden* to the cost of the *Sunset* painting, so we let the variable represent the cost of the *Sunset* painting.

$$\text{Cost of the \textit{Sunset} painting} = x; \qquad \text{Cost of \textit{The Flower Garden}} = 2x + 5000$$

(b) Form an equation.

$$\begin{array}{c}\text{Cost of the} \\ \textit{Sunset} \text{ painting}\end{array} + \begin{array}{c}\text{Cost of \textit{The}} \\ \textit{Flower Garden}\end{array} = \$35{,}000$$

$$x \quad + \quad (2x + 5000) = \$35{,}000$$

(c) Solve the equation and determine the values asked for. $3x + 5000 = \$35{,}000$

$$3x = 30{,}000$$
$$x = 10{,}000 \qquad \text{The cost of the \textit{Sunset} painting is \$10,000.}$$

Find the cost of *The Flower Garden*. $2x + 5000$ Write the variable expression, and replace x with 10,000

$$2(10{,}000) + 5000 = 25{,}000 \quad \text{The cost of \textit{The Flower Garden} is \$25,000.}$$

(d) *Check.* Is the total cost of the two paintings equal to $35,000? $\$10{,}000 + \$25{,}000 \stackrel{?}{=} \$35{,}000$

$$\$35{,}000 = \$35{,}000 \quad \checkmark$$

Chapter 7 Review Problems

Section 7.1

Solve and check your solution.

1. $x - 5 = 37$

2. $2(3x) = -36$

3. $-15 - 20 = 3x + 13 - 2x$

4. $\dfrac{y}{-2} = -8 + 3^2$

5. $-5 + 2^2 = \dfrac{x}{-5}$

6. $\dfrac{6}{7}a = 2^2 - 1$

7. $-3x + 2(4x) = \dfrac{18}{-2}$

8. $-x = -18$

9. $-y = \dfrac{3}{4}$

Section 7.2

Solve and check your solution.

10. $6x - 8 = 34$

11. $-2y + 16 = 58$

12. $-20 = 8 - 7y$

13. $14 = 10 - 4x$

14. $1 - 4y = 19$

15. $8x - 7 - 5x = 15$

16. $-6x = 9x + 36$

17. $5x = 2x + 30$

18. $3x - 8 - 5x = 6$

19. $-3 + 2y + 6 = -4y + 12$

20. $6y - 8 + 2y = -6 + 9y$

21. $8x - 9 - 5x + 18 = -3x - 2 + 2x$

22. $-3x - 2 - 9x + 5 = 2x + 8 - 7x$

Section 7.3

Solve and check your solution.

23. $-3(x + 5) = 21$

24. $4(2x + 9) + 5x = -3$

25. $7(x + 1) + 3(x + 1) = -10$

26. $-3(x - 7) = 5(x + 6) - 10$

27. $(9y^2 + 8y - 2) - (9y^2 + 5y - 1) = 14$

28. $(5x^2 + x - 2) - (5x^2 - 5) = 6x + 9$

Section 7.4

Solve and check your solution.

29. $\dfrac{x}{3} + \dfrac{x}{4} = 7$

30. $\dfrac{x}{5} - \dfrac{x}{2} = 6$

31. $y + \dfrac{y}{5} = -12$

32. $2y - \dfrac{y}{4} = 7$

33. $2x - \dfrac{3}{4} = \dfrac{1}{3}$

34. $2x - \dfrac{3}{4} = \dfrac{1}{2}$

35. $-\dfrac{1}{3} = 3y + \dfrac{1}{2}$

Section 7.5

For exercises 36 and 37:

(a) Write an equation.   *(b) Solve the equation and determine the value asked for in the problem.*

36. Four times a number decreased by nine is fifteen. What is the number?

▲ 37. Find the width of the following rectangle if the perimeter is 54 feet.

$L = 16$ ft

$W = x + 3$

▲ 38. *Geometry* Leslie has a vegetable garden in her yard that has width $= 8$ ft and length $= 15$ ft. Leslie plans to increase the length by x ft so that the new perimeter is 56 ft.

(a) By how much should the length be enlarged?

(b) What will the length of the enlarged vegetable garden be?

$L = 15$ ft $+ x$

$W = 8$ ft

For each of the following problems:

(a) Define the variable expressions.   *(b) Write an equation.*

(c) Solve the equation and determine the values asked for in the problem.   *(d) Check your answer.*

39. *Age Comparison* Sara and Tara are cousins. Sara is seven years younger than Tara. The sum of their ages is twenty-nine. How old is each?

40. *Enrollment Comparison* Last year a total of 491 students took Spanish. 85 more students took it in the spring than in the fall. 65 fewer students took it in the summer than in the fall. How many students took it during each of the three semesters?

How Am I Doing? Chapter 7 Test

Solve.

1. $5(2x) = -30$

2. $7x + 3 - 6x = -4$

3. $\dfrac{2}{3}a = 3^2 + 1$

4. $-x = 12$

ᴹᶜ **5.** $5x - 11 = -1$

6. $\dfrac{x}{5} = -2 + 4^2$

7. $4(-2y) + 5y = 12$

8. $12y + 6 - 11y - 1 = -8 + 10$

9. $-3 = 7 - 4y$

ᴹᶜ **10.** $6x + 4 - 9x = 15$

11. $14x = -2x + 16$

12. $2x + 4(3x - 6) = 4 - (6x + 2)$

Solve.

13. $3(2x + 6) + 3x = -27$

ᴹᶜ **14.** $4(x - 1) = -6(x + 2) + 48$

15. $(3x^2 - 2x + 1) + (-3x^2 - 10) = 5x + 5$

Solve.

ᴹᶜ **16.** $\dfrac{x}{5} + \dfrac{x}{2} = 7$

17. $2x + \dfrac{1}{3} = \dfrac{1}{2}$

18. $\dfrac{x}{6} + x = 14$

Write the answers. Use the following information for 19–21.

The first side of a triangle is 6 feet longer than the second. The length of the third side is triple the second. The perimeter of the triangle is 26 feet. Find the length of each side of the triangle.

19. Define the variable expressions.

20. Write an equation.

21. Solve the equation and determine the values asked for.

Write the answers. Use the following information for 22–24.

Anna is a store supervisor. The sales clerk earns $4000 per year less than Anna. The sum of Anna's annual salary and the clerk's annual salary is $61,200. Find Anna's and the clerk's annual salaries.

22. Define the variable expressions.

23. Write an equation.

24. Solve the equation and determine the values asked for.

1.	☐
2.	☐
3.	☐
4.	☐
5.	☐
6.	☐
7.	☐
8.	☐
9.	☐
10.	☐
11.	☐
12.	☐
13.	☐
14.	☐
15.	☐
16.	☐
17.	☐
18.	☐
19.	☐
20.	☐
21.	☐
22.	☐
23.	☐
24.	☐

Total Correct: ☐

MATH COACH

Mastering the skills you need to do well on the test.

Students often make the same types of errors when they do the Chapter 7 Test. Here are some helpful hints to keep you from making these common errors on test problems.

Solving Equations Using More Than One Principle of Equality—Problem 5   Solve. $5x - 11 = -1$

> **Helpful Hint** Use the addition principle *before* the multiplication or division principle. Remember that whatever you do to one side of the equation, you must do to the other side of the equation.

Examine your work for Problem 5.

As your first step, did you use the division principle and divide each term by 5?

Yes ____   No ____

If you answered Yes, your resulting equation contains fractions. To simplify calculations, use the addition principle first. Please stop now and complete this step.

Did you add 11 to *both sides* of the equation?

Yes ____   No ____

If you answered No, reread the Helpful Hint box and complete this step again.

As your last step, did you divide by 5 on both sides of the equation?

Yes ____   No ____

If you answered No, you may have used the wrong property. To undo multiplication, we must divide.

If you answered Problem 5 incorrectly, go back and rework the problem using these suggestions.

Simplifying and Solving Equations—Problem 10   Solve. $6x + 4 - 9x = 15$

> **Helpful Hint** First simplify the equation by combining like terms. Then solve the equation. Watch for sign errors.

Did you get $-3x + 4 = 15$ when you combined like terms?

Yes ____   No ____

Next did you subtract 4 from both sides of the equation?

Yes ____   No ____

If you answered No to either question, go back and complete these steps again.

As your last step, did you divide both sides by 3?

Yes ____   No ____

If you answered Yes, you solved for $-x$ not x. Stop now and look at your work. What number must you divide $-3x$ by to get x alone on one side of the equation?

Now go back and rework the problem using these suggestions.

Need help? Watch the MATH COACH videos in MyMathLab® or on You Tube™.

426

Solving Equations with Parentheses—Problem 14 Solve. $4(x - 1) = -6(x + 2) + 48$

Helpful Hint
- Did you remove parentheses?
- Did you simplify each side of the equation separately?
- Did you isolate x terms on one side of equation and then use the proper principle of equality to solve for x?

After removing parentheses, did you get
$4x - 4 = -6x - 12 + 48$?

Yes ____ No ____

If you answered No, stop and double-check all your calculations.

When simplifying the right side of the equation, did you get $-6x + 36$?

Yes ____ No ____

If you answered No, take care with signs. You may find it helpful to write out all your steps and keep your work organized.

To get all of the x terms on one side of the equation, did you get either $10x = 40$, or $-40 = -10x$?

Yes ____ No ____

If you answered No, examine your work carefully. Try to identify your error before finishing the problem.

If you answered Problem 14 incorrectly, go back and rework the problem using these suggestions.

Solving Equations Using the LCD Method   Problem 16   Solve. $\frac{x}{5} + \frac{x}{2} = 7$

Helpful Hint To make calculations easier, always clear the fractions first. Be careful not to forget to multiply *all terms* in the equation by the least common denominator (LCD).

Did you get 10 for the LCD?

Yes ____ No ____

Did you remember to multiply each term, including 7, by the LCD?

Yes ____ No ____

If you answered No to either question, stop and make these corrections.

Did you get $7x = 70$ after clearing fractions and simplifying?

Yes ____ No ____

If you answered No, examine your work closely. Check each step of the process for a calculation error.

Now go back and rework the problem using these suggestions.

Need more help? Look for section examples marked with ^{M}C to review.

CHAPTER 8

When we are seeking employment we must evaluate all the benefits and costs involved with each job offer before we determine which one will fit our needs and budget. How far must I travel? What extra expenses will the job add to my budget? Which job nets more money after expenses? The mathematics covered in this chapter will help you answer these questions.

Decimals and Percents

8.1 Understanding Decimal Fractions

① Writing Word Names for Decimal Fractions

> A **decimal fraction** is a fraction whose denominator is a power of 10: 10, 100, 1000, and so on.

$\frac{9}{10}$ is a decimal fraction. $\frac{41}{100}$ is a decimal fraction.

We can represent the shaded part of a whole as a decimal fraction in different ways (forms): using word names, as a fraction, or in decimal form.

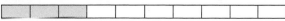

In Words	*In Fractional Form*	*In Decimal Form*
three-tenths	$\frac{3}{10}$	.3

All mean the same quantity, namely, 3 out of 10 equal parts of a whole. The "." in the decimal form .3 is called a **decimal point.** Usually a zero is placed in front of the decimal point to make sure that we don't miss seeing the decimal point.

.3 has the same value as 0.3

decimal point   extra zero used only for clarity

Since situations may require that we use different forms of decimal fractions, it is important that we understand the meaning of decimal fractions as well as how to write decimal fractions in each form. A **place-value chart** is helpful.

Hundreds	Tens	Ones	Decimal Point	Tenths	Hundredths	Thousandths	Ten Thousandths
100	10	1	"and"	$\frac{1}{10}$	$\frac{1}{100}$	$\frac{1}{1000}$	$\frac{1}{10,000}$

This place-value chart is an extension of the one we used in Chapter 1 to name whole numbers. From the chart we see the following:

1. The names for the place values to the right of the decimal point end with *"ths"* compared to those to the left of the decimal point. Therefore, we must take care when stating the name of the place value.

   "3 hundreds" names the number 300. "3 hundred*ths*" names the number $\frac{3}{100}$.
2. The word name for a decimal point is *and*.

We can use the place-value chart to help us write a word name for a decimal fraction.

EXAMPLE 1   Write a word name for the decimal 0.561.

Solution   The place value of the *last digit* is the *last word* in the word name. The last digit is 1 and is in the *thousandths* place.

We do not include 0 as part of word name. ←┐   ┌— thousandths place
                                    0.561

Five hundred sixty-one *thousandths*

Student Practice 1   Write a word name for each decimal.
(a) 0.365   **(b)** 5.32

Student Learning Objectives

After studying this section, you will be able to:

① Write word names for decimal fractions.

② Convert between decimals and fractions.

③ Compare and order decimals.

④ Round decimals.

NOTE TO STUDENT: Fully worked-out solutions to all of the Student Practice problems can be found at the back of the text starting at page SP-1.

Decimal notation is used when writing a check. Often, we write an amount that is less than 1 dollar, such as 35 cents, as $\frac{35}{100}$ dollar.

June Schultz		2882
3 Barker Road		
Placentia, CA 92870	DATE _April 6_ 20 _09_	

PAY to the
ORDER of _Jason Briggs_ $ _8.35_
Eight and ³⁵/₁₀₀ ————————————————— DOLLARS

Norwalk Central Bank
Norwalk, California

MEMO_____   _June Schultz_
⑆5800520⑆ 20550522⑈ 2882

EXAMPLE 2   Write the word name for a check written to Shandell Strong for $126.87.

Solution   One hundred twenty-six and $\dfrac{87}{100}$ Dollars

---

Student Practice 2   Write the word name for a check written to Wendy King for $245.09.   Dollars

---

② Converting Between Decimals and Fractions

Decimal fractions can be written with numerals in two ways: decimal notation or fractional notation

Decimal Notation	Fractional Notation
0.7	$\dfrac{7}{10}$
3.49	$3\dfrac{49}{100}$
0.021	$\dfrac{21}{1000}$

From the table, note the following.

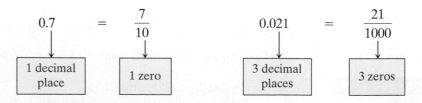

0.7 = $\dfrac{7}{10}$ 0.021 = $\dfrac{21}{1000}$

| 1 decimal place | 1 zero | 3 decimal places | 3 zeros |

PROCEDURE TO CHANGE FROM DECIMAL TO FRACTIONAL NOTATION

1. Write the whole number (if any). $9.\underbrace{7653} \to 9$

2. Count the number of decimal places.   4 decimal places

3. Write the decimal part over a denominator that has a 1 and the same number of zeros as the number of decimal places found in step 2.

$9\dfrac{7653}{10,000}$

4 zeros

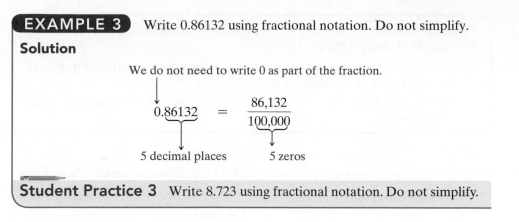

EXAMPLE 3 Write 0.86132 using fractional notation. Do not simplify.

Solution

We do not need to write 0 as part of the fraction.

$$0.86132 = \frac{86,132}{100,000}$$

5 decimal places   5 zeros

Student Practice 3 Write 8.723 using fractional notation. Do not simplify.

If a fraction has a denominator that is a power of 10 (10, 100, 1000, and so on), we use a similar procedure to change the fraction to a decimal.

PROCEDURE TO CHANGE FRACTIONAL NOTATION TO DECIMAL NOTATION WHEN THE DENOMINATOR IS A POWER OF 10

1. Count the number of zeros in the denominator. $9\frac{7653}{10,000}$ 4 zeros

2. In the numerator, move the decimal point as many places to the left as the number of zeros in step 1. If there are not enough places, add zeros until there are. Then delete the denominator. $9\frac{7653}{10,000} \rightarrow 9.7653$

      4 places

EXAMPLE 4 Write $7\frac{56}{1000}$ as a decimal.

Solution

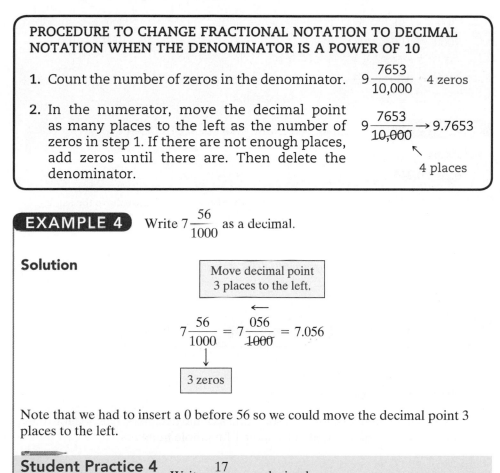

Move decimal point 3 places to the left.

$$7\frac{56}{1000} = 7\frac{056}{1000} = 7.056$$

3 zeros

Note that we had to insert a 0 before 56 so we could move the decimal point 3 places to the left.

Student Practice 4 Write $\frac{17}{1000}$ as a decimal.

In Section 8.3 we will see how to change fractions to decimals when the denominator is not a power of 10.

③ Comparing and Ordering Decimals

In Chapter 1 we studied the inequality symbols "<" and ">." Recall that

$a < b$ is read "a is less than b."

$a > b$ is read "a is greater than b."

To compare and order decimals using inequality symbols, we compare each digit.

PROCEDURE TO COMPARE TWO POSITIVE NUMBERS IN DECIMAL NOTATION

1. Start with the leftmost digit and compare corresponding digits. If the digits are the same, move one place to the right.
2. When two digits are different, the larger number is the one with the larger digit.

Note: We must write decimals such as .21 as 0.21 for the above procedure to apply.

It is easier to compare two decimals if the decimal parts of each have the same number of digits. Whenever necessary, extra zeros can be written to the right of the last digit—that is, to the *right* of the *last digit after* the decimal point—without changing the value of the decimal. To see why, let's look at $\frac{3}{10}, \frac{30}{100}, \frac{300}{1000}$.

$$\frac{300}{1000} = \frac{30}{100} = \frac{3}{10} \qquad \frac{3}{10} \text{ is the simplified form of the fractions } \frac{300}{1000} \text{ and } \frac{30}{100}.$$
$$\downarrow \qquad \downarrow \qquad \downarrow$$
$$0.300 = 0.30 = 0.3 \qquad \text{We can think of 0.3 as the simplified form of 0.300 and 0.30.}$$

EXAMPLE 5 Replace the ? with $<$ or $>$. 0.24 ? 0.244

Solution

0.24 ? 0.244

0.240 ? 0.244   Add a zero to 0.24 so that both decimal parts have the same number of digits.

0.240 ? 0.244   The tenths digits are equal and the hundredths digits are equal.

0.240 ? 0.244   The thousandths digits differ.

Since $0 < 4$, $0.240 < 0.244$.

Student Practice 5 Replace the ? with $<$ or $>$. 0.77 ? 0.771

④ Rounding Decimals

Just as with whole numbers, we must sometimes round decimals. The rule for rounding is similar to the one we used in Chapter 1 for whole numbers.

PROCEDURE TO ROUND DECIMALS

1. Identify the round-off place digit.
2. If the digit to the *right* of the round-off place digit is:
   (a) *Less than* 5, do not change the round-off place digit.
   (b) 5 *or more*, increase the round-off place digit by 1.
3. In either case, drop all digits to the right of the round-off place digit.

Thus, when rounding decimals we either *increase* the round-off place digit by 1 or *leave it the same*. We drop all digits to the right of the round-off place digit.

EXAMPLE 6   Round 237.8435 to the nearest hundredth.

Solution   The round-off place digit is in the *hundredths place*.

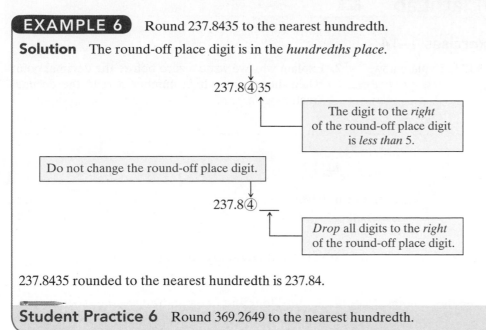

237.8④35

The digit to the *right* of the round-off place digit is *less than* 5.

Do not change the round-off place digit.

237.8④___

Drop all digits to the *right* of the round-off place digit.

237.8435 rounded to the nearest hundredth is 237.84.

Student Practice 6   Round 369.2649 to the nearest hundredth.

Remember that rounding up to the next digit in a position may result in several digits being changed.

EXAMPLE 7   Round to the nearest hundredth. Alex and Lisa used 204.9954 kilowatt-hours of electricity in their house in June.

Solution   We locate and circle the digit in the hundredths place.

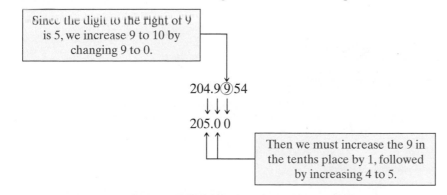

Since the digit to the right of 9 is 5, we increase 9 to 10 by changing 9 to 0.

204.9⑨54
↓↓↓
205.0 0

Then we must increase the 9 in the tenths place by 1, followed by increasing 4 to 5.

Note that we *must* include the two zeros after the decimal point because we were asked to round to the nearest hundredth.

205.0 0   When rounding to the nearest hundredth, the last digit should be in the hundredths place.

Thus 204.9954 rounds to 205.00 kilowatt-hours.

Student Practice 7   Round to the nearest tenth. Last month the college auditorium used 16,499.952 kilowatt-hours of electricity.

Verbal and Writing Skills, Exercises 1–14

1. 12.97 rounded to the nearest tenth is 13.0. Explain how rounding 9 up is similar to carrying.

2. Explain why we write a zero before the decimal point when there is no whole number part to the decimal fraction as in 0.23.

Write a word name for each decimal.

3. 5.32

4. 11.78

5. 0.428

6. 0.983

Write in decimal notation.

7. Three hundred twenty-four thousandths

8. One hundred twenty-six thousandths

9. Fifteen and three hundred forty-six ten thousandths

10. Twenty-four and one hundred seventy-six ten thousandths

Write the word name for the amount on each check.

11.
PAY to the ORDER of *Orange Coast College* $ *25.54*
_____ DOLLARS

12.
PAY to the ORDER of *Rita Smith* $ *8.75*
_____ DOLLARS

13.
PAY to the ORDER of *The Gas Company* $ *143.56*
_____ DOLLARS

14.
PAY to the ORDER of *Pep Boys Auto Parts* $ *146.32*
_____ DOLLARS

Write in fractional notation. Do not simplify.

15. 0.7

16. 0.9

17. 4.17

18. 7.21

19. 32.081

20. 0.8436

21. 0.5731

22. 30.041

Write each fraction as a decimal.

23. $6\dfrac{7}{10}$

24. $8\dfrac{1}{10}$

25. $12\dfrac{37}{1000}$

26. $63\dfrac{31}{1000}$

27. $\dfrac{1}{100}$

28. $\dfrac{3}{100}$

29. $\dfrac{1}{1000}$

30. $\dfrac{3}{1000}$

Replace the ? with < or >.

31. 0.426 ? 0.429   **32.** 0.316 ? 0.314   **33.** 0.09 ? 0.11   **34.** 0.72 ? 0.73

35. 0.36 ? 0.366   **36.** 0.12 ? 0.127   **37.** 0.7431 ? 0.743   **38.** 0.6362 ? 0.636

39. 0.3 ? 0.05   **40.** 0.6 ? 0.57   **41.** 0.502 ? 0.52   **42.** 0.703 ? 0.73

Round to the nearest hundredth.

43. 523.7235   **44.** 124.6345   **45.** 43.961   **46.** 76.996

Round to the nearest tenth.

47. 9.0546   **48.** 21.057   **49.** 462.931   **50.** 125.942

Round to the nearest thousandth.

51. 312.95144   **52.** 63.44431   **53.** 1286.3496   **54.** 2563.4895

Round to the nearest ten thousandth.

55. 0.063148   **56.** 0.043629   **57.** 0.047362   **58.** 0.095253

Applications

59. ***Mars and Earth*** Mars was only 42.5 million miles from Earth in June 2001. That was the closest it had been to Earth in 13 years. The two planets were only 34.6 million miles apart in August 2003. This was the closest they'd been in 5000 years! Round these values to the nearest million. (*Source:* mars.fpl.nasa.gov)

60. ***El Niño*** In any given year the average rainfall in Santiago Peak, California, is 33.46 inches. From July 1, 1997, to May 29, 1998, Santiago Peak had an increase in rainfall due to an El Niño condition, receiving 105.12 inches of rain. Round the average rainfall, and the rainfall during the El Niño season, to the nearest tenth. (*Source:* Orange County Register)

The menu at Emerald's Dinner House reads as follows.

Emerald's Dinner House

Grilled chicken breast . . . . $ 12.95
New York steak . . . . $ 15.25
Stuffed pork chops . . . . . . $ 13.75
Fish of the day . . . . $ 14.50
Lobster . . . . $ 18.25
Prime rib . . . . $ 15.75

61. Round the price of the New York steak to the nearest dollar.

62. Round the price of the prime rib to the nearest dollar.

63. The Austin family ordered three meals: stuffed pork chops, lobster, and prime rib. Round the price of each meal to the nearest dollar and estimate the cost of the three meals.

64. Lester, Marian, and Sean ordered three meals: grilled chicken breast, New York steak, and lobster. Round the price of each meal to the nearest dollar and estimate the cost of the three meals.

To Think About

65. Arrange in order from largest to smallest. 0.0069, 0.73, $\frac{7}{10}$, 0.007, 0.071

66. Arrange in order from smallest to largest. 0.053, 0.005, 0.52, 0.0059, $\frac{5}{100}$

Cumulative Review   *Perform the operations indicated.*

67. **[2.3.1]** $-15 - (-6)$

68. **[2.4.1]** $(356)(-28)$

69. **[2.4.4]** $-45 \div 9$

70. **[1.5.4]** $15,708 \div 231$

71. **[5.6.1]** *Mixed Nuts* Tory bought $\frac{2}{3}$ lb of cashews and $\frac{1}{2}$ lb of mixed nuts. How many total pounds of nuts did Tory buy?

72. **[5.6.1]** *Term Paper* An English instructor has specific layout requirements for the final term paper. The layout must include a top margin of $\frac{3}{4}$ inch and a bottom margin of $\frac{1}{2}$ inch. How much page length is lost because of the margins?

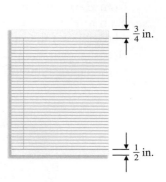

$\frac{3}{4}$ in.

$\frac{1}{2}$ in.

Quick Quiz 8.1

1. Write the word name. 7.21

2. (a) Write in fractional notation. 0.0217

(b) Write as a decimal. $4\frac{32}{1000}$

3. Round 156.748 to the nearest tenth.

4. **Concept Check** Explain how you know how many zeros to put in the denominator of your answer when you write 8.6711 as a fraction.

8.2 Adding and Subtracting Decimal Expressions

① Adding and Subtracting Decimal Expressions

We often add or subtract decimals when we work with money. For example, if we buy a drink for $1.25 and a sandwich for $3.55, how much is the total bill? It is $1.25 + $3.55, or $4.80. Now, if we pay with a $5 bill, how much change do we receive? $5.00 − $4.80, or $0.20 (20 cents).

We can relate addition and subtraction of fractions to decimals. For example,

$$2\frac{3}{10} = 2.3 \qquad\qquad 8\frac{5}{10} = 8.5$$
$$+5\frac{6}{10} = 5.6 \qquad\qquad -3\frac{2}{10} = 3.2$$
$$\overline{7\frac{9}{10} = 7.9} \qquad\qquad \overline{5\frac{3}{10} = 5.3}$$

What do you notice about the sum and difference in the examples above? Do you see that addition and subtraction of decimals are very much like those of whole numbers, except that we must consider placement of the decimal point?

Student Learning Objectives

After studying this section, you will be able to:

① Add and subtract decimal expressions.

② Combine like terms.

③ Evaluate algebraic expressions involving decimals.

④ Estimate sums and differences.

⑤ Solve applied problems involving decimals.

PROCEDURE TO ADD OR SUBTRACT DECIMALS

1. Write the numbers vertically and *line up* the decimal points. Extra zeros may be written to the right of the decimal points if needed.

2. Add or subtract all the digits with the same place value, starting with the right column and moving to the left. Use carrying or borrowing as needed.

3. Place the decimal point of the answer in line with the decimal points of all the numbers added or subtracted.

EXAMPLE 1   Add. $40 + 8.77 + 0.9$

Solution   We can write any whole number as a decimal by placing a decimal point at the end of the number: $40 = 40.$

$$
\begin{array}{r}
\overset{1}{4}0.00 \\
8.77 \\
+\ 0.90 \\
\hline
49.67
\end{array}
$$

Line up decimal points.

Add zeros so that each number has the same number of decimal places.

Place the decimal point in the answer in line with other decimal points.

Student Practice 1   Add. $50 + 4.39 + 0.7$

EXAMPLE 2   Subtract. $19.02 - 8.6$

Solution

$$
\begin{array}{r}
\overset{8}{1}\overset{10}{9}.\cancel{0}2 \\
-8.60 \\
\hline
10.42
\end{array}
$$

Line up decimal points.

Add a zero.

The decimal point in the difference is in line with the other decimal points.

Student Practice 2   Subtract.
$$26.01 - 5.7$$

When we add or subtract positive and negative decimals, we use the same rules stated in Chapter 2.

EXAMPLE 3 Perform the operation indicated. $-9.79 - (-0.68)$

Solution To subtract, we add the opposite of the second number.

$$-9.79 - (-0.68)$$
$$-9.79 + (0.68)$$
$$\downarrow \qquad \downarrow$$
Add the opposite of (-0.68).

Next, to add numbers with different signs, we "*keep the sign of the larger absolute value and subtract.*"

$$\begin{array}{r} -9.79 \\ \underline{0.68} \\ -9.11 \end{array}$$

We subtract.
The answer will be negative since $|-9.79|$ is larger than $|0.68|$.

NOTE TO STUDENT: *Fully worked-out solutions to all of the Student Practice problems can be found at the back of the text starting at page SP-1.*

Student Practice 3 Perform the operation indicated. $-3.02 - (-5.1)$

② Combining Like Terms

Recall that to combine like terms we add coefficients of like terms and the variable part stays the same.

EXAMPLE 4 Combine like terms. $11.2x + 3.6x - 7.1y$

Solution $11.2x$ and $3.6x$ are like terms, so we add them.

$$\begin{array}{r} 11.2x \\ +\ \ 3.6x \\ \hline 14.8x \end{array}$$ We line up decimal points and add.

We have $\underbrace{11.2x + 3.6x} - 7.1y$

$$= 14.8x - 7.1y$$

We cannot combine $14.8x$ and $7.1y$ since they are not like terms.

Student Practice 4 Combine like terms. $4.5y + 7.2y - 5.6x$

③ Evaluating Algebraic Expressions Involving Decimals

Recall that to evaluate an expression, we replace the variable with the given number and simplify.

EXAMPLE 5 Evaluate $x + 3.12$ for $x = 0.11$.

Solution We replace the variable with 0.11.

$$x + 3.12 = 0.11 + 3.12$$

Next we line up decimal points and add.

$$\begin{array}{r} 3.12 \\ +\ \ 0.11 \\ \hline 3.23 \end{array}$$

Student Practice 5 Evaluate $-6.2 + x$ for $x = 1.2$.

④ Estimating Sums and Differences

We can estimate a sum or difference of decimals by rounding each decimal to the nearest whole number.

EXAMPLE 6 Julie runs on her treadmill every day. She wants to run approximately 25 miles each week to prepare for a track race. She logged the distance she ran each day this week on the following chart. *Estimate* the total number of miles Julie ran this week.

Monday	Tuesday	Wednesday	Thursday	Friday	Saturday	Sunday
2.13 mi	2.79 mi	2.9 mi	3.11 mi	3.8 mi	4.12 mi	4.9 mi

Solution We round each decimal to the nearest whole number and then add.

$$2 + 3 + 3 + 3 + 4 + 4 + 5 = 24 \text{ miles}$$

Julie ran approximately 24 miles.

Student Practice 6 Allen ate at restaurants for two days while on a business trip. He logged his expenses on the chart located in the margin. *Estimate* the total cost of meals.

Student Practice 6

(4)	(5)	MEALS		(6)
DAY	BREAK-FAST	LUNCH	DINNER	
Tuesday	7.29	8.99	19.10	
Wednesday	6.99	9.20	21.76	

⑤ Solving Applied Problems Involving Decimals

EXAMPLE 7 In May 2001 Allen Iverson, a 6-foot, 165-pound guard for the Philadelphia 76ers, became the shortest and lightest player to be named the NBA Most Valuable Player. The table on the right compares some of the top players' average season statistics.

(a) How many more points did Iverson average than Bryant?

(b) Find the total of the average number of rebounds made by Iverson, Duncan, and James.

MVP	Team	MVP Season	Points	Rebounds	Assists
1. Allen Iverson	76ers	2000–2001	31.1	3.1	4.6
2. Tim Duncan	Spurs	2001–2002	25.5	12.7	3.7
3. Steve Nash	Suns	2004–2005	15.5	3.3	11.5
4. Kobe Bryant	Lakers	2007–2008	28.3	6.3	5.4
5. LeBron James	Cavaliers	2009–2010	29.7	7.3	8.6

Source: Orange County Register

Solution

(a) The key phrase "how many more" indicates the operation subtraction.

$$
\begin{array}{r}
\overset{10}{\cancel{2}}\,\overset{11}{\cancel{\emptyset}}\ \ \\
\text{Allen Iverson's points} \quad 3\cancel{1}.\cancel{1} \\
\text{Kobe Bryant's points} \quad -28.3 \\
\hline
2.8 \text{ more points}
\end{array}
$$

We borrow and then subtract.

(b) We add to find the total.

$$
\begin{array}{r}
\text{Iverson} \quad 3.1 \\
\text{Duncan} \quad 12.7 \\
\text{James} \quad 7.3 \\
\hline
23.1 \text{ rebounds}
\end{array}
$$

Student Practice 7 Refer to the table in Example 7 to answer the following.

(a) How many more assists did Nash average than Iverson?

(b) Find the total of the average number of assists for all five players listed on the chart.

Verbal and Writing Skills, Exercises 1–4 *Fill in the blanks.*

1. To add numbers in decimal notation, we _____ the decimal points.

2. When adding decimals, we place the decimal point in the answer _____ with the decimal points in the problem.

3. When subtracting decimals, we place the decimal point in the answer _____ with the decimal points in the problem.

4. When subtracting $73 - 23.4$, we rewrite 73 as _____ so that we can line up _____.

Add.

5. $0.34 + 7.21$

6. $0.56 + 3.42$

7. $1.01 + 3.46$

8. $0.63 + 8.42$

9. $63.2 + 0.2348$

10. $35.4 + 0.8759$

11. $73 + 7.54 + 0.483$

12. $59 + 1.27 + 0.345$

13. $73.1 + 0.3169$

14. $81.3 + 0.2124$

15. $15 + 2.73 + 0.423$

16. $0.562 + 13 + 4.32$

Subtract.

17. $53.783 - 2.46$

18. $48.575 - 5.44$

19. $16.54 - 3.9$

20. $125.43 - 2.8$

21. $20 - 0.36$

22. $40 - 0.72$

23. $-12.1 - 0.23$

24. $-13.6 - 0.51$

25. $-91.13 - 14.213$

26. $-88.14 - 16.315$

27. $-8.69 - (-4.12)$

28. $-7.22 - (-2.11)$

Combine like terms.

29. $2.3x + 3.9x$

30. $5.7x + 1.6x$

31. $24.8y - 9.2y$

32. $15.6y - 8.2y$

33. $3.5x + 9.1x - y$

34. $5.5x + 3.2x - 3y$

35. $1.4x + 6.2y + 3.5x$

36. $2.6x + 3.1y + 4.2x$

Mixed Problems 37–40 *Perform the operations indicated.*

37. **(a)** $-3.4 + (-2.1)$
  (b) $9.7 - (-5.4)$
  (c) $-9.2 - 4.1$

38. **(a)** $-1.13 + (-8.84)$
  (b) $8.31 - (-2.36)$
  (c) $-4.99 - 1.73$

39. **(a)** $4.6x + 2x$
  (b) $3.04y - 7.5y$
  (c) $x - 0.25x$

40. **(a)** $3.4x + 5x$
  (b) $2.06y - 1.2y$
  (c) $x - 0.44x$

Evaluate for the given value.

41. $y - 0.861$ for $y = 9$

42. $16.011 - n$ for $n = 9.7$

43. $211.2 - n$ for $n = 5.42$

44. $y - 0.12$ for $y = 7$

45. $x + 2.3$ for $x = -6.7$

46. $x - (-19.2)$ for $x = -0.09$

Applications

47. ***Payroll Deductions*** John makes $1763.24 a month. $161.96 is deducted for federal income tax, $61.23 for Social Security, and $47.82 for state taxes. *Estimate* to the nearest dollar how much money he takes home each month after the deductions are taken out.

48. ***Checking Account Balance*** Karen has $321.45 in her checking account. She makes deposits of $38.97 and $86.23. She writes checks for $23.10, $45.67, and $8.97. *Estimate* the new balance in her checking account.

Use the following bar graph to answer exercises 49–52.

The NASDAQ was founded in 1971 and is the largest U.S. electronic stock market, with about 3200 companies. The bar graph shows the closing numbers of the NASDAQ index during a selected week.

Sources: www.nasdaq.com; investor.wallstreetselect.com

49. *Estimate* the difference between closing numbers for Day 1 and Day 2.

50. *Estimate* the difference between closing numbers for Day 3 and Day 5.

51. What two days have the largest difference in closing numbers? *Estimate* this difference.

52. What two days have the smallest difference in closing numbers? *Estimate* this difference.

53. ***Change from Purchase*** Ann spent $72.31 on groceries for her family. If she gives the clerk a 100-dollar bill, how much change should she get?

54. ***Odometer Reading*** Charles checked his odometer before the summer began. It read 2301.2 miles. He traveled 1236.9 miles that summer in his car. What was the odometer reading at the end of the summer?

55. ***Olympic Record*** How much faster was the 1988 Olympic 100-meter winning time by Florence Griffith Joyner than the one by Gail Devers in the 1992 Olympics?

56. ***Olympic Record*** How much slower was the 1960 Olympic 100-meter winning time by Wilma Rudolph than the record set by Evelyn Ashford in 1984?

Women's 100-meter Winning Time

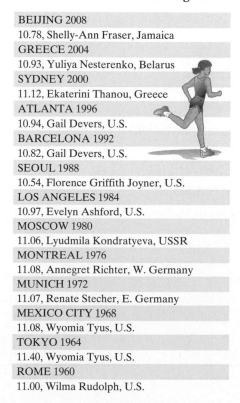

| BEIJING 2008 |
| 10.78, Shelly-Ann Fraser, Jamaica |
| GREECE 2004 |
| 10.93, Yuliya Nesterenko, Belarus |
| SYDNEY 2000 |
| 11.12, Ekaterini Thanou, Greece |
| ATLANTA 1996 |
| 10.94, Gail Devers, U.S. |
| BARCELONA 1992 |
| 10.82, Gail Devers, U.S. |
| SEOUL 1988 |
| 10.54, Florence Griffith Joyner, U.S. |
| LOS ANGELES 1984 |
| 10.97, Evelyn Ashford, U.S. |
| MOSCOW 1980 |
| 11.06, Lyudmila Kondratyeva, USSR |
| MONTREAL 1976 |
| 11.08, Annegret Richter, W. Germany |
| MUNICH 1972 |
| 11.07, Renate Stecher, E. Germany |
| MEXICO CITY 1968 |
| 11.08, Wyomia Tyus, U.S. |
| TOKYO 1964 |
| 11.40, Wyomia Tyus, U.S. |
| ROME 1960 |
| 11.00, Wilma Rudolph, U.S. |

57. *Olympic Record* What is the difference between the fastest and the slowest 100-meter Olympic winning times for the years 1960 to 2004?

58. *Olympic Record* Which pair or pairs of consecutive Olympic games had the largest difference in 100-meter winning times? Which had the smallest difference?

One Step Further *Perform the operations indicated.*

59. $-2.3 - (-0.24) + 4.6 - 9$

60. $-3.8 - (-0.46) + 8.2 - 14$

61. $\dfrac{3}{10} - 1.26 + (-2.3)$

62. $\dfrac{7}{10} - 4.36 + (-3.1)$

To Think About *Guess the next seven digits in each of the following.*

63. $5.636336333633336\ldots$

64. $6.1213314441\ldots$

65. $8.181181118\ldots$

66. $3.043004300043\ldots$

67. $12.98987987698765\ldots$

68. $7.6574839201102938\ldots$

Cumulative Review *Multiply.*

69. **[1.4.4]** $(231)(14)$

70. **[2.4.1]** $(-12)(92)$

Divide.

71. **[1.5.4]** $2940 \div 12$

72. **[2.4.4]** $3105 \div (-3)$

Quick Quiz 8.2

1. Perform the operations indicated.
   (a) $42.09 + 3.1$
   (b) $5.03 - 2.68$
   (c) $28.61 - (-4.21)$

2. Combine like terms. $5.6x + 2.13x + 9.01$

3. Evaluate $y - 18.75$ for $y = -20.96$.

4. **Concept Check** Explain how you would evaluate $x - 3.1$ for $x = 0.866$.

8.3 Multiplying and Dividing Decimal Expressions

① Multiplying Decimals

Just as with addition and subtraction, we can relate multiplication of fractions to decimals. For example:

$$\frac{5}{10} \times \frac{9}{100} = \frac{45}{1000} \quad \text{Fractional notation}$$

$$\downarrow \qquad \downarrow \qquad \downarrow$$

$$0.5 \times 0.09 = 0.045 \quad \text{Decimal notation}$$

In both cases we multiply $9 \times 5 = 45$. When we multiply using decimal notation, we must decide where to place the decimal point in the product, 45. We determine this by adding the number of decimal places in each factor.

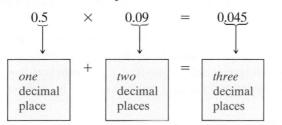

Student Learning Objectives

After studying this section, you will be able to:

① **Multiply decimals.**

② **Multiply a decimal by a power of 10.**

③ **Divide a decimal by a whole number.**

④ **Divide a decimal by a decimal.**

⑤ **Change a fraction to a decimal.**

PROCEDURE TO MULTIPLY DECIMALS

1. Multiply the numbers just as you would multiply whole numbers.
2. Find the total number of decimal places in the factors.
3. Place the decimal point in the product so that the product has the same number of decimal places as the total in step 2. To do this, you may need to write zeros to the left of the answer from step 1.

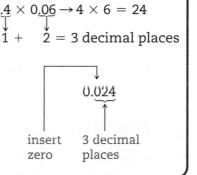

EXAMPLE 1 Multiply. 0.08×0.04

Solution

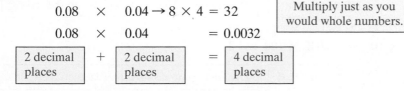

Note that we had to insert two zeros to the left of 32 in order to have 4 decimal places in the product.

Student Practice 1 Multiply. 0.05×0.07

When multiplying larger numbers, it is usually easier to perform the calculation if we multiply vertically, placing the factor with the fewer number of nonzero digits underneath the other factor.

EXAMPLE 2   Multiply. 5.33×7.2

Solution   We write the multiplication just as we would if there were no decimal points.

$$
\begin{array}{r}
5.33 \quad \text{2 decimal places} \\
\times \ 7.2 \quad \text{1 decimal place} \\
\hline
1066 \\
3731 \\
\hline
38.376 \quad \text{We need 3 decimal places } (2 + 1 = 3).
\end{array}
$$

Note that we *do not* line up the decimal points when we multiply.

Student Practice 2   Multiply. 20.1×4.32

To multiply or divide positive and negative decimals, we use the same rules stated in Chapter 2. For your convenience we summarize the rules:

The sign of the answer will be *positive* if the problem has an *even* number of negative signs and *negative* if the problem has an *odd* number of negative signs.

EXAMPLE 3   Multiply. $(-2)(4.51)$

Solution   The number of negative signs, 1, is odd so the product is negative.

$$
\begin{array}{r}
4.51 \quad \text{2 decimal places} \\
\times \ (-2) \quad \text{0 decimal places} \\
\hline
-9.02 \quad \text{We need 2 decimal places } (2 + 0 = 2).
\end{array}
$$

Student Practice 3   Multiply. $(-3)(6.22)$

② Multiplying a Decimal by a Power of 10

Observe the following pattern.

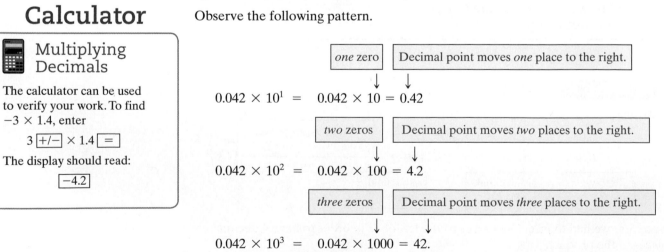

| one zero | Decimal point moves *one* place to the right. |

$0.042 \times 10^1 = \ 0.042 \times 10 = 0.42$

| two zeros | Decimal point moves *two* places to the right. |

$0.042 \times 10^2 = \ 0.042 \times 100 = 4.2$

| three zeros | Decimal point moves *three* places to the right. |

$0.042 \times 10^3 = \ 0.042 \times 1000 = 42.$

Calculator

Multiplying Decimals

The calculator can be used to verify your work. To find -3×1.4, enter

3 $\boxed{+/-}$ $\times$ 1.4 $\boxed{=}$

The display should read:

$\boxed{-4.2}$

PROCEDURE TO MULTIPLY A DECIMAL BY A POWER OF 10

To multiply a decimal by a power of 10, move the decimal point to the *right* the same number of places as the number of zeros in the power of 10. It may be necessary to add zeros at the end of the number.

EXAMPLE 4   Multiply. 0.2345×1000

Solution

three zeros

$$0.2345 \times 1000 = 234.5$$

Move decimal point *three* places to the right.

Student Practice 4   Multiply. 0.123×100

If the number that is a power of 10 is in exponent form, move the decimal point to the right the same number of places as the number that is the exponent.

EXAMPLE 5   Multiply. 15×10^4

Solution   Since 10^4 has 4 zeros ($10^4 = 10,000$), we must move the decimal point to the right 4 places.

$$(15)(10^4) \quad = \quad (15.0)(10^4) \quad = \quad 150000. \quad \text{or} \quad 150,000$$

Rewrite in decimal notation.   Add zeros and
                                    move decimal point.

Student Practice 5   Multiply. $(0.6944)(10^3)$

③ **Dividing a Decimal by a Whole Number**

Just as with addition, subtraction, and multiplication, the only new rule we must learn when dividing decimal numbers concerns the placement of the decimal point. To divide a decimal by a whole number, we place the decimal point in the quotient directly above the decimal point in the dividend.

To divide $33.6 \div 6$, we proceed as follows.

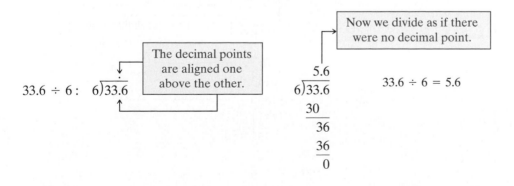

$$33.6 \div 6 : \quad 6\overline{)33.6}$$

The decimal points are aligned one above the other.

Now we divide as if there were no decimal point.

$$\begin{array}{r} 5.6 \\ 6\overline{)33.6} \\ \underline{30} \\ 36 \\ \underline{36} \\ 0 \end{array}$$

$$33.6 \div 6 = 5.6$$

PROCEDURE TO DIVIDE A DECIMAL BY A WHOLE NUMBER

1. Place the decimal point in the quotient directly above the decimal point in the dividend.
2. Divide as if there were no decimal point involved.

Often, we must add extra zeros to the right end of the dividend so that we can continue dividing.

EXAMPLE 6 Divide. $2.3 \div 5$

Solution $2.3 \div 5$: Think "5 goes into 2.3."

We place the decimal point directly above the decimal point in the dividend.

$$
2.3 \div 5 \;\longrightarrow\; 5\overline{)2.3}^{\,0.4} \qquad \text{Now we divide as if there were no decimal point.}
$$
$$
\underline{2\,0}
$$
$$
3
$$

$$
5\overline{)2.30}^{\,0.46} \qquad \text{We add a zero so that we can continue to divide.}
$$
$$
\underline{2\,0}
$$
$$
30 \qquad \text{We bring down a zero.}
$$
$$
\underline{30}
$$
$$
0
$$

$2.3 \div 5 = 0.46$

Student Practice 6   Divide. $1.3 \div 2$

Sometimes a division problem does not yield a remainder of zero, or we must carry out the division many decimal places before we get a remainder of zero. In such cases we may be asked to round the answer to a specified place.

EXAMPLE 7   Divide $-0.185 \div 13$. Round your answer to the nearest thousandth.

Solution   We must divide *one place* beyond the thousandths place—that is, to the ten thousandths place—so we can round to the nearest thousandth.

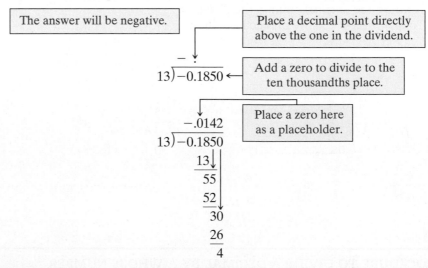

The answer will be negative.

Place a decimal point directly above the one in the dividend.

Add a zero to divide to the ten thousandths place.

Place a zero here as a placeholder.

-0.0142 rounded to the nearest thousandth is -0.014; $-0.185 \div 13 \approx -0.014$. Note that we used the symbol $\approx$ to indicate that our answer is an approximate value.

Student Practice 7   Divide $-0.3624 \div 14$. Round your answer to the nearest hundredth.

④ Dividing a Decimal by a Decimal

So far we have considered only division by whole numbers. When the *divisor is not a whole number,* we must adjust the placement of the decimal point so that we have an equivalent division problem with a whole number as the divisor.

$\mathbb{M}_\mathbb{C}$ **EXAMPLE 8**   Divide. $6.93 \div 2.2$

Solution   Since the divisor, 2.2, is *not* a whole number, let's write the division problem using fractional notation.

$$6.93 \div 2.2 = \frac{6.93}{2.2}$$

Now, if we multiply the numerator and denominator by 10, the divisor becomes a whole number.

$$\frac{(6.93)(10)}{(2.2)(10)} = \frac{69.3}{22} = 69.3 \div 22 \quad \text{The divisor is a whole number.}$$

Once the divisor is a whole number, we can divide as usual using our new equivalent fraction.

$$
\begin{array}{r}
3.15 \\
22\overline{)69.30} \\
\end{array}
\quad \text{We add a zero so that we can continue the division.}
$$

$$
\begin{array}{r}
\underline{66} \\
33 \\
\underline{22} \\
110 \\
\underline{110} \\
0
\end{array}
$$

$6.93 \div 2.2 = 3.15$

Student Practice 8   Divide.

$$14.56 \div 3.5$$

Since multiplying by a power of 10 is the same as moving the decimal point to the *right,* we could have rewritten the division statement in Example 8 by moving the decimal point to the right one place in both the divisor and dividend.

$6.93 \div 2.2$ Move the decimal point one place to the right.

or

$69.3 \div 22.0$ The divisor is a whole number.

We can summarize the process for dividing with decimals as follows.

PROCEDURE TO DIVIDE WITH DECIMALS

1. If the divisor is a decimal, change it to a whole number by moving the decimal point to the right as many places as necessary.
2. Then move the decimal point in the dividend to the right the *same* number of places.
3. Place the decimal point in the quotient directly above the decimal point in the dividend.
4. Divide until the remainder becomes zero, or the remainder repeats itself, or the desired number of decimal places is achieved.

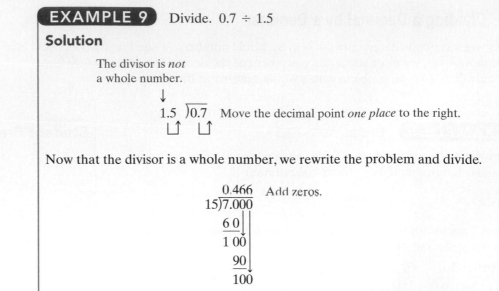

EXAMPLE 9 Divide. $0.7 \div 1.5$

Solution

The divisor is *not*
a whole number.

↓

$1.5\ \overline{)0.7}$ Move the decimal point *one place* to the right.

Now that the divisor is a whole number, we rewrite the problem and divide.

$$
\begin{array}{r}
0.466 \quad \text{Add zeros.} \\
15\overline{)7.000} \\
\underline{6\,0} \\
1\,00 \\
\underline{90} \\
100 \\
\underline{90} \\
10
\end{array}
$$

Decimals that have a digit, or a group of digits, that repeats are called **repeating decimals.** We often indicate the repeating pattern with a bar over the repeating group of digits. Thus $0.7 \div 1.5 = 0.4\overline{6}$ because if we continued dividing, the 6 would repeat.

Student Practice 9 Divide. $1.1 \div 1.8$

⑤ Changing a Fraction to a Decimal

Earlier we saw how to change fractions to decimals when the fraction has a power of 10 as a denominator: $\frac{2}{10} = 0.2$, $\frac{3}{100} = 0.03$, and so on. In this section we see how to change fractions whose denominators are not a power of 10 to decimals.

The fraction $\frac{21}{5}$ can be written as $21 \div 5$. Thus, to change a fraction to a decimal, we divide the numerator by the denominator.

Fraction		*Division*		*Decimal*
$\dfrac{21}{5}$	$=$	$21 \div 5$	$=$	4.2

PROCEDURE TO CONVERT A FRACTION TO A DECIMAL

Divide the denominator into the numerator until

(a) the remainder becomes zero, *or*

(b) the remainder repeats itself, *or*

(c) the desired number of decimal places is achieved.

EXAMPLE 10   Write as a decimal. $5\dfrac{7}{11}$

Solution

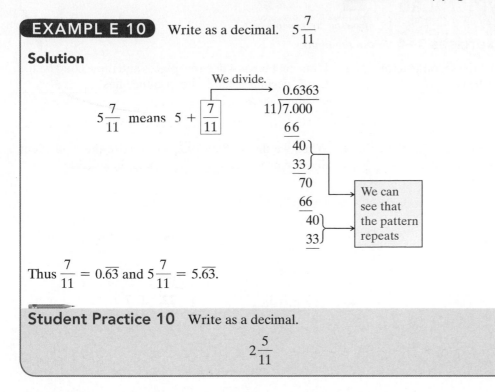

Thus $\dfrac{7}{11} = 0.\overline{63}$ and $5\dfrac{7}{11} = 5.\overline{63}$.

Student Practice 10   Write as a decimal.

$$2\dfrac{5}{11}$$

👣 STEPS TO SUCCESS Preparing for the Final Exam

To do well on the final exam, you should begin to prepare many weeks before the exam. Cramming for any test, especially the final one, often causes anxiety and fatigue and impairs your performance.

Schedule at least six study sessions now. If you come across a problem or topic during these sessions that you do not understand, even after seeking assistance, place an * beside it, then move on to another topic. When you finish reviewing all topics for the final exam, return to this topic and try again.

Making it personal: *Place a* ✔ *in the blank as you complete each of the following.*

_____ Date of your final exam: _____

_____ Find one or more study partners. Name of study

partners: _____

_____ Session 1   Date of session: _____

Review all your exams. Rework the problems you answered incorrectly.

_____ Session 2   Date of session: _____

Get help. Check with a tutor, study partners, or your instructor for those problems you cannot rework correctly.

_____ Session 3   Date of session: _____

Complete the Cumulative Test for Chapters 1–6.

_____ Session 4   Date of session: _____

Get help. Check with a tutor, study partners, or your instructor for those problems you cannot rework correctly.

_____ Session 5   Date of session: _____

Start the Practice Final Exam. Complete all problems on the exam for chapters that you have already covered in class.

Get help with those problems you do not understand.

_____ Session 6   Date of session: _____

Finish the Practice Final Exam. Do this immediately after your instructor covers the last topic for the semester. Then **revisit the * topics,** and get help with those you still do not understand.

Get a good night's sleep the night before the final exam.

Verbal and Writing Skills, Exercises 1–4 *Fill in the blanks.*

1. If one factor has 4 decimal places and the second factor has 2 decimal places, the product has _____ decimal places.

2. If one factor has 4 decimal places and the second factor has 3 decimal place, the product has _____ decimal places.

3. When we divide $4.62\overline{)12.7}$, we rewrite the equivalent division problem _____ and then divide.

4. When we divide $8.23\overline{)19.2}$, we rewrite the equivalent division problem _____ and then divide.

Multiply.

5. 0.03×0.04

6. 0.08×0.03

7. 0.05×0.07

8. 0.04×0.06

9. 7.43×8.3

10. 3.4×5.32

11. 15.2×4.3

12. 21.7×2.2

13. $(-3)(2.35)$

14. $(-7)(2.13)$

15. $(-4.23)(2.7)$

16. $(-4.15)(3.2)$

17. $(-25)(-0.613)$

18. $(-21)(-0.314)$

19. $(12.1)(-2.81)$

20. $(-11.3)(4.11)$

Multiply by powers of 10.

21. 0.1498×100

22. 0.1931×100

23. $8.554 \times 10,000$

24. $91.34 \times 10,000$

25. 41×10^4

26. 22×10^3

27. 0.6×10^4

28. 0.5×10^3

Divide.

29. $17.28 \div 8$

30. $12.6 \div 6$

31. $3.22 \div 14$

32. $5.12 \div 16$

33. $3.616 \div 64$

34. $12.6672 \div 39$

35. $82.824 \div 24$

36. $44.95 \div 31$

Divide. Round your answer to the nearest hundredth when necessary.

37. $3.25 \div 14$

38. $8.23 \div 11$

39. $-0.2988 \div 3.7$

40. $-0.2726 \div 2.9$

41. $-20.8 \div (-1.7)$

42. $-36.5 \div (-1.6)$

43. $8.343 \div 0.27$

44. $8.378 \div 0.41$

45. $13.7592 \div 5.88$

46. $15.4947 \div 4.11$

Divide. If a repeating decimal is obtained, use notation such as $0.\overline{9}$ or $0.\overline{14}$.

47. $3 \div 1.8$ **48.** $14 \div 1.5$ **49.** $0.6 \div 1.1$ **50.** $0.5 \div 3.7$

51. $11.3 \div 2.2$ **52.** $100 \div 3.3$ **53.** $200 \div 6.6$ **54.** $140 \div 1.5$

Write as a decimal. Round to the nearest hundredth.

55. $\dfrac{11}{6}$ **56.** $\dfrac{15}{7}$ **57.** $12\dfrac{7}{15}$ **58.** $14\dfrac{3}{16}$

Divide. If a repeating decimal is obtained, use notation such as $0.\overline{9}$ or $0.\overline{14}$.

59. $\dfrac{1}{3}$ **60.** $\dfrac{1}{6}$ **61.** $\dfrac{2}{15}$ **62.** $\dfrac{9}{11}$

Mixed Practice *Divide. If a repeating decimal is obtained, use notation such as $0.\overline{9}$ or $0.\overline{14}$.*

63. $20.35 \div 0.44$ **64.** $16.87 \div 0.35$

65. $\dfrac{2}{9}$ **66.** $\dfrac{5}{12}$

Multiply.

67. -3.5×4.24 **68.** -5.7×3.22

69. 0.4×0.8 **70.** 0.6×0.9

Applications

71. *Detecting Motion* A fly can detect motion in $\frac{1}{300}$ of a second, whereas the human eye detects motion in $\frac{1}{30}$ of a second.

   (a) Write $\frac{1}{300}$ of a second as a decimal.

   (b) Write $\frac{1}{30}$ of a second as a decimal.

72. *Inheritance* Erin inherited $\frac{1}{3}$ of her father's estate, and the family's favorite charity inherited $\frac{1}{60}$ of his estate.

   (a) Write $\frac{1}{3}$ as a decimal.

   (b) Write $\frac{1}{60}$ as a decimal.

Divide. Round to the nearest thousandth.

73. $-562.53 \div 13.123$

74. $-2104.03 \div 0.2346$

To Think About *Observe the pattern and then fill in the table with the missing values.*

75.

Fraction	$\dfrac{1}{9}$	$\dfrac{2}{9}$	$\dfrac{3}{9}$	$\dfrac{4}{9}$	$\dfrac{5}{9}$
Decimal	$0.11\ldots$	$0.22\ldots$	$0.33\ldots$		

76.

Fraction	$\dfrac{1}{11}$	$\dfrac{2}{11}$	$\dfrac{3}{11}$	$\dfrac{4}{11}$	$\dfrac{5}{11}$	$\dfrac{6}{11}$	$\dfrac{7}{11}$
Decimal	$0.\overline{09}$	$0.\overline{18}$	$0.\overline{27}$	$0.\overline{36}$			

Cumulative Review *Solve and check your solution.*

77. **[3.2.1]** $12x = 96$

78. **[3.1.2]** $x - 25 = -30$

79. **[3.1.2]** $x + 45 = 17$

80. **[3.2.1]** $15x = 225$

Quick Quiz 8.3

1. Multiply.
   (a) $(5.07)(3.1)$
   (b) $(4.39)(10^3)$

3. Write as a decimal. $\dfrac{29}{9}$

2. Divide. $36.54 \div 6.3$

4. **Concept Check** Marc multiplied 0.097×0.5 and obtained the answer 0.485. Is Marc's answer correct? Why or why not?

8.4 Solving Equations and Applied Problems Involving Decimals

① Solving Equations Involving Decimals

To solve equations with decimals, we use the same rule stated in Chapter 7.

Student Learning Objectives

After studying this section, you will be able to:

① Solve equations involving decimals.

② Solve applied problems involving decimals.

> **PROCEDURE TO SOLVE EQUATIONS IN THE FORM** $ax + b = c$
>
> 1. Remove any parentheses.
> 2. Combine like terms and simplify numerical work.
> 3. Add or subtract to get the variable term (ax) alone on one side of the equation.
> 4. Multiply or divide to get x alone on one side of the equation.
> 5. Check your solution.

EXAMPLE 1 Solve $x - 4.51 = 6.74$ and check your solution.

Solution We want an equation of the form $x = some\ number$.

$$
\begin{array}{rl}
x - 4.51 = & 6.74 \\
+\quad\ 4.51 & 4.51 \\
\hline
x + \ \ 0\ = & 11.25 \\
x\ = & 11.25
\end{array}
$$

We add the opposite of -4.51, or $+4.51$, to both sides of the equation. Remember to line up the decimal points.

Check.

$$
\begin{array}{r}
x - 4.51 = 6.74 \\
11.25 - 4.51 \overset{?}{=} 6.74 \\
6.74 = 6.74 \ \checkmark
\end{array}
$$

Student Practice 1 Solve $y - 2.23 = 4.69$ and check your solution.

NOTE TO STUDENT: Fully worked-out solutions to all of the Student Practice problems can be found at the back of the text starting at page SP-1.

EXAMPLE 2 Solve $-1.2x = 7.8$ and check your solution.

Solution We must *divide* both sides of the equation by -1.2 to get x alone on one side of the equation.

$$
\frac{-1.2x}{-1.2} = \frac{7.8}{-1.2}
$$

Dividing on both sides undoes the multiplication by -1.2.

$$
x = -6.5
$$

The sign of the answer is negative because the division problem has an odd number of negative signs: $7.8 \div (-1.2) = -6.5$.

Check.

$$
\begin{array}{r}
-1.2x = 7.8 \\
(-1.2)(-6.5) \overset{?}{=} 7.8 \\
7.8 = 7.8 \ \checkmark
\end{array}
$$

Student Practice 2 Solve $5.6x = -25.2$ and check your solution.

EXAMPLE 3   Solve. $2(x + 1.5) = x - 4.62$

Solution   We must first multiply each term inside the parentheses by 2 to remove the parentheses.

$$2(x + 1.5) = x - 4.62$$

$$2\,(x) + 2\,(1.5) = x - 4.62 \qquad \text{Multiply each term by 2.}$$

$$2x + 3 = x - 4.62 \qquad \text{We must add } -x \text{ to both sides of the}$$
$$\underline{+\ -x \qquad\qquad -x} \qquad\qquad \text{equation to gather terms on one side.}$$
$$x + 3 = \qquad -4.62$$

$$x + 3 = \qquad -4.62$$
$$\underline{+\qquad -3 \qquad\quad -3} \qquad \text{We add } -3 \text{ to both sides of the equation.}$$
$$x \quad = \qquad -7.62$$

We leave the check to the student.

Student Practice 3   Solve. $4(x - 2.5) = 3x + 0.6$

Sometimes it is easier to solve equations that involve decimals when we eliminate the decimals. That is, we rewrite the decimal equation as an equivalent equation that does not have decimal terms. We do this by multiplying both sides of the equation by the power of 10 necessary to eliminate the decimal from every term.

EXAMPLE 4   Solve. $0.35x + 0.3 = 1.7$

Solution   The term 0.35 has the *most* decimal places (2 decimal places), so we must multiply both sides of the equation by $10^2 = 100$ to eliminate decimals in the equation.

$$0.35x + 0.3 = 1.7$$

$$100\,(0.35x + 0.3) = 100\,(1.7) \qquad \text{Multiply } both \text{ sides of the equation by 100.}$$

$$100\,(0.35x) + 100\,(0.3) = 100\,(1.7) \qquad \text{Use the distributive property.}$$

$$35x + 30 = 170 \qquad \text{Multiply each term by 100 (by moving the decimal point to the right 2 places).}$$

$$35x = 140 \qquad \text{Subtract 30 from both sides of the equation.}$$

$$\frac{35x}{35} = \frac{140}{35} \qquad \text{Divide by 35 on both sides of the equation.}$$

$$x = 4$$

We leave the check to the student.

Student Practice 4   Solve. $0.3x + 0.15 = 2.7$

We can shorten this process if we skip the step that shows we multiplied by a power of 10. Instead, we can just move the decimal point in *each term* to the right the required number of places. The solution in Example 4 can be shortened as follows.

$$0.35x + 0.30 = 1.70$$

$$35x + 30 = 170 \qquad \text{To multiply each term by 100, move the decimal point in each term 2 places to the right.}$$

When using this method, we must remember to move the decimal point in *every term* the same number of places to the right. Note that we multiply each term by 10^2 to eliminate 2 decimal places, and by 10^3 to eliminate 3 decimal places, and so on.

② Solving Applied Problems Involving Decimals

We use the basic plan for solving applied problems that we discussed in earlier sections. Let us review how we analyze real-life situations.

> **PROCEDURE TO SOLVE APPLIED PROBLEMS**
> 1. Read the problem carefully to get an overview.
> 2. Write down formulas or draw pictures if possible.
> 3. Define the variable expressions.
> 4. Write an equation using the variable expressions you defined.
> 5. Solve the equation and determine the values asked for in the problem.
> 6. Check your answer. Ask yourself if the answers obtained are reasonable.

EXAMPLE 5 A long-distance phone carrier in New Jersey charges a base fee of $4.95 per month, plus 10 cents per minute for calls outside New Jersey and 5 cents per minute for long-distance calls within the state of New Jersey. On average, Natasha's monthly long-distance calls total 45 minutes outside the state and 220 minutes within the state. How much more or less will Natasha's average long-distance bill be than her budget of $25 per month?

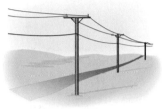

Solution *Understand the problem.* Since there is a lot of information in the problem, we fill in a Mathematics Blueprint for Problem Solving.

Mathematics Blueprint for Problem Solving

Gather the Facts	What Am I Asked to Do?	How Do I Proceed?	Key Points to Remember
Base fee: $4.95 Rates, per minute: Outside state 10¢ Inside state 5¢ Average monthly total of calls in minutes: Outside state 45 Inside state 220 Budget $25	Determine how much more or less her average calls will cost than her budget allows.	**1.** Let x = the average long-distance bill. **2.** Write equation: x = base fee + charge for calls outside the state + charge for calls outside the state. **3.** Subtract this amount from $25.	Change 10¢ and 5¢ to decimals. Follow the order of operations.

Calculate and state the answer.

Average long-distance bill	=	base fee	+	charge for calls outside state	+	charge for calls inside state

x = 4.95 + (45 min × 10¢ per min) + (220 min × 5¢ per min)

 = 4.95 + 45 × 0.10 + 220 × 0.05 10¢ = 0.10; 5¢ = 0.05

 = 4.95 + 4.50 + 11 Multiply: 45 × 0.10 = 4.50 and 220 × 0.05 = 11.

x = 20.45 Add.

$20.45 is the average cost of Natasha's long-distance calls, and this amount is less than her monthly budget of $25. We subtract: $25 − $20.45 = $4.55.

Her average monthly bill will be $4.55 less than her budget of $25.

Check. Use your calculator to verify your results.

Continued on next page

Student Practice 5 Refer to the information in Example 5 to answer the following. Natasha averages a total of 55 minutes of long-distance calls outside the state and a total of 185 minutes within the state. How much more or less will her average monthly bill be than her budget of $25?

We often deal with decimal equations when we solve problems that involve money. One type of money problem requires that we make a distinction between *how many coins* and the *value of the coins*. For example, if we let the variable d represent *how many dimes* there are, we calculate the *value of the dimes* as follows.

Each dime is worth 10 cents or $0.10.

If we had *two* dimes, they would be worth:

If we had *three* dimes, they would be worth:

Thus, if we had d dimes, they would be worth:

How Many Dimes	Coin Value	Total Value of Coins
2	0.10	$(2)(0.10) =$ **0.20**
3	0.10	$(3)(0.10) =$ **0.30**
d	0.10	$(d)(0.10) =$ **0.10d**

To solve problems involving the value of coins, we form an equation as follows:

$$\boxed{\text{total value of 1st coin}} + \boxed{\text{total value of 2nd coin}} + \cdots = \boxed{\text{total value of all coins}}$$

EXAMPLE 6 Andy must fill his vending machine with change. From past experience he knows that the number of dimes he needs to place in the machine for change is twice the number of quarters. The total value of dimes and quarters needed for change is $4.95. How many of each coin must Andy put in the vending machine?

(a) Define the variable expressions.

(b) Write an equation.

(c) Solve the equation and determine the values asked for.

(d) Check your answer.

Solution *Understand the problem.* We organize the given information in a Mathematics Blueprint for Problem Solving.

Mathematics Blueprint for Problem Solving

Gather the Facts	What Am I Asked to Do?	How Do I Proceed?	Key Points to Remember
The number of dimes is twice the number of quarters. The total value of dimes and quarters needed is $4.95.	Determine the number of dimes and the number of quarters that must be used for the $4.95 in change.	1. Define the variable expressions to find the *value* of the dimes and quarters. 2. Form an equation: value of dimes + value of quarters = $4.95. 3. Solve the equation and answer the question.	The value of the dimes is $0.10 \times$ number of dimes. The value of the quarters is $0.25 \times$ number of quarters.

Calculate and state the answer.

(a) We define our variables to find the value of each coin. Since we are comparing dimes to quarters, we let our variable represent the number of quarters.

How Many Quarters	Coin Value	Total Value of Coins
Q = number of quarters	0.25	$(0.25)Q$ = *value* of the quarters
$2Q$ = number of dimes (twice the number of quarters)	0.10	$(0.10)(2Q)$ = *value* of the dimes

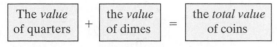

The *variable expressions* appear in the table above.

(b) We form an equation.

$$\boxed{\begin{array}{c}\text{The } value \\ \text{of quarters}\end{array}} + \boxed{\begin{array}{c}\text{the } value \\ \text{of dimes}\end{array}} = \boxed{\begin{array}{c}\text{the } total\ value \\ \text{of coins}\end{array}}$$

$$(0.25)Q + (0.10)(2Q) = \$4.95$$

(c) We solve the equation and determine the number of dimes and quarters.

$$0.25Q + 0.20Q = 4.95 \qquad \begin{array}{l}\text{Multiply: } (0.10)(2Q) \\ = (0.10 \times 2)Q = 0.20Q.\end{array}$$

$$25Q + 20Q = 495 \qquad \begin{array}{l}\text{Move the decimal point in each} \\ \text{term 2 places to the right.}\end{array}$$

$$45Q = 495$$

$$\frac{45Q}{45} = \frac{495}{45}$$

$$Q = 11 \qquad \text{There are 11 quarters.}$$

$$2Q = 2(11) = 22 \qquad \text{There are 22 dimes.}$$

Andy must put 11 quarters and 22 dimes in the vending machine for change.

(d) *Check.* Do we have twice as many dimes as quarters as stated in the problem? Yes. 22 dimes is 2×11, or 2 times the number of quarters. Is the value of the coins equal to $4.95? Yes. 22 dimes = $2.20 and 11 quarters = $2.75; $2.20 + $2.75 = $4.95. ✓

Student Practice 6 Julio must fill his vending machine with change. From past experience he knows that he must put 4 times the number of dimes as quarters in the machine for change. The total value of dimes and quarters needed for change is $3.25. How many of each coin must Julio put in the vending machine?

Solve.

1. $x + 3.7 = 9.8$

2. $x + 3.6 = 7.1$

3. $y - 2.8 = 6.95$

4. $y - 3.6 = 2.85$

5. $2.9 + x = 6$

6. $4.1 + x = 7$

7. $x + 2.5 = -9.6$

8. $x + 3.6 = -7.2$

9. $4x = 11.24$

10. $5x = 45.5$

11. $5.1x = 25.5$

12. $2.4x = 9.6$

13. $-5.6x = -19.04$

14. $-2.8x = -8.68$

15. $-5.2x - 3.3 = 22.7$

16. $-3.5x - 2.2 = 11.8$

17. $-3x - 5.3 = 11.23$

18. $1.2x + 7.15 = -4.67$

19. $0.9x + 8.7 = 15.9$

20. $0.2x - 5.9 = 3.7$

21. $2(x - 1) = 26.4$

22. $4(x - 1) = 8.9$

23. $2(x + 3.2) = x + 9.9$

24. $5(x + 1.5) = 4x + 9.6$

Eliminate the decimals by multiplying by a power of 10, and then solve.

25. $0.3x + 0.2 = 1.7$

26. $0.4x + 0.3 = 1.9$

27. $0.3x - 0.6 = 5.4$

28. $0.5x + 0.6 = 4.6$

29. $0.08x - 1.1 = 1.22$

30. $0.75x + 1.5 = 2.25$

31. $0.15x + 0.23 = 1.43$

32. $0.52x + 0.31 = 1.35$

Mixed Practice *Solve.*

33. $5.6 + x = -4.8$

34. $7.2 + x = -3.4$

35. $4.7x = 14.1$

36. $1.2x = 4.8$

37. $3(x + 1.4) = 6.9$

38. $2(x + 2.2) = 13.2$

39. $6x + 10.5 = x + 21$

40. $4x + 3.3 = x + 13.2$

Applications *Solve the following problems, which require one operation with decimals.*

41. *Gas Mileage* Sam's car travels 243.2 miles on 16 gallons of gas.

(a) How many miles per gallon does his car get?

(b) How many gallons of gasoline would he use if he drove 380 miles?

42. *Car Loan Payment* Jason's car payment is $303.12 a month.

(a) If the loan is for 60 months, how much will he pay in total for the car?

(b) If the original price of the car Jason bought was $14,297.15, how much would he have saved by paying cash?

43. *Phone Charges* The phone company charges 24 cents for each minute of phone calls during business hours and 14 cents a minute for evening calls. How much will a 17-minute phone call cost during

(a) business hours?

(b) evening hours?

(c) How much is saved by placing the 17-minute call in the evening?

44. *Bus Fare* Jerry's round-trip bus fare to work costs $1.90.

(a) If Jerry works 20 days a month, how much will it cost him to ride the bus to and from work?

(b) A bus pass for a month is $32. How much will Jerry save by buying the bus pass?

(c) If Jerry takes the bus only 15 days a month, is it still cheaper to buy the bus pass?

45. *Solution Mixture* A chemist wishes to mix 65 liters of a solution in several equal-size containers that hold 2.5 liters.

(a) How many containers will he need?

(b) If the solution costs $5.70 per liter, how much will one full container of this solution cost?

▲ **46.** *Stained Glass Dimensions* A decorative piece of stained glass measures 22.4 centimeters by 26.3 centimeters.

(a) What is the area of the stained glass in square centimeters?

(b) What is the perimeter of the stained glass?

Solve the following problems, which require several operations with decimals.

47. *Clothing Purchase* While shopping for clothes, Ruth bought 2 pairs of slacks for $22.50 each, 1 pair of jeans for $43.97, and 4 shirts for $8.88 each. How much money did she spend on her purchases?

48. *Swap Meet* Leroy bought 2 hardcover books for $1.95 each at a swap meet. He also bought 3 pairs of jeans for $19.95 each and a set of sheets for $12.95. How much money did he spend at the swap meet?

Fill in the boxes to complete each problem.

49. *Rental Car Cost* A compact car from Day One Rental Service costs $18.95 a day plus 12 cents a mile for all miles driven over 200 miles. How much will it cost Jack to rent a compact car from this company for 3 days if he drives a total of 423 miles?

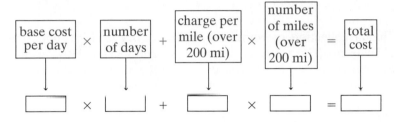

50. *Hourly Wage/Overtime* Sharon earns $8.75 an hour for the first 40 hours per week and $13.13 for overtime hours (hours worked over 40 hours). If she works 52 hours this week, how much will she earn?

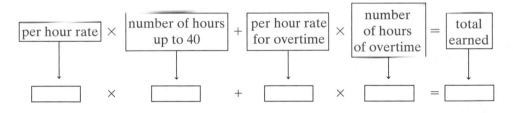

51. *Phone Charges* Jesse's phone carrier is offering the two plans in the ad. If Jesse's average monthly calls within the state total 75 minutes and outside the state total 20 minutes, how much will each plan cost him? Which plan is the better buy?

D&S Phone Service

Long Distance

Plan	Monthly Fee	Within State	Outside State
Single Rate Plan	$ 4.95	0.07 per min	0.10 per min
No Fee Plan	$ 0	0.10 per min	0.15 per min

52. *Phone Charges* If Jesse's average monthly calls within the state total 250 minutes and outside the state total 90 minutes, how much will each plan cost him? Which plan is the better buy?

53. ***Coin Problem*** Cody must place change in pay phones. From past experience he knows that the number of dimes he must place in a phone is 5 times the number of quarters. The total value of dimes and quarters needed for change is $4.50. How many of each coin must Cody put in each pay phone?

   (a) Define the variable expressions.

   (b) Write the equation.

   (c) Solve the equation and determine the values asked for.

   (d) Check your answer.

54. ***Coin Problem*** Patricia must fill her vending machine with change. From past experience she knows that the number of dimes she must place in the machine for change is 6 times the number of quarters. The total value of dimes and quarters needed for change is $4.25. How many of each coin must Patricia put in the vending machine?

   (a) Define the variable expressions.

   (b) Write the equation.

   (c) Solve the equation and determine the values asked for.

   (d) Check your answer.

55. ***Coin Problem*** Janet has a collection of coins that consists of nickels and quarters. The value of these coins is $3.50. If the number of nickels in her collection is 5 times the number of quarters, how many of each coin does she have?

56. ***Coin Problem*** Suppose that you have $10.50 in dimes and nickels. How many of each coin do you have if the number of dimes you have is twice the number of nickels?

Solve.

57. $23.098x = 103.941$

58. $42.0876x = 176.76792$

59. $x + 3.0012 = 21.566$

60. $x + 2.0011 = 33.5401$

One Step Further

61. ***Acceleration of an Object*** The average acceleration of an object is given by $a = \frac{v}{t}$, where a is the average acceleration, v is the velocity, and t is the time. Find the velocity after 3.5 seconds of an object whose acceleration is 15 feet per second squared.

62. ***Celsius Temperature*** The formula for calculating the temperature in degrees Fahrenheit when you know the temperature in degrees Celsius is $F = 1.8C + 32$. Use this formula to find the Celsius temperature when the Fahrenheit temperature is $-49°$.

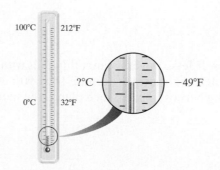

Cumulative Review

63. **[4.6.3]** Find x. $\dfrac{x}{500} = \dfrac{54}{100}$

64. **[4.6.3]** Find a. $\dfrac{a}{40} = \dfrac{13}{5}$

Change each fraction to a decimal.

65. **[8.3.5]** $\dfrac{1}{3}$

66. **[8.3.5]** $\dfrac{5}{6}$

67. **[5.6.1]** *Recipe Ingredients* A baker has a 25-pound sack of flour and uses $12\frac{1}{3}$ pounds for a recipe. How much flour is left?

68. **[5.6.1]** *Regulation Football* According to NCAA regulations, $10\frac{7}{8}$ inches is the minimum length of a regulation football, and $11\frac{7}{16}$ inches is the maximum length. What is the difference between the minimum and maximum lengths of a regulation football?

Quick Quiz 8.4

1. Solve. $0.3x + 0.2x = 2.5$

2. Solve. $3(x - 1.4) = 3.3$

3. Zach has some nickels and quarters in his pocket that total $1.05. The number of nickels in his pocket is twice the number of quarters. How many nickels and quarters does he have in his pocket?

   (a) Define the variable expressions.

   (b) Write the equation.

   (c) Solve the equation and determine the values asked for.

   (d) Check your answer.

4. **Concept Check** Fill in the table below to complete the steps necessary to solve the equation $2.2 + 2.4 = 5(x - 1) + 3x$. In the left column explain how to proceed and in the right column complete the step.

1. Add 2.2 + 2.4	$4.6 = 5(x - 1) + 3x$
2.	
3.	
4.	
5.	

How Am I Doing? Sections 8.1–8.4

1. _____

2. _____

3. _____

4. _____

5. _____

6. _____

7. _____

8. _____

9. _____

10. _____

11. _____

12. _____

13. _____

14. _____

15. _____

How are you doing with your homework assignments in Sections 8.1 to 8.4? Do you feel you have mastered the material so far? Do you understand the concepts you have covered? Before you go further in the textbook, take some time to do each of the following problems.

8.1

1. Write as a decimal. $\dfrac{2}{100}$

2. Write in fractional notation. Do not simplify. 0.027

3. Replace the ? with $<$ or $>$. 0.56 ? 0.566

4. Round to the nearest thousandth. 4212.65133

8.2

5. Add. $35 + 4.73 + 0.623$

6. Subtract. $-81.14 - 15.313$

7. Combine like terms. $2.3x + 3.1y + 4.4x$

8. Evaluate $y - 0.921$ for $y = 5.8$.

8.3

Multiply.

9. $(-3.23)(1.61)$

10. 0.2783×10^3

11. Divide. $13.806 \div 2.6$

12. Write as a decimal. $\dfrac{23}{6}$

8.4

Solve.

13. $-2.1x - 4.4 = 3.16$

14. $2(x + 4.5) = x + 9.8$

15. Lester earns $9.25 an hour for the first 40 hours per week and $13.88 for overtime (hours worked over 40 hours). If he works 53 hours this week, how much will he earn?

Now turn to page SA-15 for the answers to each of these problems. Each answer also includes a reference to the objective in which the problem is first taught. If you missed any of these problems, you should stop and review the Examples and Student Practice problems in the referenced objective. A little review now will help you master the material in the upcoming sections of the text.

8.5 Estimating with Percents

Frequently we find that we must quickly estimate a percentage of a number. For example, it is customary to leave a 15% tip at a restaurant for a meal, and discounts are generally given as percents. These are just a couple of examples of cases where one would encounter a percent situation and would benefit from the ability to quickly estimate the percentage. In this section we will look at an easy way to estimate percentages, then in Section 8.6 we will learn how to calculate percents.

① Estimating a Percentage of a Number

In previous chapters we used decimals or fractions to describe parts of a whole. Using a percent is another way to describe part of a whole. **Percents** can be described as ratios whose denominators are 100. We use the symbol % for percent. It means "parts per 100." We illustrate the meaning of percent below.

This figure has 100 squares.

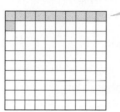

11 out of 100 squares are shaded. We write $\frac{11}{100} = 11\%$ of the squares are shaded.

If we have 50 parts out of 100 we have $\frac{50}{100}$ or 50%. To estimate 50%, we can simply think of one-half of the whole quantity. To estimate percents such as 10% or 1%, we can use the following method.

We can describe parts of a whole using a percent, fraction, or decimal.

$$10\% - \frac{10}{100} = 0.1 \qquad\qquad 1\% = \frac{1}{100} = 0.01$$

We divide $10 \div 100 = 0.1$. We divide $1 \div 100 = 0.01$.

Thus, to find 10% of a number we can multiply by 0.10, and to find 1% of a number we can multiply by 0.01.

Observe the following patterns.

10% of 2605 equals 260.5	since	$0.10 \times 2605 = 260.5$.
1% of 2605 equals 26.05	since	$0.01 \times 2605 = 26.05$.

We can see that when we find 10% of a number, we move the decimal point to the left 1 place. When we find 1% of a number, we move the decimal point to the left 2 places. We start with 2605, and then we find 10% and 1%.

Finding 10% → 260.5   We move the decimal point left 1 place to find 10%.

Finding 1% → 26.05   We move the decimal point left 2 places to find 1%.

Now, if we were only interested in *estimating* each percentage, we could just delete that last digit of 2605 to estimate 10% and delete the last two digits to estimate 1%. We start with 2605, and then we estimate 10% and 1%.

Estimating 10% → 260̸5 ≈ 260
Estimating 1% → 26̸0̸5 ≈ 26

It is easier to estimate a percentage if we round the number first. We choose the round-off place so that the rounded number is easy to work with. Note that estimated answers can vary slightly. This is because a number may be rounded to different round-off places by each student. For example, to estimate 10% of 886, one student might round 886 to the nearest ten, and another may choose to round 886 to the nearest hundred.

$$886 \rightarrow 890 \qquad 10\% \text{ of } 890 \approx 89$$
$$886 \rightarrow 900 \qquad 10\% \text{ of } 900 \approx 90$$

As you can see, answers will vary slightly but since we are estimating (looking for a ballpark figure), a slight variation is fine. You should round numbers to a place value that is easy to work with either mentally or with pencil and paper.

PROCEDURE TO ESTIMATE 10% OR 1% OF A WHOLE NUMBER

Round the number so that the rounded number is easy to work with and then estimate the percentage.

To estimate 10% of a whole number, delete the *last digit*.

To estimate 1% of a whole number, delete the *last two digits*.

EXAMPLE 1   For the number 59,040, estimate the following.

(a) 10% **(b)** 1%

Solution

(a) We estimate 10% of 59,040 as follows.

$$59,040 \rightarrow 59,000 \quad \text{First we round to the nearest thousand.}$$
$$59,00\cancel{0} \quad \text{Then we delete the last digit.}$$
$$10\% \text{ of } 59,040 \approx 5900$$

(b) We estimate 1% of 59,040 as follows.

$$59,040 \rightarrow 59,000 \quad \text{First we round to the nearest thousand.}$$
$$59,0\cancel{00} \quad \text{Then we delete the last two digits.}$$
$$1\% \text{ of } 59,040 \approx 590$$

Student Practice 1   For the number 86,205, estimate the following.

(a) 10% **(b)** 1%

NOTE TO STUDENT: *Fully worked-out solutions to all of the Student Practice problems can be found at the back of the text starting at page SP-1.*

Knowing the techniques for estimating 10% and 1% will enable you to quickly estimate other percentages.

To estimate 5%, find one-half of 10%. To estimate 15%, add 10% + 5%.

To estimate 20%, double 10%. To estimate 6%, add 5% + 1%.

EXAMPLE 2   For the number 1205, estimate the following.

(a) 5% of 1205

(b) 15% of 1205

(c) 6% of 1205

Solution   We round 1205 to the nearest hundred: 1200.

(a) To find 5%, we find $\dfrac{1}{2} \times 10\%$ of the number.

$$5\% \text{ of } 1205 \approx \frac{1}{2} \times (10\% \text{ of } 1200) = \frac{1}{2} \times 120 \qquad \text{Deleting the last digit of 1200 gives 120.}$$
$$= 60$$

(b) To find 15%, we add: (10% of 1200) + (5% of 1200).

$$15\% \text{ of } 1205 \approx 15\% \text{ of } 1200 = 120 + 60 \qquad \begin{array}{l} 10\% \text{ of } 1200 = 120; \\ 5\% \text{ of } 1200 = 60 \end{array}$$
$$= 180$$

(c) To find 6%, we add: (5% of 1200) + (1% of 1200).

$$6\% \text{ of } 1205 \approx 6\% \text{ of } 1200 = 60 + 12 \quad \text{To find 1% of 1200, we delete the last 2 digits.}$$
$$= 72$$

Student Practice 2   For the number 45,005, estimate the following.

(a) 20% of 45,005   **(b)** 11% of 45,005   **(c)** 7% of 45,005

---

② Solving Applied Problems Involving Estimating Percentages

EXAMPLE 3   Loren would like to leave a 15% tip for her dinner. If the total bill at a restaurant is $20.76, estimate the tip that Loren should leave.

Solution   We round $20.76 to the nearest ten: $20.

$$15\% \text{ of } 20.76 \approx 15\% \text{ of } 20 = (10\% \text{ of } 20) + (5\% \text{ of } 20)$$
$$= 2 + 1$$
$$= \$3$$

Student Practice 3   If the sales tax is 7%, estimate how much tax you would pay on a purchase of $199.85.

EXAMPLE 4   Everything in a store is on sale for 30% off the original price. The discount is calculated at the cash register at the time of purchase. Josh buys 2 shirts priced at $19.95 each, one pair of pants priced at $28.00, and a pair of shoes priced at $39.99.

(a) Round the original price of each item to the nearest ten and then estimate the total cost of the items before the discount.

(b) Estimate the amount of the 30% discount. Round the discount to the nearest ten.

(c) Estimate the total cost of the items after the discount is taken.

Solution

(a) We round each item to the nearest ten.

Shirt: $19.95 → $20;   Pants: $28 → $30;   Shoes: $39.99 → $40

We add these amounts: $20 + $20 + $30 + $40 = $110.

Josh bought 2 shirts.

The estimated cost of the items before the discount is $110.

(b) We find 30% of the total estimated cost.

$$10\% \text{ of } \$110 = \$11; \quad 30\% \text{ of } \$110 = \$11 + \$11 + \$11 = \$33$$

The estimated discount rounded to the nearest ten is $30.

(c) We subtract the total estimated cost minus the estimated discount.

$$\$110 - \$30 = \$80$$

The estimated cost of the items after the discount is $80.

Close out
30% off everything in the store

Student Practice 4   Josh bought 2 of each item listed in Example 4.

(a) Round the original price of each item to the nearest ten and then estimate the total cost of the items before the discount.

(b) Estimate the amount of the 30% discount. Round the discount to the nearest ten.

(c) Estimate the total cost of the items after the discount is taken.

Watch the videos
in MyMathLab

Download the
MyDashBoard App

Verbal and Writing Skills, Exercises 1–6 *Fill in the blanks with the missing phrases.*

1. To estimate 10% of a whole number, we can delete the _____.

2. To estimate 1% of a whole number, we can delete the _____.

3. We can find 15% by adding _____ and _____.

4. We can find 5% by _____.

5. We can find 6% by _____.

6. We can find 20% by _____.

For the number 701, estimate the following.

7. 10%

8. 1%

9. 5%

10. 8%

11. 20%

12. 2%

For the number 205, estimate the following.

13. 1%

14. 10%

15. 7%

16. 5%

17. 30%

18. 15%

For the number 1020, estimate the following.

19. 10%

20. 1%

21. 15%

22. 2%

23. 20%

24. 11%

For the number 3015, estimate the following.

25. 1%

26. 10%

27. 2%

28. 5%

29. 3%

30. 30%

For the number 320,050, estimate the following.

31. 10%

32. 1%

33. 5%

34. 2%

35. 15%

36. 3%

For the number 250,030, estimate the following.

37. 1%

38. 10%

39. 4%

40. 5%

41. 8%

42. 20%

Applications

43. **Sales Tax** Tobin bought a car for $22,000 in a state where the sales tax is 7%. Estimate the sales tax that Tobin paid.

44. **Late Fee** Anthony must pay a 5% late fee on his mortgage payment. If his mortgage payment is $1609, estimate the amount of the late fee.

45. *Commission* Nico paid the real estate agent who sold his home a commission of 5% of the sale price of the home. If Nico sold his home for $329,500, estimate how much commission he paid the agent.

46. *Commission* Mai Nguyen earns 11% commission on the sales she makes each month. If her total sales for the month were $25,115, estimate how much she earned in commission.

47. *Purchases* A sporting goods store is having a close-out sale with everything reduced 40% off the original price. The discount is calculated at the cash register when the purchases are made. If Leroy buys ski gloves priced at $29.99, a ski jacket priced at $199, and 2 sweaters each priced at $49.99, find the following.

   (a) Round the original price of each item to the nearest ten and then estimate the total cost of the items before the discount.

   (b) Estimate the amount of the 40% discount. If necessary, round the discount to the nearest ten.

   (c) Estimate the total cost of the purchase after the discount is taken.

48. *Purchases* A local office supply store will give Mary Ann a 20% discount off the list price on furniture for her office. Mary Ann ordered an office chair listed for $199, a desk listed for $699, a book case listed for $399, and 2 lamps listed for $59.99 each.

   (a) Round the original price of each item to the nearest ten and then estimate the total cost of the items before the discount.

   (b) Estimate the amount of the 20% discount. If necessary, round the discount to the nearest ten.

   (c) Estimate the total cost of the purchase after the discount is taken.

Cumulative Review *Solve.*

49. [4.6.3] $\dfrac{5}{18} = \dfrac{20}{n}$

50. [4.6.3] $\dfrac{n}{36} = \dfrac{7}{3}$

51. [5.6.1] *Filling a Tank* Two inlet pipes are used to fill a tank with water. After 1 hour, the smaller pipe filled $\frac{1}{4}$ of the tank, and the larger pipe filled $\frac{3}{8}$ of the tank. How much of the tank remains unfilled?

52. [5.6.1] *5-Mile Workout* Mary Beth developed a 5-mile exercise workout that consists of a combination of jogging and power walking. The first day of the workout she jogged $1\frac{3}{4}$ miles and then power walked for $2\frac{1}{2}$ miles. At this point, how far is she from completing her 5-mile workout?

Quick Quiz 8.5

1. Estimate 5% of 2995.

2. Estimate 20% of 60,015.

3. Jan earns 15% commission on the sales she makes. If her total sales for the month were $14,050, estimate how much she earned in commission.

4. **Concept Check** We can estimate 35% of 200 by finding 3 times 10% of 200, then adding $\frac{1}{2}$ times 10% of 200. Explain two other ways you can estimate 35% of 200.

8.6 Percents

Student Learning Objectives

After studying this section, you will be able to:

① Understand the meaning of percent.

② Change between decimals and percents.

③ Change between fractions, decimals, and percents.

We use percents in business, science, sports, and our everyday life: a suit you want to buy is on sale for 25% off; you receive a 5% increase in pay; and so on. What does this mean? How do we calculate percents? In this section we gain the knowledge to answer these questions.

① Understanding the Meaning of Percent

As stated in Section 8.5 we use decimals, fractions, or percents to describe parts of a whole. **Percents** can be described as ratios whose denominators are 100. The symbol % means "parts per 100." For example, 17% means 17 out of 100 parts.

It is important to know that 17% means 17 out of 100 parts. It can also be written $\frac{17}{100}$. Understanding the meaning of the notation allows you to work with percents as well as change from one notation to another.

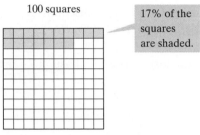

100 squares

17% of the squares are shaded.

17 of 100 squares are shaded.

EXAMPLE 1 State using percents. 13 out of 100 radios are defective.

Solution $\frac{13}{100} = 13\%$ 13% of the radios are defective.

Student Practice 1 State using percents. 42 out of 100 students in the class voted.

NOTE TO STUDENT: *Fully worked-out solutions to all of the Student Practice problems can be found at the back of the text starting at page SP-1.*

Percents can be larger than 100% or less than 1%. Consider the following situations.

EXAMPLE 2 Last year's attendance at the school's winter formal was 100 students. This year the attendance was 121. Write this year's attendance as a percent of last year's.

Solution We must write this year's attendance (121) as a percent of last year's (100).

$$\text{This year's attendance} \rightarrow \frac{121}{100} = 121\%$$
$$\text{Last year's attendance} \rightarrow$$

This year's attendance at the formal was 121% of last year's.

Note that we have 121 parts out of 100 parts, which means we have *more than* one whole amount and thus more than 100%.

Student Practice 2 Ten years ago a lawn mower cost $100. Now the average price for a lawn mower is $215. Write the present cost as a percent of the cost ten years ago.

EXAMPLE 3 There are 100 milliliters (mL) of solution in a container. Sara takes 0.3 mL of the solution. What percentage of the solution does Sara take?

Solution Sara takes 0.3 mL out of 100 mL, or

$$\frac{0.3}{100} = 0.3\% \text{ of the solution}$$

Student Practice 3 There are 100 mL of solution in a container. Julio takes 0.7 mL of the solution. What percentage of the solution does Julio take?

② Changing Between Decimals and Percents

Earlier we saw how to change between fractional and decimal notation. We review below.

Fraction → Decimal	Decimal → Fraction
$\frac{5}{8} = 5 \div 8 = 0.625$	$0.625 = \frac{625}{1000} = \frac{5}{8}$

Now we combine this skill with our knowledge of percents to see how to change between percents and decimals. Observe the pattern in the following illustrations.

We write a decimal as a percent.

Decimal → Percent

$$0.27 = \frac{27}{100} = 27\% \text{ or } 27.0\%$$

$$0.27 = 27.0\%$$

Decimal point moves 2 places to the *right*.

We reverse the process to write a percent as a decimal.

Decimal ← Percent

$$0.27 = 27.0\%$$

Decimal point moves 2 places to the *left*.

When we say we "move" the decimal point to the *right* to change a decimal to a percent, we mean we are multiplying by 100, which gives us the same result. When we say we "move" the decimal point 2 places to the *left*, we are really dividing by 100.

We summarize below.

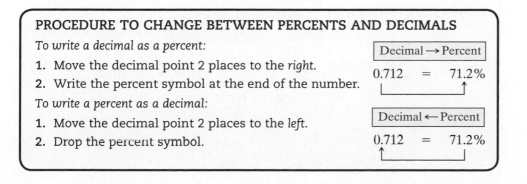

PROCEDURE TO CHANGE BETWEEN PERCENTS AND DECIMALS

To write a decimal as a percent:

1. Move the decimal point 2 places to the *right*.
2. Write the percent symbol at the end of the number.

Decimal → Percent

$$0.712 = 71.2\%$$

To write a percent as a decimal:

1. Move the decimal point 2 places to the *left*.
2. Drop the percent symbol.

Decimal ← Percent

$$0.712 = 71.2\%$$

Using the following chart can be helpful when changing between decimal and percent form since the decimal point moves the same direction as you move on the chart.

| Decimal ⟵ Percent | |
Form → Form	

EXAMPLE 4

(a) Write 3.8% as a decimal.   **(b)** Write 0.009 as a percent.

Solution

(a)

Decimal ⟵ Percent	
	3.8%
0.038 =	3.8%

We write the chart.

We move *left* on the chart, so the decimal point moves *2 places left*.

We must place an extra zero to the left of the 3.

(b)

Decimal ⟶ Percent	
0.009	
0.009 =	0.9%

We write the chart.

We move *right* on the chart, so the decimal point moves *2 places right*.

Student Practice 4

(a) Write 2.6% as a decimal.   **(b)** Write 0.001 as a percent.

EXAMPLE 5 Complete the table of equivalent notations.

Decimal Form	Percent Form
0.457	
	58.2%
	0.6%
2.9	

Solution

Decimal Form	Percent Form
0.457	45.7%
0.582	58.2%
0.006	0.6%
2.9	290%

We must insert two zeros. → (for 0.006 row)

← We must insert a zero. (for 290% row)

Student Practice 5 Complete the table of equivalent notations.

Decimal Form	Percent Form
0.511	
	84.1%
	0.2%
6.7	

③ Changing Between Fractions, Decimals, and Percents

Now that you can change between decimals and percents, you are ready to change between fractions and decimals and percents.

$$\text{Fraction} \underset{\longrightarrow}{\overset{\longleftarrow}{\;}} \text{Decimal} \underset{\longrightarrow}{\overset{\longleftarrow}{\;}} \text{Percent}$$

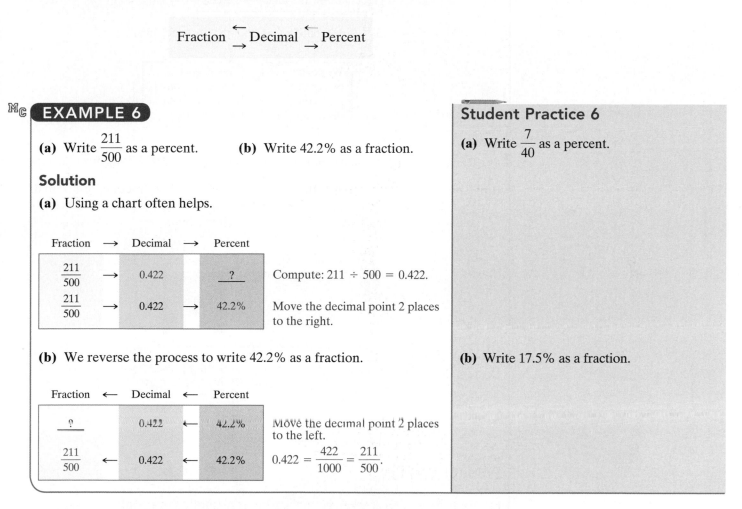

M꜀ **EXAMPLE 6**

(a) Write $\dfrac{211}{500}$ as a percent.   **(b)** Write 42.2% as a fraction.

Solution

(a) Using a chart often helps.

Fraction	→	Decimal	→	Percent	
$\dfrac{211}{500}$	→	0.422		?	Compute: $211 \div 500 = 0.422$.
$\dfrac{211}{500}$	→	0.422	→	42.2%	Move the decimal point 2 places to the right.

(b) We reverse the process to write 42.2% as a fraction.

Fraction	←	Decimal	←	Percent	
?		0.422	←	42.2%	Move the decimal point 2 places to the left.
$\dfrac{211}{500}$	←	0.422	←	42.2%	$0.422 = \dfrac{422}{1000} = \dfrac{211}{500}$.

Student Practice 6

(a) Write $\dfrac{7}{40}$ as a percent.

(b) Write 17.5% as a fraction.

Changing some fractions to decimals results in a repeating decimal. For example, $\frac{1}{3} = 0.333\ldots$. In such cases, we usually round as directed. If we are not asked to round, we use a notation such as $0.\overline{3}$.

EXAMPLE 7

(a) Write $\dfrac{5}{9}$ as a percent. Round to the nearest hundredth of a percent.

(b) Write $\dfrac{1}{4}$% as a fraction.

Solution

(a) First we change $\frac{5}{9}$ to a decimal: $5 \div 9 = 0.55555\ldots$. We must carry out the division at least *five* places beyond the decimal point so that we can move the

Continued on next page

decimal point to the right *two* places, and we then round to the nearest hundredth of a percent.

| Fraction → Decimal → Percent |

$$\frac{5}{9} = 5 \div 9 = 0.55555\ldots = 55.555\ldots\% \approx 55.56\%$$

| We need 2 places to move the decimal point. | We need 3 places to round to the nearest hundredth. |

Remember that if you are *not* asked to round, $\frac{5}{9} = 55.\overline{5}\%$.

(b) A percent can be written in the form of a whole number, fraction, or decimal. $\frac{1}{4}\%$ is a percent written in fraction form. We can rewrite the percent in its decimal form so that we can use the decimal shift method.

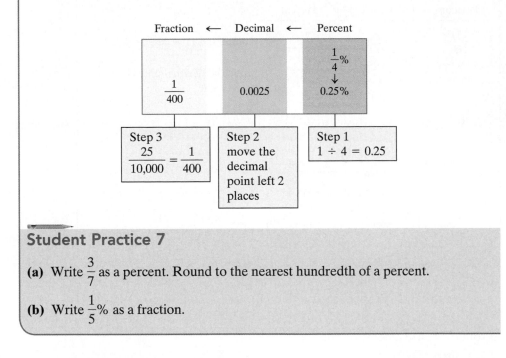

Student Practice 7

(a) Write $\dfrac{3}{7}$ as a percent. Round to the nearest hundredth of a percent.

(b) Write $\dfrac{1}{5}\%$ as a fraction.

Certain percents occur very often, especially in money matters. Here are some common equivalents that you may already know. If not, be sure to memorize them.

$$\frac{1}{4} = 0.25 = 25\% \qquad \frac{1}{3} = 0.3\overline{3} = 33\frac{1}{3}\% \qquad \frac{1}{10} = 0.10 = 10\%$$

$$\frac{1}{2} = 0.5 = 50\% \qquad \frac{2}{3} = 0.6\overline{6} = 66\frac{2}{3}\% \qquad \frac{3}{4} = 0.75 = 75\%$$

Verbal and Writing Skills, Exercises 1–2 *Fill in the blanks.*

1. To change a percent to a decimal, move the decimal point 2 places to the _____ and drop the _____.

2. To change a decimal to a percent, move the decimal point 2 places to the _____ and add the _____ to the end of the number.

State the percents.

3. 42 out of 100 students in the class voted.

4. 71 out of 100 students in the class are women.

5. *Radar Navigation* 63 out of 100 power boats had a radar navigation system.

6. *GPS Navigation* 97 out of 100 new cars have GPS navigation.

7. *Electric Toothbrushes* 28 out of 100 people use electric toothbrushes. What percentage of people use electric toothbrushes?

8. *Organized Sports* 58 out of 100 people play organized sports. What percentage of people play organized sports?

9. *School Attendance* Last year's attendance at the medical school was 100 students. This year the attendance is 113. Write this year's attendance as a percent of last year's.

10. *Club Attendance* Last year's attendance at the College Service Club was 100 students. This year the attendance is 135. Write this year's attendance as a percent of last year's.

11. *Grams of Fat* 0.9 out of 100 grams of fat is saturated fat. What percentage is saturated fat?

12. *Water Solution* 0.11 out of 100 mL of a solution is water. What percentage of the solution is water?

Complete each table of equivalent notations.

13.

Decimal Form	Percent Form
0.576	
	24.9%
	0.3%
1.546	

14.

Decimal Form	Percent Form
0.139	
	57.8%
	0.9%
5.612	

15.

Decimal Form	Percent Form
3.7	
	23.8%
	0.6%
12.882	

16.

Decimal Form	Percent Form
2.8	
	42.4%
	0.1%
13.145	

17. Write 36% as a decimal.

18. Write 51% as a decimal.

19. Write 53.8% as a decimal.

20. Write 33.4% as a decimal.

21. Write 0.075 as a percent.

22. Write 0.007 as a percent.

23. Write 2.33% as a decimal.

24. Write 7.2% as a decimal.

25. In Alaska, 0.03413 of the state is covered by water. Write the part of the state that is covered by water as a percent.

26. In Florida, 0.07689 of the state is covered by water. Write the part of the state that is covered by water as a percent.

Complete each table of equivalent notations.

27.

Fraction Form	Decimal Form	Percent Form
$\frac{4}{5}$		
	0.27	
		0.7%
$4\frac{1}{3}$		

28.

Fraction Form	Decimal Form	Percent Form
$\frac{9}{12}$		
	0.61	
		2.8%
$9\frac{5}{8}$		

29.

Fraction Form	Decimal Form	Percent Form
$\frac{5}{16}$		
	2.6	
		$\frac{1}{10}\%$
$6\frac{1}{2}$		

30.

Fraction Form	Decimal Form	Percent Form
$\frac{8}{15}$		
	3.5	
		$\frac{1}{8}\%$
$5\frac{1}{4}$		

31. Write $\dfrac{4}{32}$ as a percent.

32. Write $\dfrac{129}{250}$ as a percent.

33. Write $\dfrac{1}{5}\%$ as a fraction.

34. Write $\dfrac{1}{2}\%$ as a fraction.

35. **(a)** Write $\dfrac{14}{40}$ as a percent.

  (b) Write 22.3% as a fraction.

36. **(a)** Write $\dfrac{32}{80}$ as a percent.

  (b) Write 72.1% as a fraction.

37. *Seamstress* A seamstress wastes $1\frac{1}{4}\%$ of the material used to make a dress. Write this percent as a fraction.

38. *Photo Paper* Arran Copy Center wastes $2\frac{1}{2}\%$ of its paper supply due to poor quality of the photocopies produced. Write this percent as a fraction.

39. *Brain Size* The brain represents $\frac{1}{40}$ of an average person's weight. Express this fraction as a percent.

40. *Blinking the Eye* During waking hours a person blinks $\frac{9}{2000}$ of the time. Express the fraction as a percent.

Write each fraction as a percent. Round to the nearest hundredth of a percent.

41. $\dfrac{9}{14}$ **42.** $\dfrac{16}{35}$ **43.** $\dfrac{7}{9}$ **44.** $\dfrac{4}{9}$

Applications   *Use the bar graph to answer exercises 45–49.*

Prescription Drugs In 1980 5.5% of all personal health care spending was for prescription drugs, compared to 10.2% in 2000. Write as a fraction the percent of health care costs that was spent on prescription drugs for the following years.

45. 1980   **46.** 1990

47. 2002   **48.** 2007

49. 2010

Drug Costs as a Percentage of All Personal Health Spending

Source: National Health Care Expenditures

To Think About *Observe the pattern and then fill in the table with the missing values.*

50.

Fraction	$\frac{1}{10}$	$\frac{2}{10}$	$\frac{3}{10}$	$\frac{4}{10}$	$\frac{5}{10}$
Percent	10%	20%			50%

51.

Fraction	$\frac{1}{5}$	$\frac{2}{5}$	$\frac{3}{5}$	$\frac{4}{5}$	$\frac{5}{5}$
Percent	20%	40%			100%

52.

Fraction	2	$2\frac{1}{4}$	$2\frac{1}{2}$	$2\frac{3}{4}$	3
Percent	200%		250%		300%

53.

Fraction	4	$4\frac{1}{8}$	$4\frac{2}{8}$	$4\frac{3}{8}$	$4\frac{4}{8}$	$4\frac{5}{8}$
Percent	400%	412.5%			450%	462.5%

Cumulative Review *Translate and solve.*

54. [3.2.3] Three times what number is equal to forty-eight?

55. [3.2.3] Twice what number is equal to three hundred thirty?

56. [5.1.3] One-fourth of what number is equal to 60?

57. [5.1.3] What is one-third of sixty-nine?

58. [8.2.5] *Apartment Expenses* The utility bills for an apartment are as follows: $64.55, phone; $34.50, gas; $55.90, electricity. If the total cost of utilities is divided equally among the three roommates, how much must each contribute?

59. [8.2.5] *Land Mass* The area of the United States is 3.7 million square miles, while the area of Antarctica is 5.1 million square miles. What is the difference in area between the United States and Antarctica?

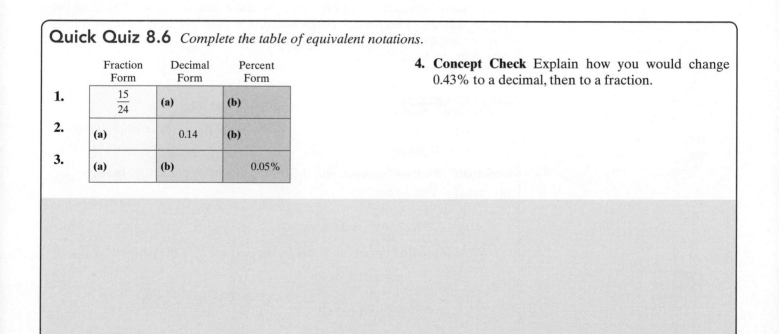

Quick Quiz 8.6 *Complete the table of equivalent notations.*

	Fraction Form	Decimal Form	Percent Form
1.	$\frac{15}{24}$	(a)	(b)
2.	(a)	0.14	(b)
3.	(a)	(b)	0.05%

4. Concept Check Explain how you would change 0.43% to a decimal, then to a fraction.

8.7 Solving Percent Problems Using Equations

Student Learning Objectives

After studying this section, you will be able to:

① Translate and solve percent problems.

② Solve applied percent problems using equations.

① Translating and Solving Percent Problems

Percents are used to describe an amount that is part of a whole base quantity. For example, 75% describes the amount *3 parts out of 4*.

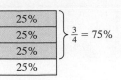

We know that the fraction $\frac{3}{4} = 75\%$. Now let's look at each of the three parts of this relationship: 3, 4, and 75%.

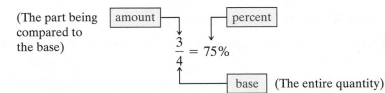

(The part being compared to the base)

amount → percent

$$\frac{3}{4} = 75\%$$

base — (The entire quantity)

We can write the relationship $\frac{3}{4} = 75\%$ as *amount = percent × base*.

$$\text{amount} = \text{percent} \times \text{base}$$
$$3 = 75\% \times 4$$

When one of the parts of the relationship (percent, amount, or base) is unknown, we can solve the equation for the unknown quantity.

It should be noted that a *percent* is used primarily for comparative and descriptive purposes. *When we perform calculations with percents, we must first change the percent to its equivalent decimal or fraction form and then perform the calculations.*

To solve applied percent problems, we must understand what each of these three parts means. Therefore, we draw a picture of each situation. Then we translate the statement by replacing "of" with ×; "is" with =; "find" with $n =$; "what" with n; "percent" with %.

EXAMPLE 1 Translate into an equation and solve.

(a) What is 25% of 40?

(b) 10 is 25% of what number?

(c) 10 is what percent of 40?

Solution We translate each into the form amount = percent × base.

(a) What is 25% of 40?

    ↓ ↓ ↓ ↓ ↓

    $n = 25\% \times 40$ Write in symbols.

We want to find the *amount,* that is, the *part of* (25% of) the *base* of 40.

$$n = 25\% \text{ of } 40$$
$$n = 0.25 \times 40 \quad \text{Change 25\% to decimal form.}$$
$$n = 10 \quad\quad\quad\quad\text{Multiply.}$$

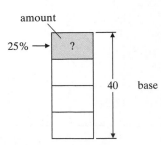

amount

25% → ?

40   base

(b) 10 is 25% of what number?

$\downarrow \quad \downarrow \quad \downarrow \quad \downarrow \qquad \qquad \downarrow$

$10 \ = 25\% \times \qquad \quad n$

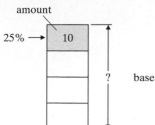

We want to find the *base* (the entire quantity).

$10 = 0.25 \times n$ Change 25% to a decimal.

$\dfrac{10}{0.25} = n$ Solve the equation for n.

$40 = n$ Divide.

10 is 25% of 40.

(c) 10 is what percent of 40?

$\downarrow \quad \downarrow \quad \downarrow \quad \downarrow \quad \downarrow \ \downarrow$

$10 = \quad n \quad \% \quad \times 40$

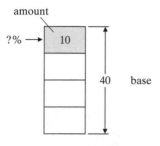

We want to find the *percent*.

$10 = n\% \times 40$

$\dfrac{10}{40} = n\%$ Solve the equation for $n\%$.

$0.25 = n\%$ The % symbol reminds us to write 0.25 as a percent.

$25 = n$

10 is 25% of 40.

    Note that when forming the equation, we used the percent symbol, %, to represent the word *percent* and remind us that the answer must be in percent form.

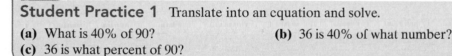

Student Practice 1   Translate into an equation and solve.

(a) What is 40% of 90?

(b) 36 is 40% of what number?

(c) 36 is what percent of 90?

NOTE TO STUDENT: Fully worked-out solutions to all of the Student Practice problems can be found at the back of the text starting at page SP-1.

    Sometimes we have more than 100%—this means that the *amount* is more than the *base*.

EXAMPLE 2   Translate into an equation and solve. 50 is what percent of 40?

Solution   We should expect to get more than 100% since 50 is more than the base 40.

50 is what percent of 40?

$\downarrow \downarrow \quad \downarrow \qquad \downarrow \qquad \downarrow \ \downarrow$

$50 = \quad n \qquad \% \qquad \times 40$

$50 = \quad n\% \times 40$

$\dfrac{50}{40} = \quad n\%$ Solve for $n\%$.

$1.25 = \quad n\%$ Divide.

$125 = \quad n$

50 is 125% of 40.

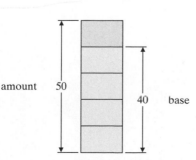

Student Practice 2   Translate into an equation and solve. 80 is what percent of 20?

CAUTION: We must remember to change the percent to its equivalent decimal or fraction form before we perform calculations.

EXAMPLE 3 Find 55% of 36.

Solution

$$\begin{array}{cccc} \text{Find} & 55\% & \text{of} & 36. \\ \downarrow & \downarrow & \downarrow & \downarrow \end{array}$$

$\begin{aligned} n &= 55\% \times 36 \quad \text{Translate "find" to } n = \text{ since it has the same} \\ & \qquad\qquad\qquad \text{meaning as "what is."} \\ &= 0.55 \times 36 \quad \text{Change 55\% to a decimal.} \\ &= 19.8 \qquad\qquad \text{Multiply.} \end{aligned}$

19.8 is 55% of 36.

Student Practice 3 Find 22% of 60.

---

② Solving Applied Percent Problems Using Equations

We can solve applied problems that involve percents by using the following three-step process.

PROCEDURE TO SOLVE APPLIED PERCENT PROBLEMS

1. Write a percent statement to represent the situation.
2. Translate the statement into an equation.
3. Solve the equation.

EXAMPLE 4 Marilyn has 850 out of 1000 points possible in her English class. What percent of the total points does Marilyn have?

Solution We must find the *percent,* so we write the statement that represents the percent situation.

$$\begin{array}{cccccc} 850 & \text{is what} & \text{percent} & \text{of} & 1000? & \text{Write the statement that represents} \\ \downarrow & \downarrow\ \downarrow & \downarrow & \downarrow & \downarrow & \text{this situation.} \end{array}$$

$\begin{aligned} 850 &= n \quad\ \% \ \times 1000 \qquad \text{Form an equation.} \\ \frac{850}{1000} &= n\% \qquad\qquad\qquad\quad \text{Solve the equation.} \\ 0.85 &= n\% \\ 85 &= n \end{aligned}$

Marilyn has 85% of the total points.

There are several ways to write equivalent statements for a percent situation. We could have written, "What percent of 1000 is 850?" Try translating and solving this statement to verify that it is equivalent.

Student Practice 4 Alisha received 35 out of 50 points on a quiz. What percent of the total points did Alisha earn on the quiz?

EXAMPLE 5   Sean's bill for his dinner at the Spaghetti House was $19.75. How much should he leave for a 15% tip? Round this amount to the nearest cent.

Solution   We must find the *amount,* that is, the *part of* (15% of) the *base* of $19.75.

What	is	15%	of	$19.75?	Write the statement for this situation.
↓	↓	↓	↓	↓	
n	=	15%	×	$19.75	Form an equation.
	=	0.15	×	$19.75	
	=	$2.9625			Solve the equation.
	≈	$2.96			Round.

The tip is $2.96.

Note that rounding to the nearest cent is the same as rounding to the nearest hundredth. Why?

Student Practice 5   Frances left a $3.50 tip for her dinner, which cost $18.55. What percent of the total bill did Frances leave for a tip? Round your answer to the nearest hundredth of a percent.

What do we mean when we say "100% of the price of an item"? We are referring to the *entire price* of the item. That is, if a sofa costs $400, then 100% of the cost of the sofa is $400. If an item costs x dollars, then 100% x = *the price of the item*. We often write x as 100% x when we are dealing with markup problems.

EXAMPLE 6   Sergio stayed in a luxury hotel on a Saturday night and paid $230 for that night. If the rate on Saturday night is 15% higher than it is on Sunday night, how much will Sergio pay for the room on Sunday night?

Solution   *Understand the problem.*

Mathematics Blueprint for Problem Solving

Gather the Facts	What Am I Asked to Do?	How Do I Proceed?	Key Points to Remember
Room rate for Saturday night is $230. The Saturday night rate is 15% *higher* than the rate on Sunday.	Find the room rate for Sunday.	Let x = the room rate on Sunday. Find the sum: The Sunday night rate *plus* the markup equals the Saturday night rate.	The markup is 15% of the rate on Sunday.

Calculate and state the answer.

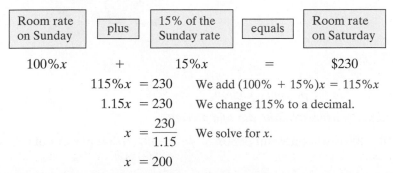

Room rate on Sunday	plus	15% of the Sunday rate	equals	Room rate on Saturday
100%x	+	15%x	=	$230

$$115\%x = 230 \qquad \text{We add } (100\% + 15\%)x = 115\%x$$

$$1.15x = 230 \qquad \text{We change 115\% to a decimal.}$$

$$x = \frac{230}{1.15} \qquad \text{We solve for } x.$$

$$x = 200$$

Sergio will pay a rate of $200 to stay Sunday night.

Student Practice 6   Refer to Example 6 to answer the following. If the rate on Saturday night is 20% higher than it is on Sunday night, how much will Sergio pay to stay Sunday? Round to the nearest cent.

Verbal and Writing Skills, Exercises 1–4

Since we use the following percents often, we should memorize them.

25% of a number is the same as $\frac{1}{4}$ of the number.   50% of a number is the same as $\frac{1}{2}$ of the number.

Knowing these facts, explain why it is obvious that there is an error in the following statements.

1. 35 is 50% of 40.

2. 50% of 30 is 40.

3. 25% of 10 is 9.

4. 200 is 25% of 60.

Each of the following requires finding the "amount" of a number in the percent equation.

Translate into an equation and solve.

5. What is 32% of 90?

6. What is 15% of 75?

7. Find 26% of 145.

8. Find 45% of 200.

9. What is 52% of 60?

10. What is 18% of 66?

11. What is 150% of 40?

12. What is 125% of 40?

Applications *Solve each applied problem by finding the "amount" of a number in the percent equation.*

13. **Restaurant Tip** The bill for Bui's dinner was $18.45. How much should he leave for a 15% tip? Round your answer to the nearest cent.

14. **Farm Acres** The Boyd farm is 250 acres. 70% of it is suitable land for farming. How many acres can be used to farm?

15. **Transfer Students** 60% of the graduates of Trinity, a two-year college, transfer to a four-year college. If the graduating class at Trinity has 650 students, how many are transferring to a four-year college?

16. **Defective Parts** H&B Manufacturing claims that no more than 0.5% of its parts are defective. If a client orders 8600 parts from H&B Manufacturing, what is the largest number of parts that could be defective?

Each of the following requires finding a "percent" in the percent equation.

Translate into an equation and solve. Round to the nearest hundredth of a percent.

17. 54 is what percent of 30?

18. 400 is what percent of 80?

19. What percent of 650 is 70?

20. What percent of 350 is 20?

21. What percent of 60 is 18?

22. What percent of 120 is 15?

Solve each applied problem by finding the "percent" in the percent equation.

23. *Calories and Fat* A snack bar has 80 calories. If 15 of those calories are from fat, what percent of the calories are from fat?

24. *Sales Tax* Wesley paid $21 tax when he bought a mountain bike for $300. What percent tax did he pay?

25. *Soccer* In a soccer game Tasha made 2 out of 5 shots on goal. What percent of shots did she make?

26. *Basketball* On the Almen High School basketball team, 9 out of the 24 players are over 6 feet tall. What percent of the players are over 6 feet tall?

Each of the following requires finding the "base" in the percent equation.

Translate into an equation and solve.

27. 56 is 70% of what number?

28. 50 is 20% of what number?

29. 24 is 40% of what number?

30. 32 is 64% of what number?

31. 70 is 20% of what number?

32. 90 is 45% of what number?

Solve each applied problem by finding the "base" in the percent equation.

33. *Employees with Flu* 24% of the employees at Jack's Sporting Goods Store called in sick with the flu. If 12 employees called in sick, how many employees are there at the company?

34. *Scholarships* 25% of the graduating class at Springdale Community College received a scholarship. If 1800 students received a scholarship, how many students graduated from the college?

35. *Nature Trail* The new 8-mile nature trail is 125% of the length of the original trail. How long was the original trail?

36. *Auditorium Capacity* The new 350-seat auditorium contains 140% of the number of seats in the original auditorium. How many seats were in the original auditorium?

37. *Golfing Cost* A small country club charges nonmembers $128 to play a game of golf at the club. If the fee to play a game of golf for nonmembers is 60% more than the fee for club members, what do club members pay to play a game of golf?

38. *Hotel Charges* A hotel charges a rate of $120 per room to individuals who are not attending a conference at the hotel. If this rate is 20% more than the rate charged to those attending the conference, how much is the room rate for a person attending a conference at the hotel?

39. *Electric Bill* Mary Beth's first electric bill for her new apartment was $95 for the month of June. When she complained to the landlord about the high cost of electricity, she was informed that summer electric bills are about 90% higher than winter bills because of the periodic summer use of air conditioners. How much can Mary Beth expect her winter electric bill to be?

40. *Restaurant Bill* Vu Nguyen and his wife have $46 to spend on dinner. What is the maximum amount they can spend on the meals and drinks so that they have enough money left to leave a 15% tip?

Mixed Practice *Translate each of these different types of percent statements into equations and then solve.*

41. 44 is 50% of what number?

42. 90 is 50% of what number?

43. 125% of 60 is what number?

44. 110% of 70 is what number?

45. What is 27% of 78?

46. What is 32% of 85?

47. 15.66 is what percent of 87?

48. 39.96 is what percent of 74?

49. 135 is 45% of what number?

50. 80 is 25% of what number?

51. 110 is what percent of 440?

52. 50 is what percent of 200?

One Step Further
When we work with large numbers such as millions, we can simplify the calculations if we write the abbreviation "mil" in place of the zeros and then perform the calculations. For example, to divide 22,000,000 by 2, we can write 22 mil ÷ 2 = 11 mil.

Roses Sold *Roses are a popular Valentine's Day flower. One year there were 130 million roses sold on Valentine's Day. Use this information to answer exercises 53 and 54.*

53. If 73% of the roses sold on Valentine's Day were red, how many red roses were sold on Valentine's Day?

54. If 48% of the cut flowers sold on Valentine's Day were roses, how many cut flowers were sold? Round your answer to the nearest tenth.

Write the statement that represents each of these different types of percent situations. Then solve the equation.

55. ***Test Score*** Robert got 86 out of the 92 questions right on his test. What percent did he get correct? Round your answer to the nearest percent.

56. ***Basketball*** A basketball player has made 25 of her 30 free throws. What percent of free throws did she make? Round your answer to the nearest percent.

Trash Recycling *About 208 million tons of residential and commercial trash are generated each year. Use the information in the illustration below to answer exercises 57 and 58.*

57. How much greater is the percentage of aluminum recycled than the percentage of glass recycled?

58. How much greater is the percentage of yard waste recycled than the percentage of plastics recycled?

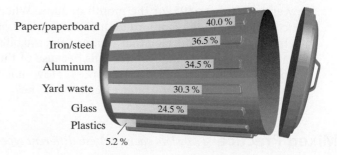

Percent of Residential and Commercial Trash Recycled Annually

Paper/paperboard	40.0 %
Iron/steel	36.5 %
Aluminum	34.5 %
Yard waste	30.3 %
Glass	24.5 %
Plastics	5.2 %

Monthly Budget *Jeremy has a monthly income of $1250. He allocates it as shown on the circle graph. Use the graph to answer exercises 59–62.*

59. What percent does he spend on recreation?

60. How much money does he spend on food?

61. How much money does he save each month?

62. After rent, food, and clothing, how much does he have left each month?

Monthly Allocation of Income

Recreation
13% Savings
10% Clothing
10% Other
27% Rent
33% Food

63. What is 67.3% of 348.9?

64. What percent of 875 is 625?

65. 368 is 20% of what number?

66. What is 18.9% of $9500?

67. ***Surfers*** 34.6% of the 1,400,000 U.S. surfers are women. How many U.S. surfers are women?

68. ***Snowboarders*** 1,501,200 U.S. snowboarders are women. What percent of the 5,400,000 U.S. snowboarders are women?

Cumulative Review *Solve.*

69. **[7.2.1]** $2x + 3 = 13$

70. **[7.2.2]** $3x - 1 = 2x + 4$

71. **[7.2.2]** $5x - 3 = 3x + 9$

72. **[7.3.1]** $2(3x + 1) = 2x - 6$

Quick Quiz 8.7

1. What is 35% of 60?

2. 15 is what percent of 50?

3. 8 is 20% of what number?

4. **Concept Check** The owner of M&R Windows determined that 0.8% of the products ordered from the manufacturer are defective. Explain how you would determine how many windows the owner should expect to be defective in a shipment from the manufacturer of 375 windows.

8.8 Solving Percent Problems Using Proportions

Student Learning Objectives

After studying this section, you will be able to:

① Identify the parts of a percent proportion.

② Use the percent proportion to solve percent problems.

③ Solve applied percent problems using proportions.

① Identifying the Parts of a Percent Proportion

In Section 8.7 we showed you how to use an equation to solve a percent problem. Some students find it easier to use proportions to solve percent problems. We will show you how to use proportions in this section. The two methods work equally well. Using percent proportions allows you to see another of the many uses of the proportions that we studied in Chapter 4.

Consider the following relationship.

$$\frac{17}{68} = 25\%$$

This can be written as follows.

$$\frac{17}{68} = \frac{25}{100}$$

In general, we can write this relationship using the following percent proportion.

$$\frac{\text{amount}}{\text{base}} = \frac{\text{percent number}}{100}$$

To use this equation effectively, we need to identify the *amount*, *base*, and *percent number* in an applied problem. The easiest of these three parts to identify is the percent number. We use the letter p (a variable) to represent the percent number.

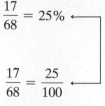

EXAMPLE 1 Identify the percent number p.

(a) Find 15% of 360.

(b) 28% of what is 25?

(c) What percent of 18 is 4.5?

Solution

(a) Find 15% of 360.   The value of p is 15.

(b) 28% of what is 25?   The value of p is 28.

(c) What percent of 18 is 4.5?
$\underbrace{\hphantom{\text{What percent}}}_{p}$

We let p represent the unknown percent number.

Student Practice 1 Identify the percent number p.

(a) Find 83% of 460.

(b) 15% of what number is 60?

(c) What percent of 45 is 9?

NOTE TO STUDENT: Fully worked-out solutions to all of the Student Practice problems can be found at the back of the text starting at page SP-1.

We use the letter b to represent the *base* number. The base is the entire quantity or the total involved. The number that is the base usually appears after the word *of*. The *amount*, which we represent by the letter a, is the *part* being compared to the whole.

EXAMPLE 2   Identify the base b and the amount a.

(a) 25% of 520 is 130.   **(b)** 19 is 50% of what?

Solution

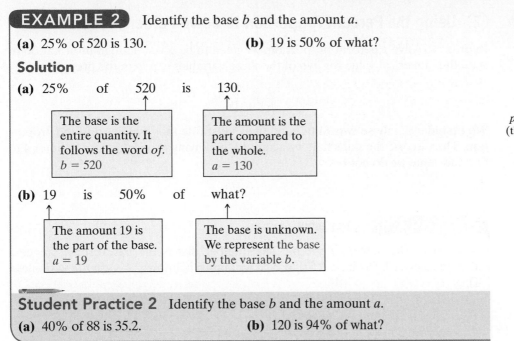

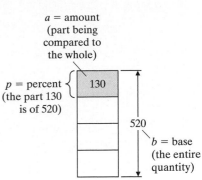

Student Practice 2   Identify the base b and the amount a.

(a) 40% of 88 is 35.2.   **(b)** 120 is 94% of what?

When identifying the percent p, base b, and amount a in a problem, it is easiest to identify p and b first. The remaining quantity or variable is then a.

EXAMPLE 3   Find the percent p, base b, and amount a.

(a) What is 77% of 210?   **(b)** What percent of 21 is 17?

Solution

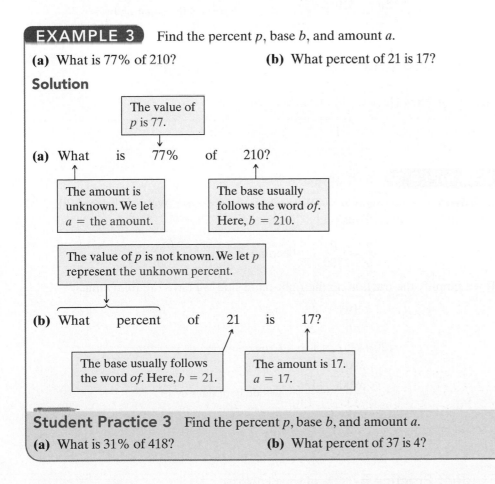

Student Practice 3   Find the percent p, base b, and amount a.

(a) What is 31% of 418?   **(b)** What percent of 37 is 4?

② Using the Percent Proportion to Solve Percent Problems

In order to solve a percent proportion, we need to be given enough information to state the numerical value for two of the three variables a, b, p in the proportion.

$$\frac{a}{b} = \frac{p}{100}$$

We first identify those two values and then substitute those values into the proportion. Then we use the skills that we acquired for solving proportions in Chapter 4 to find the value we do not know.

EXAMPLE 4 Find 260% of 40.

Solution The percent $p = 260$. The number that is the base usually appears after the word *of*. The base $b = 40$. The amount is unknown. We use the variable a. Thus

$$\frac{a}{b} = \frac{p}{100} \qquad \text{becomes} \qquad \frac{a}{40} = \frac{260}{100}$$

If we simplify the fraction on the right-hand side, we have the following.

$$\frac{a}{40} = \frac{13}{5} \qquad \frac{260}{100} = \frac{13}{5}$$

$$5a = (40)(13) \qquad \text{Cross-multiply.}$$

$$5a = 520 \qquad \text{Simplify.}$$

$$\frac{5a}{5} = \frac{520}{5} \qquad \text{Divide both sides of the equation by 5.}$$

$$a = 104$$

Thus 260% of 40 is 104.

Student Practice 4 Find 340% of 70.

EXAMPLE 5 65% of what is 195?

Solution The percent $p = 65$. The base is unknown. We use the variable b. The amount a is 195. Thus

$$\frac{a}{b} = \frac{p}{100} \qquad \text{becomes} \qquad \frac{195}{b} = \frac{65}{100}$$

If we simplify the fraction on the right-hand side, we have the following.

$$\frac{195}{b} = \frac{13}{20} \qquad \frac{65}{100} = \frac{13}{20}$$

$$(20)(195) = 13b \qquad \text{Cross-multiply.}$$

$$3900 = 13b \qquad \text{Simplify.}$$

$$\frac{3900}{13} = \frac{13b}{13} \qquad \text{Divide both sides by 13.}$$

$$300 = b$$

Thus 65% of 300 is 195.

Student Practice 5 82% of what is 246?

EXAMPLE 6   19 is what percent of 95?

Solution   The percent is unknown. We use the variable p. The base $b = 95$. The amount $a = 19$. Thus

$$\frac{a}{b} = \frac{p}{100} \quad \text{becomes} \quad \frac{19}{95} = \frac{p}{100}$$

Cross-multiplying, we have the following.

$$(100)(19) = 95p$$
$$1900 = 95p$$
$$\frac{1900}{95} = \frac{95p}{95} \quad \text{Divide both sides of the equation by 95.}$$
$$20 = p$$

Thus 19 is 20% of 95.

Student Practice 6   42 is what percent of 140?

③ Solving Applied Percent Problems Using Proportions

EXAMPLE 7   Sonia has $29.75 deducted from her weekly salary of $425 for a retirement plan.

(a) What percent of Sonia's salary is withheld for the retirement plan?

(b) What percent of Sonia's salary is *not* withheld for the retirement plan?

Solution

(a) We must find the percent p. The base $b = 425$. The amount $a = 29.75$. Thus

$$\frac{a}{b} = \frac{p}{100} \quad \text{becomes} \quad \frac{29.75}{425} = \frac{p}{100}$$

When we cross-multiply, we obtain the following.

$$100(29.75) = 425p \quad \text{Cross-multiply.}$$
$$2975 = 425p \quad (29.75)100 = 2975$$
$$\frac{2975}{425} = \frac{425p}{425} \quad \text{Divide both sides by 425.}$$
$$7 = p$$

Since p represents percent, we see that 7% of Sonia's salary is deducted for the retirement plan.

(b) We can subtract the percents to determine the percent of Sonia's salary that is *not* deducted.

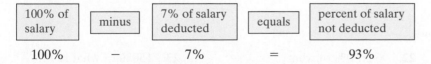

100% of salary	minus	7% of salary deducted	equals	percent of salary not deducted
100%	−	7%	=	93%

93% of Sonia's salary is not deducted for the retirement plan.

Student Practice 7   The manager of an office building has 70% of the 120 offices in the building occupied.

(a) What percent of the offices are *not* occupied?

(b) How many offices are *not* occupied?

MyMathLab®

Watch the videos
in MyMathLab

Download the
MyDashBoard App

Verbal and Writing Skills, Exercises 1–2

1. Why is it obvious that 150% of 80 is a number greater than 80?

2. Why is it obvious that 99% of 80 is equal to a number a little less than 80?

Identify the percent p, base b, and amount a. Do not solve for the unknown.

	p	b	a
3. 16% of 250 is 40.	_____	_____	_____
4. 40% of 550 is 220.	_____	_____	_____
5. What is 95% of 420?	_____	_____	_____
6. What is 50% of 600?	_____	_____	_____
7. 69% of what is 8230?	_____	_____	_____
8. 34% of what is 169?	_____	_____	_____
9. 63 is what percent of 90?	_____	_____	_____
10. 130 is what percent of 66?	_____	_____	_____
11. What percent of 47 is 10?	_____	_____	_____
12. What percent of 62 is 6?	_____	_____	_____
13. 400 is 160% of what?	_____	_____	_____
14. 800 is 225% of what?	_____	_____	_____

In exercises 15–32, solve by using the percent proportion.

$$\frac{a}{b} = \frac{p}{100}$$

In exercises 15–20, the amount a is not known.

15. 24% of 200 is what?

16. 35% of 200 is what?

17. Find 250% of 30.

18. Find 250% of 60.

19. 0.6% of 4000 is what?

20. 0.9% of 2000 is what?

In exercises 21–26, the base b is not known.

21. 82 is 50% of what?

22. 96 is 80% of what?

23. 150% of what is 75?

24. 125% of what is 75?

25. 4000 is 0.8% of what?

26. 6300 is 0.7% of what?

In exercises 27–32, the percent p is not known.

27. 70 is what percent of 280?

28. 90 is what percent of 450?

29. What percent of 140 is 11.2?

30. What percent of 170 is 3.4?

31. What percent of $5000 is $90?

32. What percent of $4000 is $64?

Mixed Practice *Solve by using the percent proportion.*

33. 26% of 350 is what?

34. 56% of 650 is what?

35. 180% of what is 720?

36. 160% of what is 320?

37. 75 is what percent of 400?

38. 88 is what percent of 500?

39. Find 0.2% of 650.

40. Find 0.5% of 500.

41. What percent of 25 is 15.2?

42. What percent of 49 is 34.3?

43. 68 is 40% of what?

44. 52 is 40% of what?

45. 94.6 is what percent of 220?

46. 83.8 is what percent of 260?

47. What is 12.5% of 380?

48. What is 20.5% of 320?

49. Find 0.05% of 5600.

50. Find 0.04% of 8700.

Applications *Solve using the percent proportion.*

51. *Apartment Renting* An apartment owner must keep 80% of the apartments rented in order to cover the costs of ownership. If there are 250 apartments, how many must be rented in order to cover the owner's costs?

52. *Midterm Grade* In a class of 25 students, 40% of the students received an A on the midterm. How many student received an A?

53. *Preschool Registration* A new preschool predicts that 60% of the 150 spaces open for registration will be filled during the first week of registration.

(a) What percent of the openings should be available *after* the first week of registration?

(b) How many spaces should be open *after* the first week of registration?

54. *Promotional Sale* The marketing division of a company predicts that 72% of the 250 special promotional sale items will sell on the first day of the sale.

(a) What percent of the promotional items should be available *after* the first day of the sale?

(b) How many special promotional items should be available *after* the first day of the sale?

55. *Vacation Savings* Owen has $42.90 deducted from his monthly salary of $1950 for a vacation savings plan.

    (a) What percent of Owen's salary is withheld for the vacation savings plan?

    (b) What percent of Owen's salary is *not* withheld for the vacation savings plan?

56. *Child Care Expenses* Dave has $120 deducted from his weekly salary of $600 to pay for the on-site child care offered by his company.

    (a) What percent of Dave's salary is deducted for child care expenses?

    (b) What percent of Dave's salary is *not* deducted for child care expenses?

57. *Office Usage* In an office building, 90 offices are currently being rented. This represents 75% of the total units. How many offices are there in the building?

58. *Mobile Homes Destroyed* Fifteen percent of the mobile homes in Senior Park Community were destroyed by a tornado. If 21 mobile homes were destroyed, how many mobile homes were in Senior Park Community?

59. *Quiz Score* Delroy got 6 out of 9 questions right on the quiz. What percent did he get correct? Round your answer to the nearest percent.

60. *Payroll Deduction* A cashier has $29.15 deducted from his weekly gross earnings of $265 for federal income taxes. What percent of the cashier's pay is withheld for federal taxes?

Solve each percent problem. Round to the nearest hundredth.

61. What is $19\frac{1}{4}\%$ of 798?

62. $140\frac{1}{2}\%$ of what number is 10,397?

63. Find 18% of 20% of $3300. (*Hint:* First find 20% of $3300.)

64. Find 42% of 16% of $5500. (*Hint:* First find 16% of $5500.)

Cumulative Review

▲ **65.** [3.3.2] Find the area of a rectangle with $L = 7$ in. and $W = 4$ in.

▲ **66.** [3.3.3] Find the volume of a cube with $L = 8$ cm, $W = 4$ cm, and $H = 6$ cm.

▲ **67.** [3.3.2] Find the area of a square with a side of 2 ft.

▲ **68.** [3.3.1] Find the perimeter of a square with a side of 3 in.

Quick Quiz 8.8 *Solve by using the percent proportion.*

1. What is 18% of 90?

2. 45 is what percent of 125?

3. 56 is 80% of what number?

4. **Concept Check** In the following percent proportion, what can you say about the *percent number* if the value of the *amount* is larger than the *base*?

$$\frac{amount}{base} = \frac{percent\,number}{100}$$

8.9 Solving Applied Problems Involving Percents

① Solving Commission Problems

If you work as a salesperson, your earnings may be in part, or in total, a certain percentage of the sales you make. This type of earnings is called a **commission.** For example, if you work on a 15% commission basis, your commission is 15% of your total sales. We call the percent, 15%, the **commission rate.** We can either write a *percent statement* or use the *formula* given below when solving commission problems.

Your commission is 15% of your total sales.

$$\text{commission} = \text{commission rate} \times \text{total sales}$$

Since the *formula* for a commission problem is equivalent to the *percent statement,* we will only write the formula. Remember, if you forget the formula, you can always write the percent statement.

EXAMPLE 1 Alex is a car salesman and earns a commission rate of 9% of the price of each car he sells. If he earned $3150 commission this month, what were his total sales for the month?

Solution

$$\text{commission} = \text{commission rate} \times \text{total sales}$$
$$\$3150 = 9\% \times n$$
$$\$3150 = 0.09 \times n$$
$$\frac{\$3150}{0.09} = \frac{0.09n}{0.09}$$
$$\$35,000 = n$$

Alex's total sales were $35,000.

Check. We will estimate to check our answer. We round 9% to 10% and then verify that 10% of his total sales ($35,000) is approximately his commission ($3150). 10% of $35,000 = $3500, which is close to his commission of $3150. ✓

Student Practice 1 You must pay your real estate agent a 6% commission on the sale price of your home. If the agent sells your home for $145,000, what amount of commission must you pay the agent?

Student Learning Objectives

After studying this section, you will be able to:

① Solve commission problems.

② Solve percent increase, decrease, and discount problems.

③ Solve simple interest problems.

NOTE TO STUDENT: Fully worked-out solutions to all of the Student Practice problems can be found at the back of the text starting at page SP-1.

② Solving Percent Increase, Decrease, and Discount Problems

There are many situations that involve increasing or decreasing an amount by a certain percent. If you receive a 5% raise or buy a CD player for 20% off the original price, you are working with a *percent increase* or a *percent decrease* situation. To find the amount of the increase or decrease, we can either write a percent statement or use a formula.

Your raise is 5% of your present salary.   The discount is 20% of the original list price.

$$\text{increase} = \text{percent increase} \times \text{original amount} \qquad \text{decrease} = \text{percent decrease} \times \text{original amount}$$

491

EXAMPLE 2   The enrollment at Laird Elementary School was 450 students in 2011. In 2012 the enrollment decreased by 36 students. What was the percent decrease?

Solution

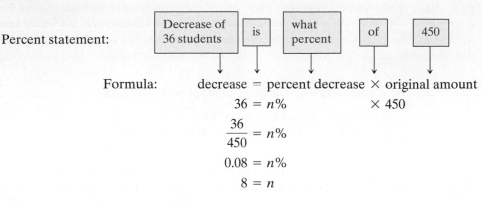

Percent statement:

| Decrease of 36 students | is | what percent | of | 450 |

Formula:

$$\text{decrease} = \text{percent decrease} \times \text{original amount}$$

$$36 = n\% \qquad \times 450$$

$$\frac{36}{450} = n\%$$

$$0.08 = n\%$$

$$8 = n$$

The enrollment decreased by 8%.

Note that we must change 0.08 to a percent since we are finding a percent.

Check. We round 8% to 10%. 10% of 450 is 45, which is close to the enrollment decrease of 36 students. ✓

Student Practice 2   A dining room set is reduced $492 from the original price of $1640. By what percent is the dining room set reduced in price?

Suppose that we want to find the new amount after the increase or decrease. We do this as follows.

$$\text{original amount} + \text{increase} = \text{new amount}$$
$$\text{original amount} - \text{decrease} = \text{new amount}$$

EXAMPLE 3   Arnold earned $26,000 a year and received a 6% raise. How much is his new yearly salary?

Solution   First, we find the amount of the raise.

$$\text{percent increase} \times \text{original amount} \qquad = \text{increase (raise)}$$

$$6\% \times 26,000 = 0.06 \times 26,000 = \$1560 \text{ raise}$$

Now we find his new yearly salary.

$$\text{original salary} + \text{raise} = \text{new salary}$$

$$\$26,000 + \$1560 = \$27,560$$

Arnold's new salary is $27,560.

Check.   6% is a little more than $\frac{1}{2}$ of 10%. Use this fact to check the answer.

Student Practice 3   Rebecca made $9.45 per hour and received a raise of 3% of her hourly rate. What is her new hourly rate?

ᴹ_c **EXAMPLE 4**   An advertisement states that all items in a department store are reduced 30% off the original list price. What is the sale price of a flat-screen television set with a list price of $2700?

Solution   First, we find the amount of the discount.

percent decrease × amount   = decrease (discount)

30% × 2700   = 0.30 × 2700 = $810 discount

Next we find the sale price.

original price − discount = sale price

$ 2700 − $ 810   = $ 1890

The sale price is $1890.

Check. A 30% discount is about $\frac{1}{3}$ of the price. Use this fact to check the answer.

Student Practice 4   1500 people attended the Westmont College Jazz Concert last year. This year the number of tickets sold is 14% less than last year. How many people have bought tickets to the concert this year?

③ **Solving Simple Interest Problems**

Interest is money paid for the use of money. If you deposit money in a bank, the bank uses that money and pays you interest. If you borrow money, you pay the bank interest for the use of that money. To solve simple interest problems, we must be familiar with the terms *interest, principal, interest rate,* and *time.*

Interest: the money earned or paid for the use of money
Principal: the amount deposited or borrowed
Interest rate: the percent used in computing the interest
Time: the period of time interest is calculated

The formula used in business to compute simple interest is as follows.

simple interest = principal × rate × time
$I = P \times R \times T$

The interest rate is assumed to be *per year* unless otherwise stated. The time T must be in years.

EXAMPLE 5   Larsen borrowed $9400 from the bank at a simple interest rate of 13%.

(a) Find the interest on the loan for 1 year.
(b) How much does Larsen pay back to the bank at the end of the year when he pays off the loan?

Solution

(a) P = principal = $9400, R = rate = 13%, T = time = 1 year

$I = P \times R \times T$

I = $9400 × 0.13 × 1 = $1222   We change 13% to 0.13, then multiply.

The interest for 1 year is $1222.

Continued on next page

(b) At the end of the year he must pay back the original amount he borrowed plus the interest.

$$\text{original loan} + \text{interest} = \text{payoff amount}$$

$$\$9400 + \$1222 = \$10{,}622$$

The total amount that Larsen must pay back is $10,622.

Student Practice 5 Damon put $2000 in a savings account that pays 3% simple interest per year.

(a) How much interest will Damon earn in 1 year?

(b) How much money will be in Damon's savings account at the end of the year?

Our formula is based on a yearly interest rate. Time periods of more than 1 year or a fractional part of a year are sometimes needed.

EXAMPLE 6 Find the interest on a loan of $2500 that is borrowed at a simple interest rate of 9% for 3 months.

Solution We must change 3 months to years since the formula requires that the time be in years: $T = 3$ months $= \frac{3}{12} = \frac{1}{4}$ year.

$$I = P \times R \times T$$

$$I = 2500 \times 0.09 \times \frac{1}{4} = 225 \times \frac{1}{4} = 56.25 \quad \text{We divide 225 by 4.}$$

The interest for 3 months is $56.25.

Student Practice 6 Find the interest on a loan of $3100 that is borrowed at a simple interest rate of 9% for 4 months.

Verbal and Writing Skills, Exercises 1–6

Fill in the blanks with the definitions of the terms used for calculating simple interest.

1. Principal: _____

2. Interest rate: _____

3. Time: _____

4. Interest: _____

Fill in the blanks to complete each formula.

5. Commission = _____ _____ × _____ _____

6. Simple interest = principal × _____ × _____

Solve. If necessary, round your answer to the nearest hundredth.

Commission

7. Becca is paid 18% commission each week based on the total dollar amount of her sales. Last week her total sales were $4200. How much did Becca earn in commission?

8. Last week Sandra sold $4400 worth of computer software. If she is paid a commission rate of 15% of her sales, how much did she earn?

9. Dalley Dodge pays its sales personnel a commission of 25% of the dealer profit on each car. The dealer made a profit of $13,500 on the cars that Brandon sold last month. What was Brandon's commission last month?

10. Latonya works as a phone solicitor and is paid a 10% commission rate based on the dollar amount of sales she makes. If Latonya earned $1150 last month, what were her total sales for the month?

Percent Increase and Decrease

11. A coat is reduced $75 from the original price of $250. What is the percent decrease?

12. A sound system is reduced $165 from the original price of $1100. What is the percent decrease?

13. A bedroom set is reduced $360 from the original price of $1800. By what percent is the bedroom set reduced?

14. The attendance at a high school homecoming game was 550 last year. Due to bad weather, the attendance at this year's homecoming game decreased by 66 people. What was the percent decrease?

15. Last year Essex Manufacturing paid $4500 for utilities. This year the company used energy-saving techniques to reduce the utility bills by 10% of last year's cost. What was the cost of utilities for Essex Manufacturing this year?

16. An appliance store is advertising that all appliances are on sale for a discount of 25% off the original list price. What is the sale price of a refrigerator with a list price of $1900?

17. Jerome earns $42,000 per year at his job. If he gets a 7% raise, what is his new salary?

18. Each year employees at LW Financial Group get a cost-of-living raise (COLA). This year the raise is 2%. How much will the new salary be for an employee making $47,000 a year?

19. Last year 760 people attended the Outdoor Concert in the Park. If 15% more tickets were sold this year than last year, how many people will attend the concert this year?

20. There were 450 seats in the Johnson School of Law's largest auditorium. This year the college remodeled the auditorium, increasing the seating 20%. How many seats does the new auditorium have?

Simple Interest

21. Pam puts $2500 into a certificate of deposit (CD) account at a simple interest of 6% per year.

   (a) How much interest will Pam earn in 1 year?

   (b) How much money will Pam have in her account at the end of the year?

22. Syed borrows $12,500 from the bank at a simple interest rate of 9%.

   (a) Find the interest on the loan after 1 year.

   (b) How much money does Syed pay back to the bank at the end of the year to pay off the loan?

23. Karen gets an emergency loan from her credit union of $500 for books and supplies at a simple interest rate of 8%. How much interest does she pay if the loan is paid back in 6 months?

24. A teachers' credit union gives $300 short-term loans to low-income students for money to buy books, supplies, and materials. If the simple interest rate is 6%, how much interest does a student pay if the loan is paid back in 3 months?

Mixed Applications *Round your answer to the nearest cent.*

25. *Income Earned* Suits Are Us pays its sales personnel a base salary of $1500 per month plus a commission of 20% of the sale price on each suit they sell each month.

   (a) If Meleena sold $4150 in suits during the month of November, what was her commission?

   (b) What was Meleena's total earnings for the month of November?

26. *Income Earned* Huy Nguyen sells major appliances at Consumer Appliance and earns a base salary of $1800 per month plus a commission of 15% of the sale price on each appliance he sells each month.

   (a) If Huy sold $6100 in appliances during September, what was his commission?

   (b) What was Huy's total earnings for the month of September?

27. *Loan Payment* Jake gets a car loan of $6600 at a simple interest rate of 9% for 2 years. If he pays back the loan in 24 equal payments, how much is each payment?

28. *Loan Payment* Alisha gets a business loan of $8500 at a simple interest rate of 9% for 1 year. If she pays back the loan in 12 equal payments, how much is each payment?

29. *Weight Loss* Sam weighed 265 pounds two years ago. After careful supervision at a weight-loss center, he reduced his weight by 15%. How much did he weigh after weight loss?

30. *Income Earned* If the cost of living went up 6% this year, how much does Bob need to make this year to have the same buying power he had last year when he earned $24,000?

Pricing Chart Use the chart below, which gives the prices for custom window shades, for exercises 31 and 32.

Window Shade Price Chart

Width to:		24"	30"	36"	42"	48"	54"	60"
Height to:	36"	$208	235	268	301	331	359	389
	42"	223	254	289	327	359	395	428
	48"	237	275	313	354	390	427	466
	54"	252	291	334	378	421	461	503
	60"	270	310	358	401	450	496	540
	66"	285	330	381	430	481	531	579
	72"	302	348	402	457	512	565	617

31. Kristen buys two shades that are 36 inches wide and 42 inches long. She also buys one shade that is 48 inches wide and 54 inches long.

(a) What is the total price of the three shades?

(b) If Kristen received a 30% discount on the shades, how much did she pay for the three shades?

32. Pam buys three shades that are 54 inches wide and 60 inches long and two shades that are 48 inches wide and 72 inches long.

(a) What is the total price of the five shades?

(b) If Pam received a 40% discount on the shades, how much did she pay for the five shades?

One Step Further

33. *Membership Cost* Marcia bought a membership in a discount warehouse for $35. She saves 5% on the purchase price of everything. What is the total amount of purchases needed in order for Marcia to save the $35 that she paid for membership?

34. *High Jump* Laird, a high jumper on the track and field team, hit the bar 58 out of 200 jump attempts. This means that he did not succeed in 29% of his jump attempts. As a result of extra practice, Laird reduced this percentage by 6%. If he attempts another 200 high-jumps, how many times can he expect to hit the bar?

Round your answer to the nearest hundredth.

35. *Apartment Rent* In a select city in Orange County, California, the average rent for an apartment was $1232 in 2010. This was a 2.2% increase from 2009.

(a) How much was the average rent for an apartment in 2009?

(b) If the same percent increase occurred in 2010, how much was the average rent in 2011?

36. *Income Earned* Audra earned $24,500 this year. This was a 5.5% increase from last year.

(a) How much did Audra earn last year?

(b) If Audra gets another 5.5% raise next year, how much will she earn?

To Think About

37. *Income Earned* Jerry works from his home selling vacuum cleaners by phone. He receives a list of 35 numbers to call. Experience has taught him to expect that about 3% will actually buy the $320 vacuum cleaner for which Jerry makes an 11% commission.

(a) What can he expect as income from calling this list of names?

(b) Jerry can buy a prescreened list of 35 names for which 9% will actually buy the vacuum cleaner. What is the maximum price he should pay for the list if he wishes to make at least $100?

Savings Account Interest Often a loan or a savings account has an interest rate that is **compounded quarterly.** *When this is the case, we must calculate the interest at the end of every quarter (3 months) and then add this amount to the principal. We use this new principal to find the interest for the next quarter.*

38. **(a)** If Sheena invests $5000 in a savings account that pays 6% compounded quarterly, how much is in the account at the end of 1 year?

(b) If Sheena invests $5000 in a savings account that pays a simple interest rate of 6%, how much is in the account at the end of 1 year?

Cumulative Review

▲ **39.** **[3.3.2]** *Geometry* Find the area of a parallelogram with a base of 6 centimeters and a height of 4 centimeters.

▲ **40.** **[3.3.3]** *Geometry* Find the volume of a rectangular solid with $L = 9$ centimeters, $W = 4$ centimeters, and $H = 7$ centimeters.

Quick Quiz 8.9

1. A ladies skirt is marked 40% off the original price of $120. How much is the skirt reduced in price?

2. Find the interest on a loan of $12,000 at a simple interest rate of 14% for 3 years.

3. Joe accepted a job for $28,000 a year with the promise of a 12% raise after 1 year. How much will he earn his second year on the job?

4. **Concept Check** Explain how to find simple interest on a loan of $6500 at an annual rate of 9% for a period of 4 months.

Did You Know...

That How Much You Get Paid at Your Job Is Not How Much You Get to Spend Because of Expenses?

DETERMINING YOUR EARNINGS

Understanding the Problem:
The real hourly wage is the amount you make per hour after taxes and other work-related expenses. When calculating real hourly wage, we cannot just look at how much a job pays an hour or how much the paycheck will be. We need to consider work-related expenses as well as extra time spent for the job.

Sharon is not currently employed. She is considering two different positions and needs to determine which job she should take. If she takes a job, she estimates she'll have the following weekly expenses: day care, $175; after-school program, $75; clothing, $25; other work-related expenses, $50; as well as $0.35 per mile for transportation.

Making a Plan:
Sharon needs to determine what the income and expenses would be for each position.

Step 1: First she calculates what her income would be for each position.

Job 1: Donivan Tech is offering her a full-time management position at a yearly salary of $50,310. Sharon estimates that she would work an average of 45 hours per week, considering the extra work commitment required for managers. The commute of 45 miles a day would be about five hours a week in driving. She estimates that payroll deductions would be 15% of her gross salary.

Job 2: J & R Financial Group is offering her a part-time position as an assistant manager. The work schedule would be 6 hours per day, 5 days a week, at an hourly rate of $20.50. The commute of 5 miles a day would be about one hour a week in driving. Due to the reduced work schedule, she calculates that payroll deductions would only be 12%, and day care, after-school program, and other work-related expenses would all be reduced by $25 per week.

Task 1: *What is the weekly pay, before deductions or expenses, at Job 1? At Job 2?*

Step 2: She needs to calculate what her expenses would be for each position.

Task 2: *How much are weekly payroll deductions at Job 1? At Job 2?*

Task 3: *What are the weekly job-related expenses at Job 1? At Job 2?*

...nings Information	Current		
...mal Gross	4,389.30		
...uctions	0.00		
...tions	0.00	**Year to Date**	
...rtime	0.00		
EARNINGS TOTAL	4,389.30	5,277.30	
...-Taxable Gross	351.14	418.18	
...able Gross	3,971.12	4,859.12	

...atutory & Other Deductions	Current	Year to Date
...eral Withholding	311.17	311.17
...itional Federal Withholding	0.00	*****
...te Withholding	135.96	135.96
...itional State Withholding	0.00	*****
...DI	0.00	55.06
	62.67	75.55
...icare	0.00	0.00
...icare Buyout	0.00	0.00
...te Disability Insurance	351.14	351.14
...RS	0.00	0.00
...RS	0.00	
...ernate Retirement	67.04	0.00

Finding a Solution:
Step 3: Using weekly pay, payroll deductions, and job-related expenses, Sharon can determine what her real hourly wage would be. Then she'll be able to compare the two jobs while taking into consideration all expenses.

Task 4: *How much would Sharon have left to spend every week after subtracting deductions and expenses from the weekly pay at Job 1? At Job 2?*

Task 5: *Considering the total time spent working and commuting, what is the real hourly wage at Job 1? At Job 2?*

Step 4: Sharon can now decide which job to take.

Task 6: *If Sharon is deciding on a job based only on real hourly wage, which job should she take?*

Task 7: *If Sharon is deciding on a job based on how much she will have left to spend each week after deductions and expenses, which job should she take?*

Applying the Situation to Your Life:
When comparing jobs, one needs to determine the real hourly wage because what you earn is not as important as what you actually get to spend. Work-related expenses and time spent commuting have an impact on the real hourly rate. Taking these into consideration will help you make an informed decision.

Chapter 8 Organizer

Topic and Procedure	Examples	✏ You Try It
Writing a decimal as a fraction, p. 430 **1.** Write the whole number (if any). **2.** Count the number of decimal places. **3.** Write the decimal part over a denominator that has a 1 and the same number of zeros as the number of decimal places found in step 2. **4.** Simplify.	**1.** Write 3.048 as a fraction. $$3.048 = 3\frac{48}{1000} = 3\frac{6}{125}$$ $\downarrow$ $\downarrow$ 3 decimal 3 zeros places	**1.** Write 20.48 as a fraction.
Comparing two positive numbers in decimal notation, pp. 431–432 **1.** Start at the left and compare corresponding digits. Write in extra zeros if needed. **2.** When two digits are different, the larger number is the one with the larger digit.	**2.** Which is larger, 0.138 or 0.13? 0.138 ? 0.130 8 > 0 So 0.138 > 0.130.	**2.** Which is larger, 1.38 or 1.308?

Rounding decimals, pp. 432–433
 	
1. Identify the round-off place digit.
2. If the digit to the *right* of the round-off place digit is:
   (a) *Less than 5,* do not change the round-off place digit.
   (b) *5 or more,* increase the round-off place digit by 1.
3. In either case, drop all digits to the right of the round-off place digit.
	3. **(a)** Round to the nearest hundredth. 0.8652
	
	0.87
	
	(b) Round to the nearest thousandth. 0.21648.
	
	0.216
		3. **(a)** Round to the nearest hundredth. 27.1048
		
		(b) Round to the nearest thousandth. 1453.1286

Adding and subtracting decimals, p. 437
 	
1. Write the numbers vertically and line up the decimal points. Extra zeros may be written to the right of the decimal points if needed.
2. Add or subtract all the digits with the same place value, starting with the right column and moving to the left. Use carrying or borrowing as needed.
3. Place the decimal point of the answer in line with the decimal points of all the numbers added or subtracted.
	4. Add. **(a)** 36.3 + 8.007 + 5.26
	       (b) −6.8 + 2.6
	(a) $\overset{1}{}$
	   36.300
	    8.007
	 + 5.260
	   49.567
	(b) −6.8
	     2.6 Keep the sign of larger
	   −4.2 absolute value and subtract.
		4. Perform the operation indicated.
		(a) 59 + 2.73 + 0.345
		
		(b) −13.6 − 0.82

Combining like terms, p. 438
 	
Add or subtract coefficients of like terms.
	5. Combine like terms. $8.1x + 0.7y + 1.6x$
	
	   8.1x We add like terms.
	 + 1.6x We line up decimal points.
	   9.7x
	
	We have $8.1x + 1.6x + 0.7y = 9.7x + 0.7y$.
		5. Combine like terms.
		$2.6y + 9.1x − 1.4y$

Multiplying decimals, p. 443
 	
1. Multiply the numbers just as you would multiply whole numbers.
2. Find the total number of the decimal places in the factors.
3. Place the decimal point in the product so that the product has the same number of decimal places as the total in step 2. You may need to insert zeros to the left of the number found in step 1.
	6. Multiply. **(a)** 3.64×2.2
	       (b) $10.2 \times (−1.3)$
	
	(a) 3.64 **(b)** 10.2
	       × 2.2 × (−1.3)
	        728 306
	        728 102
	       8.008 −13.26
	              The product is negative.
		6. Multiply.
		(a) 7.34×4.3
		
		(b) $(−3.16)(2.7)$

Multiplying a decimal by a power of 10, p. 444 Move the decimal point to the right the same number of places as there are zeros in the power of 10. (Sometimes it is necessary to write extra zeros before placing the decimal point in the answer.)	**7.** Multiply. **(a)** $0.597 \times 10^4 = 5970$ **(b)** $0.0082 \times 1000 = 8.2$	**7.** Multiply. **(a)** 85.26×10^3 **(b)** 0.00012×100

Topic and Procedure	Examples	You Try It
Dividing by a decimal, pp. 447–448 1. Make the divisor a whole number by moving the decimal point to the right. 2. Move the decimal point in the dividend to the right the same number of places. 3. Place the decimal point of the quotient directly above the decimal point in the dividend. 4. Divide as with whole numbers.	**8.** Divide. $0.003\overline{)85.8}$ $$\begin{array}{r} 28600. \\ 3\overline{)85800.} \\ \underline{6}\downarrow \\ 25 \\ \underline{24}\downarrow \\ 18 \\ \underline{18} \\ 0 \end{array}$$	**8.** Divide. $0.05\overline{)2.715}$
Converting a fraction to a decimal, pp. 448–449 Divide the denominator into the numerator until (a) the remainder is zero, *or* (b) the decimal repeats, *or* (c) the desired number of decimal places is achieved.	**9.** Find the decimal equivalent. $\dfrac{13}{22}$ $$\begin{array}{r} 0.5909 \\ 22\overline{)13.0000} \\ \underline{110} \\ 200 \\ \underline{198} \\ 200 \\ \underline{198} \\ 2 \end{array}$$ $\dfrac{13}{22} = 0.5\overline{90}$ or $0.5909090\ldots$	**9.** Find the decimal equivalent. $\dfrac{5}{12}$
Solving equations with decimals, p. 453 1. Remove any parentheses. 2. Combine like terms and simplify. 3. Add or subtract to get the *ax* term alone on one side of the equation. 4. Multiply or divide to get *x* alone on one side of the equation.	**10.** Solve. $2x - 5.6 = x - 9.87$ $$\begin{array}{rcr} 2x - 5.6 &=& x - 9.87 \\ +-x & & -x \\ \hline x - 5.6 &=& -9.87 \\ + \quad 5.6 &=& 5.6 \\ \hline x &=& -4.27 \end{array}$$	**10.** Solve. $3(x - 2) = -12.69$
Estimating with percents, pp. 463–464 To estimate 10% of a whole number, delete the last digit. To estimate 1%, delete the last 2 digits. To estimate 5%, find one-half of 10%. To estimate 20%, double 10%. To estimate 15%, add 10% + 5%. To estimate 6%, add 5% + 1%.	**11.** Sara purchased a car for \$24,000. If the sales tax is 6%, how much sales tax did Sara pay? 10% of 24,000 is 2400; thus 5% is 1200. 1% of 24,000 is 240. We add 5% + 1% = 6%: $1200 + $240 = 1440 sales tax.	**11.** Mark earns 11% commission on the sales he makes each month. If his total sales for the month were \$35,050, estimate how much he earned in commission.
Changing a decimal to a percent, p. 469 1. Move the decimal point two places to the right. 2. Add the percent sign.	**12.** Write each decimal as a percent. (a) 0.19 (b) 1.53 (c) 0.516 (d) 0.006 (e) 0.04 (a) 0.19 = 19% (b) 1.53 = 153% (c) 0.516 = 51.6% (d) 0.006 = 0.6% (e) 0.04 = 4%	**12.** Change the decimal to a percent. 1.7
Changing a percent to a decimal, p. 469 1. Move the decimal point two places to the left. 2. Drop the percent sign.	**13.** Write each percent as a decimal. (a) 49% (b) 0.5% (c) 2% (d) 196% (a) 49% = 0.49 (b) 0.5% = 0.005 (c) 2% = 0.02 (d) 196% = 1.96	**13.** Change the percent to a decimal. 0.13%
Changing a fraction to a percent, pp. 471–472 1. Divide the numerator by the denominator to obtain a decimal. 2. Change the decimal to a percent.	**14.** Write each fraction as a percent. (a) $\dfrac{3}{800}$ (b) $3\dfrac{1}{4}$ $F \rightarrow D \rightarrow P$ (a) $\dfrac{3}{800} = 0.00375 = 0.375\%$ $F \rightarrow D \rightarrow P$ (b) $3\dfrac{1}{4} = 3.25 = 325\%$	**14.** Change the fraction to a percent. $4\dfrac{1}{5}$

Topic and Procedure	Examples	You Try It
Changing a percent to a fraction, pp. 471–472 Remove the % sign and move the decimal point two places to the left to obtain a decimal. Then write the decimal as a fraction and reduce the fraction if possible.	**15.** Change 58% to a fraction. F ← D ← P $\dfrac{29}{50} = \dfrac{58}{100} \leftarrow 0.58 \leftarrow 58\%$	**15.** Change 64% to a fraction.
Solving percent problems by translating to equations, p. 476 **1.** Translate by replacing: "of" with $\times$ "find" with $n =$ "is" with $=$ "percent" with % "what" with n **2.** Solve the resulting equation.	**16.** What percent of 70 is 30? $n\%$ $\times\ 70 = 30$ $70n\% = 30$ $\dfrac{70n\%}{70} = \dfrac{30}{70}$ $n\% = 0.4285714\ldots$ n is approximately 42.86. 30 is approximately 42.86% of 70.	**16.** What percent of 75 is 15?
Solving percent problems using proportions, p. 486 **1.** Identify the parts of the percent proportion. $a =$ the amount $b =$ the base (the whole; it usually appears after the word "of") $p =$ the percent number **2.** Write the percent proportion $\dfrac{a}{b} = \dfrac{p}{100}$ using the values obtained in step 1. **3.** Solve for the unknown.	**17.** What percent of 160 is 60? The percent p is unknown. The base $b = 160$ and the amount $a = 60$. $\dfrac{a}{b} = \dfrac{p}{100}$ becomes $\dfrac{60}{160} = \dfrac{p}{100}$ If we reduce the fraction on the left side, we have $\dfrac{3}{8} = \dfrac{p}{100}$ $(100)(3) = 8p$ $300 = 8p$ $\dfrac{300}{8} = \dfrac{8p}{8}$ $37.5 = p$ Thus 60 is 37.5% of 160.	**17.** What is 18% of 60?
Solving commission problems, p. 491 commission = commission rate $\times$ total sales	**18.** A housewares salesperson gets a 16% commission on sales he makes. How much commission does he earn if he sells $12,000 in housewares? commission = (0.16)(12,000) = \$1920	**18.** Last week Tim sold \$3250 worth of computer software. If he is paid a commission rate of 14% of his sales, how much commission did he earn?
Solving percent increase, decrease, and discount problems, pp. 491–492 increase = $\dfrac{\text{percent}}{\text{increase}} \times \dfrac{\text{original}}{\text{amount}}$ $\dfrac{\text{decrease}}{\text{(discount)}} = \dfrac{\text{percent}}{\text{decrease}} \times \dfrac{\text{original}}{\text{amount}}$ original amount + increase = new amount original amount − decrease = new amount	**19.** A \$150 men's suit is discounted 20%. How much is the sale price of the suit? percent discount $\times$ original amount = discount $0.20 \times \$150 = \$30$ original amount − discount = sale price $\$150 - \$30 = \$120$	**19.** Danny earns \$42,000 per year at his job. If he gets a 7% raise, what is his new salary?
Solving simple interest problems, pp. 493–494 interest = principal $\times$ rate $\times$ time $I = P \times R \times T$	**20.** Hector borrowed \$3000 for 4 years at a simple interest rate of 12%. How much interest did he owe after 4 years? $I = P \times R \times T$ $I = (3000)(0.12)(4)$ $= (360)(4)$ $= 1440$ Hector owed \$1440 in interest.	**20.** Find the interest on a loan of \$12,000 at a simple interest rate of 14% for 3 years.

Chapter 8 Review Problems

Section 8.1

Write the word name for each decimal.

1. 0.679

2. 7.0083

Write the word name for the amount on the check.

3.

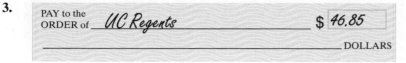

PAY to the ORDER of _UC Regents_____ $ 46.85

_____ DOLLARS

Write in fractional notation.

4. 4.267

5. 43.91

Write each fraction as a decimal.

6. $32\dfrac{761}{1000}$

7. $54\dfrac{26}{1000}$

Replace the ? with < or >.

8. 0.523 ? 0.524

9. 0.16 ? 0.168

Round to the nearest hundredth.

10. 842.8569

Round to the nearest thousandth.

11. 406.7809

Section 8.2

Add.

12. −5.2 + 0.317

13. 0.588 + 36 + 8.43

Subtract.

14. 25.98 − 2.33

15. −9.355 − 2.48

Perform the operation indicated.

16. (−9.2) + (−5.4)

17. −5.3 − (−6.67)

Fill in the blanks.

18. To subtract numbers in decimal notation, we _____ the decimal points.

19. When adding 85 + 36.5, we rewrite 85 as _____ so that we can line up _____.

20. Evaluate $x − 9.3$ for $x = 2.4$.

21. Evaluate $y + 17.2$ for $y = −2.3$.

Combine like terms.

22. $4.6x + 7.2x$

23. $8.6x + 3.9x$

24. *Gasoline Costs* Jamahl keeps a log of his gasoline receipts for his company.

| June 6, 2012 | $44.98 | June 20, 2012 | $43.77 |
| June 13, 2012 | $50.10 | June 27, 2012 | $47.05 |

Estimate Jamahl's total gasoline expenses for the month of June.

Section 8.3

Multiply.

25. 0.082×0.02

26. 5.68×7.21

27. $(3.01)(-41.25)$

28. $(-5.6)(-9.01)$

Fill in the blanks.

29. If one factor has 2 decimal places and the second factor has 3 decimal places, the product has ____ decimal places.

30. If one factor has 3 decimal places and the second factor has 4 decimal places, the product has ____ decimal places.

Multiply.

31. 0.1249×100

32. 1.6×1000

33. 41×10^5

34. *Hourly/Overtime Wages* Mark earns $9.10 per hour for the first 40 hours per week and $13.65 for overtime hours (hours worked over 40 hours). If he works 52 hours this week, how much will he earn?

35. *Lease Cost* Mark has a 5-year truck lease that allows him to buy the truck at the end of the lease for $4500 plus 12 cents a mile for each mile the truck has been driven over 36,000 miles. If at the end of the lease the mileage on Mark's truck is 44,322.6 miles, how much will it cost Mark to buy the truck?

Divide. Round your answer to the nearest hundredth when necessary.

36. $8.66 \div 12$

37. $-8.52 \div -7.2$

Divide. If a repeating decimal is obtained, use notation such as $0.\overline{8}$ or $0.\overline{16}$.

38. $21.2 \div 0.44$

Fill in the blank.

39. When we divide $7.21\overline{)25.9}$, we rewrite the equivalent division problem _____ and then divide.

40. Place the decimal point in the quotient. $12\overline{)4.349}$

_____

Write as a decimal.

41. $\dfrac{86}{8}$

42. $9\dfrac{1}{5}$

43. *Antifreeze* A mechanic wishes to pour 38.5 liters of antifreeze into several equal-size containers that hold 3.5 liters each.

   (a) How many containers will she need?

   (b) If the antifreeze costs $6.20 per liter, how much will 1 full container of antifreeze cost?

Section 8.4

Solve.

44. $x - 2.68 = 8.23$

45. $-1.6x = 5.44$

46. $2x + 2.4 = 8.7$

47. $5(x + 1.4) = x + 2.6$

48. *Coin Problem* Ernie must fill a vending machine with change. The number of nickels is 6 times the number of dimes. The total value of nickels and dimes needed for change is $12. How many of each coin must Ernie put in the vending machine?

Section 8.5

For the number 200,022, estimate the following.

49. 10%

50. 1%

51. 7%

52. 20%

Section 8.6

State using percents.

53. 85 out of 100 citizens voted.

54. *Course Enrollment* Last year's enrollment in a college Psych 100 course was 110. Now the enrollment is 132. Write this year's enrollment as a percent of last year's.

55. Write 0.016 as a percent.

56. Write 124% as a decimal.

Complete each table of equivalent notations.

57.

Decimal Form	Percent Form
0.379	
	42.8%
	0.5%
3.47	

58.

Decimal Form	Percent Form
1.2	
	3.5%
	0.25%
0.567	

59. Write $\dfrac{56}{168}$ as a percent.

60. Write 81.3% as a fraction.

Complete each table of equivalent notations.

61.

Fraction Form	Decimal Form	Percent Form
$\dfrac{3}{8}$		
	0.72	
		$\dfrac{1}{2}\%$
$7\dfrac{2}{5}$		

62.

Fraction Form	Decimal Form	Percent Form
$\dfrac{3}{4}$		
	0.56	
		$\dfrac{1}{5}\%$
$3\dfrac{6}{15}$		

63. 82 is 25% of what number?   **64.** 15 is what percent of 300?   **65.** Find 45% of 120.

66. *Restaurant Tip* José has $18 to spend on dinner. What is the maximum amount he can spend on meals and drinks so that he can leave a 15% tip? Round your answer to the nearest hundredth.

67. *Employee Car Pool* 10% of the employees at A & R Accounting Service take the company car pool van to work. If 26 employees take the company car pool van to work, how many employees work at A & R Accounting?

68. *Sales Tax* Danny bought a car for $22,000. How much sales tax did he pay if the tax is 8% of the price of the car?

69. *Restaurant Tip* Dean left a $4.25 tip for his dinner, which cost $32.40. What percent of the total bill did Dean leave for a tip? Round your answer to the nearest hundredth of a percent.

Section 8.8

Identify the percent p, the base b, and amount a. Do not solve for the unknown.

	p	b	a
70. What is 85% of 400?	_____;	_____;	_____
71. 20 is what percent of 100?	_____;	_____;	_____

Solve.

72. 9 is 45% of what?

73. 18 is what percent of 50?

Fill in the blanks.

74. *Dishwasher Price*

   (a) The original price of the dishwasher in the advertisement is _____.

   (b) The dishwasher is marked down _____.

Sale price $225
Save $75

75. *College Enrollment* A private school predicts that 70% of the 250 spaces open for registration should be *filled* during the first week of registration.

   (a) What percent of the spaces will be *available* after the first week of registration?

   (b) How many spaces should be *available* after the first week of registration?

Section 8.9

76. *Income Earned* Jason has a commission rate of 18% based on the total sales of large appliances he sells. This month Jason sold $13,250 worth of large appliances. What is his commission?

77. *Loan Interest* Find the interest on a loan of $8500 borrowed at a simple interest rate of 11% for 1 year.

How Am I Doing? Chapter 8 Test

CHAPTER Test Prep VIDEOS   MATH COACH   MyMathLab®   YouTube™

After you take this test read through the Math Coach on pages 509–510. Math Coach videos are available via MyMathLab and YouTube. Step-by-step test solutions on the Chapter Test Prep Videos are also available via MyMathLab and YouTube. (Search "BlairTobeyPrealgebra" and click on "Channels.")

1. Write the word name. 207.402

2. Write in fractional notation. 0.013

3. Write as a decimal. $\dfrac{51}{100}$

4. Replace the ? with < or >. 0.45 ? 0.412

5. Round 746.136 to the nearest hundredth.

6. Evaluate $x + 0.12$ for $x = -2.07$.

7. Combine like terms. $3.1x + 2.01y + 1.06x$

Perform the operation indicated.

8. $12.93 + 0.21$

ᴹᴄ 9. $18.8 - 6.23$

10. $(-13.2) - (-7.1)$

11. $(8.24)(1.2)$

12. $(4.72)(10^3)$

ᴹᴄ 13. $15.75 \div 3.5$

14. Write $\dfrac{19}{3}$ as a decimal.

15. Solve. $0.5x + 0.2x = 2.8$

16. Solve. $2(y + 1.3) = 7.8$

17. State as a percent. 7 out of 100 computer chips are defective.

For the number 504, estimate the following.

18. **(a)** 10%   **(b)** 5%   **(c)** 1%   **(d)** 6%   **(e)** 15%

Fill in the blanks. Complete the table of equivalent notations.

	Fraction Form	Decimal Form	Percent Form
19.	$\dfrac{76}{95}$	**(a)** ____	**(b)** ____
20.	**(a)** ____	0.10	**(b)** ____
ᴹᴄ 21.	**(a)** ____	**(b)** ____	5%

1. _____ ☐
2. _____ ☐
3. _____ ☐
4. _____ ☐
5. _____ ☐
6. _____ ☐
7. _____ ☐
8. _____ ☐
9. _____ ☐
10. _____ ☐
11. _____ ☐
12. _____ ☐
13. _____ ☐
14. _____ ☐
15. _____ ☐
16. _____ ☐
17. _____ ☐
18. (a) _____ ☐
(b) _____ ☐
(c) _____ ☐
(d) _____ ☐
(e) _____ ☐
19. (a) _____ ☐
(b) _____ ☐
20. (a) _____ ☐
(b) _____ ☐
21. (a) _____ ☐
(b) _____ ☐

22.

23.

24.

25.

26.

27.

28. (a)

(b)

29.

30.

31.

32.

Total Correct:

22. 12 is 30% of what number?

23. 5 is what percent of 250?

24. What is 20% of 48?

25. Jerald earned $150 in points on his Plus Card from The Office Discount Center. He used his point credit to purchase supplies for his business totaling $112.33. The next day the store discounted all the computer travel bags. Jerald returned to the store and bought one for $69.99 using the remainder of his points, then paid cash for the difference. How much did Jerald have to pay in cash toward the purchase of the computer travel bag?

Solve.

26. Fred has a 4-year car lease that allows him to buy the car at the end of the lease for $5500 plus 12 cents a mile for each mile the car is driven over 30,000 miles. If at the end of the lease the mileage on Fred's car is 34,100.5 miles, how much will it cost Fred to buy the car?

27. 35% of the employees at Micro Electronics Inc. contribute part of their pay each month to the company's Summer Saver Plan. If 49 employees contribute to this plan, how many employees work at Micro Electronics Inc.?

ᴹᴄ **28.** A computer is reduced 24% from the original price of $3300.

   (a) How much is the computer reduced in price?

   (b) What is the sale price?

29. Juan stayed at a resort hotel on Saturday night and paid $249 for that night. The rate on Saturday night is 20% higher than the rate on weeknights. How much will Juan pay for the room on Sunday night, if Sunday is considered a weeknight?

30. Sylvia has a commission rate of 15% based on the total sales of electronic equipment she sells. This month Sylvia sold $10,500 worth of electronic equipment. What is her commission?

31. Find the interest on a loan of $8200 at a simple interest rate of 11% for 2 years.

32. If both digital cameras in the advertisements below offer the same features and quality, which one is a better deal?

508

MATH COACH

Mastering the skills you need to do well on the test.

Students often make the same types of errors when they do the Chapter 8 Test. Here are some helpful hints to keep you from making these common errors on test problems.

Subtracting Decimal Expressions—Problem 9
Perform the operation indicated. $18.8 - 6.23$

> **Helpful Hint** Add zeros at the end of each number, if necessary, so that the same number of digits appear to the right of each decimal point. Remember to line up the decimal points.

Did you change the problem to $18.80 - 6.23$ as your first step?

Yes ▢ No ▢

If you answered No, examine your work carefully and perform this step again. This will help you avoid borrowing errors.

Did you line up the decimal points carefully when you wrote the numbers one beneath the other?

$$\begin{array}{r} 18.80 \\ -\ 6.23 \end{array}$$

Yes ▢ No ▢

If you answered No, rewrite the problem on paper. Write out your steps and show your borrowing.

If you answered Problem 9 incorrectly, go back and rework the problem using these suggestions.

Dividing a Decimal by a Decimal—Problem 13   Perform the operation indicated. $15.75 \div 3.5$

> **Helpful Hint** First, determine how many decimal places you must move the decimal point to the right in the divisor to make it a whole number. Then move the decimal point the *same number of places* to the right in the dividend. Be sure to place the decimal point in the quotient directly above the decimal point in the dividend.

Did you change 3.5 to 35, and 15.75 to 157.5 first?

Yes ▢ No ▢

If you answered No, stop and make this correction.

Did you remember to write the decimal point in the quotient directly above the decimal point in the dividend?

Yes ▢ No ▢

If you answered No, stop and make this correction.

When you performed the last step of your division, did you multiply 5×35 to obtain 175?

Yes ▢ No ▢

If you answered No, examine your work and check each step for calculation errors.

Now go back and rework the problem using these suggestions.

Need help? Watch the **MATH COACH** videos in MyMathLab® or on You.

509

Changing Between Fractions, Decimals, and Percents—Problem 21

Fill in the blanks. Complete the table of equivalent notations.

Fraction Form   Decimal Form   Percent Form

(a) _____   **(b)** _____   **(c)** 5%

> **Helpful Hint** For this problem, it is easier to convert the percent to a decimal first. To change a percent to a decimal, move the decimal point two places to the **left** and remove the % symbol. To change a decimal to a fraction, write the decimal part over a denominator that has a 1 and the same number of zeros as the number of decimal places.

Did you remember to write 5% as 5.0%, then move the decimal point two places to the left?

Yes ▢   No ▢

If you answered No, go back and complete this step again. Remember that when you change a percent to a decimal, you are dividing that number by 100. That is why the decimal point moves two places to the left.

To change 0.05 to a fraction, did you determine that the denominator of the fraction must have a 1 and 2 zeros? Then did you write 5 as the numerator of the fraction?

Yes ▢   No ▢

If you answered No, reread the Helpful Hint regarding changing a decimal to a fraction.

Did you reduce your fraction to lowest terms?

Yes ▢   No ▢

If you answered Problem 21 incorrectly, go back and rework the problem using these suggestions.

Solving Applied Problems Involving Percents—Problem 28

A computer is reduced 24% from the original price of $3300. **(a)** How much is the computer reduced in price? **(b)** What is the sale price?

> **Helpful Hint** Write a simple percent statement in your own words that describes the situation in the applied problem.

For part **(a)**, were you able to write a statement such as "The discount is 24% of the original price of $3300?"

Yes ▢   No ▢

If you answered No, reread the problem and see if you can write a similar statement.

Were you able to write the equation $n = 0.24 \times 3300$ to find the answer to part **(a)**?

Yes ▢   No ▢

If you answered No, check your calculations to be sure that you correctly changed 24% to a decimal.

For part **(b)**, once you know how much the computer is reduced in price, did you subtract as follows?

original price − discount = sale price

Yes ▢   No ▢

Remember to include the dollar sign ($) as the units for both answers.

Now go back and rework the problem using these suggestions.

Need more help? Look for section examples marked with $\mathbb{MC}$ to review.

510

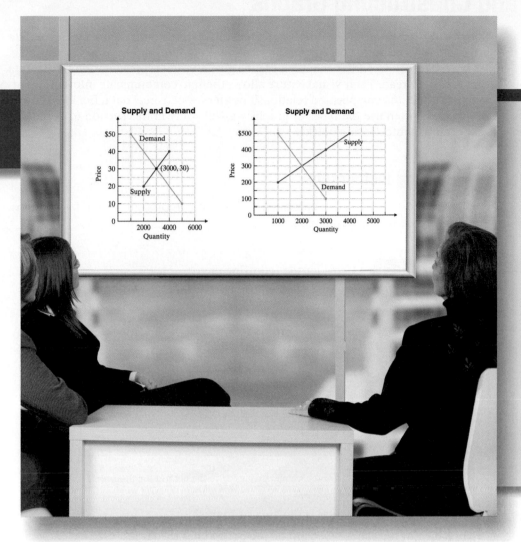

To efficiently manage a business we must be able to analyze consumers' purchase patterns and to make decisions about the sale price and demand for an item. Displaying this information on graphs allows us to see the entire financial picture in a comparative way. Many of the graphs used in business are similar to the ones studied in this chapter.

Graphing and Statistics

9.1 Interpreting and Constructing Graphs

Statistics is that branch of mathematics that collects and studies data. Once the data are collected, the data must be organized so that the information is easily readable. As we have seen in earlier chapters, graphs give a visual representation of the data that is easy to read. Their visual nature allows them to communicate information efficiently about the complicated relationships among statistical data. For this reason, newspapers often use the types of graphs we will study in this section to help their readers grasp information quickly.

① Reading Pictographs

A **pictograph** uses a visually appropriate symbol to represent a number of items. A pictograph is used in Example 1.

EXAMPLE 1 Consider the following pictograph.

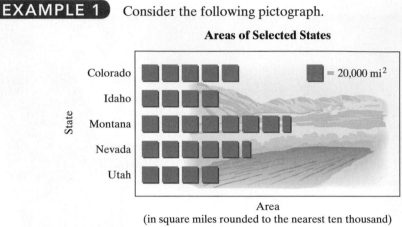

Areas of Selected States

■ = 20,000 mi^2

Area
(in square miles rounded to the nearest ten thousand)

Source: World Book Encyclopedia

(a) How many square miles is the area of Idaho?

(b) Which of the states listed on the pictograph has the largest area?

(c) How many more square miles is the area of Colorado than the area of Utah?

Solution

(a) Since there are four symbols (■) beside the state of Idaho on the pictograph, and each one equals 20,000 square miles, we have $4 \times 20{,}000 = 80{,}000$ square miles.

(b) Montana has the most symbols on the pictograph and thus has the largest area.

(c) Colorado is 20,000 square miles more in area than Utah because there is one extra symbol beside Colorado. (Each ■ equals 20,000 square miles.)

NOTE TO STUDENT: *Fully worked-out solutions to all of the Student Practice problems can be found at the back of the text starting at page SP-1.*

Student Practice 1 Consider the pictograph in Example 1.

(a) How many square miles is the area of Utah?

(b) Which of the states listed has the smallest area?

(c) How many fewer square miles is Idaho than Nevada?

② Reading Circle Graphs with Percentage Values

A **circle graph** indicates how a whole quantity is divided into parts. These graphs help us to visualize the relative sizes of the parts. Each piece of the pie or circle is called a **sector**. We sometimes refer to circle graphs as **pie graphs.**

EXAMPLE 2   Together, the Great Lakes form the largest body of freshwater in the world. The total area of these five lakes is about 290,000 square miles, almost all of which is suitable for boating. The percentage of this total area taken up by each of the Great Lakes is shown in the pie graph.

(a) What percentage of the area is taken up by Lake Michigan?

(b) What lake takes up the largest percentage of the total area?

(c) What percentage of the total area is *not* taken up by Lake Erie?

(d) How many square miles are taken up by Lake Huron and Lake Michigan together?

Area of Great Lakes

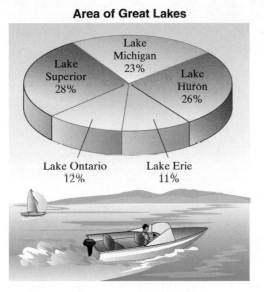

Student Practice 2   Refer to Example 2 to answer the following.

(a) What percentage of the area is taken up by Lake Superior and Lake Michigan together?

(b) What percentage of the total area is *not* taken up by Lake Superior and Lake Michigan together?

(c) How many square miles are taken up by Lake Erie and Lake Ontario together?

Solution

(a) Lake Michigan takes up 23% of the area.

(b) Lake Superior takes up the largest percentage.

(c) The entire circle represents 100%, and Lake Erie takes up 11% of the area. Thus, we subtract $100 - 11 = 89$, so 89% of the total area is *not* taken up by Lake Erie.

(d) If we add $26 + 23$, we get 49. Thus Lake Huron and Lake Michigan together take up 49% of the total area.

$$49\% \text{ of } 290,000 = (0.49)(290,000)$$
$$= 142,100 \text{ square miles}$$

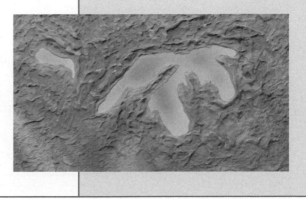

③ Reading and Interpreting Double-Bar Graphs

Double-bar graphs are useful for making comparisons. The following double-bar graph compares the percent of all traffic fatalities for each age group that involved drunk drivers. The groups are separated by gender. Use this graph for the following Examples and Student Practice problems.

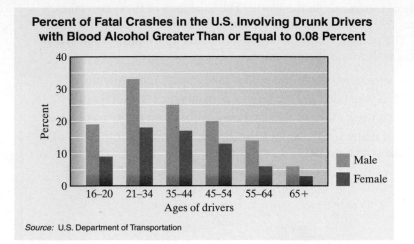

Percent of Fatal Crashes in the U.S. Involving Drunk Drivers with Blood Alcohol Greater Than or Equal to 0.08 Percent

Source: U.S. Department of Transportation

EXAMPLE 3 What percent of the traffic fatalities for males in the age range 35–44 years involved drunk drivers?

Solution The bar rises to the white line that is halfway between 20 and 30. This represents a value halfway between 20 and 30 percent.

Thus 25% of traffic fatalities for males ages 35–44 involved drunk drivers.

Student Practice 3 What percent of the traffic fatalities for males ages 45–54 years involved drunk drivers?

EXAMPLE 4 In which age group was the percent of drunk-driving fatalities the highest for males? For females?

Solution From the double-bar graph, we see that the percent of fatalities for 21–34 year olds was the highest for both males and females.

Student Practice 4 Estimate the highest percent of drunk-driving fatalities for males.

④ **Reading and Interpreting Comparison Line Graphs**

Two or more sets of data can be compared by using a **comparison line graph.** A comparison line graph shows two or more line graphs together. A different style or color for each line distinguishes them.

Note that using a blue line and a yellow line in the following graph makes it easy to read. Use this graph for the following Examples and Student Practice problems.

Grocery Shopping in the U.S.

■ Shoppers under 35 years of age   □ Shoppers over 55 years of age

Percent of people who grocery shop

Day of the week

Source: Chicago Tribune

EXAMPLE 5   Fill in the blank.

Approximately 19% of people in the _____ age group do their grocery shopping on Friday.

Solution   The blue solid line represents the under-35 age group.

Since the dot corresponding to Friday on the blue solid line is near 20%, we know that approximately 19% of the under-35 age group shops on Friday.

Student Practice 5   Fill in the blank.

Approximately 9% of people in the _____ age group do their grocery shopping on Monday.

Mᴄ **EXAMPLE 6**   On what day(s) are there more grocery shoppers from the over-55 age group than the under-35 age group?

Solution   Since the yellow solid line indicates the shopping days for the over-55 age group, we look for those days that have a yellow dot higher on the graph than a blue dot.

Thus, Tuesday, Wednesday, and Thursday are the days that have more shoppers from the over-55 age group.

Student Practice 6   Which day is the busiest shopping day for the under-35 age group? For the over-55 age group?

⑤ **Constructing Double-Bar and Line Graphs**

There are several ways to construct line and bar graphs. How you design a graph usually depends on the data you must graph and the visual appearance you want. For both types of graphs, the intervals on the horizontal and vertical lines must be equally spaced.

EXAMPLE 7   Construct a comparison line graph of the information given in the table.

Category	Activity	Hours Spent per Week
Single men	Gym	6
	Outdoor sports	4
	Dating	7
	Reading and TV	3
Single women	Gym	4
	Outdoor sports	2
	Dating	7
	Reading and TV	9

Solution   We plan our graph.

1. First, we draw a vertical and a horizontal line.

2. Since we are comparing *Hours per Week* to type of activity, we place this label on the vertical line. Mark intervals with equally spaced notches and label 0 to 9.

3. We place the label *Activity* on the horizontal line. We mark intervals with equally spaced notches and label with the name of each activity.

4. Now we place dots on the graph that correspond to the data given.

5. We choose a different color line for each of the categories we are comparing (men and women) and connect the dots with line segments.

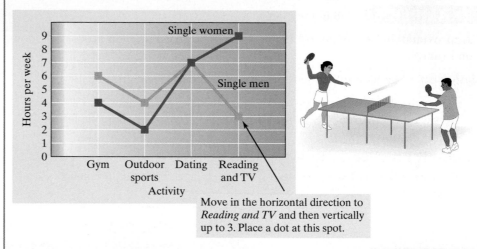

Move in the horizontal direction to *Reading and TV* and then vertically up to 3. Place a dot at this spot.

Note that instead of colored lines, we could use a solid line and a dashed line.

Student Practice 7   Construct a comparison line graph of the information given in the table.

Apartment Rent

Size	Beach Area	Inland
Studio	$800	$650
1 bedroom	$1050	$800
2 bedroom	$1400	$1100
3 bedroom	$1800	$1350

EXAMPLE 8   Construct a double-bar graph of the information given in the table.

The Window Store Profits*

Window Covering	2011	2012
Miniblinds	$12,000	$15,000
Vertical blinds	11,000	16,000
Shutters	16,000	14,000
Drapes	9,000	7,000

*Profits rounded to the nearest thousand.

Solution   We plan our graph.

1. First, we draw a vertical and a horizontal line.
2. Since we are comparing the annual profits for the items, we place the label *Profit* on the vertical line. Mark intervals with equally spaced notches and label 7,000 to 17,000.
3. We label the horizontal line with the store's products.
4. We next label a shaded and a nonshaded bar for the years we are displaying (2011 and 2012).
5. Now, we draw a bar to the appropriate height for each category of window coverings.

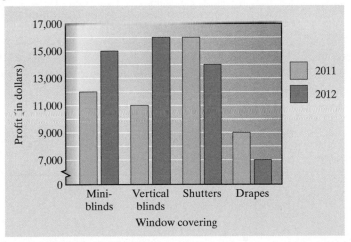

SIDELIGHT   When we construct bar graphs, sometimes the heights of many of the bars we draw do not rise exactly to one of the intervals marked on the vertical axis. When this happens we can place each bar's data above it, so that the exact number is displayed, and we can avoid estimating when reading the graph.

Student Practice 8   Construct a double-bar graph with the information given. Place the given data above each bar so that estimating is not necessary when reading the graph.

2006 Average Yearly Earnings and Highest Degree Earned*

Degree Earned	Men	Women
AA	$51,000	$40,000
BA	$77,000	$53,000
MA	$97,000	$63,000

*Rounded to the nearest thousand
Source: U.S. Census Bureau

Applications

Fastest Growing Occupations Use the pictograph to answer exercises 1–6.

1. Which profession will require the greatest number of new people?

2. Which profession will require the fewest new people?

3. How many additional computer support specialists and desktop publishers will be needed?

4. How many more systems software engineers will be needed by the end of 2014 than were needed at the start of 2004?

5. How many more new computer support specialists will be needed than new network systems and data communications analysts?

6. How many more new network and computer systems administrators will be needed than new desktop publishers?

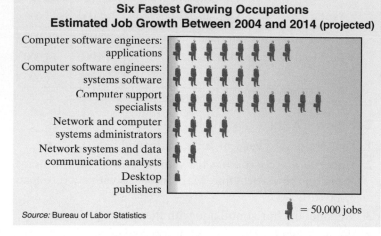

Six Fastest Growing Occupations
Estimated Job Growth Between 2004 and 2014 (projected)

Computer software engineers: applications
Computer software engineers: systems software
Computer support specialists
Network and computer systems administrators
Network systems and data communications analysts
Desktop publishers

Source: Bureau of Labor Statistics

= 50,000 jobs

Population Density Use this pictograph to answer exercises 7–10.

7. How many people per square mile are there in Asia?

8. How many people per square mile are there in Africa?

9. How many more people per square mile are there in Europe than in South America?

10. How many more people per square mile are there in Asia than in North America?

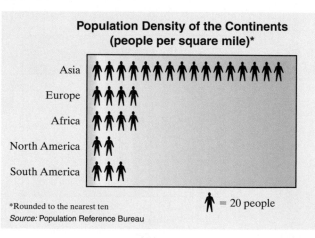

Population Density of the Continents
(people per square mile)*

Asia
Europe
Africa
North America
South America

*Rounded to the nearest ten
Source: Population Reference Bureau

= 20 people

Household Expenses In 2005 the average household expenses for a consumer under 25 years of age were approximately $27,776. The circle graph divides this average expense into basic expense categories. Use the circle graph to answer exercises 11–16.

11. What percent of the household expenses went for housing?

12. What percent of the household expenses went for food away from home?

13. (a) What are the two largest expenses?

    (b) What is the total percent of these two expenses?

14. What percent of expenses are for categories other than housing and transportation?

15. How much on the average was spent for housing?

16. How much on the average was spent for education?

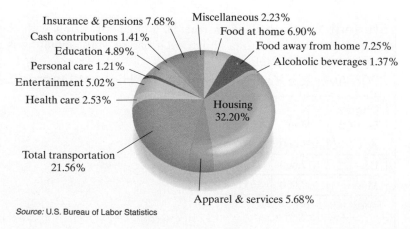

Estimated Household Expenses for People Under 25 Years of Age

Insurance & pensions 7.68%
Cash contributions 1.41%
Education 4.89%
Personal care 1.21%
Entertainment 5.02%
Health care 2.53%
Miscellaneous 2.23%
Food at home 6.90%
Food away from home 7.25%
Alcoholic beverages 1.37%
Housing 32.20%
Total transportation 21.56%
Apparel & services 5.68%

Source: U.S. Bureau of Labor Statistics

Emergency Room Visits *In the United States approximately one-quarter of all emergency room (ER) visits of people ages 5 to 24 are caused by sports. Use the circle graph to answer exercises 17–22.*

17. What percent of ER visits were caused by basketball injuries?

18. What percent of ER visits were caused by soccer injuries?

19. What percent of injuries were caused by football or cycling?

20. What percent of injuries were caused by snow sports or water sports?

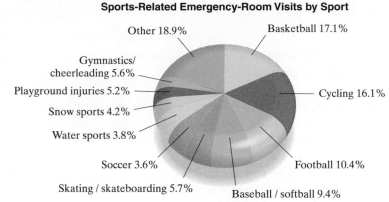

Sports-Related Emergency-Room Visits by Sport

Other 18.9%
Basketball 17.1%
Gymnastics/cheerleading 5.6%
Playground injuries 5.2%
Cycling 16.1%
Snow sports 4.2%
Water sports 3.8%
Soccer 3.6%
Football 10.4%
Skating / skateboarding 5.7%
Baseball / softball 9.4%

Source: Centers for Disease Control and Prevention

21. What percent of injuries were *not* caused by basketball?

22. What percent of injuries were *not* caused by water sports?

Snowfall *In March of a recent year, when most of the season's snow had fallen, the snowfall totals for most cities were below the average yearly totals. The double-bar graph compares the average yearly snowfall totals to the season snowfall totals as of March for selected cities in the United States. Use this graph to answer exercises 23–32.*

23. What is the average yearly snowfall total in Boston, Massachusetts?

24. As of March, what was the season snowfall total in Denver, Colorado?

25. For what cities was the average yearly snowfall less than the season snowfall as of March?

26. For which city is the average yearly snowfall the greatest?

27. For Great Falls, Montana, what was the difference between the average yearly snowfall and the season snowfall as of March?

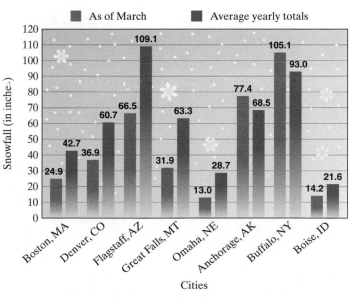

As of March   Average yearly totals

Snowfall (in inches)

Boston, MA: 24.9, 42.7
Denver, CO: 36.9, 60.7
Flagstaff, AZ: 66.5, 109.1
Great Falls, MT: 31.9, 63.3
Omaha, NE: 13.0, 28.7
Anchorage, AK: 77.4, 68.5
Buffalo, NY: 105.1, 93.0
Boise, ID: 14.2, 21.6

Cities

28. For Omaha, Nebraska, what was the difference between the average yearly snowfall and the season snowfall as of March?

29. For which city was the difference between the average yearly snowfall and the season snowfall as of March the greatest?

30. For which city was the difference between the average yearly snowfall and the season snowfall as of March the smallest?

31. In which cities were snowfalls as of March greater than the average yearly snowfall in Denver, CO?

32. In which cities were the average yearly snowfalls greater than the snowfalls as of March in Anchorage, AK?

Restaurant Customers The comparison line graph illustrates the number of customers per month coming into the Bay Shore Restaurant and the Lilly Cafe. Use this graph to answer exercises 33–38.

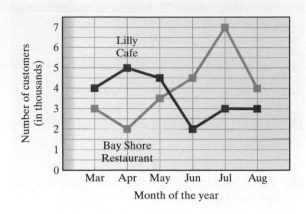

33. In which month did the Lilly Cafe have the fewest customers?

34. In which month did the Bay Shore Restaurant have the greatest number of customers?

35. (a) Approximately how many customers per month came into the Bay Shore Restaurant during the month of June?

    (b) From May to June, did the number of customers increase or decrease at the Bay Shore Restaurant?

36. (a) Approximately how many customers came into the Lilly Cafe during the month of May?

    (b) From March to April, did the number of customers increase or decrease at the Lilly Cafe?

37. Between what two months is the *increase* in customers the largest at the Bay Shore Restaurant?

38. Between what two months did the biggest *decrease* in customers occur at the Lilly Cafe?

Monthly Precipitation The comparison line graph indicates average monthly precipitation at two city airports for the thirty years between 1971 and 2000. Use this graph to answer exercises 39–44.

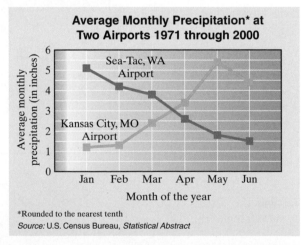

39. During which month was the average precipitation highest at the Sea-Tac (Seattle-Tacoma) airport? At the Kansas City airport?

40. During which month was the average precipitation lowest at the Sea-Tac (Seattle-Tacoma) airport? At the Kansas City airport?

41. During what months was the average precipitation at the Sea-Tac airport less than the average precipitation at the Kansas City airport?

42. During what months was the average precipitation at the Sea-Tac airport greater than the average precipitation at the Kansas City airport?

43. What was the average precipitation for June at the Sea-Tac airport?

44. At what airport was the average precipitation the highest for any month?

45. *High Low Reading* Use the table to make a double-bar graph that compares the high and low daily temperatures given in the table.

City	High °F	Low °F
Albany	50	24
Anchorage	30	16
Boise	55	30
Chicago	20	11

46. *Fat and Carbohydrates* Use the table to make a double-bar graph that compares the number of grams of fat and carbohydrates in the "fast foods" named in the following table.

Type of Sandwich	Fat (g)	Carbohydrates (g)
Wendy's Baconator™	34	43
Wendy's Ultimate Chicken Grill Sandwich	7	42
Burger King's Bacon Cheeseburger	16	28
McDonald's McChicken Sandwich	16	40

Sources: www.bk.com, www.mcdonalds.com, www.wendys.com

Life Insurance Use the information in the table to answer exercises 47 and 48.

Estimated Monthly Premiums for $200,000 5-year Term Life Insurance (Non-Nicotine User)

Age	25	30	35	40	45	50	55	60
Male	$25	$25	$30	$45	$60	$80	$130	$190
Female	$20	$20	$25	$30	$45	$60	$95	$140

47. Construct a comparison line graph that compares the male and female premiums for the ages 25–45.

48. Construct a double-bar graph that compares the male and female premiums for the ages 45–60.

Company Profit Use the information in the table for exercises 49 and 50.

Profit for Douglas Electronics

Quarter	2009	2010	2011	2012
1st	$12,000	$10,000	$15,000	$14,000
2nd	14,000	11,000	13,000	13,000
3rd	15,000	13,000	10,000	11,000
4th	10,000	15,000	12,000	14,000

49. Construct a comparison line graph that compares the profits in 2009 and 2010.

50. Construct a comparison line graph that compares the profits in 2010 and 2011.

Sport Event Tickets Use the circle graph for exercise 51 **(a)–(d)**.

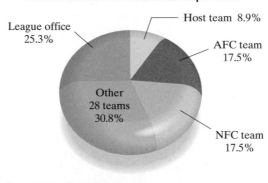

Ticket Allocation for a Selected Super Bowl

League office 25.3%
Host team 8.9%
AFC team 17.5%
Other 28 teams 30.8%
NFC team 17.5%

Source: Orange County Register

51. **(a)** What percent of the tickets were allocated to individual teams?

    (b) If 100,000 tickets were available, how many went to the host team?

    (c) If the average ticket price was $102, what was the ticket revenue for the AFC or the NFC team?

    (d) If the league office gave away 25% of its tickets as promotion, what was the value of the promotion?

Cumulative Review

Perform the operation indicated.

52. [**6.1.4**] $5x - 2(3x^2 + 1) - (-4x^2 + 6x + 3)$

53. [**6.2.3**] $(x - 2)(x + 4)$

54. [**6.4.2**] Factor. $8x^2y + 2xy + 4x$

Quick Quiz 9.1

1. Use the information in the table for exercises 49–50 to construct a double-bar graph that compares profits in 2011 and 2012.

2. Refer to the circle graph in exercise 51 to answer the following. What percent of the Super Bowl tickets were *not* allocated to individual teams?

3. Use the information in the table to construct a comparison line graph that compares the number of male and female customers in a coffee shop.

Category	Time of Day	Number of Customers
Male	5:01–6 A.M.	20
	6:01–7 A.M.	35
	7:01–8 A.M.	35
	8:01–9 A.M.	20
Female	5:01–6 A.M.	10
	6:01–7 A.M.	25
	7:01–8 A.M.	40
	8:01–9 A.M.	45

4. **Concept Check** The college administration is gathering data to determine the number of students who enroll in morning, afternoon, evening, and weekend courses. So far the following information has been gathered: 45% of students enroll in morning courses, while 10% enroll in weekend courses.

    (a) If the enrollment in the afternoon is twice the weekend enrollment, explain how you would determine the percent of students enrolled in evening and in afternoon courses.

    (b) If you create a circle graph (pie graph), describe how you would construct a pie slice that describes the percent of students enrolled in evening courses.

9.2 Mean, Median, and Mode

Student Learning Objectives

After studying this section, you will be able to:

① Find the mean of a set of numbers.

② Find the median of a set of numbers.

③ Find the mode of a set of numbers.

① Finding the Mean of a Set of Numbers

We often want to know the *middle value* of a group of numbers. In this section we learn that in statistics, there is more than one way of describing this middle value. There is the *mean* of the group of numbers, and there is the *median* of the group of numbers. In some situations it's more helpful to look at the mean, and in others it's more helpful to look at the median. We'll learn to tell which situations lend themselves to one or the other.

> The **mean** of a set of values is the sum of the values divided by the number of values. The mean is often called the **average**.

EXAMPLE 1   Find the average or mean test score of a student who has test scores of 71, 83, 87, 99, 80, and 90.

Solution   We take the sum of the six test scores and divide the sum by 6.

$$\text{Sum of test scores} \rightarrow \atop \text{Number of tests} \rightarrow \quad \frac{71 + 83 + 87 + 99 + 80 + 90}{6} = \frac{510}{6} = 85$$

The mean is 85.

Student Practice 1   Find the average or mean of the following test scores: 88, 77, 84, 97, and 89.

NOTE TO STUDENT: Fully worked-out solutions to all of the Student Practice problems can be found at the back of the text starting at page SP-1.

The mean value is often rounded to a certain decimal-place accuracy.

EXAMPLE 2   Carl and Wally each kept a log of the miles per gallon achieved by their cars for the last two months. Their results are recorded on the graph. What is the mean miles per gallon figure for the last 8 weeks for Carl? Round your answer to the nearest mile per gallon.

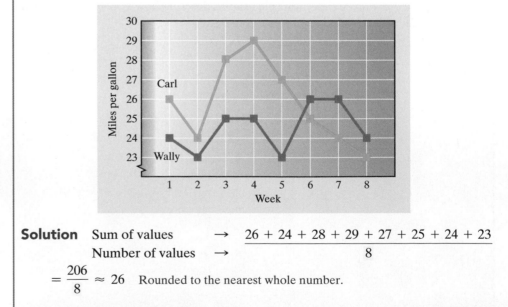

Solution
$$\text{Sum of values} \rightarrow \atop \text{Number of values} \rightarrow \quad \frac{26 + 24 + 28 + 29 + 27 + 25 + 24 + 23}{8}$$

$$= \frac{206}{8} \approx 26 \quad \text{Rounded to the nearest whole number.}$$

The mean miles per gallon figure is 26.

Student Practice 2   Use the double-line graph in Example 2 to find the mean miles per gallon figure for the last 8 weeks for Wally. Round your answer to the nearest mile per gallon.

② **Finding the Median of a Set of Numbers**

> If a set of numbers is arranged in order from smallest to largest, the **median** is that value that has the same number of values above it as below it.

EXAMPLE 3   The total minutes of daily telephone calls made by Sara and Brad during one week are indicated on the double-bar graph. Find the median value for the total minutes of Sara's daily calls.

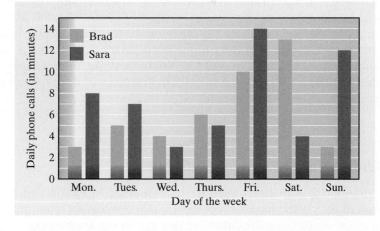

Solution   We arrange the numbers for Sara's calls in order from smallest to largest.

3, 4, 5   7   8, 12, 14

three numbers   middle number   three numbers

There are three numbers smaller than 7 and three numbers larger than 7. Thus 7 is the median.

Student Practice 3   Use the double-bar graph in Example 3 to find the median value for Brad's daily calls.

If a list of numbers contains an even number of different items, then of course there is no one middle number. In this situation we obtain the median by taking the average of the two middle numbers.

EXAMPLE 4   Find the median of the following numbers: 13, 16, 18, 26, 31, 33, 38, and 39.

Solution   13, 16, 18   26, 31   33, 38, 39

three numbers   two middle numbers   three numbers

The average (mean) of 26 and 31 is $\dfrac{26 + 31}{2} = \dfrac{57}{2} = 28.5$.

Thus the median value is 28.5.

Student Practice 4   Find the median value of the following numbers: 88, 90, 100, 105, 118, and 126.

Understanding the Concept

Use the Mean or the Median?   When would someone want to use the mean, and when would someone want to use the median? Which is more helpful? The mean (average) is used more frequently. It is most helpful when the data are distributed fairly evenly, that is, when no one value is "much larger" or "much smaller" than the rest. For example, suppose a company had employees with annual salaries of $9,000, $11,000, $14,000, $15,000, $17,000, and $20,000. All the salaries fall within a fairly limited range. The mean salary (rounded to the nearest cent),

$$\frac{9000 + 11,000 + 14,000 + 15,000 + 17,000 + 20,000}{6} \approx \$14,333.33$$

gives us a reasonable idea of the "average" salary. However, suppose that the company had six employees with salaries of $9,000, $11,000, $14,000, $15,000, $17,000, and **$90,000.** Talking about the mean salary, which is $26,000, is deceptive. No one earns a salary very close to the mean salary. The "average" worker in that company does not earn around $26,000. In this case, the median value is more appropriate. Here the median is $14,500. Problems of this type are included in Exercise Set 9.2, exercises 47 and 48.

③ Finding the Mode of a Set of Numbers

Another value that is sometimes used to describe a set of data is the **mode.** The *mode* of a set of data is the *number or numbers that occur most often*. If two values occur most often, we say that the data have two modes (or are **bimodal**).

EXAMPLE 5   Find the mode of each of the following.

(a) A student's test scores: 89, 94, 96, 89, 90.

(b) The ages of students in a calculus class: 33, 27, 28, 28, 21, 19, 18, 25, 26, 33.

Solution

(a) The mode of 89, 94, 96, 89, 90 is 89 since it occurs twice in the set of data.

(b) The data 33, 27, 28, 28, 21, 19, 18, 25, 26, 33 is bimodal since both 28 and 33 occur twice.

Student Practice 5   Find the mode of each of the following.

(a) The number of video rentals per day during a 1-week period: 121, 156, 131, 121, 142, 149, 131.

(b) Test scores: 99, 76, 79, 92, 76, 84, 83, 76.

A set of numbers may have *no mode* at all. For example, the set of numbers 10, 20, 30, 40 has *no mode* because each number occurs just once. Likewise, if all numbers occur the same number of times, there is *no mode*. The set of numbers 10, 10, 20, 20 has *no mode* because each number occurs twice.

9.2 Exercises

MyMathLab®

Watch the videos
in MyMathLab

Download the
MyDashBoard App

Verbal and Writing Skills, Exercises 1–2

1. Define the mean (average).

2. Explain the difference between the median and the mode.

Find the mean. Round to the nearest tenth when necessary.

3. *Quiz Grade* A student received grades of 84, 91, 86, 95, and 98 on math quizzes.

4. *Quiz Grade* Jessica received grades of 85, 87, 90, 92, 83, and 94 on history quizzes.

5. *Television Viewing* Sam watched television last week for the following number of hours per day.

Mon.	Tues.	Wed.	Thurs.	Fri.	Sat.	Sun.
2	3	3	4	2.5	2.5	4

6. *Mileage* Mel's car got the following miles per gallon results during the last 6 months.

Jan.	Feb.	Mar.	Apr.	May	June
24	23	25	27	28	29

7. *Calls Received* The Windy City Passport Photo Center received the following numbers of telephone calls over the last 6 days: 23, 45, 63, 34, 21, and 42.

8. *Car Rental* The local Hertz rental car office received the following numbers of inquiries over the last 7 days: 34, 57, 61, 22, 43, 80, and 39.

9. *Homes Sold* The last five houses built in town sold for the following prices: $189,000, $185,000, $162,000, $145,000, and $162,000.

10. *Sofa Price* Trang priced a sofa at six local stores. The prices were $489, $469, $590, $460, $529, and $529.

Sales Record The double-bar graph shows the number of sales made by Alex and Lisa from June through October. Use the graph to answer exercises 11 and 12.

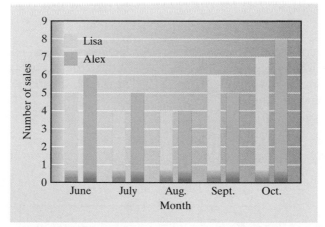

11. Find the average number of sales by Alex for the 5-month period.

12. Find the average number of sales by Lisa for the 5-month period.

13. *Baseball* The captain of the college baseball team achieved the following results.

	Game 1	Game 2	Game 3	Game 4	Game 5
Hits	0	2	3	2	2
Times at bat	5	4	6	5	4

Find his batting average by dividing his total number of hits by the total times at bat.

14. *Bowling* The captain of the college bowling team had the following results after practice.

	Practice 1	Practice 2	Practice 3	Practice 4
Score (pins)	541	561	840	422
Number of games	3	3	4	2

Find her bowling average by dividing the total number of pins scored by the total number of games.

15. *Gasoline Mileage* Frank and Wally traveled to the West Coast during the summer. The numbers of miles they drove and the numbers of gallons of gas they used each day are recorded below.

	Day 1	Day 2	Day 3	Day 4
Miles driven	276	350	391	336
Gallons of gas	12	14	17	14

Find the average miles per gallon achieved by the car on the trip by dividing the total number of miles driven by the total number of gallons used.

16. *Gasoline Mileage* Cindy and Andrea traveled to Boston this fall. The numbers of miles they drove and the numbers of gallons of gas they used each day are recorded below.

	Day 1	Day 2	Day 3	Day 4
Miles driven	260	375	408	416
Gallons of gas	10	15	17	16

Find the average miles per gallon achieved by the car on the trip by dividing the total number of miles driven by the total number of gallons used.

Find the median value.

17. 22, 42, 45, 47, 51, 50, 58

18. 30, 39, 46, 50, 57, 60, 70

19. 1052, 968, 1023, 999, 865, 1152

20. 1400, 1329, 1000, 1360, 1900, 1350

21. 0.52, 0.69, 0.71, 0.34, 0.58

22. 0.26, 0.19, 0.35, 0.43, 0.30

23. 3.7, 1.1, 3.4, 3.9, 2.8, 2.9

24. 5.3, 2.1, 4.4, 4.9, 2.9, 5.9

25. *Salaries* The annual salaries of the employees of a local cable television office are $17,000, $11,600, $23,500, $15,700, $26,700, and $31,500.

26. *Car Costs* The costs of six cars recently purchased by the Weston Company were $18,270, $11,300, $16,400, $9,100, $12,450, and $13,800.

27. *Cell Phone Usage* The numbers of cell phone minutes used on the phone per day by a San Diego teenager are 40 minutes, 108 minutes, 62 minutes, 12 minutes, 24 minutes, 31 minutes, 20 minutes, and 26 minutes.

28. *Swimmers' Ages* The ages of 10 people swimming laps at the YMCA pool one morning were 60, 18, 24, 36, 39, 32, 70, 12, 15, and 85.

Price Comparison *The comparison line graph indicates the price of the same compact disc sold at music stores and online during a 6-year period. Use the graph to answer exercises 29 and 30.*

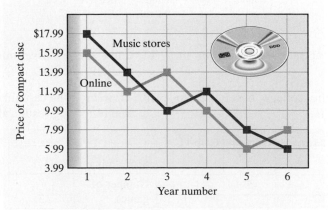

29. Find the median value of the prices at the music stores.

30. Find the median value of the prices online.

Find the median value.

31. *Actors in a Play* The numbers of potential actors who tried out for the school play at Hamilton-Wenham Regional High School over the last 10 years: 36, 48, 44, 64, 60, 71, 22, 36, 53, and 37.

32. *Football Injuries* The numbers of injuries during the high school football season for the Badgers over the last 8 years: 10, 17, 14, 29, 30, 19, 25, and 21.

Find the mode.

33. 60, 65, 68, 60, 72, 59, 80

34. 86, 84, 86, 87, 84, 83, 90

35. 121, 150, 117, 150, 121, 180, 127, 123

36. 144, 143, 140, 141, 149, 144, 141, 150

37. *Bikes Sold* The last six bicycles sold at the Skol Bike shop cost $249, $649, $439, $259, $269, and $249.

38. *TVs Sold* The last six flat-screen television sets sold at the local electronics store cost $315, $430, $515, $330, $430, and $615.

Mixed Practice *Find the mean, median, and mode for each set of numbers.*

39. 21, 82, 42, 55, 42, 45, 49
  (a) Mean
  (b) Median
  (c) Mode

40. 11, 32, 21, 74, 32, 25, 29
  (a) Mean
  (b) Median
  (c) Mode

41. 2.7, 7.1, 6.9, 7.5, 6.1
  (a) Mean
  (b) Median
  (c) Mode

42. 3.6, 7.4, 3.9, 6.2, 7.6
  (a) Mean
  (b) Median
  (c) Mode

43. *Quiz Scores* Damian's quiz scores in his history class were 97, 81, 92, 73, 86, and 81. Find the mean, median, and mode for his quiz scores.

44. *Website Inquiries* Joanna's Website received the following numbers of inquiries over the last six days: 35, 55, 22, 61, 55, and 12. Find the mean, median, and mode.

One Step Further *Use the table for exercises 45 and 46. Round to the nearest tenth if necessary.*

Inco Systems: Office Staff Positions	Inco Systems: Monthly Salary Scale
File clerk	$1350
Receptionist	$1600
Secretary	$2400
Office manager	$2800
Administrative assistant	$3200

45. (a) Find the average salary.

  (b) Find the median salary.

  (c) Find the mode.

46. All the staff at Inco Systems get a holiday bonus at the end of the year of 8.25% of their monthly salary.

  (a) What is the holiday bonus for the highest paid position?

  (b) What is the holiday bonus for the lowest paid position?

  (c) Find the mean for the holiday bonus.

  (d) Find the median for the holiday bonus.

To Think About

47. *Monthly Salaries* A local travel office has 10 employees. Their monthly salaries are $1500, $1700, $1650, $1300, $1440, $1580, $1820, $1380, $2900, and $6300.

    (a) Find the mean.

    (b) Find the median.

    (c) Which of these numbers best represents what the "average person" earns? Why?

48. *Track Meet* A college track star in California ran the 100-meter event in 8 track meets. Her times were 11.7 seconds, 11.6 seconds, 12.0 seconds, 12.1 seconds, 11.9 seconds, 18 seconds, 11.5 seconds, and 12.4 seconds.

    (a) Find the mean.

    (b) Find the median.

    (c) Which of these numbers best represents her "average running time"? Why?

Find the median value.

49. 2576; 8764; 3700; 5000; 7200; 4700; 9365; 1987

50. 15.276; 21.375; 18.90; 29.2; 14.77; 19.02

Cumulative Review *Evaluate.*

51. [2.6.2] $\dfrac{x}{2} + 4$ for $x = 26$

52. [2.6.2] $\dfrac{35}{x} - 9$ for $x = 7$

53. [2.6.2] $2x + 1$ for $x = 5$

54. [2.6.2] $3x - 7$ for $x = 0$

55. [5.1.3] *Roofing Cost* Wesley is putting a new roof on his house and must pay the contractor $\frac{3}{8}$ of the cost of the roof before she will begin the work. If the roof costs $9200, how much must Wesley give the contractor before she starts the job?

56. [2.3.3] *Mountain Altitude* The highest mountain in the United States, Mt. McKinley in Alaska, is 20,320 feet at its highest point. What is the difference in altitude between the lowest point in South America, the Valdes Peninsula, which has an altitude of 131.2 feet below sea level, and Mt. McKinley?

Quick Quiz 9.2

1. Find the average or mean quiz score for Jake if he has quiz scores of 15, 13, 14, 11, 13, 12.

2. Find the median of each of the following.

    (a) 17, 9, 5, 14, 11, 19, 3

    (b) 35, 50, 40, 20, 75, 65

3. Find the mode of each of the following.

    (a) Sara scored the following numbers of goals in the last six soccer seasons: 6, 9, 5, 7, 6, 4.

    (b) The numbers of cell phones sold each day by Peter are 5, 6, 4, 3, 5, 7, and 4.

4. **Concept Check** An Internet site had the following numbers of inquiries over the last seven days: 926, 887, 778, 887, 926, 297, 801. Would you select the mean, median, or mode to determine the most realistic estimate of how many inquires there were that particular week? Why?

How Am I Doing?　Sections 9.1–9.2

How are you doing with your homework assignments in Sections 9.1 and 9.2? Do you feel you have mastered the material so far? Do you understand the concepts you have covered? Before you go further in the textbook, take some time to do each of the following problems.

9.1

The land area of a small city on the West Coast is about 70 square miles. The percentage of the land for apartments, condominiums, homes, and business is shown on the pie graph.

1. What percentage of the city is used for business?

2. What takes up the largest percentage of land?

3. How many square miles have homes or condominiums?

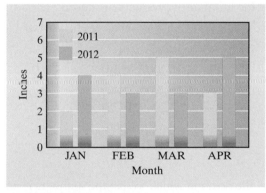

Business 26%　Homes 35%

Apartments 20%

Condominiums 19%

The double-bar graph compares the monthly rainfall total for 2 years during a city's rainy season.

4. What was the average rainfall in February 2011?

5. What was the difference in the average rainfall in March 2011 and March 2012?

6. What month and year had the least rainfall?

9.2

Refer to the double-bar graph in problems 4–6 to answer the following.

7. What was the average monthly rainfall during the rainy season in 2012?

Find the median value for each set of numbers.

8. 7, 9, 14, 19, 25, 28, 32

9. 2, 5, 9, 13, 18, 23

Find the mode.

10. Jerome's test scores: 79, 85, 81, 83, 85.

11. The numbers of cars sold at a dealership each day of one week: 8, 5, 7, 8, 4, 9, 7.

12. Find the mean, median, and mode for the following set of numbers: 2.1, 9.2, 9.6, 8.7.

Now turn to page SA-17 for the answers to each of these problems. Each answer also includes a reference to the objective in which the problem is first taught. If you missed any of these problems, you should stop and review the Examples and Student Practice problems in the referenced objective. A little review now will help you master the material in the upcoming sections of the text.

1. _____

2. _____

3. _____

4. _____

5. _____

6. _____

7. _____

8. _____

9. _____

10. _____

11. _____

12. _____

Student Learning Objectives

After studying this section, you will be able to:

① Write data as ordered pairs.

② Plot points given the coordinates.

③ Name the coordinates of points.

④ Graph points that lie on horizontal and vertical lines.

① Writing Data as Ordered Pairs

Many things in real life are clearer if we can see a picture of them. Similarly, in mathematics we often find that a drawing is helpful. We can picture relationships by drawing graphs and charts. Consider the following line graph, which shows the number of products produced by a manufacturing company over a 6-year period.

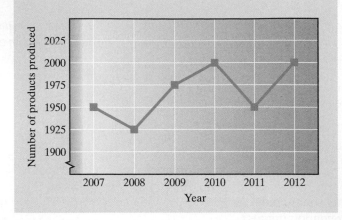

This graph indicates that 1950 products were produced in 2007, 1925 in 2008, 1975 in 2009, 2000 in 2010, 1950 in 2011, and 2000 in 2012. We could also describe the productivity without a graph. Instead, we can use **ordered pairs** of numbers stating the year followed by the number of products produced.

(year,   number of products)

↓   ↓

(2007, 1950)

(2008, 1925)

(2009, 1975)

(2010, 2000)

(2011, 1950)

(2012, 2000)

As you can see, the order in which we list the numbers is important; otherwise, we could confuse the year with the number of products produced. Since ordered pairs are pairs of numbers presented in a *specific order,* we often use ordered pairs to specify locations on a graph. We use a *dot* to represent an ordered pair on a graph.

EXAMPLE 1 Refer to the line graph and list the number of applicants to medical school in the years 2003, 2005, 2007, and 2009 using ordered pairs of the form (year, number of applicants).

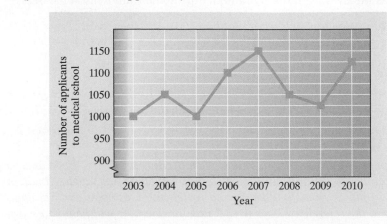

Solution

(year,   number of applicants)

(2003, 1000)
(2005, 1000)
(2007, 1150)
(2009, 1025)

Student Practice 1 Refer to the line graph in Example 1 and list the number of applicants in 2004, 2006, 2008, and 2010 using ordered pairs of the form (year, number of applicants).

NOTE TO STUDENT: *Fully worked-out solutions to all of the Student Practice problems can be found at the back of the text starting at page SP-1.*

② Plotting Points Given the Coordinates

We cannot easily use the type of line graph illustrated in Example 1 to display ordered pairs that include negative numbers. Instead, we use a **rectangular coordinate system.** We can think of a rectangular coordinate system as an extension of the line graph—we extend the horizontal line to the left and the vertical line downward to represent negative numbers.

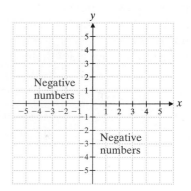

To form this coordinate system we draw two number lines, one horizontally and a second one vertically. We construct the number lines so that the zero point on each number line is at exactly the same place. We refer to this location as the **origin.** Each number line is called an **axis.** The horizontal number line is often called the **x-axis,** and the vertical number line is called the **y-axis.**

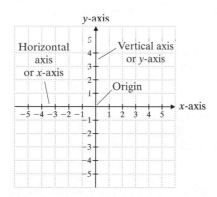

Notice on the vertical number line that the positive numbers are above the origin (positive y-direction) and the negative numbers are below the origin (negative y-direction).

Since there are two number lines in the rectangular coordinate system, we use this system to **graph** (plot) ordered pairs of numbers. The ordered pair of numbers that represents a point is often referred to as the *coordinates of the point.* The first value is called the **x-coordinate** or **x-value** and the second value the **y-coordinate** or **y-value.**

The x-value represents the distance from the origin in the x-direction.

The y-value represents the distance from the origin in the y-direction.

x-value
↑
(x, y)
↓
y-value

M꜀ **EXAMPLE 2**   Plot the ordered pair. $(-4, 3)$

Solution

The first number indicates the x-direction.

↓

$(-4, 3)$

↑

The second number indicates the y-direction.

To plot $(-4, 3)$ or $x = -4$, $y = 3$, we start at the origin and move 4 units in the negative x-direction followed by 3 units in the positive y-direction. We end up at the coordinate point $(-4, 3)$ and place a dot there.

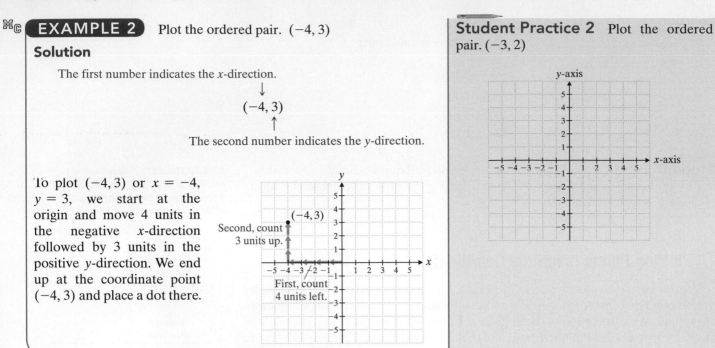

Second, count 3 units up.

$(-4, 3)$

First, count 4 units left.

Student Practice 2   Plot the ordered pair. $(-3, 2)$

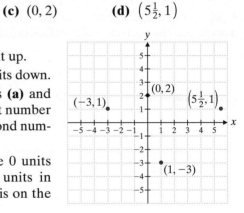

EXAMPLE 3   Plot each ordered pair.

(a) $(-3, 1)$ **(b)** $(1, -3)$ **(c)** $(0, 2)$ **(d)** $\left(5\frac{1}{2}, 1\right)$

Solution

(a) $(-3, 1)$: 3 units left followed by 1 unit up.

(b) $(1, -3)$: 1 unit right followed by 3 units down. Notice the difference between parts **(a)** and **(b)**. We must remember that the first number indicates the x-direction and the second number the y-direction.

(c) $(0, 2)$: start at the origin and move 0 units in the x-direction, followed by 2 units in the positive y-direction. The point is on the y-axis.

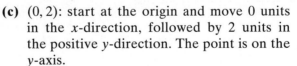

(d) $\left(5\frac{1}{2}, 1\right)$: move $5\frac{1}{2}$ units in the positive x-direction. The measure of $5\frac{1}{2}$ units is located halfway between 5 and 6. Then move 1 unit up.

Student Practice 3   Plot each ordered pair.

(a) $(-4, 2)$ **(b)** $(2, -4)$ **(c)** $\left(0, 3\frac{1}{2}\right)$ **(d)** $(1, 0)$

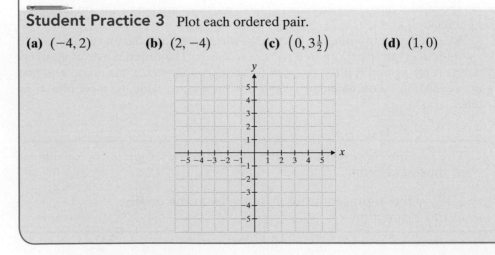

③ **Naming the Coordinates of Points**

EXAMPLE 4 Give the coordinates of each point on the graph.

Solution

(a) $S = (-5, 1)$

(b) $T = (0, 3)$

(c) $U = (1, 0)$

(d) $V = (6, 2)$

(e) $W = (-5, -1)$

(f) $X = (6, -2)$

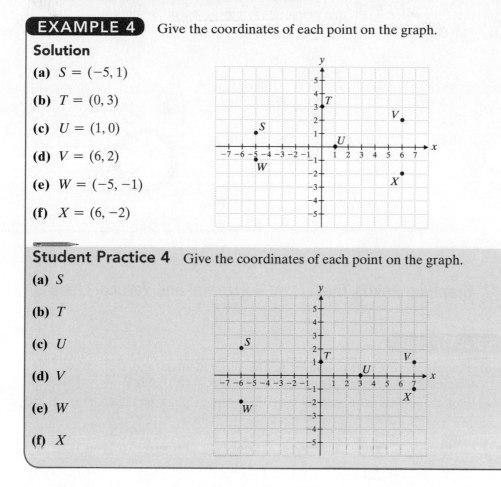

Student Practice 4 Give the coordinates of each point on the graph.

(a) S

(b) T

(c) U

(d) V

(e) W

(f) X

If the x-axis represents the directions east and west, and the y-axis represents north and south, we can indicate direction using ordered pairs.

EXAMPLE 5

(a) State the ordered pair that represents 3 miles west, 1 mile north.

(b) Plot the ordered pair.

Solution

(a) $(-3, 1)$

(b)

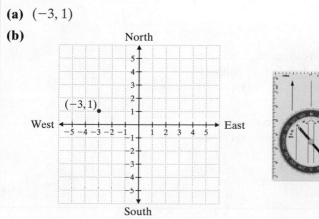

Continued on next page

Student Practice 5

(a) State the ordered pair that represents 5 miles east, 2 miles south.

(b) Plot the ordered pair.

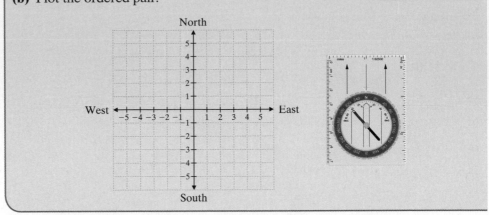

④ Graphing Points That Lie on Horizontal and Vertical Lines

EXAMPLE 6

(a) Plot the points corresponding to the ordered pairs and then draw a line connecting the coordinate points. $(2, 1), (2, -3), (2, 0), (2, 3)$

(b) Plot the points corresponding to the ordered pairs and then draw a line connecting the coordinate points. $(-3, 4), (4, 4), (1, 4), (-5, 4)$

Solution

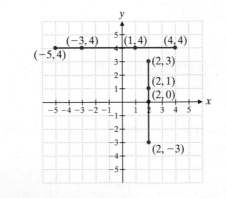

Student Practice 6

(a) Plot the points corresponding to the ordered pairs and then draw a line connecting the coordinate points. $(1, -2), (1, -3), (1, 0), (1, 2)$

(b) Plot the points corresponding to the ordered pairs, then draw a line connecting the coordinate points. $(-2, 3), (3, 3), (1, 3), (-6, 3)$

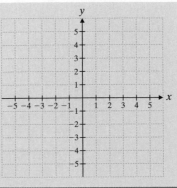

Let's take another look at the ordered pairs and points on the graph in Example 6. What do you observe about the sets of ordered pairs?

(a) $(2, 1), (2, -3), (2, 0), (2, 3)$

The x-value is the same in each ordered pair: $x = 2$.

(b) $(-3, 4), (4, 4), (1, 4), (-5, 4)$

The y-value is the same in each ordered pair: $y = 4$.

Now let's see how this affects the graphs.

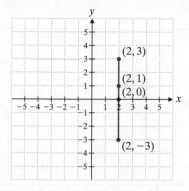

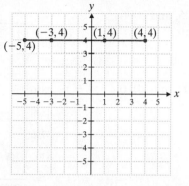

The points lie on a vertical line, since we moved the same distance in the x-direction, $x = 2$, for all the ordered pairs.

The points lie on a horizontal line, since we moved the same distance in the y-direction, $y = 4$, for all the ordered pairs.

When all x-values of a set of ordered pairs are the same number, $x = a$, the coordinate points on the graph lie on a **vertical line**.

When all y-values of a set of ordered pairs are the same number, $y = b$, the coordinate points on the graph lie on a **horizontal line**.

Verbal and Writing Skills, Exercises 1–2

1. Describe in words the directions you move on the rectangular coordinate system to plot the point $(2, -1)$.

2. Describe in words the directions you move on the rectangular coordinate system to plot the point $(-4, -2)$.

College Enrollment *Use the line graph to answer exercises 3 and 4.*

3. Represent the enrollment in 2003, 2004, 2005, and 2006 using ordered pairs (year, number of students).

4. Represent the enrollment in 2007, 2008, 2009, and 2010 using ordered pairs (year, number of students).

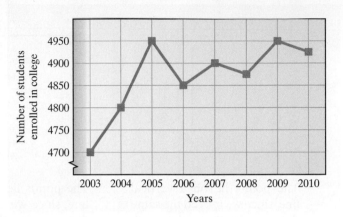

Height vs Shoe Size *Students graphed their shoe size versus their height on the graph. Use the line graph to answer exercises 5 and 6.*

5. Represent the height and corresponding shoe size for Lena, Janie, and Mark using ordered pairs (shoe size, height).

6. Represent the height and corresponding shoe size for Nho, Kelley, and Jeff using ordered pairs (shoe size, height).

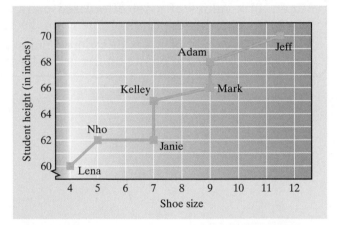

Plot and label the ordered pairs on the rectangular coordinate system.

7. $(-2, 2)$

8. $(3, 3)$

9. $(-1, 4)$

10. $(1, 5)$

11. $(3, -2)$

12. $(4, -3)$

13. $(-2, -4)$

14. $(-5, -3)$

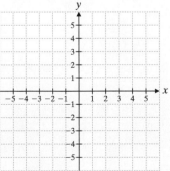

Plot and label the ordered pairs on the rectangular coordinate system.

15. $\left(4\frac{1}{2}, 3\right)$

16. $\left(2\frac{1}{2}, 2\right)$

17. $\left(-1\frac{1}{2}, -4\right)$

18. $(-3.5, -3)$

19. $\left(2\frac{1}{2}, -1\right)$

20. $\left(1\frac{1}{2}, -3\right)$

21. $\left(-3\frac{1}{2}, 2\right)$

22. $\left(-1\frac{1}{2}, 4\right)$

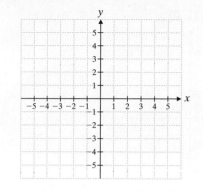

Give the coordinates of each point.

23. *K*

24. *L*

25. *M*

26. *N*

27. *O*

28. *P*

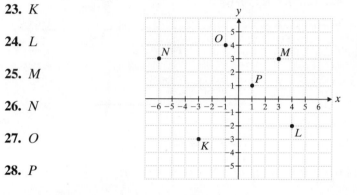

Give the coordinates of each point.

29. *Q*

30. *R*

31. *S*

32. *T*

33. *U*

34. *V*

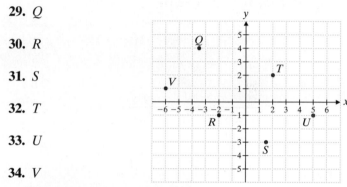

35. **(a)** State the ordered pair that represents 4 miles west, 2 miles south.
(b) Plot the ordered pair.

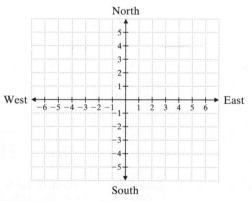

36. **(a)** State the ordered pair that represents 6 miles east, 3 miles south.
(b) Plot the ordered pair.

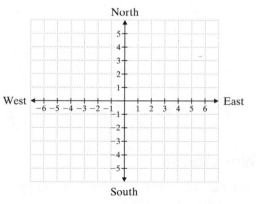

37. **(a)** State the ordered pair that represents 2 miles east, 1 mile north.
(b) Plot the ordered pair.

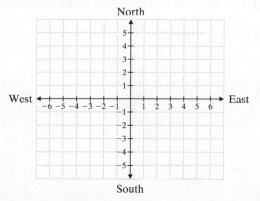

38. **(a)** State the ordered pair that represents 5 miles west, 2 miles north.
(b) Plot the ordered pair.

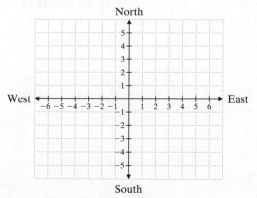

Plot the points corresponding to each set of ordered pairs and then draw a line connecting the coordinate points.

39. $(2, 4), (2, -1), (2, -3), (2, 0)$

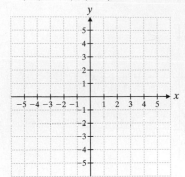

40. $(4, 2), (4, -2), (4, -1), (4, 0)$

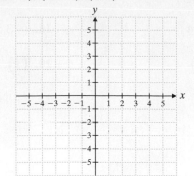

41. $(1, 3), (-5, 3), (0, 3), (-2, 3)$

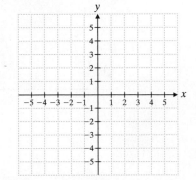

42. $(2, 5), (-3, 5), (3, 5), (-5, 5)$

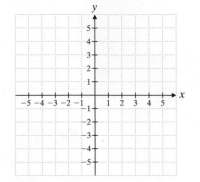

43. $(0, -1), (-2, -1), (-4, -1), (4, -1)$

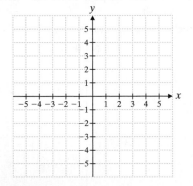

44. $(0, -2), (-3, -2), (3, -2), (2, -2)$

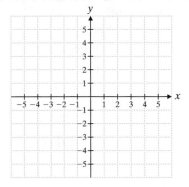

To Think About

▲ **45.** **(a)** Plot $(5, 2)$ and $(5, 6)$ and then draw a line connecting the points.

(b) Plot $(2, 2)$ and $(2, 6)$ and then draw a line connecting the points.

(c) Draw a line connecting the points $(2, 2), (5, 2)$.

(d) Draw a line connecting the points $(2, 6), (5, 6)$.

(e) Describe the figure.

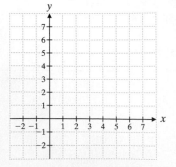

▲ **46.** **(a)** Plot $(0, 0)$ and $(4, 0)$ and then draw a line connecting the points.

(b) Name two sets of ordered pairs that you can plot so a square can be formed with the points in part **(a)**.

(c) Plot the ordered pairs from part **(b)** and draw the squares.

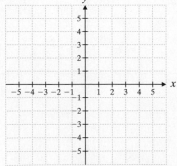

Cumulative Review

47. [7.5.2] *Pluto and the Sun* Earth is approximately 150 million kilometers from the sun. Pluto is approximately 39 times as far as Earth is from the sun. About how far is Pluto from the Sun?

48. [4.4.1] Simplify. $\dfrac{20x^6}{55x^{10}}$

49. [4.4.2] Simplify. $(2^3)^2$

50. [3.4.2] Multiply. $(2x)(y^4)(3x^2)(2y^6)$

Quick Quiz 9.3

1. Plot and label the ordered pairs on a rectangular coordinate system.

   (a) $(-1, 4)$

   (b) $\left(2\frac{1}{2}, -3\right)$

   (c) $(-2, -4)$

   (d) $(0, 1)$

2. Give the coordinates of each point.

   (a) S

   (b) T

   (c) U

   (d) V

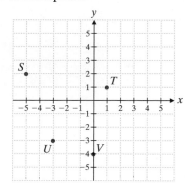

3. Plot the points corresponding to each set of ordered pairs on the graph for #2. Draw a line connecting the coordinate points. $(-2, 4), (-2, 2), (-2, 1), (-2, -2)$

4. **Concept Check** Line M passes through the points $(k, b), (n, b)$, and (a, b). Describe line M.

9.4 Linear Equations in Two Variables

Student Learning Objectives

After studying this section, you will be able to:

① Find solutions to linear equations in two variables.

② Graph linear equations in two variables.

③ Graph horizontal and vertical lines.

① Finding Solutions to Linear Equations in Two Variables

An equation such as $x + 2 = 7$ is called a **linear equation in one variable.** It is called *linear* because the exponent on x is understood to be 1.

In earlier chapters we found the solutions to linear equations with one variable. For example, the solution to $x + 2 = 7$ is 5. Now, consider the equation $x + y = 7$. This equation has *two variables* and is called a **linear equation in two variables.** A solution to $x + y = 7$ consists of a pair of numbers, one for x and one for y, whose sum is 7. When we find solutions to $x + y = 7$, we are answering the question: "The sum of what two numbers equals 7?" Since $5 + 2 = 7$, the pair of numbers $x = 5$ and $y = 2$ or $(5, 2)$ is a solution.

$$x + y = 7 \quad \text{The sum of what two numbers equals 7?}$$
$$5 + 2 = 7 \quad x = 5 \text{ and } y = 2 \text{ or } (5, 2) \text{ is a solution.}$$

Now, $3 + 4 = 7$, so the pair of numbers $x = 3$ and $y = 4$ or $(3, 4)$ is also a solution to the equation $x + y = 7$. Can you find another pair of numbers whose sum is 7? Of course you can. In fact, there are *infinitely* many pairs of numbers with a sum of 7 and thus infinitely many ordered pairs that represent solutions to $x + y = 7$. It is important that you realize that not just any ordered pair of numbers is a solution to $x + y = 7$. Only ordered pairs whose sum is 7 are solutions to $x + y = 7$. In general, we say that a solution to a linear equation in two variables is an ordered pair (x, y) whose coordinates make the equation a true statement.

EXAMPLE 1

(a) Is $(-1, 4)$ a solution to the equation $x + y = 3$?

(b) List two ordered pairs of numbers that are solutions to the equation $x + y = 3$.

Solution

(a) The ordered pair $(-1, 4)$ is a solution if, when we replace x with -1 and y with 4 in the equation $x + y = 3$, we get a true statement.

Check.
$$x = -1 \quad \text{and} \quad y = 4, \quad \text{or} \quad (-1, 4) \qquad x + y = 3$$
$$-1 + 4 = 3, \qquad 3 = 3 \checkmark$$

Since $3 = 3$ is true, $(-1, 4)$ is a solution of $x + y = 3$.

(b) There are infinitely many solutions, so answers may vary. We can choose any two numbers whose sum is 3.

$$x = 0 \quad \text{and} \quad y = 3, \quad \text{or} \quad (0, 3) \qquad x + y = 3$$
$$0 + 3 = 3, \qquad 3 = 3 \checkmark$$
$$x = 2, \quad \text{and} \quad y = 1, \quad \text{or} \quad (2, 1) \qquad x + y = 3$$
$$2 + 1 = 3, \qquad 3 = 3 \checkmark$$

Therefore $(0, 3)$ and $(2, 1)$ are solutions to $x + y = 3$.

Student Practice 1

(a) Is $(-2, 13)$ a solution to the equation $x + y = 11$?

(b) List three ordered pairs of numbers that are solutions to the equation $x + y = 11$.

NOTE TO STUDENT: *Fully worked-out solutions to all of the Student Practice problems can be found at the back of the text starting at page SP-1.*

EXAMPLE 2   A machine at a manufacturing company can seal 50 jars per minute. The equation that represents the situation is $y = 50x$, where y equals the number of jars sealed and x represents the number of minutes the machine is in operation. Determine how many minutes it takes the machine to seal the following numbers of jars. Then use your answer to write the ordered-pair solution (x, y).

(a) 200 jars   **(b)** 100 jars

Solution   We see that x represents the number of minutes and y equals the number of jars.

number of
jars sealed ———┐   ┌— number of minutes

(a) $y = 50x$

$200 = 50x$ We replace y with 200.

$\dfrac{200}{50} = x$ We solve for x.

$4 = x$

It takes the machine 4 minutes to seal 200 jars. Since $x = 4$ when $y = 200$, the ordered pair (x, y) is $(4, 200)$.

number of
jars sealed ———┐   ┌— number of minutes

(b) $y = 50x$

$100 = 50x$ We replace y with 100.

$\dfrac{100}{50} = x$ We solve for x.

$2 = x$

It takes the machine 2 minutes to seal 100 jars. Since $x = 2$ when $y = 100$, the ordered pair (x, y) is $(2, 100)$.

Student Practice 2   A machine at a manufacturing company can label 25 bottles per minute. The equation that represents the situation is $y = 25x$, where y equals the number of bottles labeled and x represents the number of minutes the machine is in operation. Determine how many minutes it takes the machine to label the following numbers of bottles. Then use your answer to write the ordered-pair solution (x, y).

(a) 125 bottles   **(b)** 200 bottles

EXAMPLE 3   Fill in the ordered pairs so that they are solutions to the equation $x + 2y = 10$.

(a) $(0, \underline{\ \ })$ **(b)** $(\underline{\ \ }, 1)$

Solution   The first number in the ordered pair is the x-value and the second number is the y-value.

(a) $(0, \underline{\ \ })$ $x = 0$, $y = ?$

We replace x with 0 and solve for y in the equation $x + 2y = 10$.

$$x + 2y = 10$$
$$0 + 2y = 10 \quad \text{Replace } x \text{ with 0.}$$
$$2y = 10 \quad \text{Solve for } y.$$
$$y = 5$$

$(0, 5)$ is a solution to $x + 2y = 10$.

Continued on next page

(b) ($\underline{\quad}$, 1) $x = ?$ $y = 1$

We replace y with 1 and solve for x in the equation $x + 2y = 10$.

$$x + 2y = 10$$
$$x + 2(1) = 10 \quad \text{Replace } y \text{ with 1.}$$
$$x + 2 = 10 \quad \text{Solve for } x.$$
$$x = 8$$

(8, 1) is a solution to $x + 2y = 10$.

It is a good idea to check your answers.

Check.

	$x + 2y = 10$		$x + 2y = 10$
(0, 5)	$0 + 2(5) \overset{?}{=} 10$	(8, 1)	$8 + 2(1) \overset{?}{=} 10$
	$10 = 10 \checkmark$		$10 = 10 \checkmark$

(0, 5) and (8, 1) are solutions to $x + 2y = 10$.

> **Student Practice 3** Fill in the ordered pairs so that they are solutions to the equation $x + 3y = 12$.
>
> **(a)** ($\underline{\quad}$, 0) **(b)** ($\underline{\quad}$, 2)

CAUTION: We must be careful when we state ordered pairs: x must be written first and y second. A common error is to reverse the numbers in the ordered pair. Consider Example 3a, which has (0, 5) as a solution. If we reverse the coordinates (in error) and state the solution as (5, 0), our proposed solution does not check in the equation $x + 2y = 10$.

(5, 0) means $x = 5$ and $y = 0$ *Check.*
$$x + 2y = 10$$
$$5 + 2(0) \overset{?}{=} 10$$
$$5 + 0 \overset{?}{=} 10$$
$$5 \neq 10$$

The ordered pair (5, 0) *does not check* and therefore is *not a solution* to the equation $x + 2y = 10$. To help avoid this error, we often organize our work in a chart, as in the following example.

EXAMPLE 4 Fill in the ordered pairs so that they are solutions to the equation $y = x + 4$. (0, $\underline{\quad}$), ($\underline{\quad}$, 5), ($\underline{\quad}$, 1)

Solution We start by organizing our ordered pairs in a chart. Then we replace x and y with the given values and solve for the unknown values.

$y = x + 4$	$y = x + 4$	$y = x + 4$
$y = 0 + 4$	$5 = x + 4$	$1 = x + 4$
$y = 4$	$1 = x$	$-3 = x$

(x,	y)
(0	)
(	5)
(	1)

We write these values in the appropriate place in the chart. (0, 4), (1, 5), (−3, 1) are solutions to $y = x + 4$. We leave the check to the student.

(x,	y)
(0,	4)
(1,	5)
(−3,	1)

Student Practice 4

(x,	y)
(	)
(	)
(	)

> **Student Practice 4** Fill in the ordered pairs so that they are solutions to the equation $y = x + 2$. (0, $\underline{\quad}$), ($\underline{\quad}$, 8), ($\underline{\quad}$, 5)

If we are not given values for either x or y, we may choose any value for x and then solve for y. Or we may choose any value for y and then solve for x.

EXAMPLE 5 Find three ordered pairs that are solutions to $y = 3x - 1$.

Solution We choose three values for x: 0, 1, -1, and write these values in a chart.

We replace the values 0, 1, and -1 for x in the equation $y = 3x - 1$ and then solve for y.

(x,	y)
(0	)
(1	)
(−1	)

$$y = 3x - 1 \qquad y = 3x - 1 \qquad y = 3x - 1$$
$$= 3(0) - 1 \qquad = 3(1) - 1 \qquad = 3(-1) - 1$$
$$= 0 - 1 \qquad = 3 - 1 \qquad = -3 - 1$$
$$y = -1 \qquad y = 2 \qquad y = -4$$

We write these values in the appropriate place in the chart. $(0, -1), (1, 2), (-1, -4)$ are solutions to $y = 3x - 1$. Answers may vary since you may start with different values.

(x,	y)
(0,	−1)
(1,	2)
(−1,	−4)

For the equation $y = 3x - 1$, the calculations are simplified if we choose three values for x and then solve for y, rather than choosing values for y and solving for x. To verify this, try choosing a few values for y in the equation $y = 3x - 1$ and solving for x.

Student Practice 5 Find three ordered pairs that are solutions to $y = 4x - 5$.

Student Practice 5

(x,	y)
(	)
(	)
(	)

Understanding the Concept

Determining Values for Ordered Pairs Why can we choose any value for either x or y when we find ordered pairs that are solutions to an equation?

The numbers for x and y that are solutions to the equation come in pairs. When we try to find a pair that fits, we have to start with some number. Usually, it is an x-value that is small and easy to work with. Then we must find the value for y so that the pair of numbers (x, y) is a solution to the given equation.

The following situation might help you understand this idea. Friends often share with each other the price they paid for some item of common interest such as a CD. Suppose Dave bought three equally priced CDs and his friend Jerry bought three equally priced CDs. The ordered pair (x, y) can represent (**CD price, total cost of three CDs**), and now Dave and Jerry can inform each other about the prices they paid for the CDs in two ways.

1. Dave can tell Jerry that his total cost for all three CDs (the y-value) is $33.24: ($x$, 33.24). Then by dividing $33.24 by 3, Jerry can compute the price per CD (the x-value) as $11.08: (11.08, 33.24).

    We have (x, y):   (CD price, $33.24)

    $\qquad\qquad\qquad\qquad\qquad$ x is the price per CD, and y is the total cost or $3x$.

    The equation is: $y = 3x$

    $\qquad\qquad\qquad$ $33.24 = 3x$ We know the y-value and solve for the x-value.

    $\qquad\qquad\qquad$ $11.08 = x$

    We end up with: $(11.08, 33.24)$

2. Jerry can inform Dave that the price he paid for each CD (the x-value) is $10.15: (10.15, y). Then by multiplying $10.15 by 3, Dave can compute the total cost for all three CDs (the y-value) as $30.45: (10.15, 30.45).

Continued on next page

We have (x, y):   ($10.15, total cost of 3 CDs)

     x is the price per CD, and y is the total cost or $3x$.

The equation is: $y = 3x$

        $y = 3(10.15)$ We know the x-value and solve for the y-value.

        $y = 30.45$

We end up with: $(10.15, 30.45)$

In other words, the ordered pairs $(11.08, 33.24)$, $(10.15, 30.45)$ can be achieved by beginning with *either* the x-value (CD price) or the y-value (total cost).

② Graphing Linear Equations in Two Variables

Now that we know how to find ordered-pair solutions to linear equations in two variables, let's see what happens when we plot these solutions on a rectangular coordinate system.

Mc EXAMPLE 6

(a) Name three ordered pairs that are solutions to $y = 2x + 4$.

(b) Plot these ordered pairs on a rectangular coordinate system.

Solution

(a) We choose three values for x.

We replace x with each of these values and solve for y.

$$y = 2x + 4 \qquad y = 2x + 4 \qquad y = 2x + 4$$
$$= 2(-1) + 4 \qquad = 2(0) + 4 \qquad = 2(1) + 4$$
$$= -2 + 4 \qquad = 0 + 4 \qquad = 2 + 4$$
$$y = 2 \qquad\qquad y = 4 \qquad\qquad y = 6$$

We place these values in the chart.

(b) Then we plot these ordered pairs.

(x,	y)
(−1	)
(0	)
(1	)

(x,	y)
(−1,	2)
(0,	4)
(1,	6)

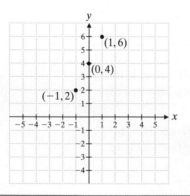

Student Practice 6

(a) Name three ordered pairs that are solutions to $y = 2x + 1$.

(b) Plot these ordered pairs on a rectangular coordinate plane.

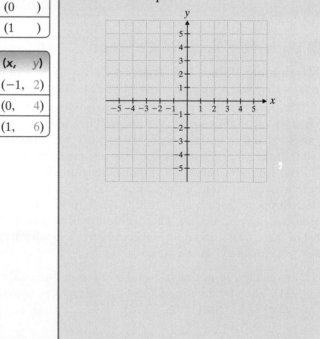

In Example 6, we found that $(-1, 2)$, $(0, 4)$, $(1, 6)$ are solutions of $y = 2x + 4$. Are these all the solutions? No. As we saw earlier there are infinitely many pairs of numbers that are solutions to a linear equation in two variables. Therefore, it is impossible to list every solution. Let's examine the points we plotted in Example 6 and look for a pattern that may help us determine how to graph the solutions. Notice that when we connect all three points, they form a straight line. If we plot other solutions to the equation $y = 2x + 4$, we find that these points also lie on this line. For example, $(-3, -2)$, and $(-4, -4)$ are solutions to $y = 2x + 4$ and lie on the line.

In fact, every ordered pair that is a solution to this equation lies on this line. Therefore, to graph the *solution set* to the equation $y = 2x + 4$, we draw a line through the points and *continue the line beyond the points*, placing an arrow on both ends to indicate that solutions lie beyond the ends of the line that we drew.

Now we state a formal definition for a linear equation in two variables. A **linear equation in two variables** is an equation that can be written in the form $Ax + By = C$ where A, B, and C are any numbers but A and B are not *both* zero.

$$\text{Examples:} \quad 2x + 5y = 3$$
$$y = 4x + 1$$
$$x = 4$$

The graph of any linear equation in two variables is a *straight line*.

We often say "graph the equation" when we mean "plot the set of ordered-pair solutions on a graph and connect the points with a line."

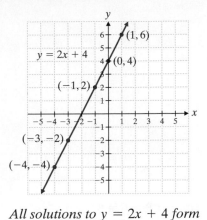

All solutions to $y = 2x + 4$ form a straight line.

PROCEDURE TO GRAPH A LINEAR EQUATION

1. Find three ordered pairs that are solutions to the equation.
2. Plot the points.
3. Draw a line through the points.
4. Continue this line beyond the points, placing an arrow at both ends.

It is important to note that *any* straight line can be determined by two points. Therefore, when two points are used to graph the equation and either of the two points is calculated incorrectly, the *wrong line* will be drawn and we will *not* be aware that an error was made. On the other hand, if we use three points to graph the equation, and if all three points do not line up, we know that a mistake has been made.

EXAMPLE 7

(a) Name three ordered pairs that are solutions to $y = -2x - 1$.

(b) Plot these ordered pairs on a rectangular coordinate system and draw a line through the points.

Solution

(a) We choose three values for x: $2, 0, -1$, and find three ordered pairs that are solutions to $y = -2x - 1$.

$$\begin{array}{lll}
y = -2x - 1 & y = -2x - 1 & y = -2x - 1 \\
= (-2)(2) - 1 & = (-2)(0) - 1 & = (-2)(-1) - 1 \\
= -4 - 1 & = 0 - 1 & = 2 - 1 \\
y = -5 & y = -1 & y = 1
\end{array}$$

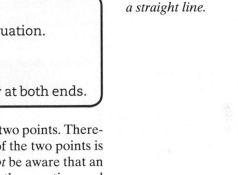

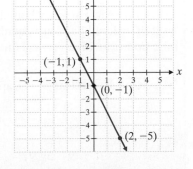

(b) Then we plot these ordered pairs and draw a straight line through the points. All solutions to $y = -2x - 1$ lie on this straight line. If any of the points we plot do not lie on this line, we made an error and must check our work.

Student Practice 7(b)

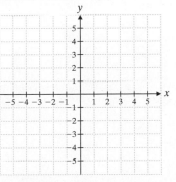

Student Practice 7

(a) Name three ordered pairs that are solutions to $y = -3x - 1$.

(b) Plot these ordered pairs on a coordinate system and draw a line through the points.

③ Graphing Horizontal and Vertical Lines

Recall that we say "graph the equation" when we mean "plot the set of ordered-pair solutions on a graph and connect the points with a line."

EXAMPLE 8 Graph. $y = -1$

Solution How do we find solutions and graph $y = -1$? A solution to $y = -1$ is any ordered pair that has y-coordinate 1. The x-coordinate can be any number, as long as y is -1.

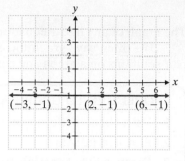

The ordered pairs $(6, -1), (2, -1),$ and $(-3, -1)$ are solutions to $y = -1$ since *all* the y-values are -1.

We plot these ordered pairs.

Recall that if all the y-values of a set of ordered pairs are equal to -1, these coordinate points lie on a horizontal line at $y = -1$.

Student Practice 8 Graph. $y = 4$

Student Practice 8

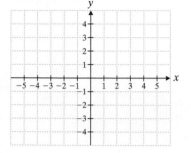

SIDELIGHT In Example 8 you could write $y = -1$ as the equation $y + 0x = -1$, and then it would be clear that for any value of x that you substitute, you will always obtain $y = -1$.

$$(6, -1) \qquad y + 0x = -1; \qquad y + 0(6) = -1; \qquad y = -1$$

$$(2, -1) \qquad y + 0x = -1; \qquad y + 0(2) = -1; \qquad y = -1$$

$$(-3, -1) \qquad y + 0x = -1; \qquad y + 0(-3) = -1; \qquad y = -1$$

EXAMPLE 9 Graph. $x = 5$

Solution A solution to $x = 5$ is any ordered pair that has x-coordinate 5. The y-coordinate can be any number as long as x is 5.

$(5, -2), (5, 3),$ and $(5, 0)$ are solutions to $x = 5$, since *all the x-values are 5.*

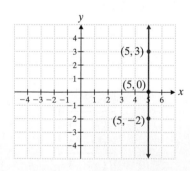

As we saw earlier, when all the x-values of a set of ordered pairs are equal to 5, the coordinate points lie on a vertical line at $x = 5$.

Student Practice 9 Graph. $x = -2$

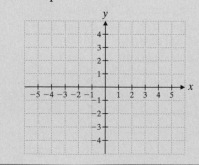

Verbal and Writing Skills, Exercises 1–2

1. Is $(2, 5)$ a solution to $x + 2y = 4$? Why or why not?

2. Is $(1, 3)$ a solution to $2x + y = 5$? Why or why not?

3. List four ordered pairs that are solutions to the equation $x + y = 4$.

4. List four ordered pairs that are solutions to the equation $x + y = 5$.

5. List four ordered pairs that are solutions to the equation $x + y = 12$.

6. List four ordered pairs that are solutions to the equation $x + y = 9$.

7. *Bottle Labeling* A machine at a manufacturing company can label 35 bottles per minute. The equation that represents the situation is $y = 35x$, where y equals the number of bottles labeled and x represents the number of minutes the machine is operating. Determine how many minutes it takes the machine to label the following numbers of bottles. Then use your answer to write an ordered-pair solution (x, y).

   (a) 140 bottles

   (b) 280 bottles

8. *Jar Sealing* A machine at a manufacturing company can seal 40 jars per minute. The equation that represents the situation is $y = 40x$, where y equals the number of jars sealed and x represents the number of minutes the machine is operating. Determine how many minutes it takes the machine to seal the following numbers of jars. Then use your answer to write an ordered-pair solution (x, y).

   (a) 160 jars

   (b) 320 jars

9. *Typing Speed* The secretary for Darwin Electronics can type 80 words per minute. The equation that represents the situation is $y = 80x$, where y equals the number of words typed and x represents the number of minutes the secretary typed. Determine how many minutes it takes the secretary to type the following numbers of words. Then use your answer to write an ordered-pair solution (x, y).

   (a) 240 words

   (b) 400 words

10. *Typing Speed* The office manager at A&L Accounting Services can type 60 words per minute. The equation that represents the situation is $y = 60x$, where y equals the number of words typed and x represents the number of minutes the manager typed. Determine how many minutes it takes the manager to type the following numbers of words. Then use your answer to write an ordered-pair solution (x, y).

   (a) 360 words

   (b) 420 words

Use a chart to organize your work for exercises 11–22.

11. Fill in the ordered pairs so that they are solutions to the equation $x + 2y = 16$.
$(0, \underline{\ \ }), (\underline{\ \ }, 0), (\underline{\ \ }, 4)$

(x,	y)

12. Fill in the ordered pairs so that they are solutions to the equation $x + 3y = 6$.
$(0, \underline{\ \ }), (\underline{\ \ }, 0), (\underline{\ \ }, 1)$

(x,	y)

13. Fill in the ordered pairs so that they are solutions to the equation $x + y = 5$.
$(\underline{\ \ }, 2), (0, \underline{\ \ }), (1, \underline{\ \ })$

14. Fill in the ordered pairs so that they are solutions to the equation $x + y = 12$.
$(\underline{\ \ }, 1), (3, \underline{\ \ }), (0, \underline{\ \ })$

15. Fill in the ordered pairs so that they are solutions to the equation $y = x + 2$.
$(-1, \underline{\ \ }), (\underline{\ \ }, 3), (\underline{\ \ }, 0)$

16. Fill in the ordered pairs so that they are solutions to the equation $y = x + 4$.
$(-1, \underline{\ \ }), (\underline{\ \ }, 4), (\underline{\ \ }, 0)$

17. Fill in the ordered pairs so that they are solutions to the equation $y = 5x + 3$.
$(0, \underline{\ \ }), (-1, \underline{\ \ }), (1, \underline{\ \ })$

18. Fill in the ordered pairs so that they are solutions to the equation $y = 3x + 4$.
$(-2, \underline{\ \ }), (-3, \underline{\ \ }), (0, \underline{\ \ })$

19. Find three ordered pairs that are solutions to $y = 5x - 3$.

20. Find three ordered pairs that are solutions to $y = 6x - 2$.

21. Find three ordered pairs that are solutions to $y = x + 6$.

22. Find three ordered pairs that are solutions to $y = x + 5$.

Plot three ordered-pair solutions of the given equation and then draw a line through the three points.

23. $y = 2x + 2$

24. $y = 3x + 2$

25. $y = -3x + 1$

26. $y = -4x - 3$

27. $y = 5x - 4$

28. $y = 3x - 6$

29. $y = 3x - 2$

30. $y = 2x - 1$

31. $y = -5x - 7$

32. $y = -6x - 4$

33. $y = 3$

34. $y = -2$

35. $x = -2$

36. $x = 3$

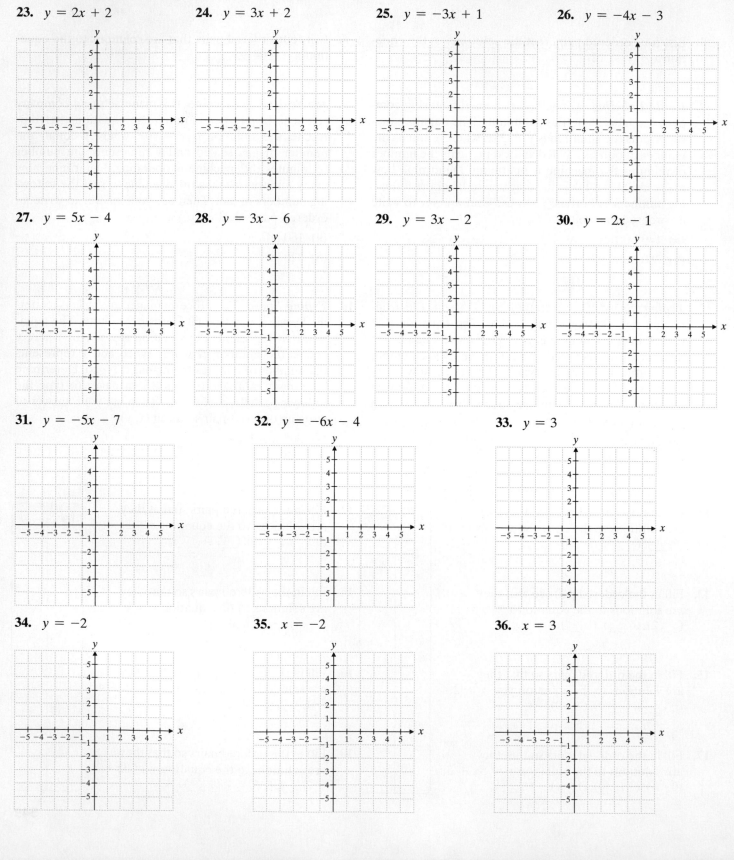

To Think About

37. All of the following ordered pairs except one are solutions to an equation $y = mx + b$. Which ordered pair is not a solution? Why?

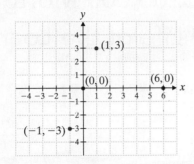

38. The ordered pairs $(-1, 3)$, $(0, 0)$, and $(1, -3)$ plotted on the rectangular coordinate system are solutions to an equation $y = mx + b$. Is $(5, 1)$ a solution to this equation? Why or why not?

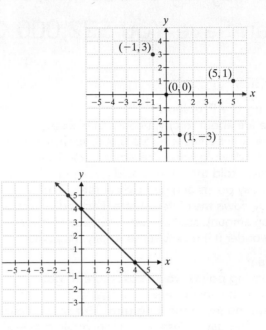

39. Three solutions to an equation are plotted on the rectangular coordinate system. Determine two more ordered pairs that are solutions to this equation.

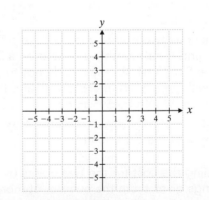

40. Find an equation that has the following solutions: $(2, 4)$, $(3, 6)$, $(4, 8)$.

Cumulative Review *Solve.*

41. **[7.2.2]** $5x + 6 - 3x - 3 = 10$

42. **[7.3.1]** $2(x + 1) = -3(x + 2) + 17$

43. **[7.2.2]** $4x - 5 + 2x = 9x + 14$

44. **[7.4.1]** $\dfrac{x}{2} + \dfrac{x}{5} = 2$

Quick Quiz 9.4

1. Fill in the ordered pairs so they are solutions to the equation $3x + y = 15$. $(-1, \underline{\hspace{1em}})$, $(\underline{\hspace{1em}}, 0)$, $(\underline{\hspace{1em}}, 3)$

2. Graph. $y = -2x + 4$

3. Graph. $y = 4$

4. **Concept Check** Professor Sanchez asked his class to name two points that lie on the graph $y = 2x - 1$. Mark answered: $(-3, 4)$ and $(0, -1)$. Did he answer the question correctly? Why or why not?

Did You Know...

That Paying $4000 in Points on a $200,000 Mortgage Can Save You $22,000 Over the Life of the Loan?

MORGAGES

Understanding the Problem:

Mike and Sue are about to buy a house that costs $220,000. They will be financing $200,000, so they are going to have a down payment of almost 10%. Their mortgage broker told them they could get a rate of 5.5% or they could pay points and get a lower rate. That is, they could buy down their interest rate. Each point costs 1% of the loan amount, so that would be $2000 for each point. They wonder if it makes sense to pay the points.

Making a Plan:

They know paying points would cost more to get the loan, but having the lower rate would lower payments and save them money in the long term. But how much money would they save? Also, how long would it take to realize the savings? They need to decide how long they plan to live in the house to determine if it's worth paying any points.

Step 1: They have enough money to pay either one or two points. Sue and Mike decide to first look at the scenario where they pay two points.

Task 1: How much would it cost to pay two points on a $200,000 mortgage?

Step 2: If they pay two points, they will get a 5% interest rate, and their monthly payment on a 30-year loan would be $1073.64. If they decide not to pay any points, they will get a 5.5% interest rate, and their monthly payment would be $1135.58.

Task 2: What would they save a month if they got the 5% loan instead of the 5.5% loan?

Step 3: Dividing the cost for the two points by the monthly savings gives them the number of months it would take to reach the break-even point, that is, when the savings of the lower payment on the 5% rate mortgage equals the cost of the points to lower the rate.

Task 3: How many months would it take to reach the break-even point?

Finding a Solution:

Step 4: The decision on whether to pay the points or not depends on how long Mike and Sue expect to stay in that home. They need to look at their long-term plans and determine how long they expect to stay in the house.

Task 4: If they expect to stay in the house for six years, should they pay the two points or not?

Task 5: If they expect to stay in the house for three years, should they pay the two points or not?

Task 6: Suppose Mike and Sue plan to stay in the house for 30 years. After paying the two points to get the lower rate, how much total would they save over the 30 years? Round your answer to the nearest hundred.

Step 5: Sue and Mike look at the scenario of paying one point. They would get a 5.25% interest rate with a monthly payment of $1104.41. It turns out the break-even point is the same. This is due to the fact that even though they would not be saving quite as much a month, they also would not pay as much upfront. They would see a benefit to paying two points over paying one point if they stayed, in the house longer than the break-even point (about 65 months). This is because at that point they would have recovered the cost of the points and would just be comparing the monthly savings.

Applying the Situation to Your Life:

When a mortgage broker is asked whether someone should pay points or not, his or her response will probably be "How long do you plan to live at this house?" It sounds like a good idea to get a mortgage at 5% instead of 5.5%, but if you are not staying in that house long enough to recoup the cost to buy down the rate, it isn't a good idea. You will need to know the break-even point and how long you intend to stay in the house in order to make your decision. If you are going to stay in the house long after the break-even point, then you should buy down the rate.

Chapter 9 Organizer

Topic and Procedure	Examples	✏ You Try It
Pictographs, p. 512		

Pictographs, p. 512

The following pictograph illustrates the oil production from a local well over a 5-year period.

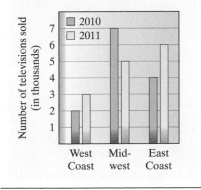

1. State the number of barrels produced each year from 2007 to 2009.

Year	Barrels Produced
2007	500
2008	1500
2009	3000

1. State the number of barrels produced each year from 2009 to 2011.

Year	Barrels Produced
2009	
2010	
2011	

Circle graphs, p. 513

The following circle graph describes the ages of the 200 men and women of a local city police force. The percentage of the 200 police officers within a given age range is illustrated.

Over age 50 12%

Under age 23 10%

Age 32–50 48%

Age 23–31 30%

2. (a) What percent of the police force is between 23 and 31 years old?

30%

(b) How many men and women in the police force are over 50 years old?

12% of 200 = (0.12)(200)
= 24 people

2. (a) What percent of the police force is between 32 and 50 years old?

(b) How many men and women in the police force are under 23 years old?

Double-bar graphs, p. 514

The following double-bar graph illustrates the sales of flat-screen television sets by a major store chain for 2010 and 2011 in three regions of the country.

Number of televisions sold (in thousands)

□ 2010
□ 2011

7, 6, 5, 4, 3, 2, 1

West Coast, Mid-west, East Coast

3. (a) How many flat-screen television sets were sold by the chain on the East Coast in 2011?

6000 sets

(b) How many more flat-screen television sets were sold in 2011 than in 2010 on the West Coast?

3000 sets were sold in 2011; 2000 sets were sold in 2010.

$$\begin{array}{r} 3000 \\ -\ 2000 \\ \hline 1000 \text{ sets more in 2011} \end{array}$$

3. (a) How many flat-screen television sets were sold by the chain on the West Coast in 2010?

(b) How many more flat-screen television sets were sold in 2010 than in 2011 in the Midwest?

Topic and Procedure	Examples	✏️ You Try It

Comparison line graphs, p. 515

The following line graph indicates the number of visitors to Wetlands State Park during a 4-month period in 2009 and 2010.

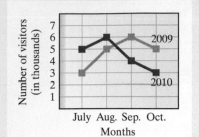

4. **(a)** How many visitors came to the park in July 2009?

3000 visitors

(b) In what months were there more visitors in 2009 than in 2010?

September and October

(c) The sharpest decrease in attendance took place between which two months?

Between August 2010 and September 2010

4. **(a)** How many visitors came to the park in September 2010?

(b) In what months were there fewer visitors in 2009 than in 2010?

(c) The sharpest increase in attendance took place between which two months?

Constructing comparison line graphs, pp. 515–516

1. Draw and label a vertical and a horizontal number line.
2. Place dots on the graph that correspond to the first category of data.
3. Connect the dots.
4. Repeat steps 2 and 3 for the second category of data.
5. Label each line.

5. The numbers of phone calls received by a hospital receptionist on the day and night shifts over five days are listed in the following table.

Hospital Calls

	M	T	W	TH	F
Night shift	125	130	120	110	115
Day shift	90	120	80	95	110

Construct a comparison line graph of the information given in the table.

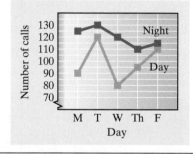

5. Use the information in the following table to construct a comparison line graph that compares the numbers of sales of the various souvenirs for week 1 and week 2 at the gift shop.

Souvenir Sales

Week	Hats	Coffee Mugs	T-Shirts	Post Cards
1	45	60	70	40
2	40	70	80	50

Constructing double-bar graphs, pp. 515–516

1. Draw and label a vertical and a horizontal number line.
2. Label a shaded and a nonshaded bar to represent each category that is being compared.
3. Draw a bar to the appropriate height for the quantity in each category.

6. Construct a double-bar graph for the number of calls received by the day and night shifts at the hospital shown in the Hospital Calls table.

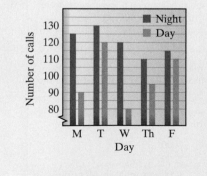

6. Use the data in the Souvenir Sales table to construct a double-bar graph that compares the numbers of sales of the various souvenirs for week 1 and week 2 at the gift shop.

Topic and Procedure	Examples	✏ You Try It
Finding the mean, p. 524 The *mean* of a set of values is the sum of the values divided by the number of values. The mean is often called the *average*.	7. Find the mean of 19, 13, 15, 25, and 18. $$\frac{19 + 13 + 15 + 25 + 18}{5} = \frac{90}{5} = 18$$ The mean is 18.	7. Find the mean of 12, 7, 9, 13, 11, and 8.

Finding the median, p. 525
 	
1. Arrange the numbers in order from smallest to largest.
2. If there is an odd number of values, the middle value is the median.
3. If there is an even number of values, the average of the two middle values is the median.
	8. **(a)** Find the median of 19, 29, 36, 15, and 20. First we arrange in order from smallest to largest: 15, 19, 20, 29, 36.
	
	  15, 19 20 29, 36
	  two middle two
	  numbers number numbers
	
	The median is 20.
	
	(b) Find the median of 67, 28, 92, 37, 81, and 75. First we arrange the numbers in order from smallest to largest: 28, 37, 67, 75, 81, 92. There is an even number of values.
	
	  28, 37 67, 75 81, 92
	         two middle
	         numbers
	
	$$\frac{67 + 75}{2} = \frac{142}{2} = 71$$
	
	The median is 71.
		8. **(a)** Find the median of 12, 16, 23, 5, 14, 25, and 8.
		
		(b) Find the median of 16, 8, 5, 12, 23, and 14.

Finding the mode, p. 526 The *mode* of a set of data is the number or numbers that occur most often.	9. Find the mode. 1, 8, 71, 18, 8, 30, 8 The number 8 is the mode since it occurs three times in the data.	9. Find the mode. 2, 6, 13, 6, 2, 8, 14
Plotting points, p. 533 To plot (x, y): 1. Begin at the origin. 2. If x is positive, move to the right along the x-axis. If x is negative, move to the left along the x-axis. 3. If y is positive, move up. If y is negative, move down. 4. Place a dot at this location and label the point.	10. Plot the ordered pair. $(-2, 3)$ 	10. Plot the ordered pair. $(3, -2)$
Graphing straight lines, p. 548 A linear equation has a graph that is a straight line. To graph such an equation, plot any three points; two points give the line and the third point checks it.	11. Graph. $3x + 2y = 6$ | (x,	y)
---	---
(0,	3)
(4,	−3)
(2,	0) | 	11. Graph. $2x - 3y = 6$

Chapter 9 Review Problems

Section 9.1

Course Enrollment *Use this pictograph to answer exercises 1–4.*

1. How many students are enrolled in prealgebra?
2. How many students are enrolled in calculus?
3. How many more students are enrolled in algebra than calculus?
4. What is the combined total enrollment in prealgebra and algebra?

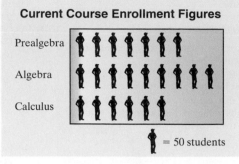

Current Course Enrollment Figures

Prealgebra

Algebra

Calculus

= 50 students

Household Budget *Nancy and Wally Worzowski's family monthly budget of $4400 is displayed in the accompanying circle graph. Use the graph to answer exercises 5–12.*

5. What percent of the budget is allotted for transportation?
6. What percent of the budget is allotted for savings?
7. What percent of the budget is used up by the food and rent categories?
8. What percent of the budget is used up by the transportation, utilities/Internet, and savings categories?

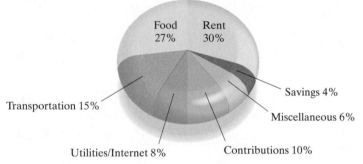

Food 27%

Rent 30%

Transportation 15%

Savings 4%

Miscellaneous 6%

Utilities/Internet 8%

Contributions 10%

9. Of the total $4400, how much money per month is budgeted for utilities/Internet?
10. Of the total $4400, how much money per month is budgeted for transportation?
11. Of the total $4400, how much money per month is budgeted for the rent and savings categories?
12. Of the total $4400, how much money per month is budgeted for the transportation and food categories?

Population Growth *The double-bar graph illustrates the predicted population explosion for the year 2050. Use this graph to answer exercises 13–20.*

13. (a) In the year 2006, which country had the largest population?
    (b) In the year 2050, which of the five countries is predicted to have the largest population?
14. (a) In the year 2006, which country had the smallest population?
    (b) In the year 2050, which of the five countries is predicted to have the smallest population?
15. In the year 2050, is the United States predicted to have the first, second, third, fourth, or fifth largest population?
16. In the year 2050, is Indonesia predicted to have the first, second, third, fourth, or fifth largest population?

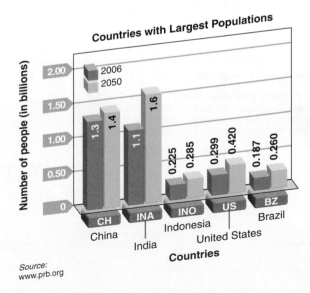

Countries with Largest Populations

Number of people (in billions)

2006
2050

2.00
1.50
1.00
0.50
0

1.3
1.4
1.6
1.1
0.225
0.285
0.299
0.420
0.187
0.260

CH — China
INA — India
INO — Indonesia
US — United States
BZ — Brazil

Countries

Source:
www.prb.org

When we work with large numbers such as millions and billions, we can simplify the calculations if we write the words million or billion in place of the zeros. For example, to subtract 6,000,000 − 2,000,000 we can write 6 million − 2 million = 4 million. Use this method as you solve exercises 17–20.

17. How much greater is the population predicted to be in China in 2050 than in 2006?

18. How much greater is the population predicted to be in the United States in 2050 than in 2006?

19. Which of the five countries is predicted to have the largest increase in population from the year 2006 to 2050?

20. How much greater is the population predicted to be in Indonesia in 2050 than in 2006?

Trade-in Values vs Private Sale *The comparison line graph indicates the suggested trade-in values for options compared with the private-sale value of those same options.*

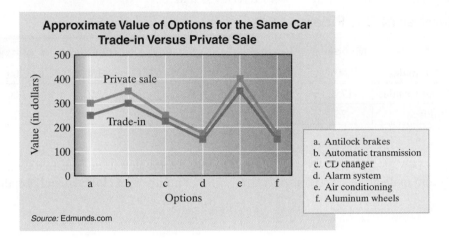

Source: Edmunds.com

21. (a) What was the value of an air conditioner on a private sale?

(b) What was the value of an air conditioner on a trade-in?

22. (a) What was the value of an automatic transmission on a private sale?

(b) What was the value of an automatic transmission on a trade-in?

23. Which three options lost the most value on a trade-in?

24. Which three options lost the least value on a trade-in?

25. What was the average value of all the options on a private sale?

26. What was the average value of all the options on a trade-in?

27. What was the median value of all the options on a private sale?

28. What was the median value of all the options on a trade-in?

Circus Souvenir *Use the information in the table to answer exercises 29 and 30.*

Circus Souvenir Sales

Week	Hats	Stuffed Animals	T-Shirts	Posters
1	55	80	75	65
2	45	60	70	40
3	40	70	80	50

29. Use the data to construct a comparison line graph that compares the numbers of sales of various souvenirs for week 1 and week 3 at the circus.

30. Use the data to construct a comparison line graph that compares the numbers of sales of the various souvenirs for week 1 and week 2 at the circus.

Bus Rides Use the information in the table to answer exercises 31 and 32.

Number of Bus Riders

	Monday	Tuesday	Wednesday	Thursday	Friday
Adult males	100	150	150	150	200
Adult females	175	150	250	250	300
Children	75	100	150	100	175

31. Construct a double-bar graph that compares the numbers of female and male adult bus riders over the 5 days.

32. Construct a double-bar graph that compares the numbers of children and female adult bus riders over the 5 days.

Section 9.2

Find the median value.

33. *Math Exam Scores* The scores on a recent mathematics exam: 69, 57, 100, 87, 93, 65, 77, 82, and 88.

34. *Student Enrollment* The numbers of students taking abnormal psychology for the fall semester for the last 9 years at Elmson College: 77, 83, 91, 104, 87, 58, 79, 81, and 88.

35. *Coffee Consumption* The numbers of cups of coffee consumed by each of the students of the 7:00 A.M. Biology III class during the last semester: 38, 19, 22, 4, 0, 1, 5, 9, 18, 36, 43, 27, 21, 19, 25, and 20.

36. *Pizza Delivery* The numbers of deliveries made each day by the Northfield House of Pizza: 21, 16, 0, 3, 19, 24, 13, 18, 9, 31, 36, 25, 28, 14, 15, and 26.

Find the mean (average).

37. **Temperature Readings** The daily maximum temperature readings in Los Angeles during the last 7 days in July were 86°F, 83°F, 88°F, 95°F, 97°F, 100°F, and 81°F.

38. **Groceries Purchased** The costs of groceries purchased by the Michael Stallard family each week for the last 7 weeks were $87, $105, $89, $120, $139, $160, and $98.

39. **Textbooks Purchased** The numbers of college textbooks purchased by each of 8 men living at Jenkins House during 4 years of college were 76, 20, 91, 57, 42, 21, 75, and 82.

40. **Engineering Student** The numbers of female students enrolled in the school of engineering at Westwood University during each of the last 10 years were 151, 140, 148, 156, 183, 201, 205, 228, 231, and 237.

Find the mode.

41. 22, 13, 18, 14, 13, 19

42. 18, 14, 28, 18, 29, 18, 14

Section 9.3

Product Sales *Use the line graph to answer exercises 43 and 44.*

43. Represent the number of products sold in the years 2002, 2003, 2004, and 2005 using ordered pairs of the form (year, number sold).

44. Represent the number of products sold in the years 2006, 2007, 2008, and 2009 using ordered pairs of the form (year, number sold).

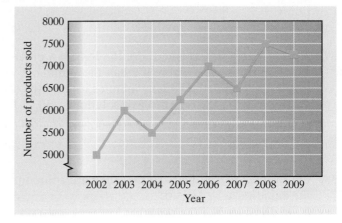

Plot and label each point on the rectangular coordinate system.

45. (3, 2)

46. $\left(2, 3\frac{1}{2}\right)$

47. (−2, 0)

48. (−3, −1)

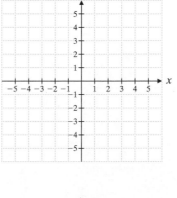

Give the coordinates of each point.

49. R

50. S

51. T

52. U

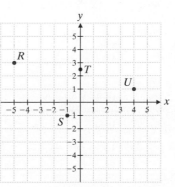

559

Plot the points corresponding to the ordered pairs and then draw a line connecting the coordinate points.

53. $(2, 0), (2, 3), (2, -1)$

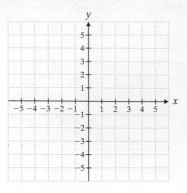

54. $(3, 1), (-4, 1), (0, 1)$

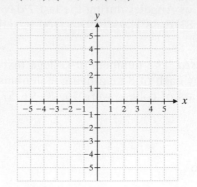

Section 9.4

55. List three ordered pairs that are solutions to the equation $x + y = 8$.

56. List three ordered pairs that are solutions to the equation $x + y = 3$.

57. ***Typing Speed*** The secretary for the J&M law offices can type 70 words per minute. The equation that represents the situation is $y = 70x$, where y equals the number of words typed and x represents the number of minutes the secretary typed. Determine how many minutes it takes the secretary to type the following numbers of words. Then use your answer to write the ordered-pair solution (x, y).

   (a) 280 words

   (b) 350 words

58. Fill in the ordered pairs so that they are solutions to the equation $y = 2x - 6$.
$(\_\_, -4), (\_\_, -6), (\_\_, -8)$

59. Fill in the ordered pairs so that they are solutions to the equation $y = -6x + 2$.
$(\_\_, 2), (\_\_, 8), (\_\_, -4)$

Graph each equation.

60. $y = 3x - 1$

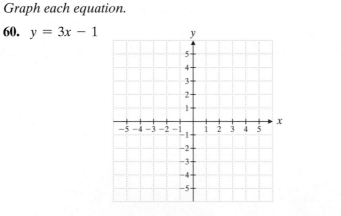

61. $y = -5x - 4$

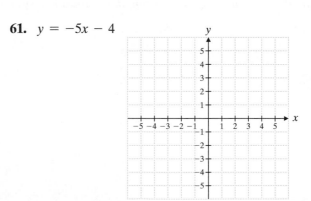

62. $y = 4x - 6$

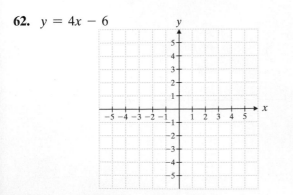

63. $y = -1$

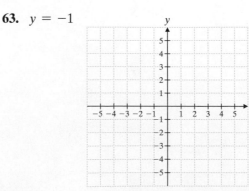

How Am I Doing?   Chapter 9 Test

After you take this test read through the Math Coach on pages 564–565. Math Coach videos are available via MyMathLab and YouTube. Step-by-step test solutions on the Chapter Test Prep Videos are also available via MyMathLab and YouTube. (Search "BlairTobeyPrealgebra" and click on "Channels.")

The ages of 5000 students on campus were recorded. The circle graph depicts the distribution. Use the graph to answer questions 1–4.

1. What age group comprises the largest percent of the student body?

Mc 2. What percent of the students are between ages 18 and 24?

3. What percent of the students are age 20 or younger?

4. If 5000 students are at the university, how many students are over age 27?

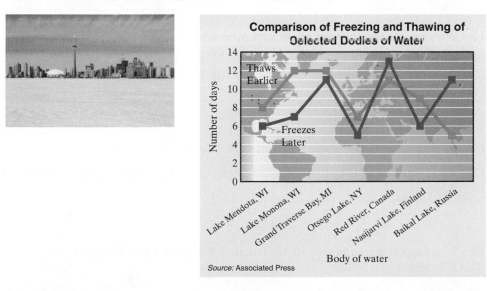

Ages of Students

Ages 25–27 10%
Over age 27 7%
Under age 18 6%
Ages 21–24 33%
Ages 18–20 44%

The water in Toronto Harbor in Canada freezes approximately 37 days later in the year than it did 100 years ago. This fact was determined in a study of the annual freezing and thawing of 26 bodies of water over the past 150 years. The outcome of the study indicates that global warming has resulted in water freezing 8.7 days later and thawing 9.8 days earlier each year. The comparison line graph gives the study results for selected bodies of water. Use this comparison line graph to answer questions 5–8.

Comparison of Freezing and Thawing of Selected Bodies of Water

Number of days

Thaws Earlier
Freezes Later

Lake Mendota, WI · Lake Monona, WI · Grand Traverse Bay, MI · Otsego Lake, NY · Red River, Canada · Nasijarvi Lake, Finland · Baikal Lake, Russia

Body of water

Source: Associated Press

Fill in the blanks.

5. In Michigan, the water in the Grand Traverse Bay freezes approximately _____ days later in the year and thaws _____ days earlier in the year than it did 100 years ago.

6. In Finland, the water in the Nasijarvi Lake freezes approximately _____ days later in the year and thaws _____ days earlier in the year than it did 100 years ago.

Mc 7. According to this graph, which body of water froze the fewest number of days later than it did 100 years ago?

8. According to this graph, the body of water where there was the least amount of change in the time of the year that the water thaws is _____.

1. _____ ☐

2. _____ ☐

3. _____ ☐

4. _____ ☐

5. _____ ☐

6. _____ ☐

7. _____ ☐

8. _____ ☐

561

The multiple bar graph indicates the number of years of various warranties for four brands of cars. Use the information on this graph to answer questions 9–14.

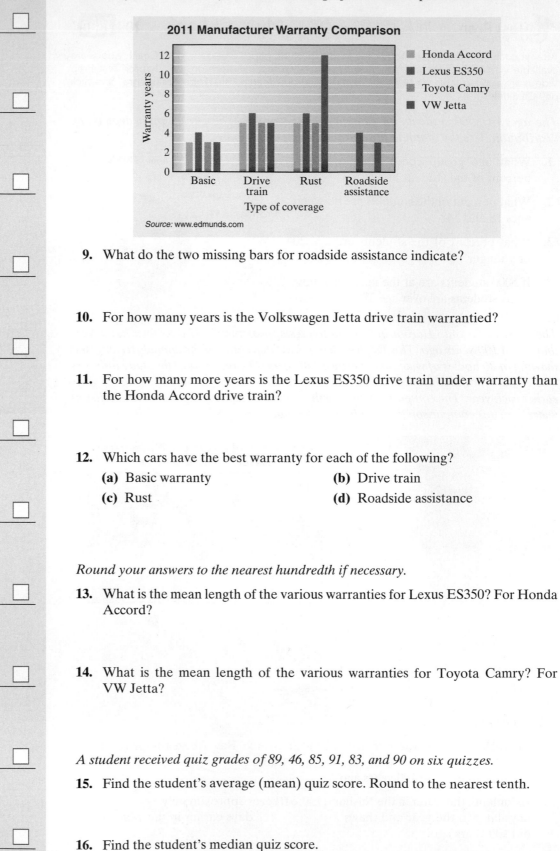

Source: www.edmunds.com

9. What do the two missing bars for roadside assistance indicate?

10. For how many years is the Volkswagen Jetta drive train warrantied?

11. For how many more years is the Lexus ES350 drive train under warranty than the Honda Accord drive train?

12. Which cars have the best warranty for each of the following?
 (a) Basic warranty   (b) Drive train
 (c) Rust   (d) Roadside assistance

Round your answers to the nearest hundredth if necessary.

13. What is the mean length of the various warranties for Lexus ES350? For Honda Accord?

14. What is the mean length of the various warranties for Toyota Camry? For VW Jetta?

A student received quiz grades of 89, 46, 85, 91, 83, and 90 on six quizzes.

15. Find the student's average (mean) quiz score. Round to the nearest tenth.

16. Find the student's median quiz score.

Plot and label each ordered pair on the rectangular coordinate system.

Ⓜ️ⓒ **17.** $(3, 5)$

18. $(0, 0)$

19. $(2, -1)$

20. $(-3, 0)$

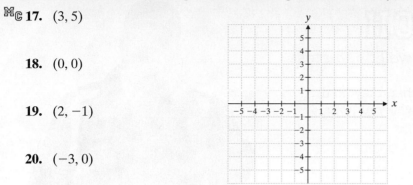

Give the coordinates of each point.

21. A

22. B

23. C

24. D

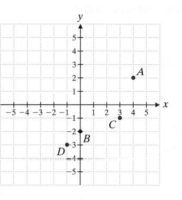

25. Plot the set of ordered pairs and then draw a line connecting the coordinate points. $(1, 2), (-3, 2), (0, 2), (4, 2)$

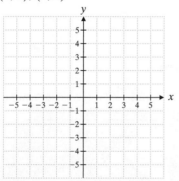

26. Graph. $y = 3$

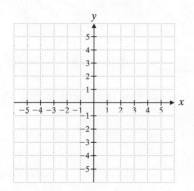

27. Graph. $y = 4x - 2$

28. Graph. $y = 3x + 1$

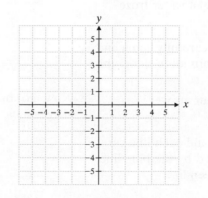

17. ☐

18. ☐

19. ☐

20. ☐

21. ☐

22. ☐

23. ☐

24. ☐

25. ☐

26. ☐

27. ☐

28. ☐

Total Correct: ☐

MATH COACH

Mastering the skills you need to do well on the test.

Students often make the same types of errors when they do the Chapter 9 Test. Here are some helpful hints to keep you from making these common errors on test problems.

Reading Circle Graphs—Problem 2

What percent of the students are between ages 18 and 24?

> **Helpful Hint** Study the labels on the circle graph very carefully. Sometimes you need to use more than one section of the circle graph in order to answer the question.

Did you get either 33% or 44% as your answer?

Yes ▭ No ▭

If you answered Yes, examine the circle graph carefully. Notice that the information for the age range 18–24 involves two sections of the graph. Please stop now and rework the problem.

Did you add 44% + 33% as your next step?

Yes ▭ No ▭

If you answered No, stop and study the graph so that you understand why the answer requires you to add these two percents.

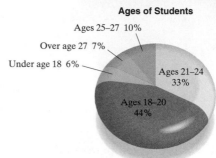

Ages of Students

Ages 25–27 10%
Over age 27 7%
Under age 18 6%
Ages 21–24 33%
Ages 18–20 44%

If you answered Problem 2 incorrectly, go back and rework the problem using these suggestions.

Reading and Interpreting a Comparison Line Graph—Problem 7

According to this graph, which body of water froze the fewest number of days later than it did 100 years ago?

> **Helpful Hint** Study the information on the comparison line graph carefully. Pay particular attention to the labels on each of the lines. Make sure you know which line represents the information you need to answer the question.

Did you realize that each point on the blue line refers to *the number of days later* that each body of water froze?

Yes ▭ No ▭

If you answered No, look at the graph carefully again and take some time to figure out what information the graph displays, then rework the problem.

Did you get Baikal Lake, Russia, as your answer?

Yes ▭ No ▭

If you answered Yes, study the labels and each line carefully. Notice that freezing is represented by the blue line, while thawing is represented by the green line.

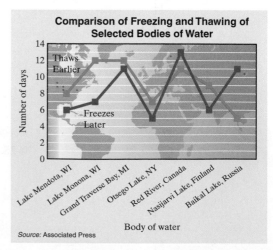

Comparison of Freezing and Thawing of Selected Bodies of Water

Thaws Earlier
Freezes Later
Number of days

Lake Mendota, WI; Lake Monona, WI; Grand Traverse Bay, MI; Otsego Lake, NY; Red River, Canada; Nasijarvi Lake, Finland; Baikal Lake, Russia

Body of water

Source: Associated Press

Now go back and rework this problem using these suggestions.

Need help? Watch the **MATH COACH** videos in MyMathLab® or on YouTube™.

564

Plotting Points Given the Coordinates—Problem 17

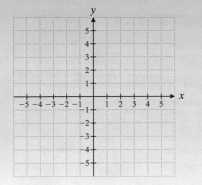

Plot and label the ordered pair on the rectangular coordinate system. $(3, 5)$

> **Helpful Hint** Remember which number in the ordered pair is x and which one is y. To plot points:
>
> - Start at the origin $(0, 0)$, where the x-axis and y-axis cross each other.
> - Move x units to the right if x is positive or to the left if x is negative.
> - Then move y units directly up if y is positive or y units directly down if y is negative.
>
> This is where your plotted point should be located.

To plot $(3, 5)$, did you start at $(0, 0)$ and move 3 units in the positive x-direction (right)?

Yes _____ No _____

If you answered No, go back and label the x-coordinate. The x-coordinate will be the first number in your ordered pair. Stop now and rework the problem.

After moving 3 units in the positive x-direction, did you then move 5 units in the positive y-direction (up)?

Yes _____ No _____

If you answered No, go back and label the y-coordinate. The y-coordinate is the second number in your ordered pair.

If you answered Problem 17 incorrectly, go back and rework the problem using these suggestions.

Graphing Linear Equations in Two Variables—Problem 28

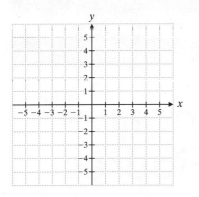

Graph. $y = 3x + 1$

> **Helpful Hint**
> - Find three ordered pairs that are solutions to the given equation.
> - Be sure to replace x and y with the correct values.
> - Verify that the three ordered pairs on the graph form a straight line.
> - Double-check to make sure that you have plotted the points correctly.
> - Complete calculations very carefully.

Did you pick three ordered pairs that are solutions to $y = 3x + 1$?

Yes _____ No _____

If you answered No, stop and complete this step. It helps to use a table of values to collect and organize the three ordered pairs.

When you connected the three points, did they form a straight line?

Yes _____ No _____

If you answered No, check your plotted points carefully to see if they are plotted correctly. Then check your calculations for any possible errors.

Now go back and rework this problem using these suggestions.

Need more help? Look for section examples marked with $\mathbb{Mc}$ to review.

565

When we plan home improvement projects, we often need to calculate perimeter, area, or volume. For example, to landscape a yard we must make decisions about the quantity of materials we must purchase. How many bricks should I buy to build a planter? How much cement must I purchase to pour a patio? If I put a spa in my yard, how much grass area will I have left? After studying the mathematics in this chapter, you will have the knowledge to perform these calculations.

Measurement and Geometric Figures

10.1 Using Unit Fractions with U.S. and Metric Units

We have worked with units of measurement throughout the book. Now we will see how to change (convert) from one unit to another. We present a table that shows the relationships between U.S. units.

Length	Weight
12 inches (in.) = 1 foot (ft)	16 ounces (oz) = 1 pound (lb)
3 feet (ft) = 1 yard (yd)	2000 pounds (lb) = 1 ton
5280 feet (ft) = 1 mile (mi)	

Volume	Time
2 cups (c) = 1 pint (pt)	60 seconds (sec) = 1 minute (min)
2 pints (pt) = 1 quart (qt)	60 minutes (min) = 1 hour (hr)
4 quarts (qt) = 1 gallon (gal)	24 hours (hr) = 1 day
	7 days = 1 week

Student Learning Objectives

After studying this section, you will be able to:

① Use unit fractions to convert between U.S. units.

② Solve applied problems involving U.S. units.

③ Convert between metric units.

④ Solve applied problems involving metric units.

① Using Unit Fractions to Convert Between U.S. Units

For many simple problems such as 24 inches = ? feet, we can easily see how to convert from inches to feet.

$$24 \text{ inches} = 12 \text{ inches} + 12 \text{ inches}$$
$$= 1 \text{ foot} + 1 \text{ foot} \qquad (12 \text{ inches} = 1 \text{ foot})$$
$$24 \text{ inches} = 2 \text{ feet}$$

For more complicated problems, for example, those with larger numbers, we need another conversion method so that the process is simple and efficient. The method we use involves multiplying by a unit fraction.

> A **unit fraction** is a fraction that shows the relationship between units and is equal to 1.

For example, since 12 in. = 1 ft, we can say that there are 12 inches per 1 foot or 1 foot per 12 inches. If we read the fraction bar as *per*, we have the following unit fractions that are equal to 1:

$$\text{per} \rightarrow \frac{12 \text{ in.}}{1 \text{ ft}} = \frac{1 \text{ ft}}{12 \text{ in.}} = 1 \qquad \frac{12 \text{ in.}}{1 \text{ ft}} \text{ and } \frac{1 \text{ ft}}{12 \text{ in.}} \quad \text{are called } \textit{unit fractions.}$$

We can multiply a quantity by a unit fraction since its value is equal to 1 and we know that multiplying by 1 does not change the value of the quantity.

EXAMPLE 1 Convert 35 yards to feet.

Solution We write the relationship between feet and yards as a unit fraction.

Since 3 ft = 1 yd, we have the unit fraction $\frac{3 \text{ ft}}{1 \text{ yd}}$.

$$35 \text{ yd} = \underline{\quad ? \quad} \text{ ft}$$

$$35 \text{ yd} \times \frac{3 \text{ ft}}{1 \text{ yd}} \qquad \text{Multiply by the unit fraction.}$$

$$= 35 \text{ y\!d} \times \frac{3 \text{ ft}}{1 \text{ y\!d}} \qquad \text{Divide out the units "yd."}$$

$$= 35 \times 3 \text{ ft} = 105 \text{ ft} \qquad \text{Multiply.}$$

Student Practice 1 Convert 420 minutes to hours.

NOTE TO STUDENT: Fully worked-out solutions to all of the Student Practice problems can be found at the back of the text starting at page SP-1.

How did we know what unit fraction to use in Example 1? We use the unit fraction that relates the units we are working with; in this case it is *feet* and *yards*. Now, to determine which unit to put in the numerator and which to put in the denominator of the fraction, we must consider what unit we want to end up with. In Example 1 we wanted to end up with feet, so we placed feet in the numerator.

$$35 \text{ yd} \times \frac{3 \text{ ft}}{1 \text{ yd}} = \underline{\quad ? \quad} \text{ ft}$$

We want to end up with feet, so we place 3 ft in the numerator.

The yards divide out, and we end up with feet.

PROCEDURE TO CONVERT FROM ONE UNIT TO ANOTHER

1. Write the relationship between the units.
2. Identify the unit you want to end up with.
3. Write a unit fraction that has the unit you want to end up with in the numerator.
4. Multiply by the unit fraction.

EXAMPLE 2   Convert 560 quarts to gallons.

Solution   We write the relationship between quarts and gallons: 4 qt = 1 gal. We want to end up with gallons, so we write *1 gal* in the numerator of the unit fraction: $\frac{1 \text{ gal}}{4 \text{ qt}}$.

$$560 \text{ qt} = \underline{\quad ? \quad} \text{ gal}$$

$$560 \text{ qt} \times \frac{1 \text{ gal}}{4 \text{ qt}} \qquad \text{We multiply by the appropriate unit fraction.}$$

$$= 560 \text{ qt} \times \frac{1 \text{ gal}}{4 \text{ qt}} \qquad \text{We divide out the units "qt."}$$

$$= 560 \times \frac{1}{4} \text{ gal} = \frac{560 \text{ gal}}{4} = 140 \text{ gal}$$

Student Practice 2   Convert 144 ounces to pounds.

---

② Solving Applied Problems Involving U.S. Units

EXAMPLE 3   A computer printout shows that a particular job took 144 seconds. How many minutes is that? (Express your answer as a decimal.)

Solution   We must change seconds to minutes.

$$144 \text{ sec} \times \frac{1 \text{ min}}{60 \text{ sec}} = \frac{144}{60} \text{ min} = 2.4 \text{ min}$$

Student Practice 3   Joe's time card read, "Hours worked today: 7.2." How many minutes are in 7.2 hours?

EXAMPLE 4   The all-night garage charges $1.50 per hour for parking both day and night. A businessman left his car there for $2\frac{1}{4}$ days. How much was he charged?

Solution   *Understand the problem.* We organize the information in a Mathematics Blueprint for Problem Solving.

Mathematics Blueprint for Problem Solving

Gather the Facts	What Am I Asked to Do?	How Do I Proceed?	Key Points to Remember
The charge for parking is $1.50 per (for each) hour. The car was in the garage $2\frac{1}{4}$ days.	Find the total parking charge for $2\frac{1}{4}$ days.	1. Change $2\frac{1}{4}$ days to hours to find the total number of hours the car was in the garage. 2. Multiply the total hours by $1.50 to find the cost for parking.	Change the number of days to a decimal so calculations are easier.

Calculate and state the answer.

1. $2\dfrac{1}{4} = 2.25$ days   We change $\dfrac{1}{4}$ to a decimal: $1 \div 4 = 0.25$.

$2.25 \text{ days} \times \dfrac{24 \text{ hr}}{1 \text{ day}} = 54 \text{ hr}$ We change days to hours.

2. Now we must find the total charge for parking.

$$54 \text{ hr} \times \dfrac{\$1.50}{1 \text{ hr}} = \$81$$

Check.   Is our answer in the desired units? Yes. The answer is in dollars, and we would expect it to be in dollars. ✓

You may want to redo the calculation or use a calculator to check. The check is up to you.

Student Practice 4   A businesswoman parked her car at a garage for $1\frac{3}{4}$ days. The garage charges $1.50 per hour. How much did she pay to park the car?

Understanding the Concept

Multiplying by a Unit Fraction   How did people first come up with the idea of multiplying by a unit fraction? What mathematical principles are involved here? Actually, this is the same as solving a proportion. Consider a situation where we change 34 quarts to 8.5 gallons by multiplying by a unit fraction.

$$34 \text{ qt} \times \dfrac{1 \text{ gal}}{4 \text{ qt}} = \dfrac{34}{4} \text{ gal} = 8.5 \text{ gal}$$

What we were actually doing is setting up the proportion 1 gal is to 4 qt as n gal is to 34 qt, and solving for n.

$$\dfrac{1 \text{ gal}}{4 \text{ qt}} = \dfrac{n \text{ gal}}{34 \text{ qt}}$$

Continued on next page

We cross-multiply: 34 qt × 1 gal = 4 qt × n gal. We divide both sides of the equation by 4 quarts.

$$\frac{34 \; \cancel{qt} \times 1 \; gal}{4 \; \cancel{qt}} = \frac{\cancel{4} \; \cancel{qt} \times n \; gal}{\cancel{4} \; \cancel{qt}}$$

$$1 \; gal \times \frac{34}{4} = n \; gal$$

$$8.5 \; gal = n \; gal$$

Thus the number of gallons is 8.5. Using proportions takes a little longer, so multiplying by a unit fraction is the more popular method.

③ Converting Between Metric Units

Now let's see how to change from one metric unit to another. We start by looking at the relationship between metric units. The basic metric units are the gram, the liter, and the meter. Units that are larger than the *basic unit* use the prefixes *kilo*, meaning 1000; *hecto*, meaning 100; and *deka*, meaning 10. For units smaller than the basic unit, we use the prefixes *deci*, meaning $\frac{1}{10}$; *centi*, meaning $\frac{1}{100}$; and *milli*, meaning $\frac{1}{1000}$.

A teaspoon can hold about 5 milliliters.

A 1-liter bottle can hold 1000 milliliters.

			gram liter meter			
Prefixes				*Prefixes*		
kilo	hecto	deka	*basic unit*	deci	centi	milli

These prefixes identify units that are larger than the basic unit.

These prefixes identify units that are smaller than the basic unit.

We list the relationships between units that are commonly used in the metric system.

1 nickel weighs about 5 grams.

200 nickels weigh about 1000 grams or 1 kilogram (kg).

COMMONLY USED METRIC MEASUREMENTS

Weight
1 kilogram (kg) = 1000 grams
1 gram (g)   the basic unit
1 milligram (mg) = 0.001 gram

Length
1 kilometer (km) = 1000 meters
1 meter (m)   the basic unit
1 centimeter (cm) = 0.01 meter
1 millimeter (mm) = 0.001 meter

Volume
1 kiloliter (kL) = 1000 liters
1 liter (L)   the basic unit
1 milliliter (mL) = 0.001 liter

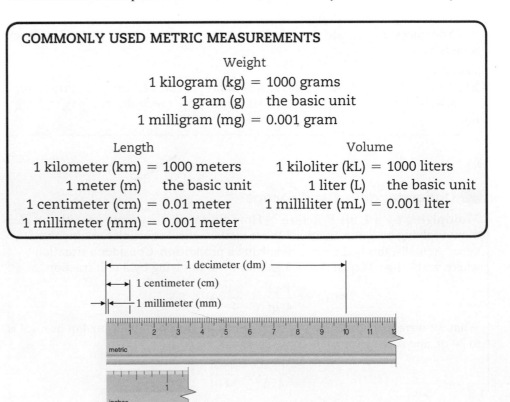

How do we convert from one metric unit to another? For example, how do we change 5 kilometers into an equivalent number of meters?

Recall from Chapter 8 that when we multiply by 10, we move the decimal point 1 place to the right. When we divide by 10, we move the decimal point 1 place to the left. Let's see how we use that idea to change from one metric unit to another.

CHANGING FROM LARGER METRIC UNITS TO SMALLER ONES

When you change from one metric prefix to another by moving to the *right* on this prefix chart, move the decimal point to the *right* the same number of places.

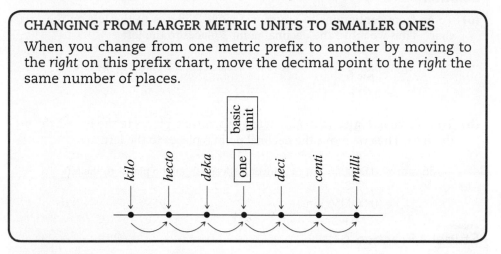

Thus 1 meter = 100 centimeters because we move two places to the right on the chart of prefixes and we also move the decimal point in 1.00 two places to the right.

EXAMPLE 5

(a) Change 7 kilometers to meters.   **(b)** Change 30 liters to centiliters.

Solution

(a) To go from kilometer to meter (basic unit), we move *3 places to the right on the prefix chart, so we move the decimal point 3 places to the right.*

$$7 \text{ km} = 7.000 \text{ m (move 3 places)} = 7000 \text{ m}$$

(b) To go from liter (basic unit) to centiliter, we move *2 places to the right on the prefix chart.* Thus we move the decimal point 2 places to the right.

$$30 \text{ L} = 30.00 \text{ cL (move 2 places)} = 3000 \text{ cL}$$

Student Practice 5

(a) Change 4 meters to centimeters.

(b) Change 30 centigrams to milligrams.

Now let us see how we can change a measurement stated in a smaller unit to an equivalent measurement in larger units.

CHANGING FROM SMALLER METRIC UNITS TO LARGER ONES

When you change from one metric prefix to another by moving to the *left* on this prefix chart, move the decimal point to the *left* the same number of places.

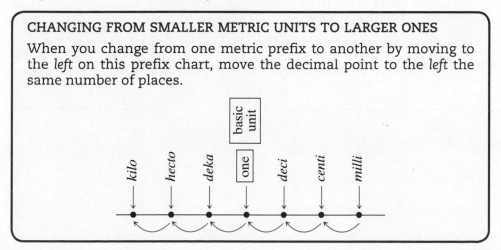

EXAMPLE 6

(a) Change 7 centigrams to grams.

(b) Change 56 millimeters to kilometers.

Solution

(a) To go from centigrams to grams, we move 2 places to the left on the prefix chart. Thus we move the decimal point 2 places to the left.

$$7 \text{ cg} = 0.07 \text{ g} \qquad \text{Move the decimal point 2 places to the left.}$$

$$= 0.07 \text{ g}$$

(b) To go from *milli*meters to *kilo*meters, we move 6 places to the left on the prefix chart. Thus we move the decimal point 6 places to the left.

$$56 \text{ mm} = 0.000056 \text{ km} \qquad \text{Move the decimal point 6 places to the left.}$$

$$= 0.000056 \text{ km}$$

Student Practice 6

(a) Change 3 milliliters to liters.

(b) Change 47 centimeters to kilometers.

EXAMPLE 7 Convert.

(a) 426 decimeters to kilometers

(b) 9.47 hectometers to meters

Solution

(a) We are converting from a smaller unit, *dm*, to a larger one, *km*. Therefore, there will be fewer kilometers than decimeters (the number we get will be smaller than 426). We move the decimal point 4 places to the left.

$$426 \text{ dm} = 0.0426 \text{ km} \qquad \text{Move the decimal point 4 places to the left.}$$

$$= 0.0426 \text{ km}$$

(b) We are converting from a larger unit, *hm,* to a smaller one, *m*. Therefore, there will be more meters than hectometers (the number will be larger than 9.47). We move the decimal point 2 places to the right.

$$9.47 \text{ hm} = 947. \text{ m} \qquad \text{Move the decimal point 2 places to the right.}$$

$$= 947 \text{ m}$$

Student Practice 7 Convert.

(a) 389 millimeters to dekameters

(b) 0.48 hectometer to centimeters

④ Solving Applied Problems Involving Metric Units

1 liter

Cleaning Fluid

1 milliliter

EXAMPLE 8 A special cleaning fluid used to rinse test tubes in a chemistry lab costs $40.00 per liter. What is the cost per milliliter?

Solution *Understand the problem.* We organize the information in a Mathematics Blueprint for Problem Solving.

Mathematics Blueprint for Problem Solving

Gather the Facts	What Am I Asked to Do?	How Do I Proceed?	Key Points to Remember
The fluid costs $40 per liter.	Find out how much 1 milliliter of fluid costs.	**1.** Change 1 liter to milliliters. **2.** Find the unit cost per milliliter.	To go from liters to milliliters we move 3 places to the right on the prefix chart. Thus we move the decimal point 3 places to the right.

Calculate and state the answer.

1. 1 L = 1000 mL   Change liters to milliliters.

2. We write $40 per liter as a rate.

$$\frac{\$40}{1\ L} = \frac{\$40}{1000\ mL}$$ Replace 1 L with 1000 mL.

$$= \$0.04\ per\ mL$$ $40 \div 1000 = 0.04$

Check. A milliliter is a very small part of a liter. Therefore it should cost much less for 1 milliliter of fluid than it does for 1 liter. $0.04 is much smaller than $40.00, so our answer seems reasonable.

Student Practice 8 A purified acid costs $110 per liter. What does it cost per milliliter?

Verbal and Writing Skills, Exercises 1–12 *From memory, write the equivalent values.*

1. 1 foot = _____ inches

2. 1 yard = _____ feet

3. _____ pints = 1 quart

4. _____ feet = 1 mile

5. 1 ton = _____ pounds

6. 1 pound = _____ ounces

7. _____ quarts = 1 gallon

8. _____ cups = 1 pint

9. 7 days = _____ week

10. 1 day = _____ hours

11. _____ seconds = 1 minute

12. _____ minutes = 1 hour

Convert.

13. 21 feet = _____ yards

14. 72 inches = _____ feet

15. 7920 feet = _____ miles

16. 31 yards = _____ feet

17. 69 inches = _____ feet

18. 192 ounces = _____ pounds

19. 13 tons = _____ pounds

20. 2.25 pounds = _____ ounces

21. 7 gallons = _____ quarts

22. 660 minutes = _____ hours

23. 11 days = _____ hours

24. 10 minutes = _____ seconds

Applications

25. *Javelin Record* Randy threw his javelin 218 feet 10 inches. How many inches did his javelin fly?

26. *Mountain Height* Mount Whitney in California is approximately 2.745 miles high. How many feet is that? Round your answer to the nearest 10 feet.

27. *Parking Costs* A stockbroker left his car in an all-night garage for $2\frac{1}{2}$ days. The garage charges $2.25 per hour. How much did he pay for parking?

28. *Wild Mushroom Sauce* Judy is making a wild mushroom sauce for a pasta dinner for a large group of friends. She bought 26 ounces of wild mushrooms at $6.00 per pound. How much did the mushrooms cost?

29. *Gourmet Cheese Purchase* Phoebe bought 24 ounces of gourmet cheese for a fondue. If the cheese sells for $4.00 per pound, how much did Phoebe pay for 24 ounces of gourmet cheese?

30. *Housesitting Expenses* Darlene's housesitter charges $1.25 per hour to stay at Darlene's home while she is out of town. How much will the housesitter charge for $2\frac{1}{4}$ days?

Memorize this chart and use it when converting between metric units. Fill in the blanks with the correct values.

31. 50 cm = _____ mm

32. 34 cm = _____ mm

basic unit

33. 3.6 km = _____ m

34. 8.4 km = _____ m

kilo | hecto | deka | one | deci | centi | milli

35. 2.43 kL = _____ mL

36. 1.76 kL = _____ mL

37. 1834 mL = _____ kL

38. 5261 mL = _____ kL

39. 0.78 g = _____ kg

40. 486 g = _____ kg

41. 5.9 kg = _____ mg

42. 5286 mg = _____ g

43. 7 mL = _____ L = _____ kL

44. 18 mL = _____ L = _____ kL

45. 413 mg = _____ g = _____ kg

46. 49 mg = _____ g = _____ kg

47. 35 mm = _____ cm = _____ m

48. 83 mm = _____ cm = _____ m

49. 3582 mm = _____ m = _____ km

50. 7812 mm = _____ m = _____ km

51. 0.32 cm = _____ m = _____ km

52. 0.81 cm = _____ m = _____ km

Applications

53. *Vaccines* A pharmaceutical firm developed a new vaccine that costs $6.00 per milliliter to produce. How much will it cost the firm to produce 1 liter of the vaccine?

54. *Anticancer Drug* A new anticancer drug costs $95.50 per gram. How much would it cost to buy 6 kilograms of the drug?

55. *Rare Flower* A very rare essence of an almost extinct flower found in the Amazon jungle of South America is extracted by a biogenetic company trying to copy and synthesize it. The company estimates that if the procedure is successful, the product will cost the company $850 per milliliter to produce. How much will it cost the company to produce 0.4 liter of the engineered essence?

56. *Longest Train Run* The world's longest train run is on the Trans-Siberian line in Russia, from Moscow to Nakhodka on the Sea of Japan. The length of the run, which makes 97 stops, measures 94,380,000 centimeters. The run takes 8 days, 4 hours, and 25 minutes.
(a) How many meters is the run?
(b) How many kilometers is the run?

57. *Highest Railroad* The highest railroad line in the world is a track on the Morococha branch of the Peruvian state railways at La Cima. The track is 4818 meters high.
(a) How many centimeters high is the track?
(b) How many kilometers high is the track?

58. *Dam Height* A dam is 335 meters high.
(a) How many kilometers high is the dam?
(b) How many centimeters high is the dam?

Mixed Practice

59. 5280 feet = _____ yards

60. 8 pints = _____ quarts

61. 4 miles = _____ feet

62. 12 feet = _____ inches

63. 9 tons = _____ pounds

64. 18 hours = _____ seconds

65. 14.6 kg = _____ g

66. 0.83 kg = _____ mg

67. 3.22 kL = _____ L

68. 3607 mL = _____ L = _____ kL

69. 7183 mg = _____ g = _____ kg

To Think About
Metric prefixes are also used for computers. A **byte** *is the amount of computer memory needed to store one alphanumeric character. In reference to computers you may hear the following words: kilobytes, megabytes, and gigabytes. Use the following chart to convert the measurements in exercises 70–73.*

70. 1.2 gigabytes = _____ bytes

71. 528 megabytes = _____ bytes

72. 78.9 kilobytes = _____ bytes

73. 24.9 gigabytes = _____ bytes

1 gigabyte (GB) = one billion bytes = 1,000,000,000 bytes
1 megabyte (MB) = one million bytes[*] = 1,000,000 bytes
1 kilobyte (KB) or K = one thousand bytes[†] = 1000 bytes

*Sometimes in computer science 1 megabyte is considered to be 1,048,576 bytes.
†Sometimes in computer science 1 kilobyte is considered to be 1024 bytes.

Cumulative Review

74. **[8.7.1]** 14 out of 70 is what percent?

75. **[8.7.1]** What is 23% of 250?

76. **[8.7.1]** What is 1.7% of $18,900?

77. **[8.7.1]** *Furniture Sold* A salesperson earns a commission of 8%. She sold furniture worth $8960. How much commission will she earn?

Quick Quiz 10.1

1. Mount Whitney is 14,496 feet high. How many miles is that? Round your answer to the nearest hundredth.

2. Convert 500 milliliters to meters.

3. Convert 9 kilograms to milligrams.

4. **Concept Check** Explain how you would convert 240 ounces to pounds.

10.2 Converting Between the U.S. and Metric Systems

Student Learning Objectives

After studying this section, you will be able to:

① Convert units of length, volume, and weight between the metric and U.S. systems.

② Convert between Fahrenheit and Celsius degrees of temperature.

① **Converting Units of Length, Volume, and Weight Between the Metric and U.S. Systems**

So far we've seen how to convert units when working *within* either the U.S. or the metric system. Many people, however, work with *both* the metric and U.S. systems. If you study such fields as chemistry, electromechanical technology, business, X-ray technology, nursing, or computers, you will probably need to convert measurements between the two systems. We learn that skill in this section.

To convert between U.S. and metric units, it is necessary to know equivalent values. The most commonly used equivalents are listed below. Most of these equivalents are approximate, denoted by $\approx$.

Equivalent Measures

	U.S. to Metric	Metric to U.S.
Units of length	1 mile $\approx$ 1.61 kilometers	1 kilometer $\approx$ 0.62 mile
	1 yard $\approx$ 0.914 meter	1 meter $\approx$ 1.09 yards
	1 foot $\approx$ 0.305 meter	1 meter $\approx$ 3.28 feet
	1 inch = 2.54 centimeters*	1 centimeter $\approx$ 0.394 inch
Units of volume	1 gallon $\approx$ 3.79 liters	1 liter $\approx$ 0.264 gallon
	1 quart $\approx$ 0.946 liter	1 liter $\approx$ 1.06 quarts
Units of weight	1 pound $\approx$ 0.454 kilogram	1 kilogram $\approx$ 2.2 pounds
	1 ounce $\approx$ 28.35 grams	1 gram $\approx$ 0.0353 ounce

*Exact value

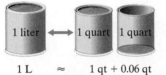

1 L $\approx$ 1 qt + 0.06 qt

Remember that to convert from one unit to another, we multiply by a unit fraction that is equivalent to 1. We create a fraction from the equivalent measures table so that the unit in the numerator is the unit we want to end up with. To change 5 miles to kilometers, we look in the table and find that 1 mile $\approx$ 1.61 kilometers.

We will use the fraction $\dfrac{1.61 \text{ km}}{1 \text{ mi}}$ because we want to have 1.61 kilometers in the numerator.

$$5 \text{ mi} \times \frac{1.61 \text{ km}}{1 \text{ mi}} = 5 \times 1.61 \text{ km} = 8.05 \text{ km}$$

Thus 5 miles is approximately 8.05 kilometers.

EXAMPLE 1

(a) Convert 26 m to yd. **(b)** Convert 1.9 km to mi. **(c)** Convert 14 gal to L.
(d) Convert 2.5 L to qt. **(e)** Convert 5.6 lb to kg. **(f)** Convert 152 g to oz.

Solution

(a) $26 \text{ m} \times \dfrac{1.09 \text{ yd}}{1 \text{ m}} = 28.34 \text{ yd}$ **(b)** $1.9 \text{ km} \times \dfrac{0.62 \text{ mi}}{1 \text{ km}} = 1.178 \text{ mi}$

(c) $14 \text{ gal} \times \dfrac{3.79 \text{ L}}{1 \text{ gal}} = 53.06 \text{ L}$ **(d)** $2.5 \text{ L} \times \dfrac{1.06 \text{ qt}}{1 \text{ L}} = 2.65 \text{ qt}$

(e) $5.6 \text{ lb} \times \dfrac{0.454 \text{ kg}}{1 \text{ lb}} = 2.5424 \text{ kg}$ **(f)** $152 \text{ g} \times \dfrac{0.0353 \text{ oz}}{1 \text{ g}} = 5.3656 \text{ oz}$

Student Practice 1

(a) Convert 17 m to yd.   **(b)** Convert 29.6 km to mi.
(c) Convert 26 gal to L.   **(d)** Convert 6.2 L to qt.
(e) Convert 16 lb to kg.   **(f)** Convert 280 g to oz.

NOTE TO STUDENT: Fully worked-out solutions to all of the Student Practice problems can be found at the back of the text starting at page SP-1.

Although the calculations in Example 1 show "=", keep in mind that most conversions are approximations and have been rounded.

Some conversions require more than one step.

EXAMPLE 2   Convert 235 cm to feet. Round your answer to the nearest hundredth of a foot.

Solution   Our first unit fraction converts centimeters to inches. Our second unit fraction converts inches to feet.

$$235 \text{ cm} \times \frac{0.394 \text{ in.}}{1 \text{ cm}} \quad \text{We convert to inches.}$$

$$235 \times \frac{0.394 \text{ in.}}{1} \times \frac{1 \text{ ft}}{12 \text{ in.}} = \frac{92.59}{12} \text{ ft} = 7.7158\overline{3} \text{ ft}$$

Rounded to the nearest hundredth we have 7.72 ft.

Student Practice 2   Convert 180 cm to feet.

The same rules can be followed to convert a rate such as 100 kilometers per hour to miles per hour.

EXAMPLE 3   Convert 100 km/hr to mi/hr.

Solution   We multiply by the unit fraction that relates mi to km.

$$\frac{100 \text{ km}}{\text{hr}} \times \frac{0.62 \text{ mi}}{1 \text{ km}} = 62 \text{ mi/hr}$$

Thus 100 km/hr is approximately equal to 62 mi/hr.

Student Practice 3   Convert 88 km/hr to mi/hr.

ᴹᴄ **EXAMPLE 4**   A camera film that is 35 mm wide is how many inches wide?

Solution   We first convert from millimeters to centimeters by moving the decimal point in the number 35 one place to the left.

$$35 \text{ mm} = 3.5 \text{ cm}$$

Then we convert to inches using a unit fraction.

$$3.5 \text{ cm} \times \frac{0.394 \text{ in.}}{1 \text{ cm}} = 1.379 \text{ in.}$$

Student Practice 4 The city police use 9-mm automatic pistols. If such a pistol fires a bullet 9 mm wide, how many inches wide is the bullet? (Round to the nearest hundredth.)

◗ Understanding the Concept

Changing the Area of a Rectangle from U.S. to Metric Units   Suppose we consider a rectangle that measures 2 yd wide by 4 yd long. The area would be 2 yd × 4 yd = 8 yd². How could you change 8 yd² to m²? Suppose that we look at 1 yd². Each side is 1 yd long, which is approximately 0.914 m.

$$\text{Area} = 1 \text{ yd} \times 1 \text{ yd} \approx 0.914 \text{ m} \times 0.914 \text{ m}$$
$$\text{Area} = 1 \text{ yd}^2 \approx 0.8354 \text{ m}^2 \text{ (rounded to the ten-thousandths place)}$$

Thus 1 yd² ≈ 0.8354 m². Therefore, we change 8 yd² to m² as follows:

$$8 \text{ yd}^2 \times \frac{0.8354 \text{ m}^2}{1 \text{ yd}^2} = 6.6832 \text{ m}^2$$

8 yd² ≈ 6.6832 m². Thus, 8 square yards is approximately 6.6832 square meters.

Fahrenheit   Celsius

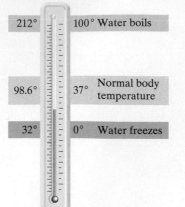

212°	100°	Water boils
98.6°	37°	Normal body temperature
32°	0°	Water freezes

② Converting Between Fahrenheit and Celsius Degrees of Temperature

In the metric system, temperature is measured on the **Celsius scale.** Water boils at 100° (100°C) and freezes at 0° (0°C) on the Celsius scale. In the **Fahrenheit system,** water boils at 212° (212°F) and freezes at 32° (32°F).

To convert a Celsius temperature to Fahrenheit, we can use the formula

$$F = 1.8 \times C + 32$$

and to convert a Fahrenheit temperature to Celsius, we can use the formula

$$C = \frac{5 \times F - 160}{9}$$

where F is the number of Fahrenheit degrees and C is the number of Celsius degrees.

EXAMPLE 5   Convert 176°F to Celsius temperature.

Solution   We use the formula that gives us Celsius degrees.

$$C = \frac{5 \times F - 160}{9}$$

$$= \frac{5 \times 176 - 160}{9} \qquad \text{We multiply, then subtract.}$$

$$= \frac{880 - 160}{9} = \frac{720}{9} = 80$$

The temperature is 80°C.

Student Practice 5   Convert 181°F to Celsius temperature. Round to the nearest degree.

Calculator

 Converting Temperature

You can use your calculator to convert temperature readings between Fahrenheit and Celsius. To convert 30°C to Fahrenheit temperature, enter:

1.8 ☒ 30 ➕ 32 ☲

The display reads:

86

The temperature is 86°F. To convert 82.4°F to Celsius temperature, enter:

5 ☒ 82.4 − 160

☲ ➗ 9 ☲

The display reads:

28

The temperature is 28°C.

EXAMPLE 6   Hester is planning a visit from his home in Rhode Island to Brazil. He checks the weather report for the part of Brazil where he will visit and finds that the temperature during the day is 37°C. If the temperature in Rhode Island is currently 87°F, what is the difference between the higher and lower temperatures in degrees Fahrenheit?

Solution   We want to convert the Celsius temperature to Fahrenheit so we use the following formula.

$$F = 1.8 \times C + 32$$
$$= 1.8 \times 37 + 32 \quad \text{We replace } C \text{ with the Celsius temperature.}$$
$$= 66.6 + 32 \qquad \text{We multiply before we add.}$$
$$= 98.6$$

It is 98.6°F in Brazil.

Now we find the difference in Fahrenheit temperatures.

$$98.6° - 87° = 11.6°F$$

Student Practice 6   On a cold winter day in England, Erin notices that the temperature reads 4°C. She calls home in Los Angeles, California, and finds out that the temperature is 79°F. What is the difference between the higher and lower temperatures in degrees Fahrenheit?

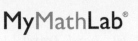 
Perform each conversion. Round your answer to the nearest hundredth when necessary.

1. 4 ft to m
2. 11 ft to m
3. 14 m to yd
4. 18 m to yd

5. 15 km to mi
6. 12 km to mi
7. 24 yd to m
8. 31 yd to m

9. 82 mi to km
10. 68 mi to km
11. 25 m to ft
12. 35 m to ft

13. 17.5 cm to in.
14. 15.2 cm to in.
15. 5 gal to L
16. 7 gal to L

17. 4.5 L to qt
18. 6.5 L to qt
19. 7 oz to g
20. 9 oz to g

21. 11 kg to lb
22. 14 kg to lb
23. 126 g to oz
24. 186 g to oz

The following problems involve a double conversion.

25. 4 kg to oz
26. 6 kg to oz

27. 230 cm to ft
28. 142 cm to ft

29. 16.5 ft to cm
30. 19.5 ft to cm

Perform each conversion. Round your answer to the nearest tenth when necessary.

31. 50 km/hr to mi/hr
32. 60 km/hr to mi/hr

33. 45 mi/hr to km/hr
34. 40 mi/hr to km/hr

Perform each conversion. Round your answer to the nearest hundredth when necessary.

35. A wire that is 13 mm wide is how many inches wide?
36. A bolt that is 7 mm wide is how many inches wide?

37. 0°C to Fahrenheit
38. 40°C to Fahrenheit

39. 85°C to Fahrenheit
40. 100°C to Fahrenheit

41. 168°F to Celsius
42. 110°F to Celsius

43. 86°F to Celsius
44. 35°F to Celsius

Mixed Practice *Perform each conversion. Round your answer to the nearest hundredth when necessary.*

45. 9 in. to cm
46. 12 in. to cm
47. 32.2 m to yd

48. 29.3 m to yd
49. 19 L to gal
50. 10 L to gal

51. 32 lb to kg
52. 27 lb to kg
53. 12°C to Fahrenheit

54. 21°C to Fahrenheit
55. 68°F to Celsius
56. 131°F to Celsius

Applications *Solve. Round your answer to the nearest hundredth when necessary.*

57. *Male's Weight* One of the heaviest human males documented in medical records weighed 635 kg in 1978. What would have been his weight in pounds?

58. *Child's Weight* The average weight for a 7-year-old girl is 22.2 kilograms. What is the average weight in pounds?

59. *Boat Travel* Mr. and Mrs. Weston have traveled 67 miles on a boat cruise from Seattle, Washington, to Victoria Island, Vancouver, B.C., Canada. They have 36 kilometers until their rendezvous point with another boat. How many kilometers in total will they have traveled?

60. *Land Travel* Marcia is traveling from Ixtapa to a beach in Zihuatenejo in Mexico. The odometer on her American car shows that the first part of her trip was 4 miles. Then she sees a sign posted: "Zihuatenejo 14 KILOMETERS." How many kilometers in total will she have traveled when she arrives at her destination?

61. *Gasoline Use* Pierre had a Jeep imported into France. During a trip from Paris to Lyon, he used 38 liters of gas. The tank, which he filled before starting the trip, holds 15 gallons of gas. How many liters of gas were left in the tank when he arrived in Lyon?

62. *Surgical Procedure* A surgeon is irrigating an abdominal cavity after a cancerous growth is removed. There is a supply of 3 gallons of distilled water in the operating room. The surgeon uses a total of 7 liters of the water during the procedure. How many liters of water are left after the operation?

63. *Temperature Comparison* Hillina is vacationing in Spain, where the temperature during the day is 32°C. She is planning her trip home, where the temperature currently is 80°F. What is the difference between the higher and lower temperatures in degrees Fahrenheit?

64. *Temperature Comparison* Jessica lives in Los Angeles, California, and is planning a trip to Germany. The temperature in the part of Germany where she will visit is 25°C. If it is currently 85°F in Los Angeles, what is the difference between the higher and lower temperatures in degrees Fahrenheit?

65. *Temperature Comparison* In central Australia at 4 o'clock in the morning, the temperature is 19°C. After 7 o'clock in the morning, the temperature can reach 45°C. What would be equivalent Fahrenheit temperatures at 4 o'clock and 7 o'clock in the morning?

66. *Roasting a Turkey* A holiday turkey in Buenos Aires, Argentina, was roasted at 200°C for 4 hours, (20 minutes per pound). What would have been the cooking temperature in Fahrenheit in Joplin, Missouri?

Round your answer to the nearest thousandth.

67. *Lead Poisoning* A pathologist found 0.768 oz of lead in the liver of a child who had died of lead poisoning. How many grams of toxic lead were in the child's liver?

68. *Gold Nugget Weight* While panning in a river in the Yukon, in Alaska, a prospector found a gold nugget that weighed 2.552 oz. How many grams did the nugget weigh?

Cumulative Review *Perform the operations indicated in the correct order.*

69. **[1.6.4]** $2^3 \times 6 - 4 + 3$

70. **[1.6.4]** $5 + 2 - 3 + 5 \times 3^2$

71. **[1.6.4]** $2^2 + 3^2 + 4^3 + 2 \times 7$

72. **[1.6.4]** $5^2 + 4^2 + 3^2 + 3 \times 8$

Quick Quiz 10.2

1. The Australian copperhead is a highly venomous snake growing to as long as 1.8 meters. How many feet is 1.8 meters? Round your answer to the nearest whole number.

2. The anaconda snake lives in the rain forests and river systems of the Amazon. This snake's average weight is 149 kg. How many pounds is this? Round your answer to the nearest whole number.

3. The center hole of a CD is 1.5 centimeters across. How many inches is this? Round your answer to the nearest hundredth.

4. **Concept Check** Explain how you would convert 50 km/hr to mi/hr.

10.3 Angles

The word **geometry** comes from the Greek words for *measure* and *earth*. This is because geometry was originally used to measure land. Today we use geometry in many fields such as physics, drafting, art, and medicine. As you have seen in previous chapters, we often use geometry in our everyday lives to measure the size of a room or to find the perimeter of a space.

In this section we will review some of the definitions and formulas studied earlier in the book and also introduce some new ones. It is important that you learn all these new terms and definitions so that you can begin working with figures and determining the relationship between figures in space.

① Understanding Angles

In geometry we symbolize a **point** by drawing a dot (•). A **line** (⟷) is one of the simplest figures and extends indefinitely. A portion of a line is called a **line segment** (•—•) and has a beginning and an end. A **ray** (•—→) starts at a point and extends indefinitely in one direction. An **angle** (∠) is formed by two rays with the same endpoint. The two rays are called the **sides** of the angle, and the point at which they meet is called the **vertex.**

We can measure the amount of the opening of an angle with **degrees.** We use the symbol ° to indicate degrees. In the figure located in the margin, the opening measures 75 degrees, or 75°. Now, if we fix one side of the angle and keep moving the other side, the angle measure will get larger and larger, forming angles that measure 90° (**right angle**), 180°, and 360°.

One-fourth revolution is 90°. We say two lines that meet at a 90° angle are **perpendicular.**

One-half revolution is 180°.

One complete revolution is 360°.

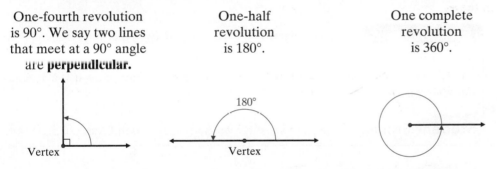

Note that we use a small □ at the vertex to indicate that an angle is 90° (a right angle). We can name an angle in different ways. For example, in the figure below we can name the angle using the letter between the rays ∠*x*; the vertex ∠*M*; or the sides and vertex ∠*LMN*, ∠*NML*.

$$\angle x, \quad \angle M, \quad \angle LMN, \quad \angle NML$$

The middle letter is the vertex.

Sometimes more than two rays meet at a point. In this case, we *do not* use the vertex alone to name the angle since this would be confusing. That is, we would not know which of the three angles was being named by the vertex. In the figure below ∠*x* = ∠*ABD*, ∠*y* = ∠*DBC*, and the largest angle is ∠*ABC*.

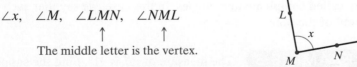

We do not use the vertex alone to name these angles.

We often encounter three types of angles. An angle that measures 180° is called a **straight angle.** An angle whose measure is between 0° and 90° is called an **acute**

Student Learning Objectives

After studying this section, you will be able to:

① Understand angles.

② Find the complement and supplement of an angle.

③ Identify adjacent and vertical angles.

④ Identify and use alternate interior and corresponding angles.

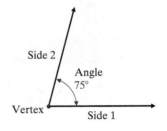

581

angle, while an angle whose measure is between 90° and 180° is called an **obtuse angle.**

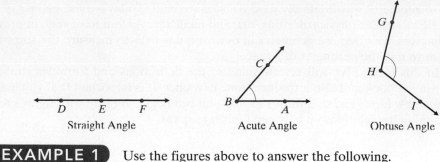

Straight Angle   Acute Angle   Obtuse Angle

EXAMPLE 1   Use the figures above to answer the following.

(a) State the measure of angle *DEF*.

(b) State the three ways we can name the obtuse angle.

Solution

(a) Since $\angle DEF$ is a straight angle, it measures 180°.

(b) We can name the obtuse angle as $\angle GHI$, $\angle IHG$, or $\angle H$.
Be sure the letter representing the vertex is the middle letter.

Student Practice 1 Use the figures above to answer the following.

(a) Using a single letter, name the angle whose measure is between 90° and 180°.

(b) State the three ways we can name the acute angle.

EXAMPLE 2   Using the figure in the margin, state two other ways to name $\angle m$.

Solution   We can name $\angle m$ as $\angle YXZ$ or $\angle ZXY$.

Note that we cannot name $\angle m$ as $\angle X$ since we would not know if $\angle X$ is naming $\angle m$, $\angle n$, or $\angle WXZ$.

Student Practice 2   State two other ways to name $\angle n$ in Example 2.

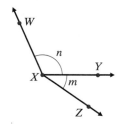

$\angle x + \angle y = 180°$

$\angle x$ and $\angle y$ are supplementary angles.

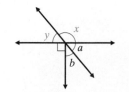

$\angle a + \angle b = 90°$

$\angle a$ and $\angle b$ are complementary angles.

② Finding the Complement and Supplement of an Angle

Recall that two angles that have a sum of 180° are called **supplementary angles.** We say that each angle is the supplement of the other. Two angles that have a sum of 90° are called **complementary angles.** In this case, we say that each angle is the complement of the other.

EXAMPLE 3   The measure of $\angle J$ is 31°. Find the supplement of $\angle J$.

Solution   If we let $\angle S$ = the supplement of $\angle J$, then we have the following.

| The supplement of $\angle J$ plus | $\angle J$ | = 180° | First, we write a statement to represent the situation. |

$$\downarrow \qquad \downarrow \quad \downarrow \qquad \downarrow$$

$$\angle S \ + \ \angle J = 180° \quad \text{Next, we translate to symbols.}$$

$$\angle S \ + \ 31° = 180° \quad \text{Then we replace } \angle J \text{ with } 31°.$$

$$\underline{+ \ - \ 31° \quad -31°}$$

$$\angle S \ = 149° \quad \text{Finally, we solve for } \angle S.$$

The supplement of $\angle J$ measures 149°.

Student Practice 3 The measure of $\angle B$ is 22°. Find the complement of $\angle B$.

EXAMPLE 4   Given that $\angle ABC$ is a right angle, find x.

Solution   A right angle has a measure of 90°. Therefore $\angle ABD$ and $\angle DBC$ are complementary, and the sum of their measures is 90°.

$$x + (x + 12°) = 90°$$ We must solve for x.
$$(x + x) + 12° = 90°$$ We regroup terms.
$$2x + 12° = 90°$$ $x + x = 2x$
$$\underline{+ \quad -12° \quad -12°}$$ We add -12 to both sides.
$$2x \qquad = 78°$$

$$\frac{2x}{2} = \frac{78°}{2}$$ We divide by 2 on both sides.

$$x = 39°$$

Student Practice 4   Given that $\angle XYZ$ is a right angle, find y.

Student Practice 4

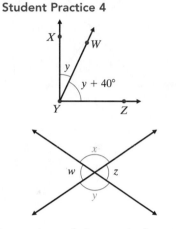

---

③ **Identifying Adjacent and Vertical Angles**

When two lines intersect, four angles are formed. Two angles that are opposite each other are called **vertical angles.** Vertical angles have the same measure. In the sketch in the margin, $\angle x$ and $\angle y$ are vertical angles, so they have the same measure. Likewise, $\angle w$ and $\angle z$ are also vertical angles since they are opposite each other. They also have the same measure.

Two angles that share a *common side* are called **adjacent angles.** For example, in the sketch in the margin, $\angle w$ and $\angle x$ are adjacent angles. Adjacent angles of intersecting lines are *supplementary*. Therefore if we know that the measure of $\angle x$ is 100°, then we also know that the measure of $\angle w$ is 80°.

$\angle x = \angle y$ and $\angle w = \angle z$ because they are vertical angles.

EXAMPLE 5

(a) Name the angles that are adjacent to $\angle a$.
(b) Find the sum of the measures of $\angle c$ and $\angle d$.

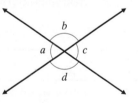

Solution

(a) Since $\angle b$ and $\angle d$ each have a side in common with $\angle a$, they are both adjacent to $\angle a$.

(b) $\angle c$ and $\angle d$ are adjacent angles of intersecting lines. Therefore, they are supplementary angles, and the sum of their measures is 180°.

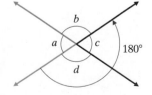

Student Practice 5

(a) Name the angles that are adjacent to $\angle y$.
(b) Find the sum of the measures of $\angle w$ and $\angle z$.

NOTE TO STUDENT: *Fully worked-out solutions to all of the Student Practice problems can be found at the back of the text starting at page SP-1.*

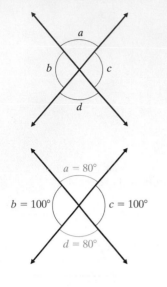

EXAMPLE 6 If $\angle a = 80°$, find the measure of each of the following.

(a) $\angle b$ **(b)** $\angle c$ **(c)** $\angle d$

Solution

(a) Since $\angle a$ and $\angle b$ are adjacent angles of intersecting lines, we know that they are supplementary angles. Thus we know that the sum of their measures is 180°. That is, $\angle a + \angle b = 180°$.

$$\angle a + \angle b = 180°$$
$$80° + \angle b = 180°$$
$$\underline{+ \; -80° \qquad\qquad -80°}$$
$$\angle b = 100°$$

(b) Since $\angle c$ and $\angle b$ are vertical angles, we know that they have the same measure. Thus we know that $\angle c$ measures 100°.

(c) $\angle d = 80°$ since $\angle a$ and $\angle d$ are vertical angles.

Student Practice 6 Refer to the figure in Example 6 to answer the following. If $\angle a = 50°$, find the measure of each of the following.

(a) $\angle d$ **(b)** $\angle c$ **(c)** $\angle b$

---

④ **Identifying and Using Alternate Interior and Corresponding Angles**

In the figure below, line p and line q are called *parallel lines*. **Parallel lines** never meet, and the distance between them is always the same. The symbol $\parallel$ means "is parallel to." For example, in the figure, $p\parallel q$.

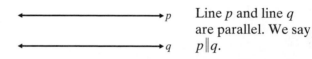

Line p and line q are parallel. We say $p\parallel q$.

A line that intersects two or more lines at different points is called a **transversal.** In the following figure, line h is a transversal that intersects line p and line q.

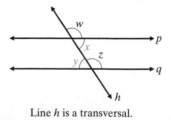

Line h is a transversal.

Alternate interior angles are two angles that are on *opposite sides* of the transversal and *between* the other two lines. In the figure, $\angle x$ and $\angle y$ are alternate interior angles. **Corresponding angles** are two angles that are on the *same side* of the transversal and are *both above* (or below) the other two lines. In the figure above, $\angle z$ and $\angle w$ are corresponding angles.

When two lines cut by a transversal are parallel, the following is true.

PARALLEL LINES CUT BY A TRANSVERSAL

If two parallel lines are cut by a transversal, then the measures of *corresponding angles are equal* and the measures of *alternate interior angles are equal.*

EXAMPLE 7   In the following figure, $a \| b$, and the measure of $\angle z$ is 56°. Find the measures of $\angle y$, $\angle x$, $\angle w$, and $\angle v$.

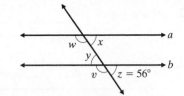

Solution   Refer to the figure in the margin.

$\angle z = \angle y = 56°$ The vertical angles z and y have the same measure.

$\angle y = \angle x = 56°$ The alternate interior angles y and x have the same measure.

To find the measure of $\angle w$, we list some facts about $\angle w$. $\angle w$ is the supplement of $\angle x$ and $\angle x = 56°$.

$\angle w + \angle x = 180°$ The sum of supplementary angles is 180°.

$\angle w + 56° = 180°$ We substitute 56° for $\angle x$.

$$\underline{+\ -56°\quad -56°}$$

$\angle w = 124°$ We solve for $\angle w$.

$\angle w = \angle v = 124°$ The corresponding angles w and v have the same measure.

Student Practice 7   In the figure in the margin, $m \| n$, and the measure of $\angle a$ is 63°. Find the measures of $\angle b$, $\angle c$, $\angle e$, and $\angle d$.

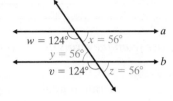

Student Practice 7

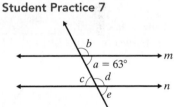

STEPS TO SUCCESS Learning Formulas and Definitions

How can I memorize all these formulas? What about all the rules and definitions I have to learn?

You should study formulas, rules, and definitions several times a week. This will allow you to rotate through the learning cycle several times, and help you learn the required information.

The Learning Cycle

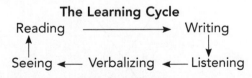

Making it personal: *Place a* ✓ *in the blank as you complete each of the following.*

1. Make 3 by 5 cards as follows.

_____ Write the name of the *new term*, or *rule*, on the front of the card. Then write the *definition* of the term or the rule on the back.

_____ Write the name and draw the picture of each *geometric shape* on the front of the card. On the back of the card write all the *formulas* that you must know for this particular geometric figure.

_____ Write *sample examples* on the front of the card and *solutions* on the back of the card.

2. Study using these cards by completing the following.

_____ Several times a week, state the definitions, rules, and formulas aloud, or read them to another student.

_____ Periodically use all these cards as flash cards and quiz yourself.

Verbal and Writing Skills, Exercises 1–6 *In your own words, give the definition of each term.*

1. Vertex: _____

_____

2. Acute angle: _____

_____

3. Obtuse angle: _____

_____

4. Complementary angles: _____

_____

5. Adjacent angles: _____

_____

_____

6. Corresponding angles: _____

_____

_____

Choose from the terms below to fill in the blanks for exercises 7–34. Each of the terms will be used at least once.

vertex	supplementary angles	corresponding angles	$\angle a$	$\angle e$
a straight angle	complementary angles	alternate interior angles	$\angle b$	$\angle f$
an obtuse angle	vertical angles	transversal	$\angle c$	$180°$
an acute angle			$\angle d$	

7. Line *u* is called a _____ since it intersects two or more lines at different points.

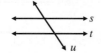

8. *C* is a point at which two rays meet and is called the _____.

Use the figure to the right to answer exercises 9–12.

9. $\angle ABC$ is called _____.

10. $\angle CBD$ is called _____.

11. $\angle ABD$ is called _____.

12. *B* is called the _____ of $\angle DBC$.

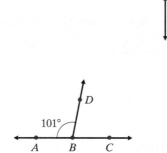

13. If the sum of two angles is 180°, these angles are called _____.

14. If the sum of two angles is 90°, these two angles are called _____.

15. An angle whose measure is between 0° and 90° is called _____.

16. An angle whose measure is between 90° and 180° is called _____.

Use the figure to the right to answer exercises 17–24.

17. $\angle a$ and $\angle c$ are called _____.

18. $\angle d$ and $\angle b$ are called _____.

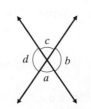

19. Name the angles that are adjacent to $\angle c$: _____ and _____.

20. Name the angles that are adjacent to $\angle b$: _____ and _____.

21. _____ is equal to $\angle d$.

22. _____ is equal to $\angle a$.

23. $\angle a + \angle b =$ _____.

24. $\angle c + \angle d =$ _____.

In the figure to the right, s∥t. Use this figure to answer exercises 25–34.

25. ∠a and ∠d are called _____.

26. ∠b and ∠c are called _____.

27. ∠b is equal to the following two angles: _____ and _____.

28. ∠a is equal to the following two angles: _____ and _____.

29. ∠f is equal to the following two angles: _____ and _____.

30. ∠e is equal to the following two angles: _____ and _____.

31. ∠e and ∠d are called _____.

32. ∠f and ∠c are called _____.

33. ∠b and ∠f are called _____.

34. ∠a and ∠e are called _____.

Use the figure to the right to answer exercises 35 and 36.

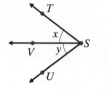

35. State two other ways to name ∠x.

36. State two other ways to name ∠y.

37. ∠M measures 29°.
  (a) Find the supplement of ∠M.
  (b) Find the complement of ∠M.

38. ∠S measures 62°.
  (a) Find the supplement of ∠S.
  (b) Find the complement of ∠S.

39. ∠X measures 45°.
  (a) Find the supplement of ∠X.
  (b) Find the complement of ∠X.

40. ∠Y measures 75°.
  (a) Find the supplement of ∠Y.
  (b) Find the complement of ∠Y.

41. Given that ∠EFG is a right angle, find x.

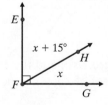

42. Given that ∠ABC is a right angle, find y.

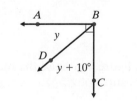

43. Given that ∠STU is a right angle, find n.

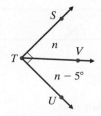

44. Given that ∠WXY is a right angle, find a.

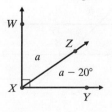

45. Given that ∠EFG is a right angle, find m.

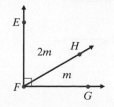

46. Given that ∠LMN is a right angle, find b.

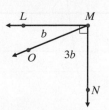

Use the figure below to answer exercises 47–54.

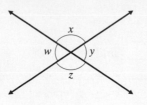

47. Name the angles that are adjacent to $\angle z$.

48. Name the angles that are adjacent to $\angle y$.

49. Find the sum of the measures of $\angle w$ and $\angle z$.

50. Find the sum of the measures of $\angle z$ and $\angle y$.

51. If $\angle x = 70°$, find $\angle z$, $\angle w$, and $\angle y$.

52. If $\angle z = 54°$, find $\angle x$, $\angle w$, and $\angle y$.

53. If $\angle w = 45°$, find $\angle y$, $\angle x$, and $\angle z$.

54. If $\angle y = 50°$, find $\angle w$, $\angle x$, and $\angle z$.

55. In the following figure, $x \| y$ and the measure of $\angle p$ is 43°. Find the measures of $\angle o$, $\angle n$, $\angle l$, and $\angle m$.

56. In the following figure, $x \| y$ and the measure of $\angle w$ is 55°. Find the measures of $\angle v$, $\angle u$, $\angle s$, and $\angle t$.

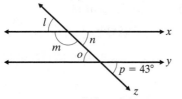

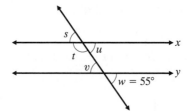

57. In the following figure, $m \| n$ and the measure of $\angle a$ is 45°. Find the measures of $\angle b$, $\angle c$, $\angle e$, and $\angle d$.

58. In the following figure, $m \| n$ and the measure of $\angle e$ is 38°. Find the measures of $\angle f$, $\angle g$, $\angle h$, and $\angle i$.

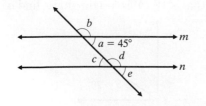

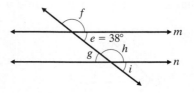

One Step Further *Find the measures of $\angle r$, $\angle s$, $\angle t$, $\angle u$, $\angle v$, $\angle w$, and $\angle x$ given that $p \| q$.*

59.

60.

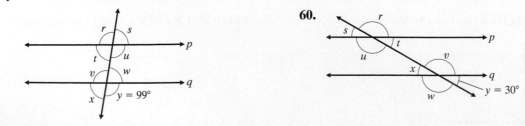

Find x.

61.

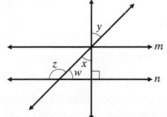

2x + 2   5x − 10

62.

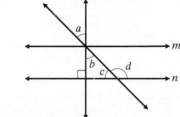

2x + 5   3x − 1

To Think About *Recall that the sum of the measures of the interior angles of a triangle equals 180°. Use this information to answer exercises 63 and 64.*

63. Given that $m \parallel n$ and $\angle y = 33°$, find the measures of $\angle x$, $\angle w$, and $\angle z$.

64. Given that $m \parallel n$ and $\angle a = 35°$, find the measures of $\angle b$, $\angle c$, and $\angle d$.

Cumulative Review

▲ **65.** **[1.9.3]** *Cross-Stitch Design* Jessica is making a cross-stitch design that is 70 squares wide and 84 squares high. There are 14 squares per inch on her material. If she needs 2 inches extra on each side of the stitchery to frame it, what are the dimensions of the material she needs?

▲ **66.** **[1.9.3]** *Earth-Sun Diameter* The approximate size of Earth's diameter is the sun's diameter divided by 109. If the sun's diameter is about 1,391,193 km, what is the diameter of Earth? Round your answer to the nearest ten.

Quick Quiz 10.3

1. $\angle A$ measures 58°.

    (a) Find the complement of $\angle A$.

    (b) Find the supplement of $\angle A$.

2. If the measure of $\angle c$ is 65°, find the measure of

    (a) $\angle a$

    (b) $\angle b$

a   b $c = 65°$ d

3. If $\angle PQR$ is a right triangle, find x.

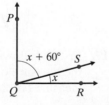

4. **Concept Check** In the figure shown for Quick Quiz 2, explain what the relationship is between angle a and angle d. If you know the measure of angle a, how can you find angle d?

1. _____

2. _____

3. _____

4. _____

5. _____

6. _____

7. _____

8. _____

9. _____

10. _____

11. _____

12. _____

13. _____

14. (a) _____

(b) _____

15. _____

16. _____

17. _____

How are you doing with your homework assignments in Sections 10.1 to 10.3? Do you feel you have mastered the material so far? Do you understand the concepts you have covered? Before you go further in the textbook, take some time to do each of the following problems.

10.1

Convert.

1. 21 ft = _____ yd

2. 6 gal = _____ qt

3. 240 min = _____ hr

4. 6.3 kg = _____ mg

Fill in the blanks with the correct values.

5. 34 mL = _____ L = _____ kL

6. Natasha must buy 63 feet of fabric to make curtains for her room. If the fabric costs $4.00 per yard, how much will it cost Natasha to buy the fabric for the curtains?

10.2

Convert. Round your answer to the nearest hundredth if necessary.

7. 10 in. to cm

8. 22 lb to kg

9. 3.5 L to qt

10. 95 km/hr to mi/hr

11. Phuong Le is planning to visit friends in Canada. He checked the weather report in Canada, and it stated that the temperature during the day is 35°C. If the temperature in his home state of Idaho is 89°F, what is the difference between the higher and lower temperatures in degrees Fahrenheit?

10.3

Fill in the blanks.

12. $\angle b$ and $\angle c$ are vertical angles. If $\angle b = 45°$, then $\angle c =$ _____.

13. $\angle x$ and $\angle y$ are alternate interior angles. If $\angle x = 32°$, then $\angle y =$ _____.

14. $\angle D$ measures 39°.

(a) Find the supplement of $\angle D$.   (b) Find the complement of $\angle D$.

15. In the figure on the right, $\angle ABC$ is a right angle. Find x.

16. If $\angle x = 55°$, find $\angle z, \angle w,$ and $\angle y$.

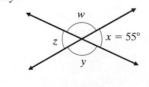

17. In the following figure, $a \parallel b$ and the measure of $\angle p$ is 40°. Find the measures of $\angle o, \angle n, \angle l,$ and $\angle m$.

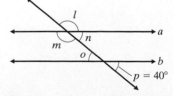

Now turn to page SA-19 for the answers to each of these problems. Each answer also includes a reference to the objective in which the problem is first taught. If you missed any of these problems, you should stop and review the Examples and Student Practice problems in the referenced objective. A little review now will help you master the material in the upcoming sections of the text.

10.4 Square Roots and the Pythagorean Theorem

① Understanding the Meaning of Square Roots

We begin our discussion of square roots by introducing some new vocabulary and mathematical symbols. When a whole number or fraction is multiplied by itself (squared), the number obtained is called a **perfect square.** The number 9 is a perfect square because 3 squared equals 9.

$$3^2 = 3 \cdot 3 = 9$$

The numbers 15 and 13 are *not* perfect squares. There is no whole number or fraction that when squared yields 15 or 13.

EXAMPLE 1 Determine if each number is a perfect square.

(a) 30 (b) 64 (c) $\dfrac{1}{4}$

Solution

(a) 30 is not a perfect square. There is no whole number or fraction that when squared equals 30.

(b) 64 is a perfect square because $8^2 = 8 \cdot 8 = 64$.

(c) $\dfrac{1}{4}$ is a perfect square because $\left(\dfrac{1}{2}\right)^2 = \dfrac{1}{2} \cdot \dfrac{1}{2} = \dfrac{1}{4}$.

Student Practice 1 Determine if each number is a perfect square.

(a) 81 (b) 17 (c) $\dfrac{1}{9}$

It is helpful to know the first 15 perfect squares for whole numbers.

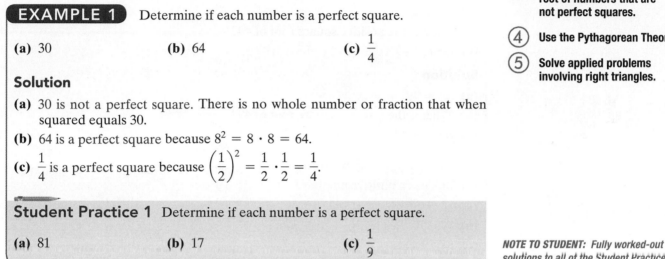

$1^2 = 1$	$6^2 = 36$	$11^2 = 121$
$2^2 = 4$	$7^2 = 49$	$12^2 = 144$
$3^2 = 9$	$8^2 = 64$	$13^2 = 169$
$4^2 = 16$	$9^2 = 81$	$14^2 = 196$
$5^2 = 25$	$10^2 = 100$	$15^2 = 225$

Suppose we are given a value, say 9, and we want to know what positive number we must multiply by itself (square) to get 9. We write this using mathematical symbols as follows.

What positive number squared equals nine?

$$\downarrow \quad \downarrow \quad \downarrow$$

$$n^2 \quad = \quad 9$$

$$n \cdot n \quad = \quad 9$$

$$n = 3$$

Note that both $3^2 = 9$ and $(-3)^2 = 9$. We were asked for a *positive number*; therefore, the answer is 3, not -3.

Another way of saying either "What positive number squared equals 9?" or "What positive number multiplied by itself equals 9?" is to ask the question: "What is the positive **square root** of 9?" We write this as $\sqrt{9}$.

Student Learning Objectives

After studying this section, you will be able to:

① Understand the meaning of square roots.

② Find the square root of perfect squares.

③ Approximate the square root of numbers that are not perfect squares.

④ Use the Pythagorean Theorem.

⑤ Solve applied problems involving right triangles.

NOTE TO STUDENT: Fully worked-out solutions to all of the Student Practice problems can be found at the back of the text starting at page SP-1.

> **SQUARE ROOT**
> We use the symbol $\sqrt{}$ to indicate that we want to find the positive **square root**. The symbol $\sqrt{}$ is called the **radical sign**.

Each of the following can be written using the radical sign.

$$n = \sqrt{36} \begin{cases} n \cdot n = 36 \\ n^2 = 36 \\ \text{What is the positive square root of 36?} \end{cases}$$

EXAMPLE 2 Write an expression for n using the radical sign. Assume n is a positive number.

(a) $n^2 = 49$

(b) What is the positive square root of 49?

(c) $n \cdot n = 49$

Solution

(a) $n^2 = 49$ $n = \sqrt{49}$

(b) What is the positive square root of 49? $n = \sqrt{49}$

(c) $n \cdot n = 49$ $n = \sqrt{49}$

Student Practice 2 Write an expression for n using the radical sign. Assume n is a positive number.

(a) $n^2 = 81$

(b) What is the positive square root of 81?

(c) $n \cdot n = 81$

② Finding the Square Root of Perfect Squares

Before we continue our discussion on square roots, let us take a closer look at the relationship between squaring and finding the square root. Consider the following.

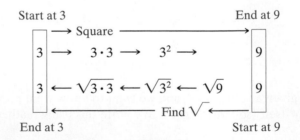

As we can see, squaring 3 gives us 9 and finding the positive square root of 9 gives us 3. Thus, *finding the square root of a number is like reversing the process of squaring a number*. We find the positive square root by asking ourselves the question, "What positive number squared equals the number under the $\sqrt{}$ symbol?"

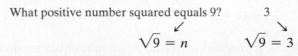

Recall that the symbol $\sqrt{}$ represents the *positive* square root. For example, 3 is the positive square root of 9, written $\sqrt{9}$, since $3 \cdot 3 = 9$. There exists a square root that is *negative*. For example, -3 is the negative square root of 9, written $-\sqrt{9}$, since $(-3)(-3) = 9$. In this book our discussion of square roots is limited to positive square roots.

> The square root of a positive number written $\sqrt{a}$ is the positive number we square to get a. In symbols:
>
> $$\text{If } \sqrt{a} = b, \quad \text{then } b^2 = a.$$

EXAMPLE 3   Find each square root. Then simplify if possible.

(a) $\sqrt{36}$ **(b)** $\sqrt{25} + \sqrt{36}$

Solution

(a) $\sqrt{36} = n$ What positive number times itself equals 36?

   $\sqrt{36} = 6$ 6, since $6 \cdot 6 = 36$.

(b) $\sqrt{25} = 5$ and $\sqrt{36} = 6$, thus $\sqrt{25} + \sqrt{36} = 5 + 6 = 11$

Student Practice 3   Find each square root. Then simplify if possible.

(a) $\sqrt{25}$ **(b)** $\sqrt{64} - \sqrt{16}$

EXAMPLE 4   Find the square root. $\sqrt{\dfrac{36}{49}}$

Solution

   $\sqrt{\dfrac{36}{49}} = n$ What positive fraction multiplied by itself equals $\dfrac{36}{49}$?

   $\sqrt{\dfrac{36}{49}} = \dfrac{6}{7}$ $\dfrac{6}{7}$, since $\left(\dfrac{6}{7}\right)\left(\dfrac{6}{7}\right) = \dfrac{36}{49}$.

Student Practice 4   Find the square root.

(a) $\sqrt{\dfrac{81}{100}}$ **(b)** $\sqrt{\dfrac{16}{49}}$

EXAMPLE 5   Find the length of the side of a square that has an area of 64 square feet.

Solution   We draw the figure and write the formula for area.

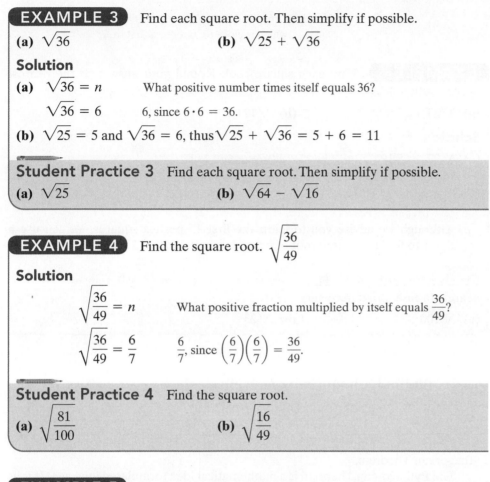

   $A = s^2$ or $s \cdot s$

   $64 = s^2$ What positive number squared equals 64?

   $\sqrt{64} = s$ Write using the radical sign.

   $8 = s$ The positive square root of 64 is 8.

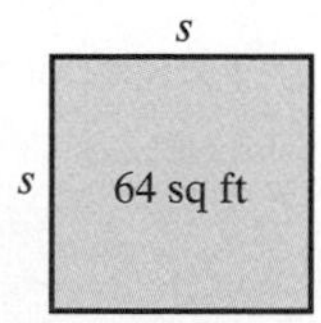

The length of the side of the square is 8 feet.

(figure: square labeled with side s on top and side s on left, interior labeled "64 sq ft")

Student Practice 5   Find the length of the side of a square that has an area of 49 square feet.

③ Approximating the Square Root of Numbers That Are Not Perfect Squares

If a number is not a perfect square, we approximate the square root. This can be done using any calculator that has a square root key. Usually, the key looks like $\boxed{\sqrt{}}$ or $\boxed{\sqrt{x}}$. To find the square root of 7 on most calculators, enter 7 and press

$\boxed{\sqrt{}}$ or $\boxed{\sqrt{x}}$. You will see displayed 2.645751. (Some calculators require that you enter the $\boxed{\sqrt{}}$ key first and then enter the 7.) Your calculator may display fewer or more digits. Remember, no matter how many digits your calculator displays, when we find $\sqrt{7}$—or the square root of any number that is not a perfect square—we have only an approximation.

$$\sqrt{7} \approx 2.645751 \quad (2.645751)(2.645751) \approx 6.9999984,$$
which is close to, but not equal to, 7.

EXAMPLE 6 Find each square root. Round your answer to the nearest thousandth if necessary.

(a) $\sqrt{27}$ **(b)** $\sqrt{121}$

Solution

(a) $\sqrt{27} \approx 5.1961524$ Use a calculator to find $\sqrt{27}$.

         ≈ 5.196 Round to the nearest thousandth.

(b) $\sqrt{121} = 11$ Use the calculator: $\sqrt{121} = 11$.

Although we advise you to learn the first 15 perfect squares, we can use a calculator to find the square root of perfect squares such as 121.

Student Practice 6 Find each square root. Round your answer to the nearest thousandth if necessary.

(a) $\sqrt{100}$ **(b)** $\sqrt{23}$

④ Using the Pythagorean Theorem

Right triangle

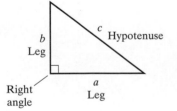

A triangle with a 90° angle is called a **right triangle.** Recall from Section 10.3 that we use a small □ to indicate that an angle measures 90°.

There are many useful properties of right triangles, including one called the Pythagorean Theorem.

The Pythagorean Theorem is a mathematical idea formulated long ago. It is as useful today as when it was discovered. The Pythagoreans, followers of the Greek philosopher and mathematician Pythagoras, lived in Italy about 2500 years ago. They studied various mathematical properties. They discovered that for any right triangle, *the square of the longest side is equal to the sum of the squares of the other two sides*. This relationship is known as the **Pythagorean Theorem.**

The square of the longest side is equal to the sum of the squares of the other two sides.

$$c^2 \qquad = \qquad a^2 + b^2$$

The shorter sides, a and b, are referred to as the **legs.** The longest side, c, which is opposite the right angle, is called the **hypotenuse** of the right triangle.

PYTHAGOREAN THEOREM

In any right triangle, if c is the length of the hypotenuse and a and b are the lengths of the two legs, then

$$c^2 = a^2 + b^2$$

In other words,

$$(\text{hypotenuse})^2 = (\text{leg})^2 + (\text{other leg})^2$$

EXAMPLE 7   Find the length of the hypotenuse of a right triangle with legs of 9 meters and 12 meters.

Solution We draw the figure and write the formula.

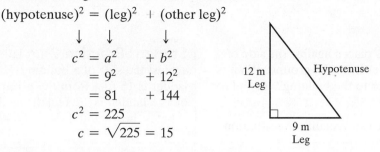

$$(\text{hypotenuse})^2 = (\text{leg})^2 + (\text{other leg})^2$$

$$
\begin{aligned}
c^2 &= a^2 + b^2 \\
&= 9^2 + 12^2 \\
&= 81 + 144 \\
c^2 &= 225 \\
c &= \sqrt{225} = 15
\end{aligned}
$$

Thus the length of the hypotenuse of the triangle is 15 meters.

Recall that the expression $c^2 = 225$ can be written as $c = \sqrt{225}$.

Student Practice 7 Find the length of the hypotenuse of a right triangle with legs of 4 meters and 3 meters.

Recall that the **height** of any triangle is the distance of a line drawn from a vertex perpendicular to the opposite side or an extension of the opposite side. The height may be one of the sides in a right triangle. The base of a triangle is perpendicular to the height.

As we saw in Chapter 5, the area of any triangle is half of the product of the base times the height of the triangle: $A = \frac{bh}{2}$.

EXAMPLE 8   Find the area of the shape made up of a square and a right triangle, shown in the margin.

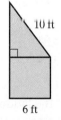

Solution We find the sum of the area of the square and the area of the triangle.

The area of the square is

$$A = s^2 = (6\ \text{ft})^2 = 36\ \text{ft}^2$$

Since all sides of a square are equal, the base of the triangle is 6 feet. We must use the Pythagorean Theorem to find the height of the triangle.

$$
\begin{aligned}
c^2 &= a^2 + b^2 \\
10^2 &= a^2 + 6^2 \\
100 &= a^2 + 36 \\
-36 &\qquad -36 \\
\hline
64 &= a^2 \\
8 &= a
\end{aligned}
$$

The height of the triangle is 8 feet.

Now we find the area of the triangle. The area of the triangle is

$$A = \frac{bh}{2} = \frac{(6\ \text{ft})(8\ \text{ft})}{2} = 24\ \text{ft}^2$$

The sum of the two areas is $36\ \text{ft}^2 + 24\ \text{ft}^2 = 60\ \text{ft}^2$.

Student Practice 8

Student Practice 8 Find the area of the shape made up of a square and a right triangle, shown in the margin.

⑤ Solving Applied Problems Involving Right Triangles

It is important to remember that the hypotenuse, c, is the side across from the right angle.

EXAMPLE 9   A 25-foot ladder is placed against the side of a building. The top of the ladder is 22 feet from the ground. What is the distance from the base of the ladder to the building? Round to the nearest tenth if necessary.

Solution   *Understand the problem.* We can visualize the situation and organize the facts by drawing a picture.

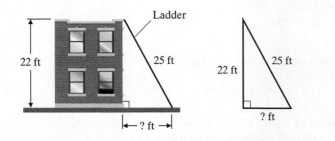

Calculate and state the answer. We can see from the picture that we must find the length of one of the legs of a right triangle.

$$c^2 = a^2 + b^2 \quad \text{We write the formula.}$$
$$25^2 = a^2 + 22^2 \quad \text{We replace } b \text{ with 22, and } c \text{ with 25.}$$
$$625 = a^2 + 484 \quad \text{Now we solve for } a.$$
$$\underline{+ \; -484 \qquad -484}$$
$$141 = a^2$$
$$\sqrt{141} = a$$
$$11.874 \approx a$$

The base of the ladder is approximately 11.9 feet from the building.

Student Practice 9   A ladder is placed against the side of a building. The top of the ladder is 15 feet from the ground. The base of the ladder is 12 feet from the building. What is the length of the ladder?

As you saw in Example 9, when solving applied problems involving geometric figures, it is helpful to organize the information in a diagram and/or draw a picture. This is because we can visualize the situation more easily when we see a picture.

10.4 Exercises · MyMathLab®

Watch the videos in MyMathLab

Download the MyDashBoard App

Verbal and Writing Skills, Exercises 1–2

1. Write each of the following as a question. Assume all variables are nonnegative.
   - (a) $\sqrt{81}$.
   - (b) $n \cdot n = 81$.
   - (c) $n^2 = 81$.

2. (a) Find $\sqrt{16}$ and then find $\sqrt{9}$. Now add the results: $\sqrt{16} + \sqrt{9} = $ _____.
   - (b) Find $16 + 9$ and then find the square root of this sum: $\sqrt{16 + 9} = $ _____.
   - (c) Based on these results, answer the following. Does $\sqrt{16} + \sqrt{9} = \sqrt{16 + 9}$?

Determine if each number is a perfect square.

3. 50 4. 64 5. 49 6. 20 7. 81 8. 85 9. $\dfrac{1}{5}$ 10. $\dfrac{1}{9}$

Write an expression for n using the radical sign $\left(\sqrt{} \right)$. Assume all variables are nonnegative.

11. (a) $n^2 = 64$
    (b) What is the positive square root of 64?
    (c) $n \cdot n = 64$

12. (a) $n^2 = 49$
    (b) What is the positive square root of 49?
    (c) $n \cdot n = 49$

Find each square root. Then simplify if possible.

13. $\sqrt{36}$

14. $\sqrt{25}$

15. $\sqrt{81}$

16. $\sqrt{121}$

17. $\sqrt{144}$

18. $\sqrt{169}$

19. $\sqrt{225} - \sqrt{16}$

20. $\sqrt{196} - \sqrt{25}$

21. $\sqrt{9} + \sqrt{49}$

22. $\sqrt{36} + \sqrt{81}$

23. $\sqrt{\dfrac{25}{49}}$

24. $\sqrt{\dfrac{16}{36}}$

25. $\sqrt{100} - \sqrt{25}$

26. $\sqrt{64} - \sqrt{49}$

27. $\sqrt{121} + \sqrt{81}$

28. $\sqrt{169} - \sqrt{36}$

29. $\sqrt{\dfrac{9}{81}}$

30. $\sqrt{\dfrac{4}{64}}$

31. $\sqrt{\dfrac{36}{49}}$

32. $\sqrt{\dfrac{25}{81}}$

33. Find the length of the side of a square that has an area of 121 square feet.

34. Find the length of the side of a square that has an area of 36 square feet.

35. Find the length of the side of a square that has an area of 144 square inches.

36. Find the length of the side of a square that has an area of 169 square inches.

Use your calculator to approximate the value. Round your answer to the nearest thousandth if necessary.

37. $\sqrt{44}$

38. $\sqrt{56}$

39. $\sqrt{69}$

40. $\sqrt{26}$

41. $\sqrt{80}$

42. $\sqrt{39}$

43. $\sqrt{90}$

44. $\sqrt{14}$

Find the unknown side of each right triangle. Use a calculator when necessary and round your answer to the nearest thousandth.

45.

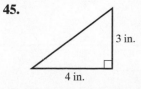

46.

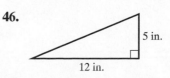

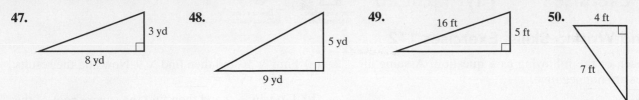

47. 3 yd, 8 yd

48. 5 yd, 9 yd

49. 16 ft, 5 ft

50. 4 ft, 7 ft

Find the unknown side of each right triangle using the information given. Round your answer to the nearest thousandth.

51. leg = 8 km, hypotenuse = 16 km

52. leg = 5 km, hypotenuse = 11 km

53. leg = 11 m, leg = 6 m

54. leg = 5 m, leg = 4 m

55. leg = 5 m, leg = 5 m

56. leg = 6 m, leg = 6 m

Find the area of each shape made up of squares and right triangles.

57.

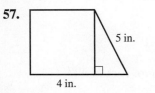

5 in., 4 in.

58.

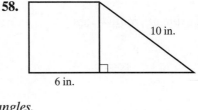

10 in., 6 in.

Find the area of each shape made up of rectangles and right triangles.

59.

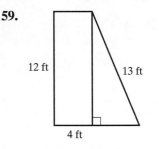

12 ft, 13 ft, 4 ft

60.

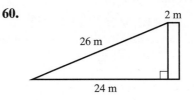

2 m, 26 m, 24 m

Applications *Solve each applied problem. Round your answer to the nearest tenth if necessary.*

61. *Guy Wire* Find the length of the guy wire supporting the telephone pole.

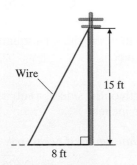

Wire, 15 ft, 8 ft

62. *Loading Dock Ramp* Find the length of this ramp to a loading dock.

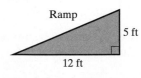

Ramp, 5 ft, 12 ft

63. *Distance Traveled* Juan runs out of gas in Los Lunas, New Mexico. He walks 4 miles west and then 3 miles south looking for a gas station. How far is he from his starting point?

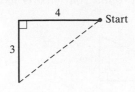

4, 3, Start

64. *Stainless Steel Plate* A construction project requires a stainless steel plate with holes drilled as shown. Find the distance between the centers of the holes in this triangular plate.

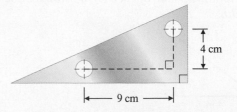

4 cm, 9 cm

65. *College Building* A 20-ft ladder is placed against a college classroom building at a point 18 ft above the ground. What is the distance from the base of the ladder to the building?

18 ft   20 ft

66. *Dragon Kite* Barbara is flying her dragon kite on 32 yd of string. The kite is directly above the edge of a pond. The edge of the pond is 30 yd from where the kite is tied to the ground. How far is the kite above the pond?

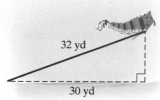

32 yd

30 yd

One Step Further *Find the total area of all the vertical areas (front, back, and both sides) of each building.*

67.

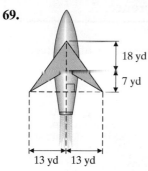

12 ft   15 ft

15 ft

30 ft

68.

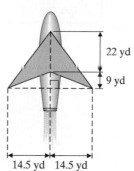

5 ft   13 ft

20 ft

45 ft

Airplane Wings *The top surface of the wings of a test plane must be coated with a special lacquer that costs $90 per square yard. Find the cost to coat the shaded wing surface of each plane.*

69.

18 yd

7 yd

13 yd   13 yd

70.

22 yd

9 yd

14.5 yd   14.5 yd

Cumulative Review

71. [5.1.1] Multiply. $\dfrac{9x}{22} \cdot \dfrac{10}{27x^2}$

72. [5.3.2] Perform the operation indicated. $\dfrac{8x}{15} - \dfrac{3x}{10}$

Quick Quiz 10.4

1. Find each square root. Then simplify if possible.

(a) $\sqrt{\dfrac{36}{49}}$ **(b)** $\sqrt{4} + \sqrt{9}$

2. Find the approximate value. Round your answer to the nearest hundredth. $\sqrt{8}$

3. Find the length of one leg of a right triangle with the other leg of length 4 cm and the hypotenuse of length 8 cm. Round your answer to the nearest hundredth.

4. Concept Check State the Pythagorean Theorem and explain when you would use it.

10.5 The Circle

Student Learning Objectives

After studying this section, you will be able to:

① Find the circumference of circles.

② Find the area of circles.

③ Solve area problems involving circles.

Every point on a circle is the same distance from the center of the circle as every other point on the circle, so the circle looks the same all around. In this section we learn how to calculate the distance around a circle as well as its area.

① Finding the Circumference of Circles

A **circle** is a figure for which all points on the figure are at an equal distance from a given point. This given point is called the **center** of the circle.

The **radius**, r, is the length of a line segment from the center to a point on the circle.

The **diameter**, d, is the length of a line segment across the circle that passes through the center.

Clearly, we can see the following relationships between the diameter and radius.

The diameter is two times the radius   or $d = 2r$.

The radius is one-half the diameter   or $r = \dfrac{d}{2}$.

> The distance around a circle is called its **circumference**, C.

The measure of the circumference is similar to the perimeter of a polygon in that they both *measure the distance around the outer edge of the figure*.

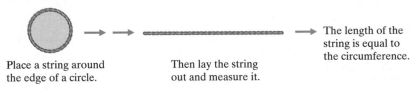

Place a string around the edge of a circle.

Then lay the string out and measure it.

The length of the string is equal to the circumference.

When we divide the circumference by the diameter, we get a special number called **pi**, denoted by the Greek lowercase letter π: $\dfrac{C}{d} = \pi$. π is approximately 3.14159265359. We can approximate π to any number of digits. For all work in this book we use the following.

> π is approximately 3.14, rounded to the nearest hundredth. $\pi \approx 3.14$

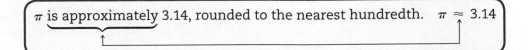

If we multiply both sides of $\dfrac{C}{d} = \pi$ by d, we get a formula for the circumference of a circle.

> The **circumference of a circle** is equal to π times the diameter.
>
> $$C = \pi d$$
>
> Since $d = 2r$, an alternative formula for the circumference is $C = \pi \cdot 2r$, or
>
> $$C = 2\pi r$$

When we approximate π with 3.14 in the following examples, the answers are *approximate values*.

EXAMPLE 1 Find the circumference of a circle if the diameter is 9.2 meters. Use $\pi \approx 3.14$.

Solution Since we are given the diameter, we use the formula for circumference that includes the diameter.

$$C = \pi d \qquad \text{Write the formula.}$$
$$= (3.14)(9.2 \text{ m}) \quad \text{Then substitute the values given.}$$
$$C \approx 28.888 \text{ m}$$

9.2 m

Remember that since π is approximately equal to 3.14, our answer is an approximate value.

Student Practice 1 Find the circumference of a circle if the diameter is 5.1 meters. Use $\pi \approx 3.14$.

NOTE TO STUDENT: *Fully worked-out solutions to all of the Student Practice problems can be found at the back of the text starting at page SP-1.*

CAUTION: When solving problems involving circles, be careful. Ask yourself, "Is the radius given or the diameter?" Then choose the appropriate formula for your calculations.

When a circular object rolls through 1 complete revolution, it rolls along the outer edge of the circle. Thus the distance the object rolls in 1 complete revolution equals the distance around the circular object, or the circumference.

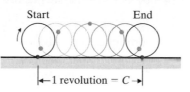

Start   End

$\leftarrow$ 1 revolution $= C \rightarrow$

EXAMPLE 2 The larger of two bicycles has a 26-inch wheel diameter, and the smaller bicycle has a 24-inch wheel diameter.

(a) If the wheels on the larger bicycle complete 12 revolutions, what distance does the larger bicycle travel?

(b) How many revolutions must each wheel on the smaller bicycle complete to travel the same distance as the larger bicycle in part **(a)**?

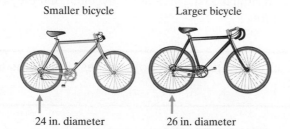

Smaller bicycle   Larger bicycle

24 in. diameter   26 in. diameter

Solution *Understand the problem for part (a).* We draw a picture of the situation:

12 revolutions of the 26-inch wheel

. . .

What is the distance traveled?

The distance traveled by the larger bicycle during 1 revolution is equal to the measure of the circumference. We must multiply the circumference by 12 to find the distance traveled in 12 revolutions.

Continued on next page

Calculate and state the answer to part (a).

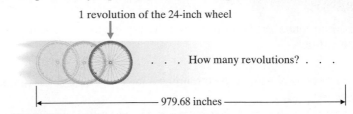

$$(12)(\pi d)$$
$$= (12)(3.14)(26 \text{ in.})$$
$$= (12)(81.64 \text{ in.}) = 979.68 \text{ in.}$$

(a) The larger bicycle travels approximately 979.68 inches when the wheels complete 12 revolutions.

Understand the problem for part (b). We draw a picture of the situation:

1 revolution of the 24-inch wheel

. . . How many revolutions? . . .

|← 979.68 inches →|

First, we find the distance traveled for 1 revolution (the circumference), and then we divide 979.68 by the circumference to find the total number of revolutions completed in this distance.

Calculate and state the answer to part (b).

$$C = \pi d$$
$$C = (3.14)(24 \text{ in.}) = 75.36 \text{ in. per revolution}$$

We divide: $979.68 \div 75.36 = 13$.

(b) The smaller bicycle must complete 13 revolutions to travel 979.68 inches.

Check. You may use your calculator to check.

Student Practice 2

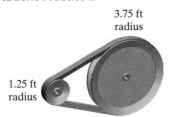

3.75 ft
radius

1.25 ft
radius

Student Practice 2   A pulley has a large wheel that has a 3.75-foot radius and a smaller wheel that has a 1.25-foot radius. If the large wheel makes 2 complete revolutions, how many revolutions does the small wheel make?

---

② Finding the Area of Circles

The area of a circle is similar to the area of a polygon in that they both measure the region inside the figure.

> The **area of a circle** is the product of π times the radius squared.
> $$A = \pi r^2$$

EXAMPLE 3   Find the area of a circle whose radius is 2.1 meters. Use $\pi \approx 3.14$. Round your answer to the nearest tenth.

Solution

$$A = \pi r^2$$
$$= (3.14)(2.1 \text{ m})^2 \quad \text{Square the radius first. Then multiply by 3.14.}$$
$$= (3.14)(4.41 \text{ m}^2)$$
$$A \approx 13.8474 \text{ m}^2$$

The area of the circle is approximately 13.8 square meters.

Student Practice 3   Find the area of a circle whose radius is 6.1 centimeters. Use $\pi \approx 3.14$. Round your answer to the nearest tenth.

③ **Solving Area Problems Involving Circles**

The formula for finding the area of a circle uses the radius. If we are given a diameter, we can use the formula $r = \dfrac{d}{2}$ to find the radius.

EXAMPLE 4 Lester wants to buy a circular braided rug that is 8 feet in diameter. Find the cost of the rug at $35 per square yard.

Solution *Understand the problem.* We organize the information in a Mathematics Blueprint for Problem Solving.

Mathematics Blueprint for Problem Solving

Gather the Facts	What Am I Asked to Do?	How Do I Proceed?	Key Points to Remember
$d = 8$ ft Cost of rug is $35 per square yard.	Find the cost of the rug.	1. Find the radius. 2. Find the area. 3. Change area to square yards. 4. Multiply area by $35 to find the cost of the rug.	$A = \pi r^2$ $r = \dfrac{d}{2}$ 1 yd = 3 ft

Solve and state the answer.

1. The diameter is 8 feet, so the radius is 4 feet.

$$r = \frac{8}{2} = 4 \text{ ft}$$

2. Find the area.

$A = \pi r^2 = (3.14)(4 \text{ ft})^2$

$\qquad = (3.14)(16 \text{ ft}^2)$ We square the radius first.

$\quad A \approx 50.24 \text{ ft}^2$ Then we multiply.

3. Change square feet to square yards.
   Since 1 yd = 3 ft, $(1 \text{ yd})^2 = (3 \text{ ft})^2$. That is, $1 \text{ yd}^2 = 9 \text{ ft}^2$.

$50.24 \text{ ft}^2 \times \dfrac{1 \text{ yd}^2}{9 \text{ ft}^2} \approx 5.58 \text{ yd}^2$ (rounded to the nearest hundredth)

1 yd = 3 ft

1 yd = 3 ft

$(1 \text{ yd})^2 = (3 \text{ ft})^2$

4. Find the cost.

$$\frac{\$35}{1 \text{ yd}^2} \times 5.58 \text{ yd}^2 = \$195.30$$

The rug costs $195.30.

Check. You may use your calculator to check.

Student Practice 4 Mary Beth wants to buy a circular crocheted tablecloth that is 60 inches in diameter. Find the cost of the tablecloth at $8 per square foot.

Verbal and Writing Skills, Exercises 1–4 *Fill in the blanks.*

1. The distance around a circle is called the _____.

2. The radius is a line segment from the _____ to a point on the circle.

3. The diameter is two times the _____ of the circle.

4. Explain in your own words how to find the area of a circle if you are given the diameter.

Find the radius of a circle if the diameter has the value given.

5. $d = 45$ yd

6. $d = 71$ yd

7. $d = 3.8$ cm

8. $d = 8.4$ cm

In exercises 9–24, use $\pi \approx 3.14$. Round your answer to the nearest tenth.

Find the circumference of each circle.

9. Diameter $= 18$ cm

10. Diameter $= 11$ cm

11. Radius $= 11$ in.

12. Radius $= 10$ in.

A bicycle wheel makes 1 revolution. Determine how far the bicycle travels in inches.

13. The diameter of the wheel is 28 in.

14. The diameter of the wheel is 26 in.

A bicycle wheel makes 5 revolutions. Determine how far the bicycle travels in inches.

15. The diameter of the wheel is 15 in.

16. The diameter of the wheel is 32 in.

Find the area of each circle.

17. Radius $= 6$ yd

18. Radius $= 5$ yd

19. Diameter $= 44$ cm

20. Diameter $= 32$ cm

A water sprinkler disperses water in a circular pattern. Determine how large an area is watered.

21. The radius of the watered area is 10 ft.

22. The radius of the watered area is 8 ft.

A radio station sends out radio waves in all directions from a tower at the center of the circle of broadcast range. Determine how large an area is reached.

23. The diameter is 90 mi.

24. The diameter is 120 mi.

Applications *In exercises 25–34, use $\pi \approx 3.14$. Round your answer to the nearest hundredth.*

25. **Porthole Window** A porthole window on a freighter ship has a diameter of 2 ft. What is the length of the insulating strip that encircles the window and keeps out wind and moisture?

26. **Manhole Cover** A manhole cover has a diameter of 3 ft. What is the length of the brass grip-strip that encircles the cover, making it easier to manage?

27. **Car Tires** Elena's car has tires with a radius of 14 in. How many feet does her car travel if the wheels make 35 revolutions?

28. **Truck Tires** Jimmy's truck has tires with a radius of 30 in. How many feet does his truck travel if the wheels make 9 revolutions?

29. **Car Tires** Carlotta's car has tires with a radius of 16 in. Carlotta drove her car down the highway 20,096 inches. How many complete revolutions did the wheels make?

30. **Car Tires** Mickey's car has tires with a radius of 15 in. He backed up his car a distance of 9891 in. How many complete revolutions did the wheels make backing up?

31. **Coca-Cola Sign** Find the area of a circular Coca-Cola sign with a diameter of 64 in.

32. **Flower Bed** Find the area of a circular flower bed with a diameter of 14 ft.

33. *Marble Tabletop* Sarah bought a circular marble tabletop to use as her dining room table. The marble is 6 ft in diameter. Find the cost of the tabletop at $72 per square yard of marble.

34. *Patio* Tom made a base for a circular patio by pouring concrete into a circular space 10 ft in diameter. Find the cost at $18 per square yard.

One Step Further *Use* $\pi \approx 3.14$ *in exercises 35 and 36. Round your answer to the nearest hundredth.*

A semicircle is one-half of a circle and has an area that is one-half of the area of a circle. Find the cost of fertilizing a playing field at $0.20 per square yard for the dimensions stated.

35. The rectangular part of the field is 120 yd long and the diameter of each semicircle is 40 yd.

36. The rectangular part of the field is 110 yd long and the diameter of each semicircle is 50 yd.

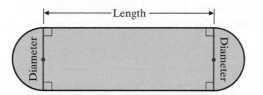

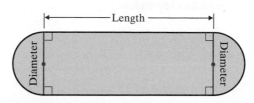

Round your answer to the nearest hundredth for exercises 37 and 38.

37. *Pizza Slices* A 15-in.-diameter pizza costs $6. A 12-in.-diameter pizza costs $4. The 12-in.-diameter pizza is cut into six slices. The 15-in.-diameter pizza is cut into eight slices.

  (a) What is the cost per slice of the 15-in.-diameter pizza? How many square inches of pizza are in one slice?

  (b) What is the cost per slice of the 12-in.-diameter pizza? How many square inches of pizza are in one slice?

  (c) If you want more value for your money, which size pizza should you buy?

38. *Pizza Slices* A 14-in.-diameter pizza costs $5.50. It is cut into eight pieces. A 12.5-in. by 12.5-in. square pizza costs $6. It is cut into nine pieces.

  (a) What is the cost of one piece of the 14-in.-diameter pizza? How many square inches of pizza are in one piece?

  (b) What is the cost of one piece of the 12.5-in. by 12.5-in. square pizza? How many square inches of pizza are in one piece?

  (c) If you want more value for your money, which size pizza should you buy?

Cumulative Review *Find the volume of each rectangular box.*

▲ **39.** [3.3.3] $L = 11$ in., $W = 5$ in., $H = 6$ in.

▲ **40.** [3.3.3] $L = 8$ in., $W = 4$ in., $H = 5$ in.

Quick Quiz 10.5 *For each of the following, use* $\pi \approx 3.14$ *and round to the nearest hundredth if necessary.*

1. Given a circle of radius 6 feet,

  (a) find the circumference.

  (b) find the area.

3. Martha wants to buy a circular rug that is 12 feet in diameter. Find the cost of the rug at $25 per square yard.

2. A pulley has a large wheel that has a 3.5-foot radius and a smaller wheel that has a 1.5-foot radius. If the large wheel makes 3 complete revolutions, how many revolutions does the small wheel make?

4. **Concept Check** Jasmine must determine the area of a semicircle with a diameter of 10 inches. Explain how you would find the area of the semicircle.

Volume

Student Learning Objectives

After studying this section, you will be able to:

① **Find the volumes of cylinders and spheres.**

② **Find the volume of cones.**

③ **Find the volume of pyramids.**

① **Finding the Volumes of Cylinders and Spheres**

In Chapter 3 we learned that the **volume** of a box is a measure of the space enclosed in the rectangular three-dimensional figure. The volume of a cylinder or a sphere is similar: it is a measure of the *amount of space enclosed in the three-dimensional figure*. Recall that volume is measured in cubic units such as cubic meters (abbreviated m^3) or cubic feet (abbreviated ft^3 or cu ft).

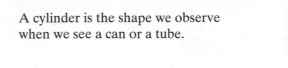

A cylinder is the shape we observe when we see a can or a tube.

A sphere is the shape we observe when we see a ball.

The **volume of a cylinder** is the area of its circular base, πr^2, times the height, h.

$$V = \pi r^2 h$$

The **volume of a sphere** is 4 times π times the radius cubed divided by 3.

$$V = \frac{4\pi r^3}{3}$$

EXAMPLE 1 Find the volume of a tin can with radius 7.62 centimeters and height 15.24 centimeters. Round your answer to the nearest hundredth.

Solution

$$V = \pi r^2 h = (3.14)(7.62 \text{ cm})^2(15.24 \text{ cm}) = 2778.59 \text{ cm}^3$$

The volume of the tin can is approximately 2778.59 cubic centimeters.

Student Practice 1 Find the volume of a tin can with radius 10.24 centimeters and height 17.78 centimeters. Round your answer to the nearest hundredth.

15.24 cm
7.62 cm

NOTE TO STUDENT: Fully worked-out solutions to all of the Student Practice problems can be found at the back of the text starting at page SP-1.

EXAMPLE 2 How much air is needed to fully inflate a soccer ball if the radius of the inner lining is 5 inches? Round your answer to the nearest hundredth.

Solution

$$V = \frac{4\pi r^3}{3} = \frac{4(3.14)(5 \text{ in.})^3}{3} = 523.33 \text{ in.}^3$$

Approximately 523.33 cubic inches of air is needed to fully inflate the ball.

Student Practice 2 How much air is needed to fully inflate a beach ball if the radius of the inner lining is 7 inches? Round your answer to the nearest hundredth.

EXAMPLE 3 A cylindrical thermos has a layer of insulation around its sides. The radius R to the outer edge of the insulation is 15 centimeters, and the radius r to the inner edge is 13 centimeters. The thermos is 27 centimeters tall. Use $\pi \approx 3.14$.

(a) What volume of coffee can the thermos hold?

(b) What is the volume of the insulated region?

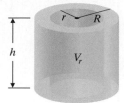

Solution *Understand the problem for part (a).* We draw a picture of the thermos. We find the volume of the inner shaded region (V_r) to determine how much coffee the thermos will hold.

Calculate and state the answer to part (a).

$$r = 13 \text{ cm} \qquad V_r = \pi r^2 h$$
$$h = 27 \text{ cm} \qquad \quad = (3.14)(13 \text{ cm})^2(27 \text{ cm})$$
$$V_r = 14{,}327.82 \text{ cm}^3$$

(a) The thermos can hold approximately 14,327.82 cubic centimeters of coffee.

Understand the problem for part (b). To determine the volume of the insulated region (blue shaded region), we find the volume of the entire cylinder (V_R) minus the volume of the inner region (V_r): $V_R - V_r$.

Calculate and state the answer to part (b).

First we find the volume of the entire cylinder V_R.

$$R = 15 \text{ cm} \qquad V_R = \pi r^2 h$$
$$h = 27 \text{ cm} \qquad \quad = (3.14)(15 \text{ cm})^2(27 \text{ cm})$$
$$V_R = 19{,}075.5 \text{ cm}^3$$

Then we subtract to find the volume of the insulated region.

$$\text{Volume of the insulated region} = V_R \quad - \quad V_r$$
$$\qquad\qquad\qquad\qquad\qquad\qquad\qquad \downarrow \qquad\qquad \downarrow$$
$$19{,}075.5 - 14{,}327.82 = 4747.68$$

(b) The volume of the insulated region is approximately 4747.68 cubic centimeters.

Student Practice 3 If $R = 12$ centimeters and $r = 10$ centimeters for the thermos in Example 3, answer the following.

(a) What volume of coffee can the thermos hold?

(b) What is the volume of the insulated region?

② Finding the Volume of Cones

We see the shape of a cone when we look at the sharpened end of a pencil or at an ice cream cone. To find the volume of a cone, we use the following formula.

> The **volume of a cone** is π times the radius squared times the height divided by 3.
>
> $$V = \frac{\pi r^2 h}{3}$$

EXAMPLE 4   Find the volume of a cone of radius 7 meters and height 9 meters. Round to the nearest tenth.

Solution

$$V = \frac{\pi r^2 h}{3}$$

$$= \frac{(3.14)(7\text{ m})^2(9\text{ m})}{3}$$

$$= \frac{(3.14)(7\text{ m})(7\text{ m})(9\text{ m})}{3}$$

$$= (3.14)(49)(3)\text{ m}^3 = (153.86)(3)\text{ m}^3 = 461.58\text{ m}^3$$

$$V \approx 461.6\text{ m}^3 \text{ rounded to the nearest tenth}$$

Student Practice 4   Find the volume of a cone of radius 5 meters and height 12 meters.

③ Finding the Volume of Pyramids

You may have seen pictures of the great pyramids of Egypt. These amazing stone structures are over 4000 years old.

> The **volume of a pyramid** is obtained by multiplying the area of the base of the pyramid B by the height h and dividing by 3.
>
> $$V = \frac{Bh}{3}$$

EXAMPLE 5   Find the volume of a pyramid with height = 6 meters, length of base = 7 meters, width of base = 5 meters.

Solution   We first find the base is a rectangle.

$$\text{area of base} = (7\text{ m})(5\text{ m}) = 35\text{ m}^2$$

Substituting the area of the base, 35 m², and the height of 6 m, we have

$$V = \frac{Bh}{3} = \frac{(35\text{ m}^2)(6\text{ m})}{3}$$

$$= (35)(2)\text{ m}^3$$

$$V = 70\text{ m}^3$$

Student Practice 5   Find the volume of each pyramid having the given dimensions.

(a) Height 10 meters, base width 6 meters, base length 6 meters

(b) Height 15 meters, base width 7 meters, base length 8 meters

10.6 Exercises

Verbal and Writing Skills, Exercises 1–6 *Match the formula for volume with the appropriate figure.*

1. $V = LWH$

2. $V = \pi r^2 h$

3. $V = \dfrac{4\pi r^3}{3}$

4. $V = \dfrac{\pi r^2 h}{3}$

5. $V = \dfrac{Bh}{3}$

6. $V = s^3$

Find each volume. Use $\pi \approx 3.14$. Round each answer to the nearest tenth when necessary.

7. A cylinder with radius 3 m and height 5 m

8. A cylinder with radius 2 m and height 5 m

9. A sphere with radius 4 m

10. A sphere with radius 3 m

11. A can with a radius of 4 in. and a height of 10 in.

12. A trash can with a radius of 2 ft and a height of 4 ft

13. How much air is needed to fill a tennis ball if the radius of the inner lining is 3.25 cm?

14. How much air is needed to fill a rubber handball if the radius of the inner lining is 3.4 cm?

Exercises 15 and 16 involve hemispheres. A hemisphere is exactly one-half of a sphere.

15. Find the volume of a hemisphere with radius = 9 m.

16. Find the volume of a hemisphere with radius = 6 m.

Find each volume. Use $\pi \approx 3.14$. Round each answer to the nearest tenth.

17. A cone with a height of 16 cm and a radius of 9 cm

18. A cone with a height of 15 cm and a radius of 8 cm

19. A cone with a height of 10 ft and a radius of 5 ft

20. A cone with a height of 12 ft and a radius of 6 ft

21. A pyramid with a height of 7 m and a square base of 3 m on a side

22. A pyramid with a height of 10 m and a square base of 7 m on a side

23. A pyramid with a height of 5 m and a rectangular base measuring 6 m by 9 m

24. A pyramid with a height of 10 m and a rectangular base measuring 8 m by 14 m

Applications
A collar of Styrofoam is made to insulate a pipe. The large radius R is to the outer rim of the insulation. The small radius r is to the inner rim of the insulation. For exercises 25–28, find the volume of the unshaded region (which represents the collar). Use π ≈ 3.14.

25. $r = 3$ in.
    $R = 5$ in.
    $h = 20$ in.

26. $r = 4$ in.
    $R = 6$ in.
    $h = 25$ in.

27. $r = 6$ in.
    $R = 8$ in.
    $h = 30$ in.

28. $r = 7$ in.
    $R = 9$ in.
    $h = 20$ in.

One Step Further

29. **Snow Cone** Find the cost of Sandy's snow cone before it melts! The top of the cone is one-half of a sphere with a diameter of 2 in. The base is the shape of a cone with a height of 6 in. If the snow cone costs $0.09 per cubic inch, how much will Sandy's snow cone cost? Round to the nearest hundredth.

30. **Gold Filling** A dentist places a gold filling in a tooth in the shape of a cylinder with a hemispherical top. The radius *r* of the filling is 1 mm. The height of the cylinder is 2 mm. If dental gold costs $95 per cubic millimeter, how much did the gold cost for the filling? Round to the nearest hundredth.

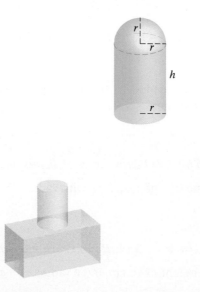

Find the volume of a shape, which consists of a rectangular box with a cylinder attached to the top.

31. The box has $L = 4$ ft, $W = 3$ ft, and $H = 2$ ft. The cylinder has a diameter of 2 ft and height of 2 ft.

32. The box has $L = 10$ in., $W = 5$ in., and $H = 2$ in. The cylinder has a diameter of 4 in. and a height of 4 in.

33. **Passenger Jet** The nose cone of a passenger jet is used to receive and send radar. It is made of a special aluminum alloy that costs $4 per cm³. The cone has a radius of 5 cm and a height of 9 cm. What is the cost of the aluminum needed to make this *solid* nose cone?

34. **TV Antenna** A special stainless steel cone sits on top of a cable television antenna. The cost of the stainless steel is $3 per cm³. The cone has a radius of 6 cm and a height of 10 cm. What is the cost of the stainless steel needed to make this *solid* steel cone?

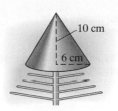

To Think About *Use the following formulas for surface area for exercises 35 and 36. Use $\pi \approx 3.14$.*

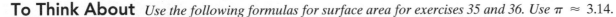

Surface area of a cylinder: $SA = 2\pi r^2 + 2\pi rh$ *Surface area of a sphere:* $SA = 4\pi r^2$

35. How much material is needed to make a beach ball that has a diameter of 14 in.?

36. What is the surface area of a cylindrical tank that has a diameter of 10 m and a height of 20 m?

Cumulative Review *Solve each proportion.*

37. **[4.6.3]** $\dfrac{21}{40} = \dfrac{x}{120}$

38. **[4.6.3]** $\dfrac{18}{x} = \dfrac{12}{10}$

39. **[5.4.2]** Multiply. $2\dfrac{1}{4} \times 3\dfrac{3}{4}$

40. **[5.4.2]** Divide. $7\dfrac{1}{2} \div 4\dfrac{1}{5}$

Quick Quiz 10.6

1. Find the volume of a tin can with a radius of 6 inches and a height of 10 inches. Round your answer to the nearest hundredth. Use $\pi \approx 3.14$.

2. Find the volume of a cone of radius 3 inches and height 5 inches. Round your answer to the nearest tenth. Use $\pi \approx 3.14$.

3. Find the volume of a pyramid with height 8 meters, length of base 6 meters, and width of base 5 meters.

4. **Concept Check** After you find the volume of the tin can described in Quick Quiz 1, you decide that you must find the volume of a similar can that has all the same measurements, except the height is 5 inches. Explain how you could do this *without* using the formula.

10.7 Similar Geometric Figures

① Finding the Corresponding Parts of Similar Triangles

In English, to say two things are *similar* means that they are, in general, alike. But in mathematics, *similar* means that the things being compared are alike in a special way—they are *alike in shape*, even though they may be different in size. So photographs that are enlarged produce images *similar* to the originals; a floor plan of a building is *similar* to the actual building; a model car is *similar* to the actual vehicle.

Two triangles with the same shape but not necessarily the same size are called **similar triangles.** Here are two pairs of similar triangles.

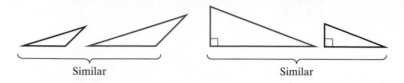

The two triangles below are similar. The smallest angle in the first triangle is angle A. The smallest angle in the second triangle is angle D. Both angles measure 36°. We say that angle A and angle D are corresponding angles in these similar triangles.

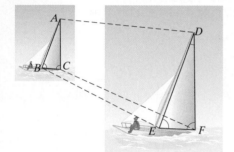

Similar Triangles

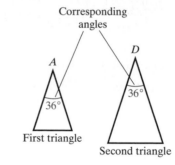

> The **corresponding angles** of similar triangles are equal.

The two triangles below are similar. Notice the corresponding sides.

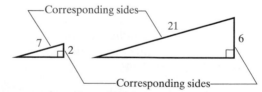

We see that the ratio of 7 to 21 is the same as the ratio of 2 to 6.

$$\frac{7}{21} = \frac{2}{6} \quad \text{is obviously true since} \quad \frac{1}{3} = \frac{1}{3}.$$

> The lengths of **corresponding sides** of similar triangles have the same ratio.

We can use the fact that corresponding sides of similar triangles have the same ratio to find the unknown lengths of sides of triangles.

EXAMPLE 1 The two triangles below are similar. Find the length of side n. Round to the nearest tenth.

Solution The ratio of 12 to 19 is the same as the ratio of 5 to n.

$$\frac{12}{19} = \frac{5}{n}$$

$12n = (19)(5)$ Cross-multiply.

$12n = 95$ Simplify.

$$\frac{12n}{12} = \frac{95}{12}$$ Divide both sides by 12.

$n = 7.91\overline{6}$

≈ 7.9 Round to the nearest tenth.

The length of side n is approximately 7.9.

NOTE TO STUDENT: *Fully worked-out solutions to all of the Student Practice problems can be found at the back of the text starting at page SP-1.*

Student Practice 1 The two triangles in the margin are similar. Find the length of side n. Round to the nearest tenth.

Student Practice 1

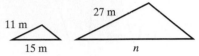

So far we have discussed the relationships between corresponding angles and corresponding sides of similar triangles. Now let's look at the relationship between perimeters of similar triangles.

The **perimeters** of similar triangles have the same ratios as the corresponding sides.

$$\frac{\text{triangle \# 1} \;\to}{\text{triangle \# 2} \;\to} \; \frac{\text{perimeter}}{\text{perimeter}} = \frac{\text{length of side}}{\text{length of side}}$$

EXAMPLE 2 Two triangles are similar. The smaller triangle has sides 5 yards, 7 yards, and 10 yards. The 7-yard side on the smaller triangle corresponds to a side of 21 yards on the larger triangle. What is the perimeter of the larger triangle? See the figures in the margin.

Solution *Understand the problem.* We are asked to find the perimeter of the larger triangle. We start by drawing the two triangles and organizing our information on the picture.

Calculate and state the answer. We know that the perimeters of similar triangles have the same ratios as the corresponding sides. Therefore, we begin by finding the perimeter of the smaller triangle.

$$5 \text{ yd} + 7 \text{ yd} + 10 \text{ yd} = 22 \text{ yd}$$

We can now write equal ratios. Let $P =$ the unknown perimeter. We will set up our ratios as: $\dfrac{\text{smaller triangle}}{\text{larger triangle}}$.

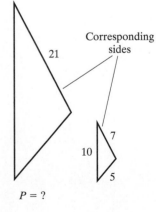

Continued on next page

Remember, once you write the first ratio, be sure to write the terms of the second ratio in the same order. We will write

perimeter of the smaller triangle → $\dfrac{22}{P} = \dfrac{7}{21}$ ← side of the smaller triangle
perimeter of the larger triangle → $\phantom{\dfrac{22}{P}}$ ← side of the larger triangle

$$7P = (21)(22)$$
$$7P = 462$$
$$\frac{7P}{7} = \frac{462}{7}$$
$$P = 66$$

The perimeter of the larger triangle is 66 yards.

Check. Are the ratios the same? Simplify each ratio to check. That is,

$$\frac{22}{66} \overset{?}{=} \frac{7}{21} \qquad \frac{\cancel{2} \cdot \cancel{11}}{\cancel{2} \cdot 3 \cdot \cancel{11}} \overset{?}{=} \frac{1 \cdot \cancel{7}}{3 \cdot \cancel{7}} \quad \text{Simplify each ratio.}$$

$$\frac{1}{3} = \frac{1}{3} \checkmark$$

Student Practice 2

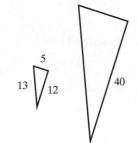

Student Practice 2 Two triangles are similar. The smaller triangle has sides of 5 yards, 12 yards, and 13 yards. The 12-yard side of the smaller triangle corresponds to a side of 40 yards on the larger triangle. What is the perimeter of the larger triangle?

Similar triangles can be used to find distances or lengths that are difficult to measure.

EXAMPLE 3 A flagpole casts a shadow of 36 feet. At the same time, a tree that is 3 feet tall has a shadow of 5 feet. How tall is the flagpole?

Solution *Understand the problem.* The shadows cast by the sun shining on vertical objects at the same time of day form similar triangles. We draw a picture and organize our information on the picture.

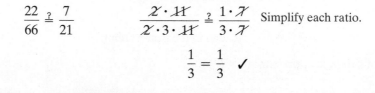

Calculate and state the answer. Let $n =$ the height of the flagpole. Thus we can say that n is to 3 as 36 is to 5.

$$\frac{n}{3} = \frac{36}{5}$$
$$5n = (3)(36)$$
$$5n = 108$$
$$\frac{5n}{5} = \frac{108}{5}$$
$$n = 21.6$$

The flagpole is 21.6 feet tall.

Check. The check is up to you.

Student Practice 3

Student Practice 3 How tall (h) is the building in the margin if the two triangles are similar?

② Finding the Corresponding Parts of Similar Geometric Figures

Geometric figures such as rectangles can also be similar figures.

> The *corresponding sides of similar geometric figures* have the same ratio.

EXAMPLE 4 The two rectangles shown here are similar because the corresponding sides of the two rectangles have the same ratio. Find the width of the larger rectangle.

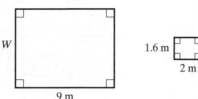

Solution Let W = the width of the larger rectangle.

$$\frac{W}{1.6} = \frac{9}{2}$$

$$2W = (1.6)(9)$$

$$2W = 14.4$$

$$\frac{2W}{2} = \frac{14.4}{2}$$

$$W = 7.2 \quad \text{The width of the larger rectangle is 7.2 meters.}$$

Student Practice 4

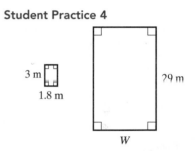

Student Practice 4 The two rectangles in the margin are similar. Find the width of the larger rectangle.

👣 STEPS TO SUCCESS Mastering Prealgebra Skills

Congratulations for successfully completing most of the prealgebra material necessary to prepare you for the final exam! There are just a few more things to do, and you will be on your way toward the next step in your educational goal. To finish the preparation for the final exam, you should complete the following in several 1- to 2-hour sessions.

Making it personal: *Place a* ✓ *in the blank as you complete each of the following.*

___ **Find a few students in your class to study with.** You will have a more productive and enjoyable study session.

___ **Ask your instructor what chapters and sections will be covered on the final exam.** Inquire about any extra review materials the instructor may have for you.

___ **Review all your exams.** Refer back to the examples in the appropriate chapters for those problems you answered incorrectly.

___ **Don't stress about topics that you cannot understand.** Move on to another topic. When you finish reviewing all the topics for the final exam, return to this topic and try again.

___ **Take the Practice Final Examination at the end of Chapter 10.** Complete this Practice Final *without assistance* from tutors and your book. Note the problems you missed and seek assistance on these problems.

You should work through the above steps at least a few days before the final. Cramming for any test, especially the final one, often causes anxiety and fatigue and impairs your performance. Finally, get a good night's sleep the night before the final.

10.7 Exercises

MyMathLab®

Watch the videos in MyMathLab · Download the MyDashBoard App

Verbal and Writing Skills, Exercises 1–4 *Fill in the blanks.*

1. The corresponding angles of _____ triangles are equal.

2. The lengths of _____ sides of similar triangles have the same ratio.

3. The _____ of similar triangles have the same ratios as the corresponding sides.

4. The corresponding sides of similar geometric figures have the same _____.

For each pair of similar triangles, find the unknown side n. Round your answer to the nearest tenth when necessary.

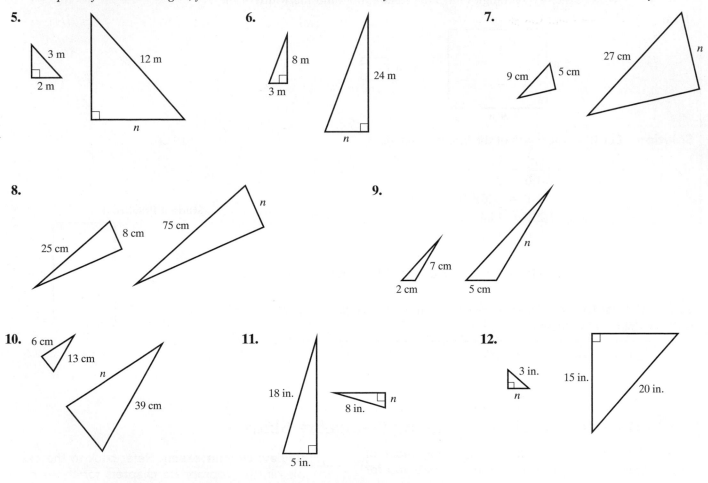

5.

6.

7.

8.

9.

10.

11.

12.

13. Two triangles are similar. The smaller triangle has sides 6 inches, 9 inches, and 10 inches. The 10-inch side on the smaller triangle corresponds to a side of 22 inches of the larger triangle. What is the perimeter of the larger triangle?

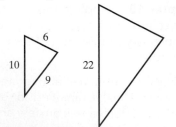

14. Two triangles are similar. The smaller triangle has sides 8 inches, 10 inches, and 12 inches. The 12-inch side on the smaller triangle corresponds to a side of 26 inches of the larger triangle. What is the perimeter of the larger triangle?

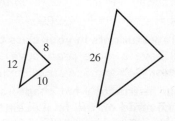

15. Two triangles are similar. The larger triangle has sides 8 ft, 10 ft, and 12 ft. The 8-ft side of the larger triangle corresponds to a side of length 6 ft of the smaller triangle. What is the perimeter of the smaller triangle?

16. Two triangles are similar. The larger triangle has sides 15 cm, 17 cm, and 24 cm. The 24-cm side on the larger triangle corresponds to a side of 9 cm of the smaller triangle. What is the perimeter of the smaller triangle?

Applications

17. ***Petting Zoo*** The zoo has hired an architect to design the triangular display case of the children's petting zoo. His scale drawing shows that the longest side of the display case is 9 cm and the shortest side is 5 cm. The longest side of the actual display case will be 90 cm. How long will the shortest side of the actual display case be?

18. ***Sculptor's Design*** A sculptor is designing her new triangular masterpiece. Her scale drawing shows that the shortest side of a triangular piece to be made measures 8 cm. The longest side of the drawing measures 25 cm. The longest side of the actual part to be sculpted must be 105 cm long. How long will the shortest side of the actual part be?

A flagpole casts a shadow. At the same time, a small tree casts a shadow. Use the sketch to find the height n of each flagpole.

19.

20.

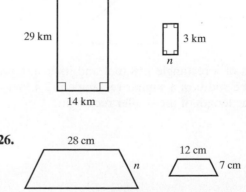

21. ***Shadow Cast*** Lola is standing outside of a department store. She is 5.5 feet tall and her shadow measures 6.5 feet long. The exterior wall of the store casts a shadow 96 feet long. How tall is the department store wall? Round your answer to the nearest foot.

22. ***Shadow Cast*** Huy is rock climbing in Utah. He is 6 ft tall and his shadow measures 8 ft long. The rock he wants to climb casts a shadow of 640 ft. How tall is the rock he is about to climb?

Each pair of figures is similar. Find the unknown side. Round to the nearest tenth when necessary.

23.

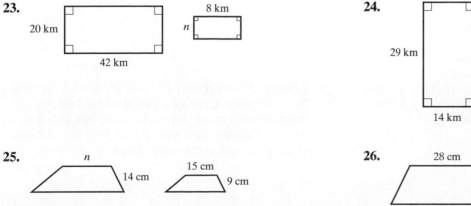

24.

25.

26.

27. ***New Kitchen*** Bob and Sue are planning a new kitchen. The old kitchen measured 9 feet by 12 feet. The new kitchen is similar in shape, but the longest dimension is 20 feet. What is the smaller dimension (width) of the new kitchen?

28. ***Family Photo*** Bertha took a great photo of the entire family at this year's reunion. She takes it to a professional photography studio and asks that the 3-in. by 5-in. photo be blown up to a larger size, which is 20 inches tall. What is the smaller dimension (width) of the enlargement?

To Think About *How would you find the relationship between the areas of two similar geometric figures? Consider the following two similar rectangles.*

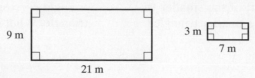

The area of the smaller rectangle is $(3 \text{ m})(7 \text{ m}) = 21 \text{ m}^2$. The area of the larger rectangle is $(9 \text{ m})(21 \text{ m}) = 189 \text{ m}^2$. How could you have predicted this result? Look at the ratio of small width to large width: $\frac{3}{9} = \frac{1}{3}$. The small rectangle has sides that are $\frac{1}{3}$ as large as the large rectangle. The ratio of the area of the small rectangle to the area of the large rectangle is $\frac{21}{189} = \frac{1}{9}$. Note that $\left(\frac{1}{3}\right)^2 = \frac{1}{9}$.

Thus we can develop the following principle:

The areas of two similar figures are in the same ratio as the square of the ratio of two corresponding sides. Use this information to complete exercises 29 and 30.

Each pair of geometric figures is similar. Find the unknown area. Round to the nearest tenth.

29.

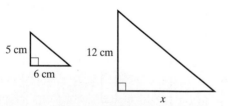

30.

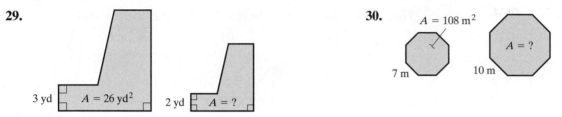

Cumulative Review *Perform the operations indicated. Use the correct order of operations.*

31. [1.6.4] $2 \times 3^2 + 4 - 2 \times 5$

32. [1.6.4] $300 \div (12 - 2 \times 3) + 2^4$

33. [1.6.4] $15 + 9 - (4^2 + 16 \div 8)$

34. [1.6.4] $108 \div 6 + 51 \div 3 + 3^3$

Quick Quiz 10.7

1. These two triangles are similar. Find the length of side x.

5 cm   12 cm
6 cm x

2. A tree that is 13.5 feet tall casts a shadow of 3 feet. At the same time, a building casts a shadow that is 10 feet. How tall is the building?

3. The width of a rectangle is 8 feet and its length is 10 feet. The width of a similar rectangle is 2.4 feet. What is the length of the similar rectangle?

4. Concept Check Refer to Quick Quiz 2 to answer the following question. The shadow cast by the building later in the day is $\frac{1}{2}$ of the length that it was earlier. Explain how you would find the shadow cast by the tree at the same time that afternoon.

Did You Know...

That You Can Save Thousands of Dollars in Fuel Costs with a Hybrid Vehicle Compared to a Less-Efficient Vehicle?

AFFORDABLE TRANSPORTATION

Understanding the Problem:

Michele wants to reduce her auto expenses. Her current vehicle is a 2003 SUV that gets only 16 miles per gallon. Different options for a replacement vehicle are

- **Option A:** A small SUV that gets 20 miles per gallon (MPG) and sells for $21,540.
- **Option B:** A hybrid model of the same SUV that gets 32 MPG and sells for $28,150.
- **Option C:** The most fuel-efficient hybrid car, which gets 60 MPG and sells for $23,770.

She wants to determine the cost and potential savings she would enjoy from the three different types of cars.

Making a Plan:

We need to calculate the cost to own the different vehicles.

Step 1: Michele drives about 15,000 miles a year. She wants to know what fuel costs would be for the three options.

Task 1: Determine the number of gallons of gasoline that each car (Options A, B, and C) would use if they are driven 15,000 miles in one year.

Task 2: If the average price of gasoline is $3.60 per gallon, how much would it cost to fuel each car for 15,000 miles?

Analyzing the Options:

Step 2: The SUV in Option B would cost Michele $6610 more to purchase than the SUV in Option A. Compare the different costs and savings of ownership.

Task 3: How much gas money would Michele save in one year with Option B compared to Option A?

Task 4: How many years would it take for Michele to save $6610 in gas money with Option B compared to Option A?

Task 5: How much gas money would Michele save after five years with Option B compared to Option A? After 10 years?

Step 3: The car in Option C would cost Michele $2230 more to purchase than the SUV in Option A. Compare the different costs of ownership.

Task 6: How much gas money would Michele save in one year with Option C compared to Option A?

Task 7: How many years would it take for Michele to save $2230 in gas money with Option C compared to Option A?

Task 8: How much gas money would Michele save after five years with Option C compared to Option A? After 10 years?

Making a Decision:

Step 4: Even though Michele enjoys the comfort of her SUV, she cannot ignore the potential savings in gas money she saw with Option C. She decides it is time to make a smart financial decision instead of one based on comfort. She expects to have her new car for at least five years. At that point, Option C will have saved her $9000 in gas money!

Applying the Situation to Your Life:

The amount of miles you drive per year as well as the price you pay for a gallon of gas is going to affect this analysis when applied to your own situation. The more miles you drive, the greater the savings from driving a fuel-efficient vehicle. Also, the higher the price of gas, the more you will save by switching.

Chapter 10 Organizer

Topic and Procedure	Examples	✏️ You Try It
Changing from one U.S. unit to another, p. 567 1. Find the equality statement that relates what you want to find and what you know. 2. Form a unit fraction. The numerator will contain the units you want to end up with. 3. Multiply by the unit fraction and simplify.	1. Convert 210 inches to feet. Use 12 inches = 1 foot. Unit fraction $= \dfrac{1\text{ ft}}{12\text{ in.}}$ $210\text{ in.} \times \dfrac{1\text{ ft}}{12\text{ in.}} = \dfrac{210}{12}\text{ ft}$ $= 17.5\text{ ft}$	1. Convert 35 feet to inches. *(handwritten)* $1\,ft = 12\,in$ $35\,ft \times \dfrac{12\,in}{1\,ft} = \dfrac{35}{12}\,ft$ $= 2.91\,ft$
Changing from one metric unit to another, p. 570 When you change from one prefix to another by moving to the *left* in the prefix guide, move the decimal point to the *left* the same number of places. kilo- hecto- deka- base unit deci- centi- milli- When you change from one prefix to another by moving to the *right* in the prefix guide, move the decimal point to the *right* the same number of places.	2. (a) Change 7.2 meters to kilometers. Move 3 decimal places to the left. $.007.2$ $7.2\text{ m} = 0.0072\text{ km}$ (b) Change 17.3 liters to milliliters. Move 3 decimal places to the right. 17.300 $17.3\text{ L} = 17{,}300\text{ mL}$	2. (a) Change 5.25 centimeters to kilometers. *(handwritten)* $0000 5.25\,km = 5.25\,cm$ (b) Change 6025 milliliters to deciliters. *(handwritten)* $6025\,mL = .006025\,L$
Changing from U.S. units to metric units, pp. 576–577 1. From the list of approximate equivalent measures, pick a statement that begins with the unit you now have and ends with the unit you want. 1 mi ≈ 1.61 km 1 gal ≈ 3.79 L 1 yd ≈ 0.914 m 1 qt ≈ 0.946 L 1 ft ≈ 0.305 m 1 lb ≈ 0.454 kg 1 in. = 2.54 cm (exact) 1 oz ≈ 28.35 g 2. Multiply by a unit fraction.	3. (a) Convert 7 gallons to liters. 1 gal ≈ 3.79 L $7\text{ gal} \times \dfrac{3.79\text{ L}}{1\text{ gal}} = 26.53\text{ L}$ (b) Convert 18 pounds to kilograms. 1 lb ≈ 0.454 kg $18\text{ lb} \times \dfrac{0.454\text{ kg}}{1\text{ lb}} = 8.172\text{ kg}$	3. (a) Convert 8 miles to kilometers. (b) Convert 25 quarts to liters.
Changing from metric units to U.S. units, pp. 576–577 1. From the list of approximate equivalent measures, pick a statement that begins with the unit you now have and ends with the unit you want. 1 km ≈ 0.62 mi 1 L ≈ 0.264 gal 1 m ≈ 1.09 yd 1 L ≈ 1.06 qt 1 m ≈ 3.28 ft 1 kg ≈ 2.2 lb 1 cm ≈ 0.394 in. 1 g ≈ 0.0353 oz 2. Multiply by a unit fraction.	4. (a) Convert 605 grams to ounces. 1 g ≈ 0.0353 oz $605\text{ g} \times \dfrac{0.0353\text{ oz}}{1\text{ g}} = 21.3565\text{ oz}$ (b) Convert 80 km/hr to mi/hr. 1 km ≈ 0.62 mi $80\,\dfrac{\text{km}}{\text{hr}} \times \dfrac{0.62\text{ mi}}{1\text{ km}} = 49.6\text{ mi/hr}$	4. (a) Convert 12 liters to gallons. (b) Convert 250 m/sec to ft/sec.
Naming angles, pp. 581–582 *Straight angle*—an angle that measures 180° *Acute angle*—an angle that measures between 0° and 90° *Obtuse angle*—an angle that measures between 90° and 180° *Right angle*—an angle that measures 90°	A B C D, 80°, E, F S, 140°, T, U 5. (a) State 3 ways we can name the acute angle. ∠DEF, or ∠FED, or ∠E (b) State 3 ways we can name the obtuse angle. ∠STU, or ∠UTS, or ∠T (c) Name the straight angle. ∠ABC or ∠CBA	X Y Z D, E, F A, 150°, O, B 5. (a) State 3 ways we can name the obtuse angle. *(handwritten)* ∠BOA, ∠AOB, O (b) State 3 ways we can name the right angle. *(handwritten)* ∠DEF, ∠FED, ∠E (c) Name the straight angle. *(handwritten)* ∠XYZ, ∠ZYX

Topic and Procedure	Examples	⟶ You Try It
Finding the complement and supplement of an angle, p. 582 *Supplementary angles*—two angles whose sum is 180° *Complementary angles*—two angles whose sum is 90°	**6.** Angle x measures 42°. Find the supplement and the complement of $\angle x$. Let $\angle s$ = the supplement of $\angle x$ and $\angle c$ = the complement of $\angle x$. $\angle s + \angle x = 180°$ $\angle c + \angle x = 90°$ $\angle s + 42° = 180°$ $\angle c + 42° = 90°$ $\angle s = 138°$ $\angle c = 48°$	**6.** Angle x measures 57°. Find the supplement and the complement of $\angle x$.
Identifying adjacent and vertical angles, p. 583 *Vertical angles*—two angles formed by intersecting lines that are opposite each other. Vertical angles have the same measure. *Adjacent angles*—two angles formed by intersecting lines that share a common side. Adjacent angles of intersecting lines are supplementary.	**7.** Find the measures of $\angle w$, $\angle y$, and $\angle z$. $\angle w + \angle x = 180°$ $\angle w$ and $\angle x$ are supplementary. $\angle w + 70° = 180°$ $\angle w = 110°$ $\angle w = \angle y = 110°$ Vertical angles are equal. $\angle x = \angle z = 70°$ Vertical angles are equal.	**7.** Find the measures of $\angle x$, $\angle y$, and $\angle z$.
Identifying alternate interior and corresponding angles, p. 584 *Transversal*—a line that intersects two or more lines at different points. *Alternate interior angles*—two angles that are on *opposite* sides of the transversal and *between* the other two lines. *Corresponding angles*—two angles that are on the *same side* of the transversal and are *both above* (or below) the other two lines. If two parallel lines are cut by a transversal, then the measures of *corresponding angles* are equal and the measures of *alternate interior angles* are equal.	**8.** $a \parallel b$ and the measure of $\angle v$ is 58°. Find the measures of $\angle x$, $\angle z$, $\angle w$, and $\angle y$. $\angle v = \angle x = 58°$ Alternate interior angles are equal. $\angle x = \angle z = 58°$ Vertical angles are equal. $\angle w = 180° - 58° = 122°$ $\angle w = \angle y = 122°$ Corresponding angles are equal.	**8.** $a \parallel b$ and the measure of $\angle y$ is 120°. Find the measures of $\angle x$, $\angle v$, $\angle w$, and $\angle z$.
Evaluating square roots of numbers that are perfect squares, pp. 592–593 The square root of a positive number written $\sqrt{a}$ is the positive number we square to get a. If $\sqrt{a} = b$, then $b^2 = a$.	**9.** Find the square root of each of the following. **(a)** $\sqrt{1}$ **(b)** $\sqrt{4}$ **(c)** $\sqrt{100}$ **(a)** $\sqrt{1} = 1$ because $(1)(1) = 1$ **(b)** $\sqrt{4} = 2$ because $(2)(2) = 4$ **(c)** $\sqrt{100} = 10$ because $(10)(10) = 100$	**9.** Evaluate. $\sqrt{49}$
Pythagorean Theorem, p. 594 In any right triangle with hypotenuse of length c and legs of lengths a and b, $c^2 = a^2 + b^2$ 	**10.** Find c to the nearest tenth if $a = 7$ and $b = 9$. $c^2 = 7^2 + 9^2 = 49 + 81 = 130$ $c = \sqrt{130} \approx 11.4$	**10.** Find a to the nearest tenth if $b = 5$ and $c = 12$.
Area of a triangle, p. 595 $A = \dfrac{bh}{2}$ b = base h = height 	**11.** Find the area of a right triangle whose base is 1.5 m and whose height is 3 m. $A = \dfrac{bh}{2} = \dfrac{(1.5 \text{ m})(3 \text{ m})}{2} = \dfrac{4.5 \text{ m}^2}{2}$ $= 2.25 \text{ m}^2$	**11.** Find the area of a right triangle whose base is 4 ft and whose height is 4.5 ft.

Topic and Procedure	Examples	✏ You Try It
Circumference of a circle, p. 600 $$C = \pi d$$ $$C = 2\pi r$$	**12.** Find the circumference of a circle with a diameter of 12 ft. $$C = \pi d = (3.14)(12 \text{ ft}) = 37.68 \text{ ft}$$ $$\approx 37.7 \text{ ft} \quad \text{(rounded to nearest tenth)}$$	**12.** Find the circumference of a circle with a radius of 5 m.
Area of a circle, p. 602 $$A = \pi r^2$$ **1.** Square the radius first. **2.** Then multiply the result by 3.14.	**13.** Find the area of a circle with radius 7 ft. $$A = \pi r^2 = (3.14)(7 \text{ ft})^2$$ $$= (3.14)(49 \text{ ft}^2)$$ $$\approx 153.9 \text{ ft}^2 \quad \text{(rounded to nearest tenth)}$$	**13.** Find the area of a circle with a radius of 5 m.
Volume of a cylinder, p. 606 $$r = \text{radius} \quad h = \text{height} \quad V = \pi r^2 h$$ **1.** Square the radius first. **2.** Then multiply the result by 3.14 and by the height.	**14.** Find the volume of a cylinder with a radius of 7 m and a height of 3 m. $$V = \pi r^2 h = (3.14)(7 \text{ m})^2(3 \text{ m})$$ $$= (3.14)(49)(3) \text{ m}^3$$ $$= (153.86)(3) \text{ m}^3 = 461.58 \text{ m}^3$$ $$\approx 461.6 \text{ m}^3 \quad \text{(rounded to nearest tenth)}$$	**14.** Find the volume of a cylinder with a radius of 4 ft and a height of 6 ft.
Volume of a sphere, p. 606 $$V = \frac{4\pi r^3}{3}$$ $$r = \text{radius}$$	**15.** Find the volume of a sphere of radius 3 m. $$V = \frac{4\pi r^3}{3} = \frac{(4)(3.14)(3 \text{ m})^3}{3}$$ $$= \frac{(4)(3.14)(\overset{9}{\cancel{27}}) \text{ m}^3}{\underset{1}{\cancel{3}}}$$ $$= (4)(3.14)(9) \text{ m}^3$$ $$= (12.56)(9) \text{ m}^3 = 113.04 \text{ m}^3$$ $$\approx 113.0 \text{ m}^3 \quad \text{(rounded to nearest tenth)}$$	**15.** Find the volume of a sphere of radius 5 ft.
Volume of a cone, p. 607 $$V = \frac{\pi r^2 h}{3}$$ $$r = \text{radius}$$ $$h = \text{height}$$	**16.** Find the volume of a cone of height 9 m and radius 7 m. $$V = \frac{\pi r^2 h}{3} = \frac{(3.14)(7 \text{ m})^2(9 \text{ m})}{3}$$ $$= \frac{(3.14)(7^2)(\overset{3}{\cancel{9}}) \text{ m}^3}{\underset{1}{\cancel{3}}}$$ $$= (3.14)(49)(3) \text{ m}^3$$ $$= (153.86)(3) \text{ m}^3 = 461.58 \text{ m}^3$$ $$\approx 461.6 \text{ m}^3 \quad \text{(rounded to nearest tenth)}$$	**16.** Find the volume of a cone of height 8 m and radius 5 m.
Volume of a pyramid, p. 608 $$V = \frac{Bh}{3}$$ $$B = \text{area of the base}$$ $$h = \text{height}$$ **1.** Find the area of the base. **2.** Multiply this area by the height and divide the result by 3.	**17.** Find the volume of a pyramid whose height is 6 m and whose rectangular base is 10 m by 12 m. $$B = (12 \text{ m})(10 \text{ m}) = 120 \text{ m}^2$$ $$V = \frac{(120)(\overset{2}{\cancel{6}}) \text{ m}^3}{\underset{1}{\cancel{3}}} = (120)(2) \text{ m}^3 = 240 \text{ m}^3$$	**17.** Find the volume of a pyramid whose height is 5 ft and whose rectangular base is 8 ft by 10 ft.
Similar figures, corresponding sides, p. 612 The corresponding sides of similar figures have the same ratio.	**18.** Find n in the following similar figures. $$\frac{n}{4} = \frac{9}{3}$$ $$3n = 36$$ $$n = 12 \text{ m}$$	**18.** Find x in the following similar figures.

Topic and Procedure	Examples	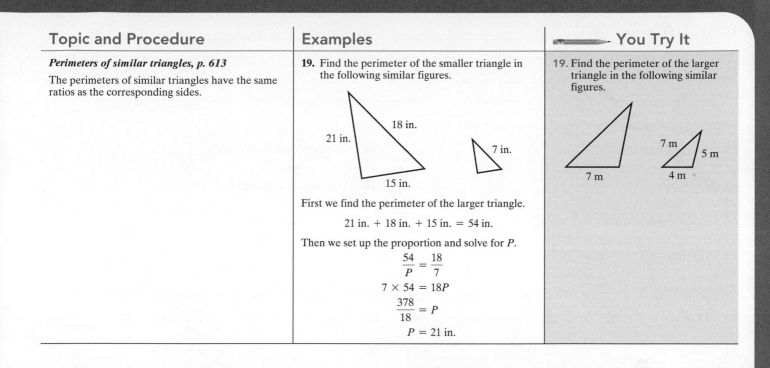 You Try It
Perimeters of similar triangles, p. 613 The perimeters of similar triangles have the same ratios as the corresponding sides.	**19.** Find the perimeter of the smaller triangle in the following similar figures. 18 in. 21 in. 7 in. 15 in. First we find the perimeter of the larger triangle. 21 in. + 18 in. + 15 in. = 54 in. Then we set up the proportion and solve for P. $$\frac{54}{P} = \frac{18}{7}$$ $$7 \times 54 = 18P$$ $$\frac{378}{18} = P$$ $$P = 21 \text{ in.}$$	**19.** Find the perimeter of the larger triangle in the following similar figures. 7 m 7 m 7 m 5 m 4 m

Chapter 10 Review Problems

Vocabulary

Fill in the blank with the definition for each word.

1. (10.3) Right angle. _____

_____

2. (10.3) Supplementary angles: _____

_____

3. (10.3) Complementary angles: _____

_____

4. (10.3) Vertical angles: _____

_____

5. (10.3) Adjacent angles: _____

_____

6. (10.3) Alternate interior angles: _____

_____

7. (10.3) Corresponding angles: _____

_____

8. (10.4) Right triangle: _____

_____

9. (10.4) Pythagorean Theorem: _____

_____

10. (10.5) Radius: _____

_____

11. (10.5) Diameter: _____

_____

12. (10.5) Circumference: _____

_____

Write each formula using symbols and numbers.

13. Area of a triangle

14. Circumference of a circle

15. Area of a circle

16. Volume of a cylinder

17. Volume of a sphere

18. Volume of a cone

19. Volume of a pyramid

Section 10.1 *Convert. When necessary, express your answer as a decimal. Round to the nearest hundredth.*

20. 33 ft = _____ yd

21. 1500 sec = _____ min

22. 78 in. = _____ ft

23. 15,840 ft = _____ mi

24. 3 tons = _____ lb

25. 15 gal = _____ qt

26. 92 oz = _____ lb

27. 31 pt = _____ qt

Convert. Do not round your answer.

28. 59 mL = _____ L

29. 8 cm = _____ mm

30. 2598 mm = _____ cm

31. 778 mg = _____ g

32. 6.3 m = _____ cm

33. 5 km = _____ m

34. 15 kL = _____ L

35. 473 m = _____ km

36. 196 kg = _____ g

37. 721 kg = _____ g

Solve each problem. Round your answer to the nearest hundredth when necessary.

38. *Chemist Solution* A chemist has a solution of 4 L. She needs to place it into 24 equal-size smaller jars. How many milliliters will be contained in each jar?

39. Find the perimeter of the triangle.
 (a) Express your answer in feet.
 (b) Express your answer in inches.

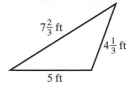

$7\frac{2}{3}$ ft $4\frac{1}{3}$ ft   5 ft

Section 10.2 *Perform each conversion. Round your answer to the nearest hundredth.*

40. 14 kg = _____ lb

41. 20 lb = _____ kg

42. 15 ft = _____ m

43. 2.4 ft = _____ cm

44. 13 oz = _____ g

45. 32°C = _____ °F

46. 14 cm = _____ in.

47. 32°F = _____ °C

48. *Distance Traveled* Keshia traveled at 90 km/hr for 3 hr. She needs to travel 200 mi. How much farther does she need to travel? State your answer in miles.

49. *Cereal Cost* The unit price on a box of cereal is $0.14 per ounce. The net weight is 450 g. How much does the cereal cost?

Section 10.3 *Use the figure to the right to answer exercises 50 and 51.*

50. State two other ways to name $\angle y$.

51. State two other ways to name $\angle x$.

52. $\angle M$ measures 39°.

    (a) Find the supplement of $\angle M$.

    (b) Find the complement of $\angle M$.

53. In the figure to the right, given that $\angle ABC$ is a right angle, find the measure of $\angle ABD$.

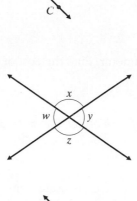

Use the figure to the right to answer exercises 54–56.

54. Name the angles that are adjacent to $\angle z$.

55. Find the sum of $\angle w$ and $\angle z$.

56. If $\angle z = 65°$, find $\angle x$, $\angle w$, and $\angle y$.

57. In the following figure, $a\|b$ and the measure of w is 45°. Find the measures of $\angle v$, $\angle u$, $\angle s$, and $\angle t$.

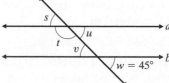

Section 10.4 *Find each square root without using a calculator.*

58. $\sqrt{144}$

59. $\sqrt{\dfrac{64}{81}}$

Approximate using a calculator. Round your answer to the nearest thousandth.

60. $\sqrt{45}$

61. $\sqrt{10}$

Find each square root and then simplify.

62. $\sqrt{64} - \sqrt{25}$

63. $\sqrt{121} + \sqrt{16}$

Find the area of the shape made up of a square and a right triangle. Round your answer to the nearest tenth.

64.

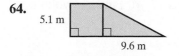

5.1 m

9.6 m

Find the unknown side of the right triangle. If the answer cannot be obtained exactly, use a calculator with a square root key. Round your answer to the nearest hundredth when necessary.

65.

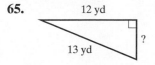

12 yd

13 yd

?

66. Find the width of a door if it is 6 ft tall and the diagonal line measures 7 ft.

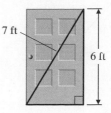

7 ft

6 ft

Section 10.5 *For exercises 67–71, use $\pi \approx 3.14$. Round your answer to the nearest hundredth if necessary.*

67. Find the circumference of a circle with diameter 12 in.

68. Find the circumference of a circle with radius 7 in.

Find the area of each circle.

69. Radius = 6 m

70. Diameter = 16 ft

71. A circular rug has a 10-foot diameter. Find the cost of the rug at $30 per square yard. ($1 \text{ yd}^2 = 9 \text{ ft}^2$.)

Section 10.6 *Find the volume of each object. Use $\pi \approx 3.14$ and round your answer to the nearest hundredth if necessary.*

72. A sphere with radius 1.2 ft

73. *Concrete Connector* A diagram of a concrete connector for the city sewer system is shown. It is shaped like a box with a hole of diameter 2 m.

    (a) Find the volume of the concrete connector.

    (b) If it is formed using concrete that costs $1.20 per cubic meter, how much will the necessary concrete cost?

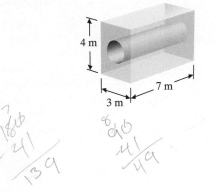

74. *Cylinder* Find the volume of the shape of a cylinder with a hemispherical top. The radius r is 2 cm and the height of the cylinder is 9 cm.

75. *Pyramid* Find the volume of a pyramid that is 18 m high and whose rectangular base measures 16 m by 18 m.

76. *Polluted Ground* A chemical has polluted a volume of ground in a cone shape. The depth of the cone is 30 yd. The radius of the cone is 17 yd. Find the volume of polluted ground.

Section 10.7

77. Find *n* in the set of similar triangles.

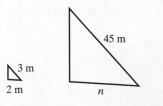

78. Find the perimeter of the larger of the two similar figures.

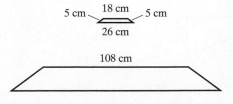

How Am I Doing? Chapter 10 Test

Test Prep
VIDEOS

MATH COACH   **MyMathLab®**   **You Tube™**

After you take this test read through the Math Coach on pages 629–630. Math Coach videos are available via MyMathLab and YouTube. Step-by-step test solutions on the Chapter Test Prep Videos are also available via MyMathLab and YouTube. (Search "BlairTobeyPrealgebra" and click on "Channels.")

Convert each of the following.

$\frac{145 \,oz \times 1\,lb}{16\,oz}$

1. **(a)** 145 ounces to pounds

(b) 5 yards to inches

$5\,yd \times \frac{3\,ft}{1\,yd} \times \frac{12\,in}{1\,ft}$

5×36

2. **(a)** 162 grams to kilograms

(b) 2.66 centimeters to millimeters

MC 3. Sharon purchased 2 kilograms of hamburger meat. $2\,kilo \times \frac{100\,g}{1\,kilo}$

(a) How many grams did she purchase?

(b) How many pounds did she purchase?

4. A recipe says to roast a turkey at 200°C for 4 hours. What temperature in Fahrenheit should be used?

5. Fred is traveling through Europe. After driving 4 miles, he sees a sign that says it is 14 kilometers to his destination. How many kilometers in total will he have traveled when he arrives at his destination?

6. $\angle T$ measures 41°.

(a) Find the supplement of $\angle T$.

(b) Find the complement of $\angle T$.

7. Given that $\angle XYZ$ is a right angle, find m.

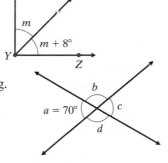

8. If $\angle a = 70°$, find the measure of the following.

(a) $\angle c$ **(b)** $\angle d$

9. In the figure to the right, $x \| y$ and the measure of $\angle w$ is 43°. Find the measures of $\angle s$, $\angle t$, $\angle u$, $\angle v$.

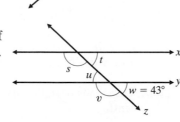

Determine if the following are perfect squares.

10. 81   11. 24   12. $\frac{1}{8}$

13. Find the length of the side of a square that has an area of 64 square feet.

Simplify.

14. $\sqrt{49}$ 15. $\sqrt{\frac{9}{16}}$

Find the approximate value. Round your answer to the nearest hundredth.

16. $\sqrt{6}$

---

1. (a) 2320 ☐

(b) 180 ☐

2. (a) ☐

(b) ☐

3. (a) 200 ☐

(b) ☐

4. ☐

5. ☐

6. (a) 139 ☐

(b) 459 ☐

7. ☐

8. (a) ☐

(b) ☐

9. ☐

10. ☐

11. ☐

12. ☐

13. ☐

14. ☐

15. ☐

16. ☐

17. _____ ☐
18. _____ ☐
19. _____ ☐
20. _____ ☐
21. _____ ☐
22. _____ ☐
23. _____ ☐
24. _____ ☐
25. _____ ☐
26. _____ ☐
27. _____ ☐
28. _____ ☐
29. _____ ☐
30. _____ ☐
31. _____ ☐
32. _____ ☐
Total Correct: ☐

Add.

17. $\sqrt{25} + \sqrt{36}$

18. Find the area of a triangle whose base is 10 centimeters and whose height is 12 centimeters.

19. Find the area of the side of this house.

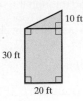

20. Find the length of one leg of a right triangle with one leg of 3 centimeters and the hypotenuse of 7 centimeters. Round to the nearest hundredth.

ᴹ𝖼 **21.** A ladder is placed against the side of a building. The top of the ladder is 14 feet from the ground. The base of the ladder is 11 feet from the building. What is the approximate length of the ladder? Round to the nearest tenth.

For questions 22–28, use $\pi \approx 3.14$.

22. Find the circumference of a circle if the diameter is 5.15 meters.

23. A pulley has a large wheel that has a 2.1-foot radius and a small wheel that has a 0.75-foot radius. If the large wheel makes 2 complete revolutions, how many revolutions does the small wheel make?

ᴹ𝖼 **24.** Find the area of a circle whose radius is 1.2 centimeters. Round your answer to the nearest tenth.

25. Joe wants to buy a circular tablecloth that is 6 feet in diameter. Find the cost of the tablecloth at $20 a square yard.

26. Find the volume of a tin can with a radius of 3 in. and a height of 7.1 in. Round your answer to the nearest hundredth.

27. How much air is needed to fully inflate a beach ball if the radius of the inner lining is 8 inches? Round your answer to the nearest hundredth.

ᴹ𝖼 **28.** Find the volume of a cone with radius 8 centimeters and height 12 centimeters.

29. Find the volume of a pyramid with height 12 meters, length of base 10 meters, and width of base 7 meters.

30. These two triangles are similar. Find the length of side n. Round to the nearest tenth.

31. A flagpole casts a shadow of 2.5 feet. At the same time, a tree that is 20 feet tall has a shadow of 2 feet. How tall is the flagpole?

32. The two rectangles shown here are similar. Find the width of the larger rectangle.

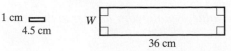

MATH COACH

Mastering the skills you need to do well on the test.

Students often make the same types of errors when they do the Chapter 10 Test. Here are some helpful hints to keep you from making these common errors on test problems.

Converting Units of Measurement—Problems 3(a) and (b)

Sharon purchased 2 kilograms of hamburger meat. (a) How many grams did she purchase? (b) How many pounds did she purchase?

> **Helpful Hint** Recall that 1 kilogram equals 1000 grams, and 1 kilogram is approximately 2.2 pounds.

For Part (a), did you understand that you have to move the decimal point 3 places?

Yes _____ No _____

Did you remember to move the decimal point to the right?

Yes _____ No _____

If you answered No to either of these questions, remember that the number of zeros in 1000 tells you how many places to move the decimal point. When converting from a larger to a smaller unit, you move the decimal point to the right.

For Part (b), did you use a unit fraction with kilograms in the denominator and pounds in the numerator?

Yes _____ No _____

Did you write $2 \text{ kg} \times \left(\dfrac{2.2 \text{ lb}}{1 \text{ kg}} \right)$?

Yes _____ No _____

If you answered No to either of these questions, consider why this is the correct unit fraction and perform this step. Be careful with your calculations.

If you answered Problem 3(a) or 3(b) incorrectly, go back and rework the problems using these suggestions.

Solving Applied Problems Involving Right Triangles—Problem 21

A ladder is placed against the side of a building. The top of the ladder is 14 feet from the ground. The base of the ladder is 11 feet from the building. What is the approximate length of the ladder? Round to the nearest tenth.

> **Helpful Hint** It is wise to first draw a picture and label each side of the figure. This will help you recognize that you must use the Pythagorean Theorem: $c^2 = a^2 + b^2$. Make sure you know which sides are the legs and which side is the hypotenuse: $\text{Hypotenuse}^2 = \text{leg}^2 + (\text{other leg})^2$.

When you substituted values, did you get $c^2 = 11^2 + 14^2$?

Yes _____ No _____

If you answered No, reread the problem and consider how to substitute these values correctly into the Pythagorean Theorem formula.

Did you simplify the equation to get $c^2 = 317$?

Yes _____ No _____

Next, did you approximate the square root of 317 and round to the nearest tenth?

Yes _____ No _____

If you answered No to either of these questions, check your calculations carefully. You can use the square root function on your calculator to approximate square roots. Remember to include the units in your final answer.

Now go back and rework this problem using these suggestions.

Need help? Watch the MATH COACH videos in MyMathLab® or on You Tube™.

629

Finding the Area of Circles—Problem 24 Find the area of a circle whose radius is 1.2 centimeters. Round your answer to the nearest tenth.

> **Helpful Hint** Try to memorize all the formulas and how to use them correctly. Take the extra time to write down the formula. This will help you to avoid errors in calculations.

Did you write down the formula $A = \pi r^2$?

Yes ____ No ____

If you answered No, go back and complete this step.

Did you perform the calculation $3.14 \times (1.2)$ to get your final answer?

Yes ____ No ____

If you answered Yes, stop and study the formula carefully. Notice that you must square the radius, 1.2 *before you multiply*. Remember to include the units in your final answer.

If you answered Problem 24 incorrectly, go back and rework the problem using these suggestions.

Finding the Volume of Pyramids—Problem 29 Find the volume of a pyramid with height 12 meters, length of base 10 meters, and width of base 7 meters.

> **Helpful Hint** It is important that you understand how to use this formula, $V = \dfrac{Bh}{3}$. The B refers to the area of the rectangular base, and the h refers to the height. You must find the area of the base first and substitute this value for B in the formula.

Did you perform the calculation $10 \text{ m} \times 7 \text{ m}$ as your first step to find B?

Yes ____ No ____

If you answered No, stop and make this correction to find the area of the base, B.

Next, did you write down $V = \dfrac{70 \text{ m}^2(12 \text{ m})}{3}$?

Yes ____ No ____

Did you get a final unit of m^3?

Yes ____ No ____

If you answered No to either of these questions, substitute values carefully and remember that the units must be multiplied too.

Now go back and rework the problem using these suggestions.

Need more help? Look for section examples marked with $\mathbb{MC}$ to review.

Practice Final Examination

This examination is based on Chapters 1–10 of the book.

Chapter 1

1. Round to the nearest hundred thousand. 7,543,876

2. Translate using numbers and symbols. A number decreased by 5.

3. Use whichever properties (associative and/or commutative) necessary to simplify the following expression. $(2 + 5 + x) + 4$

4. Evaluate $x + y$ if x is equal to 9 and y is equal to 12.

5. Find the perimeter of the shape made of two rectangles.

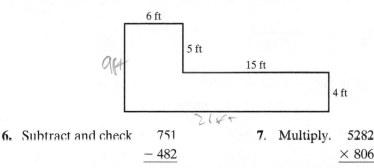

6. Subtract and check.
$$\begin{array}{r} 751 \\ -\ 482 \end{array}$$

7. Multiply.
$$\begin{array}{r} 5282 \\ \times\ 806 \end{array}$$

8. Divide. $812{,}869 \div 743$

9. Evaluate. $7 + 15 \div 3 - 3^2$

10. Simplify by combining like terms and then find the solution. $y + 4y + 3y = 16$

11. Gina must purchase supplies for her home office. The supply warehouse lists the following prices for the items she plans to purchase. Round each amount to the nearest ten and then estimate the amount of money she will pay for these supplies.

Price List	
Statistical calculator	$32
Color inkjet print cartridge	$29
Three-drawer metal file cabinet	$68
Computer chair	$189

Chapter 2

12. Replace the ? with the inequality symbol $<$ or $>$. $-5\ ?\ -3$

13. Evaluate the absolute value. $-|-13|$

14. Add. $7 + (-3) + 9 + (-4)$

15. Perform the operations indicated. $-9 - 7 + 8$

16. Multiply. $(-6)(2)(-1)(5)(-3)$ 17. Divide. $64 \div (-8)$

18. Simplify by combining like terms. $-9 + 5y + 7 - 3y$

19. Evaluate $a^2 - b$ for $a = -4$ and $b = 7$.   20. Simplify. $-3(a - 2)$

Chapter 3

In questions 21 and 22, solve and check your solution.

21. $5(3 - 7) = x - 3$ **22.** $6(3x) = 54$

23. The length (L) of a room is five times the width (W).
   (a) Write the statement as an equation.
   (b) Find the width of the room if the length is 40 feet.

24. Find the area of the following shape made up of rectangles.

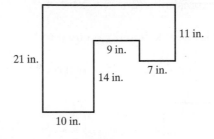

25. Find the base of the following parallelogram.

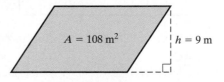

26. Find the unknown side of the following rectangular solid.

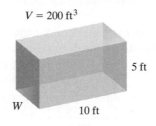

27. Daniel is purchasing carpet for his office, which measures 21 feet long and 15 feet wide. The carpet he plans to purchase is priced at $18 per square yard.
   (a) How many square yards of carpet must he purchase for the office?
   (b) How much will the carpet cost?

28. Multiply. $(5y)(z^6)(5y^4)(3z^2)$

29. Write the volume of the following rectangular solid as an algebraic expression and simplify.

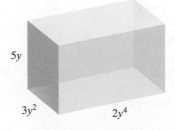

Chapter 4

30. Write the improper fraction as a mixed number. $\dfrac{57}{7}$

31. Change the mixed number to an improper fraction. $3\dfrac{1}{5}$

32. Simplify. $\dfrac{72y}{117yz}$

33. Simplify. $\dfrac{27}{-45}$

34. Simplify. $\dfrac{18x^7}{42x^{10}}$

35. Simplify. $(3^3)^2$

36. Write the ratio as a fraction in simplest form. 21 to 49

37. Write as a unit rate. 156 feet in 12 hours

38. The James and Sons law firm has 42 secretaries for every 14 lawyers and 12 paralegals for every 3 lawyers.
   (a) How many legal secretaries per lawyer does the law firm have?
   (b) How many paralegals per lawyer does the law firm have?
   (c) How many paralegals would be required if the law firm has 80 lawyers?

39. Find the value of x in the proportion and check your answer. $\dfrac{24}{x} = \dfrac{8}{5}$

40. On a map of France, 1 inch on the map represents 90 miles. How many miles do 5 inches represent?

Chapter 5

41. Multiply. $\dfrac{8y}{15} \cdot \dfrac{30}{24y^2}$

42. Divide. $\dfrac{9}{14} \div \dfrac{-72}{21}$

43. Dean wants to store $\frac{1}{3}$ pound of sugar in 4 equal-size containers. How much sugar should Dean place in each container?

44. Find the LCM. $6x, 18x, 27$

Perform the operations indicated.

45. $\dfrac{8}{y} - \dfrac{4}{y}$

46. $\dfrac{-5}{16} + \dfrac{9}{24}$

47. $\dfrac{5x}{18} - \dfrac{7x}{45}$

48. $12\dfrac{1}{7} - 7\dfrac{13}{21}$

49. $3\dfrac{3}{7} \div 6\dfrac{2}{3}$

50. Simplify. $\left(\dfrac{3}{4}\right)^2 + \dfrac{1}{9} \div \dfrac{1}{3}$

51. Solve. $\dfrac{3}{4}x = 27$

52. The Pine Ridge Country Club maintains the putting greens with a grass height of $\frac{2}{3}$ inch. The grass on the fairways is maintained at $1\frac{1}{3}$ inches. How much must the blade be lowered by a person mowing the fairways if that person will be using the same mowing machine on the putting greens?

30. _____

31. _____

32. _____

33. _____

34. _____

35. _____

36. _____

37. _____

38. (a) _____

(b) _____

(c) _____

39. _____

40. _____

41. _____

42. _____

43. _____

44. _____

45. _____

46. _____

47. _____

48. _____

49. _____

50. _____

51. _____

52. _____

53.

54.

55.

56.

57.

58.

59.

60.

61.

62.

63.

64.

65.

66.

67.

68.

69.

70. (a)

(b)

(c)

(d)

71.

72.

73.

74.

75.

634

Chapter 6

53. Perform the operations indicated. $(-4y^2 - 5) - (2y^2 + 3y + 2) + (-y^2 - 3)$

Multiply.

54. $(-3x)(x - 7)$ **55.** $(z^5)(z^6 - 3z - 5)$ **56.** $(y + 7)(4y^2 - 4y + 6)$

57. The measure of $\angle a$ is $40°$ more than the measure of $\angle b$. The measure of $\angle c$ is three times the measure of $\angle b$. Write the variable expressions for the measures of $\angle a$, $\angle b$, and $\angle c$.

Use the FOIL method to multiply.

58. $(y + 5)(y - 3)$ **59.** $(x - 4)(4x + 3)$

Factor.

60. $4z + 12x$ **61.** $9x^2y - 3xy$ **62.** $4x - 8y + 16$

Chapter 7

Solve and check your solution.

63. $-15 - 24 = 2x + 15 - 3x$ **64.** $-\dfrac{16}{4} = -3(6z) - 5z$

65. $-z = \dfrac{1}{4}$ **66.** $5y - 9 - 9y = 23$

67. $7a - 5 + 3a = -9 + 6a$

Solve.

68. $-2(y - 5) = 6(y + 5) - 12$ **69.** $\dfrac{x}{2} + \dfrac{x}{3} = 10$

For the following problem:

(a) *Define the variable expressions.*
(b) *Write an equation.*
(c) *Solve the equation.*
(d) *Check your answer.*

70. Last year 386 students took Ethnic Studies. 95 more students took it in the spring than in the fall. 75 fewer students took it in the summer than in the fall. How many students took it during each semester?

Chapter 8

71. Round to the nearest hundredth.   751.7596

72. Add. $0.588 + 75 + 8.59$ **73.** Multiply. 0.042×0.04

74. Danny earns \$9.50 per hour for the first 40 hours per week and \$13.65 for overtime hours (hours worked over 40 hours). If he works 52 hours this week, how much will he earn?

75. Divide and round your answer to the nearest hundredth. $-4.75 \div 3.2$

Complete the table of equivalent notations.

76.

Fraction Form	Decimal Form	Percent Form
$\frac{1}{4}$		
	0.56	
		350%

77. 79 is 25% of what number?

78. Find 30% of 90.

79. Amelia left a $5.75 tip for her dinner, which cost $28.50. What percent of the total bill did she leave for a tip? Round your answer to the nearest percent.

80. A private school predicts that 60% of the 350 spaces open for registration should be filled during the first week of registration. What percentage of the spaces will be available after the first week of registration?

Chapter 9

81. Use the pictograph to answer parts **(a)** and **(b)**.
   (a) How many more students are enrolled in Biology than in Chemistry?
   (b) What is the combined total enrollment in Physics and Biology?

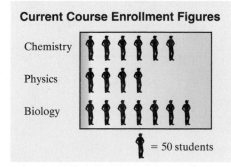

Current Course Enrollment Figures

Chemistry

Physics

Biology

= 50 students

82. The Tran family's monthly budget of $4600 is displayed in the circle graph.
   (a) What percent of the budget is taken up by the food and rent categories?
   (b) Of the total $4600, how much money per month is budgeted for utilities?

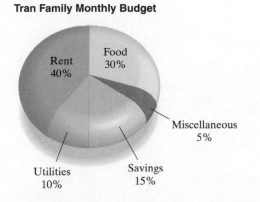

Tran Family Monthly Budget

Rent 40%
Food 30%
Miscellaneous 5%
Savings 15%
Utilities 10%

76. _____

77. _____

78. _____

79. _____

80. _____

81. (a) _____

(b) _____

82. (a) _____

(b) _____

83. Use the data below to construct a comparison line graph that compares the numbers of sales of food and snacks for Week 1 and Week 2 at the amusement park.

Amusement Park Food and Snack Sales

Week	Pizza Slices	Ice Cream	Candy
1	78	86	72
2	79	70	86

84. Find the median value of the following scores on a recent social science exam: 78, 59, 69, 77, 83, 92, 87, 95, and 72.

85. Find the mean value (average) for the following daily maximum temperature readings in Orange County during the last seven days in July: 89°F, 87°F, 92°F, 97°F, 85°F, 85°F, and 95°F.

86. Find the mode for the following set of numbers: 7, 9, 8, 9, 9, 7, 8, 10, 2.

87. Give the coordinates of each point.

    (a) R **(b)** T **(c)** S **(d)** U

88. Fill in the ordered pairs so that they are solutions to the equation $y = 2x - 4$:
($\_\_$, −6), ($\_\_$, −4), ($\_\_$, −8).

Graph.

89. $y = 3x - 2$

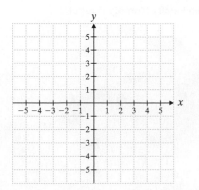

90. $y = -3$

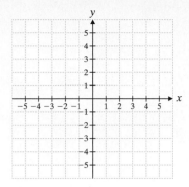

Chapter 10

91. Convert. 14 gal = _____ qt

92. Convert. 8 km = _____ m

93. Perform the conversion (1 ft ≈ 0.305 m). 13 ft = _____ m

94. Find the square root without using a calculator. $\sqrt{81}$

95. Find the sum. $\sqrt{121} + \sqrt{16}$

96. Find the unknown side of the right triangle.

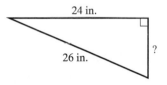

Use $\pi \approx 3.14$ *for questions 97 and 98.*

97. Find the circumference of a circle with a diameter of 13 ft.

98. Find the area of a circle with a diameter of 18 m.

99. Find the volume of a pyramid that is 9 feet high whose rectangular base measures 8 feet by 10 feet.

100. Find n in the following similar triangles.

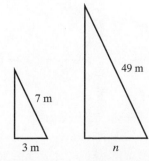

90.
_____

91.
_____

92.
_____

93.
_____

94.
_____

95.
_____

96.
_____

97.
_____

98.
_____

99.
_____

100.
_____

Appendix A Consumer Finance Applications

A.1 Balancing a Checking Account

① Calculating a Checkbook Balance

If you have a checking account, you should keep records of the checks written, ATM withdrawals, deposits, and other transactions on a check register. To find the amount of money in a checking account, you subtract debits and add credits to the balance in the account. Debits are checks written, withdrawals made, or any other amount charged to a checking account. Credits include deposits made, as well as any other money credited to the account.

Student Learning Objectives

After studying this section, you will be able to:

① Calculate a checkbook balance.

② Balance a checkbook.

EXAMPLE 1 Jesse Holm had a balance of $1254.32 in his checking account before writing five checks and making a deposit. On 9/2 Jesse wrote check #243 to the Manor Apartments for $575, check #244 to the Electric Company for $23.41, and check #245 to the Gas Company for $15.67. Then on 9/3, he wrote check #246 to Jack's Market for $125.57, check #247 to Clothing Mart for $35.85, and made a $634.51 deposit. Record the checks and deposit in Jesse's check register and then find Jesse's ending balance.

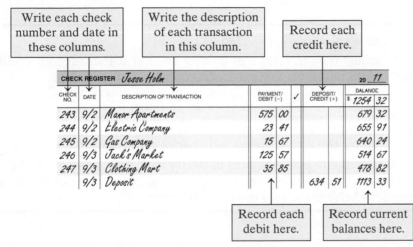

Write each check number and date in these columns.

Write the description of each transaction in this column.

Record each credit here.

Record each debit here.

Record current balances here.

CHECK NO.	DATE	DESCRIPTION OF TRANSACTION	PAYMENT/DEBIT (−)	✓	DEPOSIT/CREDIT (+)	BALANCE $ 1254	32
243	9/2	Manor Apartments	575 00			679	32
244	9/2	Electric Company	23 41			655	91
245	9/2	Gas Company	15 67			640	24
246	9/3	Jack's Market	125 57			514	67
247	9/3	Clothing Mart	35 85			478	82
	9/3	Deposit			634 51	1113	33

CHECK REGISTER Jesse Holm 20 11

Solution To find the ending balance, we subtract each check written and add the deposit to the current balance. Then we record these amounts in the check register.

$$
\begin{array}{ccccccc}
1254.32 & 679.32 & 655.91 & 640.24 & 514.67 & 478.82 \\
-\ 575.00 & -\ 23.41 & -\ 15.67 & -\ 125.57 & -\ 35.85 & +\ 634.51 \\
\hline
679.32 & 655.91 & 640.24 & 514.67 & 478.82 & 1113.33
\end{array}
$$

Jesse's balance is 1113.33.

Student Practice 1 My Chung Nguyen had a balance of $1434.52 in her checking account before writing three checks and making a deposit. On 3/1 My Chung wrote check #144 to the Leland Mortgage Company for $908 and check #145 to the Phone Company for $33.21. Then on 3/2 she wrote check #146 to Sam's Food Market for $102.37 and made a $524.41 deposit. Record the checks and deposit in My Chung's check register, and then find My Chung's ending balance.

NOTE TO STUDENT: Fully worked-out solutions to all of the Student Practice problems can be found at the back of the text starting at page SP-1.

Continued on next page

CHECK REGISTER	*My Chang Nguyen*					20 _11_
CHECK NO.	DATE	DESCRIPTION OF TRANSACTION	PAYMENT/ DEBIT (−)	✓	DEPOSIT/ CREDIT (+)	BALANCE $

② Balancing a Checkbook

The bank provides customers with bank statements each month. This statement lists the checks the bank paid, ATM withdrawals, deposits made, and all other debits and credits made to a checking account. It is very important to verify that these bank records match ours. We must make sure that we deducted all debits and added all credits in our check register. This is called **balancing a checkbook.** Balancing our checkbook allows us to make sure that the balance we think we have in our checkbook is correct. If our checkbook does not balance, we must look for any mistakes.

To balance a checkbook, proceed as follows.

1. First, *adjust the check register balance* so that it includes all credits and debits listed on the bank statement.

2. Then, *adjust the bank statement balance* so that it includes all credits and debits that may not have been received by the bank when the statement was printed. Checks that were written, but not received by the bank, are called **checks outstanding.**

3. Finally, *compare both balances* to verify that they are equal. If they are equal, the checking account balances. If they are not equal, we must find the error and make adjustments.

There are several ways to balance a checkbook. Most banks include a form you can fill out to assist you in this process.

EXAMPLE 2   Balance Jesse's checkbook using his check register and bank statement.

CHECK REGISTER	*Jesse Holm*					20 _11_
CHECK NO.	DATE	DESCRIPTION OF TRANSACTION	PAYMENT/ DEBIT (−)	✓	DEPOSIT/ CREDIT (+)	BALANCE $ 1254 32
243	9/2	Manor Apartments	575 00	✓		679 32
244	9/2	Electric Company	23 41	✓		655 91
245	9/2	Gas Company	15 67	✓		640 24
246	9/3	Jack's Market	125 57	✓		514 67
247	9/3	Clothing Mart	35 85			478 82
	9/3	Deposit		✓	634 51	1113 33
248	9/12	College Bookstore	168 96	✓		944 37
	9/18	ATM	100 00	✓		844 37
249	9/25	Telephone Company	43 29	✓		801 08
250	9/30	Sports Emporium	40 00			761 08
	10/1	Deposit			530 90	1291 98

Bank Statement: JESSE HOLM   9/1/2011 to 9/30/2011

Beginning Balance $1254.32
Ending Balance $831.68

Checks cleared by the bank

| #243 | $575.00 | #245 | $15.67 | #248 | $168.96 |
| #244 | $23.41 | #246* | $125.57 | #249 | $43.29 |

*Indicates that the next check in the sequence is outstanding (hasn't cleared).

Deposits

9/3 $634.51

Other withdrawals

9/18 ATM $100.00   Service charge $5.25

Solution   Follow steps 1–6 on the Checking Account Balance form below.

CHECKING RECONCILEMENT	This form is provided to assist you in balancing your checking account.

List checks outstanding* not charged to your checking account

CHECK NO.	AMOUNT
247	35 85
250	40 00
TOTAL	75 85

*and ATM withdrawals

Period ending   9/30 ,20 11

1. Check Register Balance		$	1291.98
Subtract any charges listed on the bank statement which you have not previously deducted from your balance.	−	$	5.25
Adjusted Check Register Balance		$	1286.73
2. Enter the ending balance shown on the bank statement.		$	831.68
3. Enter deposits made later than the ending date on the bank statement.	+	$	530.90
	+	$	
	+	$	
TOTAL (Step 2 plus Step 3)		$	1362.58
4. In your check register, **check off** all the checks paid. In the area provided to the left, **list** numbers and amounts of all outstanding checks and ATM withdrawals.			
5. Subtract the total amount in Step 4.	−	$	75.85
6. This adjusted bank balance should equal the adjusted Check Register Balance from Step 1.		$	1286.73

The balances in steps **1** and **6** are equal, so Jesse's checkbook is balanced.

Student Practice 2   Balance Anthony's checkbook using his check register, his bank statement, and the given Checking Account Balance Form.

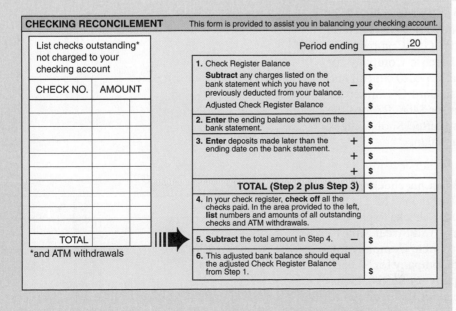

CHECK REGISTER *Anthony Maida*							20 _11_
CHECK NO.	DATE	DESCRIPTION OF TRANSACTION	PAYMENT/ DEBIT (−)	✓	DEPOSIT/ CREDIT (+)	BALANCE	
						$ 1823	00
211	7/2	Apple Apartments	985 00			838	00
	7/9	ATM	101 50			736	50
212	7/10	Leland Groceries	98 87			637	63
213	7/21	The Gas Company	45 56			592	07
	7/21	Deposit			687 10	1279	17
214	7/28	Cellular for Less	59 98			1219	19
215	7/28	The Electric Company	89 75			1129	44
216	7/28	Sports World	129 99			999	45
217	7/28	Leland Groceries	205 99			793	46
	7/30	ATM	141 50			651	96
	8/1	Deposit			398 50	1050	46

Bank Statement: ANTHONY MAIDA	7/1/2011 to 7/31/2011

Beginning Balance $1823.00
Ending Balance $934.95

Checks cleared by the bank

| #211 | $985.00 | #213 | $45.56 | #216 | $129.99 |
| #212 | $98.87 | #214* | $59.98 | | |

*Indicates that the next check in the sequence is outstanding (hasn't cleared).

Deposits

7/21 $687.10

Other withdrawals

| 7/9 ATM | $101.50 | Service charge | $3.50 |
| 7/30 ATM | $141.50 | Check purchase | $9.25 |

CHECKING RECONCILEMENT	This form is provided to assist you in balancing your checking account.

List checks outstanding* not charged to your checking account

CHECK NO.	AMOUNT
TOTAL	

*and ATM withdrawals

Period ending _____ ,20 ____

1. Check Register Balance		$	
Subtract any charges listed on the bank statement which you have not previously deducted from your balance.	−	$	
Adjusted Check Register Balance		$	
2. Enter the ending balance shown on the bank statement.		$	
3. Enter deposits made later than the ending date on the bank statement.	+	$	
	+	$	
	+	$	
TOTAL (Step 2 plus Step 3)		$	
4. In your check register, **check off** all the checks paid. In the area provided to the left, **list** numbers and amounts of all outstanding checks and ATM withdrawals.			
5. Subtract the total amount in Step 4.	−	$	
6. This adjusted bank balance should equal the adjusted Check Register Balance from Step 1.		$	

Note: To calculate the service charges, we add 3.50 + 9.25 = 12.75.

1. Shin Karasuda had a balance of $532 in his checking account before writing four checks and making a deposit. On 6/1 he wrote check #122 to the Mini Market for $124.95 and check #123 to Better Be Dry Cleaners for $41.50. Then on 6/9 he made a $384.10 deposit, wrote check #124 to Macy's Department Store for $72.98, and check #125 to Costco for $121.55. Record the checks and deposit in Shin's check register, and then find the ending balance.

CHECK REGISTER	Shin Karasuda					20 11
CHECK NO.	DATE	DESCRIPTION OF TRANSACTION	PAYMENT/ DEBIT (−)	✓	DEPOSIT/ CREDIT (+)	BALANCE $

2. Mary Beth O'Brian had a balance of $493 in her checking account before writing four checks and making a deposit. On 9/4 she wrote check #311 to Ben's Garage for $213.45 and #312 to Food Mart for $132.50. Then on 9/5 she made a $387.50 check deposit, wrote check #313 to the Shoe Pavilion for $69.98, and check #314 to the Electric Company for $92.45. Record the checks and deposit in Mary Beth's check register, and then find the ending balance.

CHECK REGISTER	Mary Beth O'Brian					20 11
CHECK NO.	DATE	DESCRIPTION OF TRANSACTION	PAYMENT/ DEBIT (−)	✓	DEPOSIT/ CREDIT (+)	BALANCE $

3. The Harbor Beauty Salon had a balance of $2498.90 in its business checking account on 3/4. The manager made a deposit on the same day for $786 and wrote check #734 to the Beauty Supply Factory for $980. Then on 3/9 he wrote check #735 to the Water Department for $131.85 and check #736 to the Electric Company for $251.50. On 3/19 he made a $2614.10 deposit and wrote two payroll checks: #737 to Ranik Ghandi for $873 and #738 to Eduardo Gomez for $750. Record the checks and deposits in the Harbor Beauty Salon's check register, and then find the ending balance.

CHECK REGISTER	Harbor Beauty Salon					20 11
CHECK NO.	DATE	DESCRIPTION OF TRANSACTION	PAYMENT/ DEBIT (−)	✓	DEPOSIT/ CREDIT (+)	BALANCE $

4. Joanna's Coffee Shop had a balance of $1108.50 in its business checking account on 1/7. The owner made a deposit on the same day for $963 and wrote check #527 to the Restaurant Supply Company for $492. Then on 1/11 she wrote check #528 to the Gas Company for $122.45 and check #529 to the Electric Company for $321.20. Then on 1/12 she made a $1518.20 deposit and wrote two payroll checks: #530 to Sara O'Conner for $579, and #531 to Nlegan Raskin for $466. Record the checks and deposits in Joanna's Coffee Shop's check register, and then find the ending balance.

CHECK REGISTER	Joanna's Coffee Shop					20 11
CHECK NO.	DATE	DESCRIPTION OF TRANSACTION	PAYMENT/ DEBIT (−)	✓	DEPOSIT/ CREDIT (+)	BALANCE $

5. On 3/3 Justin Larkin had $321.94 in his checking account before he withdrew $101.50 at the ATM. On 3/7 he made a $601.90 deposit and wrote checks to pay the following bills: check #114 to the Third Street Apartments for $550, check #115 to the Cable Company for $59.50, check #116 to the Electric Company for $43.50, and check #117 to the Gas Company for $15.90. Does Justin have enough money left in his checking account to pay $99 for his car insurance?

CHECK REGISTER	Justin Larkin					20 11
CHECK NO.	DATE	DESCRIPTION OF TRANSACTION	PAYMENT/ DEBIT (−)	✓	DEPOSIT/ CREDIT (+)	BALANCE $

6. On 5/2 Leon Jones had $423.54 in his checking account when he withdrew $51.50 at the ATM. On 5/9 he made a $601.80 deposit and wrote checks to pay the following bills: check #334 to Fasco Car Finance for $150.25, check #335 to A-1 Car Insurance for $89.20, check #336 to the Telephone Company for $33.40, and check #337 to the Apple Apartments for $615. Does Leon have enough money left in his checking account to pay $59 for his electric bill?

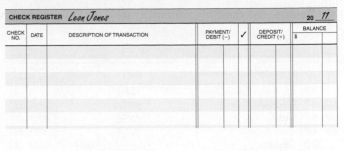

CHECK NO.	DATE	DESCRIPTION OF TRANSACTION	PAYMENT/DEBIT (−)	✓	DEPOSIT/CREDIT (+)	BALANCE
						$

CHECK REGISTER Leon Jones   20 __11__

7. Balance the monthly statement for the Carson Maid Service on the Checking Account Balance Form shown below.

CHECK REGISTER Carson Maid Service   20 __11__

CHECK NO.	DATE	DESCRIPTION OF TRANSACTION	PAYMENT/DEBIT (−)	✓	DEPOSIT/CREDIT (+)	BALANCE
						$ 1721 50
102	4/3	A&R Cleaning Supplies	422 33			1299 17
103	4/9	Allison De Julio	510 50			788 67
104	4/9	Mai Vu	320 00			468 67
105	4/9	Jon Veldez	320 00			148 67
	4/11	Deposit			1890 00	2038 67
106	4/20	Mobil Gas Company	355 35			1683 32
107	4/25	Potomac Property Mangt. warehouse rent	525 00			1158 32
108	4/29	Jack's Garage	450 10			708 22
	5/2	Deposit			540 00	1248 22

Bank Statement: CARSON MAID SERVICE　　4/1/2011 to 4/30/2011

Beginning Balance　$1721.50
Ending Balance　　$1564.82

Checks cleared by the bank

#102	$422.33	#104	$320.00	#108	$450.10
#103	$510.50	#105*	$320.00		

*Indicates that the next check in the sequence is outstanding (hasn't cleared).

Deposits

4/11　$1800.00

Other withdrawals

Service charge　$4.50
Check purchase　$19.25

CHECKING RECONCILEMENT　　This form is provided to assist you in balancing your checking account.

Period ending _____ ,20 ___

List checks outstanding* not charged to your checking account

CHECK NO.	AMOUNT
TOTAL	

*and ATM withdrawals

1. Check Register Balance　$_____

Subtract any charges listed on the bank statement which you have not previously deducted from your balance.　−　$_____

Adjusted Check Register Balance　$_____

2. Enter the ending balance shown on the bank statement.　$_____

3. Enter deposits made later than the ending date on the bank statement.　+　$_____
　+　$_____
　+　$_____

TOTAL (Step 2 plus Step 3)　$_____

4. In your check register, **check off** all the checks paid. In the area provided to the left, **list** numbers and amounts of all outstanding checks and ATM withdrawals.

5. Subtract the total amount in Step 4.　−　$_____

6. This adjusted bank balance should equal the adjusted Check Register Balance from Step 1.　$_____

8. Balance the monthly statement for The Flower Shop on the Checking Account Balance Form shown below.

CHECK REGISTER	The Flower Shop							20 _11_
CHECK NO.	DATE	DESCRIPTION OF TRANSACTION	PAYMENT/ DEBIT (−)		✓	DEPOSIT/ CREDIT (+)		BALANCE $ 3459 40
502	9/7	Whole Sale Flower Company	733	67				2725 73
503	9/7	Alexsandra Kruse	580	20				2145 53
504	9/7	Jose Sanchez	430	50				1715 03
505	9/7	Kamir Kosedag	601	90				1113 13
	9/11	Deposit				2654	00	3767 13
506	9/21	L&S Pottery	466	84				3300 29
507	9/24	Peterson Property Mangt. (warehouse rent)	985	00				2315 29
508	9/28	Barton Electric Company	525	60				1789 69
	10/2	Deposit				540	00	2329 69

Bank Statement: THE FLOWER SHOP   9/1/2011 to 9/30/2011

Beginning Balance $3459.40
Ending Balance $3220.28

Checks cleared by the bank

#502 $733.67   #504 $430.50   #508 $525.60
#503 $580.20   #505* $601.90
*Indicates that the next check in the sequence is outstanding (hasn't cleared).

Deposits

9/11 $2654.00

Other withdrawals

Service charge $4.75
Check purchase $16.50

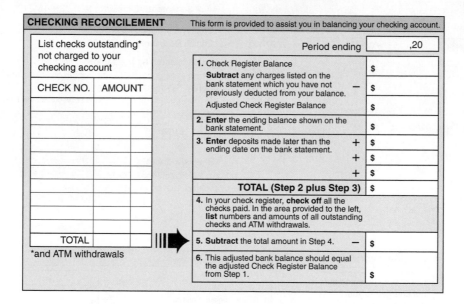

CHECKING RECONCILEMENT   This form is provided to assist you in balancing your checking account.

List checks outstanding* not charged to your checking account

CHECK NO.	AMOUNT
TOTAL	

*and ATM withdrawals

Period ending _____ ,20 ____

1. Check Register Balance — $ ____
   Subtract any charges listed on the bank statement which you have not previously deducted from your balance. — $ ____
   Adjusted Check Register Balance $ ____
2. **Enter** the ending balance shown on the bank statement. $ ____
3. **Enter** deposits made later than the ending date on the bank statement. + $ ____ + $ ____ + $ ____
   TOTAL (Step 2 plus Step 3) $ ____
4. In your check register, **check off** all the checks paid. In the area provided to the left, **list** numbers and amounts of all outstanding checks and ATM withdrawals.
5. **Subtract** the total amount in Step 4. − $ ____
6. This adjusted bank balance should equal the adjusted Check Register Balance from Step 1. $ ____

9. On 2/1 Jeremy Sirk had a balance of $672.10 in his checking account. On the same day he deposited $735 of his paycheck into his checking account and wrote check #233 to Stanton Sporting Goods for $92.99 and check #234 to the Garden Apartments for $680. Then on 2/20 he wrote check #235 to the Gas Company for $31.85, check #236 to the Cable Company for $51.50, and check #237 to Ralph's Market for $173.98, and also made an ATM withdrawal for $101.50. Then on 3/1 he made an $814.10 deposit, wrote check #238 to State Farm Insurance for $98, and made another ATM withdrawal for $41.50. Record the checks and deposits in Jeremy's check register, and then find the ending balance.

CHECK REGISTER	Jeremy Sirk							20 _11_
CHECK NO.	DATE	DESCRIPTION OF TRANSACTION	PAYMENT/ DEBIT (−)	✓	DEPOSIT/ CREDIT (+)		BALANCE $	

10. On 8/1 Shannon Mending had a balance of $525.90 in her checking account. Later that same day she deposited $588.23 into her checking account and wrote check #333 to Verizon Telephone Company for $33.20 and check #334 to Discount Car Insurance for $332.50. Then on 8/8 she wrote check #335 to the Walden Market for $21.35, made an ATM withdrawal for $81.50, and wrote check #336 to the Cable Company for $41.50. Then on 9/1 she made a $904.10 deposit, wrote check #337 to Next Day Dry Cleaners for $33.50, and check #338 to Marty's Dress Shop for $87.99. Record the checks and deposits in Shannon's check register, and then find the ending balance.

CHECK REGISTER *Shannon Mending*						20 _11_
CHECK NO.	DATE	DESCRIPTION OF TRANSACTION	PAYMENT/ DEBIT (−)	✓	DEPOSIT/ CREDIT (+)	BALANCE $

11. Refer to exercise 9 and use the bank statement and Checking Account Balance Form below to balance Jeremy's checking account.

Bank Statement: JEREMY SIRK   2/1/2011 to 2/28/2011

Beginning Balance $672.10
Ending Balance $293.88

Checks cleared by the bank

#233	$92.99	#236	$51.50
#234*	$680.00	#237	$173.98

*Indicates that the next check in the sequence is outstanding (hasn't cleared).

Deposits

2/1 $735.00

Other withdrawals

2/20 ATM $101.50   Service charge $3.50
                         Check purchase $9.75

CHECKING RECONCILEMENT   This form is provided to assist you in balancing your checking account.

List checks outstanding* not charged to your checking account

CHECK NO.	AMOUNT
TOTAL	

*and ATM withdrawals

Period ending _____ ,20 ____

1. Check Register Balance $ _____
   Subtract any charges listed on the bank statement which you have not previously deducted from your balance. − $ _____
   Adjusted Check Register Balance $ _____
2. Enter the ending balance shown on the bank statement. $ _____
3. Enter deposits made later than the ending date on the bank statement. + $ _____
   + $ _____
   + $ _____
   TOTAL (Step 2 plus Step 3) $ _____
4. In your check register, **check off** all the checks paid. In the area provided to the left, **list** numbers and amounts of all outstanding checks and ATM withdrawals.
5. **Subtract** the total amount in Step 4. − $ _____
6. This adjusted bank balance should equal the adjusted Check Register Balance from Step 1. $ _____

12. Refer to exercise 10 and use the bank statement and Checking Account Balance Form below to balance Shannon's checking account.

Bank Statement: SHANNON MENDING   8/1/2011 to 8/31/2011

Beginning Balance $525.90
Ending Balance $923.83

Checks cleared by the bank

#333*	$33.20	#336	$41.50
#335	$21.35		

*Indicates that the next check in the sequence is outstanding (hasn't cleared).

Deposits

8/1 $588.23

Other withdrawals

8/8 ATM $81.50   Service charge $3.25
                       Check purchase $9.50

CHECKING RECONCILEMENT   This form is provided to assist you in balancing your checking account.

List checks outstanding* not charged to your checking account

CHECK NO.	AMOUNT
TOTAL	

*and ATM withdrawals

Period ending _____ ,20 ____

1. Check Register Balance $ _____
   Subtract any charges listed on the bank statement which you have not previously deducted from your balance. − $ _____
   Adjusted Check Register Balance $ _____
2. Enter the ending balance shown on the bank statement. $ _____
3. Enter deposits made later than the ending date on the bank statement. + $ _____
   + $ _____
   + $ _____
   TOTAL (Step 2 plus Step 3) $ _____
4. In your check register, **check off** all the checks paid. In the area provided to the left, **list** numbers and amounts of all outstanding checks and ATM withdrawals.
5. **Subtract** the total amount in Step 4. − $ _____
6. This adjusted bank balance should equal the adjusted Check Register Balance from Step 1. $ _____

A.2 Determining the Best Deal When Purchasing a Vehicle

Student Learning Objectives

After studying this section, you will be able to:

① **Find the true purchase price of a vehicle.**

② **Find the total cost of a vehicle to determine the best deal.**

When we buy a car there are several facts to consider in order to determine which car has the best price. The sale price offered by a car dealer or seller is only one factor we must consider—others include the interest rate on the loan, sales tax, license fee, and sale promotions such as cash rebates or 0% interest. We must also consider the cost of options we choose such as extended warranties, sun roof, tinted glass, etc. In this section we will see how to calculate the total purchase price and determine the best deal when buying a vehicle.

① Finding the True Purchase Price of a Vehicle

In most states there is a sales tax and a license or title fee that must be paid on all vehicles purchased. To find the true purchase price for a vehicle, we *add* these extra costs to the sale price. In addition, we must add to the sale price the cost of any extended warranties and extra options or accessories we buy.

$$\text{purchase price} = \text{sale price} + \text{sales tax} + \text{license fee} + \text{extended warranty (and other accessories)}$$

Sometimes lenders (banks and finance companies) require a **down payment.** The amount of the down payment is usually a percent of the purchase price. We subtract the down payment from the purchase price to find the amount we must finance.

$$\text{down payment} = \text{percent} \times \text{purchase price}$$

$$\text{amount financed} = \text{purchase price} - \text{down payment}$$

EXAMPLE 1 Daniel bought a truck that was on sale for $28,999 in a city that has a 6% sales tax and a 2% license fee.

(a) Find the sales tax and license fee Daniel paid.

(b) Daniel also bought an extended warranty for $1550. Find the purchase price of the truck.

Solution

(a)
$$\begin{aligned} \text{sales tax} &= 6\% \text{ of sale price} \\ &= 0.06 \times 28{,}999 \\ \text{sales tax} &= \$1739.94 \end{aligned}$$
$$\begin{aligned} \text{license fee} &= 2\% \text{ of sale price} \\ &= 0.02 \times 28{,}999 \\ \text{license fee} &= \$579.98 \end{aligned}$$

(b)
$$\begin{aligned} \text{purchase price} &= \text{sale price} + \text{sales tax} + \text{license fee} + \text{extended warranty} \\ &= 28{,}999 + 1739.94 + 579.98 + 1550 \\ \text{purchase price} &= \$32{,}868.92 \end{aligned}$$

Student Practice 1 Huy Nguyen bought a van that was on sale for $24,999 in a city that has a 7% sales tax and a 2% license fee.

(a) Find the sales tax and license fee Huy paid.

(b) Huy also bought an extended warranty for $1275. Find the purchase price of the van.

NOTE TO STUDENT: Fully worked-out solutions to all of the Student Practice problems can be found at the back of the text starting at page SP-1.

EXAMPLE 2 The purchase price of a car Jerome plans to buy is $19,999. In order to qualify for the loan on the car, Jerome must make a down payment of 20% of the purchase price.

(a) Find the down payment.

(b) Find the amount financed.

Solution

(a) down payment = percent × purchase price

down payment = 20% × 19,999

= 0.20 × 19,999

down payment = $3999.80

(b) amount financed = purchase price − down payment

= 19,999 − 3999.80

amount financed = $15,999.20

Student Practice 2 The purchase price of a Jeep Cheryl plans to buy is $32,499. In order to qualify for the loan on the Jeep, Cheryl must make a down payment of 15% of the purchase price.

(a) Find the down payment.   **(b)** Find the amount financed.

② Finding the Total Cost of a Vehicle to Determine the Best Deal

When we borrow money to buy a car, we often pay interest on the loan and we must consider this extra cost when we calculate the **total cost** of the vehicle. If we know the amount of the car payment and the number of months it will take to pay off the loan, we can find the total of the payments on the car (the amount we borrowed plus interest) by multiplying the monthly payment amount times the number of months of the loan. Then we must add the down payment to that amount to find the total cost of the vehicle.

total cost = (monthly payment × number of months in loan) + down payment

EXAMPLE 3   Marvin went to two dealerships to find the best deal on the truck he plans to purchase. From which dealership should Marvin buy the truck so that the *total cost* of the truck is the least expensive?

Dealership 1	Dealership 2
• Purchase price: $39,999	• Purchase price: $36,499
• Financing option: 2% financing with $5000 down payment	• Financing option: 5% financing with no down payment
• Monthly payments: $743.73 per month for 48 months	• Monthly payments: $638.73 per month for 60 months

Solution   First, we find the total cost of the truck at Dealership 1.

total cost = (monthly payment × number of months in loan) + down payment

= (743.73 × 48)   + 5000   We multiply, then add.

= 40,699.04

The total cost of the truck at Dealership 1 is $40,699.04.

Next, we find the total cost of the truck at Dealership 2.

total cost = (monthly payment × number of months in loan) + down payment

= (638.73 × 60)   + 0   There is no down payment.

= 38,323.80

The total cost of the truck at Dealership 2 is $38,323.80.

We see that the best deal on the truck Marvin plans to buy is at Dealership 2.

Student Practice 3   Phoebe went to two dealerships to find the best deal on the minivan she plans to purchase. From which dealership should Phoebe buy the minivan so that the *total cost* of the minivan is the least expensive?

Dealership 1	Dealership 2
• Purchase price: $22,999	• Purchase price: $23,999
• Financing option: 4% financing with no down payment	• Financing option: 2% financing with $5500 down payment
• Monthly payments: $398.65 per month for 60 months	• Monthly payments: $393.10 per month for 48 months

1. A college student buys a car and pays a 6% sales tax on the $21,599 sale price. How much sales tax did the student pay?

2. A high school teacher buys a minivan and pays a 5% sales tax on the $26,800 sale price. How much sales tax did the teacher pay?

3. Francis is planning to buy a four-door sedan that is on sale for $18,999. He must pay a 2% license fee. Find the license fee.

4. Mai Vu saw an ad for a short-bed truck that is on sale for $17,599. If she buys the truck she must pay a 2% license fee. Find the license fee.

5. John must make a 10% down payment on the purchase price of the $42,450 sports car he is planning to buy. Find the down payment.

6. Kamir must make a 15% down payment on the purchase price of the $31,500 extended cab truck he is planning to buy. Find the down payment.

7. Tabatha bought a minivan that was on sale for $24,899 in a city that has a 5% sales tax and a 2% license fee.

   (a) Find the sales tax and license fee Tabatha paid.

   (b) Tabatha also bought an extended warranty for $1100. Find the purchase price of the minivan.

8. Dante bought a truck that was on sale for $32,499 in a city that has a 7% sales tax and a 2% license fee.

   (a) Find the sales tax and license fee Dante paid.

   (b) Dante also bought an extended warranty for $1600. Find the purchase price of the truck.

9. Jeremiah bought a sports car that was on sale for $44,799 in a city that has a 7% sales tax and a 2% license fee.

   (a) Find the sales tax and license fee Jeremiah paid.

   (b) Jeremiah also bought an extended warranty for $2100. Find the purchase price of the sports car.

10. Dawn bought a four-door sedan that was on sale for $31,899 in a city that has a 6% sales tax and a 2% license fee.

    (a) Find the sales tax and license fee Dawn paid.

    (b) Dawn also bought an extended warranty for $1300. Find the purchase price of the sedan.

11. The purchase price of an SUV that a soccer coach plans to buy is $49,999. In order to qualify for the loan on the SUV, the coach must make a down payment of 15% of the purchase price.

    (a) Find the down payment.

    (b) Find the amount financed.

12. The purchase price of a flat-bed truck a contractor plans to buy is $39,999. In order to qualify for the loan on the truck, the contractor must make a down payment of 10% of the purchase price.

    (a) Find the down payment.

    (b) Find the amount financed.

13. Tammy went to two dealerships to find the best deal on the truck she plans to purchase. From which dealership should Tammy buy the truck so that the *total cost* of the truck is the least expensive?

Dealership 1	Dealership 2
• Purchase price: $35,999	• Purchase price: $32,499
• Financing option: 0% financing with $4000 down payment	• Financing option: 4% financing with $2000 down payment
• Monthly payments: $696.65 per month for 48 months	• Monthly payments: $603.58 per month for 60 months

14. John went to two dealerships to find the best deal on a luxury SUV for his company to purchase. From which dealership should John buy the SUV so that the *total cost* of the SUV is the least expensive?

Dealership 1	Dealership 2
• Purchase price: $49,999	• Purchase price: $46,499
• Financing option: 0% financing with $5000 down payment	• Financing option: 4% financing with no down payment
• Monthly payments: $1010.39 per month for 48 months	• Monthly payments: $916.29 per month for 60 months

To Think About

Natasha went to three car dealerships to check the prices of the same Ford two-door coupe. The city where the dealerships are located has a sales tax of 5% and a license fee of 2%. All dealerships offer extended warranties that are 3 years/70,000 miles. Use the following information gathered by Natasha to answer exercises 15 and 16.

Dealership 1—Ford Coupe	Dealership 2—Ford Coupe	Dealership 3—Ford Coupe
• $24,999 plus $2000 rebate	• $23,799; dealer pays sales tax	• $23,999
• extended warranty $1350	• extended warranty $1450	• free extended warranty

15. (a) Which dealership offers the least expensive *purchase price*? State this amount.

(b) Each dealership offers a *different interest rate* on a 60-month loan without a down payment, resulting in the following monthly payments:

Dealership 1	Dealership 2	Dealership 3
$480.65/month	$485.46/month	$496.44/month

From which dealership should Natasha buy the Ford coupe so that the *total cost* of the car is the least expensive? State this amount.

(c) Compare the results of parts **(a)** and **(b)**. What conclusion can you make?

16. (a) Which dealership offers the most expensive *purchase price*? State this amount.

(b) Each dealership offers a *different interest rate* on a 48-month loan without a down payment, resulting in the following monthly payments:

Dealership 1	Dealership 2	Dealership 3
$589.27/month	$594.07/month	$603.07/month

From which dealership should Natasha buy the Ford coupe so that the *total cost* of the car is the most expensive? State this amount.

(c) Compare the results of parts **(a)** and **(b)**. What conclusion can you make?

Appendix B Introduction to U.S. and Metric Units of Measurement

The two most common systems of measurement in the world are the metric system and the U.S. system. Nearly all countries in the world use the metric system, but in the United States we use the U.S. system, except in fields such as science and medicine. It is important that you learn both systems, since you will deal with both to some extent.

In this section we introduce the U.S. and metric systems and give a general presentation of each system. That is, we answer questions such as, "How many inches are in a yard?" and "How long is a kilometer?" We also introduce the idea of "thinking metric." By this we mean the ability to estimate how big or little a metric unit is and to compare metric units to U.S. units.

① Understand U.S. Units

The table below indicates the relationships between units of measure that we often use in our daily lives. Your instructor may require that you memorize these relationships.

Length	Weight
12 inches (in.) = 1 foot (ft)	16 ounces (oz) = 1 pound (lb)
3 feet (ft) = 1 yard (yd)	2000 pounds (lb) = 1 ton
5280 feet (ft) = 1 mile (mi)	

Volume	Time
2 cups (c) = 1 pint (pt)	60 seconds (sec) = 1 minute (min)
2 pints (pt) = 1 quart (qt)	60 minutes (min) = 1 hour (hr)
4 quarts (qt) = 1 gallon (gal)	24 hours (hr) = 1 day
	7 days = 1 week

1 yard

1 foot

It is important that you become familiar with these equivalent units, since we use these units in many aspects of our daily lives.

EXAMPLE 1

(a) 3 ft = __?__ yd

(b) 1 ton = __?__ lb

Solution

(a) 3 feet has the same measure of length as 1 yard: 3 ft = 1 yd.

(b) 1 ton has the same weight as 2000 pounds: 1 ton = 2000 lb.

Student Practice 1

(a) 60 sec = __?__ min

(b) 1 gal = __?__ qt

NOTE TO STUDENT: Fully worked-out solutions to all of the Student Practice problems can be found at the back of the text starting at page SP-1.

Although we show a method to convert from one unit to another in Chapter 10, for simple conversions we can use addition facts.

EXAMPLE 2   21 days = ___?___ weeks

Solution   There are 7 days in 1 week.

$$21 \text{ days} = 7 \text{ days} + 7 \text{ days} + 7 \text{ days}$$
$$\downarrow \qquad \downarrow \qquad \downarrow$$
$$= 1 \text{ week} + 1 \text{ week} + 1 \text{ week}$$
$$21 \text{ days} = 3 \text{ weeks}$$

Student Practice 2

(a) 4 pt = ___?___ qt

(b) 2 tons = ___?___ lb

Often, we must write measurements using more than one unit. For example, 6 hours and 10 minutes can be written 6 hr + 10 min or 6 hr 10 min (we usually do not write the + sign).

EXAMPLE 3   Perform the operation indicated.

(a) 3 hr 40 min + 5 hr 30 min

(b) 5 yd 1 ft − 3 yd 2 ft

Solution   Only measurements of the same units can be added or subtracted. We line up the common units and either add or subtract, starting in the column on the right.

(a)   3 hr 40 min
       + 5 hr 30 min

       8 hr 70 min   Since 70 min = 60 min + 10 min,

       8 hr 1 hr 10 min   we can write 70 min as 1 hr 10 min.

       9 hr 10 min   We add: 8 hr + 1 hr = 9 hr.

Note that this is an example of carrying units. We carried an hour to the hours column in order to change the 70 minutes to hours and minutes. This form gives us a better idea of the size of the measurement.

(b)   5 yd 1 ft   We cannot subtract 1 ft − 2 ft.
       − 3 yd 2 ft

Since 1 yd = 3 ft we can write 5 yd = 4 yd + 1 yd = 4 yd + 3 ft

     5 yd   1 ft   4 yd + 3 ft   1 ft   4 yd   4 ft   3 ft + 1 ft = 4 ft
   − 3 yd   2 ft   − 3 yd   2 ft   − 3 yd   2 ft
                                                        1 yd   2 ft

Note that this is an example of borrowing units. We borrowed a yard from the yards column and added it to the feet column so that we could subtract.

Student Practice 3   Perform the operation indicated.

(a) 12 min 43 sec + 4 min 33 sec

(b) 4 gal 1 qt − 2 gal 3 qt

② **Understand Metric Units**

Meters, grams, and liters are the *basic units* in the metric system. All other units in the metric system are based on the number 10 and these basic units. The **meter** measures length, and the **liter** measures volume. The **gram** is a unit of measure that is related to the mass and weight of an object. We can use the basic unit gram when we are referring to weight.

Length - Meter (m)

Just as in the U.S. system, various size units have standard names. The prefixes used for these standard names are *kilo, hecto, deka, deci, centi,* and *milli.* For example, 10 *centi*meters = 1 *deci*meter. Units that are larger than the basic unit use the prefixes *kilo,* meaning 1000; *hecto,* meaning 100; and *deka,* meaning 10. Units smaller than the basic unit use the prefixes *deci,* meaning 1/10; *centi,* meaning 1/100; and *milli,* meaning 1/1000.

Weight - Gram (g)

Prefixes	Basic unit	Prefixes
kilo hecto deka	gram liter meter	deci centi milli

These prefixes identify units that are *larger* than the basic unit.   These prefixes identify units that are *smaller* than the basic unit.

Volume - Liter (L)

We say

    1 *kilo*gram for weight   1 *kilo*liter for volume   1 *kilo*meter for length

Length. We use the *basic unit* **meter** when we are dealing with length. The table in the margin indicates the abbreviations for length.

> **Abbreviations for Length**
> | | |
> |---|---|
> | 1 kilometer (km) | Larger units |
> | 1 hectometer (hm) | |
> | 1 dekameter (dam) | |
> | 1 meter (m) | |
> | 1 decimeter (dm) | |
> | 1 centimeter (cm) | |
> | 1 millimeter (mm) | Smaller units |

EXAMPLE 4   Which of the following represents a unit that is larger than the basic unit, meter: 1 millimeter or 1 dekameter?

Solution   If you memorize the prefixes in the order listed below, it is easier to determine which ones represent the larger or smaller units. The size of the units decreases as you move from left to right on the chart.

larger units ————————————————→ smaller units

Prefixes

kilo   hecto   deka   basic unit   deci   centi   milli
            ▲   (meter)   ▲

The prefix *deka* is located to the left of the meter, so it represents a unit larger than a meter. Thus 1 dekameter is larger than the basic unit, meter.

Student Practice 4   Which of the following represents a unit that is smaller than the basic unit, meter: 1 centimeter or 1 hectometer?

Thinking Metric. Since we sometimes encounter metric units in the United States, it is convenient for us to know the answers to questions such as, "Which is larger, a meter or a yard?" or "How small is a millimeter?" The following drawings show these types of relationships.

A *millimeter* is very small. The thickness of the edge of a paper clip is about 1 millimeter.

A *centimeter* is smaller than 1 inch. Approximately $2\frac{1}{2}$ centimeters equal 1 inch.

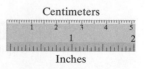

A *meter* is a little longer than a yard.

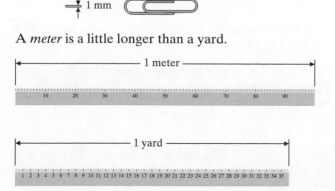

A *kilometer* is slightly more than $\frac{1}{2}$ mile.

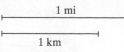

Abbreviations	
Volume	Weight
kiloliter (kL)	kilogram (kg)
hectoliter (hL)	hectogram (hg)
dekaliter (daL)	dekagram (dag)
liter (L)	gram (g)
deciliter (dL)	decigram (dg)
centiliter (cL)	centigram (cg)
milliliter (mL)	milligram (mg)

EXAMPLE 5   Place the appropriate symbol, $<$ or $>$, in the blanks.

(a) 7 in. _____ 7 cm   **(b)** 1 mi _____ 1 km

Solution

(a) Since 1 in. is larger than 1 cm, we have 7 in. $>$ 7 cm.

(b) Since 1 mi is larger than 1 km, we have 1 mi $>$ 1 km.

Student Practice 5   Place the appropriate symbol, $<$ or $>$, in the blanks.

(a) 1 mm _____ 1 in.   **(b)** 4 m _____ 4 yd

Volume and Weight. One advantage of the metric system is that the prefixes are the same for each measurement of length, volume, or weight. After each prefix we write *gram* if we are dealing with weight and *liter* for volume. The table in the margin states the abbreviations for the volume and weight units of measure.

A *milliliter* is a very small measurement and is often used in laboratories.

1 *liter* contains a little more fluid than a quart.

A *kiloliter* is used to measure large volumes.

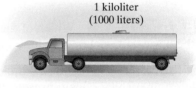

A *milligram* is a small weight. Medicine is measured in milligrams.

1 *gram* weighs about the same as two paper clips.

1 *kilogram* weighs slightly more than a 2-pound steak.

EXAMPLE 6   Place the appropriate symbol, $<$ or $>$, in the blanks.

(a) 10 lb _____ 1 g   **(b)** 5 L _____ 5 qt

Solution

(a) Since 1 gram weighs about as much as two paper clips, we have 10 lb $>$ 1 g.

(b) Since 1 liter is slightly more than 1 quart, we have 5 L $>$ 5 qt.

Student Practice 6   Place the appropriate symbol, $<$ or $>$, in the blanks.

(a) 1 L _____ 1 cup   **(b)** 1 kg _____ 1 lb

EXAMPLE 7   Select the most reasonable measurement for the weight of a backpack filled with books: 3 mg, 3 L, 3 kg, or 3 m.

Solution   We do not choose 3 L or 3 m as a reasonable weight since these units do not represent weight; 3 L is a measure of volume, and 3 m is a measure of length. Since 1 kg weighs a little more than 2 lb, the most appropriate choice is 3 kg (between 6 and 7 lb).

Student Practice 7   Select the most reasonable measurement for the volume of a bottle of soda: 1 L, 1 m, 1 g, or 1 mL.

Appendix B Exercises

MyMathLab®

Watch the videos
in MyMathLab

Download the
MyDashBoard App

Fill in the blanks.

1. 1 mi = _____ ft
2. 1 ft = _____ in.
3. 2 pt = _____ qt
4. 4 qt = _____ gal

5. 1 ton = _____ lb
6. 1 lb = _____ oz
7. 60 sec = _____ min
8. 24 hr = _____ day

9. 14 days = _____ weeks
10. 48 hr = _____ days
11. 2 gal = _____ qt
12. 2 pt = _____ c

13. 36 in. = _____ ft
14. 9 ft = _____ yd
15. 2 mi = _____ ft
16. 3 mi = _____ ft

17. 120 sec = _____ min
18. 120 min = _____ hr
19. 8 qt = _____ gal
20. 4 pt = _____ qt

21. 32 oz = _____ lb
22. 4000 lb = _____ tons
23. 28 days = _____ weeks
24. 72 hr = _____ days

Perform the operation indicated.

25. 5 ft 4 in. + 4 ft 3 in.
26. 16 ft 6 in. + 5 ft 3 in.
27. 12 ft 7 in. + 3 ft 8 in.
28. 8 ft 9 in. + 2 ft 4 in.

29. 26 yd 1 ft − 9 yd 2 ft
30. 14 yd 1 ft − 6 yd 2 ft

31. While golfing, Todd made birdie putts of 4 ft 3 in., 9 ft 10 in., 11 ft 5 in., and 12 ft 9 in. during a round. What was the total length of all the birdie putts?

32. Jessica worked out at the health spa for 2 hr 20 min on Monday, 1 hr 15 min on Wednesday, and 2 hr 45 min on Friday. What was Jessica's total workout time for the week?

33. Natalie bought a 1-pint can of tomato sauce. She used 1 cup to make a recipe. How much tomato sauce did she have left?

34. Joel has 1 quart of milk. If he uses 1 pint to prepare mashed potatoes for a family holiday gathering, how much milk does he have left?

35. It takes Isaac 2 hours to detail a car. Sean can detail the same car in 130 minutes. Who takes the lesser amount of time to detail the car?

36. A 3-pound can of Joe's Chili costs the same price as a 45-ounce can of the store brand of chili. If both brands of chili taste good, which one is the better buy?

Write the abbreviation for each metric unit.

37. decimeter _____
38. kiloliter _____
39. milligram _____

40. milliliter _____
41. centigram _____
42. centimeter _____

Place the appropriate symbol, < or >, in the blanks.

43. 1 mm _____ 1 m
44. 1 dg _____ 1 kg
45. 1 hL _____ 1 cL

46. 1 kg _____ 1 dag
47. 2 mL _____ 2 L
48. 3 km _____ 3 m

Which of the following represents a unit that is smaller than a liter?

49. mL or kL

50. hL or cL

Which of the following represents a unit that is larger than a gram?

51. dg or kg

52. hg or mg

Place the appropriate symbol, < or >, in the blanks.

53. 1 km _____ 1 mi

54. 3 yd _____ 3 m

55. 2 L _____ 2 qt

56. 3 mg _____ 3 lb

57. 2 in. _____ 2 mm

58. 1 cm _____ 1 yd

Fill in the blanks.

59. The _____ is the basic metric unit for weight.

60. The meter is the basic metric unit for _____.

61. The _____ is the basic metric unit for volume.

62. The gram is the basic metric unit for _____.

63. José bought a container of fresh-squeezed juice at the store. Choose the most appropriate unit to measure its contents.

    (a) kilometer   **(b)** gram

    (c) liter   **(d)** milliliter

64. Lisa measured the width of a doorway. Choose the most appropriate unit for this measurement.

    (a) millimeter   **(b)** centiliter

    (c) meter   **(d)** liter

Choose the unit of measure that best completes the sentence.

65. A large London broil steak weighs 2 _____.

    (a) meters   **(b)** kilograms

    (c) kiloliters   **(d)** grams

66. The thickness of the tip of a felt-tip pen is approximately 2 _____.

    (a) centigrams   **(b)** millimeters

    (c) meters   **(d)** liters

Use the figure to the right for exercises 67 and 68. This figure shows a comparison of common speed limits in miles per hour (mph) and kilometers per hour (km/h). These comparisons are approximated to the nearest kilometer per hour.

67. The speed limit posted on a road in France is 89 km/h. Sam Johnson drives an American car with a speedometer that only displays speed in miles per hour. If Sam's speedometer reads 65 mph, is he driving above or below the posted speed limit?

68. In most parts of the United States the speed limit within a school zone is 25 mph. If the speedometer reading on Edward's foreign car is 30 km/h as he drives through a school zone, is he driving above or below the speed limit?

Use the following information to answer exercises 69 and 70. Water for the Spensers' camping trip was stored in 1-quart, 1-pint, and 1-cup containers. The total quantity of water brought on the trip was 8 gallons. The Spensers recorded the amount of water used each day as shown in the table.

69. How much water was left after 4 days?

70. How much water was left after 6 days?

Water used each day			
Day 1	3 qt 3 pt	Day 4	3 qt 3 c
Day 2	4 qt 1 c	Day 5	4 qt 1 c
Day 3	3 qt 1 pt	Day 6	3 qt 3 c

Appendix C Scientific Calculators

This book does *not require* the use of a calculator. However, you may want to consider the purchase of an inexpensive scientific calculator. It is wise to ask your instructor for advice before you purchase any calculator for this course. It should be stressed that you should avoid using a calculator for any of the exercises in which the calculations can readily be done by hand. The only problems in the text that really demand the use of a scientific calculator are marked with the  symbol. Dependence on the use of the scientific calculator for regular exercises in the text will only hurt you in the long run.

The Two Types of Logic Used in Scientific Calculators

Two major styles of scientific calculators are popular today. The most common type employs a type of logic known as **algebraic logic.** Calculators manufactured by Casio, Sharp, Texas Instruments, and many other companies employ this type of logic. An example of calculation on such a calculator would be the following. To add 14 + 26 on an algebraic logic calculator, the sequence of buttons is

$$14 \boxed{+} 26 \boxed{=}$$

The second type of scientific calculator requires the entry of data in **Reverse Polish Notation (RPN).** Most calculators manufactured by Hewlett-Packard, and a few other specialized calculators, are made to use RPN. To add 14 + 26 on a RPN calculator, the sequence of buttons is

$$14 \boxed{\text{enter}} 26 \boxed{+}$$

Graphing calculators such as the TI-83 and TI-84 have a large display for viewing graphs. To perform the calculation on most graphing calculators, the sequence of buttons would be

$$14 \boxed{+} 26 \boxed{\text{enter}}$$

Mathematicians and scientists do not agree on which type of scientific calculator is superior. However, the clear majority of college students own calculators that employ *algebraic* logic. Therefore, this section of the text is explained with reference to the sequence of steps employed by an *algebraic* logic calculator. If you already own or intend to purchase a scientific calculator that uses RPN or a graphing calculator, you are encouraged to study the instruction booklet that comes with the calculator and practice the problems shown in the booklet. After this practice you will be able to solve the calculator problems discussed in this section.

Performing Simple Calculations

The following example illustrates the use of a scientific calculator in doing basic arithmetic calculations.

EXAMPLE 1 Add. 156 + 298

Solution We first key in the number 156, then press the $\boxed{+}$ key, then enter the number 298, and finally, press the $\boxed{=}$ key.

$$156 \boxed{+} 298 \boxed{=} 454$$

Student Practice 1 Add. 3792 + 5896

NOTE TO STUDENT: Fully worked-out solutions to all of the Student Practice problems can be found at the back of the text starting at page SP-1.

EXAMPLE 2   Subtract. $1508 - 963$

Solution   We first enter the number 1508, then press the $\boxed{-}$ key, then enter the number 963, and finally, press the $\boxed{=}$ key.

$$1508 \boxed{-} 963 \boxed{=} 545$$

Student Practice 2   Subtract. $7930 - 5096$

EXAMPLE 3   Multiply. 196×358

Solution $196 \boxed{\times} 358 \boxed{=} 70{,}168$

Student Practice 3   Multiply. 896×273

EXAMPLE 4   Divide. $2054 \div 13$

Solution $2054 \boxed{\div} 13 \boxed{=} 158$

Student Practice 4   Divide. $2352 \div 16$

Decimal Problems

Problems involving decimals can be done readily on a calculator. Entering numbers with a decimal point is done by pressing the $\boxed{\cdot}$ key at the appropriate time.

EXAMPLE 5   Calculate. 4.56×283

Solution   To enter 4.56, we press the $\boxed{4}$ key, the $\boxed{\cdot}$ key, then the $\boxed{5}$ key, and finally, the $\boxed{6}$ key.

$$4.56 \boxed{\times} 283 \boxed{=} 1290.48$$

The answer is 1290.48. Observe how your calculator displays the decimal point.

Student Practice 5   Calculate. 72.8×197

EXAMPLE 6   Add. $128.6 + 343.7 + 103.4 + 207.5$

Solution $128.6 \boxed{+} 343.7 \boxed{+} 103.4 \boxed{+} 207.5 \boxed{=} 783.2$

The answer is 783.2. Observe how your calculator displays the answer.

Student Practice 6   Add. $52.98 + 31.74 + 40.37 + 99.82$

Combined Operations

You must use extra caution concerning the order of mathematical operations when you are using a calculator to do a problem that involves two or more different operations.

Any scientific calculator with algebraic logic uses a priority system that has a clearly defined order of operations. It is the same order that we use in performing arithmetic operations by hand. In either situation, calculations are performed in the following order:

1. First calculations within parentheses are completed.
2. Then numbers are raised to a power, or a square root is calculated.
3. Then multiplication and division operations are performed from left to right.
4. Then addition and subtraction operations are performed from left to right.

This order is followed carefully on *scientific calculators* and graphing calculators. Small inexpensive calculators that do not have scientific functions often do not follow this order of operations.

The number of digits displayed in the answer varies from calculator to calculator. In the following examples, your calculator may display more or fewer digits than the answer we have listed.

EXAMPLE 7   Evaluate. $5.3 \times 1.62 + 1.78 \div 3.51$

Solution This problem requires that we multiply 5.3 by 1.62 and divide 1.78 by 3.51 first and then add the two results. If the numbers are entered directly into the calculator exactly as the problem is written, the calculator will perform the calculations in the correct order.

$$5.3 \; \boxed{\times} \; 1.62 \; \boxed{+} \; 1.78 \; \boxed{\div} \; 3.51 \; \boxed{=} \; 9.09312251$$

Student Practice 7   Evaluate. $0.0618 \times 19.22 - 59.38 \div 166.3$

The Use of Parentheses

To perform some calculations on a calculator, the use of parentheses is helpful. These parentheses may or may not appear in the original problem.

EXAMPLE 8   Evaluate. $5 \times (2.123 + 5.786 - 12.063)$

Solution The problem requires that the numbers in the parentheses be combined first. This will be accomplished by entering the parentheses on the calculator.

$$5 \; \boxed{\times} \; \boxed{(} \; 2.123 \; \boxed{+} \; 5.786 \; \boxed{-} \; 12.063 \; \boxed{)} \; \boxed{=} \; -20.77$$

Note: The result is a negative number.

Student Practice 8   Evaluate. $3.152 \times (0.1628 + 3.715 - 4.985)$

Negative Numbers

To enter a negative number, enter the number followed by the $\boxed{+/-}$ button.

EXAMPLE 9   Evaluate. $(-8.634)(5.821) + (1.634)(-16.082)$

Solution The products will be evaluated first by the calculator. Therefore, parentheses are not needed as we enter the data.

$$8.634 \; \boxed{+/-} \; \boxed{\times} \; 5.821 \; \boxed{+} \; 1.634 \; \boxed{\times} \; 16.082 \; \boxed{+/-} \; \boxed{=} \; -76.536502$$

Note: The result is negative.

Student Practice 9   Evaluate. $(0.5618)(-98.3) - (76.31)(-2.98)$

Scientific Notation

If you wish to enter a number in scientific notation, you should use the special scientific notation button. On most calculators it is denoted as $\boxed{\text{EXP}}$ or $\boxed{\text{EE}}$.

EXAMPLE 10   Multiply. $(9.32 \times 10^6)(3.52 \times 10^8)$

Solution   9.32 $\boxed{\text{EXP}}$ 6 $\boxed{\times}$ 3.52 $\boxed{\text{EXP}}$ 8 $\boxed{=}$ 3.28064   15

This notation means that the answer is 3.28064×10^{15}.

Student Practice 10   Divide. $(3.76 \times 10^{15}) \div (7.76 \times 10^7)$

Raising a Number to a Power

All scientific calculators have a key for finding powers of numbers. It is usually labeled $\boxed{y^x}$. (On a few calculators the notation is $\boxed{x^y}$ or sometimes $\boxed{\wedge}$.) To raise a number to a power, first you enter the base, then push the $\boxed{y^x}$ key. Then you enter the exponent, then finally, the $\boxed{=}$ button.

EXAMPLE 11   Evaluate. $(2.16)^9$

Solution   2.16 $\boxed{y^x}$ 9 $\boxed{=}$ 1023.490369

Student Practice 11   Evaluate. $(6.238)^6$

There is a special key to square a number. It is usually labeled $\boxed{x^2}$.

EXAMPLE 12   Evaluate. $(76.04)^2$

Solution   76.04 $\boxed{x^2}$ $\boxed{=}$ 5782.0816

Student Practice 12   Evaluate. $(132.56)^2$

Finding Square Roots of Numbers

To approximate square roots on a scientific calculator, use the key labeled $\boxed{\sqrt{}}$. In this example we need to use parentheses.

EXAMPLE 13   Evaluate. $\sqrt{5618 + 2734 + 3913}$

Solution $\boxed{(}$ 5618 $\boxed{+}$ 2734 $\boxed{+}$ 3913 $\boxed{)}$ $\boxed{\sqrt{}}$ $\boxed{=}$ 110.7474605

Student Practice 13   Evaluate. $\sqrt{0.0782 - 0.0132 + 0.1364}$

Note: On some scientific calculators the square root key must be entered first, followed by the left parenthesis.

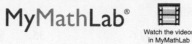

Watch the videos in MyMathLab   Download the MyDashBoard App

Use your calculator to complete each problem. Your answers may vary slightly because of the characteristics of individual calculators.

Complete the following table.

To Do This Operation	Use These Keystrokes	Record Your Answer Here
1. 8963 + 2784	8963 \[+ \] 2784 \[= \]	
2. 15,308 − 7980	15308 \[− \] 7980 \[= \]	
3. 2631 × 134	2631 \[× \] 134 \[= \]	
4. 70,221 ÷ 89	70221 \[÷ \] 89 \[= \]	
5. 5.325 − 4.031	5.325 \[− \] 4.031 \[= \]	
6. 184.68 + 73.98	184.68 \[+ \] 73.98 \[= \]	
7. 2004.06 ÷ 7.89	2004.06 \[÷ \] 7.89 \[= \]	
8. 1.34 × 0.763	1.34 \[× \] 0.763 \[= \]	

Write down the answer and then show what problem you have solved.

9. 123.45 \[+ \] 45.9876 \[+ \] 8765.3 \[= \]

10. 0.0897 \[× \] 234.56 \[× \] 2.5428 \[= \]

11. 34 \[÷ \] 8 \[+ \] 12.56 \[= \]

12. 458 \[÷ \] 4 \[− \] 16.897 \[= \]

Perform each calculation.

13. 9.467 + 0.563

14. 0.347 + 23.457

15. 34.89 + 39.6 + 214.897

16. 12.567 + 48.31 + 189.38

17. 412,899 − 34,675

18. 87,456 − 2876

19. 3,567,089 − 2,876,805

20. 8,345,802 − 4,985,004

21. 234 × 4.567

22. 1.9876 × 347

23. 0.456 × 3.48

24. 67,876 × 0.0946

25. 3458 ÷ 2.5

26. 9764 ÷ 8

27. 12.107524 ÷ 15.86

28. 16.06513 ÷ 17.98

Perform each calculation.

29.	**30.**	**31.**	**32.**
1.98	8.92	$ 103.91	$3986.21
6.34	9.31	$2653.82	$4502.89
+ 7.71	+ 7.79	+ $9804.61	+ $ 989.30

33.   368,781.5
        $\underline{-\ 283,617.8}$

34.   571,809.6
        $\underline{-\ 539,376.8}$

35. $1,393,271.80
        $\underline{-\ \$1,289,663.20}$

36. $8,571,300.70
        $\underline{-\ \$4,098,789.30}$

37.   345.34
        $\underline{\times\ \ \ 45.7}$

38.   8954.34
        $\underline{\times\ \ \ 425.4}$

39.   0.6314
        $\underline{\times\ \ \ 3.96}$

40.   0.0789
        $\underline{\times\ 12.38}$

41. $40.36\overline{)36{,}202.92}$

42. $52.98\overline{)172{,}608.84}$

43. $0.7613\overline{)17.12925}$

44. $0.9854\overline{)3.59671}$

Perform the operations in the proper order.

45. $4.567 + 87.89 - 2.45 \times 3.3$

46. $4.891 + 234.5 - 0.98 \times 23.4$

47. $7 \div 8 + 3.56$

48. $9 \div 4.5 + 0.6754$

49. $(9.34)(0.345) + 98.345$

50. $(0.628)(398) + 34.4581$

51. $\dfrac{(95.34)(0.9874)}{381.36}$

52. $\dfrac{(0.8759)(45.87)}{183.48}$

53. $2.56 + 8.98 \times 3.14$

54. $1.62 + 3.81 - 5.23 \times 6.18$

55. $(-4.23)(1.863) - 5.998$

56. $12.34 - (26.314)(-1.856)$

57. $5.62(5 \times 3.16 - 18.12)$

58. $9.356(4.8 - 7.2 - 15.94)$

59. $(3.42 \times 10^8)(0.97 \times 10^{10})$

60. $(6.27 \times 10^{20})(1.35 \times 10^3)$

61. $\dfrac{(2.16 \times 10^3)(1.37 \times 10^{14})}{6.39 \times 10^5}$

62. $\dfrac{(3.84 \times 10^{12})(1.62 \times 10^5)}{7.78 \times 10^8}$

63. $\dfrac{2.3 + 5.8 - 2.6 - 3.9}{5.3 - 8.2}$

64. $\dfrac{(2.6)(-3.2) + (5.8)(-0.9)}{2.614 + 5.832}$

65. $\sqrt{253.12}$

66. $\sqrt{0.0713}$

67. $\sqrt{5.6213 - 3.7214}$

68. $\sqrt{3417.2 - 2216.3}$

69. $(1.78)^3 + 6.342$

70. $(2.26)^8 - 3.1413$

71. $\sqrt{(6.13)^2 + (5.28)^2}$

72. $\sqrt{(0.3614)^2 + (0.9217)^2}$

73. $\sqrt{56 + 83} - \sqrt{12}$

74. $\sqrt{98 + 33} - \sqrt{17}$

Find an approximate value. Round your answer to 5 decimal places.

75. $\dfrac{7}{18} + \dfrac{9}{13}$

76. $\dfrac{5}{22} + \dfrac{1}{31}$

77. $\dfrac{7}{8} + \dfrac{3}{11}$

78. $\dfrac{9}{14} + \dfrac{5}{19}$

Appendix D Additional Arithmetic Practice

Addition Practice

1. 23
    +14

2. 42
    +33

3. 50
    +44

4. 83
    +16

5. 51
    +27

6. 16
    +13

7. 32
    +29

8. 64
    +17

9. 327
    + 42

10. 223
     + 54

11. 463
     + 28

12. 504
     + 96

13. 739
     +682

14. 567
     +485

15. 840 + 60

16. 364 + 37

17. 915 + 796

18. 420 + 899

19. 213 + 46 + 30

20. 326 + 21 + 52

21. 132 + 441 + 16

22. 671 + 204 + 12

23. 139 + 61 + 222

24. 524 + 73 + 195

25. 701 + 166 + 24 + 11

26. 439 + 365 + 45 + 81

Subtraction Practice

1. 32
    −11

2. 87
    −25

3. 56
    −34

4. 73
    −30

5. 93
    −25

6. 21
    −16

7. 40
    −11

8. 60
    −15

9. 576
    − 45

10. 294
     − 71

11. 780
     − 54

12. 208
     − 17

13. 406
     − 28

14. 100
     − 34

15. 635
     −126

16. 375
     −147

17. 500
     −244

18. 200
     −137

19. 922
     −739

20. 646
     −377

21. 1729
     − 856

22. 2382
     − 490

23. 7806
     − 327

24. 3024
     − 156

25. 8200
     −6134

26. 2004
     −1326

Multiplication Practice

1. 23
   × 3

2. 13
   × 2

3. 54
   × 7

4. 67
   × 9

5. 74
   ×21

6. 53
   ×31

7. 92
   ×40

8. 70
   ×52

9. 82
   ×95

10. 69
    ×39

11. 212
    × 43

12. 341
    × 22

13. 295
    × 41

14. 419
    × 72

15. 304
    × 68

16. 620
    × 39

17. 261
    ×144

18. 124
    ×433

19. 545
    ×522

20. 634
    ×799

21. 391
    ×609

22. 817
    ×460

23. 3844
    × 209

24. 7409
    × 106

25. 72,499(683)   26. 86,243(725)

Division Practice

1. 8)128

2. 3)168

3. 7)415

4. 6)287

5. 9)1116

6. 4)1184

7. 6)1404

8. 3)1701

9. 8)4174

10. 5)3697

11. 17)5468

12. 13)9795

13. 146)12,994

14. 163)14,833

15. 1728 ÷ 54

16. 3813 ÷ 93

17. 3701 ÷ 34

18. 6052 ÷ 49

19. 15,836 ÷ 74

20. 23,256 ÷ 68

21. 30,632 ÷ 27

22. 85,069 ÷ 79

23. 30,752 ÷ 248

24. 49,878 ÷ 326

25. 271,125 ÷ 241

26. 546,924 ÷ 357

Appendix E Congruent Triangles

Two objects are **congruent** if they can be made to coincide by placing one on top of the other. This can be done either directly or by flipping one of the objects over. In other words, two objects are congruent if they have the same size and the same shape.

Two triangles are congruent if the measures of corresponding angles are equal and the lengths of corresponding sides are equal.

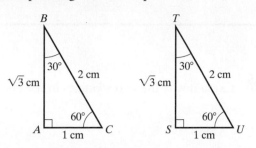

Triangle *ABC* is congruent to triangle *STU*.

We use the symbol $\angle$ to indicate that we are referring to the measure of an angle. In the triangle above, $\angle A = 90°$, $\angle B = 30°$, and $\angle C = 60°$.

Two triangles are congruent when one of the following can be shown: congruence by side-side-side (SSS), side-angle-side (SAS), or angle-side-angle (ASA).

CONGRUENCE BY SIDE-SIDE-SIDE (SSS)

If the three sides of one triangle have the same length as the corresponding three sides of a second triangle, the two triangles are congruent.

EXAMPLE 1 Explain why the two triangles in the margin are congruent.

Solution

Side *BC* and side *FD* = 12 in.
Side *AC* and side *ED* = 9 in.
Side *AB* and side *EF* = 20 in.

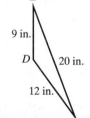

All three sides of triangle *ABC* have the same length as the corresponding sides of triangle *EFD*. The triangles satisfy SSS and are congruent.

Note: If we rotate point *B* of triangle *ABC* downward, we can lay this triangle on top of triangle *EFD* since both triangles have the same size and shape.

Student Practice 1 Explain why the two triangles below are congruent.

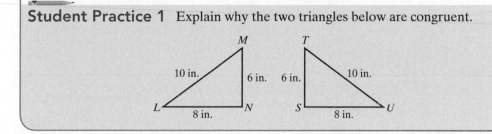

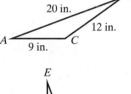

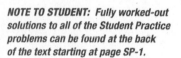

NOTE TO STUDENT: Fully worked-out solutions to all of the Student Practice problems can be found at the back of the text starting at page SP-1.

CONGRUENCE BY SIDE-ANGLE-SIDE (SAS)

Two triangles are congruent if

1. Two sides of one triangle are equal in length to two sides of the second triangle, and

2. The measure of the angle formed by these two sides in each triangle is equal.

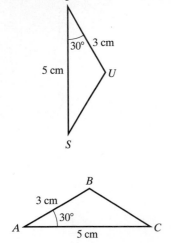

EXAMPLE 2 Explain why the two triangles in the margin are congruent.

Solution

Side ST = side AC = 5 cm.

Side TU = side AB = 3 cm.

$\angle T = \angle A = 30°$.

Two sides of triangles STU and CAB are equal in length, and the angles formed by these sides are also equal. The triangles are congruent by SAS.

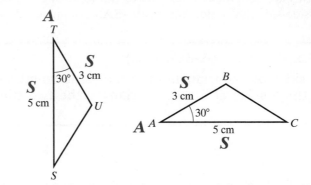

Student Practice 2 Explain why the two triangles below are congruent.

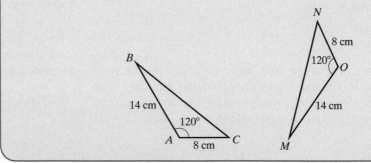

CONGRUENCE BY ANGLE-SIDE-ANGLE (ASA)

Two triangles are congruent if

1. Two angles of one triangle are equal in measure to two angles of the second triangle, and

2. The side between these two angles in each triangle is equal in length.

EXAMPLE 3   Explain why the two triangles in the margin are congruent.

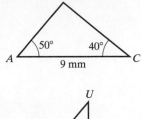

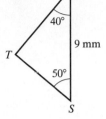

Solution

$\angle A = \angle S = 50°$.

$\angle C = \angle U = 40°$.

Side AC = side SU = 9 mm.

Two angles of triangle ABC equal two angles of triangle STU, and the sides between these angles are equal in length. The triangles are congruent by ASA.

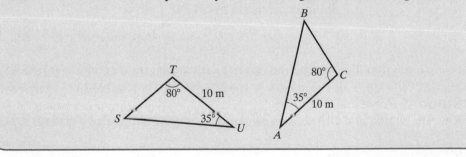

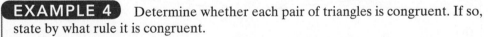

Student Practice 3   Explain why the two triangles below are congruent.

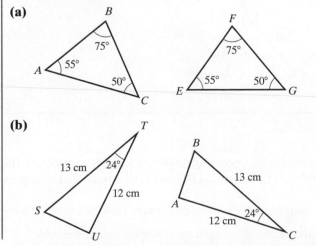

EXAMPLE 4   Determine whether each pair of triangles is congruent. If so, state by what rule it is congruent.

(a)

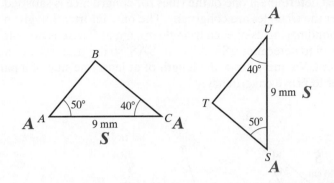

(b)

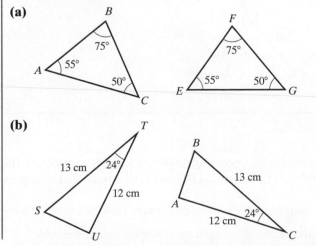

Continued on next page

Solution

(a)

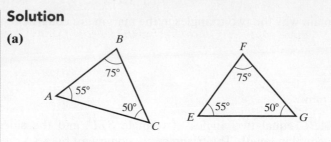

We must determine if one of the rules for congruence is satisfied to determine whether the triangles are congruent. The only information given is that all the corresponding angles of each triangle are equal. There is not enough information given to determine SSS, ASA, or SAS, so the triangles are not necessarily congruent. We must know the length of at least one side of a pair of triangles in order to prove congruency.

(b)

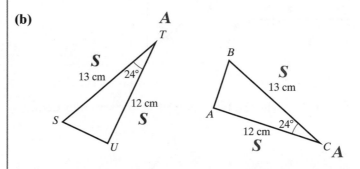

We see from the information given that the triangles are congruent by SAS.

Student Practice 4   Determine whether each pair of triangles is congruent. If so, state by what rule it is congruent.

(a)

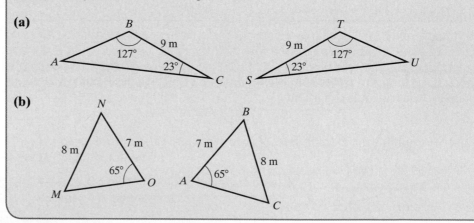

(b)

In Chapter 10 we show that if corresponding angles of triangles are equal, the triangles are similar. In Example 4(a), triangles *ABC* and *GFE* are similar triangles. Why?

The difference between similar triangles and congruent triangles is as follows.

The corresponding *sides* and *angles* of congruent triangles must be *equal.*

The corresponding *angles* of similar triangles must be *equal,* but corresponding *sides* are *not necessarily equal.*

Can you draw two triangles that are both congruent and similar?

The following pairs of triangles are congruent by SAS. Find the length of side EF.

1.

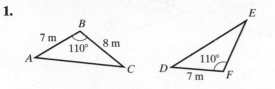

2.

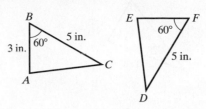

The following pairs of triangles are congruent by ASA. Name the angle in triangle DEF that is 35°.

3.

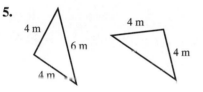

4.

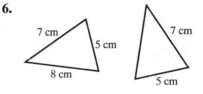

The following pairs of triangles are congruent by SSS. Find the length of the unknown side.

5.

6.

Explain why each pair of triangles is congruent.

7.

8.

9.

10.

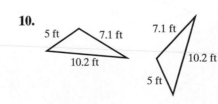

11.

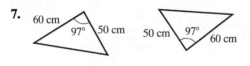

12.

13.

14.

A-31

Determine whether each pair of triangles is congruent. If so, state by what rule it is congruent.

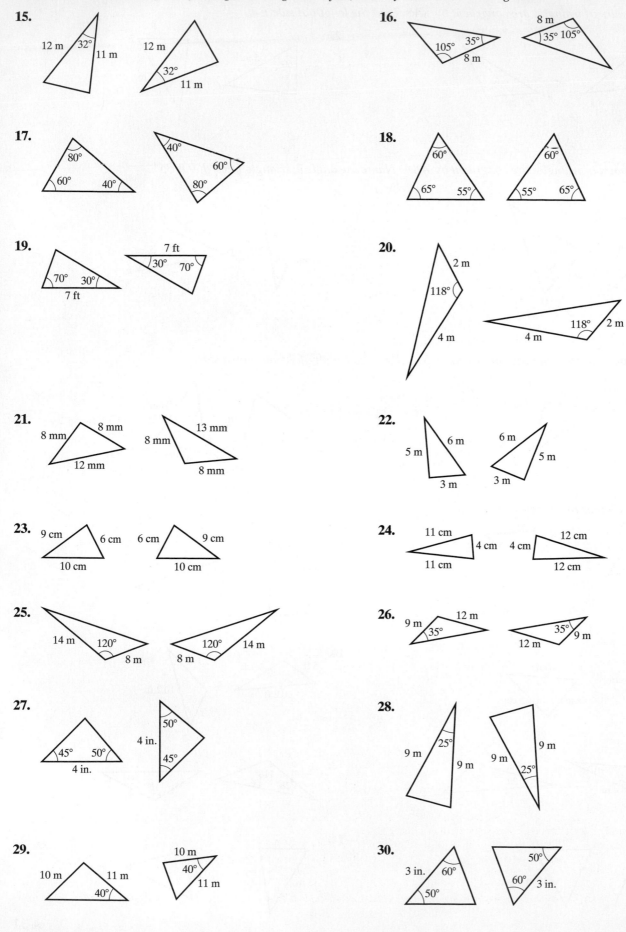

15.

12 m 32° 11 m

12 m 32° 11 m

16.

105° 35° 8 m

8 m 35° 105°

17.

80° 60° 40°

40° 60° 80°

18.

60° 65° 55°

60° 55° 65°

19.

70° 30° 7 ft

7 ft 30° 70°

20.

2 m 118° 4 m

4 m 118° 2 m

21.

8 mm 8 mm 8 mm 12 mm

13 mm 8 mm 8 mm

22.

5 m 6 m 3 m

6 m 5 m 3 m

23.

9 cm 6 cm 10 cm

6 cm 9 cm 10 cm

24.

11 cm 4 cm 11 cm

12 cm 4 cm 12 cm

25.

14 m 120° 8 m

120° 8 m 14 m

26.

9 m 12 m 35°

35° 9 m 12 m

27.

45° 50° 4 in.

50° 45° 4 in.

28.

9 m 25° 9 m

9 m 9 m 25°

29.

10 m 11 m 40°

10 m 40° 11 m

30.

3 in. 60° 50°

50° 60° 3 in.

Solutions to Student Practice Problems

Chapter 1 1.1 Student Practice

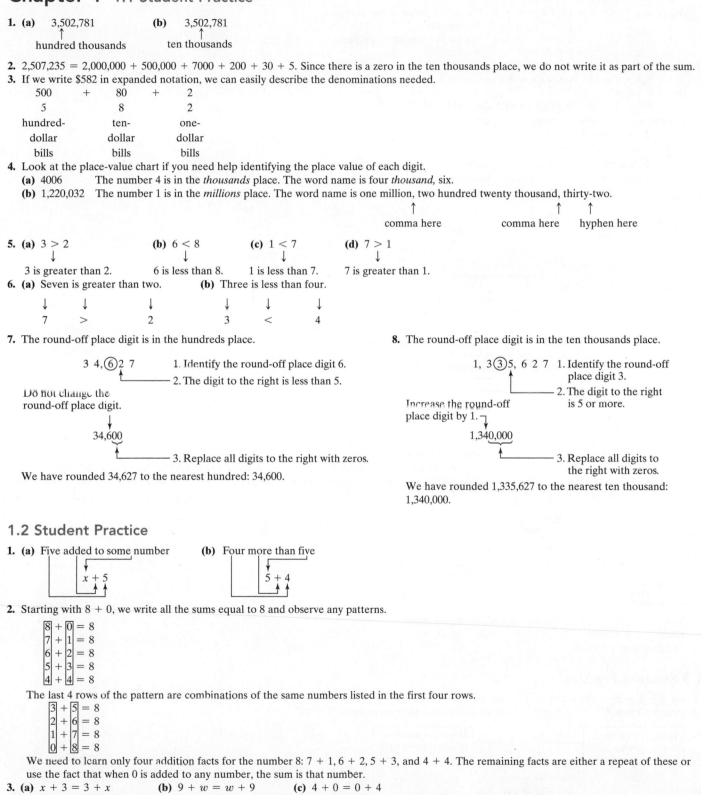

1. (a) 3,502,781   **(b)** 3,502,781

    hundred thousands   ten thousands

2. $2,507,235 = 2,000,000 + 500,000 + 7000 + 200 + 30 + 5$. Since there is a zero in the ten thousands place, we do not write it as part of the sum.

3. If we write $582 in expanded notation, we can easily describe the denominations needed.

500	+	80	+	2
5		8		2
hundred-dollar bills		ten-dollar bills		one-dollar bills

4. Look at the place-value chart if you need help identifying the place value of each digit.

    (a) 4006   The number 4 is in the *thousands* place. The word name is four *thousand,* six.

    (b) 1,220,032   The number 1 is in the *millions* place. The word name is one million, two hundred twenty thousand, thirty-two.

                                                      comma here   comma here   hyphen here

5. (a) $3 > 2$ **(b)** $6 < 8$ **(c)** $1 < 7$ **(d)** $7 > 1$

    3 is greater than 2.   6 is less than 8.   1 is less than 7.   7 is greater than 1.

6. (a) Seven is greater than two.   **(b)** Three is less than four.

7	>	2		3	<	4

7. The round-off place digit is in the hundreds place.

          3 4,⑥2 7   1. Identify the round-off place digit 6.

                       2. The digit to the right is less than 5.

Do not change the round-off place digit.

         34,600

                    3. Replace all digits to the right with zeros.

We have rounded 34,627 to the nearest hundred: 34,600.

8. The round-off place digit is in the ten thousands place.

         1, 3③5, 6 2 7   1. Identify the round-off place digit 3.

                        2. The digit to the right is 5 or more.

Increase the round-off place digit by 1.

         1,340,000

                    3. Replace all digits to the right with zeros.

We have rounded 1,335,627 to the nearest ten thousand: 1,340,000.

1.2 Student Practice

1. (a) Five added to some number   **(b)** Four more than five

          $x + 5$ $5 + 4$

2. Starting with $8 + 0$, we write all the sums equal to 8 and observe any patterns.

$$\boxed{8} + \boxed{0} = 8$$
$$\boxed{7} + \boxed{1} = 8$$
$$\boxed{6} + \boxed{2} = 8$$
$$\boxed{5} + \boxed{3} = 8$$
$$\boxed{4} + \boxed{4} = 8$$

The last 4 rows of the pattern are combinations of the same numbers listed in the first four rows.

$$\boxed{3} + \boxed{5} = 8$$
$$\boxed{2} + \boxed{6} = 8$$
$$\boxed{1} + \boxed{7} = 8$$
$$\boxed{0} + \boxed{8} = 8$$

We need to learn only four addition facts for the number 8: $7 + 1, 6 + 2, 5 + 3$, and $4 + 4$. The remaining facts are either a repeat of these or use the fact that when 0 is added to any number, the sum is that number.

3. (a) $x + 3 = 3 + x$ **(b)** $9 + w = w + 9$ **(c)** $4 + 0 = 0 + 4$

4. $y + x = 6075$ Why? The commutative property states that the order in which we add numbers doesn't affect the sum.

5. To simplify, we find the sum of the known numbers.

      $6 + 3 + x = 9 + x$ or $x + 9$

We cannot add the variable x and the number 9 because x represents an unknown quantity; we have no way of knowing what quantity to add to the number 9.

6. $(w + 1) + 4 = w + (1 + 4)$ The associative property allows us to regroup.
$\quad\quad\quad\quad\quad\quad = w + 5$ Simplify: $1 + 4 = 5$.

7. (a) $(2 + x) + 8 = (x + 2) + 8$ We use the commutative property.
$\quad\quad\quad\quad\quad\quad = x + (2 + 8)$ We use the associative property.
$\quad\quad\quad\quad\quad\quad = x + 10$ We simplify.

(b) $(4 + x + 3) + 1 = (x + 4 + 3) + 1$ We use the commutative property.
$\quad\quad\quad\quad\quad\quad = (x + 7) + 1$ We simplify.
$\quad\quad\quad\quad\quad\quad = x + (7 + 1)$ We use the associative property.
$\quad\quad\quad\quad\quad\quad = x + 8$ We simplify.

8. (a) $x + y + 6$
$\quad 9 + 3 + 6 = 18$ Replace x with 9 and y with 3, then simplify.

(b) $x + y + 6$
$\quad 1 + 7 + 6 = 14$ Replace x with 1 and y with 7, then simplify.

9. We arrange the numbers vertically and begin adding in the ones column.

$247 \rightarrow$ 2 hundreds   4 tens   7 ones
$+ 38 \rightarrow$ _____ 3 tens   8 ones
$\quad\quad\quad$ 2 hundreds   7 tens 15 ones $\rightarrow$ We cannot have two digits in the ones column so we must rename 15 as 1 ten and 5 ones.
$\quad\quad\quad$ 2 hundreds + (7 tens + 1 ten) + 5 ones = 2 hundreds + 8 tens + 5 ones = 285

10. We add whenever we must find the total amount.

$\overset{2\,1}{286}$ Milk
475 Orange Juice
$+\ \ 91$ Other
852

We add $6 + 5 + 1 = 12$. Since 12 equals 1 ten and 2 ones, we carry 1 ten placing a 1 at the top of the tens column.

We add $1 + 8 + 7 + 9 = 25$. Since 25 tens equals 2 hundreds and 5 tens, we carry 2 hundreds placing a 2 at the top of the hundreds column.

We add $2 + 2 + 4 = 8$.

$286 + 475 + 91 = 852$

11. We add the four lengths to find perimeter. Sketch a small square and label each side with a 15 ft.
$15 \text{ ft} + 15 \text{ ft} + 15 \text{ ft} + 15 \text{ ft} = 60 \text{ ft}$.   The perimeter is 60 ft.

12. We want to find the distance around the figure. We look only at the outside edges since the dashed line indicates an inside length. We must find the lengths of the two unlabeled sides. The shaded figure is a square since two adjoining sides have the same measure. Thus each side of the shaded figure has a measure of 30 feet.

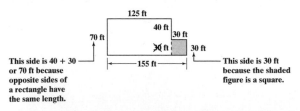

125 ft

40 ft 30 ft
70 ft
30 ft 30 ft
This side is 40 + 30
or 70 ft because
opposite sides of
a rectangle have
the same length.
←—155 ft—→
This side is 30 ft
because the shaded
figure is a square.

Now we add the lengths of the six sides to find the perimeter.
$125 \text{ ft} + 40 \text{ ft} + 30 \text{ ft} + 30 \text{ ft} + 155 \text{ ft} + 70 \text{ ft} = 450 \text{ ft}$
The perimeter is 450 feet.

1.3 Student Practice

1. (a) $5 - 2 = 3$ **(b)** $6 - 3 = 3$ **(c)** $18 - 0 = 18$ **(d)** $18 - 18 = 0$

2. Since we know $600 - 50 = 550$, we can use subtraction patterns to find $600 - 54$.

Increase numbers
in this column by 1.

Decrease numbers
in this column by 1.

$600 - 50 = 550$
$600 - 51 = 549$
$600 - 52 = 548$
$600 - 53 = 547$
$600 - 54 = 546$

3. (a) The *difference* of nine and *n*

$$9 - n$$

(b) *x minus* three

$$x - 3$$

(c) *x subtracted from* 8

$$8 - x$$

4. (a) $8 - n$

$8 - 3$ Replace *n* by 3.

5 Simplify.

When *n* is equal to 3, $8 - n$ is equal to 5.

(b) $8 - n$

$8 - 6$ Replace *n* by 6.

2 Simplify.

When *n* is equal to 6, $8 - n$ is equal to 2.

5. $\begin{array}{r} 93 \\ -46 \\ \hline \end{array}$ We cannot subtract 6 from 3, so we write 9 tens as tens and ones.

$\begin{array}{r} \overset{8}{9}\overset{13}{3} \\ -4\ 6 \\ \hline 4\ 7 \end{array}$ 9 tens and 3 ones = 8 tens and 1 ten and 3 ones =   8 tens and 13 ones

−4 tens and 6 ones

4 tens and 7 ones = 47

6. $\begin{array}{r} 603 \\ -278 \\ \hline \end{array}$ We must borrow because we cannot subtract 8 ones from 3 ones.

We cannot borrow a "ten" because there are 0 tens, so we must borrow from the 6 hundreds.

$\begin{array}{r} \overset{5}{6}\overset{\overset{9}{10}}{0}\overset{13}{3} \\ -2\ 7\ 8 \\ \hline 3\ 2\ 5 \end{array}$ = 6 hundreds + 3 ones = 5 hundreds + 10 tens + 3 ones =   5 hundreds and 9 tens and 13 ones

−2 hundreds and 7 tens and 8 ones

3 hundreds and 2 tens and 5 ones = 325

7. $\begin{array}{r} 8006 \\ -4237 \\ \hline \end{array}$ We cannot subtract 7 from 6, so we must change 800 to 799 to borrow 1 ten = 10 ones.

$\begin{array}{r} \overset{7}{8}\overset{9}{0}\overset{9}{0}\overset{16}{6} \\ -4\ 2\ 3\ 7 \\ \hline 3\ 7\ 6\ 9 \end{array}$ = 7 thousands and 9 hundreds and 9 tens and 16 ones

= −4 thousands and 2 hundreds and 3 tens and 7 ones

3 thousands and 7 hundreds and 6 tens and 9 ones = 3769

Subtraction   Check by Addition

$\begin{array}{r} 8006 \\ -4237 \\ \hline 3769 \end{array}$ $\begin{array}{r} 4237 \\ +3769 \\ \hline 8006 \end{array}$ It checks.

8. *Understand the problem.* The key phrase "how many fewer" indicates that we subtract.

Calculate and state the answer.

$$232 - 133$$

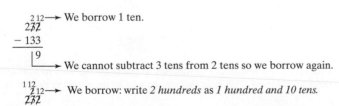

We borrow 1 ten.

$\begin{array}{r} -133 \\ \hline 9 \end{array}$

We cannot subtract 3 tens from 2 tens so we borrow again.

We borrow: write *2 hundreds* as *1 hundred and 10 tens*.

$\begin{array}{r} -133 \\ \hline 99 \end{array}$

Thus 99 fewer fish were caught on May 31 than on May 29.
We leave the check to the student.

9. To find the perimeter, we must find the distance around the figure. Therefore, we must find the measures of the unlabeled sides.

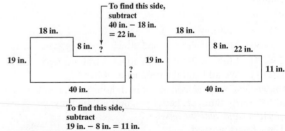

Next we add the lengths of the six sides.
18 in. + 8 in. + 22 in. + 11 in. + 40 in. + 19 in. = 118 in.
The perimeter is 118 inches.

1.4 Student Practice

1. There are two arrays consisting of 15 items that represent the multiplication 5 times 3.
One array has 3 rows and 5 columns, and the other one has 5 rows and 3 columns.

```
         ★★★
         ★★★
 ★★★★★  5 ★★★
3★★★★★    ★★★
 ★★★★★    ★★★
   5       3
15 items  15 items
```

2. (a)

	Dirt	Racer	Road
Red	Dirt; Red	Racer; Red	Road; Red
Blue	Dirt; Blue	Racer; Blue	Road; Blue
Green	Dirt; Green	Racer; Green	Road; Green
Pink	Dirt; Pink	Racer; Pink	Road; Pink
Black	Dirt; Black	Racer; Black	Road; Black

(b) There are 5 rows and 3 columns or $5 \times 3 = 15$ different bikes.

3. (a) 9 and 7 are the factors and 63 is the product.

  (b) x and y are the factors and z is the product.

4. (a) Double a number means to multiply it by 2, so $2 \cdot n = 2n$.

  (b) Two times a number also means to multiply it by 2, so $2 \cdot n = 2n$.

5. (a) $2 \cdot 6 \cdot 0 \cdot 3 = 0$ The product is 0 when one of the factors is 0.

  (b) $2 \cdot 3 \cdot 1 \cdot 5 = (3 \cdot 1) \cdot (2 \cdot 5)$ Use the commutative property to change the order of the factors so that two adjacent factors have a product of 10.

   $ = 3 \cdot 10$ $3 \cdot 1 = 3; 2 \cdot 5 = 10$

   $ = 30$ To multiply 3(10), write 3 and attach a zero at the end.

6. (a) $4(x \cdot 3) = 4 \cdot (x \cdot 3)$ Rewrite using familiar notation.

   $ = 4 \cdot (3 \cdot x)$ Change the order of factors.

   $ = (4 \cdot 3) \cdot x$ Regroup.

   $4(x \cdot 3) = 12x$ Multiply and write in standard notation: $12 \cdot x = 12x$.

  (b) $2(4)(n \cdot 5) = 2 \cdot 4 \cdot (n \cdot 5)$ Rewrite using familiar notation.

   $ = 8 \cdot (n \cdot 5)$ Multiply $2(4) = 8$.

   $ = 8 \cdot (5 \cdot n)$ Change the order of factors.

   $ = (8 \cdot 5) \cdot n$ Regroup.

   $2(4)(n \cdot 5) = 40n$ Multiply and write in standard notation: $40 \cdot n = 40n$.

7. Since the number 700 has trailing zeros, we multiply by 7 and attach the trailing zeros to the right of the product.

```
  2 4
  436
× 700
─────
305200
```
← Bring down trailing zeros.

$7(6) = 42$. Place the 2 here and carry the 4.

$7(3) = 21$. Then add the carried digit 4: $21 + 4 = 25$.

Place the 5 under the 3 and carry the 2.

$7(4) = 28$. Then add the carried digit 2: $28 + 2 = 30$.

8. To multiply 936(38), we multiply $936(30 + 8) = 936(8 + 30)$ or $936(8) + 936(30)$ using the condensed form.

```
   936
 ×  38
 ─────
  7488
 28080
 ─────
 35568
```

7488 ← Multiply 8(936).

28080 ← To find the product $30(936) = 28,080$, we multiply 3(936) and add 1 trailing zero.

35568 ← We add and the product is 35,568.

9.
```
   4651
 ×  203
 ──────
  13953
  00000
 930200
 ──────
 944153
```

13953 Multiply 3(4651).

00000 Multiply 0(4651) and attach 1 trailing zero.

930200 Multiply 2(4651) and attach 2 trailing zeros.

944153 We add and the product is 944,153.

10. *Understand the problem.* \$9 for each hour indicates that we either add \$9 thirty times or, better yet, that we multiply.

Calculate and state the answer.

```
  30    hours worked
×  9    dollars paid each hour
───
 270    total money he earned
```

Check: We have \$9 for 1 hour, \$18 for 2 hours, \$27 for 3 hours, so that would make \$270 for 30 hours.

11. *Understand the problem.* The wall has 30 bricks on the bottom row, and there are 12 rows. This indicates we can add 30 twelve times or, better yet, multiply 30(12).

Calculate and state the answer.

```
  12    rows high
×30    bricks per row
───
 360    total number of bricks
```

Check: We can use repeated addition and add 30 twelve times:
$30 + 30 + 30 + 30 + 30 + 30 + 30 + 30 + 30 + 30 + 30 + 30 = 360$.
We get the same result.

1.5 Student Practice

1. Think \$15 for the first gallon, \$15 for the second gallon, ... How many times do we add 15 to get 150? Or more efficiently, how many 15's are in 150? The division statement that corresponds to this situation is $150 \div 15$.

2. We must split \$170 into 5 equal groups. This corresponds to the division $170 \div 5$.

3. (a) The quotient of twenty-six and three

$$26 \div 3$$

  (b) The quotient of three and twenty-six

$$3 \div 26$$

4. $21 \div 7 = ?$ Think, $21 = 7 \cdot ?$
$21 \div 7 = 3$ $21 = 7 \cdot 3$

5. (a) $3 \div 3 = 1$ Any nonzero number divided by itself is 1.
(b) $3 \div 0$ Zero can never be the divisor in a division problem, so $3 \div 0$ is undefined.
(c) $0 \div 3 = 0$ 0 divided by any nonzero number is equal to 0.

6. We guess 7 because 6×7 is close to 43.

$$\begin{array}{r} 7\text{ R}1 \\ 6\overline{)43} \\ -\ 42 \\ \hline 1 \end{array}$$

To verify that this is correct, we multiply the divisor by the quotient, then add the remainder.

$43 \div 6 = 7\text{ R}1$

Multiply $6 \times 7 = 42$
$\begin{array}{r} 7\text{ R}1 \\ 6\overline{)43} \end{array}$ $\begin{array}{r} 42 \\ +\ 1 \\ \hline 43 \end{array}$ Then add the remainder.

$43 = 43$ ✓

7. Since 36 goes into 360 ten times, we guess 9.

$$\begin{array}{r} 9\text{ R}30 \\ 36\overline{)354} \\ -\ 324 \\ \hline 30 \end{array}$$

We verify the answer is correct: (divisor $\cdot$ quotient) + remainder = dividend.
$(\ 36\ \cdot\ 9\ ) +\ 30\ =\ 354$
$354 \div 36 =\ 9\text{ R}30$

8. Since 8 times $3 = 24$, we guess 3 and place 3 in the tens column.

$$\begin{array}{r} 32\text{ R}51 \\ 80\overline{)2611} \\ -240 \\ \hline 211 \\ 160 \\ \hline 51 \end{array}$$

We verify the answer is correct: (divisor $\cdot$ quotient) + remainder = dividend.
$(\ 80\ \cdot\ 32\ ) +\ 51\ =\ 2611$
$2611 \div 80 =\ 32\text{ R}51$

9. Since 3 times 4 is 12, we guess 4 and place 4 in the hundreds column.

$$\begin{array}{r} 403 \\ 37\overline{)14911} \\ -\ 148 \\ \hline 111 \\ 111 \\ \hline 0 \end{array}$$

We verify the answer is correct: (divisor $\cdot$ quotient) + remainder = dividend.
$(\ 37\ \cdot\ 403\ ) +\ 0\ =\ 14911$
$14911 \div 37 =\ 403$

10. *Understand the problem.* Since we must split 100 movie passes equally among 22 players, we divide.

Calculate and state the answer.
$$\begin{array}{r} 4\text{ R}12 \\ 22\overline{)100} \\ -\ 88 \\ \hline 12 \end{array}$$

Since 12 is the remainder, 12 movie passes are donated to a local children's home.
Check: $(22 \cdot 4) + 12 = 100$

1.6 Student Practice

1. (a) $n = n^1$ **(b)** $6 \cdot 6 \cdot y \cdot y \cdot y \cdot y = 6^2 \cdot y^4$ or $6^2 y^4$ **(c)** $5 \cdot 5 \cdot 5 \cdot 5 \cdot 5 \cdot 5 \cdot 5 \cdot 5 = 5^8$ **(d)** $x \cdot x \cdot 8 \cdot 8 \cdot 8 = x^2 \cdot 8^3$ or $8^3 x^2$

2. (a) $x^6 = x \cdot x \cdot x \cdot x \cdot x \cdot x$ **(b)** $1^7 = 1 \cdot 1 \cdot 1 \cdot 1 \cdot 1 \cdot 1 \cdot 1$ **3. (a)** $4^3 = 4 \cdot 4 \cdot 4 = 64$ **(b)** $8^1 = 8$ **(c)** $10^2 = 10 \cdot 10 = 100$

4. ┌Write 1.
$\underbrace{100{,}000}$

The exponent is 5; attach 5 trailing zeros.

$10^5 = 100{,}000$

5. $y^2 \rightarrow 8^2$ Replace y with 8 and $8(8) = 64$. When $y = 8$, $y^2 = 64$.

6. (a) Four to the sixth power $= 4^6$ **(b)** x cubed $= x^3$ **(c)** Ten squared $= 10^2$

7. $3^2 + 2 - 5 = 9 + 2 - 5$ *Identify:* The highest priority is exponents.
    Calculate: $3 \cdot 3 = 9$. *Replace:* 3^2 with 9.
$= 11 - 5$ *Identify:* Addition has the higher priority.
    Calculate: $9 + 2 = 11$. *Replace:* $9 + 2$ with 11.
$= 6$ *Identify:* Subtraction is last. *Calculate:* $11 - 5 = 6$.
    Replace: $11 - 5$ with 6.

8. $4 \cdot 2^3 = 4 \cdot 8$ *Identify:* The highest priority is exponents.
    Calculate: $2 \cdot 2 \cdot 2 = 8$. *Replace:* 2^3 with 8.
$= 32$ *Identify:* Multiplication is last. *Calculate:* $4 \cdot 8 = 32$.
    Replace $4 \cdot 8$ with 32.

9. $2 + 7(10 - 3 \cdot 2) - 4$ The product inside the parentheses is done first: $3 \cdot 2 = 6$.

$= 2 + 7(10 - 6) - 4$ We must finish all operations inside the parentheses, so we subtract: $10 - 6 = 4$.

$= 2 + 7(4) - 4$ Multiplication has the highest priority: $7 \cdot 4 = 28$.

$= 2 + 28 - 4$ Add first: $2 + 28 = 30$.

$= 30 - 4$ Subtract last: $30 - 4 = 26$.

$= 26$

10. We rewrite the problem as division and then follow the order of operations.

$(4 + 8 \div 2) \div (7 - 3)$ We perform operations inside parentheses first

$(4 + 4) \div 4$ $8 \div 2 = 4$; $7 - 3 = 4$

$8 \div 4 = 2$ Divide.

1.7 Student Practice

1. (a) Five times y plus three

$$5 \quad \cdot \quad y \quad + \quad 3 = 5y + 3$$

(b) Three times the sum of m and two

$$3 \quad \cdot \quad (m + 2) = 3(m + 2)$$

2. $\dfrac{(5y - 4)}{3} = \dfrac{(5 \cdot 2 - 4)}{3}$ We replace y with 2.

$= \dfrac{(10 - 4)}{3}$ We multiply first.

$= \dfrac{6}{3}$ Next we complete the operations within the parentheses.

$= 2$ We divide.

3. (a) $5m - 2n + 1 = 5(7) - 2(3) + 1$ We replace m with 7 and n with 3.

$= 35 - 6 + 1$ We multiply first.

$= 30$ We subtract and add last.

(b) $\dfrac{(a^3 - 2)}{b} = \dfrac{(2^3 - 2)}{3}$ We replace a with 2 and b with 3.

$= \dfrac{6}{3}$ We cube 2 first, then subtract: $2^3 = 8, 8 - 2 = 6$.

$= 2$ We divide last: $6 \div 3 = 2$.

4. (a) $2(x - 5) = 2 \cdot x - 2 \cdot 5$ The 2 "distributes" to the x and the 5 over the subtraction sign.

$= 2x - 10$

(b) $4(y + 3) = 4 \cdot y + 4 \cdot 3$ The 4 "distributes" to the y and the 3 over the addition sign.

$= 4y + 12$

5. $7(y + 3) + 2 = 7 \cdot y + 7 \cdot 3 + 2$ We use the distributive property to multiply $7(y + 3)$.

$= 7y + 21 + 2$

$= 7y + 23$ We simplify: $21 + 2 = 23$.

1.8 Student Practice

1. (a) Four n's $= 4n$ **(b)** $y + y + y = 3y$ **(c)** Eight $= 8$ **(d)** One $y = 1y$ or y

2.

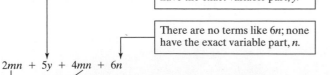

There are no terms like $5y$; none have the exact variable part, y.

There are no terms like $6n$; none have the exact variable part, n.

$2mn + 5y + 4mn + 6n$

$2mn$ and $4mn$ are like terms; the variable parts, mn, are the same.

3. (a) There are no terms like $4a$; none have the exact variable part.

$2ab + 4a + 3ab = (2ab + 3ab) + 4a$ We identify and group like terms.

$= (2 + 3)ab + 4a$ Add the numerical coefficients of like terms.

$2ab + 4a + 3ab = 5ab + 4a$ $(2 + 3)ab = 5ab$

(b) $4y + 5x + y + x = 4y + 5x + 1y + 1x$ Write the numerical coefficients of 1.

$= 4y + 1y + 5x + 1x$ We identify and group like terms.

$= (4 + 1)y + (5 + 1)x$ Add the numerical coefficients of like terms.

$4y + 5x + y + x = 5y + 6x$

(c) $7x + 3y + 3z$; There are no like terms, so no terms can be combined.

4. We add the lengths of all three sides to find the perimeter.

$(2x + y) + (x + 4y) + (3x + 2y)$ We must combine like terms.

$= (2x + x + 3x) + (y + 4y + 2y)$ We regroup.

$= 6x + 7y$ We combine like terms.

The algebraic expression for the perimeter is $6x + 7y$.

5. **(a)** Four times what number is the same as 7?

$$4 \cdot x = 7$$
$$4x = 7$$

(b) Three subtracted from what number is equal to nine?

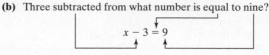

$$x - 3 = 9$$

(c) The number of cards plus 20 new cards equals 75 cards.

$$n + 20 = 75$$
$$n + 20 = 75$$

6. If 5 is a solution to $x + 8 = 11$, when we replace x with 5 we will get a true statement.
$x + 8 = 11$ "What number plus eight equals eleven?"
$5 + 8 \overset{?}{=} 11$ Replace the variable with 5 and simplify.
$13 \overset{?}{=} 11$ This is a false statement.
Since $13 = 11$ is not a true statement, 5 is not a solution to $x + 8 = 11$.

7. To solve the equation $4 + n = 9$, we answer this question:
"Four plus what number is equal to nine?"
Using addition facts we see that the answer, or solution, is 5.
Check: $4 + n = 9$ Write the equation.
$4 + 5 \overset{?}{=} 9$ Replace the variable with 5 and simplify.
$9 = 9$ ✓ Verify that we get a true statement.

8. To solve the equation $6x = 48$, we answer this question:
"Six times what number equals forty-eight?"
The answer is 8 and is written $x = 8$.
Check: $6x = 48$
$6(8) \overset{?}{=} 48$ Replace the variable with 8 and simplify.
$48 = 48$ ✓ Verify that this is a true statement.
The solution to $6x = 48$ is 8 and is written $x = 8$.

9. $\frac{x}{4} = 2$ "What number divided by 4 equals 2?"

Using division facts, we see that the answer or solution is 8 and is written $x = 8$.
Check:
$\frac{x}{4} = 2 \rightarrow \frac{8}{4} \overset{?}{=} 2$ Replace the variable with 8 and simplify.

$2 = 2$ ✓ Verify that this is a true statement.
Thus $x = 8$ is the solution.

10. First we simplify. $(3 + x) + 1 = 7$
$(x + 3) + 1 = 7$ Commutative property
$x + (3 + 1) = 7$ Associative property
$x + 4 = 7$ Simplify.
Next we solve $x + 4 = 7$.
"What number plus 4 is equal to 7?"
$x + 4 = 7$
$x = 3$
We leave the check to the student.
The solution to the equation is 3 and is written $x = 3$.

11. $n + 3n = 20$ Write the equation.
$1n + 3n = 20$ Write n as $1n$.
$(1 + 3)n = 20$ Add numerical coefficients of like terms.
$4n = 20$ Think: "Four times what number equals twenty?" 5.
$n = 5$
The solution to the equation is 5 and is written $n = 5$.

12. What number times 5 is equal to 20?

$$n \cdot 5 = 20.$$ Translate.
$$n = 4$$ Use multiplication facts to find n.

1.9 Student Practice

1. *Understand the problem.* The information we need to solve the problem is listed in the table.
   (a) *Calculate and state the answer.* To estimate, we round each number to the thousands place.

	Exact Value		Rounded Value
Price of F-150 SVT Raptor	38,020	→	38,000
Price of Taurus SEL	28,195	→	28,000

We subtract the rounded figures to estimate the difference in cost of the two vehicles.
$38,000 - 28,000 = 10,000$
The estimated difference in price is $10,000.

(b) We subtract the original figures to find the exact difference in the cost of the two vehicles.

$38,020 - 28,195 = 9825$

The exact difference in price is $9825.

The estimated difference in price, $10,000, is close to the exact difference, $9825, so our estimate is reasonable.

2. *Understand the problem.* Read the problem carefully and study the invoice. Then fill in the invoice.

Calculate and state the answer. Multiply to get the total cost per item. Then place these products on the invoice.

8	Tables	$230	1840	← 8 tables $= 8(230) = \$1840$
50	Chairs	25	1250	← 50 chairs $= 50(25) = \$1250$
3	Ovens	910	2730	← 3 ovens $= 3(910) = \$2730$.
		Total:	5820	

Now, divide the total cost by 2 to find the amount each owner paid: $5820 \div 2 = 2910$. Each owner paid $2910.

Check your answer. We estimate and compare the estimate to our calculated answer.

Tables: $8 \rightarrow 8$ $\$230 \rightarrow \$200$ $8 \cdot 200 = \$1600$

Chairs: $50 \rightarrow 50$ $\$25 \rightarrow \$30$ $50 \cdot 30 = 1500$

Ovens: $3 \rightarrow 3$ $\$910 \rightarrow 900$ $3 \cdot 900 = \underline{2700}$

                                                  $\$5800$

Divide the estimated total by 2: $\$5800 \div 2 = \$2900$.

Our estimate of $2900 per owner is close to our exact calculation of $2910. Thus our exact answer is reasonable.

3. *Understand the problem.*

2 miles + 2 miles $\cdots$ = 4500 miles   How many groups of 2 are in 4500?

3 points + 3 points $\cdots$ = ? points

We organize the information and make our plan in the Mathematics Blueprint.

Mathematics Blueprint for Problem Solving

Gather the Facts	What Am I Asked to Do?	How Do I Proceed?	Key Points to Remember
A customer is awarded 3 frequent-flyer points for every 2 miles flown.	Determine how many frequent-flyer points a customer earns after flying 4500 miles.	1. *Divide* 4500 by 2. 2. *Multiply* 3 times the number obtained in step 1.	Frequent-flyer points are determined by the number of miles flown.

Calculate and state the answer.

We divide to find how many groups of 2 are in 4500. $4500 \div 2 = 2250$

We multiply to find the total points earned. $2250 \times 3 = 6750$

The customer would earn 6750 points.

Check. If the customer earned 4 points (instead of 3) for every 2 miles traveled, we could just double the mileage to find the points earned.

2 miles earns 4 points (double 2).

4500 miles earns 9000 points (double 4500).

Since the customer earned a little less than 4 points, the total should be less than 9000. It is: $6750 < 9000$. The customer also earned more points than miles traveled (3 points for every 2 miles), so our total points should be more than the total miles traveled. It is: $6750 > 4500$. Our answer is reasonable.

4. *Understand the problem.* We organize the information in the Mathematics Blueprint.

Mathematics Blueprint for Problem Solving

Gather the Facts	What Am I Asked to Do?	How Do I Proceed?	Key Points to Remember
The commission option pays $55 per sale, and Emily averages 7 sales per week. The salary option pays $1770 per month.	Determine which option provides higher earnings.	1. Calculate the average commission per week. 2. Multiply the result of step 1 by 52 to find the yearly pay. 3. Multiply the monthly salary by 12 to find the yearly pay. 4. Compare the salaries.	I must find yearly pay: 12 months = 1 year 52 weeks = 1 year

Calculate and state the answer. From the information organized in the blueprint, we can write out a process to find the answer.

$\$55 \times 7 = \$385$ Pay for 1 week (commission)

$\$385 \times 52 = \$20,020$ Pay for 1 year (commission)

$\$1770 \times 12 = \$21,240$ Pay for 1 year (salary)

The yearly pay is $20,020 for the commission option and $21,240 for the salary option. The salary option, $1770 per month, pays more per year.

Check the answer. We estimate the commission option pay per year by rounding $55 per sale to $50 and 52 weeks to 50 weeks.

$\$50 \times 7 \text{ sales} = \$350 \text{ per week}; \$350 \times 50 \text{ weeks} = \$17,500 \text{ per year}$

We estimate the salary option pay per year by rounding $1770 per month to $2000 and 12 months to 10.

$\$2000 \times 10 \text{ months} = \$20,000 \text{ per year}$

Since $\$20,000 > \$17,500$, the salary option pays more.

Chapter 2 2.1 Student Practice

1. We draw a dot in the correct location on the number line.

$$\overset{\bullet}{-5}\ \overset{\bullet}{-4}\ -3\ \overset{\bullet}{-2}\ -1\ \ 0\ \ 1\ \ \overset{\bullet}{2}\ \ 3\ \ \overset{\bullet}{4}\ \ 5$$

3. (a) A property tax increase of \$130: $+$ \$130
(b) A dive of 7 ft below the surface of the sea: $-$ 7 ft

5. To find the opposite of a number, we change the sign of the number.
(a) The opposite of -6 is 6.   (b) The opposite of -1 is 1.
(c) The opposite of 12 is -12 (d) The opposite of 1 is -1.

7. To avoid errors involving negative signs, we can place parentheses around the variable and its replacement.
$-(-(-a)) = -(-(-(a)))$
Replace a with 6.
$\qquad = -(-(-(6)))$
The opposite of 6 is -6.
$\qquad = -(-(-6))$
The opposite of -6 is 6. $(-(-6)) = 6$.
$\qquad = -(6)$
The opposite of 6 is -6.
$\qquad = -6$

9. $|-12|\ ?\ |2|$
$\qquad\downarrow\qquad\downarrow$
$\quad$ 12 ? 2   We find the absolute values.
$\quad$ 12 > 2   We write the appropriate inequality symbol, $>$.
$|-12| > |2|$ -12 has a larger absolute value than 2.

11. On the graph, negative numbers are located below zero and positive numbers above zero.
(a) The low temperature was $-32°F$ in Fairbanks and $10°F$ in Buffalo. It was colder in Fairbanks. Note that the dot representing Fairbanks's temperature is lower on the line graph than the dot representing Buffalo's temperature.
(b) The highest point on the line graph corresponds to Boston, so Boston recorded the highest temperature. The lowest point on the line graph corresponds to Fairbanks, so Fairbanks recorded the lowest temperature.

2. (a) $-5 < 2$ -5 lies to the left of 2 on the number line.
(b) $-3 > -6$ -3 lies to the right of -6 on the number line.
(c) $-53 > -218$ -53 lies to the right of -218 on the number line.

4.

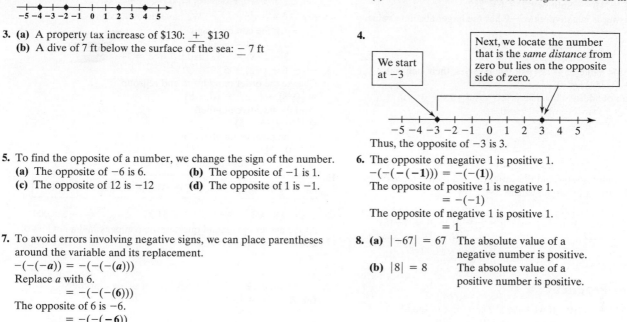

Thus, the opposite of -3 is 3.

6. The opposite of negative 1 is positive 1.
$-(-(-(-1))) = -(-(1))$
The opposite of positive 1 is negative 1.
$\qquad = -(-1)$
The opposite of negative 1 is positive 1.
$\qquad = 1$

8. (a) $|-67| = 67$ The absolute value of a negative number is positive.
(b) $|8| = 8$ The absolute value of a positive number is positive.

10. We must find the opposite of the absolute value of -1.
$-|-1|$
First we find the absolute value of -1: $|-1| = 1$.
$\qquad = -(1)$
Then we take the opposite of 1: $-(1) = -1$.
$\qquad = -1$

2.2 Student Practice

1. (a) A decrease in altitude of 100 feet followed by a decrease in altitude of 100 feet results in a decrease in altitude of 200 feet.
(b) A decrease in altitude of 100 feet followed by a decrease in altitude of 100 feet results in a decrease in altitude of 200 feet.

$$\underset{-100}{\downarrow}\qquad\underset{+}{\downarrow}\qquad\underset{(-100)}{\downarrow}\qquad\underset{=}{\downarrow}\qquad\underset{-200}{\downarrow}$$

2. (a)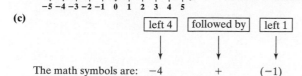

(c)
$$\boxed{\text{left 4}}\quad\boxed{\text{followed by}}\quad\boxed{\text{left 1}}$$
$$\downarrow\qquad\qquad\downarrow\qquad\qquad\downarrow$$
The math symbols are: $-4\qquad+\qquad(-1)$

(b) From the illustration, we see that the end result is in the negative region since we began at 0 and moved 4 units in the negative direction (left), followed by another 1 unit in the negative direction.

(d) We end at -5, which is the sum. $-4 + (-1) = -5$

3. (a) We are adding two numbers with the same sign, so we keep the common sign and add the absolute values.
$-2 + (-4) = -$ The answer is *negative* since the common sign is negative.
$-2 + (-4) = -6$ Add. $2 + 4 = 6$

4.   down 15 followed by   up 30
$$\qquad\downarrow\qquad\qquad\downarrow$$
(a) $-15°F$ $+$ $(+30°F)$
(b) From the chart, we see that the temperature reading is positive since it went up ($+$) more degrees than it went down ($-$).
(c) The final temperature is $+15°F$, which is the sum:
$-15°F + (+30°F) = +15°F$.

5. (a) The answer is positive, since 7 is positive and has the larger absolute value.

$-4 + 7 = +$

$-4 + 7 = +3$ Subtract: $7 - 4 = 3$.

(b) The answer is negative, since -7 has the larger absolute value.

$4 + (-7) = -$

$4 + (-7) = -3$ Subtract: $7 - 4 = 3$.

7. (a) Since 3544 and -3544 are additive inverses, their sum is 0.

$3544 + (-3544) = 0$

(b) The sum of additive inverses is 0. Thus if $-13 + y = 0$, then $y = 13$ since $-13 + 13 = 0$.

9. We place parentheses around the variables, and then replace each variable with the appropriate values.

(a) $-3 + (x) + (y)$

$= -3 + (-5) + (-11)$

$= \quad -8 \quad + (-11)$

$= -19$

(b) $-(x) + 8 + (y)$

$= -(-2) + 8 + (-15)$

$= \quad 2 \quad + \quad (-7)$

$= -5$

6. (a) $-3 + 8$ We have *different* signs.

8 is larger than 3, so the answer is positive.

$-3 + 8 = +$

$-3 + 8 = +5$ Subtract: $8 - 3 = 5$.

(b) $-3 + (-8)$ We have the *same* signs.

Keep the common sign.

$-3 + (-8) = -$

$-3 + (-8) = -11$ Add: $3 + 8 = 11$.

8. $-8 + 6 + (-2) + 5$

Change the order of addition and regroup.

$= [(-8) + (-2)] + (6 + 5)$

Add the negative numbers: $-8 + (-2) = -10$.

$= -10 + (6 + 5)$

Add the positive numbers: $6 + 5 = 11$.

$= -10 + 11$

$= 1$ Add the result: $-10 + 11 = 1$.

10.

1st quarter loss	+	2nd quarter profit	=	net loss
$-\$40{,}000$	+	$\$20{,}000$	=	$-\$20{,}000$

At the end of the second quarter, the company had a net loss of $\$20{,}000$.

2.3 Student Practice

1. (a) $\$10 - \$20 = -\$10$ If we have \$10 and want to spend \$20, we are short \$10, or $-\$10$.

(b) $5 - 7 = -2$ If we have 5 items and try to take away 7 items, we are short 2 items, or -2.

3. We replace the second number with its opposite and then add using the rules for addition.

(a)

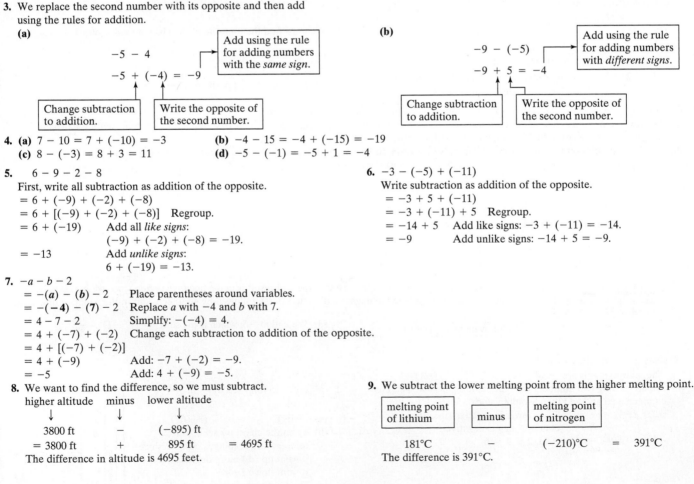

$-5 - 4$ → Add using the rule for adding numbers with the *same sign*.

$-5 + (-4) = -9$

Change subtraction to addition.   Write the opposite of the second number.

2.

Subtraction	Addition of the Opposite
(a) $20 - 10 = 10$	$20 + (-10) = 10$
(b) $5 - 2 = 3$	$5 + (-2) = 3$
(c) $20 - 5 = 15$	$20 + (-5) = 15$

(b)

$-9 - (-5)$ → Add using the rule for adding numbers with *different signs*.

$-9 + 5 = -4$

Change subtraction to addition.   Write the opposite of the second number.

4. (a) $7 - 10 = 7 + (-10) = -3$ **(b)** $-4 - 15 = -4 + (-15) = -19$

(c) $8 - (-3) = 8 + 3 = 11$ **(d)** $-5 - (-1) = -5 + 1 = -4$

5. $6 - 9 - 2 - 8$

First, write all subtraction as addition of the opposite.

$= 6 + (-9) + (-2) + (-8)$

$= 6 + [(-9) + (-2) + (-8)]$ Regroup.

$= 6 + (-19)$ Add all *like signs*:

             $(-9) + (-2) + (-8) = -19$.

$= -13$ Add *unlike signs*:

             $6 + (-19) = -13$.

6. $-3 - (-5) + (-11)$

Write subtraction as addition of the opposite.

$= -3 + 5 + (-11)$

$= -3 + (-11) + 5$ Regroup.

$= -14 + 5$ Add like signs: $-3 + (-11) = -14$.

$= -9$ Add unlike signs: $-14 + 5 = -9$.

7. $-a - b - 2$

$= -(a) - (b) - 2$ Place parentheses around variables.

$= -(-4) - (7) - 2$ Replace a with -4 and b with 7.

$= 4 - 7 - 2$ Simplify: $-(-4) = 4$.

$= 4 + (-7) + (-2)$ Change each subtraction to addition of the opposite.

$= 4 + [(-7) + (-2)]$

$= 4 + (-9)$ Add: $-7 + (-2) = -9$.

$= -5$ Add: $4 + (-9) = -5$.

8. We want to find the difference, so we must subtract.

higher altitude   minus   lower altitude

    ↓   ↓   ↓

3800 ft $-$ (-895) ft

$= 3800$ ft $+$ 895 ft $= 4695$ ft

The difference in altitude is 4695 feet.

9. We subtract the lower melting point from the higher melting point.

melting point of lithium	minus	melting point of nitrogen		
$181°C$	$-$	$(-210)°C$	$=$	$391°C$

The difference is $391°C$.

2.4 Student Practice

1. $3(-1) = (-1) + (-1) + (-1) = -3$
Therefore, $3(-1) = -3$.

2. (a) $3(8) = +24$ The number of negative signs, 0, is *even*, so the answer is *positive*. We multiply absolute values: $3 \cdot 8 = 24$.

(b) $3(-8) = -$ The number of negative signs, 1, is *odd*, so the answer is *negative*.
     $= -24$ We multiply absolute values: $3 \cdot 8 = 24$.

(c) $-3(8) = -$ The number of negative signs, 1, is *odd*, so the answer is *negative*.
     $= -24$ We multiply absolute values: $3 \cdot 8 = 24$.

(d) $-3(-8) = +$ The number of negative signs, 2, is *even*, so the answer is *positive*.
     $= +24$ We multiply absolute values: $3 \cdot 8 = 24$.

3. $(-2)(-1)(-4) = 2(-4)$ First we multiply $(-2)(-1) = 2$.
           $= -8$ Then we multiply $2(-4) = -8$.

4. $(-3)(2)(-1)(-4)(-3)$ The answer is positive since there are 4 negative signs and 4 is an even number.
   $= +[3(2)(1)(4)(3)]$ Multiply the absolute values.
   $= 72$

5. $(-4)^4$
$= (-4)(-4)(-4)(-4)$
$= 256$ The answer is positive since the exponent 4 is even.

6. (a) $(-2)^3 = -8$ The answer is negative since the exponent 3 is odd.

(b) $(-2)^6 = 64$ The answer is positive since the exponent 6 is even.

(c) $(-2)^7 = -128$ The answer is negative since the exponent 7 is odd.

7. (a) -5^2, "the opposite of five squared"
-5^2 The base is 5;
$= -(5 \cdot 5)$ We use 5 as the factor for repeated multiplication.
$= -25$ We take the opposite of the product.

(b) $(-5)^2$, "negative five squared"
$(-5)^2$ The base is (-5);
$= (-5)(-5)$ We use -5 as the factor for repeated multiplication.
$= 25$

8. (a) $42 \div 7 = 6$
(b) $42 \div (-7) = -6$
(c) $-42 \div 7 = -6$
(d) $-42 \div (-7) = 6$

9. (a) $49 \div (-7) = -7$
(b) $4(-9) = -36$
(c) $-30(-4) = 120$
(d) $\dfrac{-54}{-9} = -54 \div (-9) = 6$

10. We place parentheses around the variables and then we replace each variable with the given value.

(a) $\dfrac{-(x)}{(y)} = \dfrac{-(-22)}{(11)}$ We replace x with -22 and y with 11.
$= \dfrac{22}{11}$ $-(-22) = 22$: The opposite of negative 22 is 22.
$= 2$ $22 \div 11 = 2$

(b) $(a)^3 = (-3)^3$ We replace a with -3.
$= -27$ The answer is negative since the exponent is odd.

2.5 Student Practice

1. $-6 + 20 \div 2(-2)^2 - 5$ *Identify*: The highest priority is exponents.
                       Calculate: $(-2)^2 = 4$.   *Replace*: $(-2)^2$ with 4.

$= -6 + 20 \div 2(4) - 5$ *Identify*: The highest priority is division.
                       Calculate: $20 \div 2 = 10$.   *Replace*: $20 \div 2$ with 10.

$= -6 + 10(4) - 5$ *Identify*: The highest priority is multiplication.
                       Calculate: $10(4) = 40$.   *Replace*: $10(4)$ with 40.

$= -6 + 40 - 5$ We change subtraction to addition of the opposite.
$= -6 + 40 + (-5)$ We add: $-6 + 40 + (-5) = 29$.
$= 29$

2. We perform operations inside parentheses and brackets first.

$\dfrac{[-10 + 4(-2)]}{(11 - 20)} = \dfrac{[-10 + (-8)]}{(11 - 20)}$ We multiply: $4(-2) = -8$.
$= \dfrac{-18}{(11 - 20)}$ We add: $-10 + (-8) = -18$.
$= \dfrac{-18}{-9}$ We subtract: $11 - 20 = -9$.
$= 2$ We divide last: $-18 \div (-9) = 2$.

3. We perform the operations within the innermost grouping symbols first.

$-18 \div \{3[12 \div (-2)]\}$ We divide: $12 \div (-2) = -6$.
$= -18 \div \{3(-6)\}$ We complete operations inside the brackets: $3(-6) = -18$.
$= -18 \div (-18)$
$= 1$ Now we divide: $-18 \div (-18) = 1$.

4. We summarize the information.

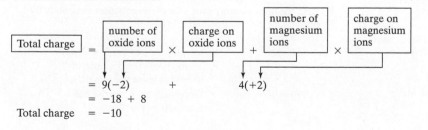

| Total charge | $=$ | number of oxide ions | $\times$ | charge on oxide ions | $+$ | number of magnesium ions | $\times$ | charge on magnesium ions |

$= 9(-2) \quad + \quad 4(+2)$
$= -18 + 8$
Total charge $= -10$

2.6 Student Practice

1. $-6y + 8x + 4y = 8x + (-6y) + 4y$ Rearrange terms.
$$= 8x + (-6 + 4)y \quad \text{Add numerical coefficients of like terms.}$$
$$= 8x + (-2)y \quad 8x \text{ and } -2y \text{ are not like terms.}$$
$$= 8x - 2y \quad \text{Rewrite addition of the opposite as subtraction.}$$

2. $7a + 4b - a = 7a + 4b + (-a)$ First we change subtraction to
     addition of the opposite.
$$= 7a + (-1a) + 4b \quad \text{Now we rearrange terms.}$$
     Note that $-a = -1a$.
$$= 6a + 4b \quad \text{We add like terms: } [7 + (-1)]a = 6a.$$

3. Perform each operation indicated.

(a) $4 - 6 + 8$ (b) $4x - 6x + 8x$
$$= 4 + (-6) + 8 \qquad\qquad = 4x + (-6x) + 8x$$
$$= -2 + 8 \qquad\qquad\qquad = -2x + 8x$$
$$= 6 \qquad\qquad\qquad\qquad\quad = 6x$$

4. $2x + 8y - 5x + 4xy - 12y$
Change subtraction to addition of the opposite.
$$= 2x + 8y + (-5x) + 4xy + (-12y)$$
Rearrange terms to group like terms.
$$= 2x + (-5x) + 8y + (-12y) + 4xy$$
Add coefficients of like terms.
$$= [2 + (-5)]x + [8 + (-12)]y + 4xy$$
$$= -3x + (-4y) + 4xy \quad \text{Rewrite addition of the opposite}$$
$$= -3x - 4y + 4xy \quad \text{as subtraction so that the answer is simplified.}$$

5. $\dfrac{[(a)^3 + (b)]}{2}$ Place parentheses around each variable.

$= \dfrac{[(-2)^3 + (-4)]}{2}$ Replace a with -2 and b with -4.

$= \dfrac{[-8 + (-4)]}{2}$ Calculate: $(-2)^3 = -8$.

$= \dfrac{-12}{2} = -6$ Simplify.

6. $-8(m - 1) = -8(m) - (-8)(1)$ Distribute the -8 over subtraction.
$$= -8m - (-8) \quad \text{Multiply: } (-8)(1) = -8.$$
$$= -8m + 8 \quad \text{Rewrite subtraction as addition of the opposite.}$$

7. We evaluate the formula for the values given: $v = -5$ and $t = 4$.
$$s = v - 32t$$
$$= -5 - 32(4)$$
$$= -5 - 128$$
$$= -133 \quad \text{Note that a negative speed means that the object is moving in a downward direction.}$$
The skydiver is falling 133 feet per second.

Chapter 3   3.1 Student Practice

1. $y - 6 + \square = y + 0 = y$

$y - 6 + 6 = y + 0 = y$

2. We want an equation of the form $x = some\ number$.
Therefore, we want to get x alone on one side of the equation.
$$x - 19 = -31$$
$$x - 19 + 19 = -31 + 19 \quad \text{Add the opposite of } -19$$
     to both sides of the equation.
$$x + 0 = -12 \qquad x - 19 + 19 = x + 0$$
     and $-31 + 19 = -12$
$$x = -12 \qquad \text{The solution is } -12.$$

3. The variable x is on the right side of the equation; therefore we want an equation of the form *some number* $= x$.
$$92 = 46 + x$$
$$92 + (-46) = 46 + (-46) + x \quad \text{Add the opposite of 46 to both}$$
     sides since $46 + (-46) = 0$.
$$46 = 0 + x$$
$$46 = x$$

Check: To check our answer, we replace x with 46 and verify we get a true statement.
$$92 = 46 + x$$
$$92 \overset{?}{=} 46 + 46$$
$$92 = 92 \ \checkmark$$

4. The variable x appears *more than once* on the left side of the equation. This means we'll need to complete an extra step, combining like terms, so that x appears only once in the equation.
$$5x - 4x - 3 = 11$$
$$x - 3 = 11 \quad \text{We combine like terms: } 5x - 4x = 1x \text{ or } x.$$
$$\underline{+ 3 \qquad 3} \quad \text{Think: "Add the opposite of } -3 \text{ to both sides of the equation."}$$
$$x + 0 = 14$$
$$x = 14$$
Check:
$$5x - 4x - 3 = 11$$
$$5(14) - 4(14) - 3 \overset{?}{=} 11$$
$$70 - 56 - 3 \overset{?}{=} 11$$
$$11 = 11 \ \checkmark$$

5. First, we must simplify each side of the equation separately by completing the addition and subtraction.
$$5 - 8 = y - 2 + 19$$
$$-3 = y + 17 \quad \text{Simplify: } 5 - 8 = -3; -2 + 19 = 17.$$
$$\underline{+ -17 \qquad - 17} \quad \text{Think: "Add the opposite of 17 to both sides of the equation."}$$
$$-20 = y \quad \text{We usually do not write the step } -20 = y + 0.$$
Check:
$$5 - 8 = y - 2 + 19$$
$$5 - 8 \overset{?}{=} -20 - 2 + 19$$
$$-3 = -3 \ \checkmark$$

6. A picture is given, so we do not need to draw one. Since $\angle x$ and $\angle y$ are supplementary angles, we know that their sum is 180°. We use this information to write an equation.

$$\angle x + \angle y = 180° \quad \text{Write an equation.}$$
$$49° + \angle y = 180° \quad \text{Replace } \angle x \text{ with } 49°.$$
$$\underline{+ -49° \qquad\qquad -49°} \quad \text{Solve the equation.}$$
$$\angle y = 131° \quad \text{The measure of } \angle y \text{ is } 131°.$$

7. (a)

Angle a measures 20° less than angle b.

$$\angle a \quad = \quad \angle b - 20°$$

(b)
$$\angle a = \angle b - 20° \quad \text{Write an equation.}$$
$$80° = \angle b - 20° \quad \text{Replace } \angle a \text{ with } 80°.$$
$$\underline{+ 20° \qquad\qquad 20°} \quad \text{Solve the equation.}$$
$$100° = \angle b \quad \text{The measure of } \angle b \text{ is } 100°.$$

3.2 Student Practice

1. $\dfrac{-9x}{\Box} = 1 \cdot x = x$ We want the quotient $(-9) \div (\Box)$ to equal 1.

$$\dfrac{-9x}{-9} = 1 \cdot x = x$$

3. First, we simplify the equation.

$$8(n \cdot 5) = \dfrac{320}{2}$$
$$8(n \cdot 5) = 160 \quad \text{Divide: } 320 \div 2 = 160.$$
$$8(5n) = 160 \quad \text{Change the order of the factors.}$$
$$(8 \cdot 5)n = 160 \quad \text{Regroup the factors.}$$
$$40n = 160 \quad \text{Simplify.}$$
$$\dfrac{40n}{40} = \dfrac{160}{40} \quad \text{Divide both sides by 40.}$$
$$n = 4$$

We leave the check to the student.

5. We are representing the measure of angle a with the letter A and the measure of angle b with the letter B. Now we translate.

The measure of angle a is four times the measure of angle b.

$$A \quad = \quad 4 \quad \cdot \quad B$$

The equation that represents the statement is $A = 4B$.

6. This statement is phrased a little differently than others we have translated. Therefore, it is helpful to write a statement that *compares* two quantities.

The number of laps Sara ran is greater than the number of laps Dave ran.

Think of a single comparison of the number of laps each ran, such as in the case when Dave only ran one lap.

If Dave ran 1 lap, then Sara ran 2 laps.

twice as many laps

Now we can rephrase the statement and translate into an equation.

The number of laps that Sara ran is twice the number of laps that Dave ran.

$$S \quad = \quad 2 \cdot \quad D$$

The equation that represents the statement is $S = 2D$.

7. (a) The number of cars (C) is twice the number of trucks (T).

The word *twice* means *two times*.

The number of cars is twice the number of trucks.

$$C \quad = \quad 2 \cdot \quad T$$

8. We know that $\angle a$ and $\angle b$ are supplementary and therefore their sum is 180°. We write the equation as follows.

$$\angle a + \angle b = 180° \quad \text{Write the equation.}$$
$$\downarrow \qquad \downarrow$$
$$55° + (x + 4°) = 180° \quad \text{Replace } \angle a \text{ with } 55° \text{ and } \angle b \text{ with } x + 4°.$$
$$59° + x = 180° \quad \text{Simplify: } 55° + 4° + x = 59° + x.$$
$$\underline{+ -59° \qquad\qquad -59°} \quad \text{Solve the equation.}$$
$$x = 121°$$

Since $\angle b = x + 4°$, we must substitute 121° for x to find the measure of $\angle b$.

$$\angle b = x + 4°$$
$$\angle b = 121° + 4° = 125°$$

Therefore, $x = 121°$ and $\angle b = 125°$.

2. We want to make $5m = -155$ into a simpler equation, $m = some\ number.$

$$5m = -155 \quad \text{The variable, } m, \text{ is multiplied by 5.}$$
$$\dfrac{5m}{5} = \dfrac{-155}{5} \quad \text{Dividing both sides by 5 undoes the multiplication by 5.}$$
$$m = -31 \qquad \dfrac{5m}{5} = 1 \cdot m = m \quad \text{and} \quad -155 \div 5 = -31$$

4. We simplify each side of the equation first.

$$30 + (-12) - 7y - 5y$$
$$18 = 7y - 5y \quad \text{Simplify the left side of the equation.}$$
$$18 = 2y \quad \text{Simplify the right side of the equation.}$$
$$\dfrac{18}{2} = \dfrac{2y}{2} \quad \text{Divide both sides of the equation by 2.}$$
$$9 = y \qquad 18 \div 2 = 9; 2 \div 2 = 1; 1 \cdot y = y$$

We leave the check to the student.

(b) Find the number of trucks if there are 150 cars on the road.

$$C = 2T \quad \text{We write the equation.}$$
$$150 = 2T \quad \text{We replace } C \text{ with } 150.$$
$$\dfrac{150}{2} = \dfrac{2T}{2} \quad \text{We divide both sides by 2.}$$
$$75 = T \quad \text{There are 75 trucks.}$$

8. *Understand the problem.* We use a Mathematics Blueprint for Problem Solving to organize the information.

Mathematics Blueprint for Problem Solving

Gather the Facts	What Am I Asked to Do?	How Do I Proceed?	Key Points to Remember
Let x = the number of shares of stock purchased. Ian paid $25 for each share. He sold each share for $45.	Find the number of shares of stock purchased.	Let $25x$ = the purchase price and $45x$ = the sale price. Find the profit: Profit = Sale price − purchase price	Profit is how much money is made.

Solve and state the answer.

$$\underset{\downarrow}{\text{profit}} = \underset{\downarrow}{\text{sale price}} - \underset{\downarrow}{\text{purchase price}}$$

$$200 = \quad 45x \quad - \quad 25x$$ We must simplify the equation.

$200 = 20x$ We combine like terms: $45x - 25x = 20x$.

$\dfrac{200}{20} = \dfrac{200x}{20}$ We divide both sides by 20.

$10 = x$ Ian purchased 10 shares of stock.

Check: We can estimate to see if our answer is reasonable. We round so that each number has 1 nonzero digit.
$200 \rightarrow 200; \quad 45 \rightarrow 50; \quad 25 \rightarrow 30$
Now we estimate the value of x.
$200 = 50x - 30x$
$200 = 20x$
$\dfrac{200}{20} = x$
$10 = x$ Our answer is reasonable.

3.3 Student Practice

1. We use the four-step process.

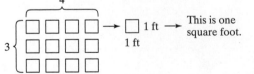

$s = 11\,\text{yd}$ *Step 1:* Draw a picture.

$P = 4s$ *Step 2:* Write the formula.
$\quad = 4(11\,\text{yd})$ *Step 3:* Replace s with 11 yd.
$P = 44\,\text{yd}$ *Step 4:* Simplify and solve for P.
The perimeter is 44 yards.

2. Since we are given the picture, we start with step 2 of the four-step process.
$P = 2L + 2W$ First, we write the formula for perimeter.
$30 = 2(2W) + 2W$ Next, we replace P with 30 and L with $2W$.
$30 = 4W + 2W$ Then, we multiply: $2(2W) = 4W$.
$30 = 6W$ We combine like terms.
$\dfrac{30}{6} = \dfrac{6W}{6}$ Now we divide each side by 6.
$5 = W$ The width of the rectangle is 5 feet.

3. Think of an array with 3 rows and 4 columns.

This is one square foot.
$\boxed{}\,1\,\text{ft}$
$1\,\text{ft}$

Just as we multiplied the number of rows times the number of columns to find the number of items in an array, we *multiply* the *length* times the *width* to find the area of the flower garden. The area is 3 feet × 4 feet = 12 square feet. Note, this means that there are 12 squares in the above illustration.

4. Follow the four-step process.

$\boxed{}$ 6 ft   *Step 1:* Draw a picture. Now we complete steps 2–4.

$A = s^2$ *Step 2:* Write the formula.
$\quad = (6\,\text{ft})^2$ *Step 3:* Replace s with the given value.
$\quad = (6 \cdot 6)(\text{ft} \cdot \text{ft})$ *Step 4:* Simplify. Multiply the units: ft *times* ft = ft^2.
$\quad = 36\,\text{ft}^2$ This is read "thirty-six square feet."

5. We are given the picture.
$A = bh$ Write the formula.
$117 = b \cdot 9$ We replace A and h with the values given.
$117 = 9 \cdot b$ Reorder.
$\dfrac{117}{9} = \dfrac{9b}{9}$ We divide both sides of the equation by 9.
$13 = b$ The base is 13 feet.

6. Divide the figure into three rectangles and then find the area of each rectangle separately. Next, add these three areas together to find the area of the figure.

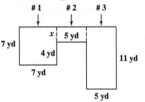

Area of rectangle #1:
$A = L \cdot W$
$\quad = 7\,\text{yd} \cdot 7\,\text{yd}$
$\quad = 49\,\text{yd}^2$ The area of rectangle #1 is 49 yd^2.
To find the area of rectangle #2, we must find the width of the rectangle. We indicate this width with the variable x in the figure. Since the width of rectangle #1 is 7 yd, we know that $x + 4 = 7$. Solving this equation, we have $x = 3$, and thus the width of rectangle #2 is 3 yd.

Area of rectangle #2:
$A = L \cdot W$
$\quad = 5\,\text{yd} \cdot 3\,\text{yd}$
$\quad = 15\,\text{yd}^2$ The area of rectangle #2 is 15 yd^2.
Next we find the area of rectangle #3.

Area of rectangle #3:
$A = L \cdot W$
$\quad = 5\,\text{yd} \cdot 11\,\text{yd}$
$\quad = 55\,\text{yd}^2$ The area of rectangle #3 is 55 yd^2.
We add the three areas to find the area of the figure.
$49\,\text{yd}^2 + 15\,\text{yd}^2 + 55\,\text{yd}^2 = 119\,\text{yd}^2$
The area of the figure is 119 yd^2.

7. We complete the four-step process. The picture is drawn for us, so we first write the formula. Then we replace the appropriate variables with the values given. Finally, we solve the equation to find the unknown side.

$V = L \cdot W \cdot H$ Write the formula.

$192 = 6 \cdot W \cdot 8$ Replace the variables with the values given.

$192 = 48W$ Simplify.

$\dfrac{192}{48} = \dfrac{48W}{48}$ Divide both sides by 48.

$4 = W$ The width of the box is 4 m.

8. *Understand the problem.* We organize the information in a Mathematics Blueprint for Problem Solving.

Mathematics Blueprint for Problem Solving

Gather the Facts	What Am I Asked to Do?	How Do I Proceed?	Key Points to Remember
The dimensions of the living room are: $L = 12\,\text{ft}$ $W = 15\,\text{ft}$ The carpet costs $11 per square yard.	**(a)** Find the area to determine the number of square yards of carpet that must be purchased. **(b)** Find the cost of the carpet for the living room.	**1.** Draw a 12-foot by 15-foot rectangle. **2.** Convert feet to yards. **3.** Find the area of the living room. **4.** Multiply the area times $11 to find the cost of the carpet.	Change feet to yards since we are asked to find the amount of carpet in square yards. There are 3 feet in 1 yard. Area = Length × Width

Solve and state the answer.

(a) To find number of *square yards,* we proceed as follows.

1. We draw a rectangle and label the sides.

2. We convert feet to yards, since the carpet is sold in square yards.

12 ft = 3 ft + 3 ft + 3 ft + 3 ft   15 ft = 3 ft + 3 ft + 3 ft + 3 ft + 3 ft
  = 4 × 3 ft   = 5 × 3 ft
  = 4 × 1 yd   = 5 × 1 yd
  = 4 yd   = 5 yd

3. We relabel the figure and find the area in square yards.

12 ft = 4 yd

15 ft = 5 yd

$A = L \times W$
  $= 4 \text{ yd} \times 5 \text{ yd}$
  $= 4 \times 5 \times \text{yd} \times \text{yd}$
  $= 20 \text{ yd}^2$

20 yd^2 of carpet must be purchased.

(b) The carpet sells for $11 per square yard and 20 square yards must be purchased: $11 \times 20 = \$220$. The carpet will cost $220.

Check: Use your calculator to verify that your calculations are correct.

3.4 Student Practice

1. 4^2 · 4^4

4 · 4 · 4 · 4 · 4 · 4 = 4^6

$4^2 \cdot 4^4 = 4^6$ Since 4 appears as a factor 6 times, the exponent is 6.

2. (a) $y^5 \cdot y = y^5 \cdot y^1 = y^{5+1} = y^6$

Note: $y = y^1$.

(b) $a^4 \cdot a^5 = a^{4+5} = a^9$

(c) $5^3 \cdot 3^4 = 5^3 \cdot 3^4$

The rule for multiplying numbers with the *same base* does not apply since the bases are *different.*

(d) $6^5 \cdot 6^6 = 6^{5+6} = 6^{11}$

3. $(4y)(5y^2)(y^5)$
  $= (4y^1)(5y^2)(1y^5)$
  $= (4 \cdot 5 \cdot 1) \cdot (y^1 \cdot y^2 \cdot y^5)$ Change the order of the factors and regroup them.
  $= 20 \cdot y^8$ $y^1 \cdot y^2 \cdot y^5 = y^{1+2+5} = y^8$.
  $= 20y^8$

4. $(4y^3)(3x^2)(5y^2) = (4 \cdot 3 \cdot 5)(y^3 \cdot x^2 \cdot y^2) = 60x^2y^5$
CAUTION: We cannot use the product rule for exponents to simplify x^2y^5 because the bases x and y are not the same.

5. (a) $(-6a)(-8a)$
  $= (-6)(-8)(a \cdot a)$
  $= 48a^2$ $a \cdot a = a^1 \cdot a^1 = a^2$

(b) $(5x^3)(-2y^5) = (5)(-2)(x^3 \cdot y^5)$ x^3 and y^5 do not have the same
  $= -10x^3y^5$ base. Thus, we cannot add exponents.

6. (a) $4x^2$ is a monomial because there is one term.

(b) $8x^2 - 9x + 1$ is a trinomial because there are three terms.

(c) $5x^3 + 8x$ is a binomial because there are two terms.

7. $4x^4(x^3 - 6x) + 3x^5$

$4x^4 \cdot x^3$

$4x^4(x^3 - 6x) + 3x^5 = 4x^4 \cdot x^3 - 4x^4 \cdot 6x + 3x^5$ Use the distributive property to multiply $4x^4(x^3 - 6x)$.

$4x^4 \cdot 6x$

$= 4x^4 \cdot 1x^3 - 4x^4 \cdot 6x^1 + 3x^5$ Write x^3 as $1x^3$ and x as x^1.

$= (4 \cdot 1) \cdot (x^4 \cdot x^3) - (4 \cdot 6) \cdot (x^4 \cdot x^1) + 3x^5$ Multiply numerical coefficients.

$= 4x^{4+3} - 24x^{4+1} + 3x^5$ Multiply variables by adding their exponents.

$4x^4(x^3 - 6x) + 3x^5 = 4x^7 - 21x^5$

8. First we write the formula for the area of a parallelogram, and then we replace b and h with the given expressions.

$A = bh$
$\quad = (7x^5 - 2x)(x^3)$ $b = 7x^5 - 2x$ and $h = x^3$
$\quad = (7x^5)(x^3) - (2x)(x^3)$ We distribute x^3.
$A = 7x^8 - 2x^4$ We simplify.

Chapter 4   4.1 Student Practice

1. (a) 975   3 and 5
Divisible by 3 because $9 + 7 + 5 = 21$, which is divisible by 3, and by 5 because 975 ends in 5.

(b) 122   2
Divisible by 2 because 122 is even.

(c) 420   2, 3, 5
Divisible by 2 because 420 is even, by 3 because $4 + 2 + 0 = 6$, which is divisible by 3, and by 5 because the last digit is 0.

(d) 11,121   3
Divisible by 3 because $1 + 1 + 1 + 2 + 1 = 6$, which is divisible by 3.

2. 0 is neither prime nor composite.
9, 16, 32, and 50 are composite.
3, 13, 19, 23, 37, and 41 are prime.

3. (a) $14 = 2 \cdot 7$
(b) $27 = 3 \cdot 3 \cdot 3$ or 3^3
Note: $3 \cdot 9$ is not correct because 9 is not a prime number.

4. Since 50 is even, it is divisible by the prime number 2. We start the division ladder by dividing 50 by 2.

$\quad\quad 25$
Step 1 $2\overline{)50}$
We must continue to divide until the quotient is a prime number.

$\quad\quad 5$
Step 2 $5\overline{)25}$
The quotient is a prime number. Thus all the factors are prime. We are finished dividing. Now we write all the divisors and the quotient as a product of prime factors.
$50 = 2 \cdot 5 \cdot 5$ or $2 \cdot 5^2$

5. We must divide by prime numbers to ensure that all factors are prime. Since 96 is even, we know that we can divide 96 by 2.

$\quad\quad\quad 3$ 3 is prime, so we are finished dividing.
Step 5 $2\overline{)6}$
Step 4 $2\overline{)12}$
Step 3 $2\overline{)24}$
Step 2 $2\overline{)48}$
Step 1 $2\overline{)96}$ We start by dividing by 2.
$96 = 2 \cdot 2 \cdot 2 \cdot 2 \cdot 2 \cdot 3$ or $2^5 \cdot 3$
We can check our answer by multiplying the prime factors.
Check:
$96 \stackrel{?}{=} 2^5 \cdot 3$
$96 \stackrel{?}{=} 32 \cdot 3$
$96 = 96$ ✓   The answer checks.

6. From the divisibility rules, we know 315 is divisible by 3 and 5. We can start with 5.

$\quad\quad\quad 7$ 7 is prime, so we are finished dividing.
Step 3 $3\overline{)21}$
Step 2 $3\overline{)63}$
Step 1 $5\overline{)315}$
$315 = 5 \cdot 3 \cdot 3 \cdot 7 = 3 \cdot 3 \cdot 5 \cdot 7$ or $3^2 \cdot 5 \cdot 7$

7. Use a factor tree to express 36 as a product of prime factors.

36
6 · 6
(2) · (3)   (2) · (3)
$36 = 2 \cdot 2 \cdot 3 \cdot 3$ or $2^2 \, 3^2$

1. Write 36 as the product of any two factors other than 1 and itself.
2. We do not circle either 6 since 6 is not prime. We must write each 6 as a product: $2 \cdot 3$.
3. Circle all factors since they are prime.
4. Write 36 as the product of prime numbers, i.e., the numbers that are circled: $2 \cdot 3 \cdot 2 \cdot 3$.

4.2 Student Practice

1. (a) Seven out of sixteen parts are shaded, or $\dfrac{7}{16}$.

(b) Three out of eight parts are shaded, or $\dfrac{3}{8}$.

(c) Two out of two parts are shaded, or $\dfrac{2}{2} = 1$.

2. (a) $y \div y = \dfrac{y}{y} = 1$ **(b)** $0 \div 18 = \dfrac{0}{18} = 0$

(c) $\dfrac{0}{65} = 0$ **(d)** $\dfrac{65}{0}$ Division by 0 is undefined.

3. First we must find the total rainfall for 1 year.
13 in. + 15 in. + 4 in. + 5 in. = 37 in.

(a) From January to March, there were 13 inches of rain out of a total of 37 inches.

$\dfrac{13}{37}$ Fractional part of rainfall that occurs from January to March.

(b) From April to June, there were 5 inches of rain out of a total of 37 inches.

37 in. − 5 in. = 32 in.   Rainfall that does not occur from April to June.

$\dfrac{32}{37}$ Fractional part of rainfall that does not occur.

4. (a) $\dfrac{6}{5}$ Improper fraction

(b) $\dfrac{x}{x}$ Improper fraction

(c) $6\dfrac{2}{9}$ Mixed number

(d) $\dfrac{1}{2}$ Proper fraction

5. The answer is in the form: quotient $\dfrac{\text{remainder}}{\text{denominator}}$.

$\dfrac{23}{6} \longrightarrow$ $6\overline{\smash{)}23}$ $3\dfrac{5}{6}$

Quotient → 3 ↓ ↓ 5 Remainder
$\dfrac{18}{5}$ Denominator from original fraction.

6. Improper fraction:
$$\dfrac{(\text{denominator} \cdot \text{whole number}) + \text{numerator}}{\text{denominator}}.$$

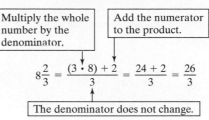

| Multiply the whole number by the denominator. | Add the numerator to the product. |

$$8\dfrac{2}{3} = \dfrac{(3 \cdot 8) + 2}{3} = \dfrac{24 + 2}{3} = \dfrac{26}{3}$$

The denominator does not change.

4.3 Student Practice

1. (a) $\dfrac{4}{7} = \dfrac{4 \cdot 3}{7 \cdot 3} = \dfrac{12}{21}$ **(b)** $\dfrac{4}{7} = \dfrac{4 \cdot 5x}{7 \cdot 5x} = \dfrac{20x}{35x}$

2. $\dfrac{2}{9} = \dfrac{\square}{36x}$

$\dfrac{2 \cdot ?}{9 \cdot ?} = \dfrac{\square}{36x}$ 9 times what number equals $36x$? $4x$

Since we must multiply the denominator by $4x$ to obtain $36x$, we must also multiply the numerator by $4x$.

$\dfrac{2 \cdot 4x}{9 \cdot 4x} = \dfrac{8x}{36x}$

3. $\dfrac{18}{54} = \dfrac{2 \cdot 3 \cdot 3}{2 \cdot 3 \cdot 3 \cdot 3}$ First, write the numerator and denominator as products of prime numbers.

$= \dfrac{\overset{1}{2} \cdot \overset{1}{3} \cdot \overset{1}{3}}{\underset{1}{2} \cdot \underset{1}{3} \cdot \underset{1}{3} \cdot 3}$ Then rewrite $\dfrac{2}{2}$ and $\dfrac{3}{3}$ as the equivalent fraction $\dfrac{1}{1}$.

$= \dfrac{1 \cdot 1 \cdot 1}{1 \cdot 1 \cdot 1 \cdot 3} = \dfrac{1}{3}$

4. If we recognize that the numerator and denominator have common factors that are not prime, we can use these factors to reduce the fraction.

$\dfrac{60}{-36} = -\dfrac{60}{36}$ Write the negative sign in front of the fraction.

$= -\dfrac{\overset{1}{12} \cdot 5}{\underset{1}{12} \cdot 3}$ Simplify.

$= -\dfrac{5}{3}$

5. Since 10 is a common factor of the numerator and denominator, we can write each as a product of 10 times some prime numbers.

$\dfrac{80x^2}{140x} = \dfrac{10 \cdot 2 \cdot 2 \cdot x \cdot x}{10 \cdot 2 \cdot 7 \cdot x}$ Write all other factors as products of prime numbers.

$= \dfrac{\overset{1}{10} \cdot \overset{1}{2} \cdot 2 \cdot \overset{1}{x} \cdot x}{\underset{1}{10} \cdot \underset{1}{2} \cdot 7 \cdot \underset{1}{x}}$

$= \dfrac{2 \cdot 2 \cdot x}{7}$

$= \dfrac{4x}{7}$

6. (a) *Understand the problem.* We use the information in the chart to find the total sales. Then we write the fraction.
Solve and state the answer.

| Condominium sales | + | Town home sales | + | Single-family home sales | = | Total sales |
| 14 | + | 21 | + | 45 | = | 80 |

$\dfrac{\text{town home sales}}{\text{total sales}} = \dfrac{21}{80}$ of the sales were town homes.

(b) *Understand the problem.* We refer to the chart and determine how many sales were not single-family homes. Then we write the fraction and simplify.
Solve and state the answer.
We want the total sales that were not single-family homes.
We find the sum: $14 + 21 = 35$.

$\dfrac{\text{sales that were not single-family homes}}{\text{total sales}} = \dfrac{35}{80}$

$= \dfrac{7 \cdot \overset{1}{5}}{16 \cdot \underset{1}{5}}$

$= \dfrac{7}{16}$ of the sales were not single-family homes.

Check:
(a) The fraction $\dfrac{21}{80}$ is a little more than $\dfrac{20}{80}$, which equals $\dfrac{1}{4}$. This means that approximately one-fourth of the total yearly sales were town homes. Our answer is reasonable.
(b) The fraction $\dfrac{7}{16}$ is a little less than $\dfrac{8}{16}$, which equals $\dfrac{1}{2}$. This means that approximately one-half of the total yearly sales were not single-family homes. Since approximately one-half of the total yearly sales were single-family homes, our answer is reasonable.

4.4 Student Practice

1. (a) $\dfrac{4^{11}}{4^7} = 4^{11-7} = 4^4$ There are more factors in the numerator.
The leftover factors are in the numerator.

(b) $\dfrac{6^9}{8^{14}}$ The bases are not the same.
We cannot divide using the rule for exponents.

(c) $\dfrac{y^5}{y^9} = \dfrac{1}{y^{9-5}} = \dfrac{1}{y^4}$ There are more factors in the denominator.
The leftover factors are in the denominator.

2. $\dfrac{25y^5x^0}{45y^8} = \dfrac{\overset{1}{\cancel{5}} \cdot 5 \cdot y^5 \cdot x^0}{\underset{1}{\cancel{5}} \cdot 3 \cdot 3 \cdot y^8}$ $x^0 = 1$

$= \dfrac{5}{9y^3}$ The leftover factors of y are in the denominator.

3. $(4^2)^3 = 4^2 \cdot 4^2 \cdot 4^2 = 4^{2+2+2} = 4^6$

4. (a) $(3^3)^4 = 3^{(3)(4)} = 3^{12}$ We multiply exponents. The base does not change when raising a power to a power.

(b) $(n^0)^7 = n^{(0)(7)} = n^0 = 1$

(c) $(3y^4)^6 = (3^1y^4)^6$ We write 3 as 3^1.
$= 3^{(1)(6)} \cdot y^{(4)(6)}$ We raise each factor to the power 6.
$= 3^6y^{24}$ We multiply exponents.

5. We must remember to raise both the numerator and the denominator to the power.

$\left(\dfrac{x}{3}\right)^3 = \left(\dfrac{x^1}{3^1}\right)^3 = \dfrac{x^{(1)(3)}}{3^{(1)(3)}} = \dfrac{x^3}{3^3} = \dfrac{x^3}{27}$

4.5 Student Practice

1. (a) 28 feet to 49 feet $= \dfrac{28 \text{ feet}}{49 \text{ feet}} = \dfrac{7 \cdot 4}{7 \cdot 7} = \dfrac{4}{7}$

Note: We treat units the same way we treat numbers and variables, i.e., $\dfrac{7}{7} = 1$ and $\dfrac{\text{feet}}{\text{feet}} = 1$.

(b) $27 : 81 = \dfrac{27}{81} = \dfrac{3 \cdot 3 \cdot 3}{3 \cdot 3 \cdot 3 \cdot 3} = \dfrac{1}{3}$

2. (a) $\dfrac{\text{men}}{\text{women}} \Rightarrow \dfrac{21}{15} = \dfrac{3 \cdot 7}{3 \cdot 5} = \dfrac{7}{5}$ **(b)** $\dfrac{\text{women}}{\text{men}} \Rightarrow \dfrac{15}{21} = \dfrac{3 \cdot 5}{3 \cdot 7} = \dfrac{5}{7}$

3. $\dfrac{\text{Midway}}{\text{Barrow}} \Rightarrow \dfrac{46 \text{ inches}}{29 \text{ inches}} = \dfrac{46}{29}$

4. $\dfrac{90 \text{ miles}}{5 \text{ gallons}}$ We divide: $5\overline{)90}\,^{18}$.

$\dfrac{90 \text{ miles}}{5 \text{ gallons}} = \dfrac{18 \text{ miles}}{1 \text{ gallon}}$ or 18 miles per gallon

5. We divide $400 \div 26$ to find the unit rate.

$\dfrac{400 \text{ pounds}}{\$26} = 15\dfrac{10}{26} = 15\dfrac{5}{13}$ pounds per dollar

6. $\dfrac{6031 \text{ tornados in June}}{70 \text{ years}}$ We divide to find the unit rate,

$70\overline{)6031}\,^{86\frac{11}{70}}$ or approximately 86 tornados per year in June.

Note that we rounded $86\dfrac{11}{70}$ to 86 because $\dfrac{11}{70}$ is closer to 0 than to 1.

7. (a) Patients per RN:
$\dfrac{\text{patients}}{\text{RN}} \Rightarrow \dfrac{40 \text{ patients}}{2 \text{ RNs}} = \dfrac{20 \text{ patients}}{1 \text{ RN}}$ or 20 patients per RN

(b) Patients per aide:
$\dfrac{\text{patients}}{\text{aide}} \Rightarrow \dfrac{30 \text{ patients}}{2 \text{ aides}} = \dfrac{15 \text{ patients}}{1 \text{ aide}}$ or 15 patients per aide

(c) Since every 15 patients require 1 aide, we divide $60 \div 15$ to find how many aides are needed for 60 patients.
$60 \div 15 = 4$ aides for 60 patients

8. (a) $\dfrac{\$78}{6} = \$13$ per towel   **(b)** 9 towels for $108 is the better buy.
$\dfrac{\$108}{9} = \$12$ per towel

4.6 Student Practice

1. 3 nails is to 6 feet as 9 nails is to 18 feet.

$\dfrac{3 \text{ nails}}{6 \text{ feet}} = \dfrac{9 \text{ nails}}{18 \text{ feet}}$

2. We can restate as follows: 4 hours is to 144 miles as 6 hours is to 216 miles. $\dfrac{4 \text{ hours}}{144 \text{ miles}} = \dfrac{6 \text{ hours}}{216 \text{ miles}}$ We write hours in the numerator. We write miles in the denominator.

3. We form the cross products to determine if the fractions are equal.

(a) $\dfrac{4}{22} \overset{?}{=} \dfrac{12}{87}$
Products are *not* equal.
$\boxed{87 \cdot 4 = 348}$ $\boxed{22 \cdot 12 = 264}$
Since $348 \neq 264$, we know that $\dfrac{4}{22} \neq \dfrac{12}{87}$.

(b) $\dfrac{84}{108} \overset{?}{=} \dfrac{7}{9}$
Products are equal.
$\boxed{9 \cdot 84 = 756}$ $\boxed{108 \cdot 7 = 756}$
Since $756 = 756$, we know that $\dfrac{84}{108} = \dfrac{7}{9}$.

4. (a) We check to see that $\dfrac{14}{45} \overset{?}{=} \dfrac{42}{135}$ by forming the two cross products.
$135 \cdot 14 = 1890;\ 45 \cdot 42 = 1890$
The two cross products are equal.
Thus $\dfrac{14 \text{ opals}}{45 \text{ diamonds}} = \dfrac{42 \text{ opals}}{135 \text{ diamonds}}$. This is a proportion.

(b) We form the two cross products.
$144 \cdot 32 = 4608;\ 72 \cdot 128 = 9216$
The two cross products are *not* equal.
Thus $\dfrac{32}{72} \neq \dfrac{128}{144}$. This is not a proportion.

5. $\dfrac{n}{18} = \dfrac{28}{72}$
$72 \cdot n = 18 \cdot 28$ Find the cross products and form an equation.
$72n = 504$ Simplify.
$\dfrac{72n}{72} = \dfrac{504}{72}$ Divide by 72 on both sides of the equation.
$n = 7$ $504 \div 72 = 7$.
Check whether the proportion is true.
$\dfrac{7}{18} \overset{?}{=} \dfrac{28}{72}$
$72 \cdot 7 \overset{?}{=} 18 \cdot 28$ We check cross products.
$504 = 504$ ✓

6. We must determine if 7 out of 8 parcels is equivalent to 10 out of 11 parcels. We have 7 of the 8 parcels $\to \dfrac{7}{8}$; 10 of the 11 parcels $\to \dfrac{10}{11}$.
We use the equality test for fractions to determine if we have a proportion. $\dfrac{7}{8} \overset{?}{=} \dfrac{10}{11}$
We form the two cross products.
$11 \cdot 7 = 77;\ 8 \cdot 10 = 80$
The cross products are *not* equal. Thus $\dfrac{7}{8} \neq \dfrac{10}{11}$ and this is not a proportion. The two parcels do not yield the same amount of land.

7. Let n represent the number of desserts for 180 people.
18 desserts is to 15 people as n desserts is to 180 people.
$$\frac{18 \text{ desserts}}{15 \text{ people}} = \frac{n \text{ desserts}}{180 \text{ people}}$$
We must solve for n.
$$\frac{18}{15} = \frac{n}{180}$$
$$\frac{6}{5} = \frac{n}{180} \qquad \frac{18}{15} = \frac{6}{5}.$$
$180 \cdot 6 = 5n$ We form the cross product.
$1080 = 5n$
$$\frac{1080}{5} = \frac{5n}{5}$$ We divide both sides by 5.
$216 = n$ We simplify.
216 desserts should be served.

9. The ratio of 3 to 5 represents Cleo's profits to Julie's profits.
$$\frac{3}{5} = \frac{\text{Cleo's profits of } \$2400}{\text{Julie's profits of } \$x}$$
$$\frac{3}{5} = \frac{2400}{x}$$
$3x = 12{,}000$
$$\frac{3x}{3} = \frac{12{,}000}{3}$$
$x = 4000$
Julie's profits would be $4000.

8. First we set up the proportion, letting the letter n represent the unknown length.
$$\frac{\text{width of garden}}{\text{length of garden}} = \frac{\text{width of yard}}{\text{length of yard}}$$
$$\frac{6 \text{ ft}}{7 \text{ ft}} = \frac{20 \text{ ft}}{n \text{ ft}}$$
$6n = 7 \cdot 20$ Form the cross product.
$6n = 140$
$$\frac{6n}{6} = \frac{140}{6}$$
$$n = 23\frac{1}{3}$$
The length of the yard is $23\frac{1}{3}$ feet.

Chapter 5 5.1 Student Practice

1. $\dfrac{5}{8}$ of $\dfrac{2}{3} = \dfrac{5}{8} \cdot \dfrac{2}{3} = \dfrac{5 \cdot 2}{8 \cdot 3} = \dfrac{5 \cdot \overset{1}{\cancel{2}}}{\underset{1}{\cancel{2}} \cdot 2 \cdot 3} = \dfrac{5}{12}$

2. $\dfrac{8}{15} \cdot \dfrac{12}{14} = \dfrac{8 \cdot 12}{15 \cdot 14} = \dfrac{2 \cdot 2 \cdot 2 \cdot 2 \cdot 2 \cdot 3}{3 \cdot 5 \cdot 2 \cdot 7} = \dfrac{\overset{1}{\cancel{2}} \cdot 2 \cdot 2 \cdot 2 \cdot 2 \cdot \overset{1}{\cancel{3}}}{\underset{1}{\cancel{3}} \cdot 5 \cdot \underset{1}{\cancel{2}} \cdot 7} = \dfrac{16}{35}$

4. $\dfrac{3x^6}{7} \cdot (14x^2) = \dfrac{3x^6}{7} \cdot \dfrac{14x^2}{1}$
$= \dfrac{3 \cdot x^6 \cdot 2 \cdot 7 \cdot x^2}{7 \cdot 1}$
$= \dfrac{3 \cdot 2 \cdot \overset{1}{\cancel{7}} \cdot x^6 \cdot x^2}{\underset{1}{\cancel{7}} \cdot 1}$
$= \dfrac{3 \cdot 2 \cdot x^{6+2}}{1}$
$= 6x^8$

3. $\dfrac{-12}{24} \cdot \dfrac{-2}{13} = (+)$ The product is positive since there are 2 negative signs and 2 is an even number.
$= \dfrac{12 \cdot 2}{24 \cdot 13}$
$= \dfrac{12 \cdot 2}{12 \cdot 2 \cdot 13}$ Factor.
$= \dfrac{\overset{1}{\cancel{12}} \cdot \overset{1}{\cancel{2}}}{\underset{1}{\cancel{12}} \cdot \underset{1}{\cancel{2}} \cdot 13}$ Factor out common factors and simplify.
$= \dfrac{1}{13}$

5. We evaluate the formula for the given values.
$A = \dfrac{1}{2}bh$
$= \dfrac{1}{2} \cdot 20 \text{ cm} \cdot 9 \text{ cm}$ We replace b with 20 cm and h with 9 cm.
$= \dfrac{1 \cdot 20 \text{ cm} \cdot 9 \text{ cm}}{2 \cdot 1 \cdot 1}$ We multiply: $\dfrac{1}{2} \cdot \dfrac{20 \text{ cm}}{1} \cdot \dfrac{9 \text{ cm}}{1}$.
$= \dfrac{1 \cdot 20 \cdot 9 \text{ cm} \cdot \text{cm}}{2}$ We factor $20 = 2 \cdot 10$. Then we simplify.
$= 90 \text{ cm}^2$ $\dfrac{\overset{1}{\cancel{2}} \cdot 10 \cdot 9}{\underset{1}{\cancel{2}}} = 10 \cdot 9 = 90; \quad \text{cm} \cdot \text{cm} = \text{cm}^2$.

6. To find the reciprocal, we invert the fraction.
(a) $\dfrac{a}{-y} \rightarrow \dfrac{-y}{a} = -\dfrac{y}{a}$ **(b)** $4 = \dfrac{4}{1} \rightarrow \dfrac{1}{4}$

8. $\dfrac{9x^6}{21} \div \left(\dfrac{-42x^4}{18}\right) = \dfrac{9x^6}{21} \cdot \left(\dfrac{18}{-42x^4}\right)$ Invert the second fraction and multiply.
$= (-)$ The product is negative since there is 1 negative sign and 1 is an odd number.
$= -\dfrac{\overset{}{\cancel{3}} \cdot \overset{}{\cancel{3}} \cdot 2 \cdot 3 \cdot 3 \cdot x^6}{\cancel{3} \cdot 7 \cdot \cancel{2} \cdot \cancel{3} \cdot 7 \cdot x^4}$ Factor and simplify.
$= -\dfrac{9x^2}{49}$ $\dfrac{x^6}{x^4} = x^{6-4} = x^2$

7. $\dfrac{-7}{8} \div \dfrac{5}{13} = \dfrac{-7}{8} \cdot \dfrac{13}{5}$ Invert the second fraction and multiply.
$= -\dfrac{7 \cdot 13}{8 \cdot 5} = -\dfrac{91}{40}$ The product of a negative number and a positive number is negative.

9. $28x^5 \div \dfrac{4x}{19} = \dfrac{28x^5}{1} \cdot \dfrac{19}{4x} = \dfrac{\overset{}{\cancel{4}} \cdot 7 \cdot 19 \cdot x^5}{1 \cdot \cancel{4} \cdot x}$
$= 133x^4$ Simplify: $\dfrac{7 \cdot 19}{1} = 133$
and $\dfrac{x^5}{x^1} = x^{5-1} = x^4$.

10. The key phrase is "$\frac{2}{13}$ *of* her income." The word *of* often indicates multiplication.

$\frac{2}{13}$ of income is placed into a savings account.

$\downarrow \quad \downarrow \quad \downarrow$

$\frac{2}{13} \cdot \$1703 = \frac{2}{13} \cdot \frac{\$1703}{1} = \$262$

$262 is placed into a savings account each week.

11. Since we are splitting $\frac{3}{4}$ pound of sugar into 2 equal parts, we divide.

$$\frac{3}{4} \div 2 = \frac{3}{4} \div \frac{2}{1} = \frac{3}{4} \cdot \frac{1}{2} = \frac{3 \cdot 1}{4 \cdot 2} = \frac{3}{8}$$

She should place $\frac{3}{8}$ pound into each container.

5.2 Student Practice

1. (a)

$$12x \cdot 1 \quad 12x \cdot 2 \quad 12x \cdot 3 \quad 12x \cdot 4 \quad 12x \cdot 5$$
$$\downarrow \qquad \downarrow \qquad \downarrow \qquad \downarrow \qquad \downarrow$$

Multiples of $12x$: $\quad 12x, \quad 24x, \quad 36x, \quad 48x, \quad 60x$

$$20x \cdot 1 \quad 20x \cdot 2 \quad 20x \cdot 3 \quad 20x \cdot 4 \quad 20x \cdot 5$$
$$\downarrow \qquad \downarrow \qquad \downarrow \qquad \downarrow \qquad \downarrow$$

Multiples of $20x$: $\quad 20x, \quad 40x, \quad 60x, \quad 80x, \quad 100x$

(b) The only multiple common to both lists is $60x$.

2. First, we list some multiples of 4: 4, 8, 12, 16, 20, 24, 28, 32, 36, 40. Next, we list some multiples of 5: 5, 10, 15, 20, 25, 30, 35, 40, 45, 50. We see that both 20 and 40 are common multiples. Since 20 is the smaller of these common multiples, we call 20 the least common multiple (LCM).

3.

Factor each number $\rightarrow$	*List requirements for factorization of LCM.* $\rightarrow$	*Build the LCM.*
$28 = 2 \cdot 2 \cdot 7 \quad \rightarrow$	must have a pair of 2's and a 7 $\rightarrow$	$\boxed{\text{LCM} = 2 \cdot 2 \cdot 7 \cdot ?}$
$36 = 2 \cdot 2 \cdot 3 \cdot 3 \quad \rightarrow$	must have a pair of 2's and a pair of 3's $\rightarrow$	$\boxed{\text{LCM} = 2 \cdot 2 \cdot 7 \cdot 3 \cdot 3 \cdot ?}$
		A pair of 2's already exists so we just multiply by a pair of 3's.
$70 = 2 \cdot 5 \cdot 7 \quad \rightarrow$	must have a 2, a 5, and a 7 $\rightarrow$	$\boxed{\text{LCM} = 2 \cdot 2 \cdot 7 \cdot 3 \cdot 3 \cdot 5}$
		2 and 7 are already factors so we just multiply by a 5.

The LCM of 28, 36, and 70 is $2 \cdot 2 \cdot 7 \cdot 3 \cdot 3 \cdot 5 = 1260$.

4.

Factor each expression $\rightarrow$	*List requirements for factorization of LCM.* $\rightarrow$	*Build the LCM.*
$4x = 2 \cdot 2 \cdot x \quad \rightarrow$	must have a pair of 2's and an x $\rightarrow$	$\boxed{\text{LCM} = 2 \cdot 2 \cdot x \cdot ?}$
$x^2 = x \cdot x \quad \rightarrow$	must have a pair of x's $\rightarrow$	$\boxed{\text{LCM} = 2 \cdot 2 \cdot x \cdot x \cdot ?}$
		One x is already a factor, so we just multiply by another x.
$10x = 2 \cdot 5 \cdot x \quad \rightarrow$	must have a 2, a 5, and an x $\rightarrow$	$\boxed{\text{LCM} = 2 \cdot 2 \cdot x \cdot x \cdot 5}$
		A 2 and an x are already factors, so we just multiply by a 5.

The LCM of $4x$, x^2, and $10x$ is $2 \cdot 2 \cdot x \cdot x \cdot 5 = 20x^2$.

5. *Understand the problem.* We organize the facts in a chart and then fill in the Mathematics Blueprint for Problem Solving.

Time	End of 1st Lap	End of 2nd Lap	. . .	
Shannon	15 min	30 min	. . .	$\leftarrow$ These are multiples of 15.
Marsha	18 min	36 min	. . .	$\leftarrow$ These are multiples of 18.

Mathematics Blueprint for Problem Solving

Gather the Facts	What Am I Asked to Do?	How Do I Proceed?	Key Points to Remember
See the chart.	Find how many minutes it takes for them to begin another lap together.	Find the LCM of 15 and 18, which is the next time they begin a new lap together.	The LCM is the least common multiple.

Solve and state the answer. We must factor 15 and 18 and find the LCM.

$\quad 15 = 3 \cdot 5 \qquad\qquad 18 = 2 \cdot 3 \cdot 3$

$\text{LCM} = 2 \cdot 3 \cdot 3 \cdot 5 = 90$

Shannon and Marsha will begin another lap together 90 minutes after they start swimming.

6. *Understand the problem.* We make a chart to help us develop a plan to solve the problem.

Tours	Number of minutes after 8 A.M. tours start
Interior tours start every 50 min (40 + 10 min break)	50, 100, 150, ...
Ground tours start every 35 min (25 + 10 min break)	35, 70, 105, ...

Mathematics Blueprint for Problem Solving

Gather the Facts	What Am I Asked to Do?	How Do I Proceed?	Key Points to Remember
See the chart.	Determine the next time when both tours start at the same time.	To find the *next* time that both tours start at the same time, we must find the LCM of 50 and 35.	60 min = 1 hr Once we have the LCM in minutes, we change the LCM to hours and minutes to find the common start time.

Solve and state the answer.

First we factor 50 and 35 and find the LCM.

$50 = 2 \cdot 5 \cdot 5$ $35 = 5 \cdot 7$

$LCM = 2 \cdot 5 \cdot 5 \cdot 7 = 350$

Both tours will start at the same time 350 minutes after 8 A.M.

Next we change minutes to hours and minutes.

8 A.M. + 5 hours and 50 minutes = 1:50 P.M.

350 minutes ÷ 60 minutes per hour = 5 hours 50 minutes after 8 A.M.

At 1:50 P.M. both tours will start at the same time.

5.3 Student Practice

1. $\dfrac{3}{13} + \dfrac{7}{13} = \dfrac{3+7}{13}$ Add numerators.

$= \dfrac{10}{13}$ The denominator stays the same.

2. $\dfrac{-7}{6} + \left(\dfrac{-21}{6}\right) = \dfrac{-7 + (-21)}{6}$ Add numerators. The denominator stays the same.

$= \dfrac{-28}{6}$

$= \dfrac{\cancel{2}(-14)}{\cancel{2}(3)}$ Factor and simpliy.

$= -\dfrac{14}{3}$

3. (a) $\dfrac{8}{x} - \dfrac{3}{x} = \dfrac{8-3}{x} = \dfrac{5}{x}$

(b) $\dfrac{y}{9} + \dfrac{5}{9} = \dfrac{y+5}{9}$; We cannot add y and 5. They are not like terms.

4. (a) $\dfrac{1}{4}, \dfrac{1}{5}$

The LCD of $\dfrac{1}{4}$ and $\dfrac{1}{5}$ is 20.

(b) We find the LCD of $\dfrac{1}{12}$ and $\dfrac{3}{28}$.

$12 = 2 \cdot 2 \cdot 3$

$28 = 2 \cdot 2 \cdot 7$

$LCD = 2 \cdot 2 \cdot 3 \cdot 7 = 84$

The LCD of $\dfrac{1}{12}$ and $\dfrac{3}{28}$ is 84.

5. (a) $\dfrac{3}{5} = \dfrac{?}{10}$ What number multiplied by the denominator, 5, yields 10? It is 2, since $5(2) = 10$.

$\dfrac{3 \cdot 2}{5 \cdot 2} = \dfrac{6}{10}$ We multiply the numerator and denominator of $\frac{3}{5}$ by 2.

$\dfrac{3}{5} = \dfrac{6}{10}$

(b) $\dfrac{1}{2} = \dfrac{?}{10}$ What number multiplied by the denominator, 2, yields 10? It is 5, since $2(5) = 10$.

$\dfrac{1 \cdot 5}{2 \cdot 5} = \dfrac{5}{10}$ We multiply the numerator and denominator of $\frac{1}{2}$ by 5.

$\dfrac{1}{2} = \dfrac{5}{10}$

6. (a) $\dfrac{-3}{8} + \dfrac{7}{9}$

Step 1: Find the LCD of $\dfrac{-3}{8}$ and $\dfrac{7}{9}$. LCD = 72

Step 2: Write equivalent fractions.

$\dfrac{-3 \cdot 9}{8 \cdot 9} = \dfrac{-27}{72}$ $\dfrac{7 \cdot 8}{9 \cdot 8} = \dfrac{56}{72}$

Step 3: Add fractions with common denominators.

$\dfrac{-3}{8} + \dfrac{7}{9} = \dfrac{-27}{72} + \dfrac{56}{72} = \dfrac{29}{72}$

(b) $\dfrac{13}{30} - \dfrac{2}{15}$

Step 1: Find the LCD of $\dfrac{13}{30}$ and $\dfrac{2}{15}$.   LCD = $2 \cdot 3 \cdot 5 = 30$

Step 2: Write equivalent fractions.

The fraction $\dfrac{13}{30}$ already has the LCD as its denominator.

$\dfrac{2 \cdot 2}{15 \cdot 2} = \dfrac{4}{30}$

Step 3: Subtract fractions with common denominators.

$\dfrac{13}{30} - \dfrac{2}{15} = \dfrac{13}{30} - \dfrac{4}{30} = \dfrac{9}{30}$

Step 4: Simplify. $\dfrac{9}{30} = \dfrac{\cancel{3} \cdot 3}{2 \cdot \cancel{3} \cdot 5} = \dfrac{3}{10}$

7. (a) $\dfrac{8}{x} + \dfrac{2}{5x}$

Step 1: Find the LCD of $\dfrac{8}{x}$ and $\dfrac{2}{5x}$. LCD = $5x$

Step 2: Write equivalent fractions.

$\dfrac{8}{x} = \dfrac{8 \cdot 5}{x \cdot 5} = \dfrac{40}{5x}$ $\dfrac{2}{5x} = \dfrac{2}{5x}$

Step 3: Add fractions with common denominators.

$\dfrac{8}{x} + \dfrac{2}{5x} = \dfrac{40}{5x} + \dfrac{2}{5x} = \dfrac{40+2}{5x} = \dfrac{42}{5x}$

(b) $\dfrac{7}{y} - \dfrac{4}{x}$

Step 1: Find the LCD of $\dfrac{7}{y}$ and $\dfrac{4}{x}$.   LCD = xy

Step 2: Write equivalent fractions.

$\dfrac{7 \cdot x}{y \cdot x} = \dfrac{7x}{xy}$ $\dfrac{4 \cdot y}{x \cdot y} = \dfrac{4y}{xy}$

Step 3: Subtract fractions with common denominators.

$\dfrac{7}{y} - \dfrac{4}{x} = \dfrac{7x}{xy} - \dfrac{4y}{xy} = \dfrac{7x - 4y}{xy}$

8. $\dfrac{8x}{15} + \dfrac{9x}{24}$

Step 1: Find the LCD of $\dfrac{8x}{15}$ and $\dfrac{9x}{24}$.   LCD = 120

Step 2: Write equivalent fractions.

$$\dfrac{8x \cdot 8}{15 \cdot 8} = \dfrac{64x}{120} \qquad \dfrac{9x \cdot 5}{24 \cdot 5} = \dfrac{45x}{120}$$

Step 3: Add fractions with common denominators.

$$\dfrac{8x}{15} + \dfrac{9x}{24} = \dfrac{64x}{120} + \dfrac{45x}{120} = \dfrac{64x + 45x}{120} = \dfrac{109x}{120}$$

9. *Understand the problem.* The phrase "how much more" indicates that we subtract. *Solve and state the answer.*

Completed Monday	Minus	Completed Tuesday
$\dfrac{1}{4}$	$-$	$\dfrac{1}{6}$

1. The LCD is 12.
2. Write equivalent fractions.

$$\dfrac{1 \cdot 3}{4 \cdot 3} = \boxed{\dfrac{3}{12}} \qquad \dfrac{1 \cdot 2}{6 \cdot 2} = \boxed{\dfrac{2}{12}}$$

3. Subtract.

$$\dfrac{1}{4} - \dfrac{1}{6} = \dfrac{3}{12} - \dfrac{2}{12} = \dfrac{1}{12}$$

Jesse completed $\dfrac{1}{12}$ more on Monday.

10. *Understand the problem.* Since we want to find how many miles Lester jogged, we must add $\dfrac{2}{3}$ and $\dfrac{4}{5}$.

To add $\dfrac{2}{3} + \dfrac{4}{5}$, we must first find the least common denominator, which is 15.

Write equivalent fractions.

$$\dfrac{2 \cdot 5}{3 \cdot 5} = \dfrac{10}{15} \qquad \dfrac{4 \cdot 3}{5 \cdot 3} = \dfrac{12}{15}$$

Add.

$$\dfrac{2}{3} + \dfrac{4}{5} = \dfrac{10}{15} + \dfrac{12}{15} = \dfrac{22}{15} \text{ or } 1\dfrac{7}{15}$$

Lester jogged $1\dfrac{7}{15}$ miles in the two-day period.

5.4 Student Practice

1. $5\dfrac{2}{9}$

$+2\dfrac{5}{9}$ Add the fractions: $\dfrac{2}{9} + \dfrac{5}{9} = \dfrac{7}{9}$.
       Add the whole numbers: $5 + 2 = 7$.
$7\dfrac{7}{9}$

2. The LCD of $\dfrac{1}{3}$ and $\dfrac{3}{5}$ is 15, so we build equivalent fractions with this denominator.

$$5\dfrac{1 \cdot 5}{3 \cdot 5} = 5\dfrac{5}{15}$$

$+6\dfrac{3 \cdot 3}{5 \cdot 3} = +6\dfrac{9}{15}$ Add the fractions: $\dfrac{5}{15} + \dfrac{9}{15} = \dfrac{14}{15}$.

$11\dfrac{14}{15}$ Add the whole numbers: $5 + 6 = 11$.

3. The LCD of $\dfrac{3}{4}$ and $\dfrac{4}{5}$ is 20.

$7\dfrac{3 \cdot 5}{4 \cdot 5} = 7\dfrac{15}{20}$

$+2\dfrac{4 \cdot 4}{5 \cdot 4} = +2\dfrac{16}{20}$ Add fractions: $\dfrac{15}{20} + \dfrac{16}{20} = \dfrac{31}{20}$.

$9\dfrac{31}{20}$ Add whole numbers: $7 + 2 = 9$.

$= 9 + 1\dfrac{11}{20}$ Change $\dfrac{31}{20}$ to $1\dfrac{11}{20}$.

$= 10\dfrac{11}{20}$ Add: $9 + 1 = 10$.

4. We cannot subtract $\dfrac{3}{12} - \dfrac{5}{12}$ without borrowing.

$5\dfrac{3}{12} = 4 + 1\dfrac{3}{12}$ We write 5 as the sum $4 + 1$ (borrowing from 5).

$= 4 + \dfrac{15}{12}$ We change $1\dfrac{3}{12}$ to $\dfrac{15}{12}$.

$5\dfrac{3}{12} = \quad 4\dfrac{15}{12}$

$-3\dfrac{5}{12} = \quad -3\dfrac{5}{12}$ Subtract fractions: $\dfrac{15}{12} - \dfrac{5}{12} = \dfrac{10}{12}$.

$1\dfrac{10}{12}$ Subtract whole numbers: $4 - 3 = 1$.

We simplify: $1\dfrac{10}{12} = 1\dfrac{5}{6}$. Thus, $5\dfrac{3}{12} - 3\dfrac{5}{12} = 1\dfrac{5}{6}$.

5. The LCD of $\dfrac{1}{7}$ and $\dfrac{3}{4}$ is 28.

$6\dfrac{1 \cdot 4}{7 \cdot 4} = 6\dfrac{4}{28}$

$-2\dfrac{3 \cdot 7}{4 \cdot 7} = -2\dfrac{21}{28}$

We cannot subtract $\dfrac{4}{28} - \dfrac{21}{28}$ without borrowing.

$6\dfrac{4}{28} = 5 + 1\dfrac{4}{28}$ We write 6 as the sum $5 + 1$ (borrowing from 6).

$= 5 + \dfrac{32}{28}$ We change $1\dfrac{4}{28}$ to $\dfrac{32}{28}$.

$6\dfrac{4}{28} = \quad 5\dfrac{32}{28}$

$-2\dfrac{21}{28} = \quad -2\dfrac{21}{28}$ Subtract fractions: $\dfrac{32}{28} - \dfrac{21}{28} = \dfrac{11}{28}$.

$3\dfrac{11}{28}$ Subtract whole numbers.

6.

$$9 = 8\frac{3}{3} \qquad 9 = 8 + 1 = 8 + \frac{3}{3} \text{ or } 8\frac{3}{3}$$

$$\begin{array}{r} -4\frac{1}{3} = -4\frac{1}{3} \\ \hline 4\frac{2}{3} \end{array}$$

When we borrowed 1 from 9 we changed the 1 to $\frac{3}{3}$ because we wanted a fraction that had the same denominator as $\frac{1}{3}$.

7. We change the mixed numbers to improper fractions and then multiply.

$$6\frac{3}{7} \cdot 1\frac{13}{15} = \frac{45}{7} \cdot \frac{28}{15} = \frac{3 \cdot \cancel{15} \cdot \cancel{7} \cdot 4}{\cancel{7} \cdot \cancel{15}} = \frac{12}{1} = 12$$

8. $1\frac{1}{4} \div (-2) = \frac{5}{4} \div \frac{(-2)}{1} = \frac{5}{4} \cdot \left(-\frac{1}{2}\right) = -\frac{5}{8}$

9. *Understand the problem.* Sara has to use $\frac{1}{2}$ tablespoon many times to measure $4\frac{1}{2}$ tablespoons of sugar.
Solve and state the answer. This is division since we want to know how many $\frac{1}{2}$'s are in $4\frac{1}{2}$.

$$4\frac{1}{2} \div \frac{1}{2} = \frac{9}{2} \div \frac{1}{2} \qquad \text{We change } 4\frac{1}{2} \text{ to an improper fraction.}$$

$$= \frac{9}{2} \cdot \frac{2}{1} \qquad \begin{array}{l}\text{We invert } \frac{1}{2} \text{ and change the division}\\ \text{to multiplication.}\end{array}$$

$$= 9 \qquad \text{We simplify.}$$

Sara must use the $\frac{1}{2}$ tablespoon 9 times to follow the recipe.

10. *Understand the problem.* Jerome needs $\frac{1}{10}$ of $32\frac{1}{2}$ feet of fencing material.
Solve and state the answer. The key word "of" ($\frac{1}{10}$ *of* this material) indicates that we multiply.

$$\frac{1}{10} \cdot 32\frac{1}{2} = \frac{1}{10} \cdot \frac{65}{2} = \frac{1 \cdot \cancel{5} \cdot 13}{2 \cdot \cancel{5} \cdot 2} = \frac{13}{4} \text{ or } 3\frac{1}{4}$$

Jerome needs $3\frac{1}{4}$ feet to repair his fence.

5.5 Student Practice

1. $\left(\frac{5}{3}\right)^2 + \frac{20}{9} \cdot \frac{1}{5} = \frac{25}{9} + \frac{20}{9} \cdot \frac{1}{5}$ First we simplify exponents:
$\left(\frac{5}{3}\right)^2 = \left(\frac{5}{3}\right)\left(\frac{5}{3}\right) = \frac{25}{9}$.

$$= \frac{25}{9} + \frac{4}{9} \qquad \begin{array}{l}\text{Then we multiply:}\\ \frac{20}{9} \cdot \frac{1}{5} = \frac{4 \cdot \cancel{5} \cdot 1}{9 \cdot \cancel{5}} = \frac{4}{9}.\end{array}$$

$$= \frac{29}{9} \text{ or } 3\frac{2}{9} \qquad \text{Now we add.}$$

2. $\dfrac{\left(\frac{4}{7}\right)}{[3^2 + (-5)]}$ We write grouping symbols in the numerator and denominator.

$$= \frac{\left(\frac{4}{7}\right)}{4} \qquad \begin{array}{l}\text{Within the brackets, we simplify exponents first,}\\ \text{and then add: } 3^2 = 9 \text{ and } 9 + (-5) = 4.\end{array}$$

$$= \frac{4}{7} \div 4 \qquad \text{The main fraction bar means divide.}$$

$$= \frac{4}{7} \cdot \frac{1}{4} \qquad \text{We invert the divisor and multiply.}$$

$$= \frac{\cancel{4} \cdot 1}{7 \cdot \cancel{4}} = \frac{1}{7} \qquad \text{We simplify.}$$

3. $\dfrac{\frac{x^2}{5}}{\frac{x}{10}} = \frac{x^2}{5} \div \frac{x}{10} = \frac{x^2}{5} \cdot \frac{10}{x} = \frac{x^2 \cdot \cancel{5} \cdot 2}{\cancel{5} \cdot x} = \frac{\cancel{x} \cdot x \cdot 2}{\cancel{x}} = 2x$

4. We write parentheses in the numerator and denominator and follow the order of operations.

$$\frac{\left(\frac{3}{5} + \frac{1}{2}\right)}{\left(\frac{5}{6} - \frac{1}{3}\right)} = \frac{\left(\frac{3 \cdot 2}{5 \cdot 2} + \frac{1 \cdot 5}{2 \cdot 5}\right)}{\left(\frac{5}{6} - \frac{1 \cdot 2}{3 \cdot 2}\right)}$$

$$= \frac{\left(\frac{6}{10} + \frac{5}{10}\right)}{\left(\frac{5}{6} - \frac{2}{6}\right)} = \frac{\frac{11}{10}}{\frac{3}{6}} \qquad \begin{array}{l}\text{Add top fractions.}\\[6pt]\text{Subtract bottom fractions.}\end{array}$$

Now we divide the top fraction by the bottom fraction.

$$\frac{11}{10} \div \frac{3}{6} = \frac{11}{10} \cdot \frac{6}{3} = \frac{11 \cdot \cancel{2} \cdot \cancel{3}}{5 \cdot \cancel{2} \cdot \cancel{3}} = \frac{11}{5}$$

$$\frac{\frac{3}{5} + \frac{1}{2}}{\frac{5}{6} - \frac{1}{3}} = \frac{11}{5}$$

5. We set up a proportion and solve for the missing number.

$$\frac{6\frac{4}{5} \text{ feet of wood}}{2 \text{ bookcases}} = \frac{x \text{ feet of wood}}{15 \text{ bookcases}}$$

$$15 \cdot 6\frac{4}{5} = 2x \quad \text{Find the cross products.}$$

$$15 \cdot \frac{34}{5} = 2x \quad \text{We change } 6\frac{4}{5} \text{ to an improper fraction.}$$

$$\frac{3 \cdot \cancel{5} \cdot 34}{\cancel{5}} = 2x \quad \text{We factor } 15 = 3 \cdot 5 \text{ and simplify.}$$

$$102 = 2x \quad \text{Now, we must solve the equation for } x.$$

$$\frac{102}{2} = \frac{2x}{2} \quad \text{We divide by 2 on both sides.}$$

$$51 = x$$

The carpenter needs 51 feet of oak wood.

5.6 Student Practice

1. *Understand the problem.* Draw a picture of the situation to help develop a plan to solve the problem.

Mathematics Blueprint for Problem Solving

Gather the Facts	What Am I Asked to Do?	How Do I Proceed?	Key Points to Remember
The room is 22 feet by 15 feet. The rug should be $2\frac{1}{2}$ feet from each wall. The binding is $2 per linear foot.	(a) Find the dimensions of the rug. (b) Calculate the cost of binding the rug.	1. Find the length and width of the rug. 2. Use the length and width to find the perimeter. 3. Multiply the perimeter by $2 to find the cost of the binding.	Use the formula $P = 2L + 2W$.

Solve and state the answer.

1. We find the length and width of the rug.

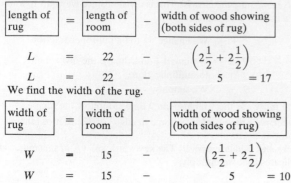

| length of rug | = | length of room | − | width of wood showing (both sides of rug) |

$$L = 22 - \left(2\tfrac{1}{2} + 2\tfrac{1}{2}\right)$$

$$L = 22 - 5 = 17$$

We find the width of the rug.

| width of rug | = | width of room | − | width of wood showing (both sides of rug) |

$$W = 15 - \left(2\tfrac{1}{2} + 2\tfrac{1}{2}\right)$$

$$W = 15 - 5 = 10$$

(a) The dimensions of the rug are 17 feet by 10 feet.

2. Now we find the perimeter.

$$P = 2L + 2W = 2(17) + 2(10) = 34 + 20 = 54$$

The perimeter is 54 feet.

3. We multiply 54 feet by \$2 to find the cost of the binding.

$$2 \cdot 54 = 108$$

(b) The binding will cost \$108.

Check.

Length: $2\tfrac{1}{2} + 17 + 2\tfrac{1}{2} = 22$ ft   Width: $2\tfrac{1}{2} + 10 + 2\tfrac{1}{2} = 15$ ft

Round the perimeter: $54 \to 50$. Then multiply to estimate the cost of the binding: $\$2 \times 50 = \$100$, which is close to our answer.   ✓

2. *Understand the problem.* You can draw a picture to help develop a plan.

Mathematics Blueprint for Problem Solving

Gather the Facts	What Am I Asked to Do?	How Do I Proceed?	Key Points to Remember
Nancy is making 2 bookcases, each having 4 shelves. Each shelf is $3\tfrac{1}{8}$ feet long. Wood is sold in 10-foot boards.	Find the number of 10-foot boards needed to make the shelves.	1. Determine the total number of shelves. 2. Determine the number of shelves that can be cut from one 10-foot board. 3. Use the information from steps 1 and 2 to find how many 10-foot boards are needed.	We must change mixed numbers to improper fractions before we perform division.

Solve and state the answer.

1. We multiply the number of bookcases by the number of shelves per bookcase to find the total number of shelves: $2 \times 4 = 8$. We need a total of 8 shelves.

2. We must divide to see how many $3\tfrac{1}{8}$-foot sections are in 10 feet.

$$10 \div 3\tfrac{1}{8} = 10 \div \frac{25}{8} = 10 \cdot \frac{8}{25} = \frac{2 \cdot \cancel{5} \cdot 8}{\cancel{5} \cdot 5} = \frac{16}{5} \text{ or } 3\tfrac{1}{5}$$

3 shelves can be cut from each board with some wood left over.

3. Now we must find how many 10-foot boards are needed.

1 board: 3 shelves
2 boards: 3 shelves + 3 shelves = 6 shelves
3 boards: 3 shelves + 3 shelves + 3 shelves = 9 shelves

Two boards are not enough to cut 8 shelves. Three boards are enough to cut 8 shelves with some left over.
Nancy must buy 3 boards.

Check. We can estimate the answer by rounding $3\tfrac{1}{8}$ to 3 and reworking the problem.
10-foot board ÷ 3 feet per shelf ≈ 3 shelves per board
8 shelves ÷ 3 shelves per board ≈ 3 boards ✓

5.7 Student Practice

1. Since we are dividing the variable a by -2, we can undo the division and get a alone by multiplying by -2.

$$\frac{a}{-2} = 17 \qquad \text{The variable } a \text{ is divided by } -2.$$

$$\frac{-2 \cdot a}{-2} = 17 \cdot (-2) \qquad \begin{array}{l}\text{We undo the division by multiplying both sides} \\ \text{by } -2.\end{array}$$

$$a = -34$$

Be sure that you check your solution.

2. We simplify each side of the equation first and then we find the solution.

$$\frac{x}{3^2} = \frac{1}{3} + \frac{1}{9}$$

$$\frac{x}{9} = \frac{1}{3} + \frac{1}{9} \qquad \text{Simplify: } 3^2 = 9$$

$$\frac{x}{9} = \frac{4}{9} \qquad \text{Add: } \frac{1}{3} + \frac{1}{9} = \frac{3}{9} + \frac{1}{9} = \frac{4}{9}$$

$$\frac{9 \cdot x}{9} = \frac{4}{9} \cdot (9) \qquad \begin{array}{l}\text{We undo the division by multiplying} \\ \text{both sides by 9.}\end{array}$$

$$x = 4 \quad \text{Multiply to find the solution.}$$

We leave the check to the student.

3.

$$-\frac{3}{8}x = 9$$

$$\left(-\frac{8}{3}\right)\left(-\frac{3}{8}\right)x = 9\left(-\frac{8}{3}\right) \qquad \begin{array}{l}\text{Multiply both sides of the equation} \\ \text{by } -\frac{8}{3}.\end{array}$$

$$1x = -\frac{\cancel{3} \cdot 3 \cdot 8}{\cancel{3}}$$

$$x = -24$$

Check. $-\frac{3}{8}x = 9 \quad \left(-\frac{3}{8}\right)(-24) \stackrel{?}{=} 9 \quad$ Replace x with -24.

$$9 = 9 \ ✓$$

Chapter 6   6.1 Student Practice

1. We include the sign in front of the term as part of the term.
   Polynomial: $y^2 - 4x^2 + 5x - 9y$
   Terms: $+y^2, -4x^2, +5x, -9y$

2. We must combine like terms.
   $$(6x^2 - 7x + 3) + (-3x^2 + 6x) = 6x^2 - 3x^2 - 7x + 6x + 3$$ We rearrange terms so that the like terms are grouped together.
   $$= 3x^2 - 1x + 3$$ We combine like terms: $6x^2 - 3x^2 = 3x^2$; $-7x + 6x = -1x$.
   $$= 3x^2 - x + 3$$ $-1x = -x$

3. Since a "$-$" sign is in front of the parentheses, we change the sign of each term.
   $$-(-4x + 6y - 2n) = 4x - 6y + 2n$$

4. $(4x^2 + 6x - 8) - (7x^2 - 6x - 5)$ A $-$ sign between the polynomials indicates we are subtracting.
   $= 4x^2 + 6x - 8 + (-7x^2) + 6x + 5$ We change the signs of terms of the second polynomial, then add.
   $= -3x^2 + 12x - 3$ We simplify by combining like terms: $4x^2 - 7x^2 = -3x^2$; $6x + 6x = +12x$;
                                            $-8 + 5 = -3$.

5. First we multiply (-2) times the binomial $(3x^2 + 1)$.
   $4x - 2(3x^2 + 1) - (-3x^2 + x - 6)$ We multiply: $-2(3x^2 + 1)$;
   $= 4x - 6x^2 - 2 - (-3x^2 + x - 6)$ $-2(3x^2) = -6x^2$; $-2(1) = -2$.
   $= 4x - 6x^2 - 2 + 3x^2 - x + 6$ We remove parentheses and change the sign of each term inside the parentheses.
   $= 3x - 3x^2 + 4$ We combine like terms: $4x - x = 3x$; $-6x^2 + 3x^2 = -3x^2$; $-2 + 6 = +4$.
   $= -3x^2 + 3x + 4$ We write the polynomial so that the powers of x decrease as we read from left to right.

6.2 Student Practice

1. We multiply each term by $-6x$.
   $-6x(3x - 8y - 2)$ $-6x(3x) = -18x^{1+1} = -18x^2$
   $= -6x(3x) - 6x(-8y) - 6x(-2)$ $-6x(-8y) = +48xy$
   $= -18x^2 + 48xy + 12x$ $-6x(-2) = +12x$

2. We move the monomial to the left side.
   $-3y^4(2y^2 + y - 5)$
   We multiply each term by $-3y^4$.
   $-3y^4(2y^2 + y - 5)$ $-3y^4(2y^2) = -6y^{4+2} = -6y^6$
                               $-3y^4(y) = -3y^{4+1} = -3y^5$
   $= -6y^6 - 3y^5 + 15y^4$ $-3y^4(-5) = +15y^4$

3. We multiply $3y$ times $4y^2 + 2y - 6$ and then -1 times $4y^2 + 2y - 6$.
   $(3y - 1)(4y^2 + 2y - 6)$
   $= 3y(4y^2 + 2y - 6) - 1(4y^2 + 2y - 6)$
   $= 3y(4y^2) + (3y)(2y) + (3y)(-6) - 1(4y^2) - 1(2y) - 1(-6)$
   $= 12y^3 + 6y^2 - 18y - 4y^2 - 2y + 6$ We multiply.
   $= 12y^3 + 2y^2 - 20y + 6$ We combine like terms.

4. We multiply $x(x + 5)$ and $+3(x + 5)$.
   $(x + 3)(x + 5) = x(x + 5) + 3(x + 5)$
                  $= x \cdot x + x \cdot 5 + 3 \cdot x + 3 \cdot 5$
                  $= x^2 + 5x + 3x + 15$
   $(x + 3)(x + 5) = x^2 + 8x + 15$ Combine like terms.

5.

	First F	Outer O	Inner I	Last L
$(x + 2)(x + 4)$	$x \cdot x$	$(+4)x$	$(+2)x$	$(+2)(+4)$
	x^2	$+4x$	$+2x$	$+8$
$(x + 2)(x + 4) = x^2 + 6x + 8$		Combine like terms.		

6.

	First F	Outer O	Inner I	Last L
$(y - 5)(y - 3)$	y^2	$-3y$	$-5y$	$+15$
$(y - 5)(y - 3) = y^2 - 8y + 15$		Combine like terms.		

7.

	First F	Outer O	Inner I	Last L
$(3x - 1)(x + 1)$	$3x^2$	$+3x$	$-1x$	-1
$(3x - 1)(x + 1) = 3x^2 + 2x - 1$		Combine like terms.		

6.3 Student Practice

1. Since we are comparing the height of the pole to the height of the building, we let the variable represent the height of the building. We can choose any variable, so we choose B.
   Let the height of the building $= B$.
   Now we write the expression for the height of the pole in terms of the height of the building by translating the statement.

   The height of the pole is one-third the height of the building.

   height of pole $= \dfrac{1}{3} \cdot B$

   We define our variable expressions as follows.
   height of building $= B$
   height of pole $= \dfrac{1}{3}B$

2. Since we are comparing all sides to the first side, we let the variable represent the length of the first side. We may choose any variable so we use the letter f.

Let the length of the first side $= f$.

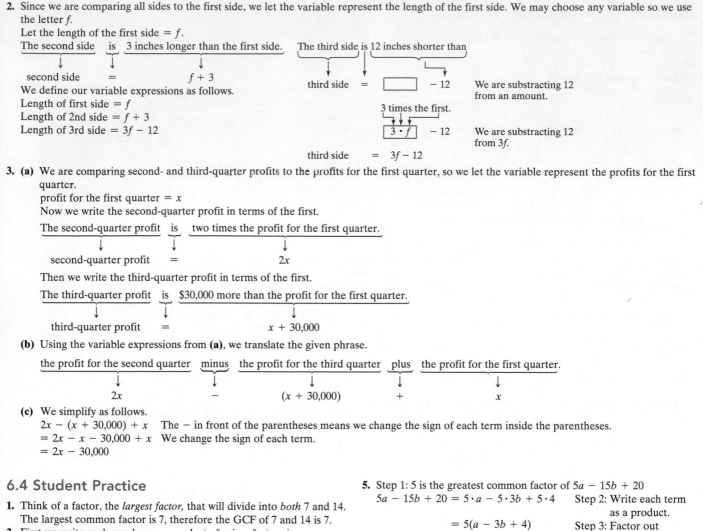

The second side ⟶ is ⟶ 3 inches longer than the first side.

second side $=$ $f + 3$

We define our variable expressions as follows.

Length of first side $= f$

Length of 2nd side $= f + 3$

Length of 3rd side $= 3f - 12$

The third side is 12 inches shorter than

third side $=$ ☐ $- 12$ We are substracting 12 from an amount.

3 times the first.

$\boxed{3 \cdot f}$ $- 12$ We are substracting 12 from $3f$.

third side $=$ $3f - 12$

3. (a) We are comparing second- and third-quarter profits to the profits for the first quarter, so we let the variable represent the profits for the first quarter.

profit for the first quarter $= x$

Now we write the second-quarter profit in terms of the first.

The second-quarter profit is two times the profit for the first quarter.

second-quarter profit $=$ $2x$

Then we write the third-quarter profit in terms of the first.

The third-quarter profit is \$30,000 more than the profit for the first quarter.

third-quarter profit $=$ $x + 30,000$

(b) Using the variable expressions from **(a)**, we translate the given phrase.

the profit for the second quarter   minus   the profit for the third quarter   plus   the profit for the first quarter.

$2x$ $-$ $(x + 30,000)$ $+$ x

(c) We simplify as follows.

$2x - (x + 30,000) + x$ The $-$ in front of the parentheses means we change the sign of each term inside the parentheses.

$= 2x - x - 30,000 + x$ We change the sign of each term.

$= 2x - 30,000$

6.4 Student Practice

1. Think of a factor, the *largest factor,* that will divide into *both* 7 and 14. The largest common factor is 7, therefore the GCF of 7 and 14 is 7.

2. First we write each number as a product of prime factors in exponent form. Then we identify the common factors and use the *smallest power* that appears on these factors to find our GCF.

$9 = \quad 3 \cdot 3 \quad = \quad 3^2$ We line up the common factors.

$21 = \quad 3 \cdot \ 7 = \quad 3 \cdot 7$ 3 is a common prime factor.

$42 = 2 \cdot 3 \cdot \ 7 = 2 \cdot 3 \cdot 7$

Notice that 3 $3^1 = 3$ The smallest power that appears

is common to   on the common factor is 1.

9, 21, and 42.

The greatest common factor of 9, 21, and 42 is 3.

3. (a) $16a^3b^4 + 12a^4b$ Rewrite each term using prime factors written in exponent form.

$16a^3b^4 = 2^4 \quad \cdot a^3 \cdot b^4$

$12a^4b = 2^2 \cdot 3^1 \cdot a^4 \cdot b^1$

$2^2 \cdot \quad a^3 \cdot b^1$ The smaller power on 2 is 2,

The GCF is $4a^3b$.   on a is 3, and on b is 1.

(b) $x^2yz^3 + xz^2$ Identify the common prime factors and use the smaller power that appears on each of these factors.

$x^2yz^3 = x^2 \cdot y \cdot z^3$

$xz^2 = x^1 \cdot \quad z^2$

$x^1 \cdot \quad z^2$ The smaller power on x is 1 and on z is 2.

The GCF is xz^2.

4. Step 1: 4 is the greatest common factor of $8y + 12$.

$8y + 12 = 4 \cdot 2y + 4 \cdot 3$ Step 2: Write each term as a product.

$8y + 12 = 4(2y + 3)$ Step 3: Factor out the GCF, 4.

Check: We multiply using the distributive property.

$4(2y + 3) \overset{?}{=} 8y + 12$

$4(2y + 3) = 4 \cdot 2y + 4 \cdot 3 = 8y + 12$ ✓

5. Step 1: 5 is the greatest common factor of $5a - 15b + 20$

$5a - 15b + 20 = 5 \cdot a - 5 \cdot 3b + 5 \cdot 4$ Step 2: Write each term as a product.

$= 5(a - 3b + 4)$ Step 3: Factor out the GCF, 5.

Check: We should always check our answer using the distributive property.

$5(a - 3b + 4) \overset{?}{=} 5a - 15b + 20$

$5a - 15b + 20 = 5a - 15b + 20$ ✓

6. We find the GCF of the expression.

$6ab^3 = 2 \cdot 3 \cdot \quad a \cdot b^3$

$18ab^4 = 2 \cdot 3 \cdot 3 \cdot a \cdot b^4$

$\text{GCF} = 2 \cdot 3 \cdot \quad a \cdot b^3 = \textbf{6ab}^3$

We factor out the GCF from the expression.

$\text{GCF}\left(\dfrac{6ab^3}{\text{GCF}} - \dfrac{18ab^4}{\text{GCF}}\right)$

$6ab^3 - 18ab^4 = \textbf{6ab}^3\left(\dfrac{6ab^3}{\textbf{6ab}^3} - \dfrac{18ab^4}{\textbf{6ab}^3}\right)$

$= 6ab^3(1 - 3b)$ $\dfrac{6ab^3}{6ab^3} = 1; \dfrac{18ab^4}{6ab^3} = 3b$

Check: We multiply $6ab^3(1 - 3b)$ using the distributive property.

$6ab^3(1 - 3b) \overset{?}{=} 6ab^3 - 18ab^4$

$6ab^3 - 18ab^4 = 6ab^3 - 18ab^4$ ✓

Chapter 7   7.1 Student Practice

1. $x - 6 = -56$ We want to get x alone.

$\underline{+\quad +6\qquad +6}$ We add the opposite of -6, or $+6$, to both sides.

$x + 0 = -50$ $-6 + 6 = 0$; $-56 + 6 = -50$

$\quad\;\; x = -50$ $x + 0 = x$

Check: $x - 6 = -56$

$-50 - 6 \overset{?}{=} -56$

$-56 = -56$ ✓

2. First we simplify each side of the equation, and then we solve for y.

$-11 - 17 = 5y + 12 - 4y - 1$ We calculate: $-11 - 17$

$= -11 + (-17) = -28.$

$-28 = 5y + 12 - 4y - 1$ We simplify: $5y - 4y = y$

and $12 - 1 = 11.$

$-28 = y + 11$

Now we solve for y.

$-28 = y + 11$ The opposite of $+11$ is -11.

$\underline{+\;\; -11\qquad -11}$ We add -11 to both sides of the equation.

$-39 = y$

We leave the check to the student.

3. First we simplify each side of the equation, and then we solve for y.

$\dfrac{y}{-3} = 2^2 - 8$

$\dfrac{y}{-3} = -4$ We simplify: $2^2 = 4$ and $4 - 8 = -4.$

$\dfrac{-3 \cdot y}{-3} = -4 \cdot (-3)$ We undo the division by multiplying by -3.

$y = 12$ We simplify: $\dfrac{-3y}{-3} = y$ and $-4 \cdot (-3) = 12.$

Check: $\dfrac{y}{-3} = 2^2 - 8$

$\dfrac{12}{-3} \overset{?}{=} 2^2 - 8$ We replace y with 12.

$-4 = -4$ ✓

4. $\dfrac{4}{5}a = 3^2 + 7$

$\dfrac{4}{5}a = 16$ Simplify: $3^2 + 7 = 9 + 7 = 16.$

$\dfrac{5}{4} \cdot \dfrac{4}{5}a = 16 \cdot \dfrac{5}{4}$ Multiply both sides by $\dfrac{5}{4}$.

$a = 20$ $16 \cdot \dfrac{5}{4} = \dfrac{4 \cdot \cancel{4} \cdot 5}{\cancel{4}} = 20$

Check: $\dfrac{4}{5}a = 3^2 + 7$ Replace a with 20, and simplify $3^2 + 7 = 16.$

$\dfrac{4}{5} \cdot 20 \overset{?}{=} 16$

$16 = 16$ ✓ $\dfrac{4}{5} \cdot 20 = \dfrac{4 \cdot 4 \cdot \cancel{5}}{\cancel{5}} = 16$

5. We begin by simplifying each side of the equation.

$\dfrac{16}{-4} = 4x + 3(-3x)$

$-4 = 4x + 3(-3x)$ Simplify: $\dfrac{16}{-4} = -4.$

$-4 = 4x - 9x$ Multiply: $3(-3x) = -9x.$

$-4 = -5x$ Add: $4x - 9x = -5x.$

$\dfrac{-4}{-5} = \dfrac{-5x}{-5}$ Dividing by -5 on both sides *undoes* the multiplication by -5.

$\dfrac{4}{5} = x$ Simplify: $\dfrac{-4}{-5} = \dfrac{4}{5}$ and $\dfrac{-5x}{-5} = 1x = x.$

Check: $\dfrac{16}{-4} = 4x + 3(-3x)$

$\dfrac{16}{-4} \overset{?}{=} 4 \cdot \dfrac{4}{5} + 3\left(-3 \cdot \dfrac{4}{5}\right)$ $4 \cdot \dfrac{4}{5} = \dfrac{16}{5}; 3\left(-3 \cdot \dfrac{4}{5}\right)$

$-4 \overset{?}{=} \dfrac{16}{5} - \dfrac{36}{5}$ $= 3\left(-\dfrac{12}{5}\right) = -\dfrac{36}{5}$

$-4 \overset{?}{=} -\dfrac{20}{5}$

$-4 = -4$ ✓

6. $-1x = 25$ Rewrite the equation: $-x$ is the same as $-1x$.

$\dfrac{-1x}{-1} = \dfrac{25}{-1}$ Dividing both sides by -1 *undoes* the multiplication by -1.

$x = -25$

7.2 Student Practice

1. First, we use the addition principle to get the variable term, $8x$, alone.

$8x + 9 = 105$

$\underline{+\qquad -9 = -9}$ We add -9 to both sides of the equation.

$8x = 96$ The variable term, $8x$, is alone.

Then we apply the division principle to get the variable, x, alone.

$\dfrac{8x}{8} = \dfrac{96}{8}$ We divide both sides by 8.

$x = 12$ The variable, x, is alone.

Check: $8x + 9 = 105$

$8(12) + 9 \overset{?}{=} 105$

$96 + 9 \overset{?}{=} 105$

$105 = 105$ ✓

2. First, we get the variable term, $-5m$, alone by using the addition principle.

$-5m - 10 = 115$

$\underline{+\qquad + 10 = +10}$ We add 10 to both sides of the equation.

$-5m \quad\;\; = 125$ The variable term, $-5m$, is alone.

Then we apply the division principle to get the variable, m, alone.

$\dfrac{-5m}{-5} = \dfrac{125}{-5}$ We divide both sides by -5.

$m = -25$ The variable, m, is alone.

We leave the check to the student.

3. We use the addition principle to get the variable term, $-5x$, alone.

$-2 = \quad 4 - 5x$

$\underline{+ \;\; -4 = -4}$ Add -4 to both sides of the equation.

$-6 = -5x$

Now we use the division principle to get x alone on the right side of the equation.

$\dfrac{-6}{-5} = \dfrac{-5x}{-5}$ Divide on both sides by -5.

$\dfrac{6}{5} = x$ $\dfrac{-6}{-5} = \dfrac{6}{5}$ and $\dfrac{-5x}{-5} = x$

Check: $-2 = 4 - 5x$

$-2 \overset{?}{=} 4 - 5\left(\dfrac{6}{5}\right)$

$-2 \overset{?}{=} 4 - 6$

$-2 = -2$ ✓

4. First we add $-4x$ to both sides of the equation so that all variable terms are on one side of the equation.

$$3x - 1 = 4x - 6$$
$$\underline{+\ -4x \qquad\quad -4x}$$
$$-x - 1 = -6$$

Then, we solve the equation.

$$-x - 1 = -6$$
$$\underline{+\quad +1 \quad +1} \qquad \text{Add 1 to both sides of the equation.}$$
$$-x \qquad = -5$$

$$\frac{-1x}{-1} = \frac{-5}{-1} \qquad \text{Write } -x \text{ as } -1x, \text{ and divide by } -1$$
$$\text{on both sides.}$$
$$x = 5$$

We leave the check to the student.

5.
$$-2 + 6y + 4 = 8y + 15$$
$$6y + 2 = 8y + 15 \qquad \text{Combine: } -2 + 4 = 2.$$
$$\underline{+\qquad -8y \qquad\quad -8y} \qquad \text{Add } -8y \text{ to both sides of the equation.}$$
$$-2y + 2 = 15$$
$$\underline{+\qquad\quad -2 = -2} \qquad \text{Add } -2 \text{ to both sides of the equation.}$$
$$-2y + 0 = 13$$
$$-2y = 13$$
$$\frac{-2y}{-2} = \frac{13}{-2} \qquad \text{Divide both sides of the equation by } -2.$$
$$y = -\frac{13}{2}$$

Check: $\quad -2 + 6y + 4 = 8y + 15$

$$-2 + \frac{6}{1}\left(\frac{-13}{2}\right) + 4 \overset{?}{=} \frac{8}{1}\left(\frac{-13}{2}\right) + 15$$
$$-2 + (-39) + 4 \overset{?}{=} -52 + 15$$
$$-41 + 4 \overset{?}{=} -37$$
$$-37 = -37 \ \checkmark$$

7.3 Student Practice

1. We use the distributive property to remove parentheses, and then we simplify and solve.

$$-5(3x + 2) - 6x = 32 \qquad \text{Multiply: } -5(3x) = -15x; -5(+2) = -10.$$
$$-15x - 10 - 6x = 32 \qquad \text{Combine like terms: } -15x - 6x = -21x.$$
$$-21x - 10 = 32$$
$$\underline{+\qquad\quad + 10 + 10}$$
$$-21x = 42$$
$$\frac{-21x}{-21} = \frac{42}{-21} \qquad \text{Divide both sides by } -21.$$
$$x = -2$$

Check: $\qquad -5(3x + 2) - 6x = 32$
$$-5[3(-2) + 2] - 6(-2) \overset{?}{=} 32$$
$$-5[-6 + 2] + 12 \overset{?}{=} 32$$
$$-5[-4] + 12 \overset{?}{=} 32$$
$$20 + 12 \overset{?}{=} 32$$
$$32 = 32 \ \checkmark$$

2.
$$2(x + 4) = -7(x - 3) + 2$$
$$2 \cdot x + 2 \cdot 4 = -7 \cdot x - 7(-3) + 2 \qquad \text{Remove parentheses.}$$
$$2x + 8 = -7x + 21 + 2$$
$$2x + 8 = -7x + 23 \qquad \text{Simplify.}$$
$$\underline{+7x \qquad\qquad +7x} \qquad \text{Add } +7x \text{ to both sides.}$$
$$9x + 8 = 23 \qquad \text{Now solve for } x.$$
$$\underline{+\quad - 8 \quad -8}$$
$$9x = 15$$
$$\frac{9x}{9} = \frac{15}{9}$$
$$x = \frac{5}{3}$$

Check: $\quad 2(x + 4) = -7(x - 3) + 2$

$$\frac{2}{1}\left(\frac{5}{3} + \frac{4}{1}\right) \overset{?}{=} \frac{-7}{1}\left(\frac{5}{3} - \frac{3}{1}\right) + \frac{2}{1}$$
$$\frac{2}{1}\left(\frac{5}{3} + \frac{12}{3}\right) \overset{?}{=} \frac{-7}{1}\left(\frac{5}{3} - \frac{9}{3}\right) + \frac{2}{1}$$
$$\frac{2}{1}\left(\frac{17}{3}\right) \overset{?}{=} \frac{-7}{1}\left(\frac{-4}{3}\right) + \frac{2}{1}$$
$$\frac{34}{3} \overset{?}{=} \frac{28}{3} + \frac{6}{3} \qquad \frac{34}{3} = \frac{34}{3} \quad \checkmark$$

3. Since there is a minus sign in front of the parentheses, $-(4x^2 + 2)$, we change the sign of each term inside the parentheses.

$$(4x^2 + 6x + 3) - (4x^2 + 2) = 3x + 1$$
$$4x^2 + 6x + 3 - 4x^2 - 2 = 3x + 1 \qquad \text{Change the sign of each term and remove parentheses.}$$
$$6x + 1 = 3x + 1 \qquad 4x^2 - 4x^2 = 0; 3 - 2 = 1.$$
$$\underline{+\ -3x \qquad\quad -3x} \qquad \text{Solve for } x.$$
$$3x + 1 = \qquad 1$$
$$\underline{+\quad -1 \qquad -1}$$
$$3x = 0$$
$$\frac{3x}{3} = \frac{0}{3} \qquad\qquad x = 0$$

We leave the check for the student.

7.4 Student Practice

1. First, we clear the fractions from the equation by multiplying each term by the LCD $= 10$. Then we solve the equation.

$$\frac{x}{5} + \frac{x}{2} = 7$$

$$10\left(\frac{x}{5}\right) + 10\left(\frac{x}{2}\right) = 10(7) \qquad \text{Multiply each term by the LCD to clear the fractions.}$$

$$2x + 5x = 70 \qquad \text{Simplify: } \frac{10x}{5} = 2x; \ \frac{10x}{2} = 5x. \text{ We cleared the fractions from the equation.}$$

$$7x = 70 \qquad \text{Combine like terms.}$$

$$\frac{7x}{7} = \frac{70}{7} \qquad \text{Divide both sides by 7.}$$

$$x = 10$$

Check: $\quad \dfrac{x}{5} + \dfrac{x}{2} = 7$

$$\frac{10}{5} + \frac{10}{2} \overset{?}{=} 7$$
$$2 + 5 = 7 \ \checkmark$$

2. The LCD is 14. We multiply each term by 14.

$$-5x + \frac{2}{7} = \frac{3}{2}$$

$$14(-5x) + 14\left(\frac{2}{7}\right) = 14\left(\frac{3}{2}\right)$$ Multiply each term by the LCD, 14.

$$-70x + 4 = 21$$ Simplify: $14 \cdot \left(\frac{2}{7}\right) = 4$; $14 \cdot \left(\frac{3}{2}\right) = 21$.

$$\begin{array}{r} + \quad -4 \quad -4 \\ \hline -70x = 17 \end{array}$$ Solve for x.

$$\frac{-70x}{-70} = \frac{17}{-70}$$

$$x = -\frac{17}{70}$$

We leave the check to the student.

3. There is only one denominator; thus the LCD is 4.

$$\frac{x}{4} + x = 5$$

$$4\left(\frac{x}{4}\right) + 4(x) = 4(5)$$ Multiply each term by the LCD, 4.

$$x + 4x = 20$$ Simplify: $\frac{4x}{4} = 1x$.

$$5x = 20$$ Combine like terms.

$$\frac{5x}{5} = \frac{20}{5}$$ Divide both sides by 5.

$$x = 4$$

7.5 Student Practice

1. *Understand the problem.*
   We organize the information in a Mathematics Blueprint for Problem Solving.

Mathematics Blueprint for Problem Solving

Gather the Facts	What Am I Asked to Do?	How Do I Proceed?	Key Points to Remember
$L = x + 6$ $W = 2$ $P = 36$ ft	**(a)** Find x: the number of feet the length should be enlarged. **(b)** Find L: the length of the enlarged planter box.	**(a)** To find x, start with $P = 2L + 2W$ and replace $P, L,$ and W with the given values. Then solve for x. **(b)** To find L, use the equation $L = x + 6$.	Place the unit "ft" in the answer.

Solve and state the answer.
(a) How much should the length be enlarged? We must find x to answer this question.

$$\begin{array}{ll} P = 2L + 2W & \text{Write the formula for the perimeter.} \\ 36 = 2(x + 6) + 2(2) & \text{Replace } P, W, \text{ and } L \text{ with the values given.} \\ 36 = 2x + 12 + 4 & \text{Multiply.} \\ 36 = 2x + 16 & \text{Simplify.} \\ \underline{+ \quad -16 \qquad -16} & \text{Solve the equation.} \\ 20 = 2x & \\ \frac{20}{2} = \frac{2x}{2} & \\ 10 = x & \text{The length should be enlarged 10 ft.} \end{array}$$

(b) What will the length of the enlarged planter box be?
   To find L, we replace x with 10 in the expression that represents the length.
   $$L = x + 6$$
   $$L = 10 + 6 = 16$$ The length of the planter box will be 16 ft.
   Check. We evaluate the formula for perimeter with our answer to check our calculations:
   $$P = 2L + 2W$$
   $$36 \overset{?}{=} 2(16) + 2(2)$$
   $$36 = 36 \checkmark$$

2. *Understand the problem.*
   To help us understand the problem, we use a Mathematics Blueprint for Problem Solving.

Mathematics Blueprint for Problem Solving

Gather the Facts	What Am I Asked to Do?	How Do I Proceed?	Key Points to Remember
The apprentice earns $7400 less than Jason. The sum of the two salaries is $83,000.	Find Jason's and the apprentice's salaries.	**(a)** Let J represent Jason's salary and write an expression for the apprentice's salary. **(b)** Form an equation by setting the sum of the expressions equal to $83,000. **(c)** Solve the equation.	I must find both Jason's and the apprentice's salaries, and check my answers.

(a) We write the variable expressions. Since we are comparing the apprentice's salary to Jason's salary, we let the variable J represent Jason's salary.
   Jason's salary $= J$
   Apprentice's salary $= J - 7400$ $7400 less than Jason's salary

(b) We form an equation.

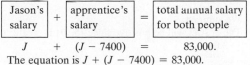

$$J + (J - 7400) = 83,000.$$
The equation is $J + (J - 7400) = 83,000$.

(c) We solve the equation.

$$J + (J - 7400) = 83{,}000$$
$$2J - 7400 = 83{,}000$$
$$\underline{+\ 7400 = \quad 7400}$$
$$2J = 90{,}400$$
$$\frac{2J}{2} = \frac{90{,}400}{2}$$
$$J = 45{,}200 \qquad \text{Jason earns \$45,200 annually.}$$

Now we find the apprentice's salary.

$$J - 7400 = \text{apprentice's salary (\$7400 less than Jason's salary)}$$
$$\$45{,}200 - \$7400 = \$37{,}800 \text{ The apprentice earns \$37,800 annually.}$$

(d) *Check:* Is the sum of their salaries \$83,000?
$$\$45{,}200 + \$37{,}800 \overset{?}{=} \$83{,}000$$
$$\$83{,}000 = \$83{,}000 \quad \checkmark$$

Chapter 8 8.1 Student Practice

1. (a) The place value of the *last digit* is the *last word* in the word name. The last digit is 5 and is in the *thousandths* place.

0.365   three hundred sixty-five thousandths

(b) The place value of the *last digit* is the *last word* in the word name. The last digit is 2 and is in the *hundredths* place.

5.32   five and thirty-two hundredths

2. Two hundred forty-five and $\frac{9}{100}$

3. The whole number is 8. There are 3 decimal places. The denominator has 3 zeros.

$$8.723 = 8\frac{723}{1000}$$

4. There are 3 zeros in the denominator. Move the decimal point 3 places to the left.

$$\frac{17}{1000} = \frac{017}{1000} = 0.017$$

Note that we had to insert a 0 before 17 so we could move the decimal point 3 places to the left.

5. 0.77   ?   0.771

0.770   ?   0.771   Add a zero to 0.77 so that both decimals have the same number of digits.

Since $0 < 1$, $0.770 < 0.771$.   The tenths digits are equal, and the hundredths digits are equal. The thousandths digits differ.

6. The round-off place digit is in the *hundredths* place.

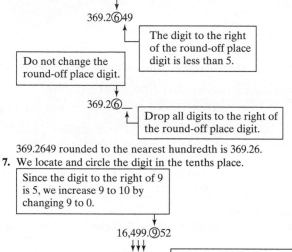

369.2649 rounded to the nearest hundredth is 369.26.

7. We locate and circle the digit in the tenths place.

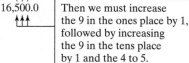

Note that we must include the zero after the decimal point because we were asked to round to the nearest tenth.

16,500.0   When rounding to the nearest tenth, the last digit should be in the tenths place.

Thus 16,499.952 rounds to 16,500.0 kilowatt-hours.

8.2 Student Practice

1. We can write any whole number as a decimal by placing a decimal point at the end of the number: 50 = 50.00.

$$\overset{1}{5}0.00$$
$$4.39 \qquad \text{Line up decimal points.}$$
$$\underline{+\ 0.70} \qquad \text{Add zeros so that each number has the same number of}$$
$$55.09 \qquad \text{decimal places. Then add.}$$

2. $2\overset{5\ 10}{6.\cancel{0}1}$ Line up decimal points.

$$\underline{-5.70} \qquad \text{Add a zero.}$$
$$20.31 \qquad \text{The decimal point in the difference is in line with the other decimal points.}$$

3. To subtract, we add the opposite of the second number.

$$-3.02 - (-5.1)$$
$$-3.02 + (5.1) \qquad \text{Add the opposite of } (-5.1).$$

Next, to add numbers with different signs, we "keep the sign of the larger absolute value and subtract."

$$5.10$$
$$\underline{-3.02} \qquad \text{We subtract.}$$
$$2.08 \qquad \text{The answer will be positive since } |5.10| \text{ is greater than } |-3.02|.$$

4. $4.5y$ and $7.2y$ are like terms, so we add them.

$$4.5y \qquad \text{We line up decimal points and add.}$$
$$\underline{+\ 7.2y}$$
$$11.7y$$

We have $4.5y + 7.2y - 5.6x = 11.7y - 5.6x$.

We cannot combine $11.7y$ and $5.6x$ since they are not like terms.

5. We replace the variable with 1.2.

$$-6.2 + x = -6.2 + 1.2$$

We line up the decimals and subtract. We keep the sign of the number with the larger absolute value.

$$-6.2$$
$$\underline{1.2}$$
$$-5.0$$

6. We round each decimal to the nearest whole number and then add.

$$7 + 9 + 19 + 7 + 9 + 22 = 73$$

The total cost of Allen's meals was approximately \$73.

7. (a) The key phrase "how many more" indicates the operation subtraction.

$$\overset{0\ \overset{10}{\cancel{8}}\ 15}{\cancel{1}\,\cancel{1}.\cancel{5}}$$
$$\underline{\quad 4.6}$$
$$\quad 6.9 \text{ assists}$$

(b) We add to find the total.
$$\begin{array}{r} 4.6 \\ 3.7 \\ 11.5 \\ 5.4 \\ \underline{8.6} \\ 33.8 \text{ assists} \end{array}$$

8.3 Student Practice

1. $0.05 \quad \times 0.07 \to 5 \times 7 = 35$
$0.05 \quad \times \quad 0.07 \quad = \quad 0.0035$

| 2 decimal + | 2 decimal | = 4 decimal |
| places | places | places |

> Multiply as if there were no decimals.

Note that we had to insert 2 zeros to the left of 35 in order to have 4 decimal places in the product.

3. The number of negative signs, 1, is odd so the product is negative.

$$\begin{array}{rl} 6.22 & \text{2 decimal places} \\ \underline{\times (-3)} & \text{0 decimal places} \\ -18.66 & \text{We need 2 decimal places. } (2 + 0 = 2) \end{array}$$

5. Since 10^3 has 3 zeros ($10^3 = 1000$), we must move the decimal point to the right 3 places.
$$(0.6944)(10^3) = 694.4$$
$$\qquad\qquad\qquad\qquad \uparrow$$
$$\qquad\qquad \text{Move decimal point.}$$

2. We write the multiplication just as we would if there were no decimal points.

$$\begin{array}{rl} 20.1 & \text{1 decimal place} \\ \underline{\times 4.32} & \text{2 decimal places} \\ 402 & \\ 6030 & \\ \underline{80400} & \\ 86.832 & \text{We need 3 decimal places. } (1 + 2 = 3) \end{array}$$

4.   *two zeros*
$$\downarrow$$
$$0.123 \times 100 = 12.3$$
$$\uparrow \qquad\qquad\quad \uparrow$$
Move the decimal point two places to the right.

6. $1.3 \div 2$: Think "2 goes into 1.3."

$1.3 \div 2 \to$
$$\begin{array}{r} 0.6 \\ 2\overline{)1.3} \\ \underline{12} \\ 1 \end{array}$$
We place the decimal point directly above the decimal point of dividend. Now we divide as if there were no decimal point.

$$\begin{array}{r} 0.65 \\ 2\overline{)1.30} \\ \underline{1\,2} \\ 10 \\ \underline{10} \\ 0 \end{array}$$
We add a zero so that we can continue to divide.
We bring down a 0.
$1.3 \div 2 = 0.65$

7. We must divide one place beyond the hundredths place—that is, to the thousandths place—so we can round to the nearest hundredth. The answer will be negative. Place a decimal point directly above the one in the dividend.

$$\begin{array}{r} -\ . \\ 14\overline{)-0.3624} \end{array}$$
Place a zero in tenths place as a placeholder.

$$\begin{array}{r} -\ .025 \\ 14\overline{)-0.3624} \\ \underline{28} \\ 82 \\ \underline{70} \\ 12 \end{array}$$
We stop because we have found the quotient to the thousandths place.

-0.025 rounded to the nearest hundredth is -0.03; $-0.3624 \div 14 \approx -0.03$.

8. Since the divisor, 3.5, is not a whole number, let's write the division problem using fractional notation.

$$14.56 \div 3.5 = \frac{14.56}{3.5}$$

Now, if we multiply the numerator and denominator by 10, the divisor becomes a whole number.

$$\frac{(14.56)(10)}{(3.5)(10)} = \frac{145.6}{35} = 145.6 \div 35$$

$$\begin{array}{r} 4.16 \\ 35\overline{)145.60} \\ \underline{140} \\ 5\,6 \\ \underline{3\,5} \\ 2\,10 \\ \underline{2\,10} \\ 0 \end{array}$$
We add a zero so we can continue dividing.

$14.56 \div 3.5 = 4.16$

9. The divisor is not a whole number. Move the decimal point one place to the right.
$$1.8\overline{)1.1} \to 18\overline{)11}$$
Now that the divisor is a whole number, we rewrite the problem and divide.

$$\begin{array}{r} 0.611 \\ 18\overline{)11.000} \\ \underline{10\,8} \\ 20 \\ \underline{18} \\ 20 \\ \underline{18} \\ 2 \end{array}$$
Add zeros.

$1.1 \div 1.8 = 0.6\overline{1}$ because if we continued dividing, the 1 would repeat.

10. $2\dfrac{5}{11}$ means $2 + \dfrac{5}{11}$.   We divide: $5 \div 11$.

$$\begin{array}{r} 0.4545 \\ 11\overline{)5.0000} \\ \underline{4\,4} \\ 60 \\ \underline{55} \\ 50 \\ \underline{44} \\ 60 \\ \underline{55} \end{array}$$

We can see that the pattern repeats.

Thus $\dfrac{5}{11} = 0.\overline{45}$ and $2\dfrac{5}{11} = 2.\overline{45}$.

8.4 Student Practice

1. We want an equation of the form $y =$ some number.

$\begin{aligned} y - 2.23 &= 4.69 \\ +2.23\ +2.23 \\ \hline y + 0 &= 6.92 \\ y &= 6.92 \end{aligned}$ We add the opposite of -2.23, or $+2.23$, to both sides of the equation. Remember to line up the decimal points.

Check:

$\begin{aligned} y - 2.23 &= 4.69 \\ 6.92 - 2.23 &\overset{?}{=} 4.69 \\ 4.69 &= 4.69 \ \checkmark \end{aligned}$

2. We must divide both sides of the equation by 5.6 to get x alone on one side of the equation.

$\dfrac{5.6x}{5.6} = \dfrac{-25.2}{5.6}$ Dividing on both sides undoes the multiplication by 5.6.

$x = -4.5$ The sign of the answer is negative because the problem has an odd number of negative signs: $-25.2 \div 5.6 = -4.5$.

Check:

$\begin{aligned} 5.6x &= -25.2 \\ (5.6)(-4.5) &\overset{?}{=} -25.2 \\ -25.2 &= -25.2 \ \checkmark \end{aligned}$

3. We multiply each term inside the parentheses by 4 to remove the parentheses.

$\begin{aligned} 4(x - 2.5) &=\ 3x + 0.6 \quad \text{Multiply each term by 4.} \\ 4(x) - 4(2.5) &=\ 3x + 0.6 \\ 4x - 10 &=\ 3x + 0.6 \\ +\ -3x \qquad\quad &\ \ -3x \qquad\qquad \text{We add } -3x \text{ to both sides of the} \\ \hline & \qquad\qquad\qquad\quad\ \text{equation.} \\ x - 10 &=\ \ \ 0.6 \\ +\qquad\quad 10 &=\ \ 10 \qquad\qquad \text{We add } +10 \text{ to both sides of the} \\ \hline x &=\ \ 10.6 \qquad\ \text{equation.} \end{aligned}$

We leave the check to the student.

4. The term 0.15 has the most decimal places (2 decimal places), so we must multiply both sides of the equation by $10^2 = 100$ to eliminate decimals in the equation.

$\begin{aligned} 0.3x + 0.15 &= 2.7 \\ 100(0.3x + 0.15) &= 100(2.7) \\ 100(0.3x) + 100(0.15) &= 100(2.7) \\ 30x + 15 &= 270 \\ +\qquad\quad -15 &\quad -15 \\ \hline 30x &= 255 \\ \dfrac{30x}{30} &= \dfrac{255}{30} \\ x &= 8.5 \end{aligned}$

Use the distributive property. Multiply each term by 100 (by moving the decimal point to the right 2 places). Subtract 15 from both sides of the equation. Divide by 30 on both sides of the equation.

We leave the check to the student.

5. *Understand the problem.* Since there is a lot of information in the problem, we fill in a Mathematics Blueprint for Problem Solving.

Mathematics Blueprint for Problem Solving

Gather the Facts	What Am I Asked to Do?	How Do I Proceed?	Key Points to Remember
Base fee: $4.95 Rates, per minute: Outside state 10¢ Inside state 5¢ Average length of calls in minutes: Outside state 55 Inside state 185 Budget $25	Determine how much more or less her average calls will cost than her budget allows.	**1.** Let x = the average long-distance bill. **2.** Write equation: x = base fee + charge for calls outside the state + charge for calls inside the state. **3.** Subtract this amount from $25.	Change 10¢ and 5¢ to decimals. Follow the order of operations.

Calculate and state the answer.

$\text{Average long-distance bill} = \text{base fee} + \text{charge for calls outside state} + \text{charge for calls inside state}$

$\begin{aligned} x &= 4.95 + (55 \times 10\text{¢ per min}) + (185 \times 5\text{¢ per min}) \\ &= 4.95 + 55 \times 0.10 + 185 \times 0.05 \\ &= 4.95 + 5.50 + 9.25 \\ x &= 19.70 \end{aligned}$

$19.70 is the average cost of Natasha's calls, and this amount is less than her monthly budget of $25. We subtract: $25 - $19.70 = $5.30. Her average monthly bill will be $5.30 less than her budget of $25.

Check: Use your calculator to verify your results.

6. *Understand the problem.* We organize information in a Mathematics Blueprint for Problem Solving.

Mathematics Blueprint for Problem Solving

Gather the Facts	What Am I Asked to Do?	How Do I Proceed?	Key Points to Remember
The number of dimes is four times the number of quarters. The total value of dimes and quarters needed is $3.25.	Determine the number of dimes and the number of quarters that must be used for the $3.25 in change.	**1.** Define the variable expressions to find the *value* of the dimes and quarters. **2.** Form an equation: value of dimes + value of quarters = $3.25. **3.** Solve the equation and answer the question.	The value of the dimes is $0.10 \times$ number of dimes. The value of the quarters is $0.25 \times$ number of quarters.

Calculate and state the answer.

(a) We define our variables to find the value of each coin. Since we are comparing dimes to quarters, we let our variable represent the number of quarters.

How Many Quarters	Coin Value	Total Value of Coins
Q = number of quarters	0.25	$(0.25)Q$ = *value* of the quarters
$4Q$ = number of dimes (four times the number of quarters)	0.10	$(0.10)(4Q)$ = *value* of the dimes

The *variable expressions* appear in the table above.

(b) We form an equation and then solve the equation.

$$\boxed{\text{The } value \text{ of quarters}} + \boxed{\text{the } value \text{ of dimes}} = \boxed{\text{the } total\ value \text{ of coins}}$$

$$(0.25)Q + (0.10)(4Q) = \$3.25$$

$$0.25Q + 0.40Q = 3.25 \qquad \text{Multiply: } (0.10)(4Q)$$
$$\qquad\qquad\qquad\qquad\qquad = (0.10 \times 4)Q = 0.40Q.$$
$$25Q + 40Q = 325 \qquad \text{Move the decimal point in each}$$
$$\qquad\qquad\qquad\qquad\qquad \text{term 2 places to the right.}$$
$$65Q = 325$$
$$\frac{65Q}{65} = \frac{325}{65}$$
$$Q = 5 \qquad \text{There are 5 quarters.}$$
$$4Q = 4(5) = 20 \quad \text{There are 20 dimes.}$$

(c) Julio must put 5 quarters and 20 dimes in the vending machine for change.

(d) *Check:* Do we have four times as many dimes as quarters as stated in the problem? Yes. 20 dimes is 4 × 5, or 4 times the number of quarters. Is the value of the coins equal to $3.25? Yes. 20 dimes = $2.00 and 5 quarters = $1.25; $2.00 + $1.25 = $3.25. ✓

8.5 Student Practice

1. (a) We estimate 10% of 86,205 as follows.

$86{,}205 \to 86{,}000$ First we round.

$86{,}00\cancel{0}$ Then we delete the last digit.

10% of 86,205 ≈ 8600

(b) We estimate 1% of 86,205 as follows.

$86{,}205 \to 86{,}000$ First we round.

$86{,}0\cancel{00}$ Then we delete the last 2 digits.

1% of 86,205 ≈ 860

3. We round $199.85 to the nearest hundred: $200.

7% of 199.85 ≈ 7% of 200 = (5% of 200) + 2(1% of 200)

$$= \tfrac{1}{2} \text{ of } (10\% \text{ of } 200) + 2(2)$$
$$= \tfrac{1}{2}(20) + 4$$
$$= 10 + 4 = \$14$$

2. We round 45,005 to the nearest thousand: 45,000.

(a) To find 20%, we add (10% of 45,000) + (10% of 45,000).

20% of 45,005 ≈ 20% of 45,000 = 4500 + 4500
$$= 9000$$

(b) To find 11%, we add (10% of 45,000) + (1% of 45,000).

11% of 45,005 ≈ 11% of 45,000 = 4500 + 450
$$= 4950$$

(c) To find 7%, we add

(5% of 45,000) + (1% of 45,000) + (1% of 45,000).

7% of 45,005 ≈ 7% of 45,000

$$= \frac{1}{2} \times (10\% \text{ of } 45{,}000) + 450 + 450$$

$$= \frac{1}{2} \times 4500 + 450 + 450$$

$$= 2250 + 450 + 450 = 3150$$

4. (a) We round each item to the nearest ten.

Shirt: $19.95 → $20   Pants: $28 → $30   Shoes: $39.99 → $40

We add these amounts:

$20 + $20 + $30 + $30 + $40 + $40 = $180.

The estimated cost of the items before the discount is $180.

(b) We find 30% of the total estimated cost.

10% of $180 = $18; 30% of $180 = $18 + $18 + $18 = $54

The estimated discount rounded to the nearest ten is $50.

(c) We subtract the total estimated cost minus the estimated discount.

$180 − $50 = $130

The estimated cost of the items after the discount is $130.

8.6 Student Practice

1. $\dfrac{42}{100} = 42\%$ 42% of the class voted.

2. We must write the present cost ($215) as a percent of the cost ten years ago ($100).

Present cost $\to \dfrac{215}{100} = 215\%$

Cost 10 years ago $\to 100$

The present cost is 215% of the cost ten years ago.

3. Julio takes 0.7 mL of 100 mL, or

$\dfrac{0.7}{100} = 0.7\%$ of the solution.

4. (a)

Decimal	←	Percent
——		2.6%
		↵
0.026	=	2.6%

We write the chart.

We move *left* on the chart, so the decimal point moves *2 places left*. We must place an extra zero to the left of the 2.

(b)

Decimal	→	Percent
0.001		——
↳		
0.001	=	0.1%

We write the chart.

We move *right* on the chart, so the decimal point moves *2 places right*.

5.

Decimal Form	Percent Form
0.511	51.1%
0.841	84.1%
0.002	0.2%
6.7	670%

We must insert two zeros. →

← We must insert a zero.

6. (a) Using a chart often helps.

Fraction	→	Decimal	→	Percent
$\frac{7}{40}$	→	0.175		?
$\frac{7}{40}$	→	0.175	→	17.5%

Compute:
$7 \div 40 = 0.175$.

Move the decimal point 2 places to the right.

(b) We reverse the process to write 17.5% as a fraction.

Fraction	←	Decimal	←	Percent
?		0.175	←	17.5%
$\frac{7}{40}$	←	0.175	←	17.5%

Move the decimal point 2 places to the left.

$0.175 = \frac{175}{1000} = \frac{7}{40}$

7. (a) First we change $\frac{3}{7}$ to a decimal: $3 \div 7 = 0.42857\ldots$.

We must carry out the division at least *five* places beyond the decimal point so that we can move the decimal point to the right *two* places, and then round to the nearest hundredth of a percent.

Fraction → Decimal → Percent

$\frac{3}{7} = 3 \div 7 = 0.42857\ldots = 42.857\ldots\% \approx 42.86\%$

(b) A percent can be written in the form of a whole number, fraction, or decimal. $\frac{1}{5}\%$ is a percent written in fraction form. We can rewrite the percent in its decimal form so that we can use the decimal shift method.

Fraction ← Decimal ← Percent

			←	$\frac{1}{5}\%$
$\frac{1}{500}$	←	0.002	←	0.2%

Step 1: $1 \div 5 = 0.2$
Step 2: Move the decimal point left 2 places.
Step 3: $\frac{2}{1000} = \frac{1}{500}$

8.7 Student Practice

1. We *translate* each into the form amount = percent × base.

(a) What is 40% of 90?
↓ ↓ ↓ ↓ ↓

$n = 40\% \times 90$ Write in symbols.
We want to find the *amount*, that is, the *part of* (40% of) the base of 90.
$n = 40\%$ of 90
$n = 0.40 \times 90$ Change 40% to decimal form.
$n = 36$ Multiply.
36 is 40% of 90.

(c) 36 is what percent of 90?
↓ ↓   ↓   ↓ ↓

$36 = n\% \times 90$
We want to find the *percent*.
$36 = n\% \times 90$

$\frac{36}{90} = n\%$ Solve the equation for $n\%$.

$0.40 = n\%$ The % symbol reminds us to write 0.40 as a percent.
$40 = n$
36 is 40% of 90.

(b) 36 is 40% of what number?
↓ ↓ ↓ ↓ ↓

$36 = 40\% \times n$
We want to find the *base* (the entire quantity).
$36 = 0.40 \times n$ Change 40% to a decimal.

$\frac{36}{0.40} = n$ Solve the equation for n.

$90 = n$ Divide.
36 is 40% of 90.

2. We should expect to get more than 100% since 80 is more than the base of 20.

80 is what percent of 20?
↓ ↓   ↓   ↓ ↓

$80 = n\% \times 20$
$80 = n\% \times 20$

$\frac{80}{20} = n\%$ Solve for $n\%$.

$4 = n\%$ Divide.
$400 = n$
80 is 400% of 20.

3. Find 22% of 60.
↓   ↓ ↓ ↓

$n = 22\% \times 60$ Translate "*find*" to "$n =$" since it has the same meaning as "*what is.*"

$= 0.22 \times 60$ Change 22% to a decimal.
$= 13.2$ Multiply.
13.2 is 22% of 60.

4. We must find the *percent*, so we write the statement that represents the percent situation.

35 is what percent of 50? Write the statement that
↓ ↓ ↓   ↓   ↓ ↓ represents this situation.

$35 = n\% \times 50$ Form an equation.
$35 = n\% \times 50$

$\frac{35}{50} = n\%$ Solve the equation.

$0.70 = n\%$
$70 = n$
Alisha earned 70% of the points.

5. We must find the percent, so we write the statement that represents the percent situation.

3.50 is what percent of 18.55? Write the statement for
↓ ↓   ↓   ↓ ↓ this situation.

$3.50 = n\% \times 18.55$ Form an equation.
$3.50 = n\% \times 18.55$

$\frac{3.50}{18.55} = n\%$ Solve the equation.

$0.18867\ldots = n\%$
$18.867\ldots = n$
$n \approx 18.87$ Round.
The tip was approximately 18.87% of the total bill.

6. *Understand the problem.*

Mathematics Blueprint for Problem Solving

Gather the Facts	What Am I Asked to Do?	How Do I Proceed?	Key Points to Remember
Room rate for Saturday night is $230. The Saturday night rate is 20% *higher* than the rate on Sunday.	Find the room rate for Sunday.	Let x = the room rate on Sunday. Find the sum: The Sunday night rate *plus* the markup equals the Saturday night rate.	The markup is 20% of the rate on Sunday.

Calculate and state the answer.

Room rate plus 20% of the equals Room rate
on Sunday   Sunday rate   on Saturday

$$100\%x \quad + \quad 20\%x \quad = \quad \$230$$

$120\%x = 230$ We add $(100\% + 20\%)x = 120\%x$.
$1.20x = 230$ We change 120% to a decimal.
$x = \dfrac{230}{1.20}$ We solve for x.
$x = 191.666\ldots$
$x \approx \$191.67$

Sergio will pay a rate of $191.67 to stay Sunday night.

8.8 Student Practice

1. (a) Find 83% of 460. The value of p is 83.
   (b) 15% of what number is 60? The value of p is 15.
   (c) What percent of 45 is 9?
         $\underset{p}{\underline{\text{What percent}}}$ We let p represent the unknown percent number.

2. (a) 40% of 88 is 35.2.
            base amount   The base is the entire quantity. It follows the word *of*. $b = 88$
            The amount is the part compared to the whole. $a = 35.2$
   (b) 120 is 94% of what?   The amount 120 is the part of the base.
      amount   base $a = 120$
          The base is unknown. We represent the base by the variable b.

3. (a) What is 31% of 418?   The amount is unknown. We let
      amount percent base a = the amount. The value of p is 31.
          The base usually follows the word *of*. Here, $b = 418$.
   (b) What percent of 37 is 4?   We let p represent the
      percent   base amount   unknown percent.
          The base usually follows the word *of*. Here, $b = 37$. The amount is 4. $a = 4$

4. The percent $p = 340$. The number that is the base usually appears after the word *of*. The base $b = 70$. The amount is the unknown. We use the variable a. Thus
$$\frac{a}{b} = \frac{p}{100} \text{ becomes } \frac{a}{70} = \frac{340}{100}$$
If we simplify the fraction on the right-hand side, we have the following.
$$\frac{a}{70} = \frac{17}{5} \qquad \frac{340}{100} = \frac{17}{5}$$
$5a = (70)(17)$ Cross-multiply.
$5a = 1190$ Simplify.
$\dfrac{5a}{5} = \dfrac{1190}{5}$ Divide both sides by 5.
$a = 238$
Thus 340% of 70 is 238.

5. The percent $p = 82$. The base is the unknown. We use the variable b. The amount a is 246. Thus
$$\frac{a}{b} = \frac{p}{100} \text{ becomes } \frac{246}{b} = \frac{82}{100}$$
If we simplify the fraction on the right hand side, we have the following.
$$\frac{246}{b} = \frac{41}{50} \qquad \frac{82}{100} = \frac{41}{50}$$
$(246)(50) = 41b$ Cross-multiply.
$12300 = 41b$ Simplify.
$\dfrac{12300}{41} = \dfrac{41b}{41}$ Divide both sides by 41.
$300 = b$ Thus 82% of 300 is 246.

6. The percent is the unknown. We use the variable p. The base $b = 140$. The amount $a = 42$. Thus
$$\frac{a}{b} = \frac{p}{100} \text{ becomes } \frac{42}{140} = \frac{p}{100}$$
Cross-multiplying, we have the following.
$(100)(42) = 140p$
$4200 = 140p$
$\dfrac{4200}{140} = \dfrac{140p}{140}$ Divide both sides by 140.
$30 = p$
Thus 42 is 30% of 140.

7. (a) We can subtract the percents to determine the percent of the offices that are not occupied.
100% of offices minus 70% occupied equals percent not occupied
     100%   −   70%   =   30%
30% of the offices are not occupied.
   (b) We must find the amount a. The base $b = 120$. The percent $p = 30$. Thus
$$\frac{a}{b} = \frac{p}{100} \text{ becomes } \frac{a}{120} = \frac{30}{100}$$
When we cross-multiply, we obtain the following.
$100a = (120)(30)$ Cross-multiply.
$100a = 3600$ $(120)(30) = 3600$
$\dfrac{100a}{100} = \dfrac{3600}{100}$ Divide both sides by 100.
$a = 36$
Thus 36 of the offices are not occupied.

8.9 Student Practice

1. commission = commission rate × total sales
$n = 6\% \times 145{,}000$
$n = 0.06 \times 145{,}000$
$n = 8700$ The real estate agent's commission is $8700.
Check: We will estimate to check our answer. We round $8700 to $9000 and $145,000 to $150,000 and then verify that $9000 is approximately 6% of $150,000: 6% of $150,000 = $9000.

2. Decrease of $492 is what percent   of $1640?
   ↓   ↓   ↓   ↓   ↓
decrease = percent decrease × original amount
    492 = $n\%$ ×   1640

$$\frac{492}{1640} = n\%$$
$$0.3 = n\%$$
$$30 = n$$

The price is reduced by 30%.
Check: We round $1640 to $1600. 30% of $1600 is $480, which is close to the price reduction of $492.

3. First, we find the amount of the raise.
percent increase × original amount = increase (raise)
   3%   × $9.45   = 0.03 × 9.45 = $0.2835
raise, which would be rounded down to $0.28 (28 cents).
Now we find the new hourly salary.
original wage + raise = new hourly wage
   $9.45   + $0.28 = $9.73
Rebecca's new hourly wage is $9.73.
Check: Rebecca receives 103% of the old wage or
    1.03(9.45) = $9.7335, which rounds to $9.73.

4. First we find the amount of decrease.
percent decrease × original amount = decrease (in attendance)
  14%   of   1500   =
         0.14(1500) = 210 fewer concertgoers
Now we find the number who bought tickets this year.
last year's attendance − decrease = this year's attendance
    1500   −   210   =   1290
1290 people have bought tickets to the concert this year.

5. (a) P = principal = $2000, R = rate = 3%, T = time = 1 year
$I = P \times R \times T$
$I = \$2000 \times 0.03 \times 1 = \$60$ We change 3% to 0.03, then multiply.
The interest for one year is $60.

  (b) At the end of the year, the interest will be added to the principal.
principal + yearly interest = new amount
$2000 + $60 = $2060
$2060 will be in Damon's savings account at the end of the year.

6. We must change 4 months to years since the formula requires that the time be in years: $T = 4$ months $= \frac{4}{12}$ or $\frac{1}{3}$ year.
$I = P \times R \times T$

$$= 3100 \times 0.09 \times \frac{1}{3}$$

$$= 279 \times \frac{1}{3} = 93$$

The interest for 4 months is $93.

Chapter 9   9.1 Student Practice

1. (a) Since there are four symbols (■) beside the state of Utah on the pictograph and each one equals 20,000 square miles, we have
    4 × 20,000 = 80,000 square miles.
  (b) Idaho and Utah have the fewest symbols on the pictograph and thus have the smallest area.
  (c) Idaho is 30,000 squares miles smaller in area than Nevada because there are $1\frac{1}{2}$ fewer symbols beside Idaho. (Each ■ equals 20,000 square miles.)

2. (a) If we add 28 + 23, we get 51. Thus Lake Superior and Lake Michigan together take up 51% of the total area.
  (b) The entire circle represents 100%, and Lake Superior and Lake Michigan together take up 51% of the area. Thus, we subtract 100 − 51 = 49% to get the percent of the total area that is not taken up by Lake Superior and Lake Michigan together.
  (c) If we add 11 + 12, we get 23. Thus Lake Erie and Lake Ontario together take up 23% of the total area.
    23% of 290,000 = (0.23)(290,000) = 66,700 square miles

3. The bar rises to the white line for 20. This represents 20 percent. Thus, 20% of traffic fatalities for males ages 45–54 years involved drunk drivers.

4. The percent of fatalities for males is the highest for 21–34 year olds. The bar rises to between 30 and 35. Thus, the highest percent of drunk driving fatalities for males in any age group is 31–34%.

5. The yellow solid line represents the over-55 age group. Since the square corresponding to Monday on the yellow solid line is near 9%, we know that approximately 9% of the over-55 group shops on Monday.

6. Since the blue solid line represents the shopping days for the under-35 age group, we look for the highest blue dot. Thus, Saturday is the busiest shopping day for the under-35 age group.
Since the yellow solid line represents the shopping days for the over-55 age group, we look for the highest yellow dot. Thus, Thursday is the busiest shopping day for the over-55 age group.

7. First, we draw a vertical and a horizontal line.
Since we are comparing *Rent amount,* we place this label on the vertical line. Mark intervals with equally spaced notches and label 400 to 2000.
We place the label *Apartment size* on the horizontal line. We mark intervals with equally spaced notches and label with each size of apartment.
Now we place dots on the graph that correspond to the given data.
We choose a different color line for each of the categories we are comparing (beach and inland) and connect the dots with line segments.

8. We plan our graph.
First we draw a vertical and a horizontal line.
Since we are comparing the average yearly earnings for the degrees, we place the label *Yearly earnings (in dollars)* on the vertical line. Mark intervals with equally spaced notches and label 40,000 to 100,000.
We label the horizontal line with *Highest degree earned* and add the degrees.
We next label a nonshaded and a shaded bar for men and women.
Now, we draw a bar to the appropriate height for each degree.

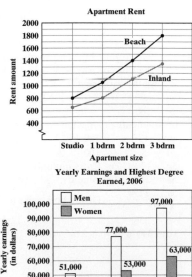

9.2 Student Practice

1. We take the sum of the five test scores and divide the sum by 5.

Sum of test scores → $\dfrac{88 + 77 + 84 + 97 + 89}{5} = \dfrac{435}{5} = 87$
Number of tests →

The mean is 87.

2. Sum of values → $\dfrac{24 + 23 + 25 + 25 + 23 + 26 + 26 + 24}{8} = \dfrac{196}{8} \approx 25$
Number of values →

The mean miles per gallon is 25.

3. We arrange the numbers for Brad's calls in order from smallest to largest:

$\underbrace{3, 3, 4}_{\text{3 numbers}} \quad \underbrace{5}_{\substack{\text{middle} \\ \text{number}}} \quad \underbrace{6, 10, 13}_{\text{3 numbers}}$

There are three numbers smaller than 5 and three numbers larger than 5. Thus 5 is the median.

4. $\underbrace{88, 90}_{\text{2 numbers}} \quad \underbrace{100, 105}_{\substack{\text{2 middle} \\ \text{numbers}}} \quad \underbrace{118, 126}_{\text{2 numbers}}$

The average (mean) of 100 and 105 is $\dfrac{100 + 105}{2} = \dfrac{205}{2} = 102.5$. Thus the median value is 102.5.

5. **(a)** The data 121, 156, 131, 121, 142, 149, and 131 are bimodal since both 121 and 131 occur twice.
   (b) The mode of 99, 76, 79, 92, 76, 84, 83, and 76 is 76 since it occurs three times in the set of data.

9.3 Student Practice

1. (year, number of applicants)
   $\downarrow \qquad \downarrow$

   (2004, 1050)
   (2006, 1100)
   (2008, 1050)
   (2010, 1125)

2. The first number indicates the x-direction.
   $\downarrow$
   $(-3, 2)$
   $\uparrow$
   The second number indicates the y-direction.

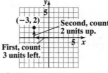

To plot $(-3, 2)$ or $x = -3$, $y = 2$, we start at the origin and move 3 units in the negative x-direction followed by 2 units in the positive y-direction. We end up at the coordinate point $(-3, 2)$ and place a dot there.

3. **(a)** $(-4, 2)$: 4 units left followed by 2 units up.   **(b)** $(2, -4)$: 2 units right followed by 4 units down.
   (c) $\left(0, 3\tfrac{1}{2}\right)$: Start at the origin and move 0 units in the x-direction,   **(d)** $(1, 0)$: 1 unit right followed by 0 units up or
   followed by $3\tfrac{1}{2}$ units in the positive y-direction. The measure of   down. The point is on the x-axis.
   $3\tfrac{1}{2}$ units is located halfway between 3 and 4. The point is on
   the y-axis.

4. **(a)** $S = (-6, 2)$ **(b)** $T = (0, 1)$ **(c)** $U = (3, 0)$ **(d)** $V = (7, 1)$ **(e)** $W = (-6, -2)$ **(f)** $X = (7, -1)$

5. **(a)** $(5, -2)$ **(b)**   **6.**

9.4 Student Practice

1. **(a)** The ordered pair $(-2, 13)$ is a solution if, when we replace x with -2 and y with 13 in the equation $x + y = 11$, we get a true statement.
   $x = -2$ and $y = 13$, or $(-2, 13)$ $x + y = 11$
   $\qquad\qquad\qquad\qquad\qquad -2 + 13 = 11, \quad 11 = 11$ ✓
   (b) There are infinitely many solutions. We can choose any two numbers whose sum is 11.
   $x = -3$ and $y = 14$, or $(-3, 14)$ $x + y = 11$
   $\qquad\qquad\qquad\qquad\qquad -3 + 14 = 11, \quad 11 = 11$ ✓
   $x = 2$ and $y = 9$, or $(2, 9)$ $x + y = 11$
   $\qquad\qquad\qquad\qquad\quad 2 + 9 = 11, \qquad 11 = 11$ ✓
   $x = 5$ and $y = 6$, or $(5, 6)$ $x + y = 11$
   $\qquad\qquad\qquad\qquad\quad 5 + 6 = 11, \qquad 11 = 11$ ✓
   Therefore $(-3, 14)$, $(2, 9)$, and $(5, 6)$ are solutions to $x + y = 11$.

2. **(a)** $\quad y = 25x$
   $125 = 25x$ We replace y with 125.
   $\dfrac{125}{25} = x$ We solve for x.
   $\qquad 5 = x$ It takes the machine 5 minutes to label 125 bottles.
   $\qquad\qquad$ Since $x = 5$ when $y = 125$, the ordered pair is $(5, 125)$.

   (b) $\quad y = 25x$
   $200 = 25x$ We replace y with 200.
   $\dfrac{200}{25} = x$ We solve for x.
   $\qquad 8 = x$ It takes the machine 8 minutes to label 200 bottles.
   $\qquad\qquad$ Since $x = 8$ when $y = 200$, the ordered pair is $(8, 200)$.

3. The first number in the ordered pair is the x-value and the
second number is the y-value.

(a) $(\underline{\quad}, 0)$ $x = ?, y = 0$
We replace y with 0 and solve for x in the equation $x + 3y = 12$.
$$x + 3y = 12$$
$x + 3(0) = 12$ Replace y with 0.
$\quad x + 0 = 12$ Solve for x.
$\qquad x = 12$
$(12, 0)$

(b) $(\underline{\quad}, 2)$ $x = ?, y = 2$
We replace y with 2 and solve for x in the equation $x + 3y = 12$.
$$x + 3y = 12$$
$x + 3(2) = 12$ Replace y with 2.
$\quad x + 6 = 12$ Solve for x.
$\qquad x = 6$
$(6, 2)$
It is a good idea to check your answers.

4. We start by organizing our ordered pairs in a chart. Then we replace x and y with the given values and solve for the unknown values.

$y = x + 2$ $y = x + 2$ $y = x + 2$
$y = 0 + 2$ $8 = x + 2$ $5 = x + 2$
$y = 2$ $\quad 6 = x$ $\quad 3 = x$

We write these values in the appropriate place in the chart. $(0, 2)$, $(6, 8)$, and $(3, 5)$
are solutions to $y = x + 2$. We leave the check to the student.

(x, y)
$(0, 2)$
$(6, 8)$
$(3, 5)$

5. We choose three values for x: 0, 1, and 2, and write these values in a chart. We replace x with the values 0, 1, and 2 in the equation $y = 4x - 5$
and then solve for y.

$y = 4x - 5$ $\quad y = 4x - 5$ $\quad y = 4x - 5$
$\quad = 4(0) - 5$ $\quad = 4(1) - 5$ $\quad = 4(2) - 5$
$\quad = 0 - 5$ $\quad = 4 - 5$ $\quad = 8 - 5$
$y = -5$ $\quad y = -1$ $\quad y = 3$

We write these values in the appropriate place in the chart. $(0, -5)$, $(1, -1)$, and $(2, 3)$ are
solutions to $y = 4x - 5$. Answers may vary since you may start with different values.

(x, y)
$(0, -5)$
$(1, -1)$
$(2, \ \ 3)$

6. (a) We choose three values for x: 0, 1, and 2. We replace x with each of these values and solve for y.

$y = 2x + 1$ $\quad y = 2x + 1$ $\quad y = 2x + 1$
$\quad = 2(0) + 1$ $\quad = 2(1) + 1$ $\quad = 2(2) + 1$
$\quad = 0 + 1$ $\quad = 2 + 1$ $\quad = 4 + 1$
$y = 1$ $\quad y = 3$ $\quad y = 5$

We write these values in the chart.

(b) Then we plot these ordered pairs.

(x, y)
$(0, 1)$
$(1, 3)$
$(2, 5)$

7. (a) We choose three values for x: -1, 0, and 1, and find three ordered pairs that are solutions to $y = -3x - 1$.

$y = -3x - 1$ $\quad y = -3x - 1$ $\quad y = -3x - 1$
$\quad = -3(-1) - 1$ $\quad = -3(0) - 1$ $\quad = -3(1) - 1$
$\quad = 3 - 1$ $\quad = 0 - 1$ $\quad = -3 - 1$
$y = 2$ $\quad y = -1$ $\quad y = -4$

(b) Then we plot these ordered pairs and draw a straight
line through the points.

(x, y)
$(-1, 2)$
$(0, -1)$
$(1, -4)$

8. A solution to $y = 4$ is any ordered pair that has y-coordinate 4. The x-coordinate can be any
number, as long as y is 4.
The ordered pairs $(-5, 4)$, $(0, 4)$, and $(3, 4)$ are solutions to $y = 4$ since all the y-values are 4.
We plot these ordered pairs. Recall that if all the y-values of a set of ordered pairs are
equal to 4, these coordinate points lie on a horizontal line at $y = 4$.

9. A solution to $x = -2$ is any ordered pair that has x-coordinate -2. The y-coordinate can be
any number, as long as x is -2. $(-2, -2)$, $(-2, 3)$, and $(-2, 0)$ are solutions to $x = -2$, since
all the x-values are -2. As we saw earlier, when all the x-values of a set of ordered pairs are
equal to -2, the coordinate points lie on a vertical line at $x = -2$.

Chapter 10   10.1 Student Practice

1. We write the relationship between minutes and hours as a unit
fraction. Since 60 min = 1 hr, we have the unit fraction $\dfrac{1 \text{ hr}}{60 \text{ min}}$.

420 min = ? hr

$420 \text{ min} \times \dfrac{1 \text{ hr}}{60 \text{ min}}$ Multiply by the unit fraction.

$= 420 \text{ m\kern-0.4em\diagdown in} \times \dfrac{1 \text{ hr}}{60 \text{ m\kern-0.4em\diagdown in}}$ Divide out the units "min."

$= 420 \times \dfrac{1}{60} \text{ hr} = 7 \text{ hr}$ Multiply.

2. We write the relationship between ounces and pounds: 1 lb = 16 oz.
We want to end up with pounds, so we write 1 lb in the numerator of
the unit fraction: $\dfrac{1 \text{ lb}}{16 \text{ oz}}$.

144 oz = ? lb

$144 \text{ oz} \times \dfrac{1 \text{ lb}}{16 \text{ oz}}$ We multiply by the appropriate fraction.

$= 144 \text{ o\kern-0.4em\diagdown z} \times \dfrac{1 \text{ lb}}{16 \text{ o\kern-0.4em\diagdown z}}$ We divide out the units "oz."

$= 144 \times \dfrac{1}{16} \text{ lb} = \dfrac{144 \text{ lb}}{16} = 9 \text{ lb}$

3. We must change hours to minutes.

$$7.2 \text{ hr} \times \frac{60 \text{ min}}{1 \text{ hr}} = 7.2 \times 60 \text{ min} = 432 \text{ min}$$

4. *Understand the problem.* We organize the information in a Mathematics Blueprint for Problem Solving.

Mathematics Blueprint for Problem Solving

Gather the Facts	What Am I Asked to Do?	How Do I Proceed?	Key Points to Remember
The charge for the parking is $1.50 per hour. The car was in the garage $1\frac{3}{4}$ days.	Find the total parking charge for $1\frac{3}{4}$ days.	1. Change $1\frac{3}{4}$ days to hours to find the total number of hours the car was in the garage. 2. Multiply the total hours by $1.50 to find the cost for parking.	Change the number of days to a decimal so calculations are easier.

Calculate and state the answer.

Step 1. $1\frac{3}{4} = 1.75$ days   We change $\frac{3}{4}$ to a decimal: $3 \div 4 = 0.75$.

$$1.75 \text{ days} \times \frac{24 \text{ hr}}{1 \text{ day}} = 42 \text{ hr} \text{ We change days to hours.}$$

Step 2. Now we must find the total charge for parking.

$$42 \text{ hr} \times \frac{\$1.50}{1 \text{ hr}} = \$63$$

Check: Is our answer in the desired units? Yes, the answer is in dollars, and we would expect it to be in dollars. ✓

5. **(a)** To go from meters to centimeters, we move *two places to the right on the prefix chart,* so we move the decimal point two places to the right.
   $4 \text{ m} = 4.00 \text{ m} = 400 \text{ cm}$

   (b) To go from centigrams to milligrams, we move *one place to the right on the prefix chart,* so we move the decimal point one place to the right.
   $30 \text{ cg} = 30.0 \text{ cg} = 300 \text{ mg}$

6. **(a)** To go from milliliters to liters, we move *three places to the left on the prefix chart,* so we move the decimal point three places to the left.
   $3 \text{ mL} = 0.003 \text{ L}$

   (b) To go from centimeters to kilometers, we move *five places to the left on the prefix chart,* so we move the decimal point five places to the left.
   $47 \text{ cm} = 0.00047 \text{ km}$

7. **(a)** We are converting the smaller unit, mm, to the larger one, dam. Therefore, there will be fewer dekameters than millimeters. We move the decimal point four places to the left. $389 \text{ mm} = 389.0 \text{ mm} = 0.0389 \text{ dam}$

   (b) We are converting the larger unit, hm, to the smaller one, cm. Therefore, there will be more centimeters than hectometers. We move the decimal point four places to the right. $0.48 \text{ hm} = 4800 \text{ cm}$

8. *Understand the problem.* We organize the information in a Mathematics Blueprint for Problem Solving.

Mathematics Blueprint for Problem Solving

Gather the Facts	What Am I Asked to Do?	How Do I Proceed?	Key Points to Remember
The acid costs $110 per liter.	Find out how much 1 milliliter of acid costs.	1. Change 1 liter to milliliters. 2. Find the unit cost per milliliter.	To go from liters to milliliters, we move 3 places to the right on the prefix chart. Thus we move the decimal point 3 places to the right.

Calculate and state the answer.

Step 1. $1 \text{ L} = 1000 \text{ mL}$ Change liters to milliliters.

Step 2. We write $110 per liter as a rate.

$$\frac{\$110}{1 \text{ L}} = \frac{\$110}{1000 \text{ mL}} \text{ Replace 1 L with 1000 mL.}$$

$$= \$0.11 \quad 110 \div 1000 = 0.11$$

Check: A milliliter is a very small part of a liter. Therefore it should cost much less for 1 milliliter of acid than it does for 1 liter. $0.11 is much less than $110, so our answer seems reasonable.

10.2 Student Practice

1. **(a)** $17 \text{ m} \times \dfrac{1.09 \text{ yd}}{1 \text{ m}} = 18.53 \text{ yd}$ **(b)** $29.6 \text{ km} \times \dfrac{0.62 \text{ mi}}{1 \text{ km}} = 18.352 \text{ mi}$ **(c)** $26 \text{ gal} \times \dfrac{3.79 \text{ L}}{1 \text{ gal}} = 98.54 \text{ L}$

   (d) $6.2 \text{ L} \times \dfrac{1.06 \text{ qt}}{1 \text{ L}} = 6.572 \text{ qt}$ **(e)** $16 \text{ lb} \times \dfrac{0.454 \text{ kg}}{1 \text{ lb}} = 7.264 \text{ kg}$ **(f)** $280 \text{ g} \times \dfrac{0.0353 \text{ oz}}{1 \text{ g}} = 9.884 \text{ oz}$

2. Our first unit fraction converts centimeters to inches. Our second unit fraction converts inches to feet.

   $$180 \text{ cm} \times \frac{0.394 \text{ in.}}{1 \text{ cm}} \times \frac{1 \text{ ft}}{12 \text{ in.}} = \frac{70.92}{12} \text{ ft} = 5.91 \text{ ft}$$

3. We multiply by the unit fraction that relates mi to km.

   $$\frac{88 \text{ km}}{\text{hr}} \times \frac{0.62 \text{ mi}}{1 \text{ km}} = 54.56 \text{ mi/hr}$$

   Thus 88 km/hr is approximately equal to 54.56 mi/hr.

4. We first convert from millimeters to centimeters by moving the decimal point in the number 9 one place to the left. 9 mm $= 0.9$ cm Then we convert to inches using a unit fraction.

$0.9 \text{ cm} \times \dfrac{0.394 \text{ in.}}{1 \text{ cm}} = 0.3546 \text{ in.} \approx 0.35 \text{ in.}$ (rounded to the nearest hundredth)

5. We use the formula that gives us Celsius degrees.

$C = \dfrac{5 \times F - 160}{9}$

$\quad = \dfrac{5 \times 181 - 160}{9}$ We multiply, then subtract.

$\quad = \dfrac{905 - 160}{9} = \dfrac{745}{9} \approx 82.778$

The temperature is about 83°C.

6. We want to convert the Celsius temperature to Fahrenheit so we use the following formula.

$F = 1.8 \times C + 32$

$\quad = 1.8 \times 4 + 32$ We replace C with the Celsius temperature.

$\quad = 7.2 + 32$ We multiply before we add.

$\quad = 39.2$

It is 39.2°F in England.

Now we find the difference in Fahrenheit temperatures.

$79 - 39.2 = 39.8$°F

The difference in temperatures is 39.8°F.

10.3 Student Practice

1. (a) We use just the vertex to name the angle: $\angle H$.

(b) We can name the acute angle as $\angle ABC$, $\angle B$, or $\angle CBA$. Be sure the letter representing the vertex is the middle letter.

2. We can name $\angle n$ as $\angle YXW$ or $\angle WXY$.

3. If we let $\angle C =$ the complement of $\angle B$, then we have the following.

The complement of $\angle B$ plus $\angle B = 90°$ First, we write a statement to represent the situation.

$\qquad\quad \downarrow \qquad\qquad \downarrow \quad \downarrow \quad \downarrow$

$\qquad \angle C \qquad\quad + \ \ \angle B = 90°$ Next, we translate to symbols.

$\angle C + 22° = 90°$ Then we replace $\angle B$ with 22°.

$\underline{+ \quad - 22° - 22°}$

$\qquad \angle C = 68°$ Finally, we solve for $\angle C$.

The complement of $\angle B$ is 68°.

4. A right angle has a measure of 90°. Therefore $\angle XYW$ and $\angle WYZ$ are complementary, and the sum of their measures is 90°.

$y + (y + 40°) = 90°$ We must solve for y.

$(y + y) + 40° = 90°$ We regroup terms.

$\quad 2y + 40° = 90°$ $y + y = 2y$

$\underline{+ \quad - 40° - 40°}$ We add −40 to both sides.

$\qquad\quad 2y = 50°$

$\qquad\quad \dfrac{2y}{2} = \dfrac{50°}{2}$ We divide by 2 on both sides.

$\qquad\quad y = 25°$

5. (a) Since $\angle x$ and $\angle z$ share a common side with $\angle y$, they are adjacent to $\angle y$.

(b) $\angle w$ and $\angle z$ are adjacent angles of intersecting lines. Therefore, they are supplementary angles and the sum of their measures is 180°.

6. (a) Since $\angle d$ and $\angle a$ are vertical angles, we know that they have the same measure. Thus we know that $\angle d$ measures 50°.

(b) Since $\angle c$ and $\angle a$ are adjacent angles of intersecting lines, we know that they are supplementary angles and the sum of their measures is 180°.

$\angle c + \angle a = 180°$

$\angle c + 50° = 180°$

$\underline{+ \quad - 50° \quad -50°}$

$\qquad\quad \angle c = 130°$

(c) $\angle b = 130°$ since $\angle b$ and $\angle c$ are vertical angles.

7. $\angle a = \angle e = 63°$ The corresponding angles a and e have the same measure.

$\angle e = \angle c = 63°$ The vertical angles e and c have the same measure.

To find the measure of $\angle b$, we list some facts about $\angle b$.

$\angle b$ is the supplement of $\angle a$ and $\angle a = 63°$.

$\angle b + \angle a = 180°$ The sum of supplementary angles is 180°.

$\angle b + 63° = 180°$ We substitute 63° for $\angle a$.

$\underline{+ \quad - 63° \quad -63°}$ We solve for $\angle b$.

$\qquad \angle b = 117°$

$\qquad \angle b = \angle d = 117°$ The corresponding angles b and d have the same measure.

10.4 Student Practice

1. (a) 81 is a perfect square because $9^2 = 9 \cdot 9 = 81$.

(b) 17 is not a perfect square. There is no whole number or fraction that when squared equals 17.

(c) $\frac{1}{9}$ is a perfect square because $\left(\frac{1}{3}\right)^2 = \frac{1}{3} \cdot \frac{1}{3} = \frac{1}{9}$.

3. (a) $\sqrt{25} = n$ What positive number times itself equals 25?

$\sqrt{25} = 5$ 5, since $5 \cdot 5 = 25$.

(b) $\sqrt{64} = 8$ and $\sqrt{16} = 4$

$\sqrt{64} - \sqrt{16} = 8 - 4 = 4$

2. (a) $n^2 = 81$ $n = \sqrt{81}$

(b) What is the positive square root of 81? $n = \sqrt{81}$

(c) $n \cdot n = 81$ $n = \sqrt{81}$

4. (a) $\sqrt{\dfrac{81}{100}} = n$ What positive fraction multiplied by itself equals $\dfrac{81}{100}$?

$\sqrt{\dfrac{81}{100}} = \dfrac{9}{10}$ $\dfrac{9}{10}$, since $\left(\dfrac{9}{10}\right)\left(\dfrac{9}{10}\right) = \dfrac{81}{100}$.

(b) $\sqrt{\dfrac{16}{49}} = n$ What positive fraction multiplied by itself equals $\dfrac{16}{49}$?

$\sqrt{\dfrac{16}{49}} = \dfrac{4}{7}$ $\dfrac{4}{7}$, since $\left(\dfrac{4}{7}\right)\left(\dfrac{4}{7}\right) = \dfrac{16}{49}$.

5. Write the formula for area.
   $A = s^2$ or $s \cdot s$
   $49 = s^2$ What positive number squared equals 49?
   $\sqrt{49} = s$ Write using the radical sign.
   $7 = s$ The positive square root of 49 is 7 feet.
   The length of the side of the square is 7 feet.

7. We write the formula.
   $(\text{hypotenuse})^2 = (\text{leg})^2 + (\text{other leg})^2$

   $$c^2 = a^2 + b^2$$
   $$= 4^2 + 3^2 \qquad \text{We evaluate the formula.}$$
   $$= 16 + 9$$
   $$c^2 = 25$$
   $$c = \sqrt{25} = 5$$

   Thus the length of the hypotenuse of the right triangle is 5 meters.

6. (a) $\sqrt{100} = 10$ Use the calculator: $\sqrt{100} = 10$.
   (b) $\sqrt{23} \approx 4.79583$ Use a calculator to find $\sqrt{23}$.
       ≈ 4.796 Round to the nearest thousandth.

8. We find the sum of the area of the square and the area of the triangle.
   The area of the square is $A = s^2 = (3\text{ ft})^2 = 9\text{ ft}^2$
   Since all sides of the square are equal, the base of the triangle is 3 feet.
   We must use the Pythagorean Theorem to find the height of the right triangle.
   $$c^2 = a^2 + b^2$$
   $$5^2 = a^2 + 3^2$$
   $$25 = a^2 + 9$$
   $$\underline{-9 \qquad\quad -9}$$
   $$16 = a^2$$
   $$4 = a \qquad \text{The height of the triangle is 4 ft.}$$
   Now we find the area of the triangle. The area of the triangle is
   $$A = \frac{bh}{2} = \frac{(3\text{ ft})(4\text{ ft})}{2} = 6\text{ ft}^2$$
   The sum of the areas is $9\text{ ft}^2 + 6\text{ ft}^2 = 15\text{ ft}^2$.

9. *Understand the problem.* We can visualize the situation and organize the facts by drawing a picture.
   Calculate and state the answer. We can see from the picture that we must find the length of the hypotenuse of a right triangle.
   $c^2 = a^2 + b^2$ We write the formula.
   $c^2 = 15^2 + 12^2$ We replace *a* with 15 and *b* with 12.
   $c^2 = 225 + 144$ Now we solve for *c*.
   $c^2 = 369$
   $c = \sqrt{369}$
   $c \approx 19.209$
   The length of the ladder is approximately 19.2 feet.

10.5 Student Practice

1. Since we are given the diameter, we use the formula for circumference that includes the diameter.
   $C = \pi d$ Write the formula.
   $= (3.14)(5.1\text{ m})$ Then substitute the values given.
   $C \approx 16.014\text{ m}$

2. *Understand the problem.* We make sketches similar to Example 2 and/or write out the facts as clearly as possible.
   The larger wheel makes 2 complete revolutions, so the belt moves a distance of $C + C = 2C$.
   The smaller wheel makes many more revolutions because its circumference C is smaller.

   Calculate and state the answer.

   First find the distance the belt moves.
   $C = 2\pi r$ Write out the formula.
   $= 2(3.14)(3.75\text{ ft})$ Then substitute the values given.
   $C = 23.55\text{ ft.}$
   $2C = 47.1\text{ ft.}$ The larger wheel rotates 47.1 ft in 2 revolutions.
   Second, find how many times this makes the smaller wheel revolve. (How many times does the smaller circumference divide into 47.1 feet?)
   $C = 2\pi r$ Find the circumference of the smaller wheel.
   $= 2(3.14)(1.25\text{ ft})$ Substitute the values given.
   $C = 7.85\text{ ft.}$
   $\dfrac{47.1}{7.85} = 6$ Divide.
   The smaller wheel makes 6 complete revolutions.
   Check: Use your calculator to check.

3. $A = \pi r^2$
   $= (3.14)(6.1\text{ cm})^2$ Square the radius first.
   $= (3.14)(37.21\text{ cm}^2)$ Then we multiply by 3.14.
   $A \approx 116.8394\text{ cm}^2$
   The area of the circle is approximately 116.8 cm^2.

4. *Understand the problem.* We organize the information in a Mathematics Blueprint for Problem Solving.

Mathematics Blueprint for Problem Solving

Gather the Facts	What Am I Asked to Do?	How Do I Proceed?	Key Points to Remember
$d = 60$ in. Cost of tablecloth is $8 per square foot.	Find the cost of the tablecloth.	**1.** Find the radius. **2.** Find the area. **3.** Change area to square feet. **4.** Multiply area by $8 to find the cost of the tablecloth.	$A = \pi r^2$ $r = \dfrac{d}{2}$ 1 ft = 12 in.

Calculate and state the answer.

1. The diameter is 60 inches, so the radius is 30 inches.

$$r = \frac{60}{2} = 30 \text{ in.}$$

3. Change square inches to square feet. Since 1 ft = 12 in., $(1 \text{ ft})^2 = (12 \text{ in.})^2$. That is, $1 \text{ ft}^2 = 144 \text{ in.}^2$.

$$2826 \text{ in.}^2 \times \frac{1 \text{ ft}^2}{144 \text{ in.}^2} = 19.625 \text{ ft}^2$$

2. Find the area.
$$A = \pi r^2 = (3.14)(30 \text{ in.})^2$$
$$= (3.14)(900 \text{ in.}^2) \quad \text{We square the radius first.}$$
$$A \approx 2826 \text{ in.}^2 \qquad \text{Then we multiply.}$$

4. Find the cost. $\dfrac{\$8}{1 \text{ ft}^2} \times 19.625 \text{ ft}^2 = \$157$

Check: You may use your calculator to check.

10.6 Student Practice

1. $V = \pi r^2 h = (3.14)(10.24 \text{ cm})^2(17.78 \text{ cm}) \approx 5854.12 \text{ cm}^3$
The volume of the tin can is approximately 5854.12 cm^3.

2. $V = \dfrac{4\pi r^3}{3} = \dfrac{4(3.14)(7 \text{ in.})^3}{3} \approx 1436.03 \text{ in.}^3$
Approximately 1436.03 cubic inches of air is needed to fully inflate the ball.

3. *Understand the problem for part (a).* We refer to the picture for Example 3. We find the volume of the inner shaded region (V_r) to determine how much coffee the thermos will hold.

Calculate and state the answer to part (a).
$r = 10$ cm $V_r = \pi r^2 h$
$h = 27$ cm $= (3.14)(10 \text{ cm})^2(27 \text{ cm})$
                 $V_r \approx 8478 \text{ cm}^3$
The thermos can hold approximately 8478 cubic centimeters of coffee.
Understand the problem for part (b). To determine the volume of the insulated region, we find the volume of the entire cylinder (V_R) minus the volume of the inner region (V_r): $V_R - V_r$.

Calculate and state the answer to part (b).
First we find the volume of the entire cylinder.
$R = 12$ cm $V_r = \pi r^2 h$
$h = 27$ cm $= (3.14)(12 \text{ cm})^2(27 \text{ cm})$
                 $V_r \approx 12{,}208.32 \text{ cm}^3$
Then we subtract to find the volume of the insulated region.
$$\text{Volume of insulated region} = \underset{\downarrow}{V_R} - \underset{\downarrow}{V_r}$$
$$12{,}208.32 - 8478 = 3730.32$$
The volume of the insulation is approximately 3730.32 cubic centimeters.

4. $V = \dfrac{\pi r^2 h}{3} = \dfrac{(3.14)(5 \text{ m})^2(12 \text{ m})}{3} \approx 314 \text{ m}^3$

5. (a) The base is a square.
area of base $= (6 \text{ m})(6 \text{ m}) = 36 \text{ m}^2$
Substituting the area of the base, 36 m^2, and the height of 10 m, we have
$$V = \frac{Bh}{3} = \frac{(36 \text{ m}^2)(10 \text{ m})}{3}$$
$$= (12)(10) \text{ m}^3$$
$$V = 120 \text{ m}^3$$

(b) The base is a rectangle.
area of base $= (7 \text{ m})(8 \text{ m}) = 56 \text{ m}^2$
Substituting the area of the base, 56 m^2, and the height of 15 m, we have
$$V = \frac{Bh}{3} = \frac{(56 \text{ m}^2)(15 \text{ m})}{3}$$
$$= (56)(5) \text{ m}^3$$
$$V = 280 \text{ m}^3$$

10.7 Student Practice

1. The ratio of 11 to 27 is the same as the ratio of 15 to n.

$$\frac{11}{27} = \frac{15}{n}$$

$11n = (15)(27)$ Cross-multiply.

$11n = 405$ Simplify.

$$\frac{11n}{11} = \frac{405}{11}$$ Divide both sides by 11.

$n = 36.\overline{81}$

≈ 36.8 Round to the nearest tenth.

The length of side n is approximately 36.8 m.

2. *Understand the problem.* We are asked to find the perimeter of the larger triangle.

Calculate and state the answer. We know that the perimeters of similar triangles have the same ratios as the corresponding sides. Therefore, we begin by finding the perimeter of the smaller triangle.

5 yd + 12 yd + 13 yd = 30 yd

We can now write equal ratios. Let P = the unknown perimter.

We will set up our ratios as $\frac{\text{smaller triangle}}{\text{larger triangle}}$.

Remember, once you write the first ratio, be sure to write the terms of the second ratio in the same order. We will write

perimeter of the smaller triangle $\rightarrow \dfrac{30}{P} = \dfrac{12}{40} \leftarrow$ side of the smaller triangle
perimeter of the larger triangle $\rightarrow$ $\leftarrow$ side of the larger triangle

$12P = (30)(40)$

$12P - 1200$

$$\frac{12P}{12} = \frac{1200}{12}$$

$P = 100$

The perimeter of the larger triangle is 100 yards.

Check: Are the ratios the same? Simplify each ratio to check.

$$\frac{30}{100} \overset{?}{=} \frac{12}{40}$$ Simplify.

$$\frac{3 \cdot \cancel{10}}{10 \cdot \cancel{10}} \overset{?}{=} \frac{3 \cdot \cancel{4}}{10 \cdot \cancel{4}}$$

$$\frac{3}{10} \overset{?}{=} \frac{3}{10} \;\checkmark$$

3. *Understand the problem.* The shadows cast by the sun shining on vertical objects at the same time of the day form similar triangles.

Calculate and state the answer.

Let h = the height of the building. Thus we can say that h is to 5 as 20 is to 2.

$$\frac{h}{5} = \frac{20}{2}$$

$2h = (20)(5)$

$2h = 100$

$$\frac{2h}{2} = \frac{100}{2}$$

$h = 50$

The building is 50 feet tall.

Check: The check is up to you.

4. Let W = the width of the larger rectangle.

$$\frac{W}{1.8} = \frac{29}{3}$$

$3W = (29)(1.8)$

$3W = 52.2$

$$\frac{3W}{3} = \frac{52.2}{3}$$

$W = 17.4$

The width W of the larger rectangle is 17.4 m.

Appendix A.1 Balancing a Checking Account   Student Practice

1.

CHECK NO.	DATE	DESCRIPTION OF TRANSACTION	PAYMENT/ DEBIT (–)	✓	DEPOSIT/ CREDIT (+)	BALANCE $ 1434 52
		CHECK REGISTER *My Chung Nguyen*				20 *11*
144	3/1	Leland Mortgage Company	908 00			526 52
145	3/1	Phone Company	33 21			493 31
146	3/2	Sam's Food Market	102 37			390 94
	3/2	Deposit			524 41	915 35

To find the ending balance we subtract each check written and add the deposit to the current balance. Then we record these amounts in the check register.

$$\begin{array}{r} 1434.52 \\ -\ 908.00 \\ \hline 526.52 \end{array} \quad \begin{array}{r} 526.52 \\ -\ 33.21 \\ \hline 493.31 \end{array} \quad \begin{array}{r} 493.31 \\ -\ 102.37 \\ \hline 390.94 \end{array} \quad \begin{array}{r} 390.94 \\ +\ 524.41 \\ \hline 915.35 \end{array}$$

My Chung's balance is $915.35.

2.

CHECKING RECONCILEMENT	This form is provided to assist you in balancing your checking account.

List checks outstanding* not charged to your checking account			Period ending	7/31 ,20 11
CHECK NO.	AMOUNT		**1.** Check Register Balance	$ 1050.46
215	89 75		**Subtract** any charges listed on the bank statement which you have not previously deducted from your balance.	− $ 12.75
217	205 99		Adjusted Check Register Balance	$ 1037.71
			2. Enter the ending balance shown on the bank statement.	$ 934.95
			3. Enter deposits made later than the ending date on the bank statement.	+ $ 398.50
				+ $
				+ $
			TOTAL (Step 2 plus Step 3)	$ 1333.45
			4. In your check register, **check off** all the checks paid. In the area provided to the left, **list** numbers and amounts of all outstanding checks and ATM withdrawals.	
TOTAL	295 74		**5. Subtract** the total amount in Step 4.	− $ 295.74
* and ATM withdrawals			**6.** This adjusted bank balance should equal the adjusted Check Register Balance from Step 1.	$ 1037.71

The balances in steps **1** and **6** are equal, so Anthony's checkbook is balanced.

Appendix A.2 Determining the Best Deal When Purchasing a Vehicle

Student Practice

1. (a) sales tax = 7% of sale price
$= 0.07 \times 24{,}999$
sales tax = $1749.93
license fee = 2% of sale price
$= 0.02 \times 24{,}999$
license fee = $499.98

(b) purchase price = sale price + sales tax + license fee + extended warranty
$= 24{,}999 + 1749.93 + 499.98 + 1275$
purchase price = $28,523.91

2. (a) down payment = percent × purchase price.
down payment = 15% × 32,499
$= 0.15 \times 32{,}499$
down payment = $4874.85

(b) amount financed = purchase price − down payment
$= 32{,}499 - 4874.85$
amount financed = $27,624.15

3. First, we find the total cost of the minivan at Dealership 1.
total cost = (monthly payment × number of months in loan) + down payment
$= (453.21 \times 60) + 0$ There is no down payment.
$= 27{,}192.60$
The total cost of the minivan at Dealership 1 is $27,192.60.
Next, we find the cost of the minivan at Dealership 2.
total cost = (monthly payment × number of months in loan) + down payment
$= (479.17 \times 48) + 3000$ We multiply, then add.
$= 26{,}000.16$
The total cost of the minivan at Dealership 2 is $26,000.16.
We see that the best deal on the minivan Phoebe plans to buy is at Dealership 2.

Appendix B   Student Practice

1. (a) 60 seconds has the same measure of time as 1 minute: 60 sec = 1 min.
(b) 1 gallon has the same volume as 4 quarts: 1 gal = 4 qt.

2. (a) There are 2 pints in 1 quart.
4 pt = 2 pt + 2 pt
$= 1 \text{ qt} + 1 \text{ qt}$
$= 2 \text{ qt}$

(b) There are 2000 pounds in 1 ton.
2 tons = 1 ton + 1 ton
$= 2000 \text{ lb} + 2000 \text{ lb}$
$= 4000 \text{ lb}$

3. (a)   12 min 43 sec
+ 4 min 33 sec
16 min 76 sec   76 sec = 60 sec + 16 sec
16 min 1 min 16 sec
17 min 16 sec

(b)   4 gal 1 qt   4 gal = 3 gal + 1 gal = 3 gal + 4 qt
−2 gal 3 qt

4 gal 1 qt   3 gal + 4 qt   1 qt   3 gal 5 qt
−2 gal 3 qt   −2 gal   3 qt   −2 gal 3 qt
                                   1 gal 2 qt

4. The prefix *centi* is located to the right of meter, so 1 centimeter is smaller than 1 meter.

5. (a) Since 1 mm is much smaller than 1 in., we have 1 mm < 1 in.
(b) Since 1 m is a little longer than 1 yd, we have 4 m > 4 yd.

6. (a) Since 1 liter is more than 1 quart, which is more than 1 cup, we have 1 L > 1 cup.
(b) Since 1 kilogram weighs more than 2 pounds, we have 1 kg > 1 lb.

7. Since neither meters nor grams measure volume, we do not choose 1 m or 1 g. Since 1 liter is a little more than 1 quart, the most appropriate choice is 1 L.

Appendix C   Student Practice

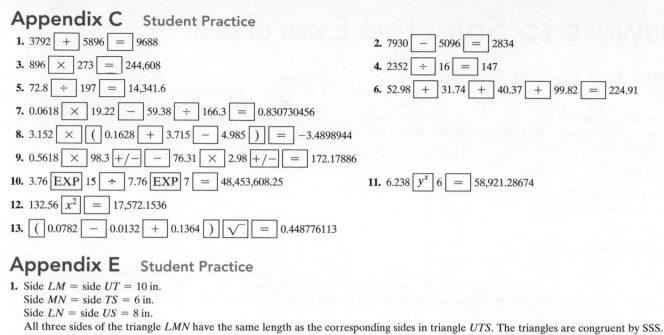

1. 3792 $\boxed{+}$ 5896 $\boxed{=}$ 9688

2. 7930 $\boxed{-}$ 5096 $\boxed{=}$ 2834

3. 896 $\boxed{\times}$ 273 $\boxed{=}$ 244,608

4. 2352 $\boxed{\div}$ 16 $\boxed{=}$ 147

5. 72.8 $\boxed{\div}$ 197 $\boxed{=}$ 14,341.6

6. 52.98 $\boxed{+}$ 31.74 $\boxed{+}$ 40.37 $\boxed{+}$ 99.82 $\boxed{=}$ 224.91

7. 0.0618 $\boxed{\times}$ 19.22 $\boxed{-}$ 59.38 $\boxed{\div}$ 166.3 $\boxed{=}$ 0.830730456

8. 3.152 $\boxed{\times}$ $\boxed{(}$ 0.1628 $\boxed{+}$ 3.715 $\boxed{-}$ 4.985 $\boxed{)}$ $\boxed{=}$ −3.4898944

9. 0.5618 $\boxed{\times}$ 98.3 $\boxed{+/-}$ $\boxed{-}$ 76.31 $\boxed{\times}$ 2.98 $\boxed{+/-}$ $\boxed{=}$ 172.17886

10. 3.76 $\boxed{\text{EXP}}$ 15 $\boxed{\div}$ 7.76 $\boxed{\text{EXP}}$ 7 $\boxed{=}$ 48,453,608.25

11. 6.238 $\boxed{y^x}$ 6 $\boxed{=}$ 58,921.28674

12. 132.56 $\boxed{x^2}$ $\boxed{=}$ 17,572.1536

13. $\boxed{(}$ 0.0782 $\boxed{-}$ 0.0132 $\boxed{+}$ 0.1364 $\boxed{)}$ $\boxed{\sqrt{}}$ $\boxed{=}$ 0.448776113

Appendix E   Student Practice

1. Side LM = side UT = 10 in.
   Side MN = side TS = 6 in.
   Side LN = side US = 8 in.
   All three sides of the triangle LMN have the same length as the corresponding sides in triangle UTS. The triangles are congruent by SSS.
2. Side AB = side OM = 14 cm.
   Side AC = side ON = 8 cm.
   $\angle A = \angle O = 120°$.
   Two sides of triangle ABC are equal in length to two sides of triangle OMN, and the angles formed by these sides are equal in measure. The triangles are congruent by SAS.
3. Side TU = side CA = 10 m.
   $\angle T = \angle C = 80°$.
   $\angle U = \angle A = 35°$.
   Two angles of triangle STU are equal in measure to two angles of triangle BCA, and the sides between these angles are equal in length. The triangles are congruent by ASA.
4. (a) We see from the information given that the two triangles are congruent by ASA.
   (b) Although two sides and one angle of triangle MNO are equal to two sides and one angle of triangle CBA, the equal angles are *not* the angles formed by the equal sides. Therefore, the SAS rule does not apply to the given information. We are also not given enough information to determine if the SSS or ASA rules apply. Therefore, we cannot determine whether these two triangles are congruent.

Answers to Selected Exercises

Chapter 1   **1.1 Exercises**

1. (a) Eight thousand two   **(b)** Eight hundred two   **(c)** Eighty-two   **(d)** One
3. (a) Hundreds   **(b)** Ones   **5. (a)** Thousands   **(b)** Hundred thousands   **7. (a)** Millions   **(b)** Hundreds   **9.** $5000 + 800 + 70 + 6$
11. $4000 + 900 + 20 + 1$ **13.** $800,000 + 60,000 + 7000 + 300 + 1$ **15.** 5 hundred-dollar bills, 6 ten-dollar bills, and 2 one-dollar bills
17. Answers may vary.   **(a)** 4 ten-dollar bills and 6 one-dollar bills   **(b)** 4 ten-dollar bills, 1 five-dollar bill, and 1 one-dollar bill
19. Six thousand, seventy-nine   **21.** Eighty-six thousand, four hundred ninety-one
23.

James Hunt 4 Platt St. Mapleville, RI 02839		2824
	DATE _____	
PAY TO THE ORDER OF *Hampton Apartments*		$ 672.00
Six hundred seventy-two and $^{00}/100$		DOLLARS
Mason Bank California		
MEMO _____	_____	
⑆5800520⑆ 55202205⑆ 2824		

25. $<$ **27.** $<$ **29.** $>$ **31.** $>$ **33.** $>$ **35.** $<$ **37.** $5 > 2$ **39.** $2 < 5$ **41.** 50   **43.** 660   **45.** 63,900
47. 823,000   **49.** 38,000   **51.** 144,000   **53.** 5,300,000   **55.** 9,000,000   **57.** 870,000 mi   **59.** $>$ **61.** $27,000   **63.** 17,000
65. Five trillion, three hundred eleven billion, one hundred ninety-two million, eight hundred nine thousand   **67.** 4 hr   **69.** 123 ft

Quick Quiz 1.1 *See Examples noted with Ex.*   **1.** $6000 + 400 + 2$ (Ex. 2)   **2. (a)** $<$ **(b)** $>$ (Ex. 5)   **3. (a)** 150,000 (Ex. 7)
(b) 154,600 (Ex. 8)   **4.** See Student Solutions Manual

Understanding the Concept: Addition Facts Made Simple   **1.** $8 + 5 = (3 + 5) + 5 = 3 + (5 + 5) = 3 + 10 = 13$
2. $6 + 8 = 6 + (6 + 2) = (6 + 6) + 2 = 12 + 2 = 14$

Understanding the Concept: Using Inductive Reasoning to Reach a Conclusion   **1.** 44   **2.** 72

1.2 Exercises

1. Ten plus a number (answers may vary)   **3.** Replace x with 9, and then add 9 and 6.   **5.** Associative property of
addition   **7.** $m + 2$ **9.** $5 + y$ or $y + 5$ **11.** $12 + m$ or $m + 12$ **13.** $m + 7$ **15.** $a + 5$ **17.** $x + 3$ **19.** 3758   **21.** 12
23. $x + 6$ **25.** $n + 12$ **27.** $x + 2$ **29.** $x + 3$ **31.** $12 + n$ **33.** $n + 11$ **35.** $x + 15$ **37.** $n + 7$ **39.** $x + 9$
41. $n + 9$ **43.** $a + 13$ **45.** $x + 16$ **47. (a)** 10   **(b)** 15   **49.** 19   **51.** 36   **53.** 83   **55. (a)** $442   **(b)** $435   **57.** 38
59. 279   **61.** 70   **63.** 344   **65.** 729   **67.** 884   **69.** 8442   **71.** 929,217   **73.** 391,850   **75. (a)** $400   **(b)** $228   **77.** $2309
79. 36 in.   **81.** 12 ft   **83.** 14 in.   **85.** 58 ft   **87.** 1120 in.   **89.** 15   **91.** 30   **93.** 52

Quick Quiz 1.2 *See Examples noted with Ex.*   **1. (a)** $a + 13$ **(b)** $x + 10$ or $10 + x$ (Ex. 7)   **2.** 46 (Ex. 8)
3. 490 in. (Ex. 12)   **4.** See Student Solutions Manual

Understanding the Concept: Money and Borrowing   **1.** We can borrow only from a place value that has a nonzero whole
number. For example, in $400 there are only hundred-dollar bills to break down (borrow from).   **2.** When we change the ten-dollar bill to
10 one-dollar bills, we have 0 ten-dollar bills and 10 one-dollar bills. This is similar to borrowing: $2\,\overset{0\ 10}{\cancel{1}\,\cancel{0}}$.

1.3 Exercises

1. Six minus x; Answers may vary   **3.** Subtraction   **5.** 3   **7.** 4   **9.** 6   **11.** 1   **13.** 15   **15.** 0   **17.** 97
19. 95   **21.** $9 - 2$ **23.** $8 - y$ **25.** $17 - 10$ **27.** $n - 1$ **29.** $a - 2$ **31.** 5   **33.** 0   **35.** 7   **37.** 1   **39.** 62   **41.** 33
43. 16   **45.** 54   **47.** 678   **49.** 457   **51.** 5065   **53.** 5116   **55.** 8679   **57.** 105,377   **59.** 36 ft   **61.** 84 ft   **63.** $1126, $989,
$920, $822, $453   **65.** 15 mph faster   **67.** 105 ft less   **69.** 862,840 mi   **71.** When the values of x and y are equal   **73.** $>$ **74.** $>$
75. 360 hr   **76.** $508

Quick Quiz 1.3 *See Examples noted with Ex.*   **1. (a)** $n - 5$ **(b)** $n - 7$ **(c)** $n - 8$ (Ex. 3)   **2. (a)** 6779   **(b)** 408,795 (Ex. 7)   **3.** $410 (Ex. 8)
4. See Student Solutions Manual

Understanding the Concept: Memorizing Multiplication Facts   **1. (a)** $2(7) + 7 = 21$ **(b)** $5(8) - 8 = 32$
(c) $5(8) + 8 = 48$ **(d)** $10(8) - 8 = 72$

1.4 Exercises

1. (a) Four times a number   **(b)** The product of a and b **3.** ★★★ ★★★ ★★★ Shapes may vary ★★ ★★ ★★
5. Associative property of multiplication   **7.** 2, 24   **9.** a, a, 24

11. (a)

	White	Pink	Blue
Brown	Brown; White	Brown; Pink	Brown; Blue
Black	Black; White	Black; Pink	Black; Blue
Gray	Gray; White	Gray; Pink	Gray; Blue
Dark Blue	Dark blue; White	Dark blue; Pink	Dark blue; Blue

(b) 12 different outfits

13. 40 **15.** 6, 3: factors; 18: product **17.** 22, x: factors; 88: product **19.** $7x$ **21.** $3x$ **23.** $6x$ **25.** 0 **27.** 40 **29.** 180 **31.** 240
33. 0 **35.** 160 **37.** $48b$ **39.** $40z$ **41.** $56a$ **43.** $14c$ **45.** $90x$ **47.** 0 **49.** $18b$ **51.** $30y$ **53.** $126x$ **55.** $90y$
57. 5733 **59.** 4214 **61.** 119,400 **63.** 475,800 **65.** 5168 **67.** 1888 **69.** 47,432 **71.** 39,130 **73.** 248,508 **75.** 176,688
77. 7,674,728 **79.** 2,516,022 **81.** 9,782,456 **83.** 61,711,000 **85.** $320 **87.** 375 trees **89.** $3924 **91.** Yes, because there are
200 rooms and 250 sets of curtains **93. (a)** 30°F **(b)** 74°F **95.** $40abc$ **97.** $240abc$ **99.** $28xyz$
101. (a) (b) (c) (d)

	0	1	2	3	4	5	6	7	8	9
0	0	0	0	0	0	0	0	0	0	0
1	0	1	2	3	4	5	6	7	8	9
2	0	2	4	6	8	10	12	14	16	18
3	0	3	6			15				
4	0	4	8			20				
5	0	5	10	15	20	25	30	35	40	45
6	0	6	12			30				
7	0	7	14			35				
8	0	8	16			40				
9	0	9	18			45				

(e) 36 **(f)** There are only a few blank spaces left on the table, so there are not many multiplication facts to learn.

102. 428,990 **103.** 6858 **104.** 827,000 **105.** 168,410,000 **106.** $23 less **107.** 465 mi

Quick Quiz 1.4 *See Examples noted with Ex.* **1.** $6n$ (Ex. 4) **2.** 169,050 (Ex. 9) **3.** $30ab$ (Ex. 6) **4.** See Student Solutions Manual

Understanding the Concept: The Commutative Property and Division **1.** When $a = b$, then $a \div b = b \div a$.

Understanding the Concept: Conclusions and Inductive Reasoning **1.** Adding consecutive whole numbers 0, 1, 2, . . . ,
we obtain 4 for the next number. Multiplying by consecutive counting numbers 1, 2, 3, . . . , we obtain 6 for the next number.

1.5 Exercises **1.** $220 \div 4$ **3.** $225 \div n$ **5. (b)** and **(c)** **7.** $27 \div x$ **9.** $42 \div 6$ **11.** $36 \div 6$ **13.** $3 \div 36$ **15.** 1 **17.** 0
19. Undefined **21.** 6 R4 **23.** 371 **25.** 448 R2 **27.** 42 R8 **29.** 21 R2 **31.** 306 R3 **33.** 48 R11 **35.** 72 R1 **37.** 703 R4
39. 340 R11 **41.** 508 R33 **43.** 515 R101 **45.** 4 tickets **47.** $17 **49.** $175 per day **51.** 62 ft **53.** 31 stamps **55.** 405
57. 52 **59.** 18 **61.** Adding the pattern 1, 3, 1, 3, . . . , we get 5. Multiplying the pattern $0 \times 0 = 0, 1 \times 1 = 1, 2 \times 2 = 4$, we get $3 \times 3 = 9$.
63. (a) 4 **(b)** 16 **(c)** The associative property does not apply to division. **65.** $7 + x = 11$ **66.** 946 **67.** 814,262 **68.** 556,000
69. 464 mi **70.** $17,899

Quick Quiz 1.5 *See Examples noted with Ex.* **1. (a)** $14 \div 7$ **(b)** $7 \div 14$ (Ex. 3) **2.** 408 R4 (Ex. 9) **3.** $1,828,000 (Ex. 10)
4. See Student Solutions Manual

1.6 Exercises **1.** What number squared is equal to 16? **3.** 2^3 **5.** a^5 **7.** 4^1 **9.** 3^4 **11.** $5^2 a^3$ **13.** $2^2 z^5$ **15.** $5^3 y^2 x^2$
17. $n^5 \cdot 9^2$ or $9^2 n^5$ **19. (a)** $7 \cdot 7 \cdot 7$ **(b)** $y \cdot y \cdot y \cdot y \cdot y$ **21.** 8 **23.** 25 **25.** 1 **27.** 49 **29.** 256 **31.** 10 **33.** 125 **35.** 1,000,000
37. 25 **39.** 1 **41.** 7^3 **43.** 9^2 **45.** 5 **47.** 51 **49.** 45 **51.** 8 **53.** 21 **55.** 13 **57.** 19 **59.** 25 **61.** 16 **63.** 6
65. 5 **67.** 1 **69.** 100 **71.** 19 **73.** 53 **75.** 19 **77.** 34 **79.** He should have multiplied 3 times 2 first and then added 4 to get 10.
81. The exponent on the 10 determines the number of trailing zeros we attach to the number. **83.** 6841 **84.** 8423 **85.** 75,852 **86.** $2x$

Quick Quiz 1.6 *See Examples noted with Ex.* **1. (a)** $9^3 x^2$ **(b)** 5^5 (Ex.1) **2. (a)** 16 **(b)** 1 (Ex. 3) **3.** 3 (Ex. 9) **4.** See Student
Solutions Manual

Use Math to Save Money **1.** $545.75 **2.** $578.06 **3.** She did not deposit enough money to cover the checks she wrote for May.
But the $300.50 she already had in the bank will help to cover her expenses for May. **4.** $268.19 **5.** Eventually Teresa will be in debt.
6. Answers will vary. **7.** Answers will vary.

How Am I Doing? Sections 1.1–1.6 **1.** $9000 + 60 + 2$ (obj. 1.1.2) **2.** < (obj. 1.1.4) **3.** 17,200,000 (obj. 1.1.5)
4. (a) $a + 9$ (obj. 1.2.2) **(b)** $x + 12$ **5.** 20 (obj. 1.2.3) **6.** 10,105 (obj. 1.2.4) **7.** 50 in. (obj. 1.2.5) **8.** $11 - x$ (obj. 1.3.2)
9. 33,222 (obj. 1.3.4) **10.** $2x$ (obj. 1.4.2) **11.** $40y$ (obj. 1.4.3) **12.** 298,746 (obj. 1.4.4) **13.** 72 rooms (obj. 1.4.5)
14. $144 \div x$ (obj. 1.5.2) **15.** 503 R1 (obj. 1.5.4) **16.** $3^3 n^4$ (obj. 1.6.1) **17.** 64 (obj. 1.6.2) **18.** 18 (obj. 1.6.2) **19.** 5 (obj. 1.6.4)

1.7 Exercises **1.** Distributive property of multiplication over addition **3. (a)** False. We only use the distributive property when the terms
inside the parentheses are being added or subtracted. **(b)** True. The terms 3 and y are separated by a + sign, so we can use the distributive property.

5. $x, 1$ **7.** $y, 3$ **9.** $6y + 2$ **11.** $7 \cdot 4 - 1$ **13.** $4(3 + 9)$ **15.** $3(y + 6)$ **17.** $8(4 - y)$ **19. (a)** $4 \cdot 2 + 7 = 15$
(b) $4(2 + 7) = 36$ **21. (a)** $4 \cdot 3 - 1 = 11$ **(b)** $4(3 - 1) = 8$ **23. (a)** $12 \cdot 1 + 3 = 15$ **(b)** $12(1 + 3) = 48$ **25.** 38 **27.** 60 **29.** 5
31. 7 **33.** 6 **35.** 2 **37.** 7 **39.** 29 **41.** 5 **43.** $4x + 4$ **45.** $3n - 15$ **47.** $3x - 18$ **49.** $4x + 16$ **51.** $2x + 17$
53. $2y + 7$ **55.** $4x + 18$ **57.** $9y + 6$ **59.** $3x + 2$ **61.** 21 **63.** 15 **65. (a)** $(x + 2) + (x + 2) + (x + 2) + (x + 2) = 4x + 8$
(b) $4(x + 2) = 4x + 8$ **(c)** The answers are the same. **67.** $64x$ **68.** 6 **69.** 8 **70.** 1538

Quick Quiz 1.7 *See Examples noted with Ex.* **1.** $2(n + 5)$ (Ex. 1) **2.** $6y + 9$ (Ex. 5) **3. (a)** 17 **(b)** 2 (Ex. 3)
4. See Student Solutions Manual

Understanding the Concept: Evaluate or Solve? **1.** Answers may vary.

1.8 Exercises **1.** Seven times x, or the product of seven and x **3.** Eight times what number equals 40? **5.** No, because the variable
parts, x and y, are not the same. **7.** variable, constant **9.** 8, coefficient **11.** $1y$ **13.** x **15.** $4xy$ **17.** $4x$ **19.** $3x$ **21.** $4a$
23. $5x$ and $2x$; $3y$ and $7y$ **25.** $2mn$ and $4mn$ **27.** $9x$ **29.** $8y$ **31.** $11x$ **33.** $11x + 5a$ **35.** $9xy + 4b$ **37.** $15xy + 3x + 9$
39. $7ab + 9$ **41.** $17xy + 10$ **43.** $18x + 14y$ **45.** $28a + 14b$ **47.** $6x + 8y$ **49.** $5 + x = 16$ **51.** $3x = 36$ **53.** $45 - x = 6$
55. $\frac{25}{n} = 5$ or $25 \div n = 5$ **57.** $J + 12 = 25$ **59.** $C - 50 = 1480$ **61.** No **63.** Yes **65.** $x = 4$ **67.** $n = 8$ **69.** $x = 6$
71. $x = 11$ **73.** $x = 5$ **75.** $x = 2$ **77.** $y = 3$ **79.** $x = 7$ **81.** $y = 15$ **83.** $x = 7$ **85.** $x = 3$ **87.** $y = 1$ **89.** $n = 2$
91. $n = 3$ **93.** $x = 1$ **95.** $x = 4$ **97.** $x = 15$ **99.** $a = 10$ **101.** $x = 5$ **103. (a)** $4 + x = 8$ **(b)** $x = 4$ **105. (a)** $3x = 9$
(b) $x = 3$ **107.** $x = 70$ ft **109.** $6x^2 + 18$ **111. (a)** $5x + 5y$ **(b)** $10xy$ **113. (a)** $7a + 6y$ **(b)** $30ay$ **115. (a)** 30 mph **(b)** 25 mph
117. d **118.** c **119.** a **120.** b

Quick Quiz 1.8 *See Examples noted with Ex.* **1.** $3ab + 4a + 1$ (Ex. 3) **2. (a)** $a = 5$ (Ex. 9) **(b)** $x = 1$ (Ex. 10) **(c)** $y = 8$ (Ex. 11)
3. (a) $3x = 18; x = 6$ **(b)** $D + 7 = 21; D = 14$ (Ex. 5) **4.** See Student Solutions Manual

1.9 Exercises **1. (a)** \$150 **(b)** \$153 **(c)** Yes **3.** 400 mi **5.** \$1185 **7.** \$53 **9.** 400 ft **11. (a)** 95 points **(b)** 44 points
13. (a) Gather the Facts: Cook earns \$8 per hour for 40 hours and \$12 per hour for overtime. Hours worked = 52. **What Am I Asked to Do?**
Calculate the cook's total pay. **How Do I Proceed?** 1. Multiply \$8 × 40 to find base pay. 2. Multiply \$12 × 12 to find overtime pay. 3. Add the
results from steps 1 and 2. **Key Points to Remember:** Overtime hours are hours worked in addition to 40 hours a week. **(b)** \$464 **15. (a)** \$54,170
(b) \$27,085 **17.** \$300 **19.** Pay each time he rides **21. (a)** \$3000 **(b)** \$1000 **23. (a)** 65 pts. **(b)** \$6 **25. (a)** 130 pts. **(b)** \$25
27. \$1300 **29.** 169
31.  **32.** 120 **33.** $x = 5$ **34.** $x = 3$

Quick Quiz 1.9 *See Examples noted with Ex.* **1. (a)** \$10,329 **(b)** \$6929 (Ex. 2) **2.** 245 gal in 1 day; 1715 gal in 1 week (Ex. 4)
3. \$4900 (Ex. 1) **4.** See Student Solutions Manual

You Try It **1.** Twenty-three million, three hundred twenty-seven thousand, four hundred fourteen **2.** $<, >$ **3.** 133,000 **4.** $x + 14$
5. 762 **6.** 54 m **7.** 41,686 **8. (a)** $2n$ **(b)** $5n$ **(c)** $x \cdot 8$ **(d)** $4 \cdot 2$ **9.** $20y$ **10.** 117,468 **11. (a)** $6 \div x$ **(b)** $x \div 6$
(c) $n \div 3$ **12.** 47 R1 **13. (a)** $8^3 n^2$ **(b)** 16 **14.** 5 **15. (a)** $4(x - 5)$ **(b)** $4x - 5$ **16.** $7n - 21$ **17.** $10mn + 2n$
18. (a) $n = 6$ **(b)** $x = 5$ **19.** $x = 10$ **20.** $x = 8$ **21. (a)** 21 **(b)** 2 **22.** \$400

Chapter 1 Review Problems **1.** A four-sided figure with adjoining sides that are perpendicular and opposite sides that are equal
2. A rectangle with all sides equal **3.** An angle that measures 90° **4.** A three-sided figure with three angles **5.** The distance around
an object **6.** The numbers or variables we multiply **7.** A number, a variable, or a product of a number and one or more variables
8. A term that has no variable **9.** The number factor in a term **10.** Terms that have identical variable parts **11.** Two expressions separated
by an equals sign **12. (a)** Ten thousands **(b)** Thousands **13.** 187.00; One hundred eighty-seven and 00/100 **14.** $7000 + 600 + 90 + 4$
15. $5000 + 800 + 30 + 1$ **16.** < **17.** > **18.** $6 > 1$ **19.** $3 < 5$ **20.** 61,300 **21.** 382,200 **22.** 6,400,000 **23.** 8,100,000
24. $x + 7$ **25.** $n + 5$ **26.** $16 + x$ **27.** $n + 11$ **28.** $7 + n$ **29.** $x + 10$ **30.** 9025 **31.** 18,651 **32.** 6018 students **33.** 66 m
34. $8 - n$ **35.** $n - 6$ **36.** $x - 10$ **37.** 5 **38.** 6 **39.** 5545 **40.** 3159 **41.** 24,116 **42.** \$378,000 more **43.** \$270,000 less
44. Factors: 4, x **45.** $3x$ **46.** Seven times what number equals 63? **47.** 0 **48.** 60 **49.** $42y$ **50.** $30x$ **51.** $24x$ **52.** 832,000
53. 1,496,352 **54.** 5,800,872 **55.** 306 mi **56.** 504 doors **57.** $300 \div 20$ **58.** $500 \div n$ **59.** $5 \div y$ **60.** $n \div 13$ **61.** Undefined
62. 1 **63.** 50 R6 **64.** 401 R35 **65.** 603 R6 **66.** \$3 **67.** \$147 **68.** $2^3 n^2$ **69.** $z^4 5^3$ or $5^3 z^4$ **70.** $x \cdot x \cdot x$ **71.** $6 \cdot 6 \cdot 6 \cdot 6 \cdot 6$
72. 1000 **73.** 16 **74.** 6^3 **75.** x^5 **76.** 5 **77.** 5 **78.** 20 **79. (a)** $3x + 2$ **(b)** $3(x + 2)$ **80. (a)** $4x - 5$ **(b)** $4(x - 5)$
81. (a) $3 \cdot 7 + 1 = 22$ **(b)** $3(7 + 1) = 24$ **82.** 13 **83.** 22 **84.** $5x + 5$ **85.** $4x - 4$ **86.** $3x + 8$ **87.** $9x$ **88.** $11x + 6y$
89. $5xy + 13y$ **90.** $10x + 10y$ **91.** $x = 7$ **92.** $n = 4$ **93.** $x = 4$ **94.** $n = 1$ **95.** $x = 3$ **96.** $x = 3$ **97.** $n = 2$ **98.** $y = 2$
99. (a) $18 - x = 3$ **(b)** $x = 15$ **100. (a)** $x + 5 = 11$ **(b)** $x = 6$ **101. (a)** $3 \cdot x = 12$ **(b)** $x = 4$ **102.** \$310 **103.** \$2746
104. (a) \$7426 **(b)** \$3713 **105.** \$468

How Am I Doing? Chapter 1 Test **1.** $1000 + 500 + 20 + 5$ (obj. 1.1.2) **2. (a)** $7 > 2$ **(b)** $5 > 0$ (obj. 1.1.4) **3. (a)** 3000
(b) 2900 (obj. 1.1.5) **4. (a)** $11 + x$ or $x + 11$ **(b)** $7 + y$ or $y + 7$ **(c)** $7 + n$ or $n + 7$ (obj. 1.2.2) **5.** 14,695 (obj. 1.2.4) **6.** 244,888,278
(obj. 1.2.4) **7. (a)** 538 **(b)** 12,279 (obj. 1.3.4) **8.** 30 ft (obj. 1.2.5) **9.** $16y$ (obj. 1.4.3) **10. (a)** 134,784 **(b)** 261,999 (obj. 1.4.4)

11. (a) 41 **(b)** 120 R3 (obj. 1.5.4) **12. (a)** $n - 7$ **(b)** $10n$ **(c)** y^4 **(d)** 7^3 **(e)** $6(x + 9)$ (obj. 1.3.2, 1.4.2) **13. (a)** $7xy + 2y - 2$
(b) $3m + 5 + 6mn$ (obj. 1.8.1) **14.** $3y + 12$ (obj. 1.7.3) **15.** $8x + 10$ (obj. 1.7.3) **16. (a)** 20 **(b)** 11 (obj. 1.7.2, 1.6.2)
17. $6^5 n^3$ (obj. 1.6.1) **18. (a)** 125 **(b)** 100,000 (obj. 1.6.2) **19.** 0 (obj. 1.6.4) **20.** 41 (obj. 1.6.4) **21.** 30 (obj. 1.6.4) **22. (a)** $x = 6$
(b) $x = 8$ **(c)** $x = 9$ **(d)** $b = 11$ **(e)** $n = 4$ (obj. 1.8.3) **23.** $B - 155 = 275$ (obj. 1.8.2) **24. (a)** $x \div 6 = 2$ **(b)** $x = 12$ (obj. 1.8.4)
25. (a) $x - 3 = 1$ **(b)** $x = 4$ (obj. 1.8.4) **26. (a)** $15,340 **(b)** $7990 (obj. 1.9.3) **27.** 12 (obj. 1.9.3) **28.** $1140 (obj. 1.9.3)
29. $1290 (obj. 1.9.3) **30. (a)** $1600 **(b)** $300 (obj. 1.9.1) **31.** 7500 points (obj. 1.9.3)

Chapter 2 2.1 Exercises **1.** The opposite of negative one **3.** $-4 - 2$ **5.** negative; positive

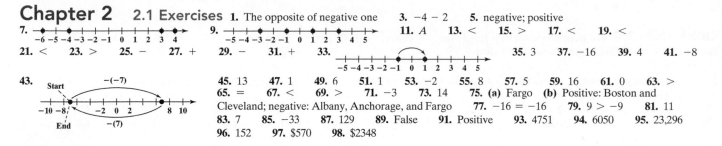

7. [number line from -6 to 4] **9.** [number line from -5 to 5] **11.** A **13.** $<$ **15.** $>$ **17.** $<$ **19.** $<$
21. $<$ **23.** $>$ **25.** $-$ **27.** $+$ **29.** $-$ **31.** $+$ **33.** [number line from -5 to 5] **35.** 3 **37.** -16 **39.** 4 **41.** -8
43. [number line Start/End from -10 to 10] **45.** 13 **47.** 1 **49.** 6 **51.** 1 **53.** -2 **55.** 8 **57.** 5 **59.** 16 **61.** 0 **63.** $>$
65. $=$ **67.** $<$ **69.** $>$ **71.** -3 **73.** 14 **75. (a)** Fargo **(b)** Positive: Boston and
Cleveland; negative: Albany, Anchorage, and Fargo **77.** $-16 = -16$ **79.** $9 > -9$ **81.** 11
83. 7 **85.** -33 **87.** 129 **89.** False **91.** Positive **93.** 4751 **94.** 6050 **95.** 23,296
96. 152 **97.** $570 **98.** $2348

Quick Quiz 2.1 *See Examples noted with Ex.* **1. (a)** $>$ (Ex. 2) **(b)** $<$ (Ex. 9) **2.** -7 (Ex. 6) **3. (a)** 3 (Ex. 8) **(b)** -5 (Ex. 10)
4. See Student Solutions Manual

2.2 Exercises **1.** Answers may vary. When we add two negative numbers, we move in the negative direction followed by another move in the negative direction. Thus, we end up in the negative region. **3.** Larger absolute value; subtract **5. (a)** $-$; When adding two numbers with the same sign, we use the common sign in the answer and then add absolute values of the numbers. **(b)** $-$; When adding two numbers with different signs, we keep the sign of the larger absolute value and subtract the absolute values. **(c)** $+$; When adding two numbers with different signs, we keep the sign of the larger absolute value and subtract the absolute values.
7. (a) [number line -2 -2] **(b)** Negative **(c)** $-2 + (-2)$ **(d)** -4
9. (a) [number line 3 2] **(b)** Positive **(c)** $3 + 2$ **(d)** 5
11. (a) Decrease of 15°F **(b)** $-10°F + (-5°F) = -15°F$ **13. (a)** Profit of $150 **(b)** $100 + $50 = $150 **15. (a)** -24 **(b)** 24
17. (a) -68 **(b)** 68 **19. (a)** 71 **(b)** 71 **21. (a)** $-3 + 10$ **(b)** Positive **(c)** 7 **23. (a)** $2 + (-4)$ **(b)** Negative **(c)** -2
25. -100 ft **27.** $100 **29. (a)** -2 **(b)** 2 **31. (a)** 4 **(b)** -4 **33. (a)** 6 **(b)** -6 **35. (a)** 2 **(b)** -4 **(c)** -2 **37. (a)** -20
(b) -2 **(c)** 2 **39.** 0 **41.** 0 **43.** 0 **45.** 0 **47.** $x = -19$ **49.** $x = 12$ **51.** 1 **53.** -6 **55.** 7 **57.** 12 **59.** -38
61. -39 **63.** 0 **65.** 4 **67.** -2 **69.** 0 **71.** -5 **73.** -38 **75.** 17 **77.** -15 **79. (a)** 3 **(b)** -2 **81. (a)** -4 **(b)** -13
83. (a) -4 **(b)** -6 **85.** 8 **87.** -3 **89.** Profit: $10,000 **91.** $79 **93.** -225 ft **95.** 5 **97.** -31 **99.** -56 **101.** -3
103. -4 **105.** 1 **107.** The sum $x + y$ equals -30; two possible values for x and y are -15 and -15 (answers may vary). **109.** There are two
solutions using six squares: $-2, 1, -4, 4, -1, 7$ and $8, -5, 2, -8, -1, 7$. **111.** $10x$ **112.** $6x$ **113.** $5x$ **114.** $3x - 12$ **115.** 24,086 miles
116. 35 people

Quick Quiz 2.2 *See Examples noted with Ex.* **1. (a)** -10 (Ex. 3) **(b)** -2 (Ex. 5) **(c)** -1 (Ex. 8) **2.** -3 (Ex. 9) **3.** Loss of $1000 (Ex. 10)
4. See Student Solutions Manual

Understanding the Concept: Another Approach to Subtracting Several Integers **1. (a)** -23 **(b)** -23
(c) Answers may vary.

2.3 Exercises **1.** Addition; opposite; add **3.** We can think of this subtraction as measuring the distance between the two numbers. -10 is 10 units below 0 on a vertical number line, so we must add these 10 units to the 25 units above 0. **5.** 3 **7.** -5 **9.** $-6, -13$ **11.** $-9, -5$
13. $+, 9$ **15.** $+, 14$ **17. (a)** $7 + (-4) = 3$ **(b)** $15 + (-7) = 8$ **(c)** $10 + (-8) = 2$ **19.** $-$15 **21.** $-$1 **23.** -10 **25.** -9
27. 7 **29.** 14 **31.** -2 **33.** 0 **35.** -5 **37.** -4 **39.** -20 **41.** -65 **43.** -10 **45.** -15 **47.** 0 **49.** -15 **51.** 0
53. -1 **55.** -14 **57.** 0 **59.** -31 **61.** 41 **63.** -7 **65.** -2 **67.** 0 **69.** -17 **71.** -14 **73.** 19 **75.** 24 **77.** -6
79. 0 **81. (a)** Brownsville, Texas **(b)** 86°F **83.** 344 ft **85.** 3706 ft **87.** 4 points **89.** Sandra **91.** -49 **93.** 17 **95.** -3902
97. -1014 **99. (a)** $-1 - 8 = n$ **(b)** $-9 = n$ **101.** Yes **103.** 2 **105.** 9 **107.** -21 **108.** 17 **109.** 3 **110.** 12 **111.** 10
112. 46 boxes **113.** 102 min, or 1 hr 42 min

Quick Quiz 2.3 *See Examples noted with Ex.* **1. (a)** -24 **(b)** 2 (Ex. 3) **2.** -7 (Ex. 5) **3.** 6719 ft (Ex. 8) **4.** See Student Solutions
Manual

Use Math to Save Money **1.** $14,970.16 **2.** $25,094.08 **3.** $25,970.16 **4.** $876.08 **5.** He will save $281.22 each month in car
payments. **6.** To get the best overall price, Louvy should buy the car since he will save $876.08 on the total price. To get a lower monthly payment,
Louvy should lease the car since he will save $281.22 each month in car payments. **7.** Answers will vary. **8.** Answers will vary.
9. Answers will vary.

How Am I Doing? Sections 2.1–2.3 **1.** $<$ (obj. 2.1.1) **2.** $>$ (obj. 2.1.1) **3.** -8 (obj. 2.1.3) **4.** -3 (obj. 2.1.2) **5.** -6 (obj. 2.1.2)
6. -16 (obj. 2.2.1) **7.** -2 (obj. 2.2.2) **8.** -2 (obj. 2.2.3) **9.** 4 (obj. 2.2.3) **10.** $20,000 profit (obj. 2.2.4) **11.** -12 (obj. 2.3.1)
12. 2 (obj. 2.3.1) **13.** -5 (obj. 2.3.2) **14.** -5 (obj. 2.3.2) **15.** -2 (obj. 2.3.2) **16.** 7783 ft (obj. 2.3.3)

2.4 Exercises
1. When we add two negative numbers the sum is negative, so the statement is only true for multiplication and division.
3. Positive   **5.** Negative   **7.** -12 **9.** -24 **11.** -6 **13. (a)** 3 **(b)** -3 **(c)** 3 **(d)** -3 **15. (a)** 4 **(b)** -4 **(c)** -4 **(d)** 4
17. (a) 18 **(b)** -18 **(c)** -18 **(d)** 18   **19. (a)** 10 **(b)** 10 **(c)** -10 **(d)** -10 **21.** 18   **23.** 6   **25.** -56 **27.** -45 **29.** Positive
31. Negative   **33.** 40   **35.** 72   **37.** 70   **39.** -40 **41.** -120 **43.** Negative   **45.** Positive   **47.** Negative   **49.** 100
51. -125 **53. (a)** 16 **(b)** -64 **55. (a)** -1 **(b)** 1   **57. (a)** -16 **(b)** 16   **59. (a)** -8 **(b)** -8 **61. (a)** -64 **(b)** -64
63. (a) 81 **(b)** -81 **65. (a)** 5 **(b)** -5 **(c)** -5 **(d)** 5   **67. (a)** 5 **(b)** -5 **(c)** -5 **(d)** 5   **69.** -6 **71.** -9 **73.** 8   **75.** 7
77. (a) -11 **(b)** -44 **79. (a)** 2 **(b)** 8   **81. (a)** -5 **(b)** -45 **83. (a)** -2 **(b)** -98 **85.** 1   **87.** -6 **89.** -10
91. (a) 8 **(b)** -16 **93.** -90 or 90 m left of 0   **95.** $-\$700$ **97.** No   **99.** -24 **101.** 18   **102.** 14   **103.** 16   **104.** 20
105. 3 seconds   **106.** <

Quick Quiz 2.4 *See Examples noted with Ex.* **1. (a)** -12 (Ex. 2) **(b)** -9 (Ex. 8) **(c)** 6 (Ex. 9)   **2.** -36 (Ex. 4)   **3. (a)** -1
(b) 5 (Ex. 10)   **4.** See Student Solutions Manual

2.5 Exercises
1. No, because we must multiply $3(-1)$ before we add.   **3.** Yes. There are no parentheses around -2, so we only square
2 not -2. Then we take the opposite of 2^2.   **5.** 10   **7.** -27 **9.** 15   **11.** 32   **13.** 54   **15.** 1   **17.** -34 **19.** -10 **21.** -36
23. -83 **25.** -24 **27.** -12 **29.** -59 **31.** 12   **33.** -2 **35.** 3   **37.** 3   **39.** -1 **41.** -2 **43.** 30,000 ft   **45.** -33
47. $+20$ **49.** 14 points   **51.** 25 points   **53.** -4 **55.** -99 **57.** $x = 9$ **59.** $2x + 6$ **60.** $3a + 6$ **61.** $4x - 8$ **62.** $7x - 7$

Quick Quiz 2.5 *See Examples noted with Ex.* **1.** 2 (Ex. 1)   **2.** 1 (Ex. 2)   **3.** 35°F (Ex. 4)   **4.** See Student Solutions Manual

2.6 Exercises
1. No, because we do not multiply variables and coefficients when combining like terms. We add the coefficients of
x: $(-2 + 5)x = 3x$.   **3.** x **5.** 3   **7.** $-$ **9.** $y, 1, +$ **11.** $-5x$ **13.** x **15.** $-12x$ **17.** $-5a$ **19.** $7y$ **21.** $-13x$
23. $7a - 9b$ **25.** $-5m - 8n$ **27.** $2x - y$ **29.** $-2a + 3b$ **31. (a)** -2 **(b)** $-2x$ **33. (a)** -1 **(b)** $-1x$ or $-x$ **35. (a)** -3
(b) $-3x$ **37.** $-6y + 4x$ **39.** $-2x + 4y$ **41.** $4x + 3y$ **43.** $-12x - y$ **45.** $-7x + 6y$ **47.** $-2x + 2y - 7$ **49.** $2 - 6ab$
51. $-4x + 6xy$ **53.** $10a - 9ab - 2$ **55.** $-2a + x + 7ax$ **57.** $-3a - 4b + 5ab$ **59.** $-3x - 2y + 6xy$ **61.** -9 **63.** 24
65. -11 **67.** -1 **69.** 81   **71.** -65 **73.** 10   **75.** -3 **77.** -4 **79.** $-3y - 3$ **81.** $-9y + 9$ **83.** $-2m + 6$
85. $-1x - 5$ or $-x - 5$ **87.** $-12 + 6y$ **89.** $-8 + 2a$ **91.** 136 ft/sec   **93.** 3 sec   **95.** -10°C   **97.** No   **98.** 18 ft
99. 28 in.   **100.** 186,000 miles in 1 sec; 11,160,000 miles in 1 min   **101.** 4380 times per hour; 105,120 times per day

Quick Quiz 2.6 *See Examples noted with Ex.* **1.** $-11x + 7y + 2$ (Ex. 4)   **2. (a)** $-3a + 3$ **(b)** $-8x - 56$ (Ex. 6)   **3.** 106 ft/sec (Ex. 7)
4. See Student Solutions Manual

You Try It 1. $-4 > -9$ **2.** $16, -12$ **3. (a)** 14 **(b)** 9   **4. (a)** Boston **(b)** Bangor   **5.** -13 **6. (a)** -4 **(b)** 7
7. (a) -15 **(b)** 12   **8. (a)** -60 **(b)** 36   **9. (a)** 36 **(b)** -36 **(c)** -8 **10. (a)** 9 **(b)** -4 **11.** -84 **12.** $-3a + 5b$
13. 45   **14.** $-3x + 6$

Chapter 2 Review Problems
1. Numbers that are less than zero   **2.** Numbers that are the same distance from zero but lie on
opposite sides of zero   **3.** Whole numbers and their opposites   **4.** The distance between a number and 0 on the number line   **5.** <
6. <   **7.** >   **8.** +   **9.** $-$ **10.** -12 **11.** -6 **12.** -11 **13.** -23 **14. (a)** May **(b)** March   **15. (a)** January, February, May
(b) March, April   **16. (a)** -59 **(b)** 59   **17. (a)** -66 **(b)** 66   **18.** Loss   **19.** Profit   **20. (a)** -10°F $+ 20$°F **(b)** Positive **(c)** 10°F
21. (a) -6 **(b)** 6 **(c)** -10 **22. (a)** 9 **(b)** -9 **(c)** -45 **23.** 4   **24.** -96 **25.** 5   **26.** -6 **27.** -290 ft   **28.** -12 **29.** -5
30. -8 **31.** 0   **32.** -11 **33.** 5   **34.** -2 **35.** -13 **36.** 2   **37.** -17 **38.** 1   **39.** \$50,000   **40.** \$40,000   **41.** 3612 ft
42. (a) 18 **(b)** -18 **(c)** -18 **(d)** 18   **43. (a)** 10 **(b)** -10 **(c)** -10 **(d)** 10   **44.** 14   **45.** -10 **46.** -12 **47.** 4   **48.** -90
49. 64   **50.** -240 **51.** 49   **52.** -81 **53.** -216 **54. (a)** 7 **(b)** -7 **55. (a)** -6 **(b)** 6   **56. (a)** 11 **(b)** -45 **(c)** 33 **(d)** -5
57. (a) -3 **(b)** -40 **(c)** 24 **(d)** -4 **58.** 1   **59.** -27 **60.** 4   **61.** -3 **62.** 7   **63.** 68   **64.** -16 **65.** 0   **66.** -5°F
67. $3x + 5y$ **68.** $-7a$ **69.** $x - 2y$ **70.** $-4 - 3z + 2yz$ **71.** -4 **72.** -3 **73.** 2   **74.** 0   **75.** 5°C   **76.** -20°C
77. $-6x - 6$ **78.** $-2a + 2$ **79.** $-8 + 4x$

How Am I Doing? Chapter 2 Test
1. < (obj. 2.1.1)   **2.** < (obj. 2.1.3)   **3.** $-$ (obj. 2.1.1)   **4.** -2 (obj. 2.1.2)
5. -10 (obj. 2.1.2)   **6. (a)** 12 **(b)** -3 (obj. 2.1.3)   **7. (a)** -10°F $+ 15$°F **(b)** 5°F (obj. 2.2.4)   **8. (a)** 2 **(b)** -2 (obj. 2.2.2)
9. -10 (obj. 2.2.1)   **10.** -19 (obj. 2.2.2)   **11.** -6 (obj. 2.3.1)   **12. (a)** -12 **(b)** 10 (obj. 2.3.1)   **13.** 13 (obj. 2.3.1)   **14.** -24 (obj. 2.3.2)
15. -21 (obj. 2.4.1)   **16.** 32 (obj. 2.4.1)   **17.** -30 (obj. 2.4.2)   **18. (a)** 25 **(b)** -125 **(c)** -25 (obj. 2.4.3)   **19. (a)** -4
(b) 4 (obj. 2.4.4)   **20.** -2 (obj. 2.4.4)   **21.** -67 (obj. 2.5.1)   **22.** 2 (obj. 2.5.1)   **23. (a)** -4 **(b)** 6 (obj. 2.6.2)   **24.** 2 (obj. 2.6.2)
25. (a) 1 **(b)** -8 (obj. 2.6.2)   **26.** -3 (obj. 2.6.2)   **27.** $-3x - 4y$ (obj. 2.6.1)   **28.** $-15x - 3y + 7xy$ (obj. 2.6.1)   **29.** $-6a - 42$
(obj. 2.6.3)   **30.** $-2x + 2$ (obj. 2.6.3)   **31.** \$15,000 profit (obj. 2.2.4)   **32.** 4292 ft (obj. 2.2.4)   **33.** 167 ft/sec (obj. 2.6.4)

Chapter 3   3.1 Exercises
1. zero   **3.** add   **5.** -3 **7.** 9   **9.** -17 **11.** 28   **13.** -5 **15.** 2   **17.** $-12, -12; 0, 4; 4$
19. $16, 16; 0, 48; 48$ **21. (a)** $x = 30$ **(b)** $x = 14$ **23. (a)** $x = -13$ **(b)** $x = -9$ **25. (a)** $-20 = x$ **(b)** $-16 = x$ **27.** $y = 15$
29. $n = -31$ **31.** $y = -55$ **33.** $x = -34$ **35.** $14 = x$ **37.** $9 = y$ **39.** $-14 = x$ **41.** $x = 11$ **43.** $y = -6$ **45.** $4 = y$
47. $x = 26$ **49.** $-18 = a$ **51.** $m = 48$ **53.** $x = -6$ **55.** $-7 = y$ **57.** $\angle a = 94$°   **59.** $\angle x = 68$°   **61.** $\angle x = 137$°
63. (a) $\angle x = \angle y + 70$° **(b)** 55° $= \angle y$ **65. (a)** $\angle a = \angle b - 40$° **(b)** 90° $= \angle b$ **67.** $x = 109$°; $\angle b = 114$°   **69.** $x = 123$°; $\angle b = 128$°
71. $-27 = x$ **73.** $3 = x$ **75.** $\angle c = 80$°; $\angle d = 40$°; $\angle f = 80$°   **77.** $7x$ **78.** $3y$ **79.** $8n = 40$ **80.** $2n = 30$ **81.** 13°F   **82.** >
83. 50 points   **84.** 46 points

Quick Quiz 3.1 *See Examples noted with Ex.* **1. (a)** $x = -27$ (Ex. 3) **(b)** $y = -13$ (Ex. 2)   **2. (a)** $x = -7$ (Ex. 4)
(b) $5 = y$ (Ex. 5)   **3. (a)** $\angle a = \angle b - 35$° (Ex. 7a) **(b)** 110° $= \angle b$ (Ex. 7b)   **4.** See Student Solutions Manual

3.2 Exercises
1. dividing; -22 **3.** 5 **5.** -2 **7.** 6 **9.** -1 **11.** $x = 12$ **13.** $x = 4$ **15.** $y = -3$ **17.** $m = -7$
19. $y = -5$ **21.** $a = -7$ **23.** $8 = x$ **25.** $8 = x$ **27.** $x = 13$ **29.** $x = 4$ **31.** $x = 9$ **33.** $x = 2$ **35.** $x = 2$ **37.** $x = 4$
39. $-4 = a$ **41.** $-1 = y$ **43.** $x = 8$ **45.** $13 = x$ **47.** $y = -11$ **49.** $x = 5$ **51.** $x = 7$ **53.** $11 = a$ **55.** $x = 2$
57. $x = -10$ **59.** $L = 3W$ **61.** $R = 2S$ **63. (a)** $R = 2B$ **(b)** $62,000 **65. (a)** $C = 3A$ **(b)** 100 adults' tickets **67. (a)** $A = 2S$
(b) 21 goals **69.** 30 shares of stock **71.** 120 miles **73. (a)** $x = 2$ **(b)** $x = -2$ **(c)** $x = 13$ **(d)** $x = 39$ **75. (a)** $x = -2$ **(b)** $x = 2$
(c) $x = -39$ **(d)** $x = -13$ **77.** $x = 30°; 5x = 150°$ **79.** $x = 36°; 4x = 144°$ **81.** $x = 37°$ **83.** 20 ft **84.** 12 in. **85.** 126 **86.** 30

Quick Quiz 3.2
See Examples noted with Ex. **1. (a)** $x = -21$ (Ex. 4) **(b)** $2 = x$ (Ex. 3) **2.** $M = 5S$ (Ex. 6) **3. (a)** $B = 2R$ (Ex. 7a)
(b) 5 red marbles (Ex. 7b) **4.** See Student Solutions Manual

How Am I Doing? Sections 3.1–3.2
1. $y = -11$ (obj. 3.1.1) **2.** $x = 11$ (obj. 3.1.2) **3.** $-16 = a$ (obj. 3.1.2) **4.** $\angle x = 65°$
(obj. 3.1.3) **5. (a)** $\angle x = \angle y + 30°$ **(b)** $\angle y = 60°$ (obj. 3.1.3) **6.** $-2 = a$ (obj. 3.2.1) **7.** $x = -3$ (obj. 3.2.1) **8.** $-29 = x$ (obj. 3.2.1)
9. $L = 4W$ (obj. 3.2.2) **10. (a)** $C = 3A$ **(b)** 50 adults' tickets (obj. 3.2.3) **11.** 53 shares of stock (obj. 3.2.3)

3.3 Exercises
1. Volume, because we want to find the amount of space inside the pool. **3. (a)** $P = 2L + 2W$ **(b)** $P = 4s$
(c) $V = LWH$ **(d)** $A = LW$ **(e)** $A = s^2$ **(f)** $A = bh$ **5. (a)** $P = 2L + 2W$ **(b)** 18 ft **7. (a)** $P = 4s$ **(b)** 44 ft **9. (a)** $P = 4s$
(b) 216 yd **11. (a)** $P = 4s$ **(b)** 10 ft **13. (a)** $P = 2L + 2W$ **(b)** 3 ft **15.** $W = 3$ ft, $L = 30$ ft **17. (a)** 50 squares **(b)** 50 in.^{2}
19. 50 **21. (a)** $A = LW$ **(b)** 396 ft^2 **23. (a)** $A = s^2$ **(b)** 100 in.^{2} **25. (a)** $A = bh$ **(b)** 108 ft^2 **27.** $x = 6$ ft **29.** $h = 11$ m
31. 259 in.^{2} **33.** 344 m^2 **35. (a)** 40 **(b)** 40 in.^{3} **37.** 36 ft^2 **39. (a)** $V = LWH$ **(b)** 825 in.^{3} **41. (a)** $V = LWH$ **(b)** 4320 yd^3
43. $W = 5$ cm **45.** 28 in. **47.** 136 m^2 **49.** 210 yd^3 **51.** $x = 5$ ft **53.** $L = 5$ m **55. (a)** 24 in. by 24 in. **(b)** 576 in.^{2}
57. (a) 4 yd by 3 yd **(b)** 12 yd^2 **59. (a)** 12 yd^2 **(b)** $96 **61. (a)** 35 yd^2 **(b)** $560 **63.** $6 **65.** 2 gal and 1 qt of paint **67.** 29 in.^{2}
69. **71.** $-7x + 14$ **72.** $-4x - 2y$ **73.** $30x$ **74.** -16

Quick Quiz 3.3
See Examples noted with Ex. **1.** $L = 12$ ft (Ex. 2) **2.** 115 in.^{2} (Ex. 6) **3.** 8 ft (Ex. 7) **4.** See Student Solutions Manual

Use Math to Save Money
1. $3 + $1 = $4 **2.** $74 - $71 = $3 **3.** $3 - $4 = -$1 **4.** It is not worth driving across town
to get gas since Mary will lose money after calculating expenses. **5.** $6 + $2 = $8 **6.** $95 - $76 = $19 **7.** $19 - $8 = $11 **8.** It is
worth driving to the mall since Mary will save $11.

Understanding the Concept: Do I Add or Multiply Coefficients?
1. (a) $8x$ **(b)** $15x^2$ **(c)** $2xy^2$ **(d)** $35x^2y^4$

3.4 Exercises
1. No, because we do not multiply the bases. **3. (a)** No. We only use the distributive property when the expression inside
the parentheses is being added or subtracted. We have $3 \cdot x^2$ inside the parentheses, not $3 + x^2$. **(b)** Yes. The parentheses between the 2 and the $3x^2$
means multiply. **(c)** No, the answer is $6 + 2x^2$. We cannot simplify this expression any further since the 6 and the $2x^2$ are not like terms.
5. coefficient or numerical coefficient **7. (a)** z^5 **(b)** z^5 **9. (a)** x^6 **(b)** x^6 **11. (a)** m^3 **(b)** m^3 **13.** $(2 \cdot 2 \cdot 2 \cdot 2) \cdot (2 \cdot 2) = 2^6$
15. $(3 \cdot 3 \cdot 3 \cdot 3 \cdot 3) \cdot (3 \cdot 3 \cdot 3) = 3^8$ **17.** x^7 **19.** u^7 **21.** 3^5 **23.** 4^6 **25.** $8^2 \cdot 7^5$ **27.** x^5y^3 **29.** x^{13} **31.** y^{14} **33.** 3^{10}
35. $2^5 \cdot 3^2 \cdot 4^7$ **37.** $24y^{12}$ **39.** $54a^{14}$ **41.** $-15x^3$ **43.** $-12y^2$ **45.** $70a^{10}$ **47.** $28x^{17}$ **49.** $-30r^2y$ **51.** $54xy^2$
53. (a) Binomial **(b)** Monomial **(c)** Trinomial **55.** $2x^3 + 10x$ **57.** $18x^5 - 6x^2$ **59.** $8y^7 - 10y^4$ **61.** $3x^5 - 8x^4$ **63.** $-12y^3 + 30y^2$
65. $6x^4 - 12x^2$ **67.** $A = 2x^7 - 5x^3$ **69.** $A = 7x^5 - 4x^3$ **71.** $V = 14x^9 - 2x^8$ **73.** $P = 8x^3 + 6x^2$ **75.** $P = 10x^3 + 10x^2 - 2x$
77. (a) $A = 6x^4 + 10x^2$ **(b)** $P = 10x^2 + 10$ **79. (a)** $13cd$ **(b)** $36c^2d^2$ **(c)** $4cd + 36$ **81. (a)** $2ab^5$ **(b)** $63a^2b^{10}$ **(c)** $9ab^5 + 63$
83. $W = L^2 + 4$ **85.** -5 **86.** -27 **87.** 65 R286 **88.** 879,844 **89.** 1 over par **90.** 1 under par

Quick Quiz 3.4
See Examples noted with Ex. **1. (a)** x^3 (Ex. 2) **(b)** 4^7 (Ex. 2) **(c)** $-42y^8$ (Ex. 3) **2. (a)** $6a^4 + 48a^2$ (Ex. 7) **(b)** $2x^7 - 6x^2$
(Ex. 7) **3.** $A = 3x^4 + 5x^2$ (Ex. 8) **4.** See Student Solutions Manual

You Try It
1. $x = 10$ **2.** $b = 125°$ **3.** $x = -4$ **4. (a)** $R = 2C$ **(b)** 6 carnations **5.** $W = 7$ ft **6.** $L = 7$ in.
7. $b = 12$ m **8.** $L = 8$ cm **9. (a)** y^9 **(b)** 6^4 **(c)** $4^2 \cdot 5^3$ **10.** $-28x^5$ **11.** $x^8 + 5x^5$ **12.** $5x^7 - 3x^4$

Chapter 3 Review Problems
1. two angles that share a common side **2.** two angles that have a sum of 180° **3.** straight lines that
are always the same distance apart **4.** variable expressions that contain terms with variable parts that have only whole number exponents
5. a number that is multiplied by a variable **6.** a polynomial with one term **7.** a polynomial with two terms **8.** a polynomial with three terms
9. $x = 27$ **10.** $x = -10$ **11.** $5 = y$ **12.** $y = 4$ **13.** $-14 = x$ **14.** $x = 12$ **15.** $\angle a = 99°$ **16.** $\angle y = 155°$
17. (a) $\angle x = \angle y + 22°$ **(b)** $\angle y = 79°$ **18.** $y = 12$ **19.** $y = -9$ **20.** $a = -7$ **21.** $x = 2$ **22.** $y = 2$ **23.** $x = -1$
24. $-4 = x$ **25.** $x = 7$ **26.** $y = 2$ **27.** $x = 10$ **28.** $x = 6$ **29. (a)** $L = 3W$ **(b)** $W = 12$ ft **30. (a)** $W = 2B$
(b) 100 blue cars **31.** $50 per hour **32.** 4 miles **33.** 8 in. **34.** $W = 4$ ft **35.** 36 ft^2 **36.** 3744 in.^{2} **37.** 99 in.^{2} **38.** 478 m^2
39. $x = 8$ cm **40.** $x = 12$ in. **41.** $H = 2$ ft **42.** $W = 3$ in. **43.** 750 ft^3 **44.** 125 in.^{3} **45.** $8 **46. (a)** 24 yd^2 **(b)** $360 **47.** 2^6
48. 7^9 **49.** a^7 **50.** $4^2 \cdot 3^3$ **51.** $15y^8$ **52.** $12x^8$ **53.** $21a^{10}$ **54.** $24y^{11}z^4$ **55.** $-12x^7$ **56.** $-35z^{10}$ **57.** $-27a^{11}$
58. $-20y^{14}$ **59.** Binomial **60.** Monomial **61.** Trinomial **62.** $x^3 + 3x$ **63.** $6x^5 - 24x^3$ **64.** $A = 3x^5 + 6x^3$ **65.** $V = 12x^7$

How Am I Doing? Chapter 3 Test
1. $x = -6$ (obj. 3.1.2) **2.** $y = 69$ (obj. 3.1.2) **3.** $-3 = a$ (obj. 3.1.2) **4.** $x = 12$ (obj. 3.1.2)
5. $4 = x$ (obj. 3.2.1) **6.** $y = -14$ (obj. 3.2.1) **7.** $x = -9$ (obj. 3.2.1) **8.** $-10 = y$ (obj. 3.2.1) **9.** $\angle y = 105°$ (obj. 3.1.3)
10. (a) $\angle x = \angle y + 10°$ **(b)** $\angle y = 85°$ (obj. 3.1.3) **11. (a)** $F = 3M$ **(b)** 7 male students (obj. 3.2.3) **12.** $W = 6$ yd (obj. 3.3.1)
13. 15 ft^2 (obj. 3.3.2) **14.** 72 in.^{2} (obj. 3.3.2) **15.** 124 cm^2 (obj. 3.3.2) **16.** $W = 3$ in. (obj. 3.3.3) **17.** $A = 4x^5 + x^4$ (obj. 3.4.4)
18. 6 yd^2 (obj. 3.3.4) **19.** 12,000 ft^3 (obj. 3.3.4) **20.** y^5 (obj. 3.4.1) **21.** z^4 (obj. 3.4.1) **22.** $2^3 \cdot 3^2$ (obj. 3.4.1) **23.** $5x^8$ (obj. 3.4.2)
24. $72x^6$ (obj. 3.4.2) **25.** $5y^5 + 40y$ (obj. 3.4.3) **26.** $18x^5 - 30x^4$ (obj. 3.4.3)

Cumulative Test for Chapters 1–3
1. 5300 **2.** 17,804 **3. (a)** $81,350 **(b)** $40,675 **4.** $84x$ **5.** $24y$ **6.** 14,003
7. 203 R2 **8. (a)** $2x = 28$ **(b)** $x + 9 = 16$ **9.** $>$ **10.** -1 **11.** -18 **12.** -25 **13.** 2 **14.** -32 **15.** -48 **16.** 22
17. (a) $-4mn + 4m$ **(b)** $10y + 8$ **18.** 54 **19.** $-2x + 6$ **20.** $y = 34$ **21.** $x = -11$ **22.** $-4 = y$ **23.** 10 in. **24.** $L = 5$ m
25. $\angle a = 75°$ **26. (a)** $C = 2A$ **(b)** 70 adults **27.** $-12x^8$ **28.** $2x^5 + 5x^3$

Chapter 4   4.1. Exercises 1. 2   3. 5, 0   5. With this method, the divisors and quotient are the factors. Therefore,

the divisor (number you divide by) must be prime for your factors to be prime.   **7.** No; it is not even.   **9.** No; the sum of the digits is not divisible
by 3.   **11.** 2, 3   **13.** 3, 5   **15.** 2, 3, 5   **17.** 3   **19.** 0, 1: neither; 17: prime; 9, 40, 8, 15, 22: composite   **21. (a)** $2 \cdot 4$ or $8 \cdot 1$
(b) $2 \cdot 2 \cdot 2$ or 2^3 **23.** 2, 2   **25.** 5, 2   **27. (a)** 5, 3   **(b)** $2 \cdot 3 \cdot 5^2$ **29. (a)** 10; 11, 2   **(b)** $2^2 \cdot 5 \cdot 11$ **31.** $3 \cdot 5$ **33.** $2 \cdot 2 \cdot 5$ or $2^2 \cdot 5$
35. $2 \cdot 2 \cdot 2 \cdot 3$ or $2^3 \cdot 3$ **37.** $2 \cdot 5 \cdot 7$ **39.** $2 \cdot 2 \cdot 2 \cdot 2 \cdot 2 \cdot 2$ or 2^6 **41.** $2 \cdot 2 \cdot 2 \cdot 2 \cdot 5$ or $2^4 \cdot 5$ **43.** $3 \cdot 5 \cdot 5$ or $3 \cdot 5^2$ **45.** $3 \cdot 3 \cdot 5$ or $3^2 \cdot 5$
47. $3 \cdot 3 \cdot 11$ or $3^2 \cdot 11$ **49.** $2 \cdot 2 \cdot 3 \cdot 5 \cdot 5$ or $2^2 \cdot 3 \cdot 5^2$ **51.** $2 \cdot 5 \cdot 11$ **53.** $2 \cdot 2 \cdot 2 \cdot 17$ or $2^3 \cdot 17$ **55.** $2 \cdot 3 \cdot 3 \cdot 5$ or $2 \cdot 3^2 \cdot 5$
57. $3 \cdot 3 \cdot 5 \cdot 5$ or $3^2 \cdot 5^2$ **59.** $7 \cdot 11 \cdot 17$ **61.** $7 \cdot 17 \cdot 23$ **63.** Answers may vary, but the number must end in the digit 5 or 0 and the sum of the
digits must be divisible by 3.   **65. (a)** 1, 3, 5, 7, 9, 11   **(b)** Third row: 1, 3, 5; Fourth row: 1, 3, 5, 7; Fifth row: 1, 3, 5, 7, 9; Sixth row: 1, 3, 5, 7, 9, 11
(c) 9, 16, 25, 36   **(d)** $3^2, 4^2, 5^2, 6^2$ **(e)** The list of numbers form a set of the first 6 perfect square whole numbers.   **(f)** The list of numbers
form a set of the first 6 whole numbers squared.   **(g)** 49, 7^2, 64, 8^2 **(h)** 12, 20   **67.** $10x^5y$ **68.** $9y^2$ **69.** $8x + 2$ **70.** $30x^2$

Quick Quiz 4.1 *See Examples noted with Ex.* **1.** 2, 3 (Ex. 1)   **2.** Composite (Ex. 2)   **3.** $3 \cdot 3 \cdot 5 \cdot 7$ or $3^2 \cdot 5 \cdot 7$ (Ex. 7)   **4.** See Student
Solutions Manual

4.2 Exercises 1. part; numerator; denominator   3. 3, 6; 2; 20, 3   5. $\frac{3}{5}$ 7. $\frac{3}{4}$ 9. Undefined   11. 0   13. 1   15. 0

17. Undefined   **19.** 1   **21.** $\frac{5}{12}$ **23.** $\frac{37}{94}$ **25.** $\frac{17}{26}$ **27.** $\frac{11}{32}$ **29.** $\frac{82}{135}$ **31.** $\frac{104}{135}$ **33.** Improper fraction   **35.** Improper fraction

37. Proper fraction   **39.** Mixed number   **41.** $1\frac{7}{8}$ **43.** $9\frac{3}{5}$ **45.** $20\frac{1}{2}$ **47.** $6\frac{2}{5}$ **49.** $9\frac{2}{5}$ **51.** 1   **53.** $\frac{59}{7}$ **55.** $\frac{97}{4}$ **57.** $\frac{47}{3}$

59. $\frac{100}{3}$ **61.** $\frac{89}{10}$ **63.** $\frac{127}{15}$ **65.** 40   **66.** -63 **67.** $3 \cdot 3 \cdot 7$ or $3^2 \cdot 7$ **68.** $2 \cdot 3 \cdot 3 \cdot 3$ or $2 \cdot 3^3$

Quick Quiz 4.2 *See Examples noted with Ex.* **1.** 0 (Ex. 2)   **2.** $\frac{23}{5}$ (Ex. 6)   **3.** $\frac{8}{15}$ (Ex. 3)   **4.** See Student Solutions Manual

4.3. Exercises 1. multiply; same   3. 2, 6   5. 3, 21   7. y, y 9. $y, y, 2y$ 11. (a) $\frac{28}{36}$ (b) $\frac{35x}{45x}$ 13. (a) $\frac{16}{44}$ (b) $\frac{20x}{55x}$ 15. $\frac{12}{32}$

17. $\frac{25}{30}$ **19.** $\frac{27}{39}$ **21.** $\frac{70}{80}$ **23.** $\frac{8y}{9y}$ **25.** $\frac{12y}{28y}$ **27.** $\frac{9a}{18a}$ **29.** $\frac{15x}{21x}$ **31.** $\frac{4}{5}$ **33.** $\frac{3}{4}$ **35.** $\frac{5}{6}$ **37.** $\frac{4}{7}$ **39.** $\frac{2}{3}$ **41.** $\frac{6}{17}$ **43.** $\frac{6}{7}$

45. $\frac{1}{2}$ **47.** $\frac{7}{5}$ or $1\frac{2}{5}$ **49.** $\frac{5}{4}$ or $1\frac{1}{4}$ **51. (a)** $-\frac{2}{3}$ **(b)** $-\frac{2}{3}$ **(c)** $-\frac{2}{3}$ **53.** $-\frac{1}{2}$ **55.** $-\frac{7}{8}$ **57.** $-\frac{5}{7}$ **59.** $-\frac{8}{9}$ **61.** $\frac{7}{8}$ **63.** $\frac{4}{7}$

65. $\frac{4y}{7}$ **67.** $\frac{1}{2x}$ **69.** $\frac{3x}{5}$ **71.** $\frac{5}{6y}$ **73.** $-\frac{6n}{7}$ **75.** $-\frac{7}{9x}$ **77.** $\frac{7}{18}$ **79.** $\frac{19}{27}$ **81.** $\frac{62}{650} = \frac{31}{325}$ **83.** $\frac{121}{650}$ **85.** $\frac{5abc^4}{11}$ **87.** $\frac{64xy^2}{75z^5}$

89. 60 in.   **90.** x^{10} **91.** 2^{10} **92.** $-6a^5$ **93. (a)** $P = 2C$ **(b)** $C = 17$ **94.** The square room has the greater area and the greater perimeter.

Quick Quiz 4.3 *See Examples noted with Ex.* **1.** $\frac{24}{56}$ (Ex. 2)   **2.** $-\frac{2}{3}$ (Ex. 4)   **3.** $\frac{3a}{4}$ (Ex. 5)   **4.** See Student Solutions Manual

Use Math to Save Money 1. Large-cap: $2500/$10,000 = 1/4; Small-cap: $2500/$10,000 = 1/4; Bonds: $2500/$10,000 = 1/4;

International: $2500/$10,000 = 1/4   **2.** Large-cap: $6000/$20,000 = 3/10; Small-cap: $5000/$20,000 = 1/4; Bonds: $4000/$20,000 = 1/5;
International: $5000/$20,000 = 1/4   **3.** Large-cap and Bonds   **4.** $5000   **5.** Large-cap, $6000 − $5000 = $1000, Jason needs to sell $1000
worth; Bonds, $5000 − $4000 = $1000, Jason needs to buy $1000 worth.

How Am I Doing?   Sections 4.1–4.3 1. 2, 3 (obj. 4.1.1)   2. $2 \cdot 2 \cdot 2 \cdot 3 \cdot 5$ or $2^3 \cdot 3 \cdot 5$ (obj. 4.1.3)   3. (a) 0   (b) Undefined

(c) 1   **(d)** 1 (obj. 4.2.1)   **4.** $7\frac{2}{7}$ (obj. 4.2.3)   **5.** $\frac{23}{4}$ (obj. 4.2.4)   **6.** $\frac{19}{36}$ (obj. 4.2.1)   **7.** $\frac{15}{35}$ (obj. 4.3.1)   **8.** $\frac{6y}{27y}$ (obj. 4.3.3)   **9.** $-\frac{10}{21}$ (obj. 4.3.2)

10. $\frac{5}{9y}$ (obj. 4.3.3)   **11.** $\frac{5}{6}$ (obj. 4.3.4)

4.4 Exercises 1. (a) We add coefficients; the variables stay the same; $20x^3$ (b) We multiply coefficients and then add exponents of like

bases; $75x^6$ **(c)** We simplify coefficients and then subtract exponents of like bases; $3x^2$ **(d)** We multiply exponents; x^6 **3.** 7   **5.** a^5
7. $\frac{1}{5}$ **9.** $\frac{1}{3^5}$ **11.** $\frac{z^4}{y^8}$ **13.** $\frac{9^3}{8^8}$ **15.** 1   **17.** $\frac{1}{y^2z^3}$ **19.** $\frac{m^2}{3}$ **21.** $\frac{a^2}{7^3}$ **23.** $\frac{b^2}{9^2}$ **25.** $\frac{1}{4^3a^3b}$ **27.** $\frac{4y^3}{7}$ **29.** $\frac{a}{3}$ **31.** $\frac{7x^6}{8}$

33. $\frac{4x^3}{5y}$ **35. (a)** z^6 **(b)** z^6 **37. (a)** x^8 **(b)** x^8 **39.** z^8 **41.** 3^{20} **43.** b^6 **45.** 1   **47.** y^9 **49.** 2^{20} **51.** 1   **53.** 6^{27} **55.** x^4

57. 3^6y^{12} **59.** 4^3x^6 **61.** 3^8a^{32} **63.** 2^6x^{15} **65.** $8^{18}n^{24}$ **67.** $\frac{4^2}{x^2}$ or $\frac{16}{x^2}$ **69.** $\frac{a^7}{b^7}$ **71.** $\frac{3^3}{x^3}$ or $\frac{27}{x^3}$ **73.** $\frac{m^4}{n^4}$ **75.** $\frac{x^2}{6^2}$ or $\frac{x^2}{36}$ **77.** $\frac{3^2}{7^2}$

or $\frac{9}{49}$ **79.** $\frac{5y^2z^4}{27x^5}$ **81.** $\frac{13b^2}{12c^9}$ **83.** $\frac{4y^2}{25x^2}$ **85.** $\frac{27a^6}{8b^9}$ **87. (a)** $20x^3$ **(b)** $75x^6$ **(c)** x^9 **(d)** $\frac{3}{x^2}$ **89. (a)** $12x^3$ **(b)** $27x^6$ **(c)** 3^3x^9 or $27x^9$

(d) $\frac{1}{3x}$ **91. (a)** $15x^4$ **(b)** $36x^8$ **(c)** 2^4x^{16} or $16x^{16}$ **(d)** $\frac{2x^3}{3}$ **93. (a)** $20y^2$ **(b)** $75y^4$ **(c)** 5^2y^4 or $25y^4$ **(d)** $\frac{1}{3y^5}$ **95.** 1, 6, 15, 20, 15, 6, 1

96. $x = 14$ **97.** $x = 3$ **98.** $x = 2$ **99.** $x = 10$ **100.** $A = 6x^3 + 6x^2$ **101.** $6960

Quick Quiz 4.4 *See Examples noted with Ex.* **1.** $\frac{a^2}{b}$ (Ex. 2)   **2.** 7^2x^6 or $49x^6$ (Ex. 4)   **3.** $\frac{y^4}{2^4}$ or $\frac{y^4}{16}$ (Ex. 5)   **4.** See Student Solutions Manual

4.5 Exercises
1. A ratio compares amounts with the same units, and a rate compares amounts with different units.   **3.** $\frac{5}{9}$ **5.** $\frac{7}{2}$ **7.** $\frac{27}{35}$

9. $\frac{17}{6}$ **11.** $\frac{2}{5}$ **13.** $\frac{17}{41}$ **15.** $\frac{121}{423}$ **17. (a)** $\frac{3}{7}$ **(b)** $\frac{7}{3}$ **19. (a)** $\frac{29}{13}$ **(b)** $\frac{13}{29}$ **21.** $\frac{8 \text{ grams}}{20 \text{ grams}} = \frac{2}{5}$ **23.** $\frac{3970}{6532} = \frac{1985}{3266}$ **25.** $21\frac{11}{19}$

calories per gram of fat   **27.** 20 miles per gal   **29.** \$8 per hour   **31.** $53\frac{1}{3}$ mph   **33.** 22 miles   **35.** \$12 per book   **37. (a)** 18 students
per instructor **(b)** 15 students per tutor **(c)** 6 tutors   **39. (a)** box of 8: \$12 per glass; box of 6: \$13 per glass **(b)** Box of 8   **41. (a)** 4 used
CDs: \$8 per CD; 6 used CDs: \$8 per CD **(b)** They both offer the same unit price.   **43.** 102 sales per day   **45.** $x = 37$ **46.** $x = 25$
47. $x = 8$ **48.** $a = 22$ **49.** 109°F   **50.** A loss of 1 yd

Quick Quiz 4.5 *See Examples noted with Ex.* **1.** $\frac{3}{5}$ (Ex. 3)   **2.** $2\frac{1}{2}$ hot dogs per minute (Ex. 5)   **3.** \$72 for 36 bars (Ex. 8)
4. See Student Solutions Manual

Understanding the Concept: Reducing a Proportion
1. A ratio is a fraction. Reducing a fraction does not change the value of the fraction.

4.6 Exercises
1. Whenever the numerator and denominator of a fraction are equal, the fraction is equal to 1. Thus both fractions are equal to 1.

3. $\frac{2}{7} = \frac{24}{84}$ **5.** $\frac{12 \text{ goals}}{7 \text{ games}} = \frac{24 \text{ goals}}{14 \text{ games}}$ **7.** $\frac{3}{8} = \frac{18}{48}$ **9.** $\frac{14 \text{ crackers}}{6 \text{ grams}} = \frac{70 \text{ crackers}}{30 \text{ grams}}$ **11.** $\frac{3\frac{1}{2} \text{ rotations}}{2 \text{ min}} = \frac{14 \text{ rotations}}{8 \text{ min}}$
13. $\frac{4 \text{ made}}{7 \text{ attempts}} = \frac{12 \text{ made}}{21 \text{ attempts}}$ **15.** $\neq$ **17.** $=$ **19.** Yes   **21.** No   **23.** No   **25.** $x = 20$ **27.** $30 = x$ **29.** $14 = x$
31. $x = 4$ **33.** $n = 6$ **35.** $n = 4$ **37.** No   **39.** Yes   **41.** No   **43.** 315 calories   **45.** 6800 calories   **47.** 112 million   **49.** 350 ft

51. 10 lb   **53.** 1360 shares   **55.** 42 mm   **57.** 20 ft wide   **59.** \$2800   **61.** $\frac{\frac{1}{3}}{\frac{1}{8}} = \frac{\frac{1}{4}}{\frac{3}{32}}$ **63.** \$240   **65.** \$192   **67.** \$1290

69. $W = 9$ in.; $H = 15$ in.   **71. (a)** 2, 3, 5, 8, 13, 21, 34 **(b)** To find the next number in the sequence, we add the two preceding numbers.
(c) 55, 89, 144   **72.** 215 points   **73.** 205 points

Quick Quiz 4.6 *See Examples noted with Ex.* **1.** No (Ex. 4)   **2.** $72 = n$ (Ex. 5)   **3.** 324 mi (Ex. 7)   **4.** See Student Solutions Manual

You Try it
1. Divisible by 3 only   **2.** $2 \cdot 7^2$ **3.** $6\frac{3}{7}$ **4.** $\frac{26}{5}$ **5.** $\frac{7}{8} = \frac{21x}{24x}$ **6.** $\frac{3x}{4}$ **7. (a)** x^5 **(b)** $\frac{1}{4^3} = \frac{1}{64}$ **8. (a)** $36x^{30}$
(b) $\frac{x^2}{9}$ **9.** The ratio of 15 to 26, 15 : 26, $\frac{15}{26}$ **10.** $\frac{2 \text{ staff}}{15 \text{ children}}$ **11.** 61 boxes per hour   **12.** $\frac{35}{55} = \frac{14}{22}$ **13.** The fractions are equal
14. This is not a proportion   **15.** $n = 23$ **16.** 16 products

Chapter 4 Review Problems
1. A whole number greater than 1 that is divisible only by itself and 1   **2.** A whole number greater
than 1 that is divisible by whole numbers other than itself and 1   **3.** A fraction that describes a quantity less than 1   **4.** A fraction that
describes a quantity greater than or equal to 1   **5.** The sum of a whole number greater than zero and a proper fraction   **6.** Fractions that look
different but have the same value   **7.** A comparison of two quantities that have the same units   **8.** A ratio that compares quantities
that have different units   **9.** A statement that two ratios or rates are equal   **10.** 2, 5   **11.** 3, 5   **12.** 0: neither; 7, 11: prime; 21, 50, 25, 51:
composite   **13.** 1: neither; 7, 13, 41: prime; 32, 12, 50, 6: composite   **14.** $2 \cdot 2 \cdot 3 \cdot 3$ or $2^2 \cdot 3^2$ **15.** $2 \cdot 2 \cdot 2 \cdot 7$ or $2^3 \cdot 7$ **16.** $5 \cdot 5 \cdot 17$ or $5^2 \cdot 17$
17. $2 \cdot 3 \cdot 3 \cdot 5$ or $2 \cdot 3^2 \cdot 5$ **18.** Undefined   **19.** 0   **20.** 1   **21.** $\frac{20}{69}$ **22.** $\frac{17}{25}$ **23.** $8\frac{3}{5}$ **24.** $9\frac{1}{6}$ **25.** 7   **26.** $\frac{7}{3}$ **27.** $\frac{33}{5}$
28. $\frac{52}{5}$ **29.** $\frac{4}{18}$ **30.** $\frac{27}{36}$ **31.** $\frac{28x}{35x}$ **32.** $\frac{18y}{33y}$ **33.** $\frac{11}{15}$ **34.** $\frac{8}{9}$ **35.** 3   **36.** $\frac{7}{3}$ or $2\frac{1}{3}$ **37.** $\frac{5}{12}$ **38.** $\frac{4}{5y}$ **39.** $-\frac{8}{9}$
40. $-\frac{2}{3}$ **41.** y^2 **42.** $\frac{1}{3}$ **43.** $\frac{8^2}{3^4}$ **44.** $\frac{x^3}{y^6}$ **45.** $\frac{1}{2^3x^9}$ or $\frac{1}{8x^9}$ **46.** $\frac{1}{3y^6}$ **47.** $\frac{4}{7x^4}$ **48.** $3y^2$ **49.** 3^3y^6 or $27y^6$ **50.** 2^8x^2
51. $\frac{3^2}{y^2}$ or $\frac{9}{y^2}$ **52.** $\frac{x^3}{2^3}$ or $\frac{x^3}{8}$ **53.** $\frac{15}{23}$ **54.** $\frac{3}{7}$ **55. (a)** $\frac{4}{11}$ **(b)** $\frac{11}{4}$ **56.** \$5 per washcloth   **57.** 4 in. per hr   **58.** 26 mi/gal
59. 103 words per minute   **60. (a)** 2 legal secretaries per lawyer **(b)** 3 paralegals per lawyer **(c)** 180 paralegals   **61. (a)** 6 for \$72, \$12 per CD;
8 for \$96, \$12 per CD **(b)** Both have the same unit price.   **62. (a)** $\frac{24}{31}$ **(b)** \$1500 per day   **63.** $\frac{3}{7} = \frac{21}{49}$ **64.** $\frac{2 \text{ teachers}}{50 \text{ students}} = \frac{6 \text{ teachers}}{150 \text{ students}}$
65. $\frac{2 \text{ in.}}{190 \text{ mi}} = \frac{6 \text{ in.}}{570 \text{ mi}}$ **66.** $\frac{234 \text{ mi}}{9 \text{ gal}} = \frac{468 \text{ mi}}{18 \text{ gal}}$ **67.** $\frac{3}{4} \neq \frac{50}{70}$ **68.** $\frac{13}{91} = \frac{12}{84}$ **69.** Yes   **70.** No   **71.** $x = 4$ **72.** $x = 45$ **73.** $34 = n$
74. 360 mi   **75.** 16 mph   **76.** 8 ft wide   **77. (a)** \$51 **(b)** \$17

How Am I Doing? Chapter 4 Test
1. 2, 5 (obj. 4.1.1)   **2. (a)** Composite **(b)** Neither **(c)** Prime (obj. 4.1.2)
3. $2 \cdot 2 \cdot 3 \cdot 7$ or $2^2 \cdot 3 \cdot 7$ (obj. 4.1.3)   **4.** $2 \cdot 2 \cdot 2 \cdot 3 \cdot 5$ or $2^3 \cdot 3 \cdot 5$ (obj. 4.1.3)   **5. (a)** 0 **(b)** 1 **(c)** Undefined (obj. 4.2.1)   **6.** $\frac{17}{36}$ (obj. 4.2.1)
7. (a) $\frac{16}{35}$ **(b)** $\frac{23}{35}$ (obj. 4.2.1)   **8.** $1\frac{3}{5}$ (obj. 4.2.3)   **9.** $\frac{43}{6}$ (obj. 4.2.4)   **10.** $\frac{35}{40}$ (obj. 4.3.1)   **11.** $\frac{12y}{27y}$ (obj. 4.3.1)   **12.** $-\frac{9}{28}$ (obj. 4.3.2)

13. $\dfrac{1}{2xy}$ (obj. 4.3.3)   **14.** $\dfrac{z^3}{y^4}$ (obj. 4.3.3)   **15.** $\dfrac{8^2}{7^3}$ (obj. 4.3.2)   **16.** $\dfrac{7x^7}{6y^3}$ (obj. 4.3.3)   **17.** 2^3y^{12} or $8y^{12}$ (obj. 4.4.2)   **18.** x^{15} (obj. 4.4.2)

19. $\dfrac{x^2}{3^2}$ or $\dfrac{x^2}{9}$ (obj. 4.4.1)   **20. (a)** $\dfrac{9}{2}$ **(b)** $\dfrac{2}{9}$ (obj. 4.5.1)   **21.** $7\dfrac{1}{2}$ cal/min (obj. 4.5.3)   **22.** 21 mi/gal (obj 4.5.2)   **23. (a)** \$3 per ream; \$2 per ream

(b) \$96 for 48 reams (obj. 4.5.3)   **24.** $\dfrac{2 \text{ in.}}{225 \text{ mi}} = \dfrac{6 \text{ in.}}{675 \text{ mi}}$ (obj. 4.6.1)   **25.** Yes (obj. 4.6.2)   **26.** $x = 30$ (obj. 4.6.3)   **27.** Yes (obj. 4.6.4)

28. 8 tablespoons (obj. 4.6.4)

Chapter 5   **5.1 Exercises 1.** We multiply numerator times numerator and denominator times denominator.   **3.** We divide because we want to split an amount into equal parts.   **5.** Answers may vary. Any applied problem that requires taking $\frac{1}{3}$ of 90 or uses repeated addition of $\frac{1}{3}$ is correct.   **7.** 14   **9.** 7; 3   **11.** 5; 6   **13.** $\cdot$; 3; 4   **15.** $\dfrac{1}{12}$ **17.** $\dfrac{5}{24}$ **19.** $\dfrac{1}{6}$ **21.** $\dfrac{2}{15}$ **23.** $\dfrac{7}{8}$ **25.** $-\dfrac{8}{11}$

27. $\dfrac{2}{27}$ **29.** $-\dfrac{1}{2}$ **31.** $\dfrac{6}{7}$ **33.** $\dfrac{2x^2}{5}$ **35.** $24x^5$ **37.** $12x^5$ **39.** $\dfrac{3}{20}$ **41.** $\dfrac{3}{10x}$ **43.** $-\dfrac{3y}{35}$ **45.** $\dfrac{9x^5}{50}$ **47.** $A = 48$ m^2

49. $A = 420$ in.^{2} **51.** $\dfrac{3}{1}$ or 3   **53.** $\dfrac{1}{5}$ **55.** $-\dfrac{5}{2}$ **57.** $-\dfrac{y}{x}$ **59.** $\dfrac{8}{7}$ **61.** $\dfrac{14}{27}$ **63.** $-\dfrac{1}{9}$ **65.** $\dfrac{1}{3}$ **67.** $\dfrac{x^4}{6}$ **69.** $-\dfrac{3x^6}{4}$

71. 49   **73.** $\dfrac{1}{44}$ **75.** $9x^3$ **77.** $12x^8$ **79. (a)** $\dfrac{5}{63}$ **(b)** $\dfrac{7}{125}$ **81. (a)** $\dfrac{7x^7}{6}$ **(b)** $\dfrac{8}{21x^3}$ **83.** $\dfrac{x^3}{12}$ **85.** $-\dfrac{15x^8}{28}$ **87.** \$180

89. 16 pipes   **91.** 8 miles   **93.** 160 bottles   **95.** $-\dfrac{3}{4}$ **97.** 6 yr   **99.** 7 pizzas   **101.** $x = 16$ **103.** $x = 6, y = 12$

105. $x = 8, y = 162$ **107.** $\dfrac{10}{15}$ **108.** $\dfrac{15}{20}$ **109.** $2 \cdot 2 \cdot 2 \cdot 3 \cdot 5$ or $2^3 \cdot 3 \cdot 5$ **110.** $5 \cdot 29$

Quick Quiz 5.1 *See Examples noted with Ex.*   **1.** $\dfrac{1}{9}$ (Ex. 1)   **2.** $-\dfrac{3x^6}{4}$ (Ex. 4)   **3.** $\dfrac{25x^2}{2}$ (Ex. 8)   **4.** See Student Solutions Manual

5.2 Exercises 1. Because $3 \cdot 4 = 12$ and there is no whole number that we can multiply by 5 to get 12.   **3. (a)** 6, 12, 18, 24; 8, 16, 24, 32 **(b)** 24   **5. (a)** 2, 4, 6, 8, 10; 5, 10, 15, 20, 25 **(b)** 10   **7. (a)** $12x, 24x, 36x, 48x$; $18x, 36x, 54x, 72x$ **(b)** $36x$ **9.** 2, 3   **11.** 7, x **13.** 15
15. 56   **17.** 60   **19.** 120   **21.** 120   **23.** 140   **25.** $36x$ **27.** $567a$ **29.** $90x^2$ **31.** $44x^3$ **33.** $60x^3$ **35.** $84x^2$ **37.** 12 min

39. 24 min   **41.** 1:15 P.M.   **43.** $40x^3y^2$ **45.** $30xyz^2$ **47.** $9\dfrac{1}{2}$ **48.** $\dfrac{22}{5}$ **49.** 0   **50.** 7

Quick Quiz 5.2 *See Examples noted with Ex.*   **1.** 45 (Ex. 3)   **2.** $36x^3$ (Ex. 4)   **3.** 90 (Ex. 3)   **4.** See Student Solutions Manual

5.3 Exercises 1. numerators; denominator   **3.** No, we must find a common denominator when we add or subtract fractions with different denominators. Also, we do not add or subtract denominators.   **5.** 2   **7.** 2   **9.** $\dfrac{15}{17}$ **11.** $\dfrac{1}{23}$ **13.** $-\dfrac{6}{7}$ **15.** $-\dfrac{20}{51}$ **17.** $\dfrac{1}{y}$ **19.** $\dfrac{39}{a}$

21. $\dfrac{x-5}{7}$ **23.** $\dfrac{y+14}{3}$ **25.** 30   **27.** 45   **29.** $\dfrac{12}{60}$ **31.** $\dfrac{50}{60}$ **33.** $\dfrac{2}{45}$ **35.** $\dfrac{13}{24}$ **37.** $\dfrac{53}{56}$ **39.** $-\dfrac{13}{20}$ **41.** $\dfrac{3}{26}$ **43.** $-\dfrac{11}{35}$

45. $\dfrac{12}{35}$ **47.** $\dfrac{19}{60}$ **49.** $\dfrac{21}{2x}$ **51.** $\dfrac{23}{7x}$ **53.** $\dfrac{7}{3x}$ **55.** $\dfrac{3y+4x}{xy}$ **57.** $\dfrac{4b-9a}{ab}$ **59.** $\dfrac{11x}{15}$ **61.** $-\dfrac{13x}{20}$ **63.** $-\dfrac{7x}{12}$ **65.** $\dfrac{17}{15}$ or $1\dfrac{2}{15}$

67. $-\dfrac{21}{80}$ **69.** $\dfrac{9x+y}{xy}$ **71.** $\dfrac{17x}{60}$ **73.** $1\dfrac{5}{8}$ lb   **75. (a)** $1\dfrac{1}{4}$ cups of sugar **(b)** $\dfrac{1}{4}$ cup   **77.** $\dfrac{1}{12}$ **79. (a)** $\dfrac{5}{24}$ **(b)** $\dfrac{1}{24}$ **81.** $\dfrac{13}{30}$ **83.** $\dfrac{1}{4}$
85. 4   **86.** 5   **87.** -2 **88.** -22 **89.** \$595   **90.** \$74

Quick Quiz 5.3 *See Examples noted with Ex.*   **1.** $\dfrac{7}{9y}$ (Ex. 7)   **2. (a)** $\dfrac{17}{30}$ **(b)** $\dfrac{1}{6}$ (Ex. 6)   **3.** $\dfrac{29x}{30}$ (Ex. 8)   **4.** See Student Solutions Manual

Understanding the Concept: Should We Change to an Improper Fraction?   **1.** With large numbers, it is easier to keep numbers as mixed numbers.

5.4 Exercises 1. (a) Marcy did not change the mixed numbers to improper fractions before she multiplied.   **(b)** $10\dfrac{2}{15}$ **3.** $21\dfrac{5}{9}$ **5.** $16\dfrac{3}{4}$

7. $13\dfrac{11}{12}$ **9.** $20\dfrac{7}{12}$ **11.** $12\dfrac{5}{24}$ **13.** $5\dfrac{3}{5}$ **15.** $3\dfrac{1}{2}$ **17.** $4\dfrac{3}{5}$ **19.** $6\dfrac{31}{60}$ **21.** $6\dfrac{3}{4}$ **23.** $2\dfrac{9}{35}$ **25.** $1\dfrac{13}{24}$ **27.** $12\dfrac{1}{12}$ **29.** $30\dfrac{7}{9}$

31. $-\dfrac{32}{5}$ or $-6\dfrac{2}{5}$ **33.** $\dfrac{39}{4}$ or $9\dfrac{3}{4}$ **35.** $-\dfrac{39}{14}$ or $-2\dfrac{11}{14}$ **37.** $-\dfrac{9}{16}$ **39.** 2   **41.** $\dfrac{26}{3}$ or $8\dfrac{2}{3}$ **43.** -24 **45.** $\dfrac{55}{12}$ or $4\dfrac{7}{12}$ **47.** $-\dfrac{21}{4}$ or $-5\dfrac{1}{4}$

49. $-\dfrac{15}{16}$ **51.** 4 pieces   **53.** $2\dfrac{2}{5}$ ft   **55.** $4\dfrac{1}{2}$ cups flour   **57.** 10 cups chocolate chips   **59.** $38\dfrac{5}{8}$ mi   **61.** $a = 5, b = 4, c = 3$

63. 74   **64.** 10   **65.** 2   **66.** 3   **67.** \$638   **68.** \$709

Quick Quiz 5.4 *See Examples noted with Ex.*   **1. (a)** $5\dfrac{7}{36}$ (Ex. 3) **(b)** $4\dfrac{11}{12}$ (Ex. 5)   **2.** -21 (Ex. 7)   **3.** -4 (Ex. 8)   **4.** See Student Solutions Manual

Use Math to Save Money **1.** $(4 \times \$8) + \$13 + (5 \times \$10) = \$95$ **2.** $10 \times \$3 = \$30$ **3.** $\$95 - \$30 = \$65$
4. $\$2275/\$65 = 35$, After 35 weeks

How Am I Doing? Sections 5.1–5.4 **1.** $-\dfrac{8}{9}$ (obj. 5.1.1) **2.** $\dfrac{3y^5}{35}$ (obj. 5.1.1) **3.** 35 (obj. 5.1.2) **4.** $\dfrac{y^2}{16}$ (obj. 5.1.2)

5. 84 parcels (obj. 5.1.3) **6.** 84 (obj. 5.2.2) **7.** $42x^2$ (obj. 5.2.2) **8.** $-\dfrac{1}{3}$ (obj. 5.3.2) **9.** $\dfrac{7}{16}$ (obj. 5.3.2) **10.** $\dfrac{23x}{56}$ (obj. 5.3.2)

11. **(a)** $\dfrac{1}{20}$ **(b)** $\dfrac{9}{20}$ (obj. 5.3.2) **12.** $7\dfrac{9}{35}$ (obj. 5.4.1) **13.** $3\dfrac{37}{60}$ (obj. 5.4.1) **14.** $\dfrac{7}{2}$ or $3\dfrac{1}{2}$ (obj. 5.4.2) **15.** $\dfrac{92}{21}$ or $4\dfrac{8}{21}$ (obj. 5.4.2)

5.5 Exercises **1.** Multiply **3.** $\dfrac{1}{5}$ **5.** $\dfrac{9}{10}$ **7.** $\dfrac{5}{6}$ **9.** $\dfrac{29}{12}$ or $2\dfrac{5}{12}$ **11.** $\dfrac{11}{12}$ **13.** $\dfrac{5}{6}$ **15.** $\dfrac{3}{4}$ **17.** $-\dfrac{1}{40}$ **19.** 18 **21.** $\dfrac{1}{28}$

23. 8 **25.** $\dfrac{4}{3}$ **27.** $2x$ **29.** $\dfrac{3}{x}$ **31.** $\dfrac{25}{18}$ or $1\dfrac{7}{18}$ **33.** $\dfrac{2}{11}$ **35.** $\dfrac{2}{x}$ **37.** $\dfrac{23}{28}$ **39.** $\dfrac{25}{62}$ **41.** 30 **43.** $4\dfrac{19}{20}$ cups of sugar

45. $22\dfrac{1}{2}$ bags **47.** 21 acres **49.** $\dfrac{10y}{21x}$ **51.** $\dfrac{y-x}{y+x}$ **53.** 2 **55.** $-\dfrac{6}{7}$ **57.** -30 **58.** 15 **59.** -10 **60.** 6

Quick Quiz 5.5 *See Examples noted with Ex.* **1.** $\dfrac{23}{45}$ (Ex. 1) **2.** $\dfrac{2}{3}$ (Ex. 3) **3.** $\dfrac{2}{5}$ (Ex. 4) **4.** See Student Solutions Manual

5.6 Exercises **1.** $1\dfrac{1}{4}$ hr **3.** $67\dfrac{7}{9}$ mph **5.** $7\dfrac{1}{2}$ cups water, $\dfrac{3}{4}$ tsp salt, $1\dfrac{1}{2}$ cups cereal **7.** $13\dfrac{3}{4}$ yd **9.** **(a)** $6\dfrac{1}{6}$ gal **(b)** 148 mi

11. $\$1500$ **13.** **(a)** $30\dfrac{1}{2}$ ft by $15\dfrac{1}{2}$ ft **(b)** $\$207$ **15.** 6 boards **17.** **(a)** 39 ft by 19 ft **(b)** $\$290$ **19.** 19 pieces of wood

21. **(a)** $53{,}333\dfrac{1}{3}$ miles per day **(b)** $2222\dfrac{2}{9}$ mph **23.** $x = 4$ **24.** $x = 9$ **25.** $x = 17$ **26.** $x = -4$

Quick Quiz 5.6 *See Examples noted with Ex.* **1.** 5 boards (Ex. 2) **2.** **(a)** $135\dfrac{3}{8}$ lb **(b)** $1\dfrac{3}{8}$ lb (Ex. 1) **3.** **(a)** $21\dfrac{1}{2}$ ft by 17 ft
(b) $\$154$ (Ex. 1) **4.** See Student Solutions Manual

5.7 Exercises **1.** 1 **3.** multiply **5.** $12; 12; -36$ **7.** $-5; -5; -20$ **9.** $\dfrac{5}{2}; \dfrac{5}{2}; -20$ **11.** $\dfrac{7}{-5}; \dfrac{7}{-5}; -14$ **13.** $y = 96$
15. $x = 217$ **17.** $m = -390$ **19.** $x = -75$ **21.** $-60 = a$ **23.** $80 = a$ **25.** $x = 36$ **27.** $x = 7$ **29.** $y = 5$ **31.** $y = 20$
33. $x = 27$ **35.** $x = -30$ **37.** $x = -48$ **39.** $a = 33$ **41.** $x = 63$ **43.** $y = 16$ **45.** $-30 = m$ **47.** **(a)** $x = 13$ **(b)** $x = 208$
49. **(a)** $x = -13$ **(b)** $x = -208$ **51.** **(a)** $x = 28$ **(b)** $x = 21$ **(c)** $x = \dfrac{3}{5}$ **(d)** $x = 55$ **53.** **(a)** $x = 5$ **(b)** $r = -\dfrac{2}{3}$ **(c)** $x = -54$

(d) $x = -45$ **55.** $\dfrac{a}{b} = \dfrac{2}{3}$ **57.** $\dfrac{a}{b} = \dfrac{3}{7}$ **59.** $-6a + 9b - 10ab$ **60.** $-8x - 2xy - 4y$ **61.** $3x^6 + 24x^3$ **62.** $-2x^5 + 12x^3$

63. $15\dfrac{3}{4}$ ft^2 **64.** 10 cups

Quick Quiz 5.7 *See Examples noted with Ex.* **1.** $x = -35$ (Ex. 2) **2.** $a = 30$ (Ex. 3) **3.** $y = -72$ (Ex. 1)
4. See Student Solutions Manual

You Try It **1.** $-\dfrac{12}{35}$ **2.** $\dfrac{21}{2x}$ **3.** 120 **4.** **(a)** $\dfrac{3}{13}$ **(b)** $\dfrac{9}{y}$ **5.** $1\dfrac{37}{90}$ **6.** $12\dfrac{2}{9}$ **7.** $4\dfrac{15}{28}$ **8.** **(a)** $\dfrac{33}{16}$ or $2\dfrac{1}{16}$ **(b)** $16\dfrac{2}{3}$ **9.** $\dfrac{3}{14}$
10. $\dfrac{22}{7}$ **11.** $x = -5$

Chapter 5 Review Problems **1.** The least common multiple is the smallest common multiple. **2.** The least common denominator is
the LCM of the denominators. **3.** A fraction that contains at least one fraction in the numerator or denominator **4.** **(a)** $\dfrac{9}{4}$ **(b)** $-\dfrac{1}{6}$

5. **(a)** $\dfrac{1}{7}$ **(b)** 11 **6.** $-\dfrac{2}{3}$ **7.** $\dfrac{1}{21}$ **8.** $-\dfrac{4}{45}$ **9.** $\dfrac{7}{10x^2}$ **10.** $-\dfrac{4x^2}{5}$ **11.** $-10x^7$ **12.** $\dfrac{6}{35}$ **13.** $-\dfrac{4}{15}$ **14.** $-\dfrac{2}{33}$ **15.** $\dfrac{11x^7}{15}$

16. $\dfrac{4}{9x^2}$ **17.** $A = 63$ m^2 **18.** $A = 110$ in.^{2} **19.** $\$1000$ withheld **20.** $\dfrac{1}{6}$ lb in each container **21.** 21 **22.** 20 **23.** 90 **24.** 36

25. $16x$ **26.** $140x^4$ **27.** $90x^2$ **28.** $100x^2$ **29.** $\dfrac{3}{17}$ **30.** $-\dfrac{34}{27}$ **31.** $\dfrac{2}{x}$ **32.** $\dfrac{x-4}{7}$ **33.** $\dfrac{11}{18}$ **34.** $\dfrac{7}{32}$ **35.** $\dfrac{1}{6}$ **36.** $\dfrac{31}{6x}$

37. $\dfrac{14x}{45}$ **38.** $\dfrac{2x}{21}$ **39.** $13\dfrac{2}{15}$ **40.** $19\dfrac{4}{9}$ **41.** $4\dfrac{19}{25}$ **42.** $8\dfrac{9}{14}$ **43.** -13 **44.** $\dfrac{11}{20}$ **45.** $\dfrac{66}{125}$ **46.** -18 **47.** $13\dfrac{1}{2}$ ft

48. $1\dfrac{7}{8}$ ft **49.** $\dfrac{19}{20}$ **50.** $\dfrac{13}{16}$ **51.** 28 **52.** 12 **53.** $\dfrac{5}{3x}$ **54.** $\dfrac{27}{20}$ **55.** $\dfrac{22}{9}$ **56.** 26 acres **57.** **(a)** $29\dfrac{1}{4}$ by 16 ft **(b)** $\$294\dfrac{1}{8}$

58. $1\dfrac{5}{8}$ in. **59.** $x = 27$ **60.** $y = -12$ **61.** $x = 24$ **62.** $y = 33$ **63.** $x = -24$ **64.** $y = 12$ **65.** $x = 104$ **66.** $y = 45$

How Am I Doing? Chapter 5 Test **1.** $\frac{3}{5}$ (obj. 5.1.1) **2.** $-\frac{1}{10}$ (obj. 5.1.1) **3.** x^3 (obj. 5.1.1) **4.** -2 (obj. 5.1.2)

5. $\frac{3x^4}{11}$ (obj. 5.1.2) **6.** $\frac{10}{9x^2}$ (obj. 5.1.1) **7.** 6, 12, 18, 24, 30, 36; 9, 18, 27, 36, 45, 54; 18 and 36 (obj. 5.2.1) **8.** 42 (obj. 5.2.2)

9. $20a^4$ (obj. 5.2.2) **10.** 70 (obj. 5.2.2) **11.** 60 (obj. 5.2.2) **12.** $\frac{15}{11}$ or $1\frac{4}{11}$ (obj. 5.3.1) **13.** $\frac{16}{5a}$ (obj. 5.3.2) **14.** $8\frac{1}{6}$ (obj. 5.4.1)

15. $\frac{41}{63}$ (obj. 5.3.2) **16.** $\frac{9}{20}$ (obj. 5.3.2) **17.** $7\frac{11}{24}$ (obj. 5.4.1) **18.** $\frac{5x}{12}$ (obj. 5.3.2) **19.** $-\frac{64}{3}$ or $-21\frac{1}{3}$ (obj. 5.4.2) **20.** $\frac{5}{9}$ (obj. 5.4.2)

21. $-\frac{25}{6}$ or $-4\frac{1}{6}$ (obj. 5.4.2) **22.** $\frac{41}{72}$ (obj. 5.5.1) **23.** 2 (obj. 5.5.2) **24.** $\frac{15}{4}$ or $3\frac{3}{4}$ (obj. 5.5.2) **25.** $x = -30$ (obj. 5.5.1)

26. $y = 49$ (obj. 5.7.1) **27.** $x = -80$ (obj. 5.7.1) **28.** 5 boards (obj. 5.6.1) **29. (a)** $100°F$ **(b)** $1\frac{1}{2}°F$ (obj. 5.6.1)

30. (a) $25\frac{1}{2}$ ft by 12 ft **(b)** $\$112\frac{1}{2}$ (obj. 5.6.1)

Chapter 6
6.1 Exercises **1.** opposite; first **3.** $+2z^2, +4z, -2y^4, +3$ **5.** $+6x^6, -3x^3, -3y, -1$ **7.** $3y + 6$
9. $2a^2 + a + 4$ **11.** $y^2 - 6y - 3$ **13.** $-5x - 2y$ **15.** $8x - 4$ **17.** $3x - 6z + 5y$ **19.** $7x + 2$ **21.** $11x - 9$ **23.** $-12a + 8$
25. $-y^2 + 10y + 3$ **27.** $-4x^2 + 10x + 3$ **29.** $-9z^2 + z + 6$ **31.** $7a^2 + 17a + 3$ **33.** $-9x^2 - 14x - 5$ **35.** $-8x^2 + 2x + 13$
37. $-5x^2 - 2x - 20$ **39.** $-5x^2 + 7x + 6$ **41.** $7x^2 - 13x - 9$ **43.** $2x^2 - x - 4$ **45.** $-3x^2 - 25x - 25$ **47.** $-11x^2 - 4x - 11$
49. $-10x^2 - 9x + 5$ **51.** $-9x^2 + 17x - 15$ **53.** $a = -23$ **55.** $a = 4, b = 3$ **57.** $-3x^6$ **58.** $-\frac{4x^4}{y^5}$ **59.** $-8x^3$
60. $-30y^3$ **61.** $1\frac{3}{14}$ more miles **62.** $\frac{5}{4}$ cups or $1\frac{1}{4}$ cups of sugar

Quick Quiz 6.1 *See Examples noted with Ex.* **1. (a)** $-8x - 1$ **(b)** $2x^2 - 5x - 5$ (Ex. 2) **2. (a)** $3a - 2$ **(b)** $7y^2 + y - 16$ (Ex. 4)
3. (a) $-8x^2 + 22x - 1$ **(b)** $6x^2 - 10x$ (Ex. 5) **4.** See Student Solutions Manual

6.2 Exercises **1.** The sign of the last term, $+4$, is wrong. It should be -4. **3. (a)** Monomial **(b)** Binomial **(c)** Trinomial
5. $-15y^2; +10y; -30; -15y^2 + 10y - 30$ **7.** $x^2 + 3x + 1, x^3 + 3x^2 + x; x^2 + 3x + 1, -x^2 - 3x - 1; x^3 + 2x^2 - 2x - 1$
9. $x^2, -3x, +1x, -3; x^2 - 2x - 3$ **11.** $16x^2 + 24x - 16$ **13.** $-8y^3 + 12y^2 - 24y$ **15.** $-3x^3 + 6x^2$ **17.** $-5y^8 - 45y^2$
19. $-3x^7 + 15x^5 + 6x^4$ **21.** $2x^3 - 5x^2 + x + 2$ **23.** $6y^3 + 14y^2 - 6y - 20$ **25.** $3x^3 + 5x^2 + x - 1$ **27.** $x^2 + 13x + 42$
29. $x^2 + 12x + 27$ **31.** $a^2 + 8a + 12$ **33.** $y^2 - 4y - 32$ **35.** $x^2 - 2x - 8$ **37.** $x^2 - 2x - 8$ **39.** $2x^2 + 5x + 2$
41. $3x^2 - 6x + 3$ **43.** $2y^2 + 3y - 2$ **45.** $2y^2 - 3y - 2$ **47.** $-10a^2 + 20ab + 30a$ **49.** $-7x^4 + 21x^3$ **51.** $x^3 - 5x^2 + 7x - 2$
53. $z^2 - 3z - 10$ **55.** $8x^3 + 8x^2 - 14x - 8$ **57.** $y^2 - 5y - 14$ **59. (a)** $z^2 + 6z + 5$ **(b)** $z^2 - 6z + 5$ **61. (a)** $x^2 + 2x - 3$
(b) $x^2 - 2x - 3$ **63.** $x^2 + 7x + 4$ **65.** $-2x^3 - 5x^2 - 3x + 6$ **67.** $a = -7$ **69. (a)** $D = 3N$ **(b)** 7 nickels **70. (a)** $L = 2W$
(b) $W = 20$ ft **71.** 625 calories **72.** $4 per hour

Quick Quiz 6.2 *See Examples noted with Ex.* **1.** $-15z^3 - 5z^2$ (Ex. 2) **2.** $4x^3 - 6x^2 + 10x - 8$ (Ex. 3) **3. (a)** $y^2 - y - 2$
(b) $2a^2 + 13a + 15$ (Ex. 7) **4.** See Student Solutions Manual

How Am I Doing? Sections 6.1–6.2 **1.** $3y^2 + 9y - 9$ (obj. 6.1.2) **2.** $-9a + 8$ (obj. 6.1.4) **3.** $-7x^2 + x - 3$ (obj. 6.1.4)
4. $-18x^2 + x - 6$ (obj. 6.1.4) **5.** $2x^2 - 13x - 20$ (obj. 6.1.4) **6.** $-16a^2 + 24a - 8$ (obj. 6.2.1) **7.** $12y^2 - 8xy + 10y$ (obj. 6.2.1)
8. $-4x^4 - 24x^2$ (obj. 6.2.1) **9.** $4y^3 + 11y^2 + 4y - 4$ (obj. 6.2.2) **10.** $y^2 - 2y - 8$ (obj. 6.2.3) **11.** $x^2 - 5x + 4$ (obj. 6.2.3)
12. $2y^2 + 11y + 12$ (obj. 6.2.3)

Understanding the Concept: Variable Expression or Equation? **1.** simplify **2.** simplify and solve **3.** simplify
and solve **4.** simplify

6.3 Exercises **1.** Mark's; Juan's age **3.** Leslie's; Alice's; Shannon's **5.** Mark drove 500 miles more than Scott. The rest of the answer
may vary. **7.** Profit for March $= x$; profit for April $= x - 8000$ **9.** Profit the second quarter $= S$; Profit the fourth quarter $= S + 31,100$
11. Height of tree $= t$; height of pole $= \frac{1}{2}t$ **13.** Width $= W$; length $= 2W$ **15.** Length $= L$; width $= 2L - 13$ **17.** Toni's DVDs $= x$;
Carl's DVDs $= 2x - 3$ **19.** Width $= W$; length $= 2W$; height $= 3W$ **21.** Number of red cars $= x$; number of blue cars $= x + 16$; number
of white cars $= x - 7$ **23.** Length of the first side $= F$; length of the second side $= F + 4$; length of the third side $= 2F - 10$ **25. (a)** Height
of tree $= t$; height of building $= 4t$ **(b)** $4t - t$ **(c)** $3t$ **27. (a)** Number of votes for: Max $= x$; Jeri $= x + 30$; Lee $= x - 45$
(b) $x - (x + 30) + (x - 45)$ **(c)** $x - 75$ **29. (a)** Sam's salary $= x$; Vu's salary $= x + 125$; Evan's salary $= x - 80$
(b) $(x + 125) + x - (x - 80)$ **(c)** $x + 205$ **31. (a)** Width $= W$; height $= W + 8$; length $= 2W - 2$ **(b)** $(W + 8) + W + (2W - 2)$
(c) $4W + 6$ **33.** False **35.** $x = 4$ **36.** $y = -83$ **37.** $m = -35$ **38.** $x = 54$ **39.** 140 in.^{2} **40.** 120 ft^3

Quick Quiz 6.3 *See Examples noted with Ex.* **1.** Mai's monthly salary $= x$; Tina's monthly salary $= 3x$ (Ex. 1) **2.** Sugar's age $= x$;
Dixie's age $= x + 4$; Pumpkin's age $= 2x - 3$ (Ex. 2) **3. (a)** Price of ring $= x$; price of watch $= x + 75$; price of bracelet $= x - 50$
(b) $(x + 75) - (x - 50) + x$ **(c)** $x + 125$ (Ex. 3) **4.** See Student Solutions Manual

Use Math to Save Money **1.** $\$440,000 - \$178,000 = \$262,000$ **2.** $\$1,007,000 - \$440,000 = \$567,000$ **3.** $\$3600 \times 10 = \$36,000$
4. $\$567,000 - \$36,000 = \$531,000$ **5.** $\$280,000 \times 3 = \$840,000$

6.4 Exercises **1.** The sign in the binomial should be $-$, not $+$. **3. (a)** 1, 2, 4 **(b)** 4 **5.** 4 **7.** 9 **9.** 3 **11.** 5 **13. (a)** a, c
(b) On a it is 3, and on c it is 1. **(c)** a^3c **15.** xy^2 **17.** a^2b^4 **19.** ac^2 **21.** xy **23.** 3 **25.** $5y$ **27.** $x; 2$ **29.** $2x; 1$
31. $6x^2; x; 3$ **33. (a)** $-$ **(b)** $+$ **35.** $3(a - 2)$ **37.** $5(y + 1)$ **39.** $2(5a + 2b)$ **41.** $3(5m + n)$ **43.** $7(x + 2y + 3)$

45. $2(4a + 9b - 3)$ **47.** $2a(a - 2)$ **49.** $b(4a - b)$ **51.** $5x(1 + 2y)$ **53.** $7xy(x - 2)$ **55.** $6a^2(2b - 1)$ **57.** $3(x^2 - 3x + 6)$
59. $4x^2(1 + 2x)$ **61.** $2xy(x + 2)$ **63.** $2(2y + 1)$ **65.** $5(3a - 4)$ **67.** $5x(1 - 2y)$ **69.** $3xy(3y^2 - 1)$ **71.** $3(2x - y + 4)$
73. $4(x^2 + 2x - 1)$ **75.** $2x^2y^2(xy - 4)$ **77.** $2x(3xy + y + 2)$ **79.** (a) $-2x + 10y$ (b) $-2x + 10y$ (c) The products are the same.

81. 12 **82.** 20 **83.** $2x$ **84.** $5x$ **85.** $1\frac{5}{12}$ in. **86.** Yes

Quick Quiz 6.4 *See Examples noted with Ex.* **1.** (a) 4 (Ex. 2) (b) x^2y (Ex. 3) **2.** $2(2x^2 - 5y + 1)$ (Ex. 5) **3.** $5ab(b - 3)$ (Ex. 6)
4. See Student Solutions Manual

You Try It **1.** $-2x^2 + 4x - 5$ **2.** $-6y + 4z - 7$ **3.** $3x^2 - 2x + 2$ **4.** $-8a^2 + 10ab - 12a$ **5.** $8x^3 + 2x^2 + 9x + 9$
6. $x^2 + 3x - 10$ **7.** $3ab(1 - 3a)$ **8.** (a) Width $= w$; height $= 3w$; length $= 2w - 3$ (b) $2w + 3$

Chapter 6 Review Problems **1.** $+2x^2$; $+5x$; $-3z^3$; $+4$ **2.** $+a^4$, $-2b^2$, $-3b$, -4 **3.** $-2a + 3$ **4.** $6x - 4y + 2$ **5.** $2x - 11$
6. $-4x + 6$ **7.** $13a^2 + 3a + 6$ **8.** $-8x^2 - 7x - 8$ **9.** $-7x^2 - 5x + 10$ **10.** $8x^2 - 16x + 12$ **11.** $-24x^2 + 32x - 20$
12. $-2y^2 + 12y$ **13.** $27x^2 - 9xy + 6x$ **14.** $20n^2 + 45mn + 35n$ **15.** $4x^6 - 16x^2$ **16.** $x^9 - 2x^5 - 3x^4$ **17.** $5z^2 + 20z$
18. $-6y^2 - 60y$ **19.** $4x^5 - 24x^3$ **20.** $2x^3 - x^2 - 7x + 2$ **21.** $3y^3 + 13y^2 - 7y + 15$ **22.** $-3y^3 + 7y^2 + y - 5$
23. $2x^3 + 9x^2 + 7x - 3$ **24.** $x^2 + 6x + 8$ **25.** $y^2 - 3y - 28$ **26.** $3x^2 - 2x - 8$ **27.** $5x^2 - 21x + 18$ **28.** First quarter $= x$;
third quarter $= x + 22{,}300$ **29.** Length $= L$; width $= L - 22$ **30.** $\angle b = x$; $\angle a = x + 30$; $\angle c = 2x$ **31.** Number of roses $= x$;
number of carnations $= 3x$; number of lilies $= x + 5$ **32.** (a) Erin's salary $= x$; Phoebe's salary $= x + 145$; Kelly's salary $= x - 60$
(b) $x + (x + 145) - (x - 60)$ (c) $x + 205$ **33.** (a) Number with blue eyes $= x$; number with brown eyes $= x + 7$; number with green
eyes $= x - 9$ (b) $x + (x + 7) - (x - 9)$ (c) $x + 16$ **34.** (a) Length of first side $= x$; length of second side $= 2x$; length of third side $= x + 10$
(b) $x + 2x + (x + 10)$ (c) $4x + 10$ **35.** (a) Width $= W$; length $= W + 7$; height $= 3W - 4$ (b) $(3W - 4) + W + (W + 7)$
(c) $5W + 3$ **36.** 7 **37.** 3 **38.** 5 **39.** 18 **40.** 2 **41.** 4 **42.** ab **43.** xy^2 **44.** $2(3x - 7)$ **45.** $5(x + 3)$
46. $4(a + 3b)$ **47.** $3(y - 3z)$ **48.** $6xy(y - 2)$ **49.** $8ab(a - 2)$ **50.** $5x^2y(2x + 1)$ **51.** $2y(2y^2 - 3y + 1)$ **52.** $3(a - 2b + 4)$
53. $2(x + 2y - 5)$

How Am I Doing? Chapter 6 Test **1.** $+x^2y$, $-2x^2$, $+3y$, $-5x$ (obj. 6.1.1) **2.** $-4x + 2y + 6$ (obj. 6.1.3) **3.** $-7x + 7$ (obj. 6.1.2)
4. $2y + 8$ (obj. 6.1.4) **5.** $-10p - 6$ (obj. 6.1.4) **6.** $13x^2 - 2x - 2$ (obj. 6.1.2) **7.** $-12m^2 - 6m - 4$ (obj. 6.1.4) **8.** $-2x^2 + 3x + 5$
(obj. 6.1.4) **9.** $-17x^2 - 9x + 4$ (obj. 6.1.4) **10.** $-14a^2 - 21ab + 28a$ (obj. 6.2.1) **11.** $-8x^5 + 6x^3$ (obj. 6.2.1) **12.** $x^2 + 14x + 45$
(obj. 6.2.3) **13.** $x^2 + x - 6$ (obj. 6.2.3) **14.** $2x^2 - 5x - 3$ (obj. 6.2.3) **15.** $-12x^7 + 4x^4$ (obj. 6.2.1) **16.** $4y^3 - 10y^2 - 12y + 18$
(obj. 6.2.2) **17.** Length $= L$; width $= L - 3$ (obj. 6.3.1) **18.** First side $= f$; second side $= f + 6$; third side $= 2f - 2$ (obj. 6.3.1)
19. (a) Votes for Lena $= x$; votes for Jason $= x - 3000$; votes for Nhan $= x + 5100$ (b) $(x + 5100) + x - (x - 3000)$ (c) $x + 8100$ (obj. 6.3.2)
20. 3 (obj. 6.4.1) **21.** 4 (obj. 6.4.1) **22.** x^2z (obj. 6.4.1) **23.** $3(x + 4)$ (obj. 6.4.2) **24.** $7(x^2 - 2x + 3)$ (obj. 6.4.2) **25.** $2xy(x - 3y)$
(obj. 6.4.2)

Cumulative Test for Chapters 1–6 **1.** No. She needs $77 **2.** $60 **3.** $3x + 5$ **4.** $-13r - 5$ **5.** 58 **6.** -5
7. -12 **8.** 25 **9.** $-50x^3$ **10.** $3x^6$ **11.** $5x^3 + 15x$ **12.** $\frac{5n}{3}$ **13.** $\frac{1}{4a^2}$ **14.** $\frac{x^3}{3^3}$ or $\frac{x^3}{27}$ **15.** (a) 144 in.^{2} (b) 150 ft^2

16. 60 yd^3 **17.** $9\frac{1}{4}$ **18.** $13\frac{1}{3}$ **19.** $2 \cdot 5 \cdot 5 \cdot 11$ or $2 \cdot 5^2 \cdot 11$ **20.** $51\frac{2}{3}$ mph **21.** 27 g **22.** 5 **23.** (a) $x = -47$ (b) $x = 43$

24. (a) $n = -48$ (b) $n = -11$ **25.** x^5 **26.** $\frac{1}{4}$ **27.** 84 **28.** $3\frac{1}{6}$ gal **29.** $3\frac{1}{2}$ pounds **30.** $-9x^2 - 2x - 6$ **31.** $3x^3 + x^2 + 4x + 6$
32. $x^2 + 8x + 12$ **33.** $2x^2 + x - 21$ **34.** Height of tree $= t$; height of building $= 2t + 4$ **35.** $9(a + 2b + 1)$ **36.** $6x^2(2y + 1)$

Chapter 7 7.1 Exercises **1.** -8 **3.** 15 **5.** simplify **7.** divide by -9 **9.** multiply by -9 **11.** add -3 **13.** $x = -13$
15. $a = -10$ **17.** $x = -10$ **19.** $y = 2$ **21.** $x = -8$ **23.** $a = 6$ **25.** $x = -3$ **27.** $y = -12$ **29.** $x = 7$ **31.** $x = -48$
33. $y = -28$ **35.** $a = -12$ **37.** $x = -8$ **39.** $y = -16$ **41.** $x = 15$ **43.** $a = -27$ **45.** $y = 18$ **47.** $x = 4$ **49.** $x = \frac{3}{5}$
51. $x = 2$ **53.** $x = 2$ **55.** $x = -\frac{5}{14}$ **57.** $x = -1$ **59.** $x = -9$ **61.** $x = 4$ **63.** (a) $x = -6$ (b) $x = -24$ (c) $x = -10$
(d) $x = -14$ **65.** (a) $y = 19$ (b) $y = -1$ (c) $y = -90$ (d) $y = -\frac{9}{10}$ **67.** $a = -20$ **69.** $x = 7$ **71.** $x = 49$ **73.** $x = -12$
75. $x = -\frac{4}{9}$ **77.** $x = 14$ **79.** (a) $x = -4$ (b) $x = -4$ (c) The answers are the same. (d) Division is defined in terms of multiplication,
therefore dividing by -2 is equivalent to multiplying by the reciprocal of -2, which is $-\frac{1}{2}$.

81.

9	4	5
2	6	10
7	8	3

83. $60°$ **85.** $36°$, $54°$ **87.** $2 \cdot 3 \cdot 5 \cdot 7$ **88.** $2 \cdot 2 \cdot 2 \cdot 2 \cdot 7$ or $2^4 \cdot 7$ **89.** 3,100,000 miles

Quick Quiz 7.1 *See Examples noted with Ex.* **1.** $y = -12$ (Ex. 2) **2.** $x = -10$ (Ex. 3) **3.** $b = 4$ (Ex. 5) **4.** See Student Solutions Manual

7.2 Exercises **1.** If 1 is added to double a number, the result is 5. What is the number? (Answers may vary.) **3.** $-3x$; $-3x$; -9; -9; 2; 2
5. $+9x$; $+9x$; -6; -6; 7; 7 **7.** -1; -2; -2; -1; -1 **9.** $x = 12$ **11.** $x = 7$ **13.** $x = 7$ **15.** $x = \frac{17}{5}$ **17.** $y = -14$

19. $m = -\frac{49}{3}$ **21.** $x = -31$ **23.** $y = -\frac{6}{5}$ **25.** $x = -1$ **27.** $y = 4$ **29.** $y = 2$ **31.** $x = \frac{8}{11}$ **33.** $x = -3$ **35.** $x = 8$

37. $x = \dfrac{7}{16}$ **39.** $x = -2$ **41.** $x = 18$ **43.** $y = -3$ **45.** $y = -\dfrac{17}{5}$ **47.** $x = \dfrac{2}{3}$ **49.** $y = -3$ **51.** $x = 5$ **53.** $x = \dfrac{10}{3}$

55. $x = 4$ **57.** $x = -8$ **59.** $x = -\dfrac{1}{5}$ **61.** $-3x + 12$ **62.** $-6 + 2y$ or $2y - 6$ **63.** $x = 105$ **64.** $x = 75$

Quick Quiz 7.2 *See Examples noted with Ex.* **1.** $a = -5$ (Ex. 4) **2.** $y = 7$ (Ex. 5) **3.** $x = 6$ (Ex. 4) **4.** See Student Solutions Manual

7.3 Exercises **1.** Use the distributive property to remove parentheses. Combine like terms. Add $+10$ to both sides of the equation. Divide both sides by -13. **3.** $x = -3$ **5.** $x = \dfrac{4}{3}$ **7.** $y = -5$ **9.** $x = -6$ **11.** $y = 1$ **13.** $x = \dfrac{1}{5}$ **15.** $y = 7$ **17.** $x = 2$ **19.** $x = 3$

21. $x = \dfrac{5}{3}$ **23.** $y = -1$ **25.** $x = 6$ **27.** $x = -\dfrac{7}{3}$ **29.** $x = \dfrac{14}{13}$ **31.** $x = -2$ **33.** $x = \dfrac{5}{2}$ **35.** $x = -\dfrac{13}{6}$ **37.** $x = -3$

39. No **41.** 12 **42.** 20 **43.** $2x$ **44.** $5x$ **45.** Overcharged **46.** $21\dfrac{5}{8}$ lb per in.^{2}

Quick Quiz 7.3 *See Examples noted with Ex.* **1.** $x = -1$ (Ex. 1) **2.** $x = 2$ (Ex. 2) **3.** $a = 3$ (Ex. 3) **4. (a)** distributive **(b)** See Student Solutions Manual

How Am I Doing? Sections 7.1–7.3 **1.** $a = -15$ (obj. 7.1.1) **2.** $x = -58$ (obj. 7.1.2) **3.** $x = -4$ (obj. 7.1.3) **4.** $x = 1$ (obj. 7.1.3) **5.** $x = 1$ (obj. 7.2.1) **6.** $y = -\dfrac{33}{5}$ (obj. 7.2.1) **7.** $y = 5$ (obj. 7.2.1) **8.** $x = \dfrac{5}{8}$ (obj. 7.2.1) **9.** $y = \dfrac{5}{12}$ (obj. 7.2.2) **10.** $x = \dfrac{4}{7}$ (obj. 7.2.2) **11.** $x = -7$ (obj. 7.3.1) **12.** $x = \dfrac{2}{5}$ (obj. 7.3.1) **13.** $y = 1$ (obj. 7.3.1)

7.4 Exercises **1.** 15 **3.** 12, 12, 12; 3, 4; 7; 12 **5.** $x = 12$ **7.** $x = 12$ **9.** $x = 20$ **11.** $x = \dfrac{23}{18}$ **13.** $x = \dfrac{1}{8}$ **15.** $x = \dfrac{1}{8}$

17. $x = \dfrac{1}{28}$ **19.** $x = \dfrac{1}{8}$ **21.** $x = 6$ **23.** $x = \dfrac{9}{2}$ **25.** $x = 4$ **27.** $x = -\dfrac{12}{7}$ **29.** $x = -\dfrac{1}{12}$ **31.** $x = 6$ **33.** $x = 10$

35. $x = -7$ **37.** $x = -13$ **39.** $x = -1$ **41.** $x = -\dfrac{11}{12}$ **43.** $x = -\dfrac{45}{44}$ **45.** $x = -\dfrac{5}{2}$ **47.** $2x + 6$ **48.** $x - 12$ **49.** $4 + x$

50. $2(5 + y)$ **51.** $\dfrac{5}{12}$ of the restaurant **52.** $3\dfrac{1}{4}$ lb

Quick Quiz 7.4 *See Examples noted with Ex.* **1.** $x = 24$ (Ex. 1) **2.** $a = \dfrac{2}{45}$ (Ex. 2) **3.** $y = 20$ (Ex. 3) **4.** See Student Solutions Manual

Understanding the Concept: Forming Equations **1.** x: 7, 8; y: 49, 64

7.5 Exercises **1. (a)** $3n + 9 = 15$ **(b)** $n = 2$ **3. (a)** $3n - 4 = 5$ **(b)** $n = 3$ **5. (a)** $2(4 + n) = 12$ **(b)** $n = 2$ **7.** $x = 3$ m
9. First side = 3 cm; second side = 4 cm; third side = 5 cm **11.** $x = 3$ in. **13. (a)** 8 ft **(b)** 20 ft **15. (a)** 6 ft **(b)** 15 ft
17. (a) Mark's salary = M; tutor's salary = $M - 11,400$ **(b)** $M + (M - 11,400) = 28,890$ **(c)** Mark's salary: \$20,145; tutor's salary: \$8745
(d) $\$20,145 + \$8745 \stackrel{?}{=} \$28,890$; $\$28,890 = \$28,890$ ✓ **19. (a)** Miles Cal drove first day = x; miles he drove the second day = $x + 165$
(b) $x + (x + 165) = 825$ **(c)** First day: 330 mi; second day: 495 mi **(d)** 330 mi $+ 495$ mi $\stackrel{?}{=} 825$ mi; 825 mi $= 825$ mi ✓

21. (a) Flight time of second flight = s; flight time of first flight = $\dfrac{1}{2}s$ **(b)** $s + \left(\dfrac{1}{2}s\right) = 15$ or $s + \dfrac{s}{2} = 15$ **(c)** Second flight: 10 hr; first flight: 5 hr
(d) 10 hr $+ 5$ hr $\stackrel{?}{=} 15$ hr; 15 hr $= 15$ hr ✓ **23. (a)** First side = L; second side = $2L$; third side = $L + 12$ **(b)** $L + 2L + (L + 12) = 120$
(c) First side: 27 m; second side: 54 m; third side: 39 m **(d)** 27 m $+ 54$ m $+ 39$ m $\stackrel{?}{=} 120$ m; 120 m $= 120$ m ✓ **25. (a)** Width = W; length = $3W - 2$
(b) $2W + 2(3W - 2) = 68$ **(c)** Width: 9 m; length: 25 m **(d)** $2(9$ m$) + 2(25$ m$) \stackrel{?}{=} 68$ m; 68 m $= 68$ m ✓ **27. (a)** Fall = x; spring = $x + 95$;
summer = $x - 75$ **(b)** $x + (x + 95) + (x - 75) = 395$ **(c)** Fall: 125; spring: 220; summer: 50 **(d)** $125 + 220 + 50 \stackrel{?}{=} 395$; $395 = 395$ ✓
29. $x = 23$ **31.** 4 units **33. (a)** 7, 8 **(b)** We multiply each x by 3 to get y. **(c)** $3x = y$ **(d)** 18, 21, 24, 90 **35. (a)** 4, 6 **(b)** We multiply
each x by 3, then add 1 to get y. **(c)** $3x + 1 = y$ **(d)** 13, 19, 136 **37.** $13\dfrac{1}{2}$ in. by $6\dfrac{3}{4}$ in. by $1\dfrac{1}{4}$ in. **38. (a)** Length: $35\dfrac{1}{4}$ in.;
width: $13\dfrac{1}{2}$ in. **(b)** $97\dfrac{1}{2}$ in.

Quick Quiz 7.5 *See Examples noted with Ex.* **1.** Third side = x; first side = $x + 5$; second side = $2x$ (Ex. 2) **2.** $x + (x + 5) + 2x = 29$
(Ex. 2) **3.** Third side: 6 m; first side: 11 m; second side: 12 m (Ex. 2) **4.** See Student Solutions Manual

Use Math to Save Money **1.** $\$315 + \$22 + \$20 = \$357$ **2.** $\$335 + \$13 = \$348$ **3.** He should buy the suit at The Men's Store.
4. $\$315 + \$22 = \$337$ **5.** Yes, he should buy the suit at the department store.

You Try It **1.** $-10 = x$ **2.** $y = 12$ **3.** $-\dfrac{3}{2} = x$ **4.** $-7 = x$ **5.** $-\dfrac{1}{2} = x$ **6.** $x = -\dfrac{1}{42}$

Chapter 7 Review Problems **1.** $x = 42$ **2.** $x = -6$ **3.** $x = -48$ **4.** $y = -2$ **5.** $x = 5$ **6.** $a = \dfrac{7}{2}$ **7.** $x = -\dfrac{9}{5}$

8. $x = 18$ **9.** $y = -\dfrac{3}{4}$ **10.** $x = 7$ **11.** $y = -21$ **12.** $y = 4$ **13.** $x = -1$ **14.** $y = -\dfrac{9}{2}$ **15.** $x = \dfrac{22}{3}$ **16.** $x = -\dfrac{12}{5}$

17. $x = 10$ **18.** $x = -7$ **19.** $y = \dfrac{3}{2}$ **20.** $y = -2$ **21.** $x = -\dfrac{11}{4}$ **22.** $x = -\dfrac{5}{7}$ **23.** $x = -12$ **24.** $x = -3$ **25.** $x = -2$

26. $x = \dfrac{1}{8}$ **27.** $y = 5$ **28.** $x = -\dfrac{6}{5}$ **29.** $x = 12$ **30.** $x = -20$ **31.** $y = -10$ **32.** $y = 4$ **33.** $x = \dfrac{13}{24}$ **34.** $x = \dfrac{5}{8}$

35. $y = -\dfrac{5}{18}$ **36. (a)** $4n - 9 = 15$ **(b)** $n = 6$ **37. (a)** $54 = 2(16) + 2(x + 3)$ **(b)** 11 ft   **38. (a)** 5 ft **(b)** 20 ft

39. (a) Tara's age $= T$; Sara's age $= T - 7$ **(b)** $T + (T - 7) = 29$ **(c)** Tara's age: 18; Sara's age: 11 **(d)** $18 + 11 \overset{?}{=} 29; 29 = 29$ ✓
40. (a) Fall students $= L$; spring students $= L + 85$; summer students $= L - 65$ **(b)** $L + (L + 85) + (L - 65) = 491$ **(c)** Fall: 157; spring: 242; summer: 92 **(d)** $157 + 242 + 92 \overset{?}{=} 491; 491 = 491$ ✓

How Am I Doing? Chapter 7 Test

1. $x = -3$ (obj. 7.1.3)   **2.** $x = -7$ (obj. 7.1.1)   **3.** $a = 15$ (obj. 7.1.2)   **4.** $x = -12$ (obj. 7.1.3)   **5.** $x = 2$ (obj. 7.2.1)   **6.** $x = 70$ (obj. 7.1.2)   **7.** $y = -4$ (obj. 7.1.3)   **8.** $y = -3$ (obj. 7.1.1)   **9.** $y = \dfrac{5}{2}$ (obj. 7.2.1)

10. $x = -\dfrac{11}{3}$ (obj. 7.2.1)   **11.** $x = 1$ (obj. 7.2.1)   **12.** $x = \dfrac{13}{10}$ (obj. 7.3.1)   **13.** $x = -5$ (obj. 7.3.1)   **14.** $x = 4$ (obj. 7.3.1)   **15.** $x = -2$

(obj. 7.3.1)   **16.** $x = 10$ (obj. 7.4.1)   **17.** $x = \dfrac{1}{12}$ (obj. 7.4.1)   **18.** $x = 12$ (obj. 7.4.1)   **19.** Length of the second side $= s$;

length of the first side $= s + 6$; length of the third side $= 3s$ (obj. 7.5.1)   **20.** $s + (s + 6) + 3s = 26$ (obj. 7.5.1)   **21.** First side: 10 ft;
second side: 4 ft; third side: 12 ft (obj. 7.5.1)   **22.** Anna's annual salary $= A$; sales clerk's annual salary $= A - 4000$ (obj. 7.5.2)
23. $A + (A - 4000) = 61{,}200$ (obj. 7.5.2)   **24.** Anna's annual salary is \$32,600; the sales clerk's salary is \$28,600. (obj. 7.5.2)

Chapter 8   8.1 Exercises

1. When we change 9 to 10, we write 0 and carry 1 to the 2, the next place value to the left. Thus the
2 changes to 3.   **3.** Five and thirty-two hundredths   **5.** Four hundred twenty-eight thousandths   **7.** 0.324   **9.** 15.0346   **11.** Twenty-five and $\dfrac{54}{100}$

13. One hundred forty-three and $\dfrac{56}{100}$ **15.** $\dfrac{7}{10}$ **17.** $4\dfrac{17}{100}$ **19.** $32\dfrac{81}{1000}$ **21.** $\dfrac{5731}{10{,}000}$ **23.** 6.7   **25.** 12.037   **27.** 0.01   **29.** 0.001

31. $<$ **33.** $<$ **35.** $<$ **37.** $>$ **39.** $>$ **41.** $<$ **43.** 523.72   **45.** 43.96   **47.** 9.1   **49.** 462.9   **51.** 312.951   **53.** 1286.350

55. 0.0631   **57.** 0.0474   **59.** 43 million mi; 35 million mi   **61.** \$15   **63.** \$14, \$18, \$16; \$48   **65.** $0.73, \dfrac{7}{10}, 0.071, 0.007, 0.0069$

67. -9 **68.** -9968 **69.** -5 **70.** 68   **71.** $\dfrac{7}{6}$ lb or $1\dfrac{1}{6}$ lb of nuts   **72.** $\dfrac{5}{4}$ in. or $1\dfrac{1}{4}$ in.

Quick Quiz 8.1 *See Examples noted with Ex.* **1.** Seven and twenty-one hundredths. (Ex. 1)   **2. (a)** $\dfrac{217}{10{,}000}$ (Ex. 3) **(b)** 4.032 (Ex. 4)
3. 156.7 (Ex. 7)   **4.** See Student Solutions Manual

8.2 Exercises

1. line up   **3.** in line   **5.** 7.55   **7.** 4.47   **9.** 63.4348   **11.** 81.023   **13.** 73.4169   **15.** 18.153   **17.** 51.323
19. 12.64   **21.** 19.64   **23.** -12.33 **25.** -105.343 **27.** -4.57 **29.** $6.2x$ **31.** $15.6y$ **33.** $12.6x - y$ **35.** $4.9x + 6.2y$
37. (a) -5.5 **(b)** 15.1 **(c)** -13.3 **39. (a)** $6.6x$ **(b)** $-4.46y$ **(c)** $0.75x$ **41.** 8.139   **43.** 205.78   **45.** -4.4 **47.** \$1492   **49.** 16
51. Day 4 and Day 5; 74   **53.** \$27.69   **55.** 0.28 sec faster   **57.** 0.86 sec   **59.** -6.46 **61.** -3.26 **63.** 3333363   **65.** 1111811
67. 9876549   **69.** 3234   **70.** -1104 **71.** 245   **72.** -1035

Quick Quiz 8.2 *See Examples noted with Ex.* **1. (a)** 45.19 (Ex. 1) **(b)** 2.35 (Ex. 2) **(c)** 32.82 (Ex. 3)   **2.** $7.73x + 9.01$ (Ex. 4)
3. -39.71 (Ex. 5)   **4.** See Student Solutions Manual

8.3 Exercises

1. 6   **3.** $462\overline{)1270}$ **5.** 0.0012   **7.** 0.0035   **9.** 61.669   **11.** 65.36   **13.** -7.05 **15.** -11.421 **17.** 15.325
19. -34.001 **21.** 14.98   **23.** 85,540   **25.** 410,000   **27.** 6000   **29.** 2.16   **31.** 0.23   **33.** 0.0565   **35.** 3.451   **37.** 0.23
39. -0.08 **41.** 12.24   **43.** 30.9   **45.** 2.34   **47.** $1.\overline{6}$ **49.** $0.\overline{54}$ **51.** $5.1\overline{36}$ **53.** $30.\overline{30}$ **55.** 1.83   **57.** 12.47   **59.** $0.\overline{3}$
61. $0.1\overline{3}$ **63.** 46.25   **65.** $0.\overline{2}$ **67.** -14.84 **69.** 0.32   **71. (a)** $0.00\overline{3}$ **(b)** $0.0\overline{3}$ **73.** -42.866 **75.** $0.44\ldots ; 0.55\ldots$ **77.** $x = 8$
78. $x = -5$ **79.** $x = -28$ **80.** $x = 15$

Quick Quiz 8.3 *See Examples noted with Ex.* **1. (a)** 15.717 (Ex. 2) **(b)** 4390 (Ex. 5)   **2.** 5.8 (Ex. 8)   **3.** $3.\overline{2}$ (Ex. 10)   **4.** See Student
Solutions Manual

8.4 Exercises

1. $x = 6.1$ **3.** $y = 9.75$ **5.** $x = 3.1$ **7.** $x = -12.1$ **9.** $x = 2.81$ **11.** $x = 5$ **13.** $x = 3.4$ **15.** $x = -5$
17. $x = -5.51$ **19.** $x = 8$ **21.** $x = 14.2$ **23.** $x = 3.5$ **25.** $x = 5$ **27.** $x = 20$ **29.** $x = 29$ **31.** $x = 8$ **33.** $x = -10.4$
35. $x = 3$ **37.** $x = 0.9$ **39.** $x = 2.1$ **41. (a)** 15.2 mi/gal **(b)** 25 gal of gas   **43. (a)** \$4.08 **(b)** \$2.38 **(c)** \$1.70   **45. (a)** 26
(b) \$14.25   **47.** \$124.49   **49.** \$18.95; 3; \$0.12; 223; \$83.61   **51.** No Fee Plan: \$10.50; Single Rate Plan: \$12.20; The No Fee Plan is the better deal.
53. (a) Q = number of quarters, $5Q$ = numbers of dimes **(b)** $(0.25)Q + (0.10)5Q = 4.50$ **(c)** 6 quarters and 30 dimes **(d)** 30 dimes is five
times the number of quarters, 6.; $0.25(6) + 0.10(30) = 4.50$ ✓   **55.** 7 quarters and 35 nickels   **57.** $x = 4.5$ **59.** $x = 18.5648$ **61.** 52.5 ft/sec

63. $x = 270$ **64.** $a = 104$ **65.** $0.\overline{3}$ **66.** $0.8\overline{3}$ **67.** $12\dfrac{2}{3}$ lb   **68.** $\dfrac{9}{16}$ in.

Quick Quiz 8.4 *See Examples noted with Ex.* **1.** $x = 5$ (Ex. 2)   **2.** $x = 2.5$ (Ex. 3)   **3. (a)** Q = number of quarters; $2Q$ = number of nickels
(b) $(0.25)Q + (0.05)2Q = 1.05$ **(c)** 3 quarters and 6 nickels **(d)** 6 nickels is twice the number of quarters, 3.; $0.25(3) + 0.05(6) = 1.05$ ✓ (Ex. 6)
4. See Student Solutions Manual

How Am I Doing? Sections 8.1–8.4

1. 0.02 (obj. 8.1.2) **2.** $\dfrac{27}{1000}$ (obj. 8.1.2) **3.** < (obj. 8.1.3) **4.** 4212.651 (obj. 8.1.4)
5. 40.353 (obj. 8.2.1) **6.** −96.453 (obj. 8.2.1) **7.** $6.7x + 3.1y$ (obj. 8.2.2) **8.** 4.879 (obj. 8.2.3) **9.** −5.2003 (obj. 8.3.1) **10.** 278.3 (obj. 8.3.2)
11. 5.31 (obj. 8.3.4) **12.** $3.8\overline{3}$ (obj. 8.3.5) **13.** $x = -3.6$ (obj. 8.4.1) **14.** $x = 0.8$ (obj. 8.4.1) **15.** \$550.44 (obj. 8.4.2)

8.5 Exercises

1. last digit **3.** 5%; 10% **5.** adding 5% and 1% **7.** 70 **9.** 35 **11.** 140 **13.** 2 **15.** 14 **17.** 60 **19.** 100
21. 150 **23.** 200 **25.** 30 **27.** 60 **29.** 90 **31.** 32,000 **33.** 16,000 **35.** 48,000 **37.** 2500 **39.** 10,000 **41.** 20,000
43. \$1540 **45.** \$16,500 **47. (a)** $30 + 200 + 50 + 50 = \$330$ **(b)** 10% of 330 = \$33; 33 × 4 = 132; \$130 **(c)** \$200 **49.** $n = 72$
50. $n = 84$ **51.** $\dfrac{3}{8}$ of the tank **52.** $\dfrac{3}{4}$ mi

Quick Quiz 8.5 *See Examples noted with Ex.* **1.** 150 (Ex. 2) **2.** 12,000 (Ex. 2) **3.** \$2100 (Ex. 3) **4.** See Student Solutions Manual

8.6 Exercises

1. left; % sign **3.** 42% **5.** 63% **7.** 28% **9.** 113% **11.** 0.9% **13.** 57.6%; 0.249; 0.003; 154.6% **15.** 370%;
0.238; 0.006; 1288.2% **17.** 0.36 **19.** 0.538 **21.** 7.5% **23.** 0.0233 **25.** 3.413% **27.** 0.8, 80%; $\dfrac{27}{100}$, 27%; $\dfrac{7}{1000}$, 0.007; $4.\overline{3}$, $433.\overline{3}$%
29. 0.3125, 31.25%; $2\dfrac{3}{5}$, 260%; $\dfrac{1}{1000}$, 0.001; 6.5, 650% **31.** 12.5% **33.** $\dfrac{1}{500}$ **35. (a)** 35% **(b)** $\dfrac{223}{1000}$ **37.** $\dfrac{1}{80}$ **39.** 2.5% **41.** 64.29%
43. 77.78% **45.** $\dfrac{11}{200}$ **47.** $\dfrac{23}{200}$ **49.** $\dfrac{159}{1000}$ **51.** 60%; 80% **53.** 425%; 437.5% **54.** $3x = 48$; 16 **55.** $2x = 330$; 165
56. $\dfrac{1}{4}x = 60$; 240 **57.** $x = \dfrac{1}{3}(69)$; 23 **58.** \$51.65 **59.** 1.4 million mi^2

Quick Quiz 8.6 *See Examples noted with Ex.* **1. (a)** 0.625 **(b)** 62.5% (Ex. 6) **2. (a)** $\dfrac{7}{50}$ **(b)** 14% (Ex. 6) **3. (a)** $\dfrac{1}{2000}$ **(b)** 0.0005 (Ex. 6)
4. See Student Solutions Manual

8.7 Exercises

1. 35 is much more than $\dfrac{1}{2}$ of 40. **3.** 9 is very close to 10, so it cannot be $\dfrac{1}{4}$ of 10. **5.** $n = 0.32 \times 90$; 28.8
7. $n = 0.26 \times 145$; 37.7 **9.** $n = 0.52 \times 60$; 31.2 **11.** $n = 1.5 \times 40$; 60 **13.** \$2.77 **15.** 390 students **17.** $54 = n\% \times 30$; 180%
19. $n\% \times 650 = 70$; 10.77% **21.** $n\% \times 60 = 18$; 30% **23.** 18.75% **25.** 40% **27.** $56 = 0.70 \times n$; 80 **29.** $24 = 0.40 \times n$; 60
31. $70 = 0.20 \times n$; 350 **33.** 50 **35.** 6.4 mi **37.** $100\%x + 60\%x = 128$; \$80 **39.** $100\%x + 90\%x = 95$; \$50 **41.** $44 = 0.5 \times n$; 88
43. $1.25 \times 60 = n$; 75 **45.** $n = 0.27 \times 78$; 21.06 **47.** $15.66 = n\% \times 87$; 18% **49.** $135 = 0.45 \times n$; 300 **51.** $110 = n\% \times 440$; 25%
53. What is 73% of 130 million? 94.9 mil **55.** 86 is what percent of 92? 93% **57.** 10% **59.** 7% **61.** \$162.50 **63.** 234.8097 **65.** 1840
67. 484,400 **69.** $x = 5$ **70.** $x = 5$ **71.** $x = 6$ **72.** $x = -2$

Quick Quiz 8.7 *See Examples noted with Ex.* **1.** 21 (Ex. 1a) **2.** 30% (Ex. 1c) **3.** 40 (Ex. 1b) **4.** See Student Solutions Manual

8.8 Exercises

1. Since 100% of any number is equal to that number, 150% of 80 must be greater than 80. **3.** 16; 250; 40 **5.** 95; 420; a
7. 69; b; 8230 **9.** p; 90; 63 **11.** p; 47; 10 **13.** 160; b; 400 **15.** $\dfrac{a}{200} = \dfrac{24}{100}$; $a = 48$ **17.** $\dfrac{a}{30} = \dfrac{250}{100}$; $a = 75$ **19.** $\dfrac{a}{4000} = \dfrac{0.6}{100}$; $a = 24$
21. $\dfrac{82}{b} = \dfrac{50}{100}$; $b = 164$ **23.** $\dfrac{75}{b} = \dfrac{150}{100}$; $b = 50$ **25.** $\dfrac{4000}{b} = \dfrac{0.8}{100}$; $b = 500,000$ **27.** $\dfrac{70}{280} = \dfrac{p}{100}$; $p = 25$; 25% **29.** $\dfrac{11.2}{140} = \dfrac{p}{100}$; $p = 8$; 8%
31. $\dfrac{90}{5000} = \dfrac{p}{100}$; $p = 1.8$; 1.8% **33.** $\dfrac{a}{350} = \dfrac{26}{100}$; $a = 91$ **35.** $\dfrac{720}{b} = \dfrac{180}{100}$; $b = 400$ **37.** $\dfrac{75}{400} = \dfrac{p}{100}$; $p = 18.75$; 18.75%
39. $\dfrac{a}{650} = \dfrac{0.2}{100}$; $a = 1.3$ **41.** $\dfrac{15.2}{25} = \dfrac{p}{100}$; $p = 60.8$; 60.8% **43.** $\dfrac{68}{b} = \dfrac{40}{100}$; $b = 170$ **45.** $\dfrac{94.6}{220} = \dfrac{p}{100}$; $p = 43$; 43%
47. $\dfrac{a}{380} = \dfrac{12.5}{100}$; $a = 47.5$ **49.** $\dfrac{a}{5600} = \dfrac{0.05}{100}$; $a = 2.8$ **51.** 200 apartments **53. (a)** 40% **(b)** 60 openings **55. (a)** 2.2% **(b)** 97.8%
57. 120 offices **59.** 67% correct **61.** $\dfrac{a}{798} = \dfrac{19.25}{100}$; $a = 153.62$ **63.** \$118.80 **65.** 28 in.^{2} **66.** 192 cm^3 **67.** 4 ft^2 **68.** 12 in.

Quick Quiz 8.8 *See Examples noted with Ex.* **1.** 16.2 (Ex. 4) **2.** 36% (Ex. 6) **3.** 70 (Ex. 5) **4.** See Student Solutions Manual

8.9 Exercises

1. The amount deposited or borrowed **3.** The period of time interest is calculated **5.** commission, rate, total, sales
7. \$756 **9.** \$3375 **11.** 30% **13.** 20% **15.** \$4050 **17.** \$44,940 **19.** 874 **21. (a)** \$150 **(b)** \$2650 **23.** \$20 **25. (a)** \$830
(b) \$2330 **27.** \$324.50 **29.** 225.25 lb **31. (a)** \$999 **(b)** \$699.30 **33.** \$700 **35. (a)** \$1205.48 **(b)** \$1259.10 **37. (a)** \$35.20
(b) The maximum price of the list must be \$5.60 **39.** 24 cm^2 **40.** 252 cm^3

Quick Quiz 8.9 *See Examples noted with Ex.* **1.** \$48 (Ex. 4) **2.** \$5040 (Ex. 5a) **3.** \$31,360 (Ex. 3) **4.** See Student Solutions Manual

Use Math to Save Money

1. Job 1: \$50,310/52 = \$967.50; Job 2: \$20.50 × 30 = \$615 **2.** Job 1: \$967.50 × 0.15 ≈ \$145.13;
Job 2: \$615 × 0.12 = \$73.80 **3.** Job 1: \$175 + \$75 + \$25 + \$50 + (\$0.35 × 45 × 5) = \$403.75; Job 2: \$150 + \$50 + \$25 + \$25 +
(\$0.35 × 5 × 5) = \$258.75 **4.** Job 1: \$967.50 − (\$145.13 + \$403.75) = \$418.62; Job 2: \$615 − (\$73.80 + \$258.75) = \$282.45
5. Job 1: \$418.62/50 ≈ \$8.37; Job 2: \$282.45/31 ≈ \$9.11 **6.** She should take Job 2. **7.** She should take Job 1.

You Try It **1.** $20\frac{12}{25}$ **2.** 1.38   **3. (a)** 27.10 **(b)** 1453.129   **4. (a)** 62.075 **(b)** −14.42   **5.** $9.1x + 1.2y$ **6. (a)** 31.562 **(b)** −8.532

7. (a) 85,260 **(b)** 0.012   **8.** 54.3   **9.** $0.41\overline{6}$ or 0.416 . . .   **10.** −2.23   **11.** $3850   **12.** 170%   **13.** 0.0013   **14.** 420%   **15.** $\frac{16}{25}$
16. 20%   **17.** 10.8   **18.** $455   **19.** $44,940   **20.** $5040

Chapter 8 Review Problems
1. Six hundred seventy-nine thousandths   **2.** Seven and eighty-three ten thousandths
3. Forty-six and $\frac{85}{100}$ **4.** $4\frac{267}{1000}$ **5.** $43\frac{91}{100}$ **6.** 32.761   **7.** 54.026   **8.** <   **9.** <   **10.** 842.86   **11.** 406.781   **12.** −4.883
13. 45.018   **14.** 23.65   **15.** −11.835   **16.** −14.6   **17.** 1.37   **18.** line up   **19.** 85.0; the decimal points   **20.** −6.9   **21.** 14.9
22. $11.8x$ **23.** $12.5x$ **24.** $186   **25.** 0.00164   **26.** 40.9528   **27.** −124.1625   **28.** 50.456   **29.** 5   **30.** 7   **31.** 12.49   **32.** 1600
33. 4,100,000   **34.** $527.80   **35.** $5498.71   **36.** 0.72   **37.** 1.18   **38.** $48.\overline{18}$ **39.** $721)\overline{2590}$ **40.** $12)\overline{4.349}$ **41.** 10.75   **42.** 9.2
43. (a) 11 **(b)** $21.70   **44.** $x = 10.91$ **45.** $x = -3.4$ **46.** $x = 3.15$ **47.** $x = -1.1$ **48.** 30 dimes; 180 nickels   **49.** 20,000
50. 2000   **51.** 14,000   **52.** 40,000   **53.** 85%   **54.** $\frac{132}{110} = 120\%$ **55.** 1.6%   **56.** 1.24   **57.** 37.9%; 0.428; 0.005; 347%   **58.** 120%;
0.035; 0.0025; 56.7%   **59.** $33.\overline{3}\%$ **60.** $\frac{813}{1000}$ **61.** 0.375, 37.5%; $\frac{18}{25}$, 72%; $\frac{1}{200}$, 0.005; 7.4, 740%   **62.** 0.75, 75%; $\frac{14}{25}$, 56%; $\frac{1}{500}$, 0.002; 3.4, 340%
63. 328   **64.** 5%   **65.** 54   **66.** $15.65   **67.** 260   **68.** $1760   **69.** 13.12%   **70.** 85; 400; a **71.** p; 100; 20   **72.** 20   **73.** 36%
74. (a) $300 **(b)** 25%   **75. (a)** 30% **(b)** 75   **76.** $2385   **77.** $935

How Am I Doing?   Chapter 8 Test
1. Two hundred seven and four hundred two thousandths (obj. 8.1.1)   **2.** $\frac{13}{1000}$ (obj. 8.1.2)
3. 0.51 (obj. 8.1.2)   **4.** > (obj. 8.1.3)   **5.** 746.14 (obj. 8.1.4)   **6.** −1.95 (obj. 8.2.3)   **7.** $4.16x + 2.01y$ (obj. 8.2.2)   **8.** 13.14 (obj. 8.2.1)
9. 12.57 (obj. 8.2.1)   **10.** −6.1 (obj. 8.2.1)   **11.** 9.888 (obj. 8.3.1)   **12.** 4720 (obj. 8.3.2)   **13.** 4.5 (obj. 8.3.4)   **14.** $6.\overline{3}$ (obj. 8.3.5)
15. $x = 4$ (obj. 8.4.1)   **16.** $y = 2.6$ (obj. 8.4.1)   **17.** 7% (obj. 8.6.1)   **18. (a)** 50 **(b)** 25 **(c)** 5 **(d)** 30 **(e)** 75 (obj. 8.5.1)   **19. (a)** 0.8
(b) 80% (obj. 8.6.3)   **20. (a)** $\frac{1}{10}$ **(b)** 10% (obj. 8.6.3)   **21. (a)** $\frac{1}{20}$ **(b)** 0.05 (obj. 8.6.3)   **22.** 40 (obj. 8.7.1)   **23.** 2% (obj. 8.7.1)
24. 9.6 (obj. 8.7.1)   **25.** $32.32 (obj. 8.2.5)   **26.** $5992.06 (obj. 8.7.2)   **27.** 140 (obj. 8.7.2)   **28. (a)** $792 **(b)** $2508 (obj. 8.7.2)
29. $207.50 (obj. 8.7.2)   **30.** $1575 (obj. 8.9.1)   **31.** $1804 (obj. 8.9.3)   **32.** One Step: $255; One Touch: $247.50 (better buy) (obj. 8.9.2)

Chapter 9   9.1 Exercises
1. Computer support specialists   **3.** 525,000   **5.** 400,000   **7.** 320   **9.** 20   **11.** 32.2%
13. (a) Transportation and housing **(b)** 53.76%   **15.** $8943.87   **17.** 17.1%   **19.** 26.5%   **21.** 82.9%   **23.** 42.7 in.   **25.** Buffalo,
NY; Anchorage, AK   **27.** 31.4 in.   **29.** Flagstaff, AZ   **31.** Flagstaff, AZ; Anchorage, AK; Buffalo, NY   **33.** June   **35. (a)** About 4500
customers **(b)** Increase   **37.** Between June and July   **39.** Sea-Tac: January; Kansas City: May   **41.** April, May, June   **43.** 1.5 in.
45. 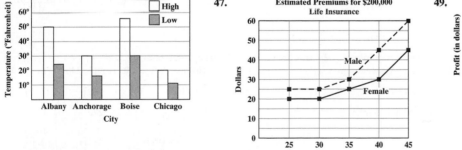   **47.** 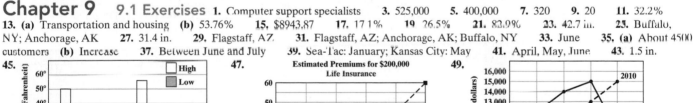   **49.**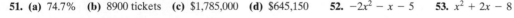

51. (a) 74.7% **(b)** 8900 tickets **(c)** $1,785,000 **(d)** $645,150   **52.** $-2x^2 - x - 5$ **53.** $x^2 + 2x - 8$ **54.** $2x(4xy + y + 2)$

Quick Quiz 9.1   *See Examples noted with Ex.*   **1.** (Ex. 8)   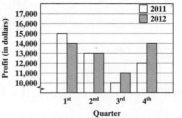   **2.** 25.3% (Ex. 2c)   **3.** (Ex. 7)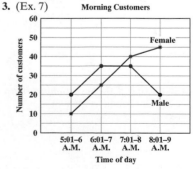

4. See Student Solutions Manual

9.2 Exercises
1. The mean of a set of values is the sum of the values divided by the number of values. **3.** 90.8 **5.** 3 hr **7.** 38
9. $168,600 **11.** 5.6 sales **13.** 0.375 **15.** 23.7 mi/gal **17.** 47 **19.** 1011 **21.** 0.58 **23.** 3.15 **25.** $20,250 **27.** 28.5 min
29. $10.99 **31.** 46 actors **33.** 60 **35.** bimodal, 121 and 150 **37.** $249 **39.** (a) 48 (b) 45 (c) 42 **41.** (a) 6.06 (b) 6.9 (c) None
43. Mean: 85; median: 83.5; mode: 81 **45.** (a) $2270 (b) $2400 (c) None **47.** (a) $2157 (b) $1615 (c) The median, because the mean
is affected by the high amount, $6300. **49.** 4850 **51.** 17 **52.** −4 **53.** 11 **54.** −7 **55.** $3450 **56.** 20,451.2 ft

Quick Quiz 9.2 *See Examples noted with Ex.* **1.** 13 (Ex. 1) **2.** (a) 11 (Ex. 3) (b) 45 (Ex. 4) **3.** (a) 6 (b) Bimodal: 4 and 5 (Ex. 5)
4. See Student Solutions Manual

How Am I Doing? Sections 9.1–9.2 **1.** 26% (obj. 9.1.2) **2.** Homes (obj. 9.1.2) **3.** 37.8 mi² (obj. 9.1.2) **4.** 4 in. (obj. 9.1.3)
5. 2 in. (obj. 9.1.3) **6.** January 2011 (obj. 9.1.3) **7.** 3.75 in. (obj. 9.2.1) **8.** 19 (obj. 9.2.2) **9.** 11 (obj. 9.2.2) **10.** 85 (obj. 9.2.3)
11. Bimodal: 8 and 7 (obj. 9.2.3) **12.** Mean: 7.4; median: 8.95; mode: none (obj. 9.2.3)

9.3 Exercises
1. Starting at the origin, move 2 units right followed by 1 unit down. **3.** (2003, 4700), (2004, 4800), (2005, 4950), (2006, 4850)
5. Lena (4, 60); Janie (7, 62); Mark (9, 66) **7–13.** **15–21.** **23.** (−3, −3) **25.** (3, 3)
27. (−1, 4) **29.** $\left(-3\frac{1}{2}, 4\right)$

31. $\left(1\frac{1}{2}, -3\right)$ **33.** (5, −1) **35.** (a) (−4, −2) (b) **37.** (a) (2, 1) (b)

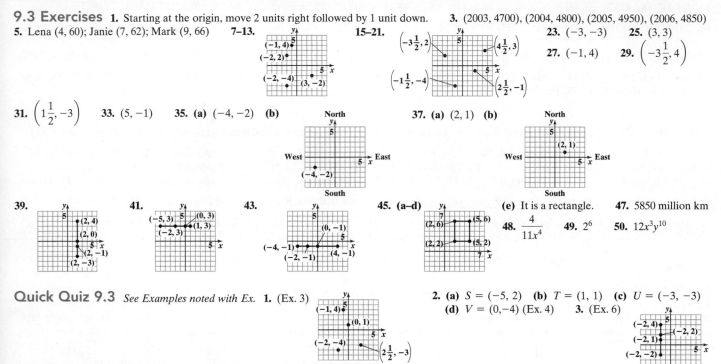

39. **41.** **43.** **45.** (a–d) **(e)** It is a rectangle. **47.** 5850 million km
48. $\dfrac{4}{11x^4}$ **49.** 2^6 **50.** $12x^3y^{10}$

Quick Quiz 9.3 *See Examples noted with Ex.* **1.** (Ex. 3) **2.** (a) $S = (-5, 2)$ (b) $T = (1, 1)$ (c) $U = (-3, -3)$
(d) $V = (0, -4)$ (Ex. 4) **3.** (Ex. 6)

4. See Student Solutions Manual

9.4 Exercises
1. No, because substituting $x = 2$ and $y = 5$ into $x + 2y = 4$ does not yield a true statement. **3.** (0, 4), (4, 0), (1, 3), (−1, 5)
(Answers may vary.) **5.** (0, 12), (12, 0), (1, 11), (−1, 13) (Answers may vary.) **7.** (a) (4, 140) (b) (8, 280) **9.** (a) (3, 240) (b) (5, 400)
11. (0, 8), (16, 0), (8, 4) **13.** (3, 2), (0, 5), (1, 4) **15.** (−1, 1), (1, 3), (−2, 0) **17.** (0, 3), (−1, −2), (1, 8) **19.** $(0, -3), \left(\dfrac{3}{5}, 0\right), (1, 2)$
(Answers may vary.) **21.** (0, 6), (−6, 0), (1, 7) (Answers may vary.) **23.** **25.** **27.**

29. **31.** **33.** **35.** **37.** (6, 0); because this is not on the line formed by
the other points.
39. Any ordered pair that is on the line formed
by the points on the graph is a solution. For
example, (2, 2) and (3, 1) are solutions.

41. $x = \dfrac{7}{2}$ **42.** $x = \dfrac{9}{5}$ **43.** $x = -\dfrac{19}{3}$ **44.** $x = \dfrac{20}{7}$

Quick Quiz 9.4 *See Examples noted with Ex.* **1.** (−1, 18), (5, 0), (4, 3) (Ex. 4) **2.** (Ex. 7) **3.** (Ex. 8)
4. See Student Solutions Manual

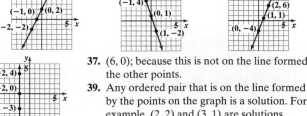

Use Math to Save Money **1.** $2000 × 2 = $4000 **2.** $1135.58 − $1073.64 = $61.94 **3.** $4000/$61.94 ≈ 64.58 or about 65 months
4. They should pay the points. **5.** They should not pay the points. **6.** $18,300

You Try It **1.**

Year	Barrels Produced
2009	3000
2010	2000
2011	1000

2. (a) 48% **(b)** 20   **3. (a)** 2000 **(b)** 2000
4. (a) 4000 visitors   **(b)** July and August
(c) Between July 2009 and August 2009   **5.**

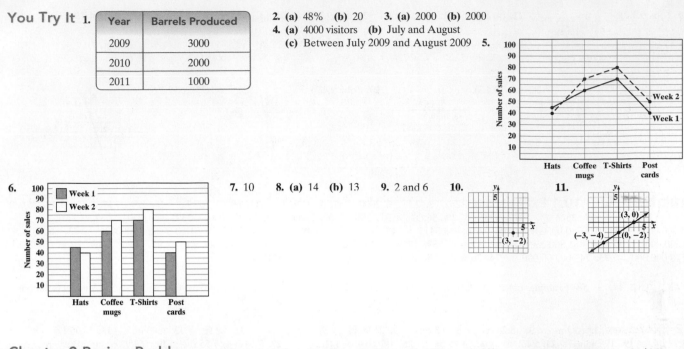

6.   **7.** 10   **8. (a)** 14 **(b)** 13   **9.** 2 and 6   **10.**   **11.**

Chapter 9 Review Problems **1.** 350   **2.** 300   **3.** 150   **4.** 800   **5.** 15%   **6.** 4%   **7.** 57%   **8.** 27%   **9.** $352
10. $660   **11.** $1496   **12.** $1848   **13. (a)** China **(b)** India   **14. (a)** Brazil **(b)** Brazil   **15.** Third largest population
16. Fourth largest population   **17.** 0.1 billion more people   **18.** 0.121 billion more people   **19.** India   **20.** 0.06 billion more people
21. (a) $400 **(b)** $350   **22. (a)** $350 **(b)** $300   **23.** Antilock brakes, automatic transmission, and air conditioning   **24.** CD changer,
alarm system, and aluminum wheels   **25.** $275   **26.** $237.50   **27.** $275   **28.** $237.50
29.   **30.**   **31.**

33. 82   **34.** 83   **35.** 19.5   **36.** 18.5   **37.** 90°F   **38.** $114   **39.** 58   **40.** 188
41. 13   **42.** 18   **43.** (2002, 5000), (2003, 6000), (2004, 5500), (2005, 6250)
44. (2006, 7000), (2007, 6500), (2008, 7500), (2009, 7250)

32.   **45–48.**   **49.** (−5, 3)   **50.** (−1, −1)   **51.** $\left(0, 2\frac{1}{2}\right)$ **52.** (4, 1)

53.   **54.**   **55.** (6, 2), (8, 0), (7, 1) (Answers may vary.)   **56.** (0, 3), (2, 1), (3, 0) (Answers may vary.)
57. (a) (4, 280) **(b)** (5, 350)   **58.** (1, −4), (0, −6), (−1, −8)   **59.** (0, 2), (−1, 8), (1, −4)

60.   **61.**   **62.**   **63.**

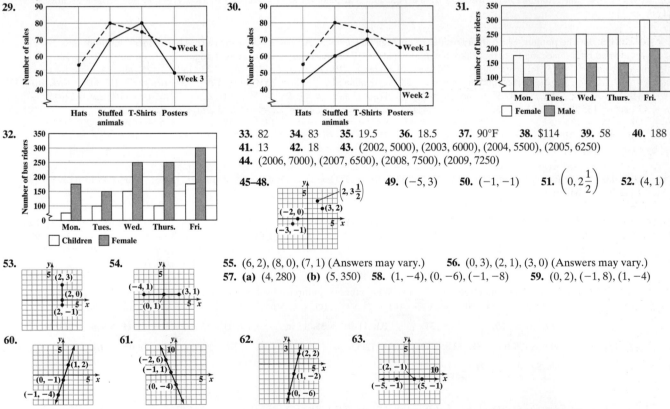

How Am I Doing? Chapter 9 Test **1.** Ages 18–20 (obj. 9.1.2)   **2.** 77% (obj. 9.1.2)   **3.** 50% (obj. 9.1.2)   **4.** 350 students
(obj. 9.1.2)   **5.** 11; 12 (obj. 9.1.4)   **6.** 6; 9 (obj. 9.1.4)   **7.** Otsego Lake, NY (obj. 9.1.4)   **8.** Baikal Lake, Russia (obj. 9.1.4)
9. Honda Accord and Toyota Camry do not have a roadside assistance plan. (obj. 9.1.3)   **10.** 5 yr (obj. 9.1.3)   **11.** 1 yr (obj. 9.1.3)
12. (a) Lexus ES350 **(b)** Lexus ES350 **(c)** VW Jetta **(d)** Lexus ES350 (obj. 9.1.3)   **13.** 5 yr for Lexus ES350; 4.33 yr for Honda
Accord (obj. 9.1.3)   **14.** 3.25 yr for Toyota Camry; 5.75 yr for VW Jetta (obj. 9.1.3)   **15.** 80.7 (obj. 9.2.1)   **16.** 87 (obj. 9.2.2)

17–20. (obj. 9.3.2) 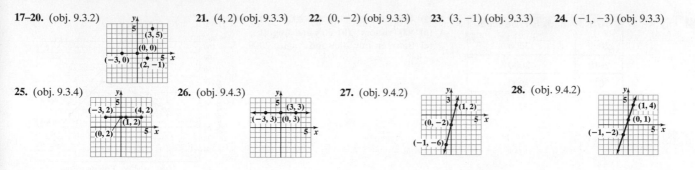 **21.** (4, 2) (obj. 9.3.3)   **22.** (0, −2) (obj. 9.3.3)   **23.** (3, −1) (obj. 9.3.3)   **24.** (−1, −3) (obj. 9.3.3)

25. (obj. 9.3.4)   **26.** (obj. 9.4.3)   **27.** (obj. 9.4.2)   **28.** (obj. 9.4.2)

Chapter 10   10.1 Exercises
1. 12   **3.** 2   **5.** 2000   **7.** 4   **9.** 1   **11.** 60   **13.** 7   **15.** 1.5   **17.** 5.75   **19.** 26,000
21. 28   **23.** 264   **25.** 2626 in.   **27.** $135   **29.** $6.00   **31.** 500   **33.** 3600   **35.** 2,430,000   **37.** 0.001834   **39.** 0.00078
41. 5,900,000   **43.** 0.007; 0.000007   **45.** 0.413; 0.000413   **47.** 3.5; 0.035   **49.** 3.582; 0.003582   **51.** 0.0032; 0.0000032   **53.** $6000
55. $340,000   **57. (a)** 481,800 cm   **(b)** 4.818 km   **59.** 1760   **61.** 21,120   **63.** 18,000   **65.** 14,600   **67.** 3220   **69.** 7.183; 0.007183
71. 528,000,000   **73.** 24,900,000,000   **74.** 20%   **75.** 57.5   **76.** $321.30   **77.** $716.80

Quick Quiz 10.1 *See Examples noted with Ex.* **1.** 2.75 mi (Ex. 1)   **2.** 0.5 m (Ex. 6)   **3.** 9,000,000 mg (Ex. 5)   **4.** See Student Solutions Manual

10.2 Exercises **1.** 1.22 m   **3.** 15.26 yd   **5.** 9.3 mi   **7.** 21.94 m   **9.** 132.02 km   **11.** 82 ft   **13.** 6.90 in.   **15.** 18.95 L
17. 4.77 qt   **19.** 198.45 g   **21.** 24.2 lb   **23.** 4.45 oz   **25.** 140.8 oz or 141.2 oz   **27.** 7.54 ft or 7.55 ft   **29.** 502.92 cm or 503.25 cm
31. 31 mph   **33.** 72.5 km/hr   **35.** 0.51 in.   **37.** 32°F   **39.** 185°F   **41.** 75.56°C   **43.** 30°C   **45.** 22.86 cm   **47.** 35.10 yd
49. 5.02 gal   **51.** 14.53 kg   **53.** 53.6°F   **55.** 20°C   **57.** 1397 lb   **59.** 143.87 km   **61.** 18.85 L   **63.** 9.6°F   **65.** 66.2°F; 113°F
67. 21.773 g   **69.** 47   **70.** 49   **71.** 91   **72.** 74

Quick Quiz 10.2 *See Examples noted with Ex.* **1.** 6 ft (Ex. 2)   **2.** 328 lb (Ex. 1)   **3.** 0.59 in. (Ex. 4)   **4.** See Student Solutions Manual

10.3 Exercises **1.** The point at which the two sides of an angle meet   **3.** An angle whose measure is between 90° and 180°
5. Two angles formed by intersecting lines that share a common side   **7.** transversal   **9.** a straight angle   **11.** an obtuse angle
13. supplementary angles   **15.** an acute angle   **17.** vertical angles   **19.** $\angle d, \angle b$ **21.** $\angle b$ **23.** 180°   **25.** alternate interior angles
27. $\angle c, \angle f$ **29.** $\angle b, \angle c$ **31.** corresponding angles   **33.** vertical angles   **35.** $\angle TSV$ or $\angle VST$ **37. (a)** 151°   **(b)** 61°   **39. (a)** 135°
(b) 45°   **41.** $x = 37.5°$ **43.** $n = 47.5°$ **45.** $m = 30°$ **47.** $\angle w$ and $\angle y$ **49.** 180°   **51.** $\angle z = 70°; \angle w = 110°; \angle y = 110°$
53. $\angle y = 45°; \angle x = 135°; \angle z = 135°$ **55.** $\angle p = \angle o = 43°; \angle o = \angle n = 43°; \angle n = \angle l = 43°;$ and $\angle m = 180° - 43° = 137°$
57. $\angle b = 180° - 45° = 135°; \angle a = \angle c = 45°; \angle c = \angle e = 45°;$ and $\angle d = 180° - 45° = 135°$ **59.** $\angle y = \angle v = \angle u = \angle r = 99°;$
$\angle x = \angle w = \angle t = \angle s = 81°$ **61.** $x = 4$ **63.** $\angle y = \angle x = 33°; \angle w + \angle x = 90°, \angle w = 90° - 33° = 57°; \angle z = 180° - 57° = 123°$
65. 9 in. by 10 in.   **66.** 12,760 km

Quick Quiz 10.3 *See Examples noted with Ex.* **1. (a)** 32°   **(b)** 122° (Ex. 3)   **2. (a)** 65°   **(b)** 115° (Ex. 6)   **3.** $x = 15°$ (Ex. 4)
4. See Student Solutions Manual

How Am I Doing? **Sections 10.1–10.3** **1.** 7 (obj. 10.1.1)   **2.** 24 (obj. 10.1.1)   **3.** 4 (obj. 10.1.1)   **4.** 6,300,000 (obj. 10.1.3)
5. 0.034; 0.000034 (obj. 10.1.3)   **6.** $84 (obj. 10.1.2)   **7.** 25.4 cm (obj. 10.2.1)   **8.** 9.99 kg (obj. 10.2.1)   **9.** 3.71 qt (obj. 10.2.1)   **10.** 58.9 mi/hr
(obj. 10.2.1)   **11.** 6°F (obj. 10.2.2)   **12.** 45° (obj. 10.3.3)   **13.** 32° (obj. 10.3.4)   **14. (a)** 141°   **(b)** 51° (obj. 10.3.2)   **15.** $x = 25°$
(obj. 10.3.2)   **16.** $\angle z = 55°; \angle w = \angle y = 125°$ (obj. 10.3.3)   **17.** $\angle o = \angle n = 40°; \angle l = \angle m = 140°$ (obj. 10.3.4)

10.4 Exercises **1. (a)** What is the square root of 81?   **(b)** What number multiplied by itself equals 81?   **(c)** What number squared equals 81?
3. No   **5.** Yes   **7.** Yes   **9.** No   **11. (a)** $n = \sqrt{64}$ **(b)** $n = \sqrt{64}$ **(c)** $n = \sqrt{64}$ **13.** 6   **15.** 9   **17.** 12   **19.** 11

21. 10   **23.** $\dfrac{5}{7}$ **25.** 5   **27.** 20   **29.** $\dfrac{3}{9} = \dfrac{1}{3}$ **31.** $\dfrac{6}{7}$ **33.** 11 ft   **35.** 12 in.   **37.** 6.633   **39.** 8.307   **41.** 8.944   **43.** 9.487

45. 5 in.   **47.** 8.544 yd   **49.** 15.199 ft   **51.** 13.856 km   **53.** 12.530 m   **55.** 7.071 m   **57.** 22 in.^{2} **59.** 78 ft^2 **61.** 17 ft   **63.** 5 mi

65. 8.7 ft   **67.** 1656 ft^2 **69.** $21,060   **71.** $\dfrac{5}{33x}$ **72.** $\dfrac{7x}{30}$

Quick Quiz 10.4 *See Examples noted with Ex.* **1. (a)** $\dfrac{6}{7}$ (Ex. 4)   **(b)** 5 (Ex. 3)   **2.** 2.83 (Ex. 6)   **3.** 6.93 cm (Ex. 9)   **4.** See Student Solutions Manual

10.5 Exercises **1.** circumference   **3.** radius   **5.** 22.5 yd   **7.** 1.9 cm   **9.** 56.5 cm   **11.** 69.1 in.   **13.** 87.9 in.   **15.** 235.5 in.
17. 113.0 yd^2 **19.** 1519.8 cm^2 **21.** 314 ft^2 **23.** 6358.5 mi^2 **25.** 6.28 ft   **27.** 256.43 ft   **29.** 200   **31.** 3215.36 in.^{2} **33.** $226.08
35. $1211.20   **37. (a)** $0.75 per slice, 22.08 in.^{2} **(b)** $0.67 per slice, 18.84 in.^{2} **(c)** 15-in. pizza is a better value.   **39.** 330 in.^{3} **40.** 160 in.^{3}

Quick Quiz 10.5 *See Examples noted with Ex.* **1. (a)** 37.68 ft (Ex. 1)   **(b)** 113.04 ft^2 (Ex. 3)   **2.** 7 (Ex. 2b)   **3.** $314 (Ex. 4)
4. See Student Solutions Manual

10.6 Exercises
1. Box **3.** Sphere **5.** Pyramid **7.** 141.3 m³ **9.** 267.9 m³ **11.** 502.4 in.³ **13.** 143.7 cm³ **15.** 1526.0 m³
17. 1356.5 cm³ **19.** 261.7 ft³ **21.** 21 m³ **23.** 90 m³ **25.** 1004.8 in.³ **27.** 2637.6 in.³ **29.** $0.75 **31.** 30.28 ft³ **33.** $942

35. 615.44 in.² **37.** $x = 63$ **38.** $x = 15$ **39.** $\frac{135}{16}$ or $8\frac{7}{16}$ **40.** $\frac{25}{14}$ or $1\frac{11}{14}$

Quick Quiz 10.6 *See Examples noted with Ex.* **1.** 1130.40 in.³ (Ex. 1) **2.** 47.1 in.³ (Ex. 4) **3.** 80 m³ (Ex. 5) **4.** See Student Solutions Manual

10.7 Exercises
1. similar **3.** perimeters **5.** 8 m **7.** 15 cm **9.** 17.5 cm **11.** 2.2 in. **13.** $P = 55$ in. **15.** $P = 22.5$ ft
17. 50 cm **19.** 36 ft **21.** 81 ft **23.** $n = 3.8$ km **25.** $n = 23.3$ cm **27.** 15 ft **29.** $A \approx 11.6$ yd² **31.** 12 **32.** 66 **33.** 6
34. 62

Quick Quiz 10.7 *See Examples noted with Ex.* **1.** 14.4 cm (Ex. 1) **2.** 45 ft (Ex. 3) **3.** 3 ft (Ex. 4) **4.** See Student Solutions Manual

Use Math to Save Money
1. 750, 468.75, and 250 gallons respectively **2.** $2700.00, $1687.50, and $900.00 respectively
3. $1012.50 **4.** About six and a half years **5.** $5062.50, $10,125.00 **6.** $1800 **7.** About one year and three months
8. $9000, $18,000

You Try It
1. 420 in. **2. (a)** 0.0000525 km **(b)** 60.25 dL **3. (a)** 12.88 km **(b)** 23.65 L **4. (a)** 3.17 gal **(b)** 820 ft/sec
5. (a) $\angle AOB$, or $\angle BOA$, or $\angle O$ **(b)** $\angle DEF$, or $\angle FED$, or $\angle E$ **(c)** $\angle XYZ$, or $\angle ZYX$ **6.** Supplement of $\angle x$: 123°; Complement of $\angle x$: 33°
7. $\angle x = 75°$; $\angle y = 105°$; $\angle z = 75°$ **8.** $\angle x = 60°$; $\angle v = 60°$; $\angle w = 120°$; $\angle z = 60°$ **9.** 7 **10.** 10.9 **11.** 9 ft² **12.** 31.4 m
13. 78.5 m² **14.** 301.44 ft³ **15.** 523.33 ft³ **16.** 209.33 m³ **17.** 133.33 ft³ **18.** 9 ft **19.** 28 m

Chapter 10 Review Problems
1. An angle that measures 90° **2.** Two angles whose sum is 180° **3.** Two angles whose sum is 90°
4. Two angles formed by intersecting lines that are opposite each other **5.** Two angles formed by intersecting lines that share a common side
6. Two angles that are on opposite sides of the transversal and between the other two lines **7.** Two angles that are on the same side of the
transversal and are both above or below the other two lines **8.** A triangle with a 90° angle **9.** In a right triangle the square of the longest side
is equal to the sum of the squares of the other two sides **10.** The length of a line segment from the center to a point on the circle **11.** The length
of a line segment across the circle that passes through the center **12.** The distance around a circle **13.** $A = \frac{bh}{2}$ **14.** $C = \pi d$ or $C = 2\pi r$

15. $A = \pi r^2$ **16.** $V - \pi r^2 h$ **17.** $V = \frac{4\pi r^3}{3}$ **18.** $V = \frac{\pi r^2 h}{3}$ **19.** $V = \frac{Bh}{3}$ **20.** 11 **21.** 25 **22.** 6.5 **23.** 3 **24.** 6000

25. 60 **26.** 5.75 **27.** 15.5 **28.** 0.059 **29.** 80 **30.** 259.8 **31.** 0.778 **32.** 630 **33.** 5000 **34.** 15,000 **35.** 0.473 **36.** 196,000
37. 721,000 **38.** 166.67 mL per jar **39. (a)** 17 ft **(b)** 204 in. **40.** 30.8 **41.** 9.08 **42.** 4.58 **43.** 73.20 **44.** 368.55 **45.** 89.6
46. 5.52 **47.** 0 **48.** 32.6 mi **49.** $2.22 **50.** $\angle MNO$ or $\angle ONM$ **51.** $\angle PNO$ or $\angle ONP$ **52. (a)** 141° **(b)** 51° **53.** 40°

54. $\angle w$ and $\angle y$ **55.** 180° **56.** $\angle x = 65°$; $\angle w = 115°$; $\angle y = 115°$ **57.** $\angle u = \angle v = \angle s = 45°$; $\angle t = 135°$ **58.** 12 **59.** $\frac{8}{9}$ **60.** 6.708

61. 3.162 **62.** 3 **63.** 15 **64.** 50.5 m² **65.** 5 yd **66.** 3.61 ft **67.** 37.68 in. **68.** 43.96 in. **69.** 113.04 m² **70.** 200.96 ft²
71. $261.67 **72.** 7.23 ft³ **73. (a)** 62.02 m³ **(b)** $74.42 **74.** 129.79 cm³ **75.** 1728 m³ **76.** 9074.6 yd³ **77.** 30 m **78.** 324 cm

How Am I Doing? Chapter 10 Test
1. (a) 9.0625 lb **(b)** 180 in. (obj. 10.1.1) **2. (a)** 0.162 kg **(b)** 26.6 mm (obj. 10.1.3)
3. (a) 2000 g **(b)** 4.4 lb (obj. 10.1.4) **4.** 392°F (obj. 10.2.2) **5.** 20.44 km (obj. 10.2.1) **6. (a)** 139° **(b)** 49° (obj. 10.3.2) **7.** $m = 41°$
(obj. 10.3.2) **8. (a)** $\angle c = 70°$ **(b)** $\angle d = 110°$ (obj. 10.3.3) **9.** $\angle s = \angle v = 137°$; $\angle w = \angle u = \angle t = 43°$ (obj. 10.3.4) **10.** Yes (obj. 10.4.2)

11. No (obj. 10.4.2) **12.** No (obj. 10.4.2) **13.** 8 ft (obj. 10.4.2) **14.** 7 (obj. 10.4.2) **15.** $\frac{3}{4}$ (obj. 10.4.2) **16.** 2.45 (obj. 10.4.3)

17. 11 (obj. 10.4.2) **18.** 60 cm² (obj. 10.4.5) **19.** 700 ft² (obj. 10.4.5) **20.** 6.32 cm (obj. 10.4.4) **21.** 17.8 ft (obj. 10.4.5) **22.** 16.171 m
(obj. 10.5.1) **23.** 5.6 revolutions (obj. 10.5.1) **24.** 4.5 cm² (obj. 10.5.2) **25.** $62.80 (obj. 10.5.3) **26.** 200.65 in.³ (obj. 10.6.1)
27. 2143.57 in.³ (obj. 10.6.1) **28.** 803.84 cm³ (obj. 10.6.2) **29.** 280 m³ (obj. 10.6.3) **30.** 58.5 in. (obj. 10.7.1) **31.** 25 ft (obj. 10.7.1)
32. 8 cm (obj. 10.7.2)

Practice Final Examination
1. 7,500,000 **2.** $x - 5$ **3.** $11 + x$ **4.** 21 **5.** 60 ft **6.** 269 **7.** 4,257,292 **8.** 1094 R27
9. 3 **10.** $y = 2$ **11.** $320 **12.** $<$ **13.** -13 **14.** 9 **15.** -8 **16.** -180 **17.** -8 **18.** $2y - 2$ **19.** 9 **20.** $-3a + 6$
21. $x = -17$ **22.** $x = 3$ **23. (a)** $L = 5W$ **(b)** $W = 8$ ft **24.** 350 in.² **25.** 12 m **26.** $W = 4$ ft **27. (a)** 35 yd² **(b)** $630

28. $75y^5 z^8$ **29.** $(5y)(3y^2)(2y^4)$; $30y^7$ **30.** $8\frac{1}{7}$ **31.** $\frac{16}{5}$ **32.** $\frac{8}{13z}$ **33.** $-\frac{3}{5}$ **34.** $\frac{3}{7x^3}$ **35.** 3^6 or 729 **36.** $\frac{3}{7}$ **37.** 13 ft per hr

38. (a) 3 secretaries **(b)** 4 paralegals **(c)** 320 paralegals **39.** $x = 15$ **40.** 450 mi **41.** $\frac{2}{3y}$ **42.** $-\frac{3}{16}$ **43.** $\frac{1}{12}$ lb in each **44.** $54x$

45. $\frac{4}{y}$ **46.** $\frac{1}{16}$ **47.** $\frac{11x}{90}$ **48.** $4\frac{11}{21}$ **49.** $\frac{18}{35}$ **50.** $\frac{43}{48}$ **51.** $x = 36$ **52.** $\frac{2}{3}$ in. **53.** $-7y^2 - 3y - 10$ **54.** $-3x^2 + 21x$
55. $z^{11} - 3z^6 - 5z^5$ **56.** $4y^3 + 24y^2 - 22y + 42$ **57.** $\angle b = x$; $\angle a = x + 40°$; $\angle c = 3x$ **58.** $y^2 + 2y - 15$ **59.** $4x^2 - 13x - 12$

60. $4(z + 3x)$ **61.** $3xy(3x - 1)$ **62.** $4(x - 2y + 4)$ **63.** $x = 54$ **64.** $z = \frac{4}{23}$ **65.** $z = -\frac{1}{4}$ **66.** $y = -8$ **67.** $a = -1$

68. $y = -1$ **69.** $x = 12$ **70. (a)** $L =$ fall students; $L + 95 =$ spring students; $L - 75 =$ summer students **(b)** $L + (L + 95) + (L - 75) = 386$
(c) Fall = 122; spring = 217; summer = 47 **(d)** $122 + 217 + 47 = 386$ ✓ **71.** 751.76 **72.** 84.178 **73.** 0.00168 **74.** $543.80 **75.** -1.48

76.

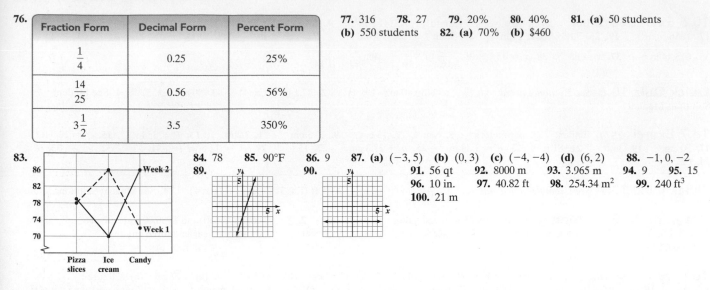

Fraction Form	Decimal Form	Percent Form
$\frac{1}{4}$	0.25	25%
$\frac{14}{25}$	0.56	56%
$3\frac{1}{2}$	3.5	350%

77. 316 **78.** 27 **79.** 20% **80.** 40% **81. (a)** 50 students **(b)** 550 students **82. (a)** 70% **(b)** $460

83.

84. 78 **85.** 90°F **86.** 9 **87. (a)** $(-3, 5)$ **(b)** $(0, 3)$ **(c)** $(-4, -4)$ **(d)** $(6, 2)$ **88.** $-1, 0, -2$
89. **90.** **91.** 56 qt **92.** 8000 m **93.** 3.965 m **94.** 9 **95.** 15
96. 10 in. **97.** 40.82 ft **98.** 254.34 m² **99.** 240 ft³
100. 21 m

Appendix A.1 Balancing a Checking Account Exercises 1. $555.12 3. $2912.65
5. $153.44; Yes, Justin can pay his car insurance. **7.** The account balances. **9.** $949.88 **11.** Jeremy's account balances.

Appendix A.2 Determining the Best Deal When Purchasing a Vehicle
Exercises **1.** $1295.94 **3.** $379.98 **5.** $4245 **7. (a)** $1244.95; $497.98 **(b)** $27,741.93 **9. (a)** $3135.93; $895.98
(b) $50,930.91 **11. (a)** $7499.85 **(b)** $42,499.15 **13.** Dealership 1 **15. (a)** Dealership 3; $25,678.93 **(b)** Dealership 1; $28,839 **(c)** The
least expensive purchase price does not guarantee the least expensive total cost. Many factors need to be considered to determine the best deal.

Appendix B 1. 5280 3. 1 5. 2000 7. 1 9. 2 11. 8 13. 3 15. 10,560 17. 2 19. 2 21. 2 23. 4
25. 9 ft 7 in. **27.** 16 ft 3 in. **29.** 16 yd 2 ft **31.** 38 ft 3 in. **33.** 1 c **35.** Isaac **37.** dm **39.** mg **41.** cg **43.** < **45.** >
47. < **49.** mL **51.** kg **53.** < **55.** > **57.** > **59.** gram **61.** liter **63. (c)** liter **65. (b)** kilograms
67. Above **69.** 4 gal

Appendix C 1. 11,747 3. 352,554 5. 1.294 7. 254 9. 8934.7376; 123.45 + 45.9876 + 8765.3 11. 16.81; $\frac{34}{8}$ + 12.56
13. 10.03 **15.** 289.387 **17.** 378,224 **19.** 690,284 **21.** 1068.678 **23.** 1.58688 **25.** 1383.2 **27.** 0.7634 **29.** 16.03
31. $12,562.34 **33.** 85,163.7 **35.** $103,608.60 **37.** 15,782.038 **39.** 2.500344 **41.** 897 **43.** 22.5 **45.** 84.372 **47.** 4.435
49. 101.5673 **51.** 0.24685 **53.** 30.7572 **55.** −13.87849 **57.** −13.0384 **59.** 3.3174×10^{18} **61.** $4.630985915 \times 10^{11}$
63. −0.5517241379 **65.** 15.90974544 **67.** 1.378368601 **69.** 11.981752 **71.** 8.090444982 **73.** 8.325724507 **75.** 1.08120 **77.** 1.14773

Appendix D Addition Practice 1. 37 2. 75 3. 94 4. 99 5. 78 6. 29 7. 61 8. 81 9. 369
10. 277 **11.** 491 **12.** 600 **13.** 1421 **14.** 1052 **15.** 900 **16.** 401 **17.** 1711 **18.** 1319 **19.** 289 **20.** 399 **21.** 589
22. 887 **23.** 422 **24.** 792 **25.** 902 **26.** 930

Subtraction Practice **1.** 21 **2.** 62 **3.** 22 **4.** 43 **5.** 68 **6.** 5 **7.** 29 **8.** 45 **9.** 531 **10.** 223 **11.** 726
12. 191 **13.** 378 **14.** 66 **15.** 509 **16.** 228 **17.** 256 **18.** 63 **19.** 183 **20.** 269 **21.** 873 **22.** 1892 **23.** 7479
24. 2868 **25.** 2066 **26.** 678

Multiplication Practice **1.** 69 **2.** 26 **3.** 378 **4.** 603 **5.** 1554 **6.** 1643 **7.** 3680 **8.** 3640 **9.** 7790 **10.** 2691
11. 9116 **12.** 7502 **13.** 12,095 **14.** 30,168 **15.** 20,672 **16.** 24,180 **17.** 37,584 **18.** 53,692 **19.** 284,490 **20.** 506,566
21. 238,119 **22.** 375,820 **23.** 803,396 **24.** 785,354 **25.** 49,516,817 **26.** 62,526,175

Division Practice **1.** 16 **2.** 56 **3.** 59 R2 **4.** 47 R5 **5.** 124 **6.** 296 **7.** 234 **8.** 567 **9.** 521 R6 **10.** 739 R2
11. 321 R11 **12.** 753 R6 **13.** 89 **14.** 91 **15.** 32 **16.** 41 **17.** 108 R29 **18.** 123 R25 **19.** 214 **20.** 342 **21.** 1134 R14
22. 1076 R65 **23.** 124 **24.** 153 **25.** 1125 **26.** 1532

Appendix E 1. 8 m 3. ∠F 5. 6 m 7. SAS 9. SSS 11. ASA 13. SAS 15. SAS 17. Not necessarily
congruent **19.** ASA **21.** Not congruent **23.** SSS **25.** SAS **27.** ASA **29.** Not necessarily congruent

Subject Index

Photo Credits

CHAPTER 1   CO Radius/SuperStock   **p. 25** Photodisc/Photo Library New York   **p. 29** Michael Stokes/Shutterstock   **p. 39** Jupiterimages/Thinkstock   **p. 53** Alberto Pomares/iStockphoto   **p. 62** Creatas/Thinkstock   **p. 75** Andy Z./Shutterstock   **p. 95** David Freund/iStockphoto   **p. 101** Courtesy of authors

CHAPTER 2   CO p. 103 Ryan Mcvay/Getty Images   **p. 129** Nico Schmidt/Still Pictures/Specialist Stock   **p. 134** Eugene Choi/iStockphoto   **p. 148** Robert Brenner/PhotoEdit, Inc.   **p. 163** Courtesy of authors

CHAPTER 3   CO p. 165 Photos.com   **p. 174** Gheorge Roman/Fotolia   **p. 197** Andreblais/Dreamstime   **p. 206** Stockbyte   **p. 214** Courtesy of authors

CHAPTER 4   CO p. 218 Lisa Kloosterhof/iStockphoto   **p. 225** Roberto A. Sanchez/iStockphoto   **p. 267** Marcusarm/Dreamstime   **p. 279** Courtesy of authors

CHAPTER 5   CO p. 281 Kbfmedia/Dreamstime   **p. 299** Graphixel/iStockphoto; Exactostock/SuperStock   **p. 318** Amy Etra/Photo Edit, Inc.   **p. 319** Exactostock/SuperStock   **p. 337** Pxlxl/Dreamstime   **p. 344** Courtesy of authors

CHAPTER 6   CO p. 346 Robert Crandall/Alamy   **p. 361** John Neubauer/Photo Edit   **p. 369** Tdiscpicture/Shutterstock   **p. 382** Courtesy of authors

CHAPTER 7   CO p. 386 Andresr/Shutterstock   **p. 420** Goruppa/Dreamstime   **p. 426** Courtesy of authors

CHAPTER 8   CO p. 428 GeoM/Shutterstock   **p. 433** Brandon Seidel/Shutterstock   **p. 456** Doring Kindersley, Ltd.   **p. 482** Losevsky Pavol/Shutterstock   **p. 499** Josh Randall/iStockphoto   **p. 509** Courtesy of authors

CHAPTER 9   CO p. 511 Andresr/iStockphoto   **p. 513** Design Pics Inc./Photo Library New York   **p. 552** Wavebreakmedia, Ltd./Shutterstock   **p. 561** Stephanie Barbary/Shutterstock   **p. 564** Courtesy of authors

CHAPTER 10   CO p. 566 Jupiterimages/Thinkstock   **p. 610** SW Productions/Thinkstock   **p. 619** Getty Images/Thinkstock   **p. 629** Courtesy of authors

MEASUREMENT ABBREVIATIONS

U.S. System

LENGTH		CAPACITY		WEIGHT		AREA		TIME	
in.	inches	oz	ounces	oz	ounces	in^2	square inches	h	hours
ft	feet	c	cups	lb	pounds	ft^2	square feet	min	minutes
yd	yards	qt	quarts			yd^2	square yards	s	seconds
mi	miles	gal	gallons			mi^2	square miles		

Metric System

LENGTH		CAPACITY		WEIGHT/MASS		AREA	
mm	millimeters	mL	milliliters	mg	milligrams	cm^2	square centimeters
cm	centimeters	cL	centiliters	cg	centigrams	m^2	square meters
m	meters	L	liters	g	grams	km^2	square kilometers
km	kilometers	kL	kiloliters	kg	kilograms		

USEFUL FORMULAS FROM GEOMETRY

Perimeters, Areas, and Volumes

Rectangle

L = length W = width
Perimeter: $P = 2L + 2W$
Area: $A = LW$

Rectangular Solid

L = length W = width
H = height
Volume: $V = LWH$
Surface Area:
$S = 2LW + 2LH + 2WH$

Square

s = length of each side
Perimeter: $P = 4s$
Area: $A = s^2$

Cylinder

r = radius h = height
$\pi \approx 3.14$
Volume: $V = \pi r^2 h$

Parallelogram

b = base h = height
Perimeter: P = sum of all sides
Area: $A = bh$

Sphere

r = radius
$\pi \approx 3.14$
Volume: $V = \dfrac{4\pi r^3}{3}$

Triangle

b = base h = height
Perimeter: P = sum of all sides
Area: $A = \dfrac{bh}{2}$
Angles: Sum of the measure of all 3 angles of any triangle is 180°

Cone

r = radius h = height
$\pi \approx 3.14$
Volume: $V = \dfrac{\pi r^2 h}{3}$

Circle

r = radius
d = diameter
$\pi \approx 3.14$ $d = 2r$
Circumference: $C = \pi d = 2\pi r$
Area: $A = \pi r^2$

Pyramid

B = Area of the base
h = height
Volume: $V = \dfrac{Bh}{3}$